Tables for the hydraulic design of pipes, sewers and channels

Sixth edition   –   Volume II

Tables for the hydraulic design of pipes, sewers and channels

Sixth edition – Volume II

HR Wallingford and D.I.H. Barr

 Thomas Telford, London

Published by Thomas Telford Services Ltd, Thomas Telford House,
1 Heron Quay, London E14 4JD, UK

Distributors for Thomas Telford books are
USA: American Society of Civil Engineers, Publications Sales
Department, 345 East 47th Street, New York, NY 10017-2398
Japan: Maruzen Co Ltd, Book Department, 3-10 Nihonbashi
2-chome, Chuo-ku, Tokyo 103
Australia: DA Information Services, 648 Whitehorse Road, Mitcham,
Victoria 3132

First published 1963
Sixth edition 1994

A catalogue record for this book is available from the British Library

ISBN: 0 7277 2004 X

Set in Helvetica by D. I. H. Barr
Printed and bound in Great Britain by Redwood Books,
Trowbridge, Wiltshire

Helvetica™ is a trademark of Linotype AG and its subsidiaries in the UK and other countries

Foreword to First Edition

Hydraulics Research Papers Nos 1 and 2 were published in 1958 under the titles *Resistance of fluids flowing in channels and pipes* and *Charts for the hydraulic design of channels and pipes*. These dealt with the application of the Colebrook-White equation for turbulent-transitional flow in determining the discharge capacity of channels and pipes. The Wallingford Charts have achieved wide circulation, but there have been requests for the design data to be made available in tabular form.

With the collaboration of the Road Research Laboratory of the Department of Scientific and Industrial research, the present publication has been prepared, as part of the programme of the Hydraulics Research Board, to meet this demand. It is hoped that it will be of particular value to civil engineers engaged on the design of urban drainage systems.

F H ALLEN
Director of Hydraulics Research

Hydraulics Research Station
Wallingford, Berks

March 1963

Foreword to Sixth Edition

The *Tables for the hydraulic design of pipes, sewers and channels* continue to provide a valuable reference for civil engineers working in the field of hydraulics.

The range, and therefore usefulness, of this approach was recently extended by the production of *Additional Tables* which increased the range of diameters considered and presented a new approach to the solution of problems involving non-circular flow cross-sections. This present edition is intended to consolidate these improvements within the main body of the Tables. It is hoped that, by doing this, the usefulness of these Tables to the industry will be enhanced.

Since Hydraulics Research Paper No 2 was published in 1958, there have been both developments in pipe materials and jointing methods and continuing research into roughness coefficient and the resistance of flow. These processes continue and it is expected that, as developments occur, they will be incorporated into future editions of these Tables

Dr W R White
Research Director

HR Wallingford
Wallingford, Oxfordshire

February 1994

Preface

The Sixth edition of the Wallingford Tables is published in two volumes, which are designed both to be mutually supportive and to be individually free-standing in use. To the intended amalgamation of the 5th edition of *Tables for the hydraulic design of pipes and sewers* with the Additional Tables, there has been added an alternative route for table-aided solution of the Colebrook-White equation. This has allowed of considerable increase in coverage of standard pipe diameters with the established form of Colebrook-White solution tables in Volume I, which draws also on the material and methods first presented in the Additional Tables.

Volume II uses the new, alternative, route to support the application of the unit size method. For this route, Manning equation tables also act as a carrier for the obtaining of Colebrook-White equation based solutions, as required. A wide range of conduit and channel shapes is covered by tables of properties based on unit size, with key examples of these tables also included in Volume I. This gives illustration of solutions supported by the established form of Colebrook-White tables, as is possible for most conduit and smaller channel circumstances, when the two volumes are used in conjunction.

In both volumes, there is extension of the unit properties tables to provide aid for rapidly varied flow problems, this augmenting the use of these tables in treatment of gradually varied flows.

An Annexure to Volume II enables solutions of the Hazen-Williams formula to be obtained from the Manning equation tables.

The authors acknowledge the contribution of Ronald Baron, Computer Officer, Department of Civil Engineering, University of Strathclyde, to the production of the various forms of table.

Users of these Tables are invited to provide comments or corrections, particularly on conduit or channel shapes which are in common use but which are not covered. The authors are grateful for various comments which have been received already, many of which have influenced the content of this 6th edition.

Contents

Contents (continued)

Contents (continued)

Tables C (continued)
Tables of properties of unit sections

(continued)

Contents (continued)

Tables C (continued)
Tables of properties of unit sections

Contents (continued)

(continued)

Contents (continued)

Tables H (continued)
Values of m_H deriving from the Hazen-Williams equation

The Wallingford Charts and the Wallingford Tables

The first editions of the Wallingford Charts[1] and the Wallingford Tables[2] were published in 1958 and 1963 respectively. Corresponding 6th and 5th editions were published in 1990. The continuing availability of these Charts and Tables has greatly facilitated the more general adoption of the Colebrook-White resistance equation[3] in the U.K. and elsewhere. This equation relates specifically to steady turbulent flow in circular pipes flowing full, but more and more it is adopted also for flow calculation within other shapes of pipes or channels. Hitherto, the Tables have concentrated on circular pipe flow, with coverage of part-full flow in such pipes and of flow in two forms of egg-shape.

The 3rd edition of the Tables, as published in 1977, introduced a selection of 36 metric pipe diameters from 50 to 2100 mm for which Colebrook-White equation assessed flow data was given directly in Tables 1-33. This pattern was continued in the 4th and 5th editions.

The Additional Tables

Published in 1993, the Additional Tables[2a] were designed as a companion to Tables 1-33 of the 3rd, 4th or 5th editions of the Wallingford Tables. The Additional Tables introduced a new approach to calculation of flows in non-circular conduits and channels. This approach uses tables which give geometrical and hydraulic properties for a wide range of conduit and channel shapes. Each table is based on a conduit or channel of unit size, and the data is presented for appropriate increments of depth. A linear multiplying factor, M, can then be used to give values of the properties for any size of conduit or channel of the given geometry.

This 6th Edition of the Wallingford Tables in two volumes

For the mid-nineties and beyond, it seemed desirable to enlarge the range of diameters with the established form of tables of Colebrook-White solutions. A selection of 65 diameters from 20 to 4000 mm has been made which includes both the 1977 selection of standard diameters and most of the differing diameters which are now standard for newer pipe materials. This new selection includes also all the diameters specified in both appropriate existing and draft European Standards. The resulting tables (Tables A) form the main element of Volume I, which thus succeeds the 5th edition.

It also seemed desirable to provide smaller increments of diameter than had been feasible in the Additional Tables, especially in the range relating primarily to open channels. The enlargement of Tables A in Volume I does go some way towards meeting this objective. An alternative procedure for table-aided solution of the Colebrook-White equation has been developed which leads to significant reduction in the bulk of tables required. This has allowed the main coverage of conduit and channel shapes to be presented in a linked but essentially independent Volume II, which thus succeeds the Additional Tables.

Arrangement and functions of Volume II

Table 1 provides a summary of the main capacities of the tabular approach now available in the two volumes.

In this volume, Tables C give a range of properties for a wide variety of unit size conduits and channels. Tables D and Tables E provide a method of solving both Manning's equation and the Colebrook-White equation.

Tables D have designed to provide a method for solving Manning's equation. The tables show values of mV and mQ for wide ranges of pipe diameters and gradients, in the manner of Tables A of Volume I, with V and Q indicating mean velocity and discharge respectively. The coefficient m is $100 \times n$, with n indicating the general case of the well known Manning[4] coefficient. The diameters treated are from 20 mm to 60·00 m, with the purpose of applications to open channel flows as equivalent diameters in the case of the larger values.

Tables E are designed to enable one to solve the Colebrook-White equation using the values of mV or mQ contained in Tables D. The tables show values of divisor m_C calculated from the Colebrook-White equation.

Values of m are broadly centred on one. This makes Tables D more compatible with Tables A of Volume I, and greatly simplifies the presentation of both Tables D and Tables E in comparison with basing the system on $n = m/100$. Solutions of the Manning[4] equation are obtained from Tables D by using $m_M = 100 n_M$ as divisor of mV and/or mQ. Here the subscript M indicates dependence on past experience, as summarised in Appendix 2.

Solutions to the Hazen-Williams formula can also be obtained, using the Annexure. This gives values of m_H to be applied as divisor to the values in Tables D to provide the Hazen-Williams solutions.

Review of hydraulic resistance

The Colebrook-White equation

In 1939, Colebrook[3] published this equation for turbulent flow in circular pipes flowing full. It followed from the smooth and rough turbulent logarithmic resistance laws for circular tubes. These had been evaluated experimentally by Nikuradse[5,6], after work by Prandtl and von Karman. The equations describing smooth and rough turbulent flow are

$$\frac{1}{\sqrt{\lambda}} = 2 \log \left\{ \frac{R\sqrt{\lambda}}{2 \cdot 51} \right\} \qquad \text{(Smooth turbulent flow)} \qquad (1)$$

$$\frac{1}{\sqrt{\lambda}} = 2 \log \left\{ \frac{3 \cdot 71 D}{k_s} \right\} \qquad \text{(Rough turbulent flow)} \qquad (2)$$

TABLE 1 : Overall solution paths for uniform flow problems

Applications	Resistance equations	Table sequences (Volume shown I or II)					Numerical (Fig. 2; Table 2)
		A (I)	B (I)	C (I&II)	D (II)	E (II)	
Water or sewage at normal temperature in circular pipe, full bore flow.	C-W equation:-	●	○				
	or:-				♦	✳	
	Manning eq:-				♦		
As above but any fluid.	C-W equation:-						▲
Water or sewage in circular pipe flowing part-full. Solution for discharge or for proportional depth.	C-W equation:-	●	○	C1(a)			
	or:-			C1(a)	♦	✳	
Water or sewage in non-circular flow sections, including part-full circular pipes. All solutions covered.	C-W equation:-	●	○	■			
	or:-			■	♦	✳	
	Manning eq:-			■	♦		
As above but any fluid.	C-W equation:-			■			▲

● Tables A in Volume I give Colebrook-White solutions for circular pipes flowing full. The diameter range covered is 20 mm to 4000 mm. Use for water at normal temperature (15°C), and with enhancement of range of water temperature using Appendix 2.

○ Tables B in Volume I give a full range of proportioning exponents for use in interpolations of Tables A, if and as required.

■ Tables C give details of geometric and hydraulic properties of cross-sections on a unit size basis. Key examples are included in Volume I and the full coverage is given in Volume II. Figure 3 outlines the combination of these tables with table aided, graphical or numerical solution of the Colebrook-White equation or the Manning equation.

♦ Tables D in Volume II allow of table-aided solution of the Manning equation, for circular pipes flowing full, in a parallel mode to that of the Colebrook-White equation with Tables A. Use for water or sewage at normal temperature: the unqualified Manning equation is less appropriate for smoother surfaces, but see Tables E below. The range of diameters covered is 20 mm to 60 m, with emphasis on the equivalent diameter function of the larger values.

✳ Tables E in Volume II allow selection of value of the Manning coefficient so as to give Colebrook-White solutions from Tables D.

▲ Figure 2 with Table 2 provides for non-tabular solution of the Colebrook-White equation for full bore flow of fluid of any viscosity. Note that roughness size, k_s, is entered into the equations in metres. Thereby, the general method for non-circular flow sections, using Tables C, can be applied widely.

Note:- The use of these tables for both gradually and rapidly varied flow assessments, and in treatment of different roughnesses around the wetted perimeter, is covered in the text.

The following combination provides a transition between the individual laws. It is known as the Colebrook-White equation to indicate the influence of C. M. White, Colebrook's collaborator and former research supervisor.

$$\frac{1}{\sqrt{\lambda}} = -2 \log \left\{ \frac{k_s}{3 \cdot 71 D} + \frac{2 \cdot 51}{R\sqrt{\lambda}} \right\} \qquad (3)$$

Here :

Friction factor, $\lambda = 2(Sg)D/V^2 = 1 \cdot 2337 (Sg)D^5/Q^2$ (4)

Reynolds number, $R = VD/v = 4Q/\pi v D$ (5)

Mean velocity, $V = Q/(\pi/4)D^2$ (6)

where D, k_s, Q, (Sg) and v are diameter, roughness size, discharge, product of piezometric gradient and acceleration due to gravity and kinematic viscosity respectively.

If Eqs (4), (5) and (6) are substituted in Eq. (3), one obtains

$$\frac{V}{\sqrt{(2SgD)}} = \frac{0 \cdot 9003\, Q}{\sqrt{(Sg)D^{2 \cdot 5}}} = -2 \log \left\{ \frac{k_s}{3 \cdot 71 D} + \frac{1 \cdot 775\, v}{\sqrt{(Sg)D^{1 \cdot 5}}} \right\} \qquad (7)$$

The linear measure of surface roughness

In his experiments, Nikuradse[6] used pipes which were roughened by a uniformly graded sand applied to the semi-dry lacquered surface and then re-lacquered. The original diameter of those grains has provided a standard of comparison against which other surfaces may be evaluated. Thus the roughness of another surface is given as the size of sand which, as a so applied uniform coating, would give the same resistance under rough turbulent flow. Such k_s values are available for most of the surfaces likely to be used in drainage works. Many of these values are listed in Appendix 1, which is based on recommendations made by Colebrook[3], Rouse[7], King[8], Lamont[9] and Perkins and Gardiner[10].

The Colebrook-White equation is most appropriate where the roughness consists of separate protuberances of random height and spacing. Some classes of pipe, not having this form of roughness, do not follow the same resistance function in the transition zone between smooth-turbulence and rough-turbulence. However in such cases flow is often rough turbulent, or close to this, in the circumstances of most concern.

Simplified forms of the Colebrook-White equation

The SU (simplified usage) equations of Fig. 2 (page 6) derive from the non-dimensional equations[3,11,12] of Fig. 1 and apply to fresh water at 15°C and hence of kinematic viscosity $1 \cdot 141 \times 10^{-6}$ m²s⁻¹. Where the kinematic viscosity differs significantly from this value, a simple correction detailed in Table 2 (page 7) may be applied.

Figure 1 : Colebrook-White equation and direct solution approximations

Solution for Q (or V) (i.e. Colebrook-White equation)

$$\frac{V}{\sqrt{(2SgD)}} = \frac{0 \cdot 9003 Q}{\sqrt{(Sg)} D^{2 \cdot 5}} = -2 \log \left\{ \frac{k_s}{3 \cdot 71 D} + \frac{1 \cdot 775 \, \nu}{\sqrt{(Sg)} D^{1 \cdot 5}} \right\} \qquad (7)$$

Solution for S (Q as input variable for flow) (Barr[11] approximation)

$$\frac{0 \cdot 9003 Q}{\sqrt{(Sg)} D^{2 \cdot 5}} = -1 \cdot 9 \log \left\{ \left(\frac{k_s}{3 \cdot 71 D} \right)^{1 \cdot 053} + \left(\frac{4 \cdot 932 \, \nu \, D}{Q} \right)^{0 \cdot 937} \right\}$$

Solution for D (Q as input variable for flow) (Pham[12] approximation)

$$\frac{0 \cdot 9003 Q}{\sqrt{(Sg)} D^{2 \cdot 5}} = -1 \cdot 8844 \log \left\{ \frac{0 \cdot 365 (Sg)^{0 \cdot 2} k_s}{Q^{0 \cdot 4}} + \frac{3 \cdot 55 \, \nu}{Q^{0 \cdot 6} (Sg)^{0 \cdot 2}} \right\}$$

D is diameter of circular pipe flowing full
V is mean velocity
Q is discharge
S is piezometric gradient
g is acceleration due to gravity
ν is kinematic viscosity
k_s is equivalent sand roughness size

The foregoing variables are in any coherent system of units such as pure SI (kg-m-s units)

Figure 2 : Colebrook-White equations in simplified usage mode (SU)

For SI units (D and k_s in m, Q in m^3s^{-1}); the fluid flowing in the first instance is water at 15°C (kinematic viscosity $1{\cdot}141 \times 10^{-6}$ m^2s^{-1}).

Solution for Q

$$\frac{Q}{\sqrt{S}D^{2{\cdot}5}} = -6{\cdot}957 \log \left\{ \frac{k_s}{3{\cdot}71D} + \frac{(0{\cdot}647 \times 10^{-6})^{*}}{\sqrt{S}D^{1{\cdot}5}} \right\}$$

Solution for S

$$\frac{Q}{\sqrt{S}D^{2{\cdot}5}} = -6{\cdot}61 \log \left\{ \left(\frac{k_s}{3{\cdot}71D} \right)^{1{\cdot}053} + \left(\frac{(5{\cdot}63 \times 10^{-6})^{*}D}{Q} \right)^{0{\cdot}937} \right\}$$

Solution for D (Q given as flow variable)

$$\frac{Q}{\sqrt{S}D^{2{\cdot}5}} = -6{\cdot}555 \log \left\{ \frac{0{\cdot}576 \, S^{0{\cdot}2} \, k_s}{Q^{0{\cdot}4}} + \frac{(2{\cdot}566 \times 10^{-6})^{*}}{Q^{0{\cdot}6} \, S^{0{\cdot}2}} \right\}$$

For other viscosities, multiply at * by the factor :-

$$\frac{\text{actual kinematic viscosity in } m^2s^{-1}}{1{\cdot}141 \times 10^{-6} \, m^2s^{-1}}$$

For other than circular pipes flowing full, D and Q are the values found for the equivalent pipe.

TABLE 2 : Values of multiplying factor for SU Colebrook-White equations

Fluid	Density, ρ (kgm^{-3})	Kin. visc., ν (m^{2}s$^{-1}\times10^6$)	Factor
Water (0°C)	999·9	1·787	1·566
Water (10°C)	999·6	1·307	1·145
Water (15°C)	999·1	1·141	**1·00**
Water (20°C)	998·2	1·004	0·880
Water (30°C)	995·7	0·801	0·702
Water (40°C)	992·9	0·658	0·579
Water (50°C)	988·8	0·553	0·485
Water (60°C)	983·2	0·475	0·416
Water (80°C)	974·5	0·364	0·319
Water (100°C)	958·4	0·294	0·258
Salt water (0°C)	c 1027	c 1·923	c 1·69
Salt water (20°C)	c 1025	c 1·085	c 0·95
Salt water (40°C)	c 1019	c 0·881	c 0·77
Jet fuel (JR4)(15·6°C)	773	1·125	0·986
Paraffin oil (20°C)	800	2·375	2·08
Turpentine (26°C)	868	1·58	1·39
Kerosene (26°C)	820	2·00	1·75
Ethyl alcohol	785	1·40	0·22
Methyl alcohol	787	0·71	0·62
Propyl alcohol	800	2·40	2·10
Ether (26°C)	714	0·31	0·27
Linseed oil (26°C)	929	35·6	31·2
Crude oil(−10°C)	925	2000	1753
Crude oil (20°C)	855	74	64·9
Petrol (0°C)	c 716	0·8	0·70
Petrol (20°C)	c 716	0·59	0·52
Petrol (60°C)	c 716	0·4	0·35
Fuel oil (20°C)	940	1200	1052
Glycerin (20°C)	1258	1188	1041
SAE 10 oil (20°C)	918	89·3	78·3
SAE 30 oil (20°C)	918	479	420
Mercury (20°C)	13546	0·114	0·10
Free flowing lava (c 1140°C)	-	cc 50000	cc 44000
Air (Atmos. 20°C)	1·205	14·9	13·1
Hydrogen (Atmos. 20°C)	0·0839	107	93·8

Note :- SU Colebrook-White equations are given in Fig. 2. Then the above values are used as multiplier at * in these equations, for fluids other than water at normal temperature (15°C).

7

The Manning equation arranged for parallel usage

Some engineers still prefer to work with the earlier Manning[4] equation. Strictly, this applies to rough turbulent flow only. For SI (metre-second) units, the Manning equation is stated as

$$V = (1/n_M) R^{2/3} S^{1/2} \qquad (8)$$

where n_M is the Manning coefficient and R is hydraulic mean depth (or hydraulic radius). Alternatively

$$n_M V = 0.3969 (D_{ep})^{2/3} S^{1/2} \qquad (9)$$

where D_{ep} is the diameter of the equivalent pipe ($D_{ep} = 4R$).

Williamson[13,13a] linked the Nikuradse data for rough turbulent flow in uniform sand roughened pipes[6] with a non-dimensional Manning type equation. The resulting Manning-Williamson equation is

$$\lambda = 0.18 (k_s/D)^{1/3} \qquad (10)$$

which can be re-arranged as

$$(k_s)^{1/6} V = 3.333 \sqrt{g} (D_{ep})^{2/3} S^{1/2} \qquad (11)$$

or

$$[(k_s)^{1/6}/8.398 \sqrt{g}] V = 0.3969 (D_{ep})^{2/3} S^{1/2} \qquad (12)$$

Thus for SI basic units and standard gravity

$$n_W V = 0.3969 (D_{ep})^{2/3} S^{1/2} \qquad (13)$$

where $n_W = (k_s)^{1/6}/26.3$, i.e. n_W is the value of coefficient which derives from the Manning-Williamson equation.

Formation of Tables D and Tables E

For a system for table-aided solutions, it is more convenient to arrange the tables, and thus to calculate, in terms of $m = 100n$ than in terms of n. Thus the generating equations for Tables D are

$$mV = 39.69 (D_{ep})^{2/3} S^{1/2} \qquad (14)$$

and correspondingly with $Q = V \times (\pi/4) D^2$

$$mQ = 31.17 (D_{ep})^{8/3} S^{1/2} \qquad (15)$$

In the main, it is not $m_W = 100 n_W$ which is of concern but $m_C = 100 n_C$, the value of m to give the Colebrook-White solution from Tables D. Thus Tables E show values of $m_C = mV/V_C$ where mV is given by Eq. (14) and the Colebrook-White mean velocity V_C is given by Eq. (7). Then Tables E can effectively be used to move from a solution of Manning's equation to a solution of the Colebrook-White equation.

For Reynolds numbers below 4000, values of m_C are omitted, but below 2000 there is substituted $m_P = mV/V_P$ where laminar flow mean velocity V_P is given by the Poiseiulle equation, i.e. $\lambda = 64/\boldsymbol{R}$, arranged as

$$V_P = SgD^2/32\nu \qquad (16)$$

Proportioning exponents arising from the structure of the Manning equation

In many cases, linear interpolation can be used where a value is required between entries in a table. If in some applications this is not considered sufficiently accurate, then non-linear interpolation can be used. In this case the interpolation should be based on the following proportioning equations and values of the exponents.

In these proportioning equations (Eqs (17)-(19)), given values, table values and required values are indicated by subscripts G, T and R respectively. The undifferentiated m indicates that the exponents apply equally with m_M, m_W or m_C.

$$mQ_R = mQ_T \left[\frac{D_G}{D_T}\right]^{2\cdot667} \left[\frac{S_G}{S_T}\right]^{1/2\cdot00} \qquad (17)$$

For mV_R, mQ_T is replaced by mV_T and the exponent 2·667 by 0·667

$$S_R = S_T \left[\frac{mQ_G}{mQ_T}\right]^{2\cdot00} \left[\frac{D_T}{D_G}\right]^{5\cdot333} \qquad (18)$$

$$D_R = D_T \left[\frac{mQ_G}{mQ_T}\right]^{1/2\cdot667} \left[\frac{S_T}{S_G}\right]^{1/5\cdot333} \qquad (19)$$

These basic proportioning procedures are equally applicable, as found necessary, in Manning, Colebrook-White or Hazen-Williams based solutions.

Arrangement and use of Tables D and Tables E

Tables D are compressed parallels to the Colebrook-White solution tables used in Volume I. There is only one table for each diameter range, allowing some additional diameters to be added in the range up to 4000 mm, and continuing up to 60·00 m with quite small increments of (equivalent) diameter. With the adoption of m as coefficient in the variables mV and mQ of Tables D, the change from litres per second to m^3s^{-1} for specification of product mQ occurs at a similar diameter in Tables D to that for Q in Tables A.

Each of the Tables E is for a standard value of k_s, and gives values of m_C for a range of diameters and gradients, all for water of standard temperature 15°C, i.e. kinematic viscosity $1\cdot141 \times 10^{-6}$ m^2s^{-1}.

For convenience, Tables E also contain the values of the Ackers[1] parameter θ, which is used, as explained later, in determining conditions in part-full pipes. This parameter applies to turbulent flow, and the values can be found below the values of m_C. Null entries are shown for laminar flow and for the laminar to turbulent transition.

In using the tables, the sole function of m_C, as obtained from Tables E, is as a divisor of product mV or of product mQ, as obtained from Tables D. Thus it is convenient to suppress the units of m in the tables and illustrative solutions.

The coefficient $m_W = 100n_W$, where n_W is defined by Eq. (13), is effective for approximating the corresponding Colebrook-White derived coefficient m_C for rough turbulent flow, but increasingly underestimates values of m_C as conditions move towards smooth turbulent flow. Table 3 shows values of m_W for the values of uniform sand roughness size k_s, which are adopted as standard.

Table 4 shows values of m_C for smooth turbulent flow for a range of values of D and S, and so represent extracts from Tables E.

Thus Tables 3 and 4 can be used to obtain preliminary estimations of likely values of coefficients for Colebrook-White solutions without reference to Tables E. The limitations on the possible relevance of m_W are well illustrated by comparison of these two tables, the values from which must be used only in the context of the situation in hand.

Again to simplify the use of Tables D, typical values of the Manning coefficient n_M are given in Appendix 2. As already detailed, direct application to Tables D is in terms of $m_M = 100n_M$. More comprehensive lists of values of n_M are available, especially from Ven Te Chow's[14] 'Open Channel Hydraulics', and from Barnes[15].

General

Throughout this volume, pipe diameters are specified in millimetres to 4000 mm, corresponding with Volume I and with standard practice, and thereafter in metres. Roughness sizes are specified in millimetres. Ratio values, D/k_s and the like, must be obtained with consistent units. However, the basic unit adopted for the unit size method is the metre, and calculations are illustrated in terms of (equivalent) diameters sized in metres, where appropriate.

In Tables E, values which correspond to values of D/k_s less than 5 are excluded. This follows the practice adopted for the established form of tables of Colebrook-White solutions.

Design of circular section pipelines and sewers

Use of the Tables D and E to find velocity, *V*, and discharge, *Q*

Tables D relate values of pipe diameter and gradient to the products mV and mQ. Tables E give values of the coefficient m_C for combinations of diameters, gradients and roughness sizes k_s, so as to provide Colebrook-White solutions for water at 15°C when these values are used to isolate V or Q from mV and mQ in Tables D.

> *What is the mean velocity and the discharge in a 1000 mm diameter pipe, with roughness size 0·060 mm, flowing full under a piezometric gradient of 0·00050? (Numerical solution of the Colebrook-White equation gives 0·8607 ms^{-1} and 0·6760 m^{3}s^{-1} respectively)*

TABLE 3 : Values of m_W from Manning-Williamson equation

k_s (mm)	n_W	m_W	k_s (mm)	n_W	m_W
0·060	0·00752	0·752	30·00	0·02120	2·120
0·150	0·00876	0·876	60·00	0·02379	2·379
0·300	0·00984	0·984	150·0	0·02772	2·772
0·600	0·01104	1·104	300·0	0·03111	3·111
1·500	0·01286	1·286	600·0	0·03492	3·492
3·000	0·01444	1·444	1500	0·04068	4·068
6·000	0·01621	1·621	3000	0·04566	4·566
15·00	0·01888	1·888	6000	0·05125	5·125

Note :- Values of n_W and m_W are given by $n_W = (k_s)^{1/6}/26\cdot3$, deriving from Eqs (10) to (13), together with $m_W = 100\,n_W$. Thus these values apply specifically to rough turbulent flows in circular pipes, or in conceptual circular cross-sections relating to equivalent diameters. Also note the presence of some values of n_W and m_W, for small values of k_s, which may be less than correspond to real flow conditions for water, as delineated by the values of m_C for smooth turbulent flow given in Table 4.

TABLE 4 : Values of m_C for smooth turbulent flow

Gradient	(Equivalent) Pipe diameters in mm					
	30	100	300	1000	3000	10000
0·00003	-	1·257	1·166	1·141	1·159	1·212
0·00010	-	1·135	1·077	1·070	1·097	1·156
0·00030	-	1·043	1·006	1·012	1·046	1·109
0·00100	1·038	0·957	0·939	0·955	0·996	1·063
0·00300	0·945	0·891	0·885	0·909	0·953	1·023
0·01000	0·860	0·827	0·832	0·864	0·911	0·983
0·03000	0·794	0·777	0·789	0·825	0·876	0·949
0·10000	0·733	0·739	0·747	0·788	0·840	0·915
0·30000	0·686	0·689	0·713	0·756	0·810	0·885
1·00000	0·640	0·651	0·678	0·724	0·779	0·855

Note :- Values of m_C derive from the smooth turbulent law; i.e. Eq. (1), in conjunction with the Manning equation.

From Table D4, read the values of mV and mQ for the stated combination of diameter and gradient, i.e. $0.887\,\text{ms}^{-1}$ and $0.6969\,\text{m}^3\text{s}^{-1}$. Merely from past experience, one might estimate m_M as 1·10 (i.e. 100×0.011, where 0·011 is the estimate of the Manning coefficient n_M), thus giving Manning equation based estimates of velocity and discharge as $0.887/1.10 = 0.807\,\text{ms}^{-1}$ and $0.6969/1.10 = 0.634\,\text{m}^3\text{s}^{-1}$ respectively.

However, inspection of Appendix 2 suggests adjustment of m_M to 1·05, then giving estimates of $0.845\,\text{ms}^{-1}$ and $0.6637\,\text{m}^3\text{s}^{-1}$.

To obtain, as thought necessary, accurate solution of the Colebrook-White equation, Table E5, for roughness size $k_s=0.06$ mm, gives, for diameter 1000 mm, values of m_C of 1·039 and 1·025 for gradients S of 0·00040 and 0·00060 respectively. This gives, by linear proportioning, $m_C = 1.032$ for the case in hand. On application of this value the solutions are $V = 0.887/1.032 = 0.8595\,\text{ms}^{-1}$ and $Q = 0.6969/1.032 = 0.6753\,\text{m}^3\text{s}^{-1}$ respectively (cf. $0.8607\,\text{ms}^{-1}$ and $0.6760\,\text{m}^3\text{s}^{-1}$ by numerical solution of the Colebrook-White equation).

Solutions for gradient and for pipe diameter

In the same basic case, to find the gradient using the Colebrook-White equation, given the discharge as $0.6760\,\text{m}^3\text{s}^{-1}$, a trial value of m_C is taken from Table E5 for roughness size 0·06 mm. For the stipulated diameter of 1000 m, a suitable medial value of m_C is 1·0. This in turn combines with the discharge of $0.6760\,\text{m}^3\text{s}^{-1}$ to give a trial value for mQ of $0.6760\,\text{m}^3\text{s}^{-1}$.

On Table D4, this indicates a required gradient of approximately 0·00048, and thus, on returning to Table E5, a corrected m_C value of 1·033. With the given value of mQ now adjusted to $0.6983\,\text{m}^3\text{s}^{-1}$, the mQ value found on Table D4 for gradient 0·00050 is $0.6969\,\text{m}^3\text{s}^{-1}$. This indicates a gradient of 0·00050 as closely representing the Colebrook-White solution.

Had entry values of 1·10 or 0·95 for m_C been chosen, these corresponding to the limits of the gradient range on Table E5, the final results would have differed only fractionally, if at all.

If the diameter is treated as unknown, the same pattern applies. By trying an entry value for m_C of 1·0, with required discharge as $0.6760\,\text{m}^3\text{s}^{-1}$, there is indicated, in the 0·00050 gradient line of Table D4, a solution of approximately 1000 mm. Hence, from Table E5, the corrected value of m_C is approximately 1·032. In Table D4, on the 0·00050 gradient line, an mQ value of $0.6969\,\text{m}^3\text{s}^{-1}$ appears in the 1000 mm column. Then $0.6969/1.032 = 0.6753\,\text{m}^3\text{s}^{-1}$ (cf. $0.6760\,\text{m}^3\text{s}^{-1}$) demonstrates that the required diameter is 1000 mm.

Interpolation between entries in Tables D

Interpolation should be used where intermediate values between the entries in Tables D are required. Visual or simple linear interpolation may be sufficient in many cases, because the increments in diameter are small throughout Tables D.

In the preceding example, the pipe size is increased to 1025 mm and the gradient to 0·000525, roughness size remaining the same. Obtain the new flow values. (Numerical solution of the Colebrook-White equation now gives $0·8972\,ms^{-1}$ and $0·7404\,m^3s^{-1}$)

On Table D4, linear proportioning as between the values for 1000 and 1050 mm diameter and for 0·00050 and 0·00055 gradients gives $mV = 0·924\,ms^{-1}$ and $mQ = 0·7636\,m^3s^{-1}$. Continuing to use $m_C = 1·032$, velocity and discharge are obtained as $0·895\,ms^{-1}$ and $0·7399\,m^3s^{-1}$ respectively.

Where linear interpolation is not considered sufficiently accurate in respect of assessment of mV or mQ, or for convenience, exponential interpolation can be adopted. In this case the proportioning equations (17) to (19) described earlier should be used. To demonstrate, the modified case as already treated, can be solved for discharge, gradient and then diameter as unknowns. Substituting in Eq. (17) to estimate the required mQ value, starting from the table value for 1000 mm and 0·00050 gradient

$$mQ_R \;=\; 0·6969 \left[\frac{1025}{1000} \right]^{2·667} \left[\frac{0·000525}{0·000500} \right]^{1/2·00}$$

$$=\; 0·7627\,m^3s^{-1}$$

Again adopting the divisor $m_C = 1·032$, the estimate of discharge is then given as $0·7391\,m^3s^{-1}$ (cf. $0·7404\,m^3s^{-1}$).

To determine the gradient or diameter, with m_C again as 1·032, the given value of mQ becomes $0·7404 \times 1·032 = 0·7641\,m^3s^{-1}$, compared with table value $0·6969\,m^3s^{-1}$ for 1000 mm diameter and 0·00050 gradient. Substituting as appropriate in Eqs (18) and (19) gives gradient 0·000527 and diameter 1025·7 mm respectively (cf. 0·00525 and 1025 mm).

Solution involving an intermediate roughness size

For water at normal temperature, determine the discharge in a 575 mm diameter pipe of 1 mm roughness size under a piezometric gradient of 0·00525 (Numerical solution of the Colebrook-White equation gives $0·4183\,m^3s^{-1}$).

From Table D3, mQ for a 550 mm diameter pipe under 0·0050 gradient is read as $447·54 \times 10^{-3}\,m^3s^{-1}$ (i.e. mQ_T is given directly as 447·54 litres per second). Substituting in Eq. (17), as before, to relate to the tabulated values, gives $mQ = 516·31 \times 10^{-3}\,m^3s^{-1}$, for 575 mm pipe size and 0·00525 gradient.

From examination of Tables E8 for $k_s = 0·60\,mm$ and E9 for $k_s = 1·50\,mm$, it can be seen that in these cases m_C varies quite slowly with changes of diameter and gradient, but is strongly dependent on roughness size. This indicates conditions close to rough turbulent. The m_C values for 0·60 mm and 1·5 mm are estimated as 1·161 and 1·302 respectively.

13

Linear interpolation between these values gives an estimate of $m_C = 1\cdot224$ for $1\cdot00\,mm$ roughness size at the stipulated diameter and gradient. Hence the estimate of discharge is $0\cdot4218\,m^3s^{-1}$ (cf. $0\cdot4182\,m^3s^{-1}$).

Alternatively, the local proportioning exponent r can be obtained:

$$r = \log\left(m_{C(1\cdot5)}/m_{C(0\cdot60)}\right)/\log(1\cdot5/0\cdot60) = 0\cdot125$$

Then $m_{C(1\cdot0)} = 1\cdot161\times\left(1\cdot0/0\cdot60\right)^{0\cdot125} = 1\cdot238$, this value giving discharge as $0\cdot4171\,m^3s^{-1}$ (cf. $0\cdot4182\,m^3s^{-1}$ by numerical solution of the Colebrook-White equation).

Multiplying factors, on tabulated values of *mQ*, for standard but non-tabulated diameters

Estimates may be required for pipe-full discharges in a pipe (or tunnel) of a standard diameter which is not included in Tables D. A list of multiplying factors is provided as Appendix 4. Values of *mQ* so obtained are then operated on in the usual way by the appropriate value of m_C from Tables E, or by an otherwise determined value of *m*.

> *Estimate the discharge in a 3658 mm (144 in) diameter tunnel of roughness size 0·60 mm, under a gradient of 0·0010. (The Colebrook-White numerical solution is 24·359 m³s⁻¹)*

From Table D8 read *mQ* as $30\cdot005\,m^3s^{-1}$ for $3600\,mm$ diameter and gradient $0\cdot0010$. From Appendix 4, read factor $1\cdot04$ for the adjustment to $3658\,mm$ from $3600\,mm$ diameter. Hence, *mQ* is adjusted to $31\cdot205\,m^3s^{-1}$. From Table E8, for k_s of $0\cdot60\,mm$, interpolate between $3000\,mm$ and $4000\,m$ diameter to obtain m_C as $1\cdot283$. The estimate of discharge is then

$$31\cdot205 \,/\, 1\cdot283 = 24\cdot322\,m^3s^{-1} \text{ (cf. } 24\cdot359\,m^3s^{-1}\text{)}$$

Perimeters involving dissimilar roughness

The wetted perimeter of a channel or pipe may be composed of surfaces of dissimilar roughness. This occurs, for example, where sliming has taken place only below the usual water level or where the invert is of concrete and the walls of brick. Then there are two ways[1,2] of computing discharges using these Tables.

The first method is based on the concept of an equivalent grain roughness for the whole perimeter. It utilises the expression

$$k_s = p_1\,k_{s1} + p_2\,k_{s2} \qquad\qquad (20)$$

where p_1, p_2 denote the proportions of the total perimeter occupied by surfaces 1 and 2, and k_{s1}, k_{s2} denote the equivalent sand grain roughness of surfaces 1 and 2.

This method may be used in situations where the difference in roughness values is not excessive and where the two surfaces occupy similar proportions of the total wetted perimeter. It will also give approximate answers outside these ranges. These ranges can be defined (somewhat arbitrarily) as

$$0 \cdot 05 < k_{s1}/k_{s2} < 20 \quad \text{and} \quad 0 \cdot 33 < p_1/p_2 < 3 \cdot 0$$

This method provides a direct solution to the problem of composite roughness. The tabular system can be used without the need to resort to a successive approximation technique.

The second method is based on the concept of an equivalent friction factor for the whole wetted perimeter. It utilises the expression

$$\lambda_s = p_1 \lambda_1 + p_2 \lambda_2 \tag{21}$$

where the suffices denote surfaces 1 and 2 as before. The method uses the following approximation[16] to the Colebrook-White equation

$$\frac{1}{\sqrt{\lambda}} = -2 \log \left\{ \frac{k_s}{3 \cdot 71 D} + \frac{5 \cdot 1286}{R^{0 \cdot 89}} \right\} \tag{22}$$

Later the use of an equivalent pipe diameter is explained, and on this basis the RHS of Eq. (19) can be written

$$\left\{ \frac{k_s}{3 \cdot 71 D} + \frac{5 \cdot 1286 \, v^{0 \cdot 89}}{(D_{ep} V)^{0 \cdot 89}} \right\} \tag{22a}$$

Or, for water at normal temperature

$$\left\{ \frac{k_s}{3 \cdot 71 D} + \frac{2 \cdot 63 \times 10^{-5}}{(D_{ep} V)^{0 \cdot 89}} \right\} \tag{22b}$$

This second method is not so restricted as the first method, but errors will occur where there are large differences in the grain roughness of the two surfaces, say $k_{s1}/k_{s2} < 0 \cdot 01$ or $k_{s1}/k_{s2} > 100$. The method requires a successive approximation approach as follows

(i) Approximate to the composite value of k_s using $k_s = p_1 k_{s1} + p_2 k_{s2}$.

(ii) Use the appropriate tables to determine the flow, Q, and also the mean velocity, V, for the actual pipe flowing full, or for the equivalent pipe.

(iii) Determine values C_1 and C_2 of the expression (13b) by adopting k_{s1} and k_{s2} respectively.

(iv) Determine C from $\log C = \dfrac{\log C_1}{\sqrt{\{ p_1 + (\log C_1 / \log C_2)^2 \, p_2 \}}}$

(v) Calculate the composite k_s value from

$$k_s = 3 \cdot 71 \, D_{ep} \left\{ C - \frac{2 \cdot 63 \times 10^{-5}}{(D_{ep} V)^{0 \cdot 89}} \right\}$$

15

(vi)   Use the appropriate tables to determine the flow, Q.

(vii)   Repeat (iii) to (vi) until Q changes by less than the required tolerance.

Non-circular cross-sections of flow

Calculation of discharge and velocity in part-full circular pipes
Ackers[1] defined a transitioning parameter θ. The value of θ determines the pattern of proportional flow variation in circular tubes of widely varying degrees of roughness. With SI units and water at standard temperature 15°C, θ is evaluated as follows

$$\theta = \left\{ \frac{k_s}{D} + \frac{1}{3600\,D\,S^{1/3}} \right\}^{-1} \tag{23}$$

In Tables E, the value of θ is shown below each value of m_C. Table C1(a) shows values of proportional discharge against proportional depth for a range of values of θ, for circular pipes. The following example shows how data from the tables are combined.

> *Estimate the discharge at 0·66 proportional depth in a 3400 mm diameter conduit at a gradient of 0·0008. Roughness size is 1·5 mm. (Numerical solution with the Colebrook-White equation gives 12·574 m³s⁻¹)*

In Table D8 the value of mQ read from the 3400 mm diameter column at 0·0008 gradient is 23·043 m³s⁻¹. From Table E9, m_C is interpolated linearly as 1·404, and θ as 750. The full pipe discharge is then 16·412 m³s⁻¹. On Table C1(a), the proportional discharge interpolates as 0·7665, giving the part-full discharge as 12·580 m³s⁻¹ (cf. 12·574 m³s⁻¹) Table C1 gives unit sectional areas, i.e. 0·5499 m² at 0·66 proportional depth. Then, if an estimate of mean velocity is required as well, the discharge of 12·580 m³s⁻¹ should be divided by $(0·5499 \times 3·40^2)$ m².

Calculation of depth in part-full circular pipes
The converse problem of estimating depth, given discharge, also can be solved using Table C1(a). Full pipe discharge is assessed as before, together with θ. Then the proportional discharge is applied to the table. Where necessary, interpolation can be used between table entries. However, for direct assessment of gradient for a part-full circular pipe, or of size of circular pipe to run at a stipulated proportional depth, a different approach should be adopted. This utilises the concept of 'unit size' and is now described.

Hydraulic equivalence
For any steady flow in a non-circular section one can find an equivalent circular cross-section where the flow characteristics are the same, in terms of hydraulic gradient and mean velocity, because the hydraulic mean depth (hydraulic radius) is the same. This includes part-full flow in a circular pipe. According to Chezy[17], the hydraulic mean depth is A/P, which is one quarter of the diameter of the equivalent pipe. Following Johnson[18] and Ackers[19], one can make

16

the following comparison between a non-circular flow section and its equivalent pipe. Since

$$V = V_{ep}$$

$$Q/A = Q_{ep} / [(\pi/4)(4A/P)^2]$$

Thus

$$Q_{ep} = Q \times 4\pi A/P^2 \qquad (24)$$

Here A and P are cross-sectional flow area and wetted perimeter of the non-circular flow section, V and Q represent mean velocity and discharge and subscript ep indicates equivalent pipe in the sense of providing equivalent (i.e. the same) mean velocity, though not, of course, equivalent conveyance. The discharge conversion factor $J = 4\pi A/P^2$ must apply also to the relation between flow areas, because discharge is mean velocity times flow area.

Therefore the core of solving flow in a non-circular cross-section reduces to finding the equivalent circular pipe and then solving for flow in this equivalent pipe, using the methods already presented.

'Unit size' measures for shapes of conduits and channels

For any geometrically similar series of conduit or channel shapes, a 'unit size' can be defined in terms of a key dimension. Then, a multiplying factor M specifies the size of an actual example of the shape within the geometrically similar series. Also for each proportional or relative depth value, there is a 'unit' case wetted perimeter value P_u and a 'unit' case cross-sectional area of flow A_u.

Hence for an actual example, MP_u and $M^2 A_u$ are the size of the wetted perimeter and of the area, respectively, at the specified proportional or relative depth. It follows that there is a 'unit' case value of hydraulic mean depth (hydraulic radius), i.e. $R_u = A_u/P_u$ and hence a 'unit' case value of diameter of equivalent pipe, i.e. $D_{ep(u)} = 4R_u$.

In these Tables, the sizes of $D_{ep(u)}$ and P_u etc. are quoted in metres, while the values of the other variables in the tables of sequence C are quoted in metre-second units as appropriate. Thus with values of $D_{ep(u)}$ made available in table sequence C

$$D_{ep} = MD_{ep(u)} \qquad (25)$$

In using the Tables, Eqs (24) and (25) provide the essential relationships between a given non-circular flow circumstance and the corresponding equivalent pipe, whether a free surface is involved or not. Normally, in design calculations, the gradient is taken as the same for both circumstances, as is the roughness size and the kinematic viscosity. Thus the completing relationship for the calculations is normally

$$S_{ep} = S \qquad (26)$$

Tables of properties of unit sections (Tables C)

Details of the shape are given at the start of each table. The cross-sections fall into groups as follows.

(i) Conduits running full or part-full. Here, overall height is the natural choice for unit size, leading to the conventional definition of proportional depth of flow. Many of the section shapes most commonly used in the U.K. and in North America for drainage and sewerage conduits are covered. For the U.K., there is included the coverage of shapes by Wallingford Software's WALLRUS suite of programs[20] and by the Water Research Centre's Sewerage Rehabilitation Manual[21]. Also, the effect of relining of egg-shapes is treated. Past practices in North American sewerage have been illustrated by Metcalf and Eddy[22] and summarised into key shapes both by Metcalf and Eddy Inc.[23] and by Babbitt and Baumann[24]. These summarising shapes are included in the Tables, as are a range of standard corrugated metal conduit shapes[25,26].

There are four tables, C33-C36, which treat circular pipes with bottom deposits of 7%, 14%, 21% and 28% of the pipe diameter at the centre line, with the deposit having a curvature with radius 3 times that of the pipe. Then Table C32 effectively provides a further case in the sequence, with a centre line deposit of 39% of pipe diameter. In these cases, the unit size is that of the obstructed pipe, from the deposited surface to soffit. Thus a 2000 mm pipe with 14% deposit has a centre line clearance from depoisited surface to soffit of 1·72 m and the multiplying factor M is 1·72. Then proportional depths as given in the tables apply to depths above the deposited surface.

(ii) Narrow channels and related conduits. For narrow channels, breadth is the natural choice for unit size, and relative depth of flow is the ratio of depth of flow to this chosen breadth. Limitations on space require that conduit shapes such as closed rectangles and, particularly, arch culverts and ovals be grouped with rectangular and U-shaped channels respectively. Thus only one table needs to be allocated to each such case, with the table dealing with a range of relative geometries but always for conduit full flow. An example of this device is the relating of Tables C12 and C13 for U-shaped capped and oval respectively, to Table C11 for U-shaped free surface. Then for part-full arch culverts and ovals and for box-culverts, the effect of the soffit arching, or of the upper splays, is not included for nearly full conditions. But in all cases there is coverage of a wide range of relative proportions for the full condition.

(iii) Wide open channels. To deal with wide, shallower flowing, open channels, it is more expedient to define arbitrarily a unit depth in relation to the specified unit geometry. This applies to canals and rivers in Tables C65 to C96 and accommodates sufficient detail within the standard table arrangement for cases of small relative depths in terms of channel breadth. There are three series of cross-sections. Firstly, there are wider trapezoidal sections termed 'Regime' in Tables C65-C70. Secondly, there is a series with a semi-circular bed between notional slope change points between bed and banks in Tables C74-C86. This concave element of the cross-section always occupies 0·2 of the depth range covered in the table. There is a range of ratios of the depth of the concave section to its width. For example '10% concave bed river' dips one tenth of the distance between the slope change points. Thirdly, there is a series with tangential meetings of side slopes and a circularly curved element through the centre line in Tables C87-C96. Hence '1·0 to 1 tangent river', where the side-slope meeting the curve is the defining factor.

18

(iv) Triangular open channels. For the case of triangular channels, determination of depth of flow is the same as determination of size. Table C97 is modified to show side-slope instead of proportional depth. In consequence, there is in this case a direct solution for side-slope, given discharge and required depth of flow, as well as the alternative solutions for discharge or for depth of flow.

(v) Flow in ducts. Table C98 links the 'unit size' approach for non-circular flow sections with some standard cases of flow in ducts. Applications are restricted to flows that are clearly turbulent. For refinement of estimates in cases (a), (c) and (d) of Table C98, material from ESDU International[27] should be consulted, with respect to adjustment of value of S_{ep}.

If a direct solution for depth of flow in a part-full conduit, or in a channel, is required where shape, size, gradient and roughness is known, then this involves comparison of the given discharge with that occurring at a defined depth in the same conduit or channel. Tables C include discharge ratios for this purpose. The appropriate proportional depth is shown at the start of each table. Such ratios are invariant for any given shape when the Manning equation is adopted, but may vary slightly with change of values of Reynolds number and relative roughness when the Colebrook-White equation is used. Thus, the adopted medial conditions are given for which the calculation of the tabulated ratios has been made. Selection of these medial conditions has been made taking into account typical usage of the tables. By comparing the values of discharge ratios using both the Colebrook-White equation and the Manning equation, it can be seen that, in general, they provide very similar results in such assessments.

Finding discharge in a rectangular open channel
The steps for the 'unit size' method of finding a discharge are as follows.

1) From table sequence C determine, for the given shape, $D_{ep(u)}$ and J.
2) Determine the multiplying factor M between the 'unit' size used in Table C and the real size.
3) Determine the equivalent diameter of circular pipe, D_{ep} using Eq. (25) (i.e. $D_{ep} = MD_{ep(u)}$).
4) Use the Tables D and, as appropriate, Tables E to determine the discharge for pipe with diameter D_{ep} and given gradient and roughness size, i.e. Q_{ep}.
5) Determine the required discharge, i.e. $Q = Q_{ep}/J$.

Estimate the uniform flow discharge at 1·20 m depth in a 2·40 m wide rectangular channel where the gradient is 0·0020 and the roughness size is 1·50 mm. (Numerical solution with the Colebrook-White equation gives $6·649 m^3 s^{-1}$)

1) The relative depth is 0·50 and, from Table C14, $D_{ep(u)}$ and J are 1·00 m and 1·5708 respectively.
2) From the diagram at the head of Table C14 the value of M is 2·40.

3) Then the diameter of the equivalent circular pipe, D_{ep}, is $2{\cdot}40 \times 1{\cdot}00 = 2{\cdot}40\,m$ (2400 mm).

4) From Table D7, mQ for 2400 mm diameter and 0·0020 gradient is $14{\cdot}392\,m^3s^{-1}$, and from Table E9, for k_s of 1·50 mm, m_C interpolates linearly as 1·376 giving $Q_{ep} = 14{\cdot}392/1{\cdot}376$, i.e. $10{\cdot}459\,m^3s^{-1}$.

5) The estimated discharge in the channel is Q_{ep}/J, i.e. $10{\cdot}459/1{\cdot}5708 = 6{\cdot}659\,m^3s^{-1}$ (cf. $6{\cdot}649\,m^3s^{-1}$), or in practical terms, $6{\cdot}6\,m^3s^{-1}$.

This solution route is shown schematically in Fig. 3, together with the solutions for gradient and for the size of conduit or channel (of stipulated proportional depth of flow where there is a free surface).

Solutions for egg-shape sewer

Finding (i) discharge, or (ii) gradient, or (iii) size where proportional depth is stipulated

In the case of the rectangular channel just treated, both the gradient and the diameter of the equivalent pipe corresponded to table values. Typically this is often true for gradient but rarely so when equivalent diameter is the unknown.

To illustrate solution procedures more generally, consider a Form 1 egg-shape (see Table C2). The overall height is 1·800 m, so M is 1·8. At a depth of flow of 0·540 m (i.e. a proportional depth (Y_N) of 0·30) with roughness size of 15 mm (k_s) and at gradient 0·001 (S), the numerical solution for discharge is $0{\cdot}2223\,m^3s^{-1}$ (compared with $1{\cdot}3845\,m^3s^{-1}$ for just full flow).

For 0·30 proportional depth, Table C2 gives the values of $D_{ep(u)}$ and J as 0·5123 m and 1·9042 respectively. From Fig. 3, these values are seen both to be needed for each of the three examples as follows.

(i) To estimate the discharge Q, given gradient, size of conduit, roughness size and proportional depth

Substituting in Eq. (25)

$$D_{ep} = 1{\cdot}8 \times 0{\cdot}5123 = 0{\cdot}9222\,m \; (922{\cdot}2\,mm)$$

$$S_{ep} = 0{\cdot}001 \; (given)$$

From Table D4, mQ for 900 mm diameter and 0·001 gradient is $0{\cdot}7442\,m^3s^{-1}$. Then with exponential proportioning to 922·2 mm diameter (exponent 2·667), the required mQ_{ep} is obtained as $0{\cdot}7942\,m^3s^{-1}$. Linear interpolation in Table E12, for k_s of 15 mm, gives m_C as 1·877, and hence

$$Q_{ep} = [(0{\cdot}7942)\,/\,1{\cdot}877\,]\,m^3s^{-1}$$

$$= 0{\cdot}4231\,m^3s^{-1} = Q\,J \qquad \text{(i.e. from Eq. (24))}$$

Then $\quad Q = 0{\cdot}4231\,/\,1{\cdot}9042 = 0{\cdot}2222\,m^3s^{-1}$

Figure 3 : Solution routes for uniform flow in non-circular cross-sections

For the three explicit problems of finding (i) discharge (Q), (ii) friction gradient (S_f) or (iii) size (factor M), the proportional or the relative depth (Y_N) is known. Hence the values of equivalent diameter for the unit case ($D_{ep(u)}$) and of the equivalent discharge factor (J) can be read from the appropriate Table C. Very often, it is expedient to treat residual shape effect as assimilable in evaluation of roughness size, with the gradient factor $C = 1$, and thus $S_f = S_{ep}$

Then (i) *Find Q* :

$$M \times D_{ep(u)} = D_{ep} \qquad \searrow \qquad \begin{matrix} k_s \\ \downarrow \end{matrix}$$
$$\boxed{\text{C-W}} \;\longrightarrow\; Q_{ep} = Q \times J$$
$$S_f \times C = S_{ep} \qquad \nearrow \qquad\qquad \text{Hence } Q$$

(ii) *Find S_f* :

$$M \times D_{ep(u)} = D_{ep} \qquad \searrow \qquad \begin{matrix} k_s \\ \downarrow \end{matrix}$$
$$\boxed{\text{C-W}} \;\longrightarrow\; S_{ep} = S_f \times C$$
$$Q \times J = Q_{ep} \qquad \nearrow \qquad\qquad \text{Hence } S_f$$

(iii) *Find size (i.e. factor M)* :

$$Q \times J = Q_{ep} \qquad \searrow \qquad \begin{matrix} k_s \\ \downarrow \end{matrix}$$
$$\boxed{\text{C-W}} \;\longrightarrow\; D_{ep} = M \times D_{ep(u)}$$
$$S_f \times C = S_{ep} \qquad \nearrow \qquad\qquad \text{Hence } M$$

For the inherently implicit problem of finding (iv) normal depth (y_N) in a channel of known shape and size, find Q_s by route (i) above, where Q_s corresponds to that proportional, or that relative, depth which is specified for the medial condition on the appropriate Table C, and using the gradient and roughness size stipulated for the problem. Evaluate Q/Q_s where Q is the stipulated discharge.

Then (iv) *Find y_N* : $Q/Q_s \rightarrow$ value of Y_N on the appropriate Table C.

$$\text{Hence } y_N = M \times Y_N$$

Note : The Colebrook-White element of a solution, is indicated by the block containing 'C-W'. This may be aided by Tables, by Charts or be accomplished by solution of an equation. For the Manning equation, the same overall routes apply, with tabular or numerical solution, and with the Manning discharge ratios from Tables C for problem (iv).

(ii) To estimate gradient S, now given that $Q = 0.2223\,m^3s^{-1}$

$$D_{ep} = 1.8 \times 0.5123 = 0.9222\,m\ (922.2\,mm)\ \text{(as for (i))}$$

$$Q_{ep} = 0.2223 \times J = 0.4233\,m^3s^{-1}\ \text{(from Eq. (24))}$$

Table E12, for k_s of 15 mm, shows that the value of m_C is insensitive to gradient about 900 mm diameter, and a medial trial value for this diameter is 1.878. With this value, $mQ_{ep} = 0.7950$. Inspection of Table D4 shows $mQ = 0.7442\,m^3s^{-1}$ for 900 mm diameter and 0.0010 gradient. Thus there is no need for any further adjustment of the value of m_C.

Then substituting these values in Eq. (18)

$$S_{ep} = 0.0010 \left[\frac{0.7950}{0.7442} \right]^{2.00} \left[\frac{900}{922.2} \right]^{5.333}$$

$$= 0.00100 = S$$

(iii) To find size of Form 1 egg-shape, $k_s = 15$ mm, which runs at 0.30 proportional depth with $Q = 0.2223\,m^3s^{-1}$ and $S = 0.001$

$$Q_{ep} = 0.2223 \times J = 0.4233\,m^3s^{-1}\ \text{(as for (ii))}$$

$$S_{ep} = S = 0.001$$

The corresponding value of D_{ep} is to be found by applying the appropriate value of m_C to a value of mQ_{ep} from the 0.001 gradient line of a Table D, with interpolation as required. In the 0.001 gradient line in Table E12 for k_s of 15 mm, the m_C values vary slowly but significantly with diameter. The first value of m_C that appears in Table E12 for a gradient 0.001 is 2.28, corresponding to an equivalent pipe diameter of only 80 mm.

Adopting this value leads to a trial value of $mQ_{ep} = 0.9651\,m^3s^{-1}$. In Table D4, with gradient 0.001, this value of mQ_{ep} points to a diameter rather less than 1000 mm, for which diameter the value of mQ_{ep} is given as $0.9856\,m^3s^{-1}$. In Table E12, use of a diameter value less than 1000 mm gives a revised value for m_C of 1.875. With the value of mQ_{ep} therefore adjusted to $0.4233 \times 1.875 = 0.7937\,m^3s^{-1}$ and substituting in Eq. (19)

$$D_{ep} = 1000 \times [0.7937 / 0.9856]^{1/2.667} = 922.0\,mm\ (0.9220\,m)$$

However

$$D_{ep} = M\,D_{ep(u)} = M \times 0.5132\ \text{(from Eq. (25))}$$

This gives the scale factor M as $0.922/0.5132 = 1.797$ on the basis of 1.000 m unit height egg-shape. Then the required egg-shape is estimated to be 1.797 m in overall height (cf. 1.800 m). Note that the accuracy of the solution is little affected by the poor initial estimate of the diameter and hence of m_C. The net result was merely that the table diameter on which the proportioning for the required D_{ep} was based (i.e. 1000 mm) was slightly further from the actual solution of 922 mm than would have obtained with a more realistic initial estimate of diameter.

Finding depth of flow in a conduit of specified boundary shape and size, with discharge, gradient and roughness size fixed.

Consider again the egg-shape sewer example. We wish to find the depth corresponding to a discharge of $0 \cdot 2223 \, \text{m}^3\text{s}^{-1}$. Using the unit size method for estimation of discharge, as already illustrated, the discharge at $0 \cdot 60$ proportional depth in this particular conduit is estimated as $0 \cdot 8068 \, \text{m}^3\text{s}^{-1}$. Thus the discharge of $0 \cdot 2223 \, \text{m}^3\text{s}^{-1}$ as stipulated gives a $Q/Q_{0 \cdot 60}$ ratio of $0 \cdot 2754$. From Table C2, this corresponds closely with a proportional depth of $0 \cdot 30$ (by interpolation $0 \cdot 298$) and hence to a depth of flow of $1 \cdot 80 \times 0 \cdot 30 = 0 \cdot 540 \, \text{m}$.

The values of the Colebrook-White discharge ratio in Table C2 were calculated for the case of a $1 \cdot 50 \, \text{m}$ high egg-shape at gradient of $0 \cdot 001$ and with roughness size of $3 \, \text{mm}$. For changes in conduit size, gradient or roughness size, the corresponding changes in discharge ratio are small. Fig. 3 on page 21 includes an abstract of the foregoing route for determination of normal depth.

Table C1(b), which applies directly to circular pipes, provides a means of final corrections to estimates of proportional depth when conditions vary from medial as defined by variation of the obtaining value of θ. The pattern of corrections shown serves to emphasise the small changes in proportional flows caused by large changes in conduit size, gradient and roughness size. Indeed, because of its non-dimensional nature, the values given on Table C1(b) can be applied to quite different basic shapes as in Table 5 on page 24. This demonstrates that the methods for estimation of depth of flow give accuracy of solution which is well within normal design requirements.

Solutions for trapezoidal open channel

> *Find the size (i.e. bottom width) of a 45° side slopes (1 to 1) open channel to convey $15 \, \text{m}^3\text{s}^{-1}$ at a surface breadth to depth ratio of 3. The gradient is $0 \cdot 00095$ and the roughness size is $1 \cdot 5$ mm. Then find the depth of flow with $10 \, \text{m}^3\text{s}^{-1}$ in the same channel (Numerical solutions adopting the Colebrook-White equation are $4 \cdot 0365 \, \text{m}$ and $1 \cdot 060 \, \text{m}$ respectively).*

The geometric parameters for the appropriate trapezoidal channel can be found in Table C58. For $0 \cdot 333$ relative depth, linear interpolation gives $D_{ep(u)}$ and J values of $0 \cdot 9143 \, \text{m}$ and $1 \cdot 4750$ respectively.

$$Q_{ep} = Q \, J = 15 \times 1 \cdot 4750 = 22 \cdot 125 \, \text{m}^3\text{s}^{-1}$$

Table E9, for k_s of $1 \cdot 5 \, \text{mm}$, indicates $1 \cdot 5$ as trial value of m_C. This gives a trial value of $mQ_{ep} = 33 \cdot 188 \, \text{m}^3\text{s}^{-1}$. On Table D8, this points to a diameter somewhat under $3800 \, \text{mm}$ (for which mQ is given as $33 \cdot 781 \, \text{m}^3\text{s}^{-1}$) and thus, from Table E9, adjustment of m_C value to $1 \cdot 412$. With mQ_{ep} therefore adjusted to $31 \cdot 241 \, \text{m}^3\text{s}^{-1}$ and substituting as relevant in Eq. (19)

$$D_{ep} = 3800 \times [31 \cdot 241 / 33 \cdot 781]^{1/2 \cdot 667} = 3690 \cdot 2 \, \text{mm}$$

$$\text{i.e.} \quad MD_{ep(u)} = 3 \cdot 6902 \, \text{m}$$

TABLE 5 : Predictions of proportional depth in Form 1 egg-shape (egg-s.) with range of extreme combinations of conditions.

Height of egg-s. (mm)	Grad't S	R'ness size k_s (mm)	First assess't for prop. depth of 0·150	First assess't for prop. depth of 0·850	θ / θ_{med}	Corrected assess't for prop. depth of 0·150	Corrected assess't for prop. depth of 0·850
500	0·00015	15·0	0·143	0·859	0·089	0·149	0·852
500	0·00015	0·006	0·147	0·854	0·346	0·150	0·851
500	0·075	15·0	0·143	0·859	0·116	0·149	0·853
500	0·075	0·006	0·151	0·848	2·721	0·150	0·851
4000	0·00015	15·0	0·149	0·851	0·715	0·150	0·850
4000	0·00015	0·006	0·151	0·849	2·765	0·149	0·850
4000	0·075	15·0	0·149	0·841	0·924	0·149	0·851
4000	0·075	0·006	0·154	0·845	21·77	0·150	0·849

Note :- The first assessments were made using discharge ratio values from Table C2; the corrections were made using Table C1(b).

With the $D_{ep(u)}$ value known, $M = 3 \cdot 6902/0 \cdot 9143 = 4 \cdot 0361$. Thus the bottom width of the channel is estimated as $4 \cdot 0361$ m (cf. $4 \cdot 0365$ m).

Then to determine the depth of flow for a discharge of $10 \, m^3 s^{-1}$ in this same channel, we first calculate the discharge at the prescribed relative depth on Table C58, i.e. $Q_{0 \cdot 50}$. At relative depth $0 \cdot 50$ on this table, $D_{ep(u)}$ and J are read as $1 \cdot 2426$ m and $1 \cdot 6170$ respectively.

$$D_{ep(0 \cdot 50)} = 4 \cdot 0361 \times 1 \cdot 2426 = 5 \cdot 0153 \text{ m}$$

From Table D9, for a diameter of $5 \cdot 000$ m and a gradient of $0 \cdot 0095$, the value of mQ is $70 \cdot 226 \, m^3 s^{-1}$. Table E9, for k_s of $1 \cdot 5$ mm, gives the value of m_C as $1 \cdot 434$, by linear interpolation.

$$mQ_{ep(0 \cdot 50)} = 70 \cdot 226 \times [5 \cdot 0153/5 \cdot 000]^{2 \cdot 667} \qquad \text{(from Eq. 17)}$$

$$= 70 \cdot 801 \, m^3 s^{-1}$$

Thus
$$Q_{ep(0 \cdot 50)} = 70 \cdot 801/1 \cdot 434$$

$$= 49 \cdot 373 = Q_{0 \cdot 50} \, J$$

$$Q_{0 \cdot 50} = 49 \cdot 373/1 \cdot 6170 = 30 \cdot 534 \, m^3 s^{-1}$$

$$Q/Q_{0 \cdot 50} = 10/30 \cdot 534 = 0 \cdot 3275$$

With this discharge ratio, linear interpolation on Table C58 gives the relative depth as $0 \cdot 2631$.

Then normal depth $y_N = 0 \cdot 2631 \times 4 \cdot 0354 \text{ m} = 1 \cdot 0617 \text{ m}$ (cf. $1 \cdot 060$ m).

Manning-Williamson solutions for preceding examples

In the case of the egg-shape sewer example already treated, flow is effectively rough turbulent. For a roughness size k_s of 15 mm ($0 \cdot 015$ m) the Manning-Williamson equation (Eq. (10) *et seq.*) gives m_W as $1 \cdot 888$, and the results are

Find discharge	: $0 \cdot 221 \, m^3 s^{-1}$, cf. $0 \cdot 2222$
Find gradient	: $0 \cdot 00101$, cf. $0 \cdot 00100$
Find size	: $1 \cdot 804$ m, cf. $1 \cdot 800$ m
Find depth of flow	: $0 \cdot 301$ proportional depth, cf. $0 \cdot 300$

The corresponding results for the trapezoidal channel both show slight underestimation of the linear size because the flow conditions are actually in the transition turbulent region, and this is not taken into account where the Manning-Williamson conversion is adopted. A roughness size k_s of $1 \cdot 5$ mm ($0 \cdot 0015$ m) gives $m_W = 1 \cdot 286$ and with this value the bottom breadth is estimated as $3 \cdot 901$ m (cf. $4 \cdot 0365$ m), while the depth of flow is estimated as $1 \cdot 030$ m (cf. $1 \cdot 060$ m).

Other sources of resistance

When designing a drainage pipe-line on the basis of its full-bore capacity, allowances must be made for head losses which will occur at bends, manholes or other appurtenances involving changes in cross-section. Similar allowances should also be made under conditions of free-surface flow. Hydraulic handbooks, manuals and other technical publications give head-loss coefficients, $\varepsilon_1, \varepsilon_2, \ldots$ for use in the expressions $h_1 = \varepsilon_1 V^2/2g$ etc., with $h_1, h_2, \ldots$ contributing cumulatively to the total head difference Σh necessary to drive the system. Here $V^2/2g$ is the kinetic head based simply on the mean velocity. Head losses at straight through open-channel manholes are generally small, the head loss coefficients being of the following approximate magnitudes.

	Part-full	*Full-bore*
Open-channel manhole	< 0·1	0·05 - 0·25
Open-channel manhole with bend	~ 0·3	~ 1·5
Open-channel manhole with pipe bend beyond manhole	~ 0·3	~ 0·3

If a manhole incorporates a junction, losses are much higher and depend on the relative magnitudes of the branches and the geometry of the junction.

There is some loss of head at the entry to a pipe, which will depend upon the sharpness of the arris. Furthermore, the kinetic energy of flow $\alpha(V^2/2g)$, where α is the Coriolis coefficient, is generally not recovered at the exit. These factors may well prove important if the pipe or conduit is relatively short. The head loss with a sharp-edged re-entrant inlet is approximately $V^2/2g$. With a flush headwall the loss coefficient drops to about 0·4, whilst a rounding as little as a sixth or seventh of the pipe diameter will almost eliminate the entrance loss, but not the need to allow for a surface elevation including kinetic head. Much information on losses at features and bends is given by Miller [28] and by Fried and Idelchik [29].

Calculating with additional head losses present

Particularly for pipe-lines, a common practice is to take into account losses over and above those for uniform flow as an equivalent additional length of pipe. From the original definition of the Darcy-Weisbach friction factor λ, the value of its reciprocal, i.e. $1/\lambda$, is the number of diameters to give the length of pipe in which the losses are equivalent to the kinetic head $V^2/2g$. Consequently, if an additional loss is expressed as $\varepsilon (V^2/2g)$, ε/λ is the number of pipe diameters to give the equivalent length for the feature. This is constant in rough turbulent flow but otherwise varies with Reynolds number. For metre-second units, $1/\lambda$ is given by $0·051(V^2/SD)$, or the length in metres per unit ε is $0·051(V^2/S)$. This allows estimation of the required total head for a selected discharge through a pipe system, by adding the length $\Sigma \varepsilon \times [0·051(V^2/S)]$ to the actual pipe length before multiplying by the already calculated gradient for the simple pipe flow, to obtain the required total head difference. However, it follows that information provided simply in terms of equivalent length of pipe may not sufficiently allow for variation in Reynolds number in smooth pipe flows.

Alternatively, we can adopt values of the coefficient

$$\varepsilon_f = SL/(V^2/2g) \tag{28}$$

where L is the pipe length, to allow for the pipe flow resistance element in the following total head equation.

$$\Sigma h = (\varepsilon_1 + \varepsilon_2 \ldots + \varepsilon_f)(V^2/2g) \tag{29}$$

This approach is preferable where discharge must be estimated by iteration, especially in smooth pipe systems.

Solutions for successive trial values of mean velocity corresponding to the assessed pipe friction gradient S_f, i.e. $V = mV/m_C$, can be obtained using Tables D to give values of mV and Tables E to give values of m_C. In the following table, column two is for the foregoing step and column three for application of Eq. (28). Then column four shows the trial summations of total head loss coefficient and column five shows the check values of velocity obtained from Eq. (29) rearranged, using these summations. In the illustrative solution, the starting assumption for the pipe friction gradient S_f is the available head, and then pipe friction gradients are adjusted by the square of the velocity balance quotients. As can be seen as follows, convergence is normally rapid.

A 200 mm pipe, roughness size 0·03 mm, runs for 100 m between two tanks where the surface levels differs by 7·5 m. Allowing for entry, exit, valve and bend effects, the combined additional loss coefficient ε_c is 4·0. Estimate the discharge (Numerical solution using the Colebrook-White equation gives $0·1133 m^3 s^{-1}$ with the pipe resistance gradient S_f at 0·04848).

S_f (trial)	V (by Tables) (i.e. mV/m_C)	$\varepsilon_f = \dfrac{(2gL)S}{V^2}$	$\Sigma \varepsilon$ ($\varepsilon_c + \varepsilon_f$)	$V = \sqrt{\left\{\dfrac{2g \times \Sigma h}{\Sigma \varepsilon}\right\}}$ (check)
0·07500	4·527 (3·717/0·821)	7·178	11·178	3·628

[adjust gradient - 0·07500 × (3·628 / 4·527)² = 0·0481]

S_f (trial)	V (by Tables)	ε_f	$\Sigma \varepsilon$	V (check)
0·04800	3·586 (2·973/0·829)	7·321	11·321	3·605

[adjust gradient - 0·04800 × (3·605 / 3·586)² = 0·0485]

S_f (trial)	V (by Tables)	ε_f	$\Sigma \varepsilon$	V (check)
0·0485	3·605 (2·9885/0·829)	7·312	11·312	3·605

The third assessment requires a linear interpolation for mV in Table D2 but would be unnecessary in practice. From the second assessment the average of the already close velocities is $3·596 ms^{-1}$, which gives a discharge of $0·1130 m^3 s^{-1}$ (cf. $0·1133 m^3 s^{-1}$).

In short culverts and conduits at hydraulically steep slopes, separation may occur at the inlet if it is square-edged. Then the inlet acts as a controlling section and precludes full-bore operation. It cannot be overemphasised that for very short culverts and conduits, friction loss may not be significant. Of primary concern is the relationship of inlet and outlet water levels to surrounding features, with the conduit tending to act as an orifice. For intermediate cases, these Tables may aid assessments of both gradually varied and rapidly varied flow conditions, as is shown in following sections.

Checks on mean velocity, Reynolds number and Froude number

With the tabular method now introduced (i.e. for solution for discharge, for gradient or for size), mean velocity in a section is the same as that for the flow in the equivalent pipe. Thus mean velocities may be read directly from the Colebrook-White solution tables.

Reynolds number is DV/v where the variables D, V and v are in consistent units. The value is then given by $DV/(1 \cdot 141 \times 10^{-6})$ for water at normal temperature (15°C) with kinematic viscosity value $1 \cdot 141 \times 10^{-6} \, \mathrm{m^2 s^{-1}}$, where D is the diameter for pipe-full flows and is the diameter of the equivalent pipe for cases other than pipe full.

Froude number is $V/\sqrt{(gy_{mean})}$ where mean depth y_{mean} is the cross-sectional area of flow divided by the free surface width. This can be evaluated using Tables C.

Viscosities other than that of water at 15°C

In the previous editions of the Wallingford Tables[2] there is provided a system to determine factors to be applied to velocity (and hence discharge) for change in temperature of water over the normal range of variation of temperature in the civil engineering context. A so derived table of velocity corrections is given here as Appendix 3. Alternatively, the equations of Fig. 2 can be used. This means that one can no longer use the tables to determine the solution. However, it does allow that the tables of sequence C be applied to the solution of flow problems involving fluids of any known viscosity, provided that checks are made that the Reynolds number, in terms of D_{ep}, remains above 2000.

Critical depth and critical discharge

The tables of sequence C include values of critical discharge for the unit case, $Q_{c(u)}$, this being the product of unit area and the critical velocity for unit mean depth. Division of discharge in a conduit or channel by $M^{2 \cdot 5}$ gives the corresponding value of unit critical discharge. Hence, the corresponding proportional or relative depth is defined on the table and then multiplication of this by M gives the critical depth.

Alternatively, in a given shape of channel for which the scale factor M is known, division of a selected depth by M gives the proportional or relative depth. Then the corresponding unit critical discharge can be read from the appropriate table and multiplied by $M^{2.5}$ to obtain the critical discharge for the chosen depth.

Use of Tables with data from natural channels

A reach of cobbled bed river has an average observed surface gradient of 0·00034 along its centre line. At a section chosen as representative, under low flow conditions, the centre line depth is 1·5 m and the surface breadth is 220 m. Bank slopes of 1 in 3 phase into a gently concave bed. Gauging upriver gives the discharge as 153 m³s⁻¹. Estimate the capacity of this section of the river at bank full conditions, where the centre line depth at the representative section will be 5·0 m, and also the corresponding Froude number.

The surface breadth over centre line depth quotient is $220/1·5 = 146·7$. In Table C83, for 0·50% concave bed river with 1 to 3 side-slopes, proportional depth 0·28 gives the quotient of unit surface breadth over proportional depth as $40·48/0·28 = 144·6$. Adopting this, as sufficiently close agreement in the light of the data, gives multiplication factor M as $1·5/0·28 = 5·36$.

From Table C83, the unit equivalent diameter at a proportional depth of 0·28 is 0·8445 and the equivalent discharge factor J is 0·0665. Then the equivalent diameter for the river at the observed stage is $0·8445 \times 5·36 = 4·53$ m, and the discharge in the equivalent pipe is $153 \times 0·0665 = 10·17$ m³s⁻¹ (Q).

In Table D9, for an equivalent diameter 4·50 m, a gradient of 0·00034 gives a mQ value of 31·72 m³s⁻¹. Thus $31·72/10·17 = 3·12$ (i.e. mQ/Q) gives the required value of m, in this case of m_C because the roughness size measure k_s is to be adopted to quantify the channel boundary resistance. On examination of Tables E, the m_C value of 3·12 places the flow into rough turbulent conditions in Table E13, in which table one can determine by interpolation that the corresponding value of roughness size k_s is approximately 240 mm.

For the projected bank full condition with 5 m centre line depth, the proportional depth is $0·28 \times (5/1·5)$ or 0·933 and, from Table C83, the unit equivalent diameter is then interpolated as 3·25, and the equivalent discharge factor J as 0·229.

With an M value of 5·36, the diameter of the equivalent pipe is now $5·36 \times 3·25 = 17·4$ m. With a gradient of 0·00034 in Table D11, proportioning for this diameter from the value of mQ for table diameter 17·0 m (i.e. 1098·1 m³s⁻¹), with exponent 2·667, gives mQ as 1168 m³s⁻¹. From Table E13, for $k_s = 240$ mm and diameter 17·4 m, the value of m_C is approximately 3·00. Then the equivalent pipe discharge is $1168/3·00 = 389$ m³s⁻¹, and the estimated bank full discharge in the river is $389/0·229 = 1700$ m³s⁻¹.

Again from Table C83, unit critical discharge at 0·933 proportional depth is approximately 102·5 m³s⁻¹, and multiplication by 5·36²·⁵ gives 6818 m³s⁻¹ as the critical discharge. For a given configuration, the Froude number is predicted by the estimated discharge over critical discharge, so is assessed as 1700/6818 = 0·25.

Gradually varied flow in prismatic channels

For gradually varied flows, the relative values of normal and of critical depths define the circumstances for the occurrence of flow profiles. This is described in appropriate texts. Then profile details can be estimated using the direct step method, adapted for use with Tables C.

This level of application fits well with the use of a brief routine on a programable hand calculator, but even this is not essential. With or without the use of such a routine, profile calculations to be undertaken immediately for any shape of conduit or channel for which a Table C is available, using Eq. 30. The increments of unit proportional or relative depth, as adopted for Tables C, are likely to be suitable as the basis for a series of steps. Then for each successive step of that series, the only new information needed is the new end-of-step value of unit cross-sectional area A_u (subscript 2) and the new end-of-step value of unit wetted perimeter P_u. These are operated on by M^2 and M respectively, as is unit Δy. Here it is assumed that the brief routine includes transfer of the end-of-step values for one step to be start of step values (subscript 1) for the next.

$$\Delta x = \left[\Delta y + \frac{\alpha}{2g} \left(V_2^2 - V_1^2 \right) \right] \Big/ \left(S_o - S_f \right) \qquad (30)$$

Here Δx is the resulting distance along the conduit or channel, positive indicating the direction of flow; Δy is the direct step value, M times Δy_u, or $M \times (Y_2 - Y_1)$; α is the Coriolis coefficient; g is acceleration due to gravity; V is mean velocity, $Q/M^2 A_u$, with subscripts 1 and 2 corresponding to successive end of step positions; S_o is channel gradient, positive when sustaining flow; S_f is the friction gradient, which for the Manning equation is given by

$$S_f = \left[\frac{m_M Q_{ep}}{31 \cdot 17 (D_{ep})^{2 \cdot 667}} \right]^2 \qquad (15b)$$

Alternatively, the friction gradient can be assessed using the second equation of Fig. 2, rearranged as shown in Volume I, where the Colebrook-White equation is preferred.

In comparing the procedure outlined here with existing solutions, the average value of unit area and wetted perimeter at the beginning and end of each step has been used before applying M^2 and M, respectively.

Solution for gradually varied flow in a trapezoidal channel
Consider the application of the system to the following problem[30].

> *A 1:1 side slopes trapezoidal channel of 3 m bed width carries 19 m³s⁻¹ at a bed slope of 0·0015. Trace the water surface profile back from a control giving 4 m depth, to the point where the depth is 1·8 m. Take Manning's coefficient n_M as 0·017 (i.e. $m_M = 1.70$) and the Coriolis coefficient α as 1·1.*

With multiplying factor $M = 3$, unit critical discharge is $19/3^{2.5} = 1\cdot2188\,\text{m}^3\text{s}^{-1}$. On Table C58, this interpolates to give a relative depth of 0·455, giving the estimate of critical depth as $3 \times 0\cdot455 = 1\cdot365\,\text{m}$.

To find normal depth directly, $Q_{0.50}$ is required. From Table C58, $D_{ep(u)}$ and J at 0·50 relative depth, are 1·2426 m and 1·6170 respectively.

$$D_{ep(0\cdot50)} = 3 \times 1\cdot2426 = 3\cdot7278\,\text{m}\ (3727\cdot8\,\text{mm})$$

From Table D8, at $D = 3600\,\text{mm}$; $S = 0\cdot0015$, mQ is $36\cdot748\,\text{m}^3\text{s}^{-1}$.

Using Eq. (17), $36\cdot748 \times [3727\cdot8/3600]^{2\cdot667} = 40\cdot331\,\text{m}^3\text{s}^{-1}$

(i.e. $mQ_{ep(0\cdot50)}$).

Then with $J = 1\cdot6170$, $mQ_{0\cdot50} = 40\cdot331/1\cdot6170 = 24\cdot942\,\text{m}^3\text{s}^{-1}$

and with $m = 1\cdot70$, $Q_{0\cdot50} = 24\cdot942/1\cdot70 = 14\cdot672\,\text{m}^3\text{s}^{-1}$

and hence
$$Q/Q_{0\cdot50} = 19/14\cdot672 = 1\cdot295$$

On Table C58, this value corresponds to a Y value of 0·5750, giving normal depth as $0\cdot5750 \times 3 = 1\cdot725\,\text{m}$. Thus flow is subcritical throughout and the profile is of the M1 form.

The adoption of 19 steps is imposed by the relative depth increments used in Table C58. A calculator aided solution is given in Table 6 on page 32, showing the result as 1797 m upstream. Clearly, it is simple to amplify such a calculator routine to cover this or any other trapezoidal channel shape, without recourse to the values from the appropriate Table C. But this does not hold for many more complex shapes. A basic routine for use with these Tables, once prepared, equally can be applied to any of the shapes treated in Tables C.

Rapidly varied flow

> *There is a discharge of 40 m³s⁻¹ at a depth of 1·12 m in a horizontal 1·6 pipe arch with a rise of 4 m (i.e. M = 4). Establish the conjugate depth.*

The value of $M^{2\cdot5}$ is 32·5, giving Q_u as 1·25 m³s⁻¹ A depth of 1·12 m gives a proportional depth of 0·28, and on Table C43 this shows $Q_{c(u)}$ as 0·4806 m³s⁻¹, and thus flow Q_u to be supercritical.

31

TABLE 6 : Computation of M1 flow profile in trapezoidal channel

Given : 45° (1 to 1) side-slopes with multiplying factor $M = 3$; Discharge $Q = 19$ m³s⁻¹ ; Gradient $S_o = 0.0015$; Manning coefficient $n_M = 0.017$ (i.e. $m_M = 1.70$); Coriolis coefficient $\alpha = 1.1$; Normal depth $y_N = 1.725$ m and Critical depth $y_c = 1.364$ m.

Y (m)	Δy_u (m)	P_u (m)	A_u (m²)	Δy (m)	ΔK (m)	J (calc.)	D_{ep} (m)	Q_{ep} (m³s⁻¹)	S_f	$S_o\text{-}S_f$	Δx (m)	x (m)	y (m)
1·3333	--	4·7618	3·0992	--	--	--	--	--	--	--	--	--	4·00
1·32	– 0·0133	4·7335	3·0624	– 0·04	0·00063	1·7175	7·7869	32·634	$5·57\times10^{-5}$	0·00144	– 27·19	– 27·26	3·96
1·28	– 0·04	4·6204	2·9184	– 0·12	0·00270	1·7178	7·6727	32·641	$6·03\times10^{-5}$	0·00144	– 81·48	– 108·67	3·84
1·24	– 0·04	4·5072	2·7776	– 0·12	0·00347	1·7183	7·4885	32·648	$6·87\times10^{-5}$	0·00143	– 81·71	– 190·38	3·72
1·20	– 0·04	4·3941	2·6400	– 0·12	0·00305	1·7185	7·3036	32·651	$7·85\times10^{-5}$	0·00142	– 81·98	– 272·36	3·60
1·16	– 0·04	4·2810	2·5056	– 0·12	0·00395	1·7184	7·1176	32·650	$9·01\times10^{-5}$	0·00141	– 82·31	– 354·68	3·48
1·12	– 0·04	4·1678	2·3744	– 0·12	0·00452	1·7182	6·9312	32·645	$1·04\times10^{-4}$	0·00140	– 82·71	– 437·39	3·36
1·08	– 0·04	4·0547	2·2464	– 0·12	0·00520	1·7177	6·7436	32·636	$1·20\times10^{-4}$	0·00138	– 83·20	– 520·58	3·24
1·04	– 0·04	3·9416	2·1216	– 0·12	0·00600	1·7169	6·5550	32·621	$1·40\times10^{-4}$	0·00136	– 83·80	– 604·38	3·12
1·00	– 0·04	3·8284	2·0000	– 0·12	0·00696	1·7158	6·3654	32·600	$1·63\times10^{-4}$	0·00134	– 84·55	– 688·93	3·00
0·96	– 0·04	3·7153	1·8816	– 0·12	0·00811	1·7151	6·1761	32·587	$1·91\times10^{-4}$	0·00131	– 85·50	– 774·43	2·88
0·92	– 0·04	3·6022	1·7664	– 0·12	0·00951	1·7131	5·9838	32·549	$2·26\times10^{-4}$	0·00127	– 86·73	– 861·16	2·76
0·88	– 0·04	3·4890	1·6544	– 0·12	0·01121	1·7097	5·7888	32·485	$2·69\times10^{-4}$	0·00123	– 88·34	– 949·50	2·64
0·84	– 0·04	3·3759	1·5456	– 0·12	0·01331	1·7066	5·5937	32·425	$3·21\times10^{-4}$	0·00118	– 90·52	– 1040·02	2·52
0·80	– 0·04	3·2627	1·4400	– 0·12	0·01591	1·7026	5·3968	32·350	$3·87\times10^{-4}$	0·00111	– 93·54	– 1133·56	2·40
0·76	– 0·04	3·1496	1·3376	– 0·12	0·01916	1·6978	5·1980	32·258	$4·70\times10^{-4}$	0·00103	– 97·93	– 1231·49	2·28
0·72	– 0·04	3·0365	1·2384	– 0·12	0·02328	1·6918	4·9970	32·144	$5·76\times10^{-4}$	0·00092	– 104·72	– 1336·22	2·16
0·68	– 0·04	2·9233	1·1424	– 0·12	0·02854	1·6846	4·7937	32·008	$7·13\times10^{-4}$	0·00079	– 116·24	– 1452·46	2·04
0·64	– 0·04	2·8102	1·0496	– 0·12	0·03536	1·6759	4·5878	31·842	$8·92\times10^{-4}$	0·00061	– 139·23	– 1591·40	1·92
0·60	– 0·04	2·6971	0·9600	– 0·12	0·04433	1·6652	4·3788	31·639	0·001130	0·00037	– 204·26	– 1795·28	1·80

Notes : (a) For assessment of change in kinetic head, ΔK, velocities are $Q/M^2 A_u$.

(b) Assessments of J, D_{ep}, Q_{ep} and S_f are for mid-step averaged values of P_u and A_u. The latter values are read directly from Table C54, except for the starting values. Here the linear interpolation values have been assessed from the adjoining table values.

(c) In practice, only the first four and last three columns need be recorded.

Following Daugherty and Franzini[31], $(Q^2/Ag) + Ay_d$ is constant between conjugate depths. But under free surface conditions, $Q = M^{2.5}Q_u$, $A = M^2A_u$ and $y_d = My_{d(u)}$. Thus for conjugate depths

$$M^3 \left[(Q_u^2/A_u g) + A_u y_{d(u)} \right] = \text{a constant} \qquad (31)$$

Then the operation of finding conjugate depth can be done in terms of table values. In this case, the value within the square bracket for the given supercritical condition is 0·5176 and this value is also found at a proportional depth just under 0·75. Thus the upper conjugate depth is about 3·0 m.

The values from Tables C overcome the problems arising from the section geometry. Methods for allowance for channel slope are treated in hydraulics texts, where illustration of solutions is usually confined to rectangular channel circumstances.

Review

Most illustrations are to a greater accuracy than is physically significant. This has been done in order that the user can appreciate the small degree of numerical variability that is intrinsic in the methods provided. The intention is also to enable first estimates to be made as readily as possible. This has led to the adoption of quite small increments of (equivalent) diameter throughout Tables D.

Eq. 26 shows $S_{ep} = S$, this being the normal assumption in the design processes for uniform flows in simple free-surface cross-sections. There is the implicit assumption that the evaluation of the adopted roughness measure has been in the context of its further use. This applies to the values given in Appendices 1 and 2, for cross-sectional shapes other than circular flowing full. All the assessments of n_M given by Chow[14] were for free surface conditions, including those for smooth pipes where part full flow was stipulated. In Appendix 2, values are given also for pipe full flows, but these are drawn from Tables E and are conditioned by approximate gradients. Essentially, these values lead towards Colebrook-White solutions, and this provision is occasioned by the newly available Tables D.

References

1. HYDRAULICS RESEARCH LTD., WALLINGFORD. *Charts for the hydraulic design of channels and pipes*, 6th edition. Thomas Telford, London, 1990. (Note: the earlier editions of the Wallingford Charts were published by H.M.S.O. in 1958, 1963, 1969 and 1978, in the earlier cases under the authorship of P. Ackers, and by Hydraulics Research Ltd. in 1983).

2. HYDRAULICS RESEARCH LTD., WALLINGFORD. *Tables for the hydraulic design of pipes and sewers*, 5th edition. Thomas Telford, London, 1990. (Note: the earlier editions of the Wallingford Tables were published by H.M.S.O. in 1963, 1969 and 1977, in the first two cases under the authorship of P. Ackers and by Hydraulics Research Ltd. in 1983).

2a. BARR, D.I.H. and HR WALLINGFORD *Additional tables for the hydraulic design of pipes and sewers*. Thomas Telford Services Ltd., London, 1993.

3. COLEBROOK, C.F. Turbulent flow in pipes, with particular reference to the transition region between the smooth and the rough pipe laws. *J. Instn. Civ. Engrs*, 1939, Vol. 11, pp 133-156.

4. MANNING, R. On the flow of water in open channels and pipes. *Proc. Instn Civ. Engrs*, Ire., 1891, Vol. 20, p161; 1895, Vol. 24, p 179.

5. NIKURADSE, J. Gesetzmäßigkeit der turbulenten Strömung in glatten Rohren. *Forsch. Arb. Ing.-Wes.* No.356 (1932).

6. NIKURADSE, J. Strömungsgesetze in rauhen Rohren. *Forsch. Arb. Ing.-Wes.* No.361 (1933).

7. ROUSE, H. Evaluation of boundary roughness. *Proc. 2nd Hydraulics Conf.*, University of Iowa, 1943, Bulletin 27.

8. KING, H.W. *Handbook of Hydraulics*, McGraw-Hill, N.York, 1954, Section 6.

9. LAMONT, P.A. A review of pipe friction data and formulae, with a proposed set of exponential formulae based on the theory of roughness. *Proc.Instn Civ.Engrs*, Part 3, 1954, 3, p 248.

10. PERKINS, J.A. and GARDINER, I.M. *The effect of sewage slime on the hydraulic roughness of pipes.* Report IT 218, Hydraulics Research Station, Wallingford , 1982.

11. BARR, D.I.H. Explicit Colebrook-White solutions. *Civil Engineering*, Sept. 1986, pp 19-31.

12. PHAM, Q.T. Explicit equations for the solution of turbulent pipe-flow problems. *Trans. Instn Chem. Engrs*, 1979, Vol. 57, pp 281-283.

13. WILLIAMSON, J. Correspondence on "Turbulent flow in pipes, with particular reference to the transition region between the smooth and rough pipe laws," by C. F. Colebrook. *J. Instn Civ. Engrs,* Vol. 11 (1938-39), pp 419-422 (October, 1939)

13a. WILLIAMSON, J. The laws of flow in rough pipes, Strickler, Manning, Nikuradse and drag-velocity. *La Houille Blanche,* 1951, Vol. 6, No. 5, pp 738-57.

14. CHOW, V-T. *Open channel hydraulics.* McGraw-Hill, New York, 1959.

15. BARNES, H.H., JR Roughness characteristics of natural channels. *Water-Supply Paper 1849.* U.S. Geological Survey, Washington, D.C.

16. BARR, D.I.H. Two additional methods of direct solution of the Colebrook-White function. TN 128, *Proc. Instn Civ. Engrs,* Part 2, Dec. 1975, 3, p 827.

17. MOURET, G. Antoine Chezy, histoire d'une formule d'hydraulique. *Annales des Ponts et Chaussees,* 1921-II. (See *History of Hydraulics* by Hunter Rouse and Simon Ince, Dover N.Y., 1963).

18. JOHNSON, S.P. *A survey of flow calculation methods.* Pre-printed programme for June 19-21, 1934, meeting of the Amer. Soc. Mech. Engrs., University of California, Berkeley.

19. ACKERS, P. *Resistance of fluids flowing in channels and pipes.* Hydraulics Research Paper No. 1, H.M.S.O., London, 1958.

20. HYDRAULICS RESEARCH LTD., WALLINGFORD. *Wallingford Software User Manual for WALLRUS,* 4th Edition, 1991.

21. WATER RESEARCH CENTRE. *Sewerage Rehabilitation Manual.* 2nd Edition, Swindon, U.K., 1986.

22. METCALF, L. and EDDY, H.P. *American Sewerage Practice,* 2nd Edition, Vol. I, Design of sewers. McGraw-Hill, N.Y., 1928.

23. METCALF and EDDY INC. *Wastewater engineering.* McGraw-Hill, N.Y., 1972.

24. BABBITT, H.E. and BAUMANN, E.R. *Sewerage and sewage treatment,* 8th Edition. Wiley, N.Y. 1958.

25. ARMCO INT. CORP. *Handbook of drainage and construction products.* Middleton, Ohio, USA, 1958.

26. ASSET INT. LTD. *Design Manual,* Newport, U.K., 1989.

27. ESDU International, Item No 66027. *Friction losses for fully developed flow in straight pipes.* 1966 and amended 1971.

28. MILLER, D.S. *Internal flow systems* - 2nd Edition. BHRA (Information Services), Bedford, 1990.

29. FRIED, E. and IDELCHIK, I.E. *Flow resistance - A design guide for engineers.* Hemisphere Publishing, New York, 1989.

30. FEATHERSTONE, R.E. and NALLURI, C. *Civil engineering hydraulics - essential theory with worked examples.* 2nd Edn, BSP Prof. Books, Oxford, 1988.

31. DAUGHERTY, R.L. and FRANZINI, J.B. *Fluid mechanics with engineering applications.* 6th Edn, McGraw-Hill, New York, 1965.

32. YEN, B.C. Hydraulic resistance in open channels *Channel flow resistance: Centennial of Manning's formula* B.C. Yen, Editor, Water Resources Publications, Colorado, 1992, pp 1-135

A Area of cross-section of flow.

B Width of channel; horizontal size used in specifying relative depth.

C Gradient factor.

D Diameter. When used in resistance formula there is the implication that flow is pipe-full, or that the equivalent diameter measure is being utilised. The diameter of a full pipe, or that of the equivalent pipe to a specified non-circular section of flow, is used in ratio D/k_s. Values are given in both millimetres and metres, and must be used consistently in respect of the ratio D/k_s.

g Acceleration due to gravity, $9.80665\,\mathrm{ms}^{-2}$ in SI.

h Head loss in uniform flow.

J Discharge conversion factor, $4\pi A/P^2$ where A and P relate to a particular flow circumstance.

l Length over which head loss is assessed.

L Specific pipe length.

k_s Nikuradse equivalent sand roughness size; the linear measure of roughness size. Stated in millimetres, and to be used with consistency of units in evaluating D/k_s and the like.

log   Common logarithm (base 10).

M Multiplying factor on unit measure of conduit or channel size to specify size of given example.

m $100 \times n$ as defined below; adopted to simplify conjointly Tables D, for values of mV and mQ, and Tables E, for values of m_C.

n General case of friction coefficient in the mode of the Manning coefficient n_M.

p A proportion of total wetted perimeter.

P Wetted perimeter of flow section.

Q Discharge (volume per unit time).

R Hydraulic mean depth (hydraulic radius) of flow section, A/P, i.e. $(\pi/4)D^2/\pi D$ or $(1/4)D$.

$\boldsymbol{R}$ Reynolds number, VD/ν or $4Q/\pi\,\nu D$.

S	Hydraulic (piezometric) gradient; head loss per unit length of uniform flow, h/l.

V Mean velocity of flow through cross-section.

Δx Distance along channel in gradually varied flow, corresponding to Δy.

y a depth of flow; vertical measure of size.

Y a non-dimensional depth of flow in terms of either vertical or horizontal measure of conduit or channel size. For tables applying only to conduit full flow, $Y = y_f/B$.

Δy Direct step value in gradually varied flow.

α Coriolis coefficient; $\alpha V^2/2g$ is kinetic energy head.

ε Head loss coefficient, normally applied to $V^2/2g$.

θ Proportional flow parameter for circular pipes.

λ Darcy-Weisbach friction factor, $2(Sg)D/V^2$.

ν Kinematic viscosity, i.e. dynamic viscosity divided by mass density. This is $1 \cdot 141 \times 10^{-6}\ \text{m}^2\text{s}^{-1}$ for water at $15°C$.

Σ Summation.

Subscripts

c   Combined head loss coefficient.

C   Indicating derivation from solution of the Colebrook-White equation in respect of friction coefficient m (or $n = m/100$).

d   Depth to centroid.

ep   Relating to equivalent pipe - i.e. circular pipe flowing full at same gradient and with same hydraulic mean depth (hydraulic radius).

ep(u)   Relating to equivalent pipe for unit case.

f   Relating to vertical measure of conduit size for tables dealing only with conduit full flow. Also, indicating friction gradient in gradually varied flow.

G   Given value between table values in Colebrook-White (or Manning) solutions.

m   Medial: typical case without necessarily implying specific numerical basis.

M	Specifically linked with Manning equation in relation to use of friction coefficient n (or $m = 100n$).
N	Relating to normal depth.
o	Relating to overall height measure in closed conduit, or to depth corresponding to upper limit of tabulated values in certain open channel flows i.e. y_N/y_o is (normal) proportional depth Y_N for flow which has y_N as normal depth. Also, channel gradient in gradually varied flow.
P	Indicating derivation from solution of Poiseuille equation.
R	Required value between table values in Colebrook-White (or Manning) solutions.
T	Table value in Colebrook-White (or Manning) solutions.
u, (u)	Relating to unit section.
c(u)	Critical flow in unit section.
d(u)	Depth to centroid in unit section.
m(u)	Mean depth in unit section.
W	Indicating derivation from Manning-Williamson equation in relation to use of friction coefficient n (or $m = 100n$).
0·50	(for example) - A proportional or a relative depth so defining a flow condition in a given conduit or channel.

APPENDIX 1 : Recommended roughness values

Classification (assumed clean and new unless otherwise stated)	Suitable values of k_s (mm)		
	Good	Normal	Poor
Smooth materials (pipes)			
Drawn non-ferrous pipes of aluminium, brass, copper, lead etc, and non-metallic pipes of Alkathene, glass, perspex etc	--	0·003	--
Asbestos-cement	0·015	0·03	--
Metal			
Spun bitumen or concrete lined	--	0·03	--
Wrought iron	0·03	0·06	0·15
Rusty wrought iron	0·15	0·6	3·0
Uncoated steel	0·015	0·03	0·06
Coated steel	0·03	0·06	0·15
Galvanised iron, coated cast iron	0·06	0·15	0·3
Uncoated cast iron	0·15	0·3	0·6
Tate relined pipes	0·15	0·3	0·6
Old tuberculated water mains as follows:			
Slight degree of attack	0·6	1·5	3·0
Moderate degree of attack	1·5	3·0	6·0
Appreciable degree of attack	6·0	15	30
Severe degree of attack	15	30	60
(Good: up to 20 years use; Normal: 40 to 50 years use; Poor: 80 to 100 years use)			
Wood			
Wood stave pipes, planed plank conduits	0·3	0·6	1·5
Concrete			
Precast concrete pipes with "O" ring joints	0·06	0·15	0·6
Spun precast concrete pipes with "O" ring joints	0·06	0·15	0·3
Monolithic construction against steel forms	0·3	0·6	1·5
Monolithic construction against rough forms	0·6	1·5	--
Clayware			
Glazed or unglazed pipe:			
With sleeve joints	0·03	0·06	0·15
With spigot and socket joints and "O" ring seals			
-- dia < 0·150 m	--	0·03	--
-- dia > 0·150 m	--	0·06	--
Pitch fibre			
(lower value refers to full bore flow)	0·003	0·03	
Glass fibre	--	0·06	--
uPVC			
With chemically cemented joints	--	0·03	--
With spigot and socket joints, "O" ring seals at 6 to 9 metre intervals	--	0·06	--

Classification (assumed clean and new unless otherwise stated)	Suitable values of k_s (mm)		
	Good	Normal	Poor

Brickwork

Glazed	0·6	1·5	3·0
Well pointed	1·5	3·0	6·0
Old, in need of pointing	--	15	30

Slimed sewers [*]

Sewers slimed to about half depth; velocity, when flowing half full, approximately 0·75 ms^{-1}:

Concrete, spun or vertically cast	--	3·0	6·0
Asbestos cement	--	3·0	6·0
Clayware	--	1·5	3·0
uPVC	--	0·6	1·5

Sewers slimed to about half depth; velocity, when flowing half full, approximately 1·2 ms^{-1}:

Concrete, spun or vertically cast	--	1·5	3·0
Asbestos cement	--	0·6	1·5
Clayware	--	0·3	0·6
uPVC	--	0·15	0·3

Sewer rising mains

All materials, operating as follows			
Mean velocity 1 ms^{-1}	0·15	0·3	0·6
Mean velocity 1·5 ms^{-1}	0·06	0·15	0·30
Mean velocity 2 ms^{-1}	0·03	0·06	0·15

Unlined rock tunnels

Granite and other homogeneous rocks	60	150	300
Diagonally bedded slates	--	300	600
(values to be used with *design* diameter)			

Earth channels

Straight uniform artificial channels	15	60	150
Straight natural channels, free from shoals, boulders and weeds	150	300	600

[*] The roughness of a slimed sewer varies considerably during any year. The normal value is that roughness which is exceeded for approximately half of the time. The poor value is that which is exceeded, generally on a continuous basis, for one month of the year. The value of k_s should be interpolated for velocities between 0·75 and 1·2 ms^{-1}.

The lists of values in Appendices 1 and 2 should not be taken as absolving the engineer of the responsibility for checking the actual surface roughness achieved in particular projects by precise, in context, hydraulic tests whenever possible. Where such direct evidence is available from comparable projects it should clearly take precedence over the values quoted here.

APPENDIX 2 : Values of coefficient m_M, for use with Tables D

The coefficient $m_M = 100n_M$, where n_M is the Manning coefficient. The adoption of m, as divisor coefficient in general for Tables D, is for convenience in tabulation and in calculation.

This listing is in two parts. Below there are values of m_M which apply to circular or near circular pipes flowing full. These values are conditioned by size range and gradient range, and derive from Tables E in conjunction with Appendix 1. Essentially, when these values are applied directly to Tables D, the solutions obtained are first approximations to those obtained with full involvement of Tables E by the user, the latter being accurate solutions of the Colebrook-White equation for water at 15°C.

The second part of the listing of m_M values, opposite, is based on Chow's[14] comprehensive listing of values of the Manning coefficient n_M, recently re-listed in modified form by Yen[32]. Throughout it is stipulated that free-surface flow is involved. **Boldface** values are those recommended by Chow[14] for design.

For general guidance, the second part of the listing includes also the directly equivalent value of roughness size k_s, as calculated from the Manning-Williamson[13,13a] equation (Eq. (10)) For m_M greater than 4 (i.e. n_M greater than 0·04), this conversion is omitted as inappropriate.

It is reiterated that the lists of values in Appendices 1 and 2 should not be taken as absolving the engineer of the responsibility for checking the actual surface roughness achieved in particular projects by precise, in context, hydraulic tests whenever possible. Where such direct evidence is available from comparable projects it should clearly take precedence over the values quoted here.

Pipes and near circular conduits running full	Medial values of m_M for gradients:			
	c 0·0001	c 0·0010	c 0·0100	c 0·1000
Smooth pipes - drawn non-ferrous metal, plastic, pitch fibre, and the like: k_s = 0·003 and 0·006 mm: to 400 mm diameter	1·12	0·96	0·85	0·77
Ditto; 400 mm to 1000 mm diameter	1·07	0·96	0·86	0·81
Good asbestos cement; good uncoated steel: k_s = 0·015 mm: to 400 mm diameter	1·12	0·96	0·85	0·78
Ditto; 400 mm to 1000 mm diameter	1·08	0·97	0·88	0·85
Good wrought iron & coated steel; normal asb. cement; normal uncoated steel; spun lined metal pipes; smooth jointed uPVC: k_s = 0·030 mm: to 400 mm diameter	1·12	0·97	0·87	0·80
Ditto; 400 mm to 1000 mm diameter	1·08	0·98	0·91	0·87
Ditto; 1000 mm to 3000 mm diameter	1·10	1·02	0·97	0·95
Good galv. iron & concrete pipes; normal W.I. & coated steel; normal clayware; poor uncoated steel; glass fibre; S. & S. jointed uPVC: k_s = 0·060 mm: to 400 mm diameter	1·12	0·99	0·90	0·85
Ditto; 400 mm to 1000 mm diameter	1·09	1·00	0·95	0·93
Ditto; 1000 mm to 3000 mm diameter	1·12	1·03	1·01	0·99
Good uncoated cast iron; normal G.I. & concrete pipes; poor W.I. & coated steel; etc: k_s = 0·15 mm: to 400 mm diameter	1·14	1·03	0·96	0·93
Ditto; 400 mm to 1000 mm diameter	1·12	1·05	1·02	1·01
Ditto; 1000 mm to 3000 mm diameter	1·15	1·10	1·08	1·08
Normal uncoated C.I.; poor G.I. & concrete pipes; slow flowing rising main sewer; etc: k_s = 0·30 mm: to 400 mm diameter	1·17	1·08	1·03	1·02
Ditto; 400 mm to 1000 mm diameter	1·16	1·11	1·09	1·08
Ditto; 1000 mm to 3000 mm diameter	1·20	1·17	1·15	1·15
Good brickwork; normal rusty W.I.; poor slow flowing rising main sewer; poor uncoated C.I.; very poor concrete pipes; etc: k_s = 0·60 mm: to 400 mm diameter	1·23	1·15	1·12	1·11
Ditto; 400 mm to 1000 mm diameter	1·22	1·19	1·18	1·17
Ditto; 1000 mm to 3000 mm diameter	1·26	1·24	1·23	1·23

APPENDIX 2 : Values of coefficient m_M, for use with Tables D (continued)

	Minimum		Normal		Maximum	
	m_M	$(k_s(mm))$	m_M	$(k_s(mm))$	m_M	$(k_s(mm))$
Closed conduits, partially full						
Lucite® or Perspex®	0·8	(0·09)	0·9	(0·18)	1·0	(0·33)
Smooth pipe, brass or glass	0·9	(0·18)	**1·0**	(0·33)	1·3	(1·60)
Riveted or spiral steel	1·3	(1·60)	1·6	(5·55)	1·7	(8·00)
Coated cast iron	1·0	(0·33)	1·3	(1·60)	1·4	(2·50)
Uncoated cast iron	1·1	(0·59)	1·4	(2·50)	1·6	(5·55)
Black wrought iron	1·2	(1·00)	1·4	(2·50)	1·5	(3·77)
Galvanised wrought iron	1·3	(1·60)	1·6	(5·55)	1·7	(8·00)
Corrugated metal storm drain	2·1	(28·4)	**2·4**	(63·2)	3·0	(240)
Cement mortar	1·1	(0·59)	1·3	(1·60)	1·5	(3·77)
Straight clean conc. culvert	1·0	(0·33)	1·1	(0·59)	1·3	(1·60)
Conc. culv. with bends etc. and debris	1·1	(0·59)	**1·3**	(1·60)	1·4	(2·49)
Finished concrete	1·1	(0·59)	1·2	(1·00)	1·4	(2·49)
Conc., smooth wood formwork, unfinished	1·2	(1·00)	**1·4**	(2·50)	1·6	(5·55)
Conc., poor wood formwork, unfinished	1·5	(3·77)	1·7	(8·00)	2·0	(21·2)
Clay drainage tile	1·1	(0·59)	**1·3**	(1·60)	1·7	(8·00)
Vitrified clay sewer with features	1·3	(1·60)	1·5	(3·77)	1·7	(8·00)
Vitrified subdrain with open joints	1·4	(2·50)	**1·6**	(5·55)	**1·8**	(11·3)
Used sanitary sewers with features	1·2	(1·00)	1·3	(1·60)	1·6	(5·55)
Rubble masonry, cemented	1·8	(11·3)	2·5	(80·8)	3·0	(240)
Lined or built-up channels						
Corrugated metal lining	2·1	(28·4)	2·5	(80·8)	3·0	(240)
Trowel finish concrete	1·1	(0·59)	**1·3**	(1·60)	1·5	(3·77)
Unfinished concrete	1·4	(2·50)	1·7	(8·00)	2·0	(21·2)
Gunite, poor	1·8	(11·3)	2·2	(37·5)	2·5	(80·8)
Concrete bottom, dressed stone sides	1·5	(3·77)	1·7	(8·00)	2·0	(21·2)
Concrete bottom, dry rubble sides	2·0	(21·2)	3·0	(240)	3·5	(610)
Gravel bottom, formed concrete sides	1·7	(8·00)	2·0	(21·2)	2·5	(80·8)
Gravel bottom, rubble or riprap sides	2·3	(50·0)	3·3	(427)	3·6	(720)
Glazed brick	1·1	(0·59)	**1·3**	(1·60)	1·5	(3·77)
Brick in cement mortar	1·2	(1·00)	**1·5**	(3·77)	1·8	(11·3)
Cemented rubble masonry	1·7	(8·00)	2·5	(80·8)	3·0	(240)
Dry rubble masonry	2·3	(50·0)	3·2	(355)	3·5	(610)
Dressed ashlar	1·3	(1·60)	1·5	(3·77)	1·7	(8·00)
Asphalt (good)	1·3	(1·60)	1·3	(1·60)	-	-
Asphalt (rough)	1·6	(5·55)	1·6	(5·55)	-	-
Excavated or dredged channels						
Straight and uniform channel, new earth	1·6	(5·55)	1·8	(11·3)	2·0	(21·2)
Ditto after weathering	1·8	(11·3)	**2·2**	(37·5)	2·5	(80·8)
Ditto gravel, well maintained	2·2	(37·5)	2·5	(80·8)	3·0	(240)
Ditto earth, some grass and weeds	2·2	(37·5)	2·7	(128)	3·3	(427)
Winding and sluggish, no vegetation	2·3	(50·0)	2·5	(80·8)	3·0	(240)
Ditto, dense weeds	3·0	(240)	3·5	(608)	4·0	(1360)
Dredged, with light brush on banks	3·5	(610)	5·0	-	6·0	-
Non-maintained, vegetated to flow depth	5·0	-	8·0	-	12	-
Smooth and uniform rock cut	2·5	(80·8)	3·5	(610)	4·0	(1360)
Jagged and irregular rock cut	3·5	(610)	4·0	(1360)	5·0	-
Small natural streams; less than 30m wide at flood stage						
Clean, straight, full stage without features	2·5	(80·8)	**3·0**	(240)	3·3	(427)
Clean, winding, with some features	3·3	(427)	4·0	(1360)	4·5	-
Very poor conditions with much vegetation	7·5	-	10	-	15	-
Clean mountain stream, mainly gravel bed	3·0	(240)	4·0	(1360)	5·0	-
Ditto, cobble and boulder bottom	4·0	(1360)	5·0	-	7·0	-

APPENDIX 3 : Velocity correction for variation in temperature

$$V = V_{15} + at_{15} \quad \text{(in ms}^{-1}) \quad \text{where } t_{15} = t - 15 \quad \text{(Celsius)}$$

D S (mm)	Roughness values, k_s in mm							
	0·006	0·015	0·03	0·06	0·15	0·30	0·60	1·5
2000	0·0327	0·0151	0·0080	0·0041	0·0017	0·0008	0·0004	0·0002
800	0·0290	0·0142	0·0077	0·0040	0·0017	0·0008	0·0004	0·0002
400	0·0257	0·0134	0·0075	0·0040	0·0016	0·0008	0·0004	0·0002
200	0·0222	0·0124	0·0071	0·0039	0·0016	0·0008	0·0004	0·0002
80	0·0174	0·0107	0·0065	0·0037	0·0016	0·0008	0·0004	0·0002
40	0·0140	0·0093	0·0060	0·0035	0·0016	0·0008	0·0004	0·0002
20	0·0110	0·0079	0·0054	0·0033	0·0015	0·0008	0·0004	0·0002
8	0·0077	0·0060	0·0044	0·0029	0·0014	0·0008	0·0004	0·0002
4	0·0057	0·0048	0·0037	0·0026	0·0013	0·0008	0·0004	0·0002
2	0·0042	0·0037	0·0030	0·0022	0·0012	0·0007	0·0004	0·0002
0·8	0·0028	0·0025	0·0022	0·0017	0·0011	0·0007	0·0004	0·0002
0·4	0·0020	0·0019	0·0017	0·0014	0·0009	0·0006	0·0004	0·0002
0·2	0·0014	0·0014	0·0013	0·0011	0·0008	0·0005	0·0003	0·0002
0·08	0·0009	0·0009	0·0008	0·0008	0·0006	0·0004	0·0003	0·0002
0·04	0·0007	0·0006	0·0006	0·0006	0·0005	0·0004	0·0003	0·0002

The values of 'a' tabulated above show the increase in mean velocity in ms^{-1} per 1°C increase in temperature. (This table derives from Table 34 of the 5th edition of the Wallingford Tables[2].)

APPENDIX 4: Multiplying factors for discharges in pipes and lined tunnels

D_G (mm)	D_T (mm)	x	Fact.	Notes	D_G (mm)	D_T (mm)	x	Fact.	Notes
12.7	20	2.72	0.29	0.5in	1070	1050	2.61	1.05	
15.875	20	2.72	0.53	0.625in	1117.6	1125	2.61	0.98	44in
19.05	20	2.72	0.88	0.75in	1143	1150	2.61	0.98	45in
22.225	20	2.71	1.33	0.875in	1207	1200	2.61	1.01	
25.4	25	2.71	1.04	1.0in	1219.2	1200	2.61	1.04	48in, *
28.575	30	2.70	0.88	1.125in	1264	1275	2.61	0.98	
31.75	30	2.70	1.17	1.25in	1295.4	1300	2.61	0.99	51in
38.10	40	2.70	0.88	1.5in	1320.8	1300	2.61	1.04	52in
44.45	40	2.70	1.33	1.75in	1333	1350	2.61	0.97	52.5in
50.8	50	2.69	1.04	2.0in	1371.6	1350	2.61	1.04	54in
63.5	65	2.69	0.94	2.5in	1380	1400	2.61	0.96	*
67.0	65	2.69	1.08	2.64in	1422.4	1425	2.61	0.99	56in
76.2	75	2.68	1.04	3.0in	1447.8	1450	2.61	0.99	57in
88.9	90	2.68	0.97	3.5in	1511	1500	2.61	1.02	59.5in
101.6	100	2.67	1.04	4.0in	1520	1500	2.61	1.03	
108.0	110	2.67	0.95	4.25in	1524	1500	2.61	1.04	60in
127.0	125	2.66	1.04	5.0in	1530	1550	2.61	0.97	*
133.4	135	2.66	0.97	5.25in	1625.6	1600	2.60	1.04	64in
152.4	150	2.65	1.04	6.0in	1638	1650	2.60	0.98	64.5in
159.0	150	2.65	1.17	6.26in	1660	1650	2.60	1.02	*
160	150	2.65	1.19		1676.4	1650	2.60	1.04	66in
177.8	175	2.65	1.04	7.0in	1680	1700	2.60	0.97	
203.2	200	2.64	1.04	8.0in	1727.2	1725	2.60	1.00	68in
219.2	225	2.64	0.93	8.63in	1753	1750	2.60	1.00	69in
228.6	225	2.64	1.04	9.0in	1803	1800	2.60	1.00	71in
254.0	250	2.63	1.04	10.0in	1828.8	1800	2.60	1.04	72in, *
267	275	2.63	0.93	10.51in	1867	1875	2.60	0.99	73.5in
279.4	275	2.63	1.04	11.0in	1880	1875	2.60	1.01	74in
304.8	300	2.63	1.04	12.0in	1905	1900	2.60	1.01	75in
323.9	330	2.63	0.95	12.75in	1910	1900	2.60	1.01	*
355.6	350	2.63	1.04	14.0in	1930	1950	2.60	0.97	76in
368	375	2.63	0.95	14.49in	1943	1950	2.60	0.99	76.5in
381.0	375	2.63	1.04	15.0in	1981.2	2000	2.60	0.98	78in
406.4	400	2.63	1.04	16.0in	1990	2000	2.60	0.99	*
419	425	2.63	0.96	16.5in	2133.6	2150	2.60	0.98	84in
440	450	2.63	0.94		2140	2150	2.60	0.99	*
457.2	450	2.63	1.04	18.0in	2280	2300	2.60	0.98	*
508	500	2.63	1.04	20.0in	2286.0	2300	2.60	0.98	90in
515	525	2.63	0.95		2438.4	2450	2.60	0.99	96in, *
533.4	525	2.62	1.04	21.0in	2590.8	2600	2.60	0.99	102in
558.8	550	2.62	1.04	22in	2743.2	2750	2.60	0.99	108in
560	550	2.62	1.05		2895.6	2900	2.60	1.00	114in
609.6	600	2.62	1.04	24in	3048.0	3000	2.60	1.04	120in, *
635.0	630	2.62	1.02	25in	3200.4	3200	2.60	1.00	126in
660.4	675	2.62	0.94	26in	3352.8	3400	2.60	0.96	132in, *
685.8	675	2.62	1.04	27in	3505.2	3500	2.60	1.00	138in
711.2	700	2.62	1.04	28in	3657.6	3600	2.60	1.04	144in
730	750	2.62	0.93		3670	3600	2.60	1.05	*
762	750	2.62	1.04	30in	3810	3800	2.60	1.01	*
812.8	800	2.61	1.04	32in	3962.4	4000	2.60	0.97	156in
838.2	825	2.61	1.04	33in	4120	4000	2.60	1.08	*
914	900	2.61	1.04	36in	4267.2	4250	2.60	1.01	168in, *
990.6	1000	2.61	0.97	39in	4572.0	4500	2.60	1.04	180in
1016	1000	2.61	1.04	40in					
1066.8	1050	2.61	1.04	42in					

x is exponent used in **Fact.** $= (D_G/D_T)^x$

* indicates tunnel size; one-pass or grano. lined

To estimate pipe full discharge for diameter D_G, obtain discharge for diameter D_T, for same gradient and roughness size, from the appropriate tables and multiply by the factor given above.

C1

Circular pipe

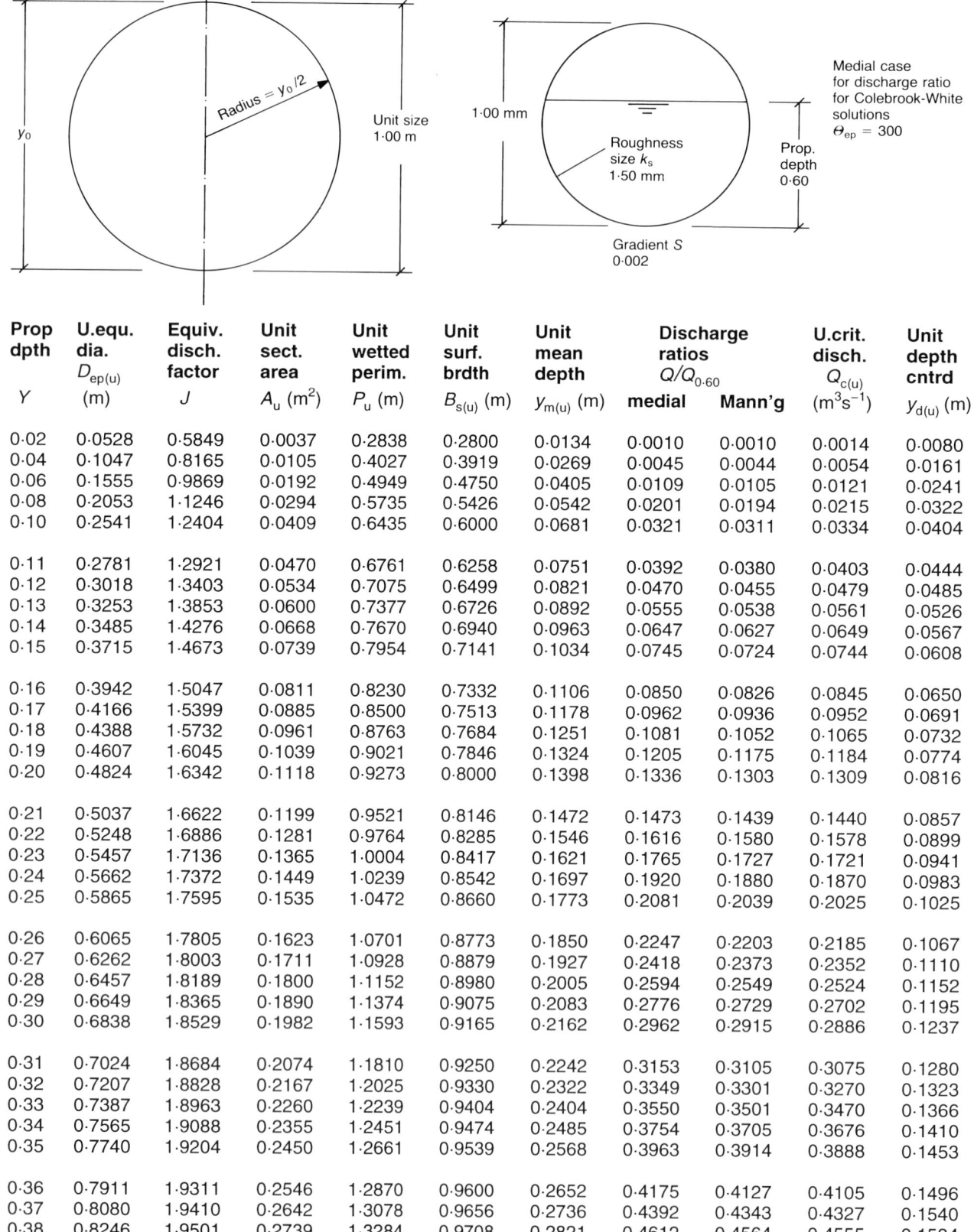

Radius = $y_0/2$

Unit size 1·00 m

y_0

1·00 mm

Roughness size k_s 1·50 mm

Prop. depth 0·60

Medial case for discharge ratio for Colebrook-White solutions $\Theta_{ep} = 300$

Gradient S 0·002

Prop dpth Y	U.equ. dia. $D_{ep(u)}$ (m)	Equiv. disch. factor J	Unit sect. area A_u (m²)	Unit wetted perim. P_u (m)	Unit surf. brdth $B_{s(u)}$ (m)	Unit mean depth $y_{m(u)}$ (m)	Discharge ratios $Q/Q_{0.60}$ medial	Mann'g	U.crit. disch. $Q_{c(u)}$ (m³s⁻¹)	Unit depth cntrd $y_{d(u)}$ (m)
0·02	0·0528	0·5849	0·0037	0·2838	0·2800	0·0134	0·0010	0·0010	0·0014	0·0080
0·04	0·1047	0·8165	0·0105	0·4027	0·3919	0·0269	0·0045	0·0044	0·0054	0·0161
0·06	0·1555	0·9869	0·0192	0·4949	0·4750	0·0405	0·0109	0·0105	0·0121	0·0241
0·08	0·2053	1·1246	0·0294	0·5735	0·5426	0·0542	0·0201	0·0194	0·0215	0·0322
0·10	0·2541	1·2404	0·0409	0·6435	0·6000	0·0681	0·0321	0·0311	0·0334	0·0404
0·11	0·2781	1·2921	0·0470	0·6761	0·6258	0·0751	0·0392	0·0380	0·0403	0·0444
0·12	0·3018	1·3403	0·0534	0·7075	0·6499	0·0821	0·0470	0·0455	0·0479	0·0485
0·13	0·3253	1·3853	0·0600	0·7377	0·6726	0·0892	0·0555	0·0538	0·0561	0·0526
0·14	0·3485	1·4276	0·0668	0·7670	0·6940	0·0963	0·0647	0·0627	0·0649	0·0567
0·15	0·3715	1·4673	0·0739	0·7954	0·7141	0·1034	0·0745	0·0724	0·0744	0·0608
0·16	0·3942	1·5047	0·0811	0·8230	0·7332	0·1106	0·0850	0·0826	0·0845	0·0650
0·17	0·4166	1·5399	0·0885	0·8500	0·7513	0·1178	0·0962	0·0936	0·0952	0·0691
0·18	0·4388	1·5732	0·0961	0·8763	0·7684	0·1251	0·1081	0·1052	0·1065	0·0732
0·19	0·4607	1·6045	0·1039	0·9021	0·7846	0·1324	0·1205	0·1175	0·1184	0·0774
0·20	0·4824	1·6342	0·1118	0·9273	0·8000	0·1398	0·1336	0·1303	0·1309	0·0816
0·21	0·5037	1·6622	0·1199	0·9521	0·8146	0·1472	0·1473	0·1439	0·1440	0·0857
0·22	0·5248	1·6886	0·1281	0·9764	0·8285	0·1546	0·1616	0·1580	0·1578	0·0899
0·23	0·5457	1·7136	0·1365	1·0004	0·8417	0·1621	0·1765	0·1727	0·1721	0·0941
0·24	0·5662	1·7372	0·1449	1·0239	0·8542	0·1697	0·1920	0·1880	0·1870	0·0983
0·25	0·5865	1·7595	0·1535	1·0472	0·8660	0·1773	0·2081	0·2039	0·2025	0·1025
0·26	0·6065	1·7805	0·1623	1·0701	0·8773	0·1850	0·2247	0·2203	0·2185	0·1067
0·27	0·6262	1·8003	0·1711	1·0928	0·8879	0·1927	0·2418	0·2373	0·2352	0·1110
0·28	0·6457	1·8189	0·1800	1·1152	0·8980	0·2005	0·2594	0·2549	0·2524	0·1152
0·29	0·6649	1·8365	0·1890	1·1374	0·9075	0·2083	0·2776	0·2729	0·2702	0·1195
0·30	0·6838	1·8529	0·1982	1·1593	0·9165	0·2162	0·2962	0·2915	0·2886	0·1237
0·31	0·7024	1·8684	0·2074	1·1810	0·9250	0·2242	0·3153	0·3105	0·3075	0·1280
0·32	0·7207	1·8828	0·2167	1·2025	0·9330	0·2322	0·3349	0·3301	0·3270	0·1323
0·33	0·7387	1·8963	0·2260	1·2239	0·9404	0·2404	0·3550	0·3501	0·3470	0·1366
0·34	0·7565	1·9088	0·2355	1·2451	0·9474	0·2485	0·3754	0·3705	0·3676	0·1410
0·35	0·7740	1·9204	0·2450	1·2661	0·9539	0·2568	0·3963	0·3914	0·3888	0·1453
0·36	0·7911	1·9311	0·2546	1·2870	0·9600	0·2652	0·4175	0·4127	0·4105	0·1496
0·37	0·8080	1·9410	0·2642	1·3078	0·9656	0·2736	0·4392	0·4343	0·4327	0·1540
0·38	0·8246	1·9501	0·2739	1·3284	0·9708	0·2821	0·4612	0·4564	0·4555	0·1584
0·39	0·8409	1·9583	0·2836	1·3490	0·9755	0·2907	0·4835	0·4788	0·4788	0·1628
0·40	0·8569	1·9658	0·2934	1·3694	0·9798	0·2994	0·5062	0·5016	0·5027	0·1672
0·41	0·8726	1·9724	0·3032	1·3898	0·9837	0·3082	0·5292	0·5247	0·5271	0·1716
0·42	0·8880	1·9783	0·3130	1·4101	0·9871	0·3171	0·5525	0·5481	0·5521	0·1760
0·43	0·9031	1·9835	0·3229	1·4303	0·9902	0·3261	0·5761	0·5718	0·5775	0·1805
0·44	0·9179	1·9879	0·3328	1·4505	0·9928	0·3353	0·5999	0·5958	0·6035	0·1850
0·45	0·9323	1·9916	0·3428	1·4706	0·9950	0·3445	0·6239	0·6200	0·6301	0·1895

Circular pipe

Prop dpth Y	U.equ. dia. $D_{ep(u)}$ (m)	Equiv. disch. factor J	Unit sect. area A_u (m²)	Unit wetted perim. P_u (m)	Unit surf. brdth $B_{s(u)}$ (m)	Unit mean depth $y_{m(u)}$ (m)	Discharge ratios $Q/Q_{0.60}$ medial	Mann'g	U.crit. disch. $Q_{c(u)}$ (m³s⁻¹)	Unit depth cntrd $y_{d(u)}$ (m)
0·46	0·9465	1·9947	0·3527	1·4907	0·9968	0·3539	0·6482	0·6444	0·6571	0·1940
0·47	0·9604	1·9970	0·3627	1·5108	0·9982	0·3634	0·6727	0·6691	0·6847	0·1985
0·48	0·9739	1·9986	0·3727	1·5308	0·9992	0·3730	0·6973	0·6940	0·7128	0·2031
0·49	0·9871	1·9996	0·3827	1·5508	0·9998	0·3828	0·7222	0·7190	0·7415	0·2076
0·50	1·0000	2·0000	0·3927	1·5708	1·0000	0·3927	0·7471	0·7442	0·7706	0·2122
0·51	1·0126	1·9996	0·4027	1·5908	0·9998	0·4028	0·7722	0·7696	0·8003	0·2168
0·52	1·0248	1·9987	0·4127	1·6108	0·9992	0·4130	0·7974	0·7950	0·8306	0·2214
0·53	1·0367	1·9971	0·4227	1·6308	0·9982	0·4234	0·8227	0·8205	0·8613	0·2261
0·54	1·0483	1·9949	0·4327	1·6509	0·9968	0·4340	0·8480	0·8461	0·8926	0·2308
0·55	1·0595	1·9920	0·4426	1·6710	0·9950	0·4448	0·8734	0·8718	0·9245	0·2355
0·56	1·0704	1·9886	0·4526	1·6911	0·9928	0·4558	0·8987	0·8975	0·9568	0·2402
0·57	1·0810	1·9845	0·4625	1·7113	0·9902	0·4671	0·9241	0·9232	0·9898	0·2449
0·58	1·0912	1·9798	0·4724	1·7315	0·9871	0·4785	0·9495	0·9488	1·0232	0·2497
0·59	1·1011	1·9746	0·4822	1·7518	0·9837	0·4902	0·9748	0·9744	1·0573	0·2545
0·60	1·1106	1·9687	0·4920	1·7722	0·9798	0·5022	1·0000	1·0000	1·0919	0·2593
0·61	1·1197	1·9623	0·5018	1·7926	0·9755	0·5144	1·0252	1·0255	1·1271	0·2642
0·62	1·1285	1·9553	0·5115	1·8132	0·9708	0·5269	1·0502	1·0508	1·1628	0·2690
0·63	1·1369	1·9476	0·5212	1·8338	0·9656	0·5398	1·0751	1·0760	1·1992	0·2739
0·64	1·1449	1·9394	0·5308	1·8546	0·9600	0·5530	1·0998	1·1010	1·2362	0·2789
0·65	1·1526	1·9306	0·5404	1·8755	0·9539	0·5665	1·1243	1·1259	1·2738	0·2839
0·66	1·1599	1·9213	0·5499	1·8965	0·9474	0·5804	1·1486	1·1505	1·3120	0·2889
0·67	1·1667	1·9113	0·5594	1·9177	0·9404	0·5948	1·1727	1·1749	1·3510	0·2939
0·68	1·1732	1·9007	0·5687	1·9391	0·9330	0·6096	1·1965	1·1990	1·3906	0·2990
0·69	1·1793	1·8896	0·5780	1·9606	0·9250	0·6249	1·2200	1·2227	1·4309	0·3041
0·70	1·1849	1·8779	0·5872	1·9823	0·9165	0·6407	1·2431	1·2462	1·4720	0·3093
0·71	1·1902	1·8655	0·5964	2·0042	0·9075	0·6571	1·2659	1·2693	1·5139	0·3144
0·72	1·1950	1·8526	0·6054	2·0264	0·8980	0·6741	1·2884	1·2920	1·5566	0·3197
0·73	1·1994	1·8390	0·6143	2·0488	0·8879	0·6919	1·3104	1·3142	1·6001	0·3250
0·74	1·2033	1·8249	0·6231	2·0715	0·8773	0·7103	1·3319	1·3360	1·6446	0·3303
0·75	1·2067	1·8101	0·6319	2·0944	0·8660	0·7296	1·3530	1·3573	1·6901	0·3357
0·76	1·2097	1·7947	0·6405	2·1176	0·8542	0·7498	1·3735	1·3780	1·7367	0·3411
0·77	1·2123	1·7786	0·6489	2·1412	0·8417	0·7710	1·3935	1·3982	1·7844	0·3466
0·78	1·2143	1·7618	0·6573	2·1652	0·8285	0·7933	1·4130	1·4178	1·8334	0·3521
0·79	1·2158	1·7444	0·6655	2·1895	0·8146	0·8169	1·4317	1·4367	1·8837	0·3577
0·80	1·2168	1·7263	0·6736	2·2143	0·8000	0·8420	1·4498	1·4549	1·9355	0·3633
0·81	1·2172	1·7074	0·6815	2·2395	0·7846	0·8686	1·4672	1·4724	1·9890	0·3691
0·82	1·2171	1·6879	0·6893	2·2653	0·7684	0·8970	1·4838	1·4891	2·0443	0·3748
0·83	1·2164	1·6675	0·6969	2·2916	0·7513	0·9276	1·4996	1·5049	2·1018	0·3807
0·84	1·2150	1·6463	0·7043	2·3186	0·7332	0·9605	1·5146	1·5198	2·1616	0·3866
0·85	1·2131	1·6243	0·7115	2·3462	0·7141	0·9963	1·5286	1·5338	2·2241	0·3927
0·86	1·2104	1·6013	0·7186	2·3746	0·6940	1·0354	1·5416	1·5467	2·2897	0·3988
0·87	1·2071	1·5775	0·7254	2·4039	0·6726	1·0785	1·5535	1·5585	2·3591	0·4050
0·88	1·2029	1·5525	0·7320	2·4341	0·6499	1·1263	1·5643	1·5691	2·4328	0·4113
0·89	1·1980	1·5265	0·7384	2·4655	0·6258	1·1800	1·5739	1·5784	2·5118	0·4177
0·90	1·1921	1·4992	0·7445	2·4981	0·6000	1·2409	1·5821	1·5864	2·5972	0·4242
0·91	1·1853	1·4706	0·7504	2·5322	0·5724	1·3110	1·5889	1·5928	2·6906	0·4308
0·92	1·1775	1·4404	0·7560	2·5681	0·5426	1·3933	1·5940	1·5975	2·7943	0·4376
0·93	1·1684	1·4085	0·7612	2·6061	0·5103	1·4917	1·5973	1·6004	2·9115	0·4445
0·94	1·1579	1·3744	0·7662	2·6467	0·4750	1·6131	1·5986	1·6011	3·0472	0·4517
0·95	1·1458	1·3378	0·7707	2·6906	0·4359	1·7681	1·5975	1·5994	3·2093	0·4590
0·96	1·1316	1·2980	0·7749	2·7389	0·3919	1·9771	1·5936	1·5947	3·4119	0·4665
0·97	1·1148	1·2537	0·7785	2·7934	0·3412	2·2819	1·5861	1·5863	3·6829	0·4743
0·98	1·0941	1·2027	0·7816	2·8578	0·2800	2·7916	1·5738	1·5728	4·0898	0·4823
0·99	1·0663	1·1389	0·7841	2·9413	0·1990	3·9401	1·5533	1·5509	4·8738	0·4908
1·00	1·0000	1·0000	0·7854	3·1416	0·0000	-	1·4942	1·4884	-	0·5000

C1(a)   Proportional discharges in part-full circular pipes

Prop. depth   Coefficient θ for full pipe = $[\{ k_s/D \} + \{ 1/(3600DS^{1/3}) \}]^{-1}$
(Water at $15°C$ - k_s and D in m)

Prop. depth	5	10	20	50	100	200	500	1000	2000	5000	10000	20000
0·02	0·0000	0·0002	0·0004	0·0005	0·0006	0·0007	0·0007	0·0008	0·0008	0·0008	0·0008	0·0008
0·04	0·0011	0·0017	0·0021	0·0025	0·0028	0·0030	0·0032	0·0033	0·0034	0·0035	0·0036	0·0036
0·06	0·0036	0·0048	0·0056	0·0064	0·0068	0·0071	0·0075	0·0077	0·0079	0·0081	0·0082	0·0083
0·08	0·0080	0·0097	0·0109	0·0121	0·0127	0·0132	0·0137	0·0141	0·0144	0·0147	0·0149	0·0150
0·10	0·0142	0·0165	0·0182	0·0196	0·0205	0·0212	0·0219	0·0224	0·0228	0·0232	0·0234	0·0236
0·12	0·0224	0·0253	0·0273	0·0291	0·0302	0·0311	0·0320	0·0325	0·0330	0·0335	0·0338	0·0340
0·14	0·0325	0·0360	0·0383	0·0405	0·0418	0·0428	0·0439	0·0446	0·0451	0·0457	0·0461	0·0463
0·16	0·0446	0·0486	0·0513	0·0538	0·0552	0·0564	0·0576	0·0584	0·0590	0·0597	0·0601	0·0604
0·18	0·0587	0·0631	0·0661	0·0689	0·0704	0·0717	0·0731	0·0739	0·0747	0·0754	0·0759	0·0762
0·20	0·0748	0·0795	0·0827	0·0857	0·0874	0·0888	0·0902	0·0912	0·0920	0·0928	0·0933	0·0936
0·22	0·0927	0·0977	0·1011	0·1042	0·1060	0·1075	0·1090	0·1100	0·1109	0·1117	0·1122	0·1126
0·24	0·1124	0·1176	0·1211	0·1244	0·1263	0·1278	0·1294	0·1304	0·1313	0·1322	0·1327	0·1331
0·26	0·1340	0·1392	0·1428	0·1462	0·1481	0·1496	0·1513	0·1525	0·1532	0·1541	0·1546	0·1550
0·28	0·1572	0·1625	0·1661	0·1694	0·1713	0·1729	0·1745	0·1756	0·1765	0·1774	0·1779	0·1783
0·30	0·1821	0·1873	0·1908	0·1941	0·1960	0·1975	0·1992	0·2002	0·2011	0·2020	0·2025	0·2029
0·32	0·2085	0·2135	0·2170	0·2202	0·2220	0·2234	0·2250	0·2260	0·2269	0·2278	0·2283	0·2286
0·34	0·2364	0·2412	0·2444	0·2475	0·2492	0·2506	0·2521	0·2530	0·2538	0·2547	0·2551	0·2555
0·36	0·2657	0·2701	0·2731	0·2759	0·2775	0·2788	0·2802	0·2811	0·2818	0·2826	0·2831	0·2834
0·38	0·2962	0·3002	0·3029	0·3055	0·3069	0·3081	0·3093	0·3101	0·3108	0·3115	0·3119	0·3122
0·40	0·3279	0·3314	0·3338	0·3360	0·3373	0·3383	0·3394	0·3401	0·3407	0·3413	0·3417	0·3419
0·42	0·3607	0·3636	0·3656	0·3674	0·3685	0·3693	0·3703	0·3708	0·3713	0·3719	0·3722	0·3724
0·44	0·3944	0·3967	0·3982	0·3997	0·4005	0·4011	0·4019	0·4023	0·4027	0·4031	0·4033	0·4035
0·46	0·4289	0·4305	0·4316	0·4326	0·4331	0·4336	0·4341	0·4344	0·4347	0·4349	0·4351	0·4352
0·48	0·4642	0·4650	0·4655	0·4661	0·4663	0·4666	0·4668	0·4670	0·4671	0·4673	0·4674	0·4674
0·50	0·5000	0·5000	0·5000	0·5000	0·5000	0·5000	0·5000	0·5000	0·5000	0·5000	0·5000	0·5000
0·52	0·5363	0·5355	0·5349	0·5343	0·5340	0·5338	0·5335	0·5333	0·5332	0·5330	0·5330	0·5329
0·54	0·5729	0·5712	0·5700	0·5689	0·5683	0·5678	0·5672	0·5669	0·5666	0·5663	0·5661	0·5660
0·56	0·6097	0·6071	0·6053	0·6036	0·6026	0·6019	0·6010	0·6005	0·6000	0·5996	0·5993	0·5991
0·58	0·6466	0·6430	0·6406	0·6383	0·6370	0·6360	0·6348	0·6341	0·6335	0·6329	0·6325	0·6322
0·60	0·6834	0·6788	0·6757	0·6729	0·6712	0·6699	0·6685	0·6676	0·6668	0·6660	0·6655	0·6652
0·62	0·7199	0·7144	0·7107	0·7072	0·7052	0·7036	0·7019	0·7008	0·6999	0·6989	0·6984	0·6979
0·64	0·7560	0·7496	0·7452	0·7411	0·7388	0·7370	0·7349	0·7337	0·7326	0·7314	0·7308	0·7303
0·66	0·7915	0·7842	0·7792	0·7745	0·7719	0·7698	0·7675	0·7660	0·7648	0·7635	0·7627	0·7622
0·68	0·8263	0·8181	0·8125	0·8073	0·8043	0·8019	0·7993	0·7977	0·7963	0·7948	0·7940	0·7934
0·70	0·8601	0·8511	0·8449	0·8391	0·8359	0·8333	0·8304	0·8286	0·8271	0·8255	0·8245	0·8239
0·72	0·8928	0·8830	0·8763	0·8700	0·8665	0·8636	0·8605	0·8586	0·8569	0·8552	0·8542	0·8534
0·74	0·9242	0·9137	0·9065	0·8998	0·8960	0·8929	0·8896	0·8875	0·8857	0·8838	0·8827	0·8819
0·76	0·9540	0·9429	0·9352	0·9281	0·9241	0·9208	0·9173	0·9151	0·9132	0·9112	0·9101	0·9092
0·78	0·9821	0·9704	0·9624	0·9549	0·9507	0·9473	0·9436	0·9413	0·9393	0·9372	0·9360	0·9351
0·80	1·0081	0·9960	0·9877	0·9800	0·9756	0·9720	0·9682	0·9658	0·9637	0·9616	0·9603	0·9594

Prop. depth	Coefficient θ for full pipe $= [\{\,k_s/D\,\} + \{\,1/(3600\,D\,S^{1/3}\,)\}]^{-1}$ (Water at 15°C - k_s and D in m)											
	5	10	20	50	100	200	500	1000	2000	5000	10000	20000
0·82	1·0318	1·0194	1·0109	1·0030	0·9985	0·9948	0·9909	0·9884	0·9863	0·9841	0·9828	0·9819
0·84	1·0529	1·0403	1·0317	1·0237	1·0191	1·0154	1·0115	1·0090	1·0068	1·0046	1·0033	1·0023
0·86	1·0709	1·0583	1·0497	1·0417	1·0372	1·0335	1·0295	1·0270	0·0249	1·0227	1·0214	1·0204
0·88	1·0854	1·0731	1·0646	1·0568	1·0523	1·0487	1·0448	1·0423	1·0403	1·0380	1·0368	1·0359
0·90	1·0959	1·0840	1·0759	1·0683	1·0640	1·0605	1·0568	1·0544	1·0524	1·0503	1·0491	1·0482
0·92	1·1015	1·0904	1·0828	1·0757	1·0716	1·0684	1·0648	1·0626	1·0607	1·0587	1·0576	1·0568
0·94	1·1012	1·0912	1·0843	1·0779	1·0742	1·0713	0·0681	1·0661	1·0644	1·0626	1·0616	1·0608
0·96	1·0929	1·0845	1·0787	1·0733	1·0702	1·0677	1·0650	1·0634	1·0619	1·0604	1·0595	1·0589
0·98	1·0723	1·0662	1·0620	1·0581	1·0559	1·0541	1·0522	1·0510	1·0499	1·0488	1·0482	1·0478
1·00	1·0000	1·0000	1·0000	1·0000	1·0000	1·0000	1·0000	1·0000	1·0000	1·0000	1·0000	1·0000

C1(b)

Corrections to assessed proportional depths for circular pipes, as based on shift of θ ratio.

Prop. Depth	θ/θ_{medial} at 0·60 proportional depth								
	0·05	0·10	0·20	0·50	1	2	5	10	20
0·05	+ 0·0075	+ 0·005	+ 0·003	+ 0·001	--	− 0·001	− 0·0015	− 0·002	− 0·0025
0·10	+ 0·0085	+ 0·0055	+ 0·0035	+ 0·0015	--	− 0·0015	− 0·002	− 0·003	− 0·0035
0·15	+ 0·009	+ 0·006	+ 0·004	+ 0·0015	--	− 0·0015	− 0·0025	− 0·0035	− 0·004
0·20	+ 0·009	+ 0·006	+ 0·004	+ 0·0015	--	− 0·0015	− 0·0025	− 0·0035	− 0·004
0·25	+ 0·0085	+ 0·006	+ 0·0035	+ 0·0015	--	− 0·0015	− 0·0025	− 0·003	− 0·004
0·30	+ 0·008	+ 0·005	+ 0·003	+ 0·001	--	− 0·001	− 0·0025	− 0·003	− 0·0035
0·35	+ 0·007	+ 0·0045	+ 0·0025	+ 0·001	--	− 0·001	− 0·002	− 0·0025	− 0·003
0·40	+ 0·0055	+ 0·004	+ 0·0025	+ 0·001	--	− 0·001	− 0·0015	− 0·002	− 0·0025
0·45	+ 0·0045	+ 0·003	+ 0·002	+ 0·0005	--	− 0·0005	− 0·0015	− 0·0015	− 0·002
0·50	+ 0·003	+ 0·002	+ 0·0015	+ 0·0005	--	− 0·0005	− 0·001	− 0·001	− 0·0015
0·55	+ 0·0015	+ 0·001	+ 0·0005	--	--	--	− 0·0005	− 0·0005	− 0·001
0·60	--	--	--	--	--	--	--	--	--
0·65	− 0·0015	− 0·001	− 0·0005	--	--	--	+ 0·0005	+ 0·001	+ 0·001
0·70	− 0·003	− 0·002	− 0·0015	− 0·0005	--	+ 0·0005	+ 0·001	+ 0·0015	+ 0·0015
0·75	− 0·005	− 0·0035	− 0·0025	− 0·001	--	+ 0·001	+ 0·0015	+ 0·002	+ 0·002
0·80	− 0·007	− 0·005	− 0·003	− 0·001	--	+ 0·001	+ 0·002	+ 0·0025	+ 0·003
0·85	− 0·009	− 0·006	− 0·004	− 0·0015	--	+ 0·0015	+ 0·0025	+ 0·0035	+ 0·004
0·90	− 0·0135	− 0·009	− 0·006	− 0·0025	--	+ 0·002	+ 0·004	+ 0·005	+ 0·0065

Note : These values derive from θ_{medial} for 1·50 m diameter pipe at 0·60 proportional depth with gradient 0·002 and roughness size k_s of 1·50 mm.

Form 1 egg-shape (3:2 old type)

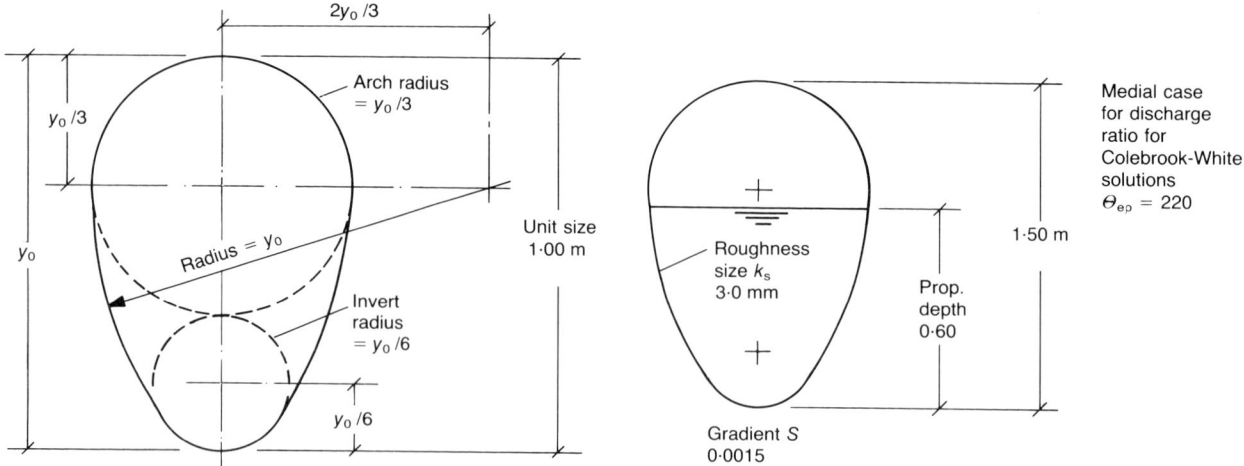

Prop dpth	U.equ. dia. $D_{ep(u)}$	Equiv. disch. factor	Unit sect. area	Unit wetted perim.	Unit surf. brdth	Unit mean depth	Discharge ratios $Q/Q_{0.60}$		U.crit. disch. $Q_{c(u)}$	Unit depth cntrd
Y	(m)	J	A_u (m²)	P_u (m)	$B_{s(u)}$ (m)	$y_{m(u)}$ (m)	medial	Mann'g	(m³s⁻¹)	$y_{d(u)}$ (m)
0·02	0·0518	0·9869	0·0021	0·1650	0·1583	0·0135	0·0012	0·0012	0·0008	0·0080
0·04	0·1006	1·3403	0·0059	0·2358	0·2166	0·0274	0·0051	0·0051	0·0031	0·0162
0·06	0·1463	1·5732	0·0107	0·2921	0·2561	0·0417	0·0120	0·0118	0·0068	0·0244
0·08	0·1883	1·7287	0·0161	0·3422	0·2863	0·0563	0·0215	0·0211	0·0120	0·0328
0·10	0·2262	1·8167	0·0221	0·3912	0·3146	0·0703	0·0333	0·0328	0·0184	0·0411
0·11	0·2440	1·8454	0·0253	0·4154	0·3281	0·0772	0·0402	0·0395	0·0220	0·0452
0·12	0·2612	1·8674	0·0287	0·4393	0·3414	0·0840	0·0476	0·0467	0·0260	0·0494
0·13	0·2778	1·8843	0·0322	0·4631	0·3543	0·0908	0·0556	0·0546	0·0303	0·0535
0·14	0·2939	1·8971	0·0358	0·4868	0·3668	0·0975	0·0641	0·0631	0·0350	0·0576
0·15	0·3097	1·9068	0·0395	0·5102	0·3790	0·1042	0·0733	0·0721	0·0399	0·0617
0·16	0·3250	1·9140	0·0433	0·5335	0·3910	0·1109	0·0831	0·0817	0·0452	0·0658
0·17	0·3400	1·9193	0·0473	0·5566	0·4025	0·1175	0·0934	0·0919	0·0508	0·0698
0·18	0·3547	1·9229	0·0514	0·5796	0·4138	0·1242	0·1043	0·1027	0·0567	0·0739
0·19	0·3692	1·9252	0·0556	0·6024	0·4248	0·1309	0·1158	0·1141	0·0630	0·0779
0·20	0·3833	1·9264	0·0599	0·6251	0·4355	0·1375	0·1279	0·1260	0·0696	0·0820
0·21	0·3972	1·9266	0·0643	0·6476	0·4459	0·1442	0·1406	0·1386	0·0765	0·0860
0·22	0·4108	1·9261	0·0688	0·6700	0·4561	0·1509	0·1538	0·1517	0·0837	0·0900
0·23	0·4242	1·9249	0·0734	0·6923	0·4659	0·1576	0·1676	0·1653	0·0913	0·0941
0·24	0·4374	1·9232	0·0781	0·7145	0·4755	0·1643	0·1819	0·1796	0·0992	0·0981
0·25	0·4504	1·9209	0·0829	0·7365	0·4848	0·1711	0·1968	0·1943	0·1074	0·1021
0·26	0·4632	1·9183	0·0878	0·7585	0·4938	0·1778	0·2122	0·2097	0·1160	0·1062
0·27	0·4757	1·9152	0·0928	0·7803	0·5026	0·1847	0·2282	0·2256	0·1249	0·1102
0·28	0·4881	1·9118	0·0979	0·8021	0·5111	0·1915	0·2447	0·2420	0·1341	0·1142
0·29	0·5003	1·9081	0·1030	0·8237	0·5194	0·1984	0·2618	0·2590	0·1437	0·1183
0·30	0·5123	1·9042	0·1083	0·8453	0·5274	0·2053	0·2793	0·2765	0·1536	0·1223
0·31	0·5242	1·8999	0·1136	0·8667	0·5351	0·2122	0·2974	0·2945	0·1639	0·1263
0·32	0·5358	1·8955	0·1190	0·8881	0·5426	0·2192	0·3160	0·3130	0·1744	0·1304
0·33	0·5473	1·8908	0·1244	0·9094	0·5499	0·2263	0·3351	0·3320	0·1853	0·1345
0·34	0·5586	1·8860	0·1300	0·9306	0·5569	0·2333	0·3547	0·3516	0·1966	0·1385
0·35	0·5698	1·8809	0·1356	0·9517	0·5637	0·2405	0·3747	0·3716	0·2082	0·1426
0·36	0·5808	1·8757	0·1412	0·9727	0·5703	0·2477	0·3952	0·3921	0·2201	0·1467
0·37	0·5916	1·8703	0·1470	0·9937	0·5766	0·2549	0·4162	0·4131	0·2324	0·1507
0·38	0·6023	1·8648	0·1528	1·0146	0·5827	0·2622	0·4376	0·4345	0·2449	0·1548
0·39	0·6128	1·8591	0·1586	1·0355	0·5886	0·2695	0·4595	0·4564	0·2579	0·1589
0·40	0·6231	1·8533	0·1645	1·0562	0·5942	0·2769	0·4818	0·4787	0·2711	0·1630
0·41	0·6333	1·8473	0·1705	1·0770	0·5997	0·2843	0·5045	0·5015	0·2847	0·1671
0·42	0·6433	1·8413	0·1765	1·0976	0·6049	0·2919	0·5276	0·5247	0·2986	0·1713
0·43	0·6532	1·8350	0·1826	1·1182	0·6098	0·2994	0·5511	0·5483	0·3129	0·1754
0·44	0·6629	1·8287	0·1887	1·1388	0·6146	0·3071	0·5750	0·5723	0·3275	0·1796
0·45	0·6725	1·8223	0·1949	1·1593	0·6192	0·3148	0·5993	0·5966	0·3424	0·1837

Form 1 egg-shape (3:2 old type)

Prop dpth Y	U.equ. dia. $D_{ep(u)}$ (m)	Equiv. disch. factor J	Unit sect. area A_u (m²)	Unit wetted perim. P_u (m)	Unit surf. brdth $B_{s(u)}$ (m)	Unit mean depth $y_{m(u)}$ (m)	Discharge ratios $Q/Q_{0.60}$ medial	Mann'g	U.crit. disch. $Q_{c(u)}$ (m³s⁻¹)	Unit depth cntrd $y_{d(u)}$ (m)
0·46	0·6819	1·8157	0·2011	1·1798	0·6235	0·3226	0·6240	0·6214	0·3577	0·1879
0·47	0·6911	1·8090	0·2074	1·2002	0·6276	0·3304	0·6490	0·6465	0·3733	0·1921
0·48	0·7002	1·8022	0·2137	1·2206	0·6315	0·3383	0·6743	0·6719	0·3892	0·1963
0·49	0·7091	1·7953	0·2200	1·2409	0·6352	0·3463	0·7000	0·6977	0·4054	0·2005
0·50	0·7179	1·7883	0·2264	1·2612	0·6387	0·3544	0·7259	0·7239	0·4220	0·2047
0·51	0·7266	1·7812	0·2328	1·2815	0·6420	0·3626	0·7522	0·7503	0·4389	0·2089
0·52	0·7350	1·7740	0·2392	1·3017	0·6450	0·3708	0·7788	0·7770	0·4562	0·2132
0·53	0·7434	1·7666	0·2457	1·3219	0·6479	0·3792	0·8056	0·8040	0·4737	0·2174
0·54	0·7515	1·7592	0·2522	1·3421	0·6506	0·3876	0·8327	0·8313	0·4916	0·2217
0·55	0·7596	1·7517	0·2587	1·3622	0·6530	0·3961	0·8601	0·8589	0·5098	0·2260
0·56	0·7674	1·7441	0·2652	1·3824	0·6553	0·4048	0·8876	0·8867	0·5284	0·2303
0·57	0·7752	1·7364	0·2718	1·4025	0·6573	0·4135	0·9154	0·9147	0·5473	0·2346
0·58	0·7827	1·7285	0·2784	1·4225	0·6591	0·4223	0·9434	0·9429	0·5665	0·2389
0·59	0·7901	1·7206	0·2850	1·4426	0·6608	0·4313	0·9716	0·9714	0·5860	0·2433
0·60	0·7974	1·7126	0·2916	1·4627	0·6622	0·4403	1·0000	1·0000	0·6059	0·2477
0·61	0·8045	1·7046	0·2982	1·4827	0·6635	0·4495	1·0286	1·0288	0·6261	0·2520
0·62	0·8115	1·6964	0·3048	1·5027	0·6645	0·4588	1·0573	1·0578	0·6466	0·2564
0·63	0·8183	1·6881	0·3115	1·5228	0·6653	0·4682	1·0861	1·0869	0·6675	0·2609
0·64	0·8249	1·6797	0·3182	1·5428	0·6660	0·4777	1·1150	1·1161	0·6886	0·2653
0·65	0·8314	1·6713	0·3248	1·5628	0·6664	0·4874	1·1441	1·1454	0·7102	0·2698
0·66	0·8377	1·6627	0·3315	1·5828	0·6666	0·4973	1·1732	1·1749	0·7320	0·2742
0·67	0·8439	1·6541	0·3381	1·6028	0·6666	0·5072	1·2025	1·2044	0·7542	0·2787
0·68	0·8499	1·6454	0·3448	1·6228	0·6661	0·5176	1·2317	1·2340	0·7769	0·2832
0·69	0·8558	1·6365	0·3515	1·6428	0·6650	0·5285	1·2610	1·2635	0·8001	0·2878
0·70	0·8614	1·6274	0·3581	1·6629	0·6633	0·5399	1·2902	1·2931	0·8240	0·2923
0·71	0·8669	1·6181	0·3647	1·6830	0·6610	0·5518	1·3194	1·3225	0·8484	0·2970
0·72	0·8721	1·6085	0·3713	1·7032	0·6581	0·5643	1·3484	1·3518	0·8735	0·3016
0·73	0·8770	1·5985	0·3779	1·7235	0·6545	0·5774	1·3771	1·3809	0·8992	0·3063
0·74	0·8817	1·5883	0·3844	1·7440	0·6503	0·5911	1·4057	1·4098	0·9255	0·3110
0·75	0·8861	1·5776	0·3909	1·7646	0·6455	0·6056	1·4339	1·4383	0·9526	0·3157
0·76	0·8902	1·5665	0·3973	1·7853	0·6400	0·6208	1·4618	1·4665	0·9804	0·3206
0·77	0·8940	1·5549	0·4037	1·8062	0·6338	0·6369	1·4893	1·4942	1·0089	0·3254
0·78	0·8975	1·5429	0·4100	1·8274	0·6270	0·6540	1·5162	1·5214	1·0383	0·3303
0·79	0·9006	1·5303	0·4162	1·8488	0·6194	0·6720	1·5427	1·5481	1·0685	0·3353
0·80	0·9033	1·5171	0·4224	1·8704	0·6110	0·6913	1·5685	1·5742	1·0998	0·3404
0·81	0·9056	1·5033	0·4284	1·8924	0·6019	0·7118	1·5936	1·5995	1·1320	0·3455
0·82	0·9075	1·4889	0·4344	1·9148	0·5919	0·7339	1·6180	1·6241	1·1654	0·3507
0·83	0·9090	1·4738	0·4403	1·9375	0·5811	0·7576	1·6415	1·6478	1·2001	0·3559
0·84	0·9100	1·4580	0·4460	1·9607	0·5694	0·7833	1·6641	1·6705	1·2362	0·3613
0·85	0·9105	1·4414	0·4517	1·9843	0·5568	0·8112	1·6857	1·6922	1·2740	0·3667
0·86	0·9104	1·4240	0·4572	2·0086	0·5431	0·8418	1·7062	1·7128	1·3136	0·3722
0·87	0·9098	1·4056	0·4625	2·0335	0·5283	0·8756	1·7255	1·7321	1·3553	0·3779
0·88	0·9086	1·3863	0·4677	2·0591	0·5122	0·9131	1·7434	1·7500	1·3996	0·3836
0·89	0·9067	1·3658	0·4728	2·0856	0·4949	0·9553	1·7599	1·7664	1·4470	0·3895
0·90	0·9042	1·3442	0·4776	2·1130	0·4761	1·0032	1·7747	1·7812	1·4981	0·3954
0·91	0·9008	1·3213	0·4823	2·1416	0·4556	1·0585	1·7877	1·7941	1·5539	0·4016
0·92	0·8965	1·2969	0·4867	2·1716	0·4333	1·1234	1·7988	1·8049	1·6155	0·4079
0·93	0·8913	1·2708	0·4909	2·2033	0·4087	1·2011	1·8076	1·8134	1·6849	0·4143
0·94	0·8849	1·2427	0·4949	2·2370	0·3816	1·2970	1·8139	1·8193	1·7650	0·4210
0·95	0·8772	1·2122	0·4986	2·2734	0·3512	1·4197	1·8172	1·8221	1·8603	0·4278
0·96	0·8678	1·1785	0·5019	2·3133	0·3166	1·5851	1·8169	1·8213	1·9788	0·4350
0·97	0·8563	1·1408	0·5049	2·3583	0·2764	1·8266	1·8122	1·8159	2·1368	0·4424
0·98	0·8418	1·0967	0·5074	2·4112	0·2274	2·2309	1·8014	1·8042	2·3733	0·4501
0·99	0·8217	1·0411	0·5094	2·4796	0·1621	3·1429	1·7808	1·7823	2·8279	0·4584
1·00	0·7725	0·9181	0·5105	2·6433	0·0000	-	1·7155	1·7140	-	0·4674

Form 1 egg-shape with 5% lining

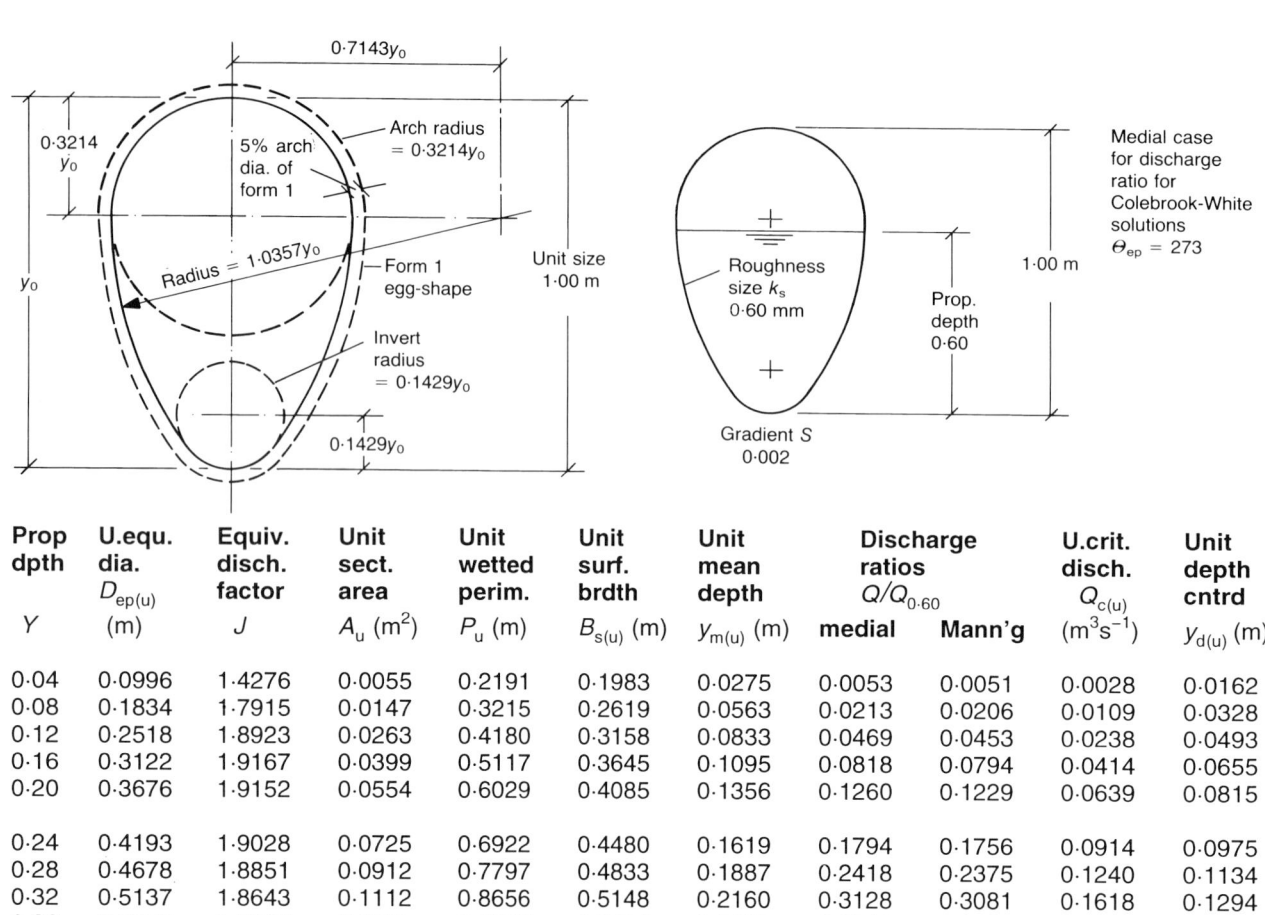

Prop dpth Y	U.equ. dia. $D_{ep(u)}$ (m)	Equiv. disch. factor J	Unit sect. area A_u (m²)	Unit wetted perim. P_u (m)	Unit surf. brdth $B_{s(u)}$ (m)	Unit mean depth $y_{m(u)}$ (m)	Discharge ratios $Q/Q_{0.60}$ medial	Mann'g	U.crit. disch. $Q_{c(u)}$ (m³s⁻¹)	Unit depth cntrd $y_{d(u)}$ (m)
0.04	0.0996	1.4276	0.0055	0.2191	0.1983	0.0275	0.0053	0.0051	0.0028	0.0162
0.08	0.1834	1.7915	0.0147	0.3215	0.2619	0.0563	0.0213	0.0206	0.0109	0.0328
0.12	0.2518	1.8923	0.0263	0.4180	0.3158	0.0833	0.0469	0.0453	0.0238	0.0493
0.16	0.3122	1.9167	0.0399	0.5117	0.3645	0.1095	0.0818	0.0794	0.0414	0.0655
0.20	0.3676	1.9152	0.0554	0.6029	0.4085	0.1356	0.1260	0.1229	0.0639	0.0815
0.24	0.4193	1.9028	0.0725	0.6922	0.4480	0.1619	0.1794	0.1756	0.0914	0.0975
0.28	0.4678	1.8851	0.0912	0.7797	0.4833	0.1887	0.2418	0.2375	0.1240	0.1134
0.32	0.5137	1.8643	0.1112	0.8656	0.5148	0.2160	0.3128	0.3081	0.1618	0.1294
0.36	0.5570	1.8413	0.1323	0.9503	0.5424	0.2439	0.3919	0.3871	0.2047	0.1455
0.40	0.5978	1.8166	0.1545	1.0338	0.5665	0.2727	0.4785	0.4738	0.2527	0.1617
0.42	0.6174	1.8037	0.1660	1.0752	0.5773	0.2875	0.5244	0.5199	0.2786	0.1699
0.44	0.6363	1.7904	0.1776	1.1164	0.5872	0.3025	0.5720	0.5677	0.3059	0.1781
0.46	0.6546	1.7768	0.1894	1.1575	0.5962	0.3177	0.6211	0.6172	0.3344	0.1863
0.48	0.6724	1.7629	0.2014	1.1983	0.6044	0.3333	0.6717	0.6681	0.3642	0.1946
0.50	0.6896	1.7486	0.2136	1.2390	0.6118	0.3491	0.7236	0.7205	0.3952	0.2030
0.52	0.7062	1.7340	0.2259	1.2795	0.6184	0.3653	0.7768	0.7742	0.4276	0.2114
0.54	0.7223	1.7190	0.2383	1.3199	0.6242	0.3818	0.8311	0.8291	0.4612	0.2198
0.56	0.7377	1.7038	0.2509	1.3602	0.6292	0.3987	0.8865	0.8851	0.4961	0.2283
0.58	0.7526	1.6882	0.2635	1.4005	0.6335	0.4160	0.9429	0.9421	0.5322	0.2369
0.60	0.7669	1.6724	0.2762	1.4406	0.6369	0.4337	1.0000	1.0000	0.5696	0.2455
0.62	0.7806	1.6562	0.2890	1.4807	0.6395	0.4518	1.0579	1.0587	0.6083	0.2543
0.64	0.7938	1.6397	0.3018	1.5207	0.6414	0.4705	1.1164	1.1180	0.6482	0.2630
0.66	0.8063	1.6230	0.3146	1.5608	0.6425	0.4897	1.1753	1.1778	0.6894	0.2719
0.68	0.8183	1.6059	0.3275	1.6008	0.6429	0.5094	1.2347	1.2380	0.7319	0.2808
0.70	0.8297	1.5885	0.3403	1.6408	0.6414	0.5306	1.2942	1.2985	0.7763	0.2899
0.72	0.8403	1.5703	0.3531	1.6810	0.6375	0.5539	1.3535	1.3587	0.8230	0.2990
0.74	0.8500	1.5510	0.3658	1.7215	0.6310	0.5797	1.4122	1.4184	0.8722	0.3083
0.76	0.8586	1.5304	0.3783	1.7626	0.6219	0.6084	1.4699	1.4769	0.9241	0.3177
0.78	0.8661	1.5080	0.3907	1.8043	0.6100	0.6404	1.5259	1.5338	0.9790	0.3274
0.80	0.8722	1.4835	0.4027	1.8469	0.5952	0.6766	1.5799	1.5886	1.0374	0.3373
0.82	0.8768	1.4567	0.4145	1.8908	0.5773	0.7179	1.6313	1.6406	1.0997	0.3475
0.84	0.8797	1.4273	0.4258	1.9362	0.5559	0.7659	1.6793	1.6892	1.1669	0.3579
0.86	0.8806	1.3947	0.4367	1.9835	0.5307	0.8229	1.7234	1.7336	1.2404	0.3688
0.88	0.8793	1.3586	0.4470	2.0333	0.5010	0.8922	1.7625	1.7728	1.3222	0.3800
0.90	0.8755	1.3182	0.4567	2.0865	0.4660	0.9800	1.7958	1.8060	1.4157	0.3918
0.92	0.8686	1.2726	0.4656	2.1442	0.4244	1.0970	1.8218	1.8315	1.5271	0.4041
0.94	0.8577	1.2201	0.4736	2.2085	0.3740	1.2662	1.8387	1.8475	1.6688	0.4171
0.96	0.8416	1.1578	0.4805	2.2836	0.3106	1.5470	1.8434	1.8507	1.8714	0.4310
0.98	0.8166	1.0781	0.4859	2.3797	0.2232	2.1765	1.8295	1.8343	2.2446	0.4461
1.00	0.7498	0.9033	0.4888	2.6077	0.0000	-	1.7452	1.7435	-	0.4633

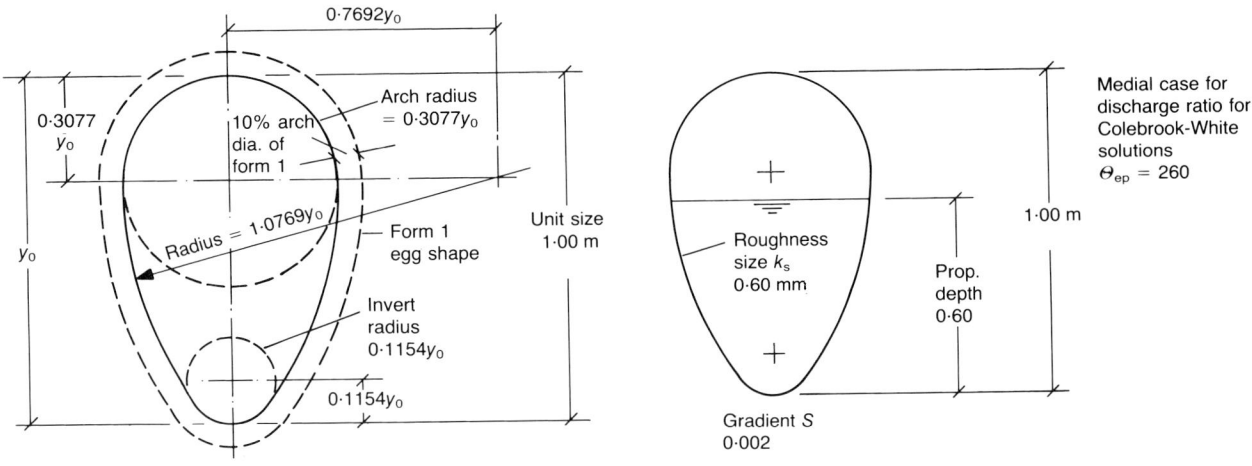

Prop dpth	U.equ. dia. $D_{ep(u)}$	Equiv. disch. factor	Unit sect. area	Unit wetted perim.	Unit surf. brdth	Unit mean depth	Discharge ratios $Q/Q_{0.60}$		U.crit. disch. $Q_{c(u)}$	Unit depth cntrd
Y	(m)	J	A_u (m²)	P_u (m)	$B_{s(u)}$ (m)	$y_{m(u)}$ (m)	medial	Mann'g	(m³s⁻¹)	$y_{d(u)}$ (m)
0·04	0·0979	1·5512	0·0048	0·1982	0·1747	0·0278	0·0051	0·0049	0·0025	0·0163
0·08	0·1754	1·8524	0·0130	0·2974	0·2334	0·0559	0·0202	0·0195	0·0097	0·0329
0·12	0·2385	1·9054	0·0234	0·3932	0·2861	0·0820	0·0445	0·0431	0·0210	0·0491
0·16	0·2949	1·9045	0·0359	0·4864	0·3339	0·1074	0·0780	0·0759	0·0368	0·0650
0·20	0·3470	1·8882	0·0501	0·5774	0·3772	0·1328	0·1211	0·1181	0·0572	0·0807
0·24	0·3960	1·8668	0·0660	0·6664	0·4162	0·1585	0·1735	0·1699	0·0823	0·0964
0·28	0·4422	1·8432	0·0833	0·7538	0·4513	0·1847	0·2352	0·2310	0·1122	0·1121
0·32	0·4860	1·8184	0·1020	0·8397	0·4826	0·2114	0·3057	0·3011	0·1469	0·1279
0·36	0·5275	1·7927	0·1219	0·9244	0·5103	0·2389	0·3846	0·3800	0·1866	0·1437
0·40	0·5667	1·7663	0·1428	1·0080	0·5345	0·2672	0·4715	0·4669	0·2311	0·1597
0·42	0·5855	1·7528	0·1536	1·0494	0·5454	0·2816	0·5177	0·5133	0·2553	0·1678
0·44	0·6037	1·7390	0·1646	1·0906	0·5554	0·2964	0·5656	0·5614	0·2806	0·1759
0·46	0·6214	1·7250	0·1758	1·1317	0·5647	0·3114	0·6152	0·6113	0·3072	0·1841
0·48	0·6386	1·7109	0·1872	1·1726	0·5731	0·3266	0·6663	0·6628	0·3350	0·1923
0·50	0·6552	1·6964	0·1987	1·2133	0·5808	0·3422	0·7189	0·7158	0·3641	0·2005
0·52	0·6713	1·6818	0·2104	1·2539	0·5876	0·3581	0·7728	0·7702	0·3943	0·2088
0·54	0·6868	1·6669	0·2222	1·2944	0·5937	0·3743	0·8280	0·8260	0·4258	0·2172
0·56	0·7018	1·6518	0·2342	1·3347	0·5991	0·3909	0·8844	0·8829	0·4585	0·2256
0·58	0·7162	1·6364	0·2462	1·3750	0·6036	0·4078	0·9417	0·9410	0·4924	0·2341
0·60	0·7301	1·6208	0·2583	1·4152	0·6075	0·4252	1·0000	1·0000	0·5275	0·2427
0·62	0·7435	1·6049	0·2705	1·4553	0·6105	0·4430	1·0591	1·0599	0·5638	0·2513
0·64	0·7563	1·5888	0·2827	1·4953	0·6128	0·4613	1·1189	1·1205	0·6013	0·2600
0·66	0·7685	1·5725	0·2950	1·5354	0·6144	0·4801	1·1793	1·1818	0·6401	0·2687
0·68	0·7802	1·5559	0·3073	1·5754	0·6152	0·4995	1·2401	1·2435	0·6801	0·2776
0·70	0·7914	1·5391	0·3196	1·6154	0·6152	0·5195	1·3013	1·3056	0·7214	0·2865
0·72	0·8019	1·5218	0·3319	1·6555	0·6129	0·5415	1·3625	1·3678	0·7648	0·2955
0·74	0·8117	1·5037	0·3441	1·6958	0·6079	0·5660	1·4233	1·4296	0·8107	0·3047
0·76	0·8205	1·4843	0·3562	1·7365	0·6003	0·5933	1·4833	1·4905	0·8592	0·3140
0·78	0·8282	1·4634	0·3681	1·7778	0·5899	0·6240	1·5418	1·5499	0·9106	0·3235
0·80	0·8346	1·4406	0·3798	1·8200	0·5765	0·6588	1·5984	1·6073	0·9653	0·3333
0·82	0·8396	1·4156	0·3911	1·8633	0·5599	0·6986	1·6524	1·6621	1·0237	0·3433
0·84	0·8430	1·3879	0·4021	1·9081	0·5399	0·7449	1·7031	1·7134	1·0869	0·3536
0·86	0·8445	1·3573	0·4127	1·9547	0·5160	0·7999	1·7498	1·7605	1·1558	0·3643
0·88	0·8439	1·3231	0·4227	2·0037	0·4876	0·8669	1·7916	1·8025	1·2326	0·3754
0·90	0·8408	1·2847	0·4322	2·0560	0·4540	0·9518	1·8274	1·8382	1·3204	0·3870
0·92	0·8347	1·2412	0·4409	2·1127	0·4139	1·0651	1·8557	1·8661	1·4248	0·3992
0·94	0·8248	1·1909	0·4487	2·1758	0·3651	1·2289	1·8746	1·8841	1·5576	0·4121
0·96	0·8098	1·1310	0·4554	2·2493	0·3034	1·5008	1·8809	1·8890	1·7470	0·4259
0·98	0·7863	1·0540	0·4607	2·3435	0·2182	2·1107	1·8679	1·8736	2·0958	0·4409
1·00	0·7225	0·8843	0·4636	2·5666	0·0000	-	1·7830	1·7821	-	0·4581

C5

Form 2 egg-shape (3:2 new type)

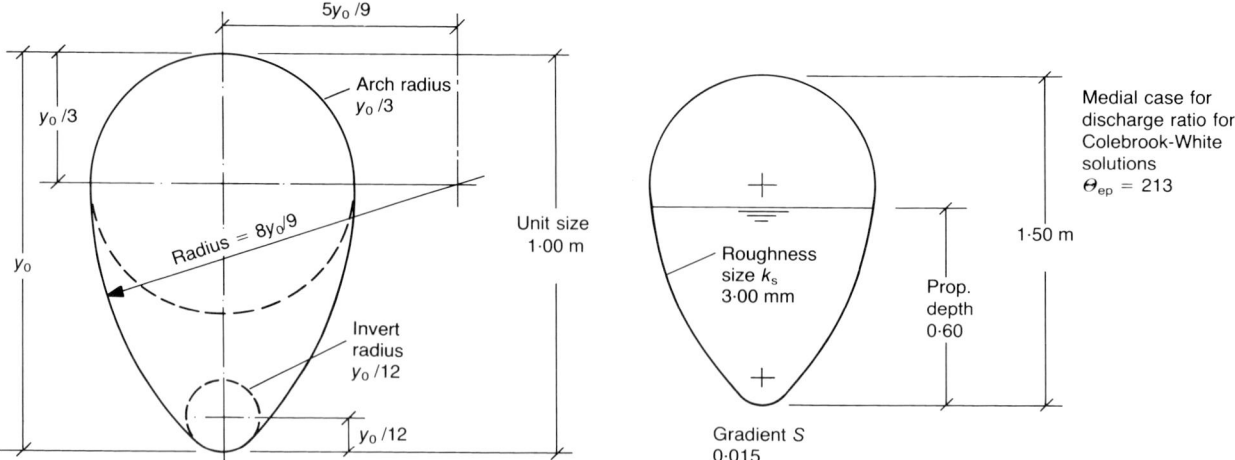

Prop dpth Y	U.equ. dia. $D_{ep(u)}$ (m)	Equiv. disch. factor J	Unit sect. area A_u (m²)	Unit wetted perim. P_u (m)	Unit surf. brdth $B_{s(u)}$ (m)	Unit mean depth $y_{m(u)}$ (m)	Discharge ratios $Q/Q_{0.60}$ medial	Mann'g	U.crit. disch. $Q_{c(u)}$ (m³s⁻¹)	Unit depth cntrd $y_{d(u)}$ (m)
0·02	0·0503	1·3403	0·0015	0·1179	0·1083	0·0137	0·0008	0·0009	0·0005	0·0081
0·04	0·0928	1·6611	0·0041	0·1755	0·1497	0·0272	0·0036	0·0036	0·0021	0·0163
0·06	0·1291	1·7550	0·0075	0·2310	0·1882	0·0396	0·0083	0·0082	0·0046	0·0242
0·08	0·1626	1·7923	0·0116	0·2850	0·2245	0·0516	0·0150	0·0148	0·0082	0·0319
0·10	0·1946	1·8106	0·0164	0·3376	0·2586	0·0635	0·0240	0·0237	0·0130	0·0395
0·11	0·2101	1·8165	0·0191	0·3634	0·2749	0·0694	0·0294	0·0290	0·0157	0·0433
0·12	0·2254	1·8210	0·0219	0·3889	0·2907	0·0754	0·0354	0·0348	0·0188	0·0470
0·13	0·2405	1·8246	0·0249	0·4141	0·3061	0·0813	0·0420	0·0413	0·0222	0·0508
0·14	0·2554	1·8275	0·0280	0·4391	0·3210	0·0873	0·0493	0·0484	0·0259	0·0545
0·15	0·2701	1·8298	0·0313	0·4638	0·3355	0·0933	0·0571	0·0562	0·0300	0·0583
0·16	0·2847	1·8317	0·0347	0·4882	0·3496	0·0994	0·0656	0·0645	0·0343	0·0620
0·17	0·2990	1·8331	0·0383	0·5124	0·3633	0·1055	0·0747	0·0735	0·0390	0·0658
0·18	0·3132	1·8343	0·0420	0·5365	0·3765	0·1116	0·0845	0·0832	0·0439	0·0696
0·19	0·3273	1·8351	0·0458	0·5602	0·3894	0·1177	0·0949	0·0935	0·0492	0·0733
0·20	0·3412	1·8357	0·0498	0·5838	0·4020	0·1239	0·1060	0·1044	0·0549	0·0771
0·21	0·3549	1·8359	0·0539	0·6073	0·4141	0·1301	0·1177	0·1159	0·0609	0·0809
0·22	0·3685	1·8360	0·0581	0·6305	0·4259	0·1364	0·1300	0·1281	0·0672	0·0847
0·23	0·3819	1·8358	0·0624	0·6535	0·4374	0·1427	0·1430	0·1410	0·0738	0·0885
0·24	0·3952	1·8353	0·0668	0·6764	0·4485	0·1490	0·1566	0·1545	0·0808	0·0923
0·25	0·4083	1·8347	0·0714	0·6991	0·4593	0·1554	0·1708	0·1686	0·0881	0·0961
0·26	0·4213	1·8338	0·0760	0·7217	0·4697	0·1618	0·1857	0·1834	0·0957	0·0999
0·27	0·4341	1·8327	0·0808	0·7441	0·4798	0·1683	0·2012	0·1988	0·1037	0·1037
0·28	0·4468	1·8314	0·0856	0·7664	0·4897	0·1748	0·2173	0·2148	0·1121	0·1076
0·29	0·4593	1·8299	0·0905	0·7885	0·4992	0·1814	0·2341	0·2314	0·1208	0·1114
0·30	0·4717	1·8282	0·0956	0·8105	0·5084	0·1880	0·2514	0·2487	0·1298	0·1153
0·31	0·4839	1·8263	0·1007	0·8324	0·5173	0·1947	0·2693	0·2665	0·1392	0·1192
0·32	0·4960	1·8242	0·1059	0·8542	0·5259	0·2014	0·2878	0·2850	0·1489	0·1230
0·33	0·5080	1·8219	0·1112	0·8759	0·5342	0·2082	0·3069	0·3040	0·1589	0·1269
0·34	0·5198	1·8194	0·1166	0·8974	0·5423	0·2150	0·3266	0·3236	0·1693	0·1309
0·35	0·5314	1·8167	0·1221	0·9189	0·5500	0·2219	0·3469	0·3438	0·1801	0·1348
0·36	0·5429	1·8139	0·1276	0·9402	0·5575	0·2289	0·3676	0·3646	0·1912	0·1387
0·37	0·5542	1·8108	0·1332	0·9615	0·5647	0·2359	0·3890	0·3859	0·2026	0·1427
0·38	0·5654	1·8076	0·1389	0·9827	0·5717	0·2430	0·4108	0·4077	0·2144	0·1466
0·39	0·5765	1·8041	0·1447	1·0038	0·5784	0·2501	0·4332	0·4301	0·2266	0·1506
0·40	0·5873	1·8005	0·1505	1·0248	0·5848	0·2573	0·4561	0·4530	0·2390	0·1546
0·41	0·5981	1·7967	0·1564	1·0457	0·5909	0·2646	0·4795	0·4765	0·2518	0·1586
0·42	0·6087	1·7928	0·1623	1·0666	0·5968	0·2719	0·5034	0·5004	0·2650	0·1626
0·43	0·6191	1·7886	0·1683	1·0873	0·6025	0·2793	0·5277	0·5248	0·2785	0·1666
0·44	0·6294	1·7843	0·1743	1·1081	0·6079	0·2868	0·5525	0·5496	0·2924	0·1707
0·45	0·6395	1·7798	0·1804	1·1287	0·6130	0·2943	0·5778	0·5750	0·3066	0·1747

Form 2 egg-shape (3:2 new type)

Prop dpth Y	U.equ. dia. $D_{ep(u)}$ (m)	Equiv. disch. factor J	Unit sect. area A_u (m²)	Unit wetted perim. P_u (m)	Unit surf. brdth $B_{s(u)}$ (m)	Unit mean depth $y_{m(u)}$ (m)	Discharge ratios $Q/Q_{0.60}$ medial	Mann'g	U.crit. disch. $Q_{c(u)}$ (m³s⁻¹)	Unit depth cntrd $y_{d(u)}$ (m)
0·46	0·6494	1·7752	0·1866	1·1493	0·6179	0·3020	0·6034	0·6007	0·3211	0·1788
0·47	0·6592	1·7704	0·1928	1·1698	0·6226	0·3097	0·6295	0·6270	0·3360	0·1829
0·48	0·6689	1·7654	0·1991	1·1903	0·6270	0·3175	0·6560	0·6536	0·3512	0·1870
0·49	0·6784	1·7602	0·2053	1·2107	0·6312	0·3253	0·6829	0·6806	0·3668	0·1911
0·50	0·6877	1·7549	0·2117	1·2311	0·6351	0·3333	0·7102	0·7080	0·3827	0·1952
0·51	0·6969	1·7495	0·2180	1·2515	0·6388	0·3413	0·7378	0·7358	0·3989	0·1994
0·52	0·7059	1·7438	0·2244	1·2718	0·6423	0·3494	0·7658	0·7639	0·4155	0·2035
0·53	0·7148	1·7380	0·2309	1·2920	0·6455	0·3577	0·7941	0·7924	0·4324	0·2077
0·54	0·7235	1·7321	0·2374	1·3123	0·6485	0·3660	0·8227	0·8212	0·4497	0·2119
0·55	0·7321	1·7260	0·2439	1·3324	0·6513	0·3744	0·8516	0·8503	0·4673	0·2161
0·56	0·7405	1·7198	0·2504	1·3526	0·6538	0·3830	0·8808	0·8798	0·4852	0·2204
0·57	0·7487	1·7134	0·2569	1·3727	0·6561	0·3916	0·9102	0·9094	0·5035	0·2246
0·58	0·7567	1·7068	0·2635	1·3928	0·6582	0·4003	0·9399	0·9394	0·5221	0·2289
0·59	0·7646	1·7001	0·2701	1·4129	0·6600	0·4092	0·9699	0·9696	0·5411	0·2332
0·60	0·7724	1·6933	0·2767	1·4330	0·6617	0·4182	1·0000	1·0000	0·5604	0·2375
0·61	0·7800	1·6863	0·2833	1·4530	0·6631	0·4273	1·0304	1·0306	0·5800	0·2418
0·62	0·7874	1·6792	0·2900	1·4731	0·6642	0·4366	1·0609	1·0614	0·6000	0·2462
0·63	0·7946	1·6719	0·2966	1·4931	0·6652	0·4459	1·0916	1·0924	0·6203	0·2506
0·64	0·8017	1·6645	0·3033	1·5131	0·6659	0·4555	1·1224	1·1236	0·6409	0·2550
0·65	0·8086	1·6570	0·3099	1·5331	0·6664	0·4651	1·1534	1·1548	0·6619	0·2594
0·66	0·8154	1·6493	0·3166	1·5531	0·6666	0·4749	1·1845	1·1862	0·6832	0·2638
0·67	0·8220	1·6415	0·3233	1·5731	0·6666	0·4849	1·2156	1·2177	0·7049	0·2683
0·68	0·8284	1·6335	0·3299	1·5931	0·6661	0·4953	1·2469	1·2493	0·7271	0·2727
0·69	0·8346	1·6253	0·3366	1·6132	0·6650	0·5061	1·2781	1·2809	0·7499	0·2772
0·70	0·8406	1·6169	0·3432	1·6332	0·6633	0·5174	1·3093	1·3124	0·7732	0·2818
0·71	0·8464	1·6082	0·3498	1·6534	0·6610	0·5293	1·3404	1·3438	0·7970	0·2864
0·72	0·8519	1·5992	0·3564	1·6736	0·6581	0·5416	1·3714	1·3751	0·8215	0·2910
0·73	0·8572	1·5898	0·3630	1·6939	0·6545	0·5546	1·4021	1·4063	0·8466	0·2956
0·74	0·8622	1·5800	0·3695	1·7143	0·6503	0·5682	1·4326	1·4371	0·8723	0·3003
0·75	0·8669	1·5698	0·3760	1·7349	0·6455	0·5825	1·4628	1·4676	0·8987	0·3050
0·76	0·8713	1·5591	0·3824	1·7556	0·6400	0·5976	1·4927	1·4978	0·9258	0·3098
0·77	0·8754	1·5480	0·3888	1·7766	0·6338	0·6134	1·5220	1·5275	0·9536	0·3147
0·78	0·8791	1·5363	0·3951	1·7977	0·6270	0·6302	1·5509	1·5566	0·9823	0·3196
0·79	0·8825	1·5240	0·4013	1·8191	0·6194	0·6480	1·5792	1·5852	1·0117	0·3245
0·80	0·8855	1·5112	0·4075	1·8408	0·6110	0·6669	1·6068	1·6131	1·0421	0·3296
0·81	0·8881	1·4977	0·4136	1·8628	0·6019	0·6871	1·6338	1·6403	1·0735	0·3347
0·82	0·8902	1·4835	0·4195	1·8851	0·5919	0·7087	1·6599	1·6667	1·1060	0·3398
0·83	0·8919	1·4686	0·4254	1·9078	0·5811	0·7320	1·6852	1·6921	1·1398	0·3451
0·84	0·8931	1·4530	0·4312	1·9310	0·5694	0·7571	1·7094	1·7166	1·1748	0·3504
0·85	0·8938	1·4365	0·4368	1·9547	0·5568	0·7845	1·7326	1·7399	1·2115	0·3558
0·86	0·8940	1·4192	0·4423	1·9789	0·5431	0·8144	1·7547	1·7620	1·2499	0·3613
0·87	0·8936	1·4009	0·4476	2·0038	0·5283	0·8474	1·7754	1·7828	1·2904	0·3669
0·88	0·8925	1·3816	0·4528	2·0294	0·5122	0·8840	1·7947	1·8022	1·3334	0·3727
0·89	0·8909	1·3613	0·4579	2·0559	0·4949	0·9252	1·8125	1·8199	1·3792	0·3785
0·90	0·8884	1·3397	0·4627	2·0834	0·4761	0·9719	1·8286	1·8359	1·4286	0·3845
0·91	0·8852	1·3167	0·4674	2·1120	0·4556	1·0258	1·8427	1·8499	1·4825	0·3906
0·92	0·8811	1·2923	0·4718	2·1420	0·4333	1·0890	1·8548	1·8617	1·5420	0·3969
0·93	0·8760	1·2661	0·4761	2·1737	0·4087	1·1647	1·8644	1·8711	1·6089	0·4033
0·94	0·8698	1·2379	0·4800	2·2074	0·3816	1·2580	1·8714	1·8777	1·6860	0·4100
0·95	0·8623	1·2073	0·4837	2·2438	0·3512	1·3773	1·8752	1·8811	1·7776	0·4168
0·96	0·8530	1·1735	0·4870	2·2837	0·3166	1·5380	1·8753	1·8806	1·8914	0·4239
0·97	0·8417	1·1355	0·4900	2·3286	0·2764	1·7727	1·8707	1·8752	2·0430	0·4313
0·98	0·8272	1·0912	0·4925	2·3815	0·2274	2·1654	1·8597	1·8632	2·2696	0·4391
0·99	0·8073	1·0353	0·4945	2·4499	0·1621	3·0511	1·8384	1·8406	2·7049	0·4473
1·00	0·7584	0·9116	0·4956	2·6136	0·0000	-	1·7703	1·7693	-	0·4563

C6

Form 2 egg-shape with 3% lining

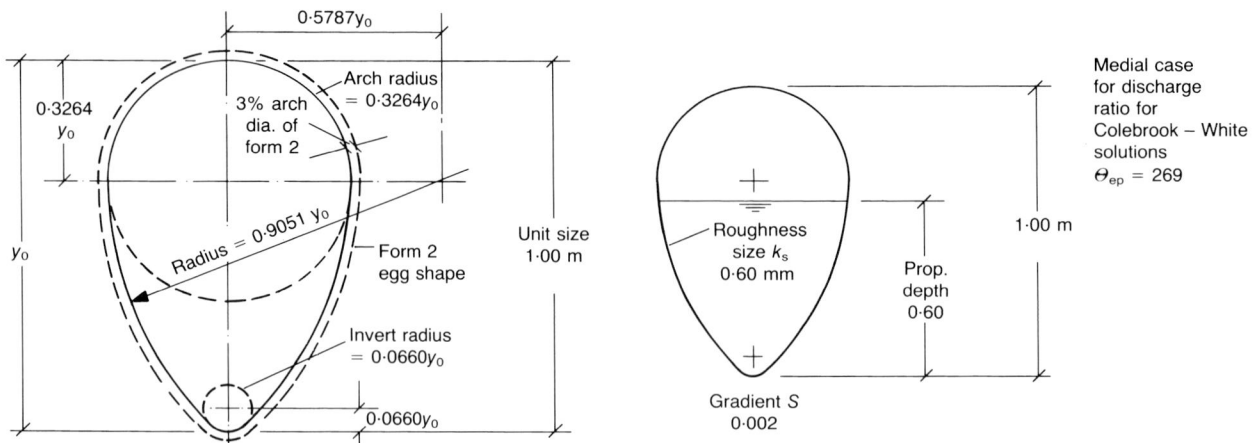

Prop dpth Y	U.equ. dia. $D_{ep(u)}$ (m)	Equiv. disch. factor J	Unit sect. area A_u (m²)	Unit wetted perim. P_u (m)	Unit surf. brdth $B_{s(u)}$ (m)	Unit mean depth $y_{m(u)}$ (m)	Discharge ratios $Q/Q_{0.60}$ medial	Mann'g	U.crit. disch. $Q_{c(u)}$ (m³s⁻¹)	Unit depth cntrd $y_{d(u)}$ (m)
0·04	0·0888	1·7174	0·0036	0·1625	0·1352	0·0267	0·0034	0·0033	0·0018	0·0162
0·08	0·1551	1·7956	0·0105	0·2714	0·2091	0·0503	0·0143	0·0137	0·0074	0·0315
0·12	0·2158	1·8086	0·0202	0·3749	0·2747	0·0736	0·0341	0·0329	0·0172	0·0464
0·16	0·2735	1·8127	0·0324	0·4740	0·3331	0·0973	0·0637	0·0617	0·0316	0·0612
0·20	0·3287	1·8133	0·0468	0·5694	0·3852	0·1215	0·1036	0·1008	0·0511	0·0761
0·24	0·3816	1·8111	0·0631	0·6619	0·4315	0·1463	0·1538	0·1502	0·0756	0·0911
0·28	0·4322	1·8060	0·0812	0·7518	0·4726	0·1719	0·2143	0·2100	0·1055	0·1063
0·32	0·4806	1·7980	0·1009	0·8397	0·5089	0·1982	0·2846	0·2799	0·1407	0·1216
0·36	0·5267	1·7872	0·1219	0·9258	0·5406	0·2255	0·3644	0·3595	0·1812	0·1372
0·40	0·5704	1·7736	0·1441	1·0103	0·5681	0·2536	0·4530	0·4482	0·2272	0·1530
0·42	0·5914	1·7658	0·1556	1·0522	0·5803	0·2681	0·5004	0·4957	0·2522	0·1609
0·44	0·6118	1·7574	0·1673	1·0937	0·5914	0·2828	0·5497	0·5452	0·2786	0·1690
0·46	0·6316	1·7483	0·1792	1·1350	0·6016	0·2979	0·6008	0·5967	0·3063	0·1770
0·48	0·6508	1·7385	0·1913	1·1760	0·6109	0·3132	0·6537	0·6499	0·3354	0·1852
0·50	0·6694	1·7281	0·2036	1·2169	0·6192	0·3289	0·7081	0·7048	0·3657	0·1934
0·52	0·6874	1·7172	0·2161	1·2576	0·6265	0·3449	0·7640	0·7612	0·3975	0·2017
0·54	0·7047	1·7056	0·2287	1·2981	0·6329	0·3613	0·8213	0·8191	0·4305	0·2100
0·56	0·7215	1·6934	0·2414	1·3384	0·6385	0·3781	0·8798	0·8783	0·4649	0·2184
0·58	0·7376	1·6807	0·2542	1·3787	0·6431	0·3953	0·9394	0·9386	0·5006	0·2269
0·60	0·7531	1·6674	0·2671	1·4189	0·6468	0·4130	1·0000	1·0000	0·5376	0·2355
0·62	0·7679	1·6535	0·2801	1·4590	0·6496	0·4312	1·0614	1·0623	0·5760	0·2441
0·64	0·7821	1·6391	0·2931	1·4990	0·6515	0·4499	1·1235	1·1253	0·6157	0·2528
0·66	0·7957	1·6242	0·3062	1·5390	0·6526	0·4691	1·1862	1·1889	0·6567	0·2616
0·68	0·8086	1·6087	0·3192	1·5790	0·6527	0·4891	1·2493	1·2530	0·6991	0·2705
0·70	0·8208	1·5926	0·3322	1·6191	0·6506	0·5106	1·3125	1·3172	0·7435	0·2795
0·72	0·8322	1·5755	0·3452	1·6594	0·6462	0·5343	1·3755	1·3813	0·7902	0·2886
0·74	0·8425	1·5570	0·3581	1·7000	0·6391	0·5603	1·4378	1·4446	0·8393	0·2979
0·76	0·8518	1·5369	0·3708	1·7411	0·6295	0·5890	1·4989	1·5067	0·8911	0·3074
0·78	0·8598	1·5148	0·3832	1·7830	0·6171	0·6210	1·5583	1·5671	0·9457	0·3170
0·80	0·8663	1·4906	0·3954	1·8258	0·6018	0·6570	1·6155	1·6252	1·0037	0·3270
0·82	0·8713	1·4638	0·4073	1·8699	0·5834	0·6981	1·6699	1·6803	1·0657	0·3371
0·84	0·8745	1·4342	0·4187	1·9155	0·5616	0·7456	1·7208	1·7318	1·1323	0·3477
0·86	0·8756	1·4013	0·4297	1·9630	0·5359	0·8019	1·7674	1·7788	1·2051	0·3585
0·88	0·8746	1·3647	0·4402	2·0131	0·5057	0·8704	1·8089	1·8204	1·2859	0·3698
0·90	0·8708	1·3238	0·4499	2·0666	0·4702	0·9568	1·8441	1·8555	1·3782	0·3815
0·92	0·8640	1·2774	0·4589	2·1247	0·4281	1·0719	1·8717	1·8827	1·4879	0·3939
0·94	0·8531	1·2241	0·4670	2·1895	0·3772	1·2381	1·8898	1·8997	1·6272	0·4069
0·96	0·8369	1·1607	0·4739	2·2651	0·3131	1·5135	1·8950	1·9034	1·8258	0·4208
0·98	0·8118	1·0797	0·4794	2·3619	0·2250	2·1305	1·8807	1·8865	2·1911	0·4359
1·00	0·7445	0·9025	0·4824	2·5917	0·0000	-	1·7929	1·7920	-	0·4531

Form 2 egg-shape with 6% lining

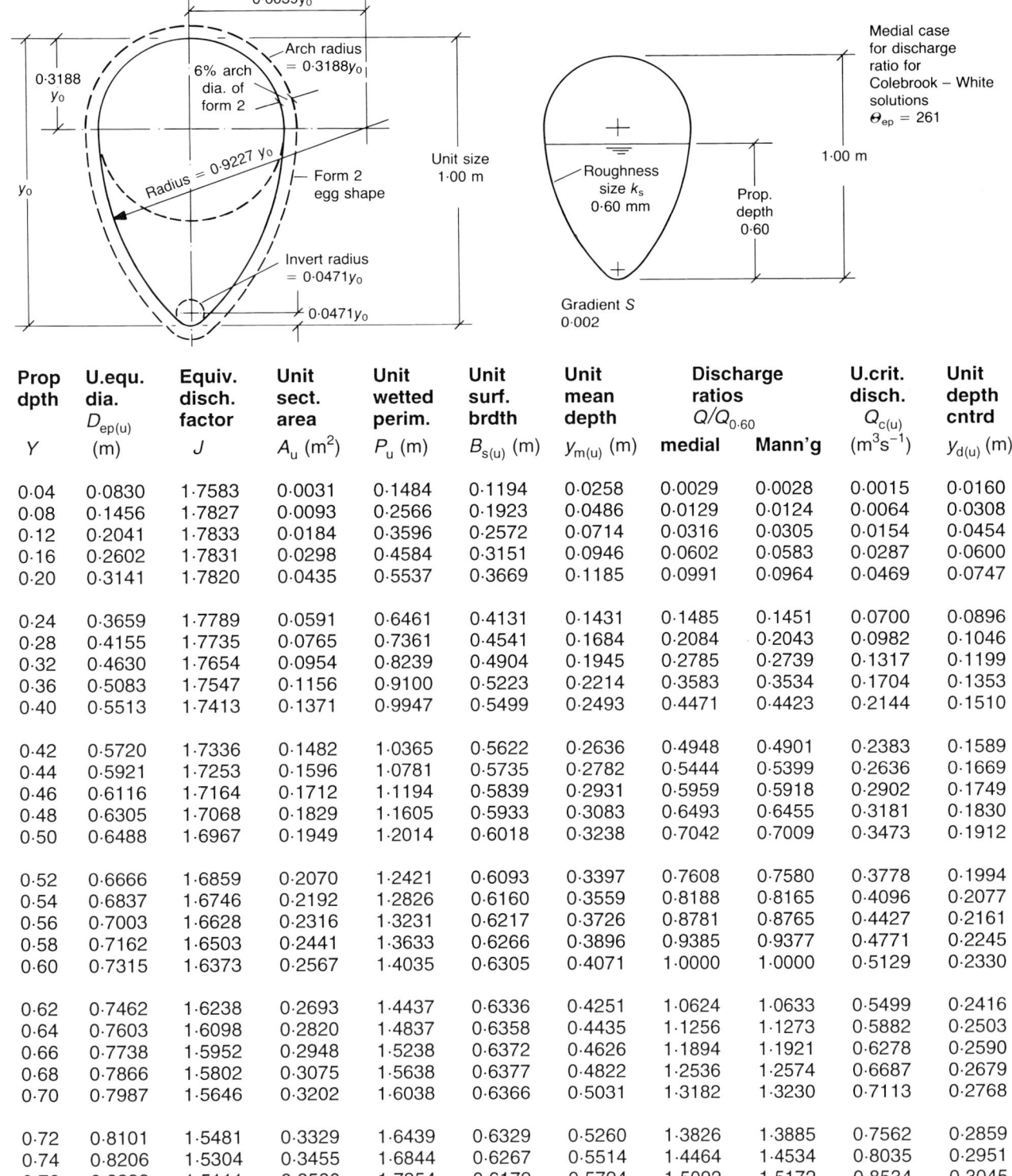

Prop dpth Y	U.equ. dia. $D_{ep(u)}$ (m)	Equiv. disch. factor J	Unit sect. area A_u (m²)	Unit wetted perim. P_u (m)	Unit surf. brdth $B_{s(u)}$ (m)	Unit mean depth $y_{m(u)}$ (m)	Discharge ratios $Q/Q_{0.60}$ medial	Mann'g	U.crit. disch. $Q_{c(u)}$ (m³s⁻¹)	Unit depth cntrd $y_{d(u)}$ (m)
0·04	0·0830	1·7583	0·0031	0·1484	0·1194	0·0258	0·0029	0·0028	0·0015	0·0160
0·08	0·1456	1·7827	0·0093	0·2566	0·1923	0·0486	0·0129	0·0124	0·0064	0·0308
0·12	0·2041	1·7833	0·0184	0·3596	0·2572	0·0714	0·0316	0·0305	0·0154	0·0454
0·16	0·2602	1·7831	0·0298	0·4584	0·3151	0·0946	0·0602	0·0583	0·0287	0·0600
0·20	0·3141	1·7820	0·0435	0·5537	0·3669	0·1185	0·0991	0·0964	0·0469	0·0747
0·24	0·3659	1·7789	0·0591	0·6461	0·4131	0·1431	0·1485	0·1451	0·0700	0·0896
0·28	0·4155	1·7735	0·0765	0·7361	0·4541	0·1684	0·2084	0·2043	0·0982	0·1046
0·32	0·4630	1·7654	0·0954	0·8239	0·4904	0·1945	0·2785	0·2739	0·1317	0·1199
0·36	0·5083	1·7547	0·1156	0·9100	0·5223	0·2214	0·3583	0·3534	0·1704	0·1353
0·40	0·5513	1·7413	0·1371	0·9947	0·5499	0·2493	0·4471	0·4423	0·2144	0·1510
0·42	0·5720	1·7336	0·1482	1·0365	0·5622	0·2636	0·4948	0·4901	0·2383	0·1589
0·44	0·5921	1·7253	0·1596	1·0781	0·5735	0·2782	0·5444	0·5399	0·2636	0·1669
0·46	0·6116	1·7164	0·1712	1·1194	0·5839	0·2931	0·5959	0·5918	0·2902	0·1749
0·48	0·6305	1·7068	0·1829	1·1605	0·5933	0·3083	0·6493	0·6455	0·3181	0·1830
0·50	0·6488	1·6967	0·1949	1·2014	0·6018	0·3238	0·7042	0·7009	0·3473	0·1912
0·52	0·6666	1·6859	0·2070	1·2421	0·6093	0·3397	0·7608	0·7580	0·3778	0·1994
0·54	0·6837	1·6746	0·2192	1·2826	0·6160	0·3559	0·8188	0·8165	0·4096	0·2077
0·56	0·7003	1·6628	0·2316	1·3231	0·6217	0·3726	0·8781	0·8765	0·4427	0·2161
0·58	0·7162	1·6503	0·2441	1·3633	0·6266	0·3896	0·9385	0·9377	0·4771	0·2245
0·60	0·7315	1·6373	0·2567	1·4035	0·6305	0·4071	1·0000	1·0000	0·5129	0·2330
0·62	0·7462	1·6238	0·2693	1·4437	0·6336	0·4251	1·0624	1·0633	0·5499	0·2416
0·64	0·7603	1·6098	0·2820	1·4837	0·6358	0·4435	1·1256	1·1273	0·5882	0·2503
0·66	0·7738	1·5952	0·2948	1·5238	0·6372	0·4626	1·1894	1·1921	0·6278	0·2590
0·68	0·7866	1·5802	0·3075	1·5638	0·6377	0·4822	1·2536	1·2574	0·6687	0·2679
0·70	0·7987	1·5646	0·3202	1·6038	0·6366	0·5031	1·3182	1·3230	0·7113	0·2768
0·72	0·8101	1·5481	0·3329	1·6439	0·6329	0·5260	1·3826	1·3885	0·7562	0·2859
0·74	0·8206	1·5304	0·3455	1·6844	0·6267	0·5514	1·4464	1·4534	0·8035	0·2951
0·76	0·8299	1·5111	0·3580	1·7254	0·6179	0·5794	1·5092	1·5172	0·8534	0·3045
0·78	0·8381	1·4900	0·3702	1·7671	0·6063	0·6107	1·5704	1·5793	0·9061	0·3141
0·80	0·8449	1·4667	0·3822	1·8096	0·5917	0·6460	1·6294	1·6392	0·9620	0·3239
0·82	0·8501	1·4410	0·3939	1·8534	0·5740	0·6862	1·6856	1·6962	1·0218	0·3340
0·84	0·8536	1·4124	0·4052	1·8986	0·5529	0·7328	1·7383	1·7496	1·0861	0·3445
0·86	0·8551	1·3806	0·4160	1·9458	0·5279	0·7880	1·7867	1·7984	1·1564	0·3552
0·88	0·8544	1·3452	0·4263	1·9955	0·4985	0·8551	1·8299	1·8418	1·2344	0·3664
0·90	0·8512	1·3053	0·4359	2·0485	0·4638	0·9399	1·8667	1·8786	1·3234	0·3781
0·92	0·8448	1·2601	0·4448	2·1060	0·4224	1·0528	1·8958	1·9072	1·4291	0·3904
0·94	0·8345	1·2080	0·4527	2·1701	0·3723	1·2159	1·9151	1·9256	1·5633	0·4033
0·96	0·8189	1·1459	0·4596	2·2449	0·3092	1·4861	1·9213	1·9303	1·7545	0·4172
0·98	0·7945	1·0664	0·4649	2·3407	0·2223	2·0916	1·9075	1·9140	2·1057	0·4323
1·00	0·7289	0·8918	0·4679	2·5677	0·0000	-	1·8190	1·8187	-	0·4495

C8

4:3 egg-shape (WRc)

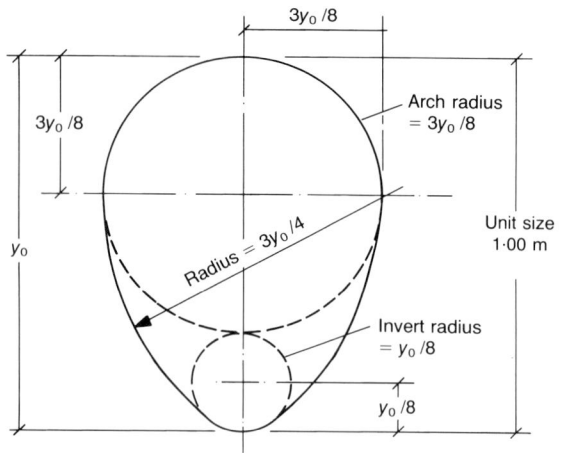

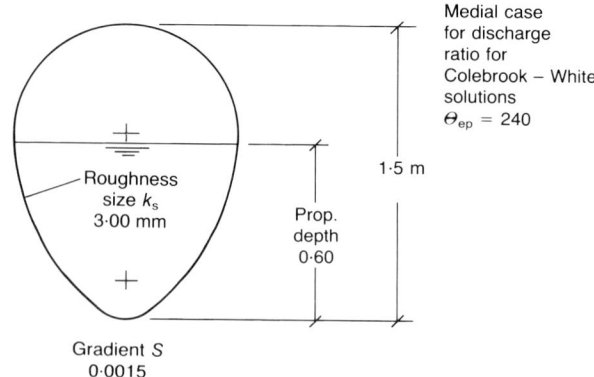

Prop dpth Y	U.equ. dia. $D_{ep(u)}$ (m)	Equiv. disch. factor J	Unit sect. area A_u (m²)	Unit wetted perim. P_u (m)	Unit surf. brdth $B_{s(u)}$ (m)	Unit mean depth $y_{m(u)}$ (m)	Discharge ratios $Q/Q_{0.60}$ medial	Mann'g	U.crit. disch. $Q_{c(u)}$ (m³s⁻¹)	Unit depth cntrd $y_{d(u)}$ (m)
0·02	0·0513	1·1246	0·0018	0·1434	0·1356	0·0136	0·0008	0·0009	0·0007	0·0081
0·04	0·0972	1·4551	0·0051	0·2098	0·1887	0·0270	0·0037	0·0036	0·0026	0·0162
0·06	0·1375	1·5874	0·0094	0·2721	0·2365	0·0396	0·0085	0·0084	0·0058	0·0241
0·08	0·1753	1·6603	0·0145	0·3316	0·2805	0·0518	0·0157	0·0154	0·0104	0·0318
0·10	0·2115	1·7093	0·0206	0·3887	0·3212	0·0640	0·0251	0·0247	0·0163	0·0395
0·11	0·2292	1·7287	0·0239	0·4165	0·3405	0·0701	0·0308	0·0302	0·0198	0·0433
0·12	0·2466	1·7460	0·0274	0·4438	0·3590	0·0762	0·0371	0·0364	0·0237	0·0472
0·13	0·2638	1·7613	0·0310	0·4706	0·3769	0·0824	0·0440	0·0432	0·0279	0·0510
0·14	0·2808	1·7752	0·0349	0·4970	0·3942	0·0885	0·0516	0·0506	0·0325	0·0548
0·15	0·2977	1·7878	0·0389	0·5230	0·4108	0·0947	0·0598	0·0587	0·0375	0·0586
0·16	0·3143	1·7993	0·0431	0·5487	0·4269	0·1010	0·0687	0·0674	0·0429	0·0624
0·17	0·3307	1·8098	0·0475	0·5740	0·4424	0·1073	0·0782	0·0767	0·0487	0·0662
0·18	0·3469	1·8195	0·0520	0·5990	0·4574	0·1136	0·0883	0·0867	0·0548	0·0701
0·19	0·3630	1·8284	0·0566	0·6237	0·4719	0·1199	0·0991	0·0974	0·0614	0·0739
0·20	0·3789	1·8365	0·0614	0·6481	0·4859	0·1263	0·1106	0·1087	0·0683	0·0777
0·21	0·3946	1·8440	0·0663	0·6723	0·4994	0·1328	0·1227	0·1206	0·0757	0·0816
0·22	0·4101	1·8508	0·0714	0·6962	0·5125	0·1393	0·1354	0·1332	0·0834	0·0855
0·23	0·4255	1·8571	0·0766	0·7198	0·5251	0·1458	0·1488	0·1464	0·0916	0·0893
0·24	0·4407	1·8627	0·0819	0·7432	0·5373	0·1524	0·1628	0·1603	0·1001	0·0932
0·25	0·4557	1·8679	0·0873	0·7664	0·5490	0·1590	0·1774	0·1748	0·1090	0·0971
0·26	0·4705	1·8725	0·0929	0·7894	0·5604	0·1657	0·1926	0·1899	0·1184	0·1010
0·27	0·4852	1·8767	0·0985	0·8122	0·5713	0·1724	0·2085	0·2056	0·1281	0·1049
0·28	0·4997	1·8803	0·1043	0·8348	0·5819	0·1792	0·2249	0·2220	0·1383	0·1088
0·29	0·5140	1·8836	0·1102	0·8573	0·5921	0·1861	0·2420	0·2389	0·1488	0·1128
0·30	0·5281	1·8864	0·1161	0·8795	0·6019	0·1929	0·2596	0·2565	0·1597	0·1167
0·31	0·5421	1·8887	0·1222	0·9017	0·6113	0·1999	0·2778	0·2746	0·1711	0·1207
0·32	0·5559	1·8907	0·1284	0·9236	0·6204	0·2069	0·2966	0·2933	0·1828	0·1246
0·33	0·5695	1·8922	0·1346	0·9454	0·6291	0·2140	0·3160	0·3126	0·1950	0·1286
0·34	0·5829	1·8934	0·1409	0·9671	0·6375	0·2211	0·3358	0·3324	0·2075	0·1326
0·35	0·5961	1·8942	0·1473	0·9887	0·6455	0·2283	0·3563	0·3528	0·2205	0·1366
0·36	0·6092	1·8946	0·1538	1·0101	0·6532	0·2355	0·3772	0·3737	0·2338	0·1406
0·37	0·6221	1·8947	0·1604	1·0314	0·6606	0·2428	0·3986	0·3952	0·2475	0·1447
0·38	0·6348	1·8944	0·1671	1·0527	0·6677	0·2502	0·4206	0·4171	0·2617	0·1487
0·39	0·6473	1·8938	0·1738	1·0738	0·6745	0·2576	0·4430	0·4395	0·2762	0·1528
0·40	0·6596	1·8929	0·1805	1·0948	0·6809	0·2651	0·4659	0·4625	0·2911	0·1569
0·41	0·6718	1·8916	0·1874	1·1157	0·6870	0·2727	0·4893	0·4859	0·3064	0·1610
0·42	0·6838	1·8900	0·1943	1·1365	0·6929	0·2804	0·5131	0·5097	0·3222	0·1651
0·43	0·6956	1·8881	0·2012	1·1573	0·6984	0·2881	0·5373	0·5340	0·3383	0·1692
0·44	0·7072	1·8859	0·2082	1·1780	0·7037	0·2960	0·5619	0·5588	0·3548	0·1733
0·45	0·7186	1·8834	0·2153	1·1986	0·7086	0·3039	0·5870	0·5839	0·3717	0·1775

Prop dpth	U.equ. dia. $D_{ep(u)}$	Equiv. disch. factor	Unit sect. area	Unit wetted perim.	Unit surf. brdth	Unit mean depth	Discharge ratios $Q/Q_{0.60}$		U.crit. disch. $Q_{c(u)}$	Unit depth cntrd
Y	(m)	J	A_u (m^2)	P_u (m)	$B_{s(u)}$ (m)	$y_{m(u)}$ (m)	medial	Mann'g	(m^{3}s^{-1})	$y_{d(u)}$ (m)
0.46	0.7298	1.8806	0.2224	1.2191	0.7132	0.3118	0.6124	0.6095	0.3890	0.1816
0.47	0.7408	1.8775	0.2296	1.2396	0.7176	0.3199	0.6382	0.6354	0.4066	0.1858
0.48	0.7517	1.8741	0.2368	1.2600	0.7217	0.3281	0.6644	0.6617	0.4247	0.1900
0.49	0.7623	1.8705	0.2440	1.2803	0.7255	0.3363	0.6909	0.6883	0.4431	0.1942
0.50	0.7728	1.8665	0.2513	1.3006	0.7290	0.3447	0.7177	0.7153	0.4620	0.1985
0.51	0.7831	1.8623	0.2586	1.3209	0.7323	0.3531	0.7448	0.7426	0.4812	0.2027
0.52	0.7931	1.8579	0.2659	1.3411	0.7352	0.3617	0.7722	0.7702	0.5008	0.2070
0.53	0.8030	1.8532	0.2733	1.3613	0.7379	0.3704	0.7999	0.7981	0.5208	0.2113
0.54	0.8127	1.8482	0.2807	1.3815	0.7403	0.3791	0.8279	0.8263	0.5412	0.2156
0.55	0.8222	1.8429	0.2881	1.4016	0.7425	0.3880	0.8561	0.8547	0.5620	0.2199
0.56	0.8315	1.8374	0.2955	1.4217	0.7444	0.3970	0.8845	0.8834	0.5831	0.2242
0.57	0.8406	1.8317	0.3030	1.4417	0.7460	0.4062	0.9131	0.9123	0.6047	0.2286
0.58	0.8495	1.8257	0.3104	1.4618	0.7473	0.4154	0.9419	0.9413	0.6266	0.2330
0.59	0.8582	1.8195	0.3179	1.4818	0.7484	0.4248	0.9709	0.9706	0.6489	0.2374
0.60	0.8667	1.8130	0.3254	1.5018	0.7492	0.4344	1.0000	1.0000	0.6716	0.2418
0.61	0.8750	1.8063	0.3329	1.5218	0.7497	0.4441	1.0293	1.0296	0.6947	0.2463
0.62	0.8831	1.7994	0.3404	1.5418	0.7500	0.4539	1.0587	1.0592	0.7182	0.2507
0.63	0.8910	1.7923	0.3479	1.5618	0.7499	0.4639	1.0881	1.0890	0.7421	0.2552
0.64	0.8987	1.7849	0.3554	1.5818	0.7494	0.4743	1.1177	1.1189	0.7665	0.2597
0.65	0.9062	1.7772	0.3629	1.6019	0.7483	0.4849	1.1473	1.1488	0.7914	0.2643
0.66	0.9134	1.7692	0.3704	1.6219	0.7467	0.4960	1.1768	1.1787	0.8168	0.2688
0.67	0.9204	1.7609	0.3778	1.6420	0.7446	0.5074	1.2063	1.2085	0.8428	0.2734
0.68	0.9271	1.7522	0.3853	1.6622	0.7419	0.5193	1.2358	1.2383	0.8694	0.2780
0.69	0.9335	1.7431	0.3927	1.6825	0.7386	0.5316	1.2651	1.2679	0.8965	0.2827
0.70	0.9397	1.7336	0.4000	1.7028	0.7348	0.5444	1.2942	1.2973	0.9243	0.2874
0.71	0.9455	1.7237	0.4074	1.7233	0.7305	0.5577	1.3231	1.3266	0.9526	0.2922
0.72	0.9511	1.7132	0.4146	1.7439	0.7255	0.5715	1.3517	1.3555	0.9816	0.2969
0.73	0.9563	1.7024	0.4219	1.7647	0.7200	0.5859	1.3801	1.3842	1.0112	0.3018
0.74	0.9611	1.6910	0.4290	1.7856	0.7139	0.6010	1.4081	1.4125	1.0416	0.3066
0.75	0.9656	1.6790	0.4361	1.8067	0.7071	0.6168	1.4356	1.4404	1.0727	0.3116
0.76	0.9697	1.6665	0.4432	1.8280	0.6997	0.6334	1.4627	1.4678	1.1045	0.3165
0.77	0.9735	1.6535	0.4501	1.8496	0.6917	0.6508	1.4893	1.4946	1.1372	0.3216
0.78	0.9768	1.6398	0.4570	1.8714	0.6829	0.6692	1.5154	1.5209	1.1707	0.3267
0.79	0.9797	1.6255	0.4638	1.8935	0.6735	0.6886	1.5408	1.5466	1.2052	0.3318
0.80	0.9822	1.6105	0.4705	1.9160	0.6633	0.7093	1.5655	1.5715	1.2408	0.3370
0.81	0.9842	1.5949	0.4771	1.9388	0.6524	0.7313	1.5895	1.5957	1.2775	0.3423
0.82	0.9858	1.5785	0.4835	1.9620	0.6406	0.7548	1.6126	1.6190	1.3155	0.3477
0.83	0.9868	1.5613	0.4899	1.9856	0.6280	0.7800	1.6349	1.6414	1.3548	0.3531
0.84	0.9873	1.5434	0.4961	2.0097	0.6145	0.8073	1.6561	1.6628	1.3958	0.3586
0.85	0.9873	1.5246	0.5022	2.0344	0.6000	0.8369	1.6764	1.6831	1.4386	0.3642
0.86	0.9867	1.5048	0.5081	2.0598	0.5845	0.8693	1.6955	1.7022	1.4834	0.3699
0.87	0.9854	1.4842	0.5138	2.0858	0.5678	0.9050	1.7133	1.7200	1.5307	0.3757
0.88	0.9835	1.4624	0.5194	2.1126	0.5499	0.9446	1.7298	1.7365	1.5809	0.3816
0.89	0.9808	1.4396	0.5248	2.1404	0.5307	0.9890	1.7448	1.7514	1.6345	0.3876
0.90	0.9774	1.4154	0.5300	2.1692	0.5099	1.0395	1.7581	1.7646	1.6923	0.3938
0.91	0.9731	1.3900	0.5350	2.1993	0.4874	1.0976	1.7697	1.7760	1.7553	0.4000
0.92	0.9678	1.3629	0.5398	2.2309	0.4630	1.1657	1.7793	1.7853	1.8251	0.4065
0.93	0.9615	1.3341	0.5443	2.2642	0.4363	1.2473	1.7867	1.7924	1.9036	0.4131
0.94	0.9540	1.3032	0.5485	2.2998	0.4069	1.3479	1.7916	1.7968	1.9941	0.4199
0.95	0.9450	1.2697	0.5524	2.3382	0.3742	1.4764	1.7936	1.7983	2.1019	0.4269
0.96	0.9343	1.2330	0.5560	2.3803	0.3370	1.6495	1.7921	1.7961	2.2361	0.4341
0.97	0.9212	1.1919	0.5591	2.4279	0.2939	1.9022	1.7862	1.7894	2.4149	0.4416
0.98	0.9047	1.1443	0.5618	2.4839	0.2417	2.3248	1.7743	1.7766	2.6825	0.4495
0.99	0.8824	1.0844	0.5639	2.5563	0.1720	3.2776	1.7528	1.7536	3.1970	0.4578
1.00	0.8279	0.9528	0.5651	2.7299	0.0000	-	1.6865	1.6842	-	0.4669

C9

4:3 egg-shape with 3% lining

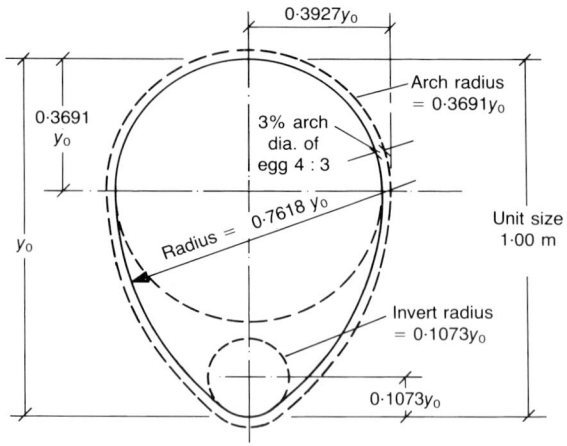

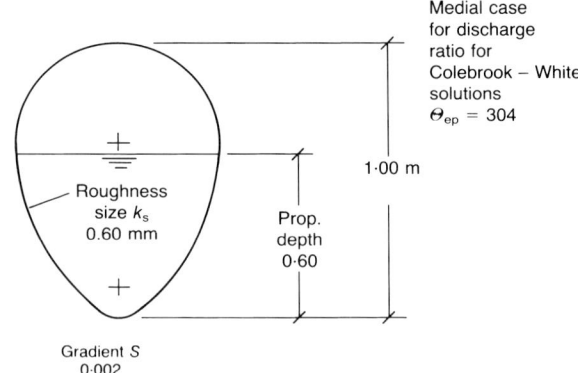

Medial case for discharge ratio for Colebrook – White solutions $\Theta_{ep} = 304$

Prop dpth Y	U.equ. dia. $D_{ep(u)}$ (m)	Equiv. disch. factor J	Unit sect. area A_u (m²)	Unit wetted perim. P_u (m)	Unit surf. brdth $B_{s(u)}$ (m)	Unit mean depth $y_{m(u)}$ (m)	Discharge ratios $Q/Q_{0.60}$ medial	Discharge ratios $Q/Q_{0.60}$ Mann'g	U.crit. disch. $Q_{c(u)}$ (m³s⁻¹)	Unit depth cntrd $y_{d(u)}$ (m)
0·04	0·0951	1·5065	0·0047	0·1983	0·1762	0·0268	0·0036	0·0034	0·0024	0·0161
0·08	0·1707	1·6795	0·0136	0·3193	0·2670	0·0511	0·0154	0·0147	0·0096	0·0316
0·12	0·2404	1·7524	0·0259	0·4310	0·3448	0·0751	0·0365	0·0351	0·0222	0·0468
0·16	0·3067	1·7989	0·0411	0·5357	0·4123	0·0996	0·0678	0·0655	0·0406	0·0619
0·20	0·3702	1·8317	0·0588	0·6350	0·4711	0·1248	0·1095	0·1063	0·0650	0·0771
0·24	0·4310	1·8549	0·0787	0·7300	0·5224	0·1506	0·1615	0·1575	0·0956	0·0925
0·28	0·4891	1·8702	0·1005	0·8216	0·5670	0·1772	0·2235	0·2188	0·1324	0·1080
0·32	0·5445	1·8789	0·1239	0·9104	0·6056	0·2047	0·2951	0·2900	0·1756	0·1237
0·36	0·5971	1·8815	0·1488	0·9970	0·6386	0·2331	0·3756	0·3703	0·2250	0·1396
0·40	0·6469	1·8788	0·1750	1·0818	0·6666	0·2625	0·4644	0·4592	0·2807	0·1558
0·42	0·6708	1·8755	0·1884	1·1236	0·6787	0·2776	0·5116	0·5066	0·3109	0·1639
0·44	0·6939	1·8711	0·2021	1·1650	0·6896	0·2931	0·5605	0·5558	0·3426	0·1722
0·46	0·7163	1·8655	0·2160	1·2062	0·6994	0·3088	0·6111	0·6067	0·3759	0·1804
0·48	0·7379	1·8588	0·2301	1·2471	0·7080	0·3249	0·6632	0·6592	0·4107	0·1888
0·50	0·7588	1·8510	0·2443	1·2878	0·7156	0·3414	0·7166	0·7131	0·4470	0·1972
0·52	0·7790	1·8422	0·2587	1·3284	0·7220	0·3583	0·7713	0·7684	0·4849	0·2057
0·54	0·7983	1·8324	0·2732	1·3687	0·7273	0·3756	0·8272	0·8248	0·5243	0·2142
0·56	0·8170	1·8216	0·2878	1·4089	0·7316	0·3933	0·8840	0·8824	0·5652	0·2229
0·58	0·8348	1·8099	0·3024	1·4491	0·7348	0·4116	0·9416	0·9408	0·6076	0·2316
0·60	0·8519	1·7972	0·3172	1·4891	0·7370	0·4303	1·0000	1·0000	0·6515	0·2404
0·62	0·8682	1·7837	0·3319	1·5291	0·7381	0·4497	1·0590	1·0598	0·6970	0·2492
0·64	0·8837	1·7692	0·3467	1·5691	0·7380	0·4697	1·1183	1·1201	0·7441	0·2582
0·66	0·8984	1·7538	0·3614	1·6092	0·7359	0·4911	1·1779	1·1806	0·7931	0·2673
0·68	0·9120	1·7371	0·3761	1·6494	0·7317	0·5140	1·2373	1·2410	0·8444	0·2764
0·70	0·9247	1·7189	0·3907	1·6900	0·7252	0·5387	1·2963	1·3009	0·8979	0·2858
0·72	0·9361	1·6990	0·4051	1·7309	0·7164	0·5655	1·3544	1·3601	0·9539	0·2952
0·74	0·9463	1·6772	0·4193	1·7725	0·7052	0·5946	1·4114	1·4180	1·0125	0·3049
0·76	0·9550	1·6533	0·4333	1·8147	0·6916	0·6265	1·4668	1·4742	1·0739	0·3147
0·78	0·9623	1·6271	0·4469	1·8579	0·6753	0·6618	1·5202	1·5284	1·1387	0·3248
0·80	0·9678	1·5984	0·4603	1·9023	0·6562	0·7014	1·5712	1·5801	1·2072	0·3351
0·82	0·9716	1·5669	0·4732	1·9480	0·6340	0·7464	1·6191	1·6286	1·2801	0·3457
0·84	0·9734	1·5324	0·4856	1·9955	0·6083	0·7983	1·6635	1·6734	1·3587	0·3566
0·86	0·9729	1·4944	0·4975	2·0453	0·5788	0·8595	1·7037	1·7138	1·4443	0·3678
0·88	0·9700	1·4526	0·5087	2·0978	0·5447	0·9339	1·7390	1·7490	1·5395	0·3795
0·90	0·9642	1·4062	0·5192	2·1540	0·5053	1·0277	1·7683	1·7781	1·6484	0·3916
0·92	0·9550	1·3543	0·5289	2·2152	0·4589	1·1524	1·7905	1·7996	1·7780	0·4043
0·94	0·9415	1·2952	0·5375	2·2837	0·4035	1·3323	1·8037	1·8117	1·9430	0·4176
0·96	0·9222	1·2257	0·5449	2·3637	0·3342	1·6304	1·8052	1·8115	2·1790	0·4318
0·98	0·8932	1·1377	0·5507	2·4664	0·2397	2·2976	1·7885	1·7921	2·6142	0·4472
1·00	0·8175	0·9475	0·5540	2·7105	0·0000	-	1·7023	1·6993	-	0·4645

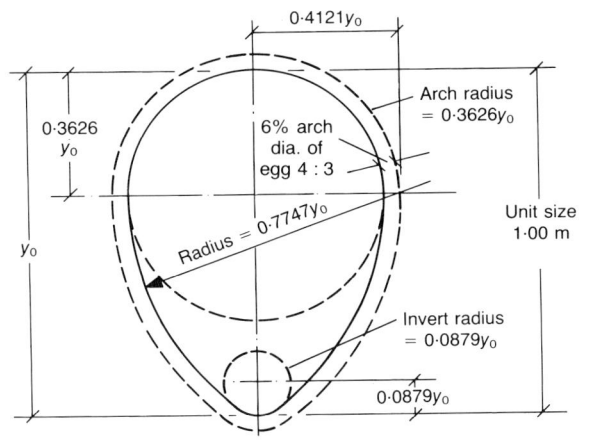

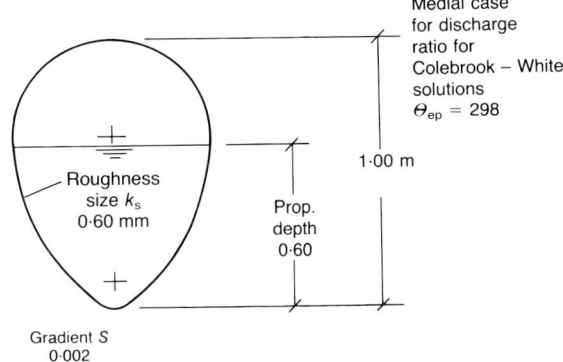

Prop dpth	U.equ. dia. $D_{ep(u)}$	Equiv. disch. factor	Unit sect. area	Unit wetted perim.	Unit surf. brdth	Unit mean depth	Discharge ratios $Q/Q_{0.60}$		U.crit. disch. $Q_{c(u)}$	Unit depth cntrd
Y	(m)	J	A_u (m²)	P_u (m)	$B_{s(u)}$ (m)	$y_{m(u)}$ (m)	medial	Mann'g	(m³s⁻¹)	$y_{d(u)}$ (m)
0·04	0·0922	1·5597	0·0043	0·1856	0·1624	0·0263	0·0033	0·0032	0·0022	0·0161
0·08	0·1650	1·6954	0·0126	0·3058	0·2520	0·0500	0·0145	0·0139	0·0088	0·0313
0·12	0·2329	1·7549	0·0243	0·4169	0·3291	0·0737	0·0349	0·0336	0·0206	0·0462
0·16	0·2978	1·7946	0·0388	0·5213	0·3962	0·0980	0·0655	0·0633	0·0380	0·0612
0·20	0·3601	1·8232	0·0559	0·6205	0·4548	0·1228	0·1066	0·1035	0·0613	0·0763
0·24	0·4198	1·8434	0·0751	0·7155	0·5059	0·1484	0·1580	0·1541	0·0906	0·0915
0·28	0·4770	1·8567	0·0962	0·8071	0·5506	0·1748	0·2197	0·2151	0·1260	0·1069
0·32	0·5315	1·8637	0·1191	0·8960	0·5893	0·2020	0·2910	0·2860	0·1676	0·1226
0·36	0·5834	1·8652	0·1433	0·9826	0·6226	0·2302	0·3715	0·3663	0·2153	0·1384
0·40	0·6325	1·8615	0·1688	1·0675	0·6508	0·2594	0·4604	0·4553	0·2692	0·1545
0·42	0·6560	1·8579	0·1819	1·1093	0·6630	0·2744	0·5078	0·5028	0·2985	0·1626
0·44	0·6789	1·8532	0·1953	1·1508	0·6742	0·2897	0·5570	0·5522	0·3292	0·1708
0·46	0·7010	1·8473	0·2089	1·1920	0·6841	0·3053	0·6078	0·6034	0·3615	0·1790
0·48	0·7224	1·8405	0·2227	1·2330	0·6930	0·3213	0·6602	0·6562	0·3953	0·1873
0·50	0·7430	1·8326	0·2366	1·2738	0·7007	0·3377	0·7140	0·7105	0·4306	0·1957
0·52	0·7630	1·8237	0·2507	1·3143	0·7074	0·3544	0·7691	0·7662	0·4673	0·2041
0·54	0·7822	1·8138	0·2649	1·3547	0·7130	0·3715	0·8254	0·8231	0·5056	0·2127
0·56	0·8006	1·8030	0·2792	1·3950	0·7175	0·3891	0·8828	0·8812	0·5454	0·2212
0·58	0·8183	1·7913	0·2936	1·4351	0·7210	0·4072	0·9410	0·9402	0·5867	0·2299
0·60	0·8353	1·7787	0·3080	1·4752	0·7235	0·4258	1·0000	1·0000	0·6294	0·2387
0·62	0·8514	1·7653	0·3225	1·5152	0·7249	0·4449	1·0596	1·0605	0·6737	0·2475
0·64	0·8668	1·7510	0·3370	1·5552	0·7253	0·4647	1·1197	1·1215	0·7195	0·2564
0·66	0·8814	1·7358	0·3515	1·5952	0·7239	0·4856	1·1801	1·1828	0·7671	0·2654
0·68	0·8951	1·7194	0·3660	1·6354	0·7202	0·5081	1·2404	1·2442	0·8169	0·2746
0·70	0·9078	1·7017	0·3803	1·6758	0·7144	0·5324	1·3004	1·3051	0·8690	0·2838
0·72	0·9193	1·6823	0·3945	1·7167	0·7062	0·5587	1·3596	1·3653	0·9234	0·2932
0·74	0·9295	1·6610	0·4085	1·7581	0·6956	0·5873	1·4177	1·4243	0·9805	0·3028
0·76	0·9384	1·6377	0·4223	1·8001	0·6825	0·6188	1·4742	1·4818	1·0403	0·3126
0·78	0·9458	1·6121	0·4358	1·8431	0·6668	0·6536	1·5288	1·5372	1·1034	0·3226
0·80	0·9516	1·5841	0·4490	1·8872	0·6482	0·6926	1·5809	1·5900	1·1702	0·3329
0·82	0·9556	1·5533	0·4617	1·9327	0·6266	0·7369	1·6300	1·6397	1·2413	0·3434
0·84	0·9576	1·5194	0·4740	1·9800	0·6015	0·7881	1·6756	1·6857	1·3178	0·3543
0·86	0·9575	1·4822	0·4858	2·0294	0·5725	0·8485	1·7169	1·7273	1·4013	0·3655
0·88	0·9549	1·4411	0·4969	2·0816	0·5390	0·9219	1·7533	1·7636	1·4940	0·3771
0·90	0·9494	1·3954	0·5073	2·1374	0·5001	1·0144	1·7836	1·7937	1·6000	0·3891
0·92	0·9405	1·3442	0·5169	2·1981	0·4544	1·1374	1·8067	1·8161	1·7262	0·4017
0·94	0·9275	1·2858	0·5254	2·2660	0·3996	1·3149	1·8207	1·8290	1·8867	0·4150
0·96	0·9086	1·2171	0·5327	2·3453	0·3311	1·6089	1·8226	1·8293	2·1162	0·4292
0·98	0·8802	1·1299	0·5385	2·4472	0·2375	2·2670	1·8062	1·8103	2·5391	0·4445
1·00	0·8057	0·9412	0·5417	2·6892	0·0000	-	1·7195	1·7168	-	0·4618

C11

U-shaped (free surface)

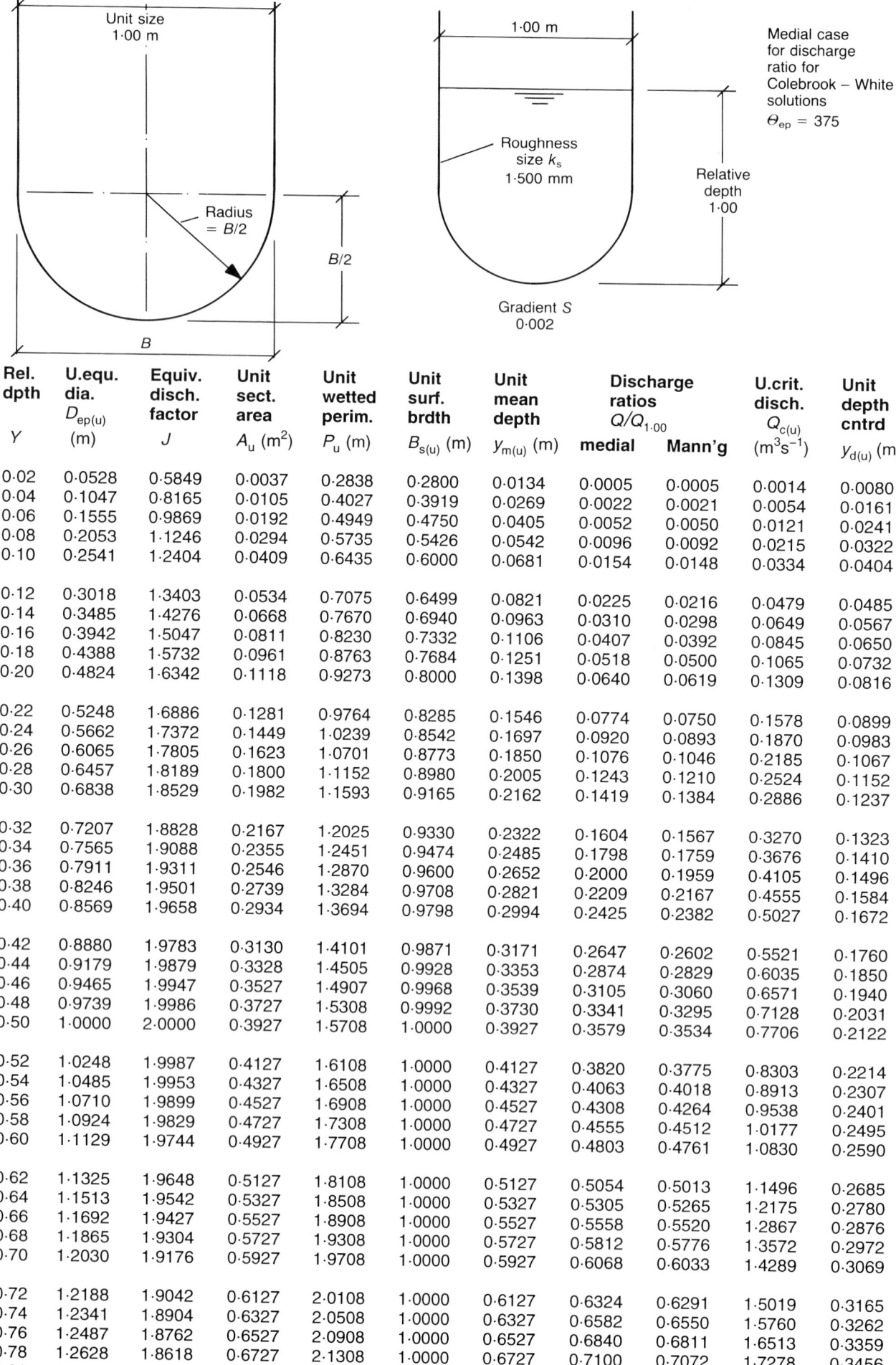

Unit size 1·00 m

Radius = B/2

B/2

B

1·00 m

Roughness size k_s 1·500 mm

Medial case for discharge ratio for Colebrook – White solutions $\Theta_{ep} = 375$

Relative depth 1·00

Gradient S 0·002

Rel. dpth Y	U.equ. dia. $D_{ep(u)}$ (m)	Equiv. disch. factor J	Unit sect. area A_u (m²)	Unit wetted perim. P_u (m)	Unit surf. brdth $B_{s(u)}$ (m)	Unit mean depth $y_{m(u)}$ (m)	Discharge ratios $Q/Q_{1·00}$ medial	Mann'g	U.crit. disch. $Q_{c(u)}$ (m³s⁻¹)	Unit depth cntrd $y_{d(u)}$ (m)
0·02	0·0528	0·5849	0·0037	0·2838	0·2800	0·0134	0·0005	0·0005	0·0014	0·0080
0·04	0·1047	0·8165	0·0105	0·4027	0·3919	0·0269	0·0022	0·0021	0·0054	0·0161
0·06	0·1555	0·9869	0·0192	0·4949	0·4750	0·0405	0·0052	0·0050	0·0121	0·0241
0·08	0·2053	1·1246	0·0294	0·5735	0·5426	0·0542	0·0096	0·0092	0·0215	0·0322
0·10	0·2541	1·2404	0·0409	0·6435	0·6000	0·0681	0·0154	0·0148	0·0334	0·0404
0·12	0·3018	1·3403	0·0534	0·7075	0·6499	0·0821	0·0225	0·0216	0·0479	0·0485
0·14	0·3485	1·4276	0·0668	0·7670	0·6940	0·0963	0·0310	0·0298	0·0649	0·0567
0·16	0·3942	1·5047	0·0811	0·8230	0·7332	0·1106	0·0407	0·0392	0·0845	0·0650
0·18	0·4388	1·5732	0·0961	0·8763	0·7684	0·1251	0·0518	0·0500	0·1065	0·0732
0·20	0·4824	1·6342	0·1118	0·9273	0·8000	0·1398	0·0640	0·0619	0·1309	0·0816
0·22	0·5248	1·6886	0·1281	0·9764	0·8285	0·1546	0·0774	0·0750	0·1578	0·0899
0·24	0·5662	1·7372	0·1449	1·0239	0·8542	0·1697	0·0920	0·0893	0·1870	0·0983
0·26	0·6065	1·7805	0·1623	1·0701	0·8773	0·1850	0·1076	0·1046	0·2185	0·1067
0·28	0·6457	1·8189	0·1800	1·1152	0·8980	0·2005	0·1243	0·1210	0·2524	0·1152
0·30	0·6838	1·8529	0·1982	1·1593	0·9165	0·2162	0·1419	0·1384	0·2886	0·1237
0·32	0·7207	1·8828	0·2167	1·2025	0·9330	0·2322	0·1604	0·1567	0·3270	0·1323
0·34	0·7565	1·9088	0·2355	1·2451	0·9474	0·2485	0·1798	0·1759	0·3676	0·1410
0·36	0·7911	1·9311	0·2546	1·2870	0·9600	0·2652	0·2000	0·1959	0·4105	0·1496
0·38	0·8246	1·9501	0·2739	1·3284	0·9708	0·2821	0·2209	0·2167	0·4555	0·1584
0·40	0·8569	1·9658	0·2934	1·3694	0·9798	0·2994	0·2425	0·2382	0·5027	0·1672
0·42	0·8880	1·9783	0·3130	1·4101	0·9871	0·3171	0·2647	0·2602	0·5521	0·1760
0·44	0·9179	1·9879	0·3328	1·4505	0·9928	0·3353	0·2874	0·2829	0·6035	0·1850
0·46	0·9465	1·9947	0·3527	1·4907	0·9968	0·3539	0·3105	0·3060	0·6571	0·1940
0·48	0·9739	1·9986	0·3727	1·5308	0·9992	0·3730	0·3341	0·3295	0·7128	0·2031
0·50	1·0000	2·0000	0·3927	1·5708	1·0000	0·3927	0·3579	0·3534	0·7706	0·2122
0·52	1·0248	1·9987	0·4127	1·6108	1·0000	0·4127	0·3820	0·3775	0·8303	0·2214
0·54	1·0485	1·9953	0·4327	1·6508	1·0000	0·4327	0·4063	0·4018	0·8913	0·2307
0·56	1·0710	1·9899	0·4527	1·6908	1·0000	0·4527	0·4308	0·4264	0·9538	0·2401
0·58	1·0924	1·9829	0·4727	1·7308	1·0000	0·4727	0·4555	0·4512	1·0177	0·2495
0·60	1·1129	1·9744	0·4927	1·7708	1·0000	0·4927	0·4803	0·4761	1·0830	0·2590
0·62	1·1325	1·9648	0·5127	1·8108	1·0000	0·5127	0·5054	0·5013	1·1496	0·2685
0·64	1·1513	1·9542	0·5327	1·8508	1·0000	0·5327	0·5305	0·5265	1·2175	0·2780
0·66	1·1692	1·9427	0·5527	1·8908	1·0000	0·5527	0·5558	0·5520	1·2867	0·2876
0·68	1·1865	1·9304	0·5727	1·9308	1·0000	0·5727	0·5812	0·5776	1·3572	0·2972
0·70	1·2030	1·9176	0·5927	1·9708	1·0000	0·5927	0·6068	0·6033	1·4289	0·3069
0·72	1·2188	1·9042	0·6127	2·0108	1·0000	0·6127	0·6324	0·6291	1·5019	0·3165
0·74	1·2341	1·8904	0·6327	2·0508	1·0000	0·6327	0·6582	0·6550	1·5760	0·3262
0·76	1·2487	1·8762	0·6527	2·0908	1·0000	0·6527	0·6840	0·6811	1·6513	0·3359
0·78	1·2628	1·8618	0·6727	2·1308	1·0000	0·6727	0·7100	0·7072	1·7278	0·3456
0·80	1·2764	1·8472	0·6927	2·1708	1·0000	0·6927	0·7360	0·7334	1·8054	0·3553

Rel. dpth Y	U.equ. dia. $D_{ep(u)}$ (m)	Equiv. disch. factor J	Unit sect. area A_u (m²)	Unit wetted perim. P_u (m)	Unit surf. brdth $B_{s(u)}$ (m)	Unit mean depth $y_{m(u)}$ (m)	Discharge ratios $Q/Q_{1.00}$ medial	Mann'g	U.crit. disch. $Q_{c(u)}$ (m³s⁻¹)	Unit depth cntrd $y_{d(u)}$ (m)
0·82	1·2895	1·8323	0·7127	2·2108	1·0000	0·7127	0·7621	0·7598	1·8842	0·3651
0·84	1·3021	1·8174	0·7327	2·2508	1·0000	0·7327	0·7883	0·7862	1·9640	0·3748
0·86	1·3143	1·8024	0·7527	2·2908	1·0000	0·7527	0·8145	0·8127	2·0450	0·3846
0·88	1·3261	1·7873	0·7727	2·3308	1·0000	0·7727	0·8409	0·8392	2·1270	0·3944
0·90	1·3374	1·7722	0·7927	2·3708	1·0000	0·7927	0·8672	0·8659	2·2102	0·4042
0·92	1·3484	1·7571	0·8127	2·4108	1·0000	0·8127	0·8937	0·8926	2·2943	0·4140
0·94	1·3591	1·7421	0·8327	2·4508	1·0000	0·8327	0·9202	0·9193	2·3795	0·4238
0·96	1·3694	1·7271	0·8527	2·4908	1·0000	0·8527	0·9468	0·9462	2·4658	0·4337
0·98	1·3793	1·7122	0·8727	2·5308	1·0000	0·8727	0·9734	0·9731	2·5530	0·4435
1·00	1·3890	1·6973	0·8927	2·5708	1·0000	0·8927	1·0000	1·0000	2·6413	0·4533
1·02	1·3983	1·6826	0·9127	2·6108	1·0000	0·9127	1·0267	1·0270	2·7306	0·4632
1·04	1·4074	1·6680	0·9327	2·6508	1·0000	0·9327	1·0535	1·0540	2·8208	0·4730
1·06	1·4162	1·6535	0·9527	2·6908	1·0000	0·9527	1·0803	1·0811	2·9120	0·4829
1·08	1·4248	1·6391	0·9727	2·7308	1·0000	0·9727	1·1071	1·1083	3·0042	0·4928
1·10	1·4331	1·6248	0·9927	2·7708	1·0000	0·9927	1·1340	1·1354	3·0973	0·5026
1·12	1·4412	1·6107	1·0127	2·8108	1·0000	1·0127	1·1609	1·1627	3·1914	0·5125
1·14	1·4490	1·5968	1·0327	2·8508	1·0000	1·0327	1·1878	1·1899	3·2864	0·5224
1·16	1·4566	1·5830	1·0527	2·8908	1·0000	1·0527	1·2148	1·2172	3·3823	0·5323
1·18	1·4640	1·5693	1·0727	2·9308	1·0000	1·0727	1·2418	1·2445	3·4792	0·5422
1·20	1·4713	1·5558	1·0927	2·9708	1·0000	1·0927	1·2689	1·2719	3·5769	0·5520
1·24	1·4851	1·5293	1·1327	3·0508	1·0000	1·1327	1·3230	1·3267	3·7751	0·5718
1·28	1·4983	1·5034	1·1727	3·1308	1·0000	1·1727	1·3773	1·3817	3·9769	0·5917
1·32	1·5108	1·4782	1·2127	3·2108	1·0000	1·2127	1·4317	1·4368	4·1821	0·6115
1·36	1·5227	1·4536	1·2527	3·2908	1·0000	1·2527	1·4862	1·4919	4·3907	0·6313
1·40	1·5340	1·4297	1·2927	3·3708	1·0000	1·2927	1·5407	1·5472	4·6026	0·6512
1·44	1·5448	1·4063	1·3327	3·4508	1·0000	1·3327	1·5954	1·6025	4·8179	0·6710
1·48	1·5551	1·3837	1·3727	3·5308	1·0000	1·3727	1·6501	1·6580	5·0364	0·6909
1·52	1·5650	1·3616	1·4127	3·6108	1·0000	1·4127	1·7049	1·7135	5·2582	0·7108
1·56	1·5744	1·3401	1·4527	3·6908	1·0000	1·4527	1·7597	1·7691	5·4831	0·7306
1·60	1·5834	1·3192	1·4927	3·7708	1·0000	1·4927	1·8146	1·8247	5·7111	0·7505
1·64	1·5921	1·2988	1·5327	3·8508	1·0000	1·5327	1·8696	1·8805	5·9422	0·7704
1·68	1·6004	1·2790	1·5727	3·9308	1·0000	1·5727	1·9246	1·9362	6·1763	0·7903
1·72	1·6084	1·2598	1·6127	4·0108	1·0000	1·6127	1·9797	1·9921	6·4134	0·8102
1·76	1·6160	1·2410	1·6527	4·0908	1·0000	1·6527	2·0348	2·0480	6·6535	0·8301
1·80	1·6234	1·2228	1·6927	4·1708	1·0000	1·6927	2·0899	2·1039	6·8965	0·8500
1·84	1·6305	1·2050	1·7327	4·2508	1·0000	1·7327	2·1451	2·1599	7·1424	0·8699
1·88	1·6373	1·1877	1·7727	4·3308	1·0000	1·7727	2·2004	2·2159	7·3912	0·8899
1·92	1·6439	1·1708	1·8127	4·4108	1·0000	1·8127	2·2557	2·2720	7·6427	0·9098
1·96	1·6502	1·1544	1·8527	4·4908	1·0000	1·8527	2·3110	2·3281	7·8971	0·9297
2·00	1·6563	1·1384	1·8927	4·5708	1·0000	1·8927	2·3663	2·3842	8·1542	0·9496
2·05	1·6637	1·1190	1·9427	4·6708	1·0000	1·9427	2·4355	2·4544	8·4795	0·9746
2·10	1·6707	1·1002	1·9927	4·7708	1·0000	1·9927	2·5048	2·5247	8·8089	0·9995
2·15	1·6775	1·0819	2·0427	4·8708	1·0000	2·0427	2·5741	2·5950	9·1425	1·0244
2·20	1·6840	1·0643	2·0927	4·9708	1·0000	2·0927	2·6434	2·6654	9·4803	1·0493
2·25	1·6902	1·0471	2·1427	5·0708	1·0000	2·1427	2·7128	2·7358	9·8220	1·0743
2·30	1·6962	1·0305	2·1927	5·1708	1·0000	2·1927	2·7822	2·8063	10·168	1·0992
2·35	1·7020	1·0144	2·2427	5·2708	1·0000	2·2427	2·8517	2·8768	10·518	1·1241
2·40	1·7075	0·9988	2·2927	5·3708	1·0000	2·2927	2·9212	2·9473	10·871	1·1491
2·45	1·7129	0·9836	2·3427	5·4708	1·0000	2·3427	2·9907	3·0178	11·229	1·1740
2·50	1·7180	0·9688	2·3927	5·5708	1·0000	2·3927	3·0602	3·0884	11·590	1·1990
2·55	1·7230	0·9545	2·4427	5·6708	1·0000	2·4427	3·1298	3·1591	11·955	1·2239
2·60	1·7278	0·9406	2·4927	5·7708	1·0000	2·4927	3·1994	3·2297	12·324	1·2488
2·65	1·7324	0·9270	2·5427	5·8708	1·0000	2·5427	3·2690	3·3004	12·697	1·2738
2·70	1·7369	0·9139	2·5927	5·9708	1·0000	2·5927	3·3386	3·3711	13·073	1·2988
2·75	1·7413	0·9011	2·6427	6·0708	1·0000	2·6427	3·4083	3·4418	13·453	1·3237

C12 U-shaped capped (running full - modified table)

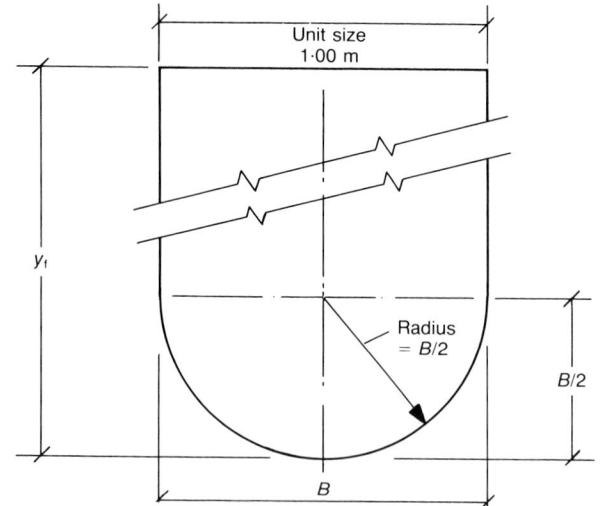

Unit size 1·00 m

Radius = B/2

B/2

B

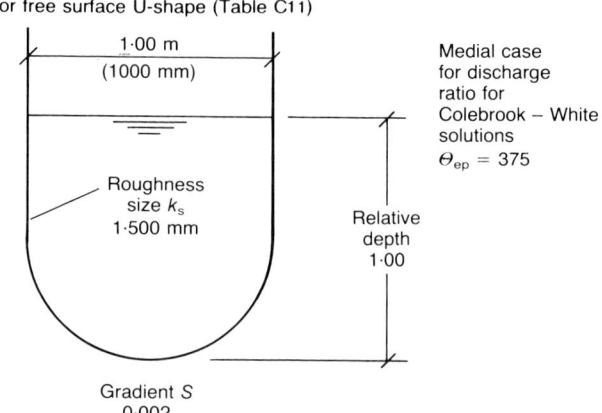

Reference discharges for discharge ratios are for free surface U-shape (Table C11)

1·00 m (1000 mm)

Roughness size k_s 1·500 mm

Relative depth 1·00

Medial case for discharge ratio for Colebrook – White solutions $\Theta_{ep} = 375$

Gradient S 0·002

Rel. dpth Y	U.equ. dia. $D_{ep(u)}$ (m)	Equiv. disch. factor J	Unit sect. area A_u (m²)	Unit wetted perim. P_u (m)	Discharge ratios $Q/Q_{1.00*}$ medial	Mann'g	Unit dist. cntrd $y_{d(u)}$ (m)
0·50	0·6110	0·7467	0·3927	2·5708	0·2617	0·2544	0·2122
0·54	0·6529	0·7738	0·4327	2·6508	0·3009	0·2930	0·2307
0·58	0·6924	0·7965	0·4727	2·7308	0·3412	0·3329	0·2495
0·62	0·7296	0·8155	0·5127	2·8108	0·3826	0·3739	0·2685
0·66	0·7648	0·8311	0·5527	2·8908	0·4250	0·4159	0·2876
0·70	0·7980	0·8439	0·5927	2·9708	0·4683	0·4589	0·3069
0·74	0·8296	0·8542	0·6327	3·0508	0·5123	0·5026	0·3262
0·78	0·8595	0·8624	0·6727	3·1308	0·5571	0·5472	0·3456
0·82	0·8879	0·8687	0·7127	3·2108	0·6025	0·5924	0·3651
0·86	0·9149	0·8734	0·7527	3·2908	0·6485	0·6383	0·3846
0·90	0·9407	0·8767	0·7927	3·3708	0·6951	0·6848	0·4042
0·94	0·9652	0·8787	0·8327	3·4508	0·7421	0·7318	0·4238
0·96	0·9771	0·8793	0·8527	3·4908	0·7658	0·7555	0·4337
0·98	0·9887	0·8797	0·8727	3·5308	0·7897	0·7793	0·4337
1·00	1·0000	0·8798	0·8927	3·5708	0·8136	0·8033	0·4533
1·04	1·0219	0·8794	0·9327	3·6508	0·8617	0·8515	0·4730
1·08	1·0429	0·8782	0·9727	3·7308	0·9103	0·9001	0·4928
1·12	1·0630	0·8763	1·0127	3·8108	0·9592	0·9491	0·5125
1·16	1·0822	0·8738	1·0527	3·8908	1·0084	0·9985	0·5323
1·20	1·1007	0·8709	1·0927	3·9708	1·0579	1·0482	0·5520
1·24	1·1185	0·8674	1·1327	4·0508	1·1078	1·0982	0·5718
1·28	1·1356	0·8636	1·1727	4·1308	1·1579	1·1486	0·5917
1·32	1·1520	0·8595	1·2127	4·2108	1·2082	1·1992	0·6115
1·36	1·1678	0·8550	1·2527	4·2908	1·2588	1·2500	0·6313
1·40	1·1830	0·8503	1·2927	4·3708	1·3096	1·3011	0·6512
1·44	1·1977	0·8454	1·3327	4·4508	1·3606	1·3525	0·6710
1·48	1·2119	0·8403	1·3727	4·5308	1·4118	1·4040	0·6909
1·52	1·2256	0·8350	1·4127	4·6108	1·4632	1·4558	0·7108
1·56	1·2388	0·8296	1·4527	4·6908	1·5148	1·5078	0·7306
1·60	1·2515	0·8241	1·4927	4·7708	1·5666	1·5599	0·7505
1·64	1·2639	0·8185	1·5327	4·8508	1·6185	1·6122	0·7704
1·68	1·2758	0·8129	1·5727	4·9308	1·6705	1·6647	0·7903
1·72	1·2874	0·8071	1·6127	5·0108	1·7227	1·7173	0·8102
1·76	1·2986	0·8013	1·6527	5·0908	1·7750	1·7701	0·8301
1·80	1·3094	0·7955	1·6927	5·1708	1·8275	1·8230	0·8500
1·84	1·3200	0·7897	1·7327	5·2508	1·8801	1·8761	0·8699
1·88	1·3302	0·7839	1·7727	5·3308	1·9328	1·9293	0·8899
1·92	1·3401	0·7780	1·8127	5·4108	1·9856	1·9826	0·9098
1·96	1·3497	0·7722	1·8527	5·4908	2·0385	2·0360	0·9297
2·00	1·3590	0·7664	1·8927	5·5708	2·0915	2·0896	0·9496

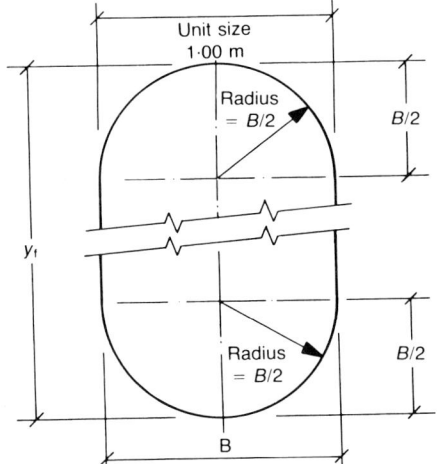

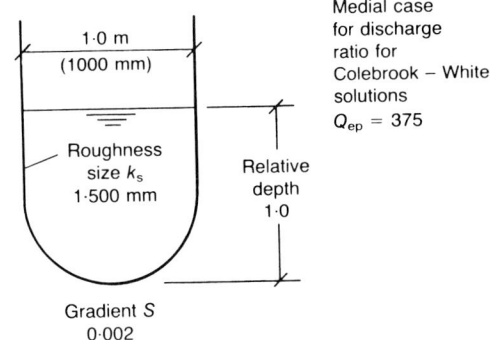

Reference discharges for discharge ratios
are for free surface U-shape (Table C11)

Medial case for discharge ratio for Colebrook – White solutions $Q_{ep} = 375$

Rel. dpth Y	U.equ. dia. $D_{ep(u)}$ (m)	Equiv. disch. factor J	Unit sect. area A_u (m²)	Unit wetted perim. P_u (m)	Discharge ratios $Q/Q_{1.00^*}$ medial	Mann'g	Unit dist. cntrd $y_{d(u)}$ (m)
1·04	1·0248	0·9994	0·8254	3·2216	0·7640	0·7550	0·5200
1·08	1·0485	0·9976	0·8654	3·3016	0·8126	0·8037	0·5400
1·12	1·0710	0·9949	0·9054	3·3816	0·8616	0·8528	0·5600
1·16	1·0924	0·9914	0·9454	3·4616	0·9110	0·9024	0·5800
1·20	1·1129	0·9872	0·9854	3·5416	0·9607	0·9523	0·6000
1·24	1·1325	0·9824	1·0254	3·6216	1·0107	1·0025	0·6200
1·28	1·1513	0·9771	1·0654	3·7016	1·0610	1·0531	0·6400
1·32	1·1692	0·9713	1·1054	3·7816	1·1116	1·1040	0·6600
1·36	1·1865	0·9652	1·1454	3·8616	1·1625	1·1551	0·6800
1·40	1·2030	0·9588	1·1854	3·9416	1·2136	1·2065	0·7000
1·44	1·2188	0·9521	1·2254	4·0216	1·2649	1·2582	0·7200
1·48	1·2341	0·9452	1·2654	4·1016	1·3164	1·3100	0·7400
1·52	1·2487	0·9381	1·3054	4·1816	1·3681	1·3621	0·7600
1·56	1·2628	0·9309	1·3454	4·2616	1·4199	1·4144	0·7800
1·60	1·2764	0·9236	1·3854	4·3416	1·4720	1·4669	0·8000
1·64	1·2895	0·9162	1·4254	4·4216	1·5242	1·5195	0·8200
1·68	1·3021	0·9087	1·4654	4·5016	1·5766	1·5724	0·8400
1·72	1·3143	0·9012	1·5054	4·5816	1·6291	1·6253	0·8600
1·76	1·3261	0·8937	1·5454	4·6616	1·6817	1·6785	0·8800
1·80	1·3374	0·8861	1·5854	4·7416	1·7345	1·7317	0·9000
1·84	1·3484	0·8786	1·6254	4·8216	1·7874	1·7852	0·9200
1·88	1·3591	0·8711	1·6654	4·9016	1·8404	1·8387	0·9400
1·92	1·3694	0·8636	1·7054	4·9816	1·8935	1·8923	0·9600
1·96	1·3793	0·8561	1·7454	5·0616	1·9467	1·9461	0·9800
2·00	1·3890	0·8487	1·7854	5·1416	2·0000	2·0000	1·0000
2·04	1·3983	0·8413	1·8254	5·2216	2·0535	2·0540	1·0200
2·08	1·4074	0·8340	1·8654	5·3016	2·1070	2·1081	1·0400
2·12	1·4162	0·8267	1·9054	5·3816	2·1605	2·1622	1·0600
2·16	1·4248	0·8195	1·9454	5·4616	2·2142	2·2165	1·0800
2·20	1·4331	0·8124	1·9854	5·5416	2·2680	2·2709	1·1000
2·24	1·4412	0·8054	2·0254	5·6216	2·3218	2·3253	1·1200
2·28	1·4490	0·7984	2·0654	5·7016	2·3757	2·3798	1·1400
2·32	1·4566	0·7915	2·1054	5·7816	2·4296	2·4344	1·1600
2·36	1·4640	0·7847	2·1454	5·8616	2·4836	2·4891	1·1800
2·40	1·4713	0·7779	2·1854	5·9416	2·5377	2·5438	1·2000
2·44	1·4783	0·7712	2·2254	6·0216	2·5919	2·5986	1·2200
2·48	1·4851	0·7646	2·2654	6·1016	2·6461	2·6535	1·2400
2·52	1·4918	0·7581	2·3054	6·1816	2·7003	2·7084	1·2600
2·56	1·4983	0·7517	2·3454	6·2616	2·7546	2·7634	1·2800
2·60	1·5046	0·7454	2·3854	6·3416	2·8090	2·8184	1·3000

C14

Rectangular (free surface)

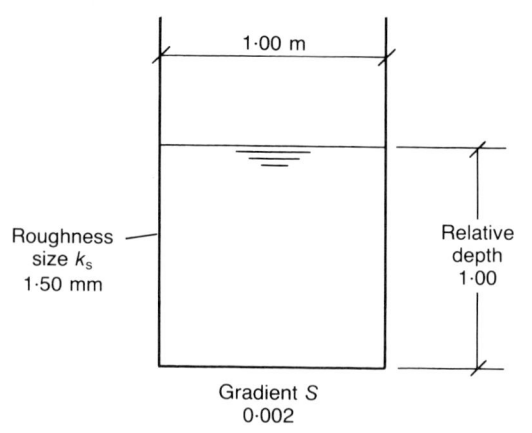

Unit size 1·00 m

1·00 m

Medial case for discharge ratio for Colebrook – White solutions $\Theta_{ep} = 360$

Roughness size k_s 1·50 mm

Relative depth 1·00

Gradient S 0·002

B

Rel. dpth Y	U.equ. dia. $D_{ep(u)}$ (m)	Equiv. disch. factor J	Unit sect. area A_u (m²)	Unit wetted perim. P_u (m)	Unit surf. brdth $B_{s(u)}$ (m)	Unit mean depth $y_{m(u)}$ (m)	Discharge ratios $Q/Q_{1.00}$ medial	Mann'g	U.crit. disch. $Q_{c(u)}$ (m³s⁻¹)	Unit depth cntrd $y_{d(u)}$ (m)
0·02	0·0769	0·2324	0·0200	1·0400	1·0000	0·0200	0·0031	0·0030	0·0089	0·0100
0·04	0·1481	0·4309	0·0400	1·0800	1·0000	0·0400	0·0096	0·0092	0·0251	0·0200
0·06	0·2143	0·6011	0·0600	1·1200	1·0000	0·0600	0·0185	0·0177	0·0460	0·0300
0·08	0·2759	0·7471	0·0800	1·1600	1·0000	0·0800	0·0291	0·0280	0·0709	0·0400
0·10	0·3333	0·8726	0·1000	1·2000	1·0000	0·1000	0·0412	0·0397	0·0990	0·0500
0·12	0·3871	0·9807	0·1200	1·2400	1·0000	0·1200	0·0545	0·0526	0·1302	0·0600
0·14	0·4375	1·0738	0·1400	1·2800	1·0000	0·1400	0·0689	0·0666	0·1640	0·0700
0·16	0·4848	1·1539	0·1600	1·3200	1·0000	0·1600	0·0842	0·0815	0·2004	0·0800
0·18	0·5294	1·2229	0·1800	1·3600	1·0000	0·1800	0·1002	0·0972	0·2391	0·0900
0·20	0·5714	1·2823	0·2000	1·4000	1·0000	0·2000	0·1169	0·1137	0·2801	0·1000
0·22	0·6111	1·3332	0·2200	1·4400	1·0000	0·2200	0·1343	0·1308	0·3231	0·1100
0·24	0·6486	1·3769	0·2400	1·4800	1·0000	0·2400	0·1522	0·1485	0·3682	0·1200
0·26	0·6842	1·4141	0·2600	1·5200	1·0000	0·2600	0·1706	0·1667	0·4152	0·1300
0·28	0·7179	1·4458	0·2800	1·5600	1·0000	0·2800	0·1894	0·1853	0·4640	0·1400
0·30	0·7500	1·4726	0·3000	1·6000	1·0000	0·3000	0·2087	0·2044	0·5146	0·1500
0·32	0·7805	1·4951	0·3200	1·6400	1·0000	0·3200	0·2283	0·2239	0·5669	0·1600
0·34	0·8095	1·5138	0·3400	1·6800	1·0000	0·3400	0·2483	0·2438	0·6208	0·1700
0·36	0·8372	1·5291	0·3600	1·7200	1·0000	0·3600	0·2685	0·2640	0·6764	0·1800
0·38	0·8636	1·5416	0·3800	1·7600	1·0000	0·3800	0·2891	0·2845	0·7336	0·1900
0·40	0·8889	1·5514	0·4000	1·8000	1·0000	0·4000	0·3099	0·3053	0·7922	0·2000
0·42	0·9130	1·5589	0·4200	1·8400	1·0000	0·4200	0·3310	0·3263	0·8524	0·2100
0·44	0·9362	1·5644	0·4400	1·8800	1·0000	0·4400	0·3523	0·3476	0·9140	0·2200
0·46	0·9583	1·5680	0·4600	1·9200	1·0000	0·4600	0·3738	0·3691	0·9770	0·2300
0·48	0·9796	1·5701	0·4800	1·9600	1·0000	0·4800	0·3955	0·3908	1·0414	0·2400
0·50	1·0000	1·5708	0·5000	2·0000	1·0000	0·5000	0·4173	0·4127	1·1072	0·2500
0·52	1·0196	1·5702	0·5200	2·0400	1·0000	0·5200	0·4394	0·4348	1·1743	0·2600
0·54	1·0385	1·5684	0·5400	2·0800	1·0000	0·5400	0·4616	0·4571	1·2427	0·2700
0·56	1·0566	1·5657	0·5600	2·1200	1·0000	0·5600	0·4839	0·4796	1·3123	0·2800
0·58	1·0741	1·5621	0·5800	2·1600	1·0000	0·5800	0·5064	0·5021	1·3833	0·2900
0·60	1·0909	1·5578	0·6000	2·2000	1·0000	0·6000	0·5290	0·5249	1·4554	0·3000
0·62	1·1071	1·5527	0·6200	2·2400	1·0000	0·6200	0·5518	0·5477	1·5288	0·3100
0·64	1·1228	1·5471	0·6400	2·2800	1·0000	0·6400	0·5746	0·5707	1·6034	0·3200
0·66	1·1379	1·5409	0·6600	2·3200	1·0000	0·6600	0·5976	0·5938	1·6791	0·3300
0·68	1·1525	1·5342	0·6800	2·3600	1·0000	0·6800	0·6206	0·6171	1·7560	0·3400
0·70	1·1667	1·5271	0·7000	2·4000	1·0000	0·7000	0·6438	0·6404	1·8340	0·3500
0·72	1·1803	1·5197	0·7200	2·4400	1·0000	0·7200	0·6671	0·6638	1·9132	0·3600
0·74	1·1935	1·5119	0·7400	2·4800	1·0000	0·7400	0·6904	0·6873	1·9935	0·3700
0·76	1·2063	1·5039	0·7600	2·5200	1·0000	0·7600	0·7138	0·7109	2·0748	0·3800
0·78	1·2187	1·4956	0·7800	2·5600	1·0000	0·7800	0·7373	0·7346	2·1573	0·3900
0·80	1·2308	1·4871	0·8000	2·6000	1·0000	0·8000	0·7609	0·7584	2·2408	0·4000

Rel. dpth Y	U.equ. dia. $D_{ep(u)}$ (m)	Equiv. disch. factor J	Unit sect. area A_u (m²)	Unit wetted perim. P_u (m)	Unit surf. brdth $B_{s(u)}$ (m)	Unit mean depth $y_{m(u)}$ (m)	Discharge ratios $Q/Q_{1.00}$ medial	Mann'g	U.crit. disch. $Q_{c(u)}$ (m³s⁻¹)	Unit depth cntrd $y_{d(u)}$ (m)
0·82	1·2424	1·4784	0·8200	2·6400	1·0000	0·8200	0·7845	0·7823	2·3253	0·4100
0·84	1·2537	1·4696	0·8400	2·6800	1·0000	0·8400	0·8082	0·8062	2·4109	0·4200
0·86	1·2647	1·4607	0·8600	2·7200	1·0000	0·8600	0·8320	0·8302	2·4975	0·4300
0·88	1·2754	1·4517	0·8800	2·7600	1·0000	0·8800	0·8559	0·8543	2·5851	0·4400
0·90	1·2857	1·4425	0·9000	2·8000	1·0000	0·9000	0·8798	0·8784	2·6738	0·4500
0·92	1·2958	1·4333	0·9200	2·8400	1·0000	0·9200	0·9037	0·9026	2·7634	0·4600
0·94	1·3056	1·4241	0·9400	2·8800	1·0000	0·9400	0·9277	0·9269	2·8540	0·4700
0·96	1·3151	1·4148	0·9600	2·9200	1·0000	0·9600	0·9518	0·9512	2·9456	0·4800
0·98	1·3243	1·4055	0·9800	2·9600	1·0000	0·9800	0·9759	0·9756	3·0381	0·4900
1·00	1·3333	1·3962	1·0000	3·0000	1·0000	1·0000	1·0000	1·0000	3·1316	0·5000
1·02	1·3421	1·3869	1·0200	3·0400	1·0000	1·0200	1·0242	1·0245	3·2260	0·5100
1·04	1·3506	1·3776	1·0400	3·0800	1·0000	1·0400	1·0485	1·0490	3·3213	0·5200
1·06	1·3590	1·3683	1·0600	3·1200	1·0000	1·0600	1·0727	1·0735	3·4176	0·5300
1·08	1·3671	1·3591	1·0800	3·1600	1·0000	1·0800	1·0971	1·0982	3·5148	0·5400
1·10	1·3750	1·3499	1·1000	3·2000	1·0000	1·1000	1·1214	1·1228	3·6128	0·5500
1·12	1·3827	1·3407	1·1200	3·2400	1·0000	1·1200	1·1458	1·1475	3·7118	0·5600
1·14	1·3902	1·3315	1·1400	3·2800	1·0000	1·1400	1·1702	1·1722	3·8117	0·5700
1·16	1·3976	1·3225	1·1600	3·3200	1·0000	1·1600	1·1947	1·1970	3·9124	0·5800
1·18	1·4048	1·3134	1·1800	3·3600	1·0000	1·1800	1·2192	1·2218	4·0141	0·5900
1·20	1·4118	1·3044	1·2000	3·4000	1·0000	1·2000	1·2437	1·2466	4·1165	0·6000
1·24	1·4253	1·2867	1·2400	3·4800	1·0000	1·2400	1·2928	1·2964	4·3241	0·6200
1·28	1·4382	1·2691	1·2800	3·5600	1·0000	1·2800	1·3421	1·3463	4·5350	0·6400
1·32	1·4505	1·2519	1·3200	3·6400	1·0000	1·3200	1·3914	1·3963	4·7492	0·6600
1·36	1·4624	1·2350	1·3600	3·7200	1·0000	1·3600	1·4409	1·4464	4·9667	0·6800
1·40	1·4737	1·2183	1·4000	3·8000	1·0000	1·4000	1·4904	1·4966	5·1874	0·7000
1·44	1·4845	1·2020	1·4400	3·8800	1·0000	1·4400	1·5400	1·5469	5·4113	0·7200
1·48	1·4949	1·1860	1·4800	3·9600	1·0000	1·4800	1·5897	1·5973	5·6384	0·7400
1·52	1·5050	1·1703	1·5200	4·0400	1·0000	1·5200	1·6395	1·6478	5·8685	0·7600
1·56	1·5146	1·1549	1·5600	4·1200	1·0000	1·5600	1·6893	1·6983	6·1016	0·7800
1·60	1·5238	1·1398	1·6000	4·2000	1·0000	1·6000	1·7392	1·7490	6·3378	0·8000
1·64	1·5327	1·1250	1·6400	4·2800	1·0000	1·6400	1·7892	1·7997	6·5770	0·8200
1·68	1·5413	1·1105	1·6800	4·3600	1·0000	1·6800	1·8392	1·8504	6·8191	0·8400
1·72	1·5495	1·0964	1·7200	4·4400	1·0000	1·7200	1·8893	1·9013	7·0640	0·8600
1·76	1·5575	1·0825	1·7600	4·5200	1·0000	1·7600	1·9394	1·9521	7·3119	0·8800
1·80	1·5652	1·0689	1·8000	4·6000	1·0000	1·8000	1·9896	2·0031	7·5626	0·9000
1·84	1·5726	1·0557	1·8400	4·6800	1·0000	1·8400	2·0398	2·0541	7·8160	0·9200
1·88	1·5798	1·0427	1·8800	4·7600	1·0000	1·8800	2·0901	2·1051	8·0723	0·9400
1·92	1·5868	1·0299	1·9200	4·8400	1·0000	1·9200	2·1404	2·1562	8·3313	0·9600
1·96	1·5935	1·0175	1·9600	4·9200	1·0000	1·9600	2·1908	2·2073	8·5930	0·9800
2·00	1·6000	1·0053	2·0000	5·0000	1·0000	2·0000	2·2412	2·2585	8·8574	1·0000
2·05	1·6078	0·9904	2·0500	5·1000	1·0000	2·0500	2·3042	2·3225	9·1916	1·0250
2·10	1·6154	0·9759	2·1000	5·2000	1·0000	2·1000	2·3673	2·3866	9·5299	1·0500
2·15	1·6226	0·9618	2·1500	5·3000	1·0000	2·1500	2·4304	2·4507	9·8723	1·0750
2·20	1·6296	0·9481	2·2000	5·4000	1·0000	2·2000	2·4936	2·5149	10·219	1·1000
2·25	1·6364	0·9347	2·2500	5·5000	1·0000	2·2500	2·5568	2·5791	10·569	1·1250
2·30	1·6429	0·9216	2·3000	5·6000	1·0000	2·3000	2·6201	2·6434	10·923	1·1500
2·35	1·6491	0·9089	2·3500	5·7000	1·0000	2·3500	2·6834	2·7078	11·281	1·1750
2·40	1·6552	0·8965	2·4000	5·8000	1·0000	2·4000	2·7468	2·7721	11·643	1·2000
2·45	1·6610	0·8844	2·4500	5·9000	1·0000	2·4500	2·8101	2·8365	12·009	1·2250
2·50	1·6667	0·8726	2·5000	6·0000	1·0000	2·5000	2·8736	2·9010	12·379	1·2500
2·55	1·6721	0·8612	2·5500	6·1000	1·0000	2·5500	2·9370	2·9655	12·752	1·2750
2·60	1·6774	0·8499	2·6000	6·2000	1·0000	2·6000	3·0005	3·0300	13·129	1·3000
2·65	1·6825	0·8390	2·6500	6·3000	1·0000	2·6500	3·0640	3·0945	13·509	1·3250
2·70	1·6875	0·8283	2·7000	6·4000	1·0000	2·7000	3·1275	3·1591	13·893	1·3500
2·75	1·6923	0·8179	2·7500	6·5000	1·0000	2·7500	3·1911	3·2237	14·281	1·3750

C15 Rectangular capped (running full - modified table)

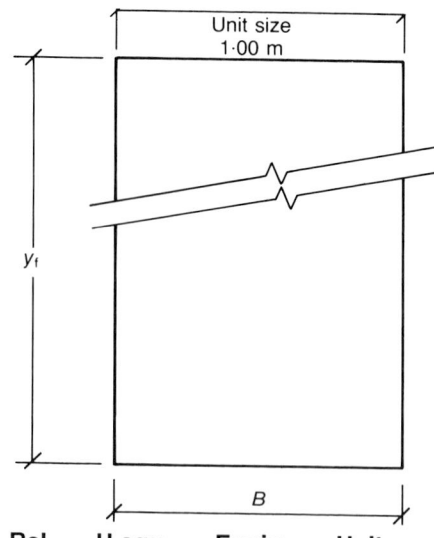

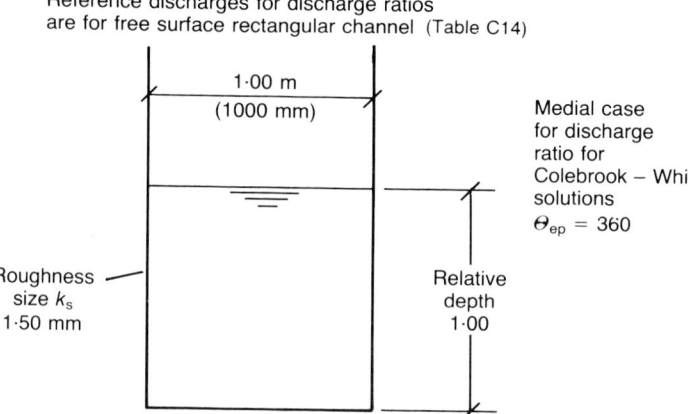

Reference discharges for discharge ratios
are for free surface rectangular channel (Table C14)

1·00 m
(1000 mm)

Medial case
for discharge
ratio for
Colebrook – White
solutions
$\Theta_{ep} = 360$

Roughness
size k_s
1·50 mm

Relative
depth
1·00

Gradient S
0·002

Rel. dpth Y	U.equ. dia. $D_{ep(u)}$ (m)	Equiv. disch. factor J	Unit sect. area A_u (m²)	Unit wetted perim. P_u (m)	Discharge ratios $Q/Q_{1.00}*$ medial	Mann'g	Unit dist. cntrd $y_{d(u)}$ (m)
0·02	0·0392	0·0604	0·0200	2·0400	0·0019	0·0019	0·0100
0·04	0·0769	0·1162	0·0400	2·0800	0·0061	0·0060	0·0200
0·06	0·1132	0·1678	0·0600	2·1200	0·0120	0·0116	0·0300
0·08	0·1481	0·2155	0·0800	2·1600	0·0192	0·0185	0·0400
0·10	0·1818	0·2596	0·1000	2·2000	0·0276	0·0265	0·0500
0·12	0·2143	0·3005	0·1200	2·2400	0·0369	0·0355	0·0600
0·14	0·2456	0·3384	0·1400	2·2800	0·0472	0·0453	0·0700
0·16	0·2759	0·3735	0·1600	2·3200	0·0582	0·0560	0·0800
0·18	0·3051	0·4061	0·1800	2·3600	0·0700	0·0673	0·0900
0·20	0·3333	0·4363	0·2000	2·4000	0·0825	0·0794	0·1000
0·22	0·3607	0·4643	0·2200	2·4400	0·0955	0·0920	0·1100
0·24	0·3871	0·4904	0·2400	2·4800	0·1091	0·1052	0·1200
0·26	0·4127	0·5145	0·2600	2·5200	0·1232	0·1190	0·1300
0·28	0·4375	0·5369	0·2800	2·5600	0·1378	0·1332	0·1400
0·30	0·4615	0·5577	0·3000	2·6000	0·1528	0·1479	0·1500
0·32	0·4848	0·5770	0·3200	2·6400	0·1683	0·1630	0·1600
0·34	0·5075	0·5949	0·3400	2·6800	0·1842	0·1786	0·1700
0·36	0·5294	0·6115	0·3600	2·7200	0·2004	0·1945	0·1800
0·38	0·5507	0·6269	0·3800	2·7600	0·2170	0·2108	0·1900
0·40	0·5714	0·6411	0·4000	2·8000	0·2339	0·2274	0·2000
0·42	0·5915	0·6544	0·4200	2·8400	0·2511	0·2443	0·2100
0·44	0·6111	0·6666	0·4400	2·8800	0·2686	0·2616	0·2200
0·46	0·6301	0·6779	0·4600	2·9200	0·2863	0·2791	0·2300
0·48	0·6486	0·6884	0·4800	2·9600	0·3044	0·2969	0·2400
0·50	0·6667	0·6981	0·5000	3·0000	0·3226	0·3150	0·2500
0·52	0·6842	0·7071	0·5200	3·0400	0·3412	0·3333	0·2600
0·54	0·7013	0·7153	0·5400	3·0800	0·3599	0·3519	0·2700
0·56	0·7179	0·7229	0·5600	3·1200	0·3789	0·3706	0·2800
0·58	0·7342	0·7299	0·5800	3·1600	0·3980	0·3896	0·2900
0·60	0·7500	0·7363	0·6000	3·2000	0·4174	0·4089	0·3000
0·62	0·7654	0·7422	0·6200	3·2400	0·4369	0·4283	0·3100
0·64	0·7805	0·7475	0·6400	3·2800	0·4566	0·4478	0·3200
0·66	0·7952	0·7524	0·6600	3·3200	0·4765	0·4676	0·3300
0·68	0·8095	0·7569	0·6800	3·3600	0·4965	0·4876	0·3400
0·70	0·8235	0·7609	0·7000	3·4000	0·5167	0·5077	0·3500
0·72	0·8372	0·7646	0·7200	3·4400	0·5371	0·5280	0·3600
0·74	0·8506	0·7678	0·7400	3·4800	0·5576	0·5484	0·3700
0·76	0·8636	0·7708	0·7600	3·5200	0·5782	0·5690	0·3800
0·78	0·8764	0·7734	0·7800	3·5600	0·5990	0·5897	0·3900
0·80	0·8889	0·7757	0·8000	3·6000	0·6198	0·6105	0·4000

Rel. dpth	U.equ. dia.	Equiv. disch. factor	Unit sect. area	Unit wetted perim.	Discharge ratios $Q/Q_{1\cdot00}*$		Unit dist. cntrd
	$D_{ep(u)}$						
Y	(m)	J	A_u (m²)	P_u (m)	medial	Mann'g	$y_{d(u)}$ (m)
0·82	0·9011	0·7777	0·8200	3·6400	0·6408	0·6315	0·4100
0·84	0·9130	0·7794	0·8400	3·6800	0·6620	0·6526	0·4200
0·86	0·9247	0·7809	0·8600	3·7200	0·6832	0·6738	0·4300
0·88	0·9362	0·7822	0·8800	3·7600	0·7045	0·6952	0·4400
0·90	0·9474	0·7832	0·9000	3·8000	0·7260	0·7166	0·4500
0·92	0·9583	0·7840	0·9200	3·8400	0·7475	0·7382	0·4600
0·94	0·9691	0·7846	0·9400	3·8800	0·7692	0·7599	0·4700
0·96	0·9796	0·7851	0·9600	3·9200	0·7909	0·7816	0·4800
0·98	0·9899	0·7853	0·9800	3·9600	0·8128	0·8035	0·4900
1·00	1·0000	0·7854	1·0000	4·0000	0·8347	0·8255	0·5000
1·02	1·0099	0·7853	1·0200	4·0400	0·8567	0·8475	0·5100
1·04	1·0196	0·7851	1·0400	4·0800	0·8788	0·8697	0·5200
1·06	1·0291	0·7847	1·0600	4·1200	0·9009	0·8919	0·5300
1·08	1·0385	0·7842	1·0800	4·1600	0·9231	0·9142	0·5400
1·10	1·0476	0·7836	1·1000	4·2000	0·9455	0·9366	0·5500
1·12	1·0566	0·7829	1·1200	4·2400	0·9678	0·9591	0·5600
1·14	1·0654	0·7820	1·1400	4·2800	0·9903	0·9817	0·5700
1·16	1·0741	0·7811	1·1600	4·3200	1·0128	1·0043	0·5800
1·18	1·0826	0·7800	1·1800	4·3600	1·0354	1·0270	0·5900
1·20	1·0909	0·7789	1·2000	4·4000	1·0580	1·0497	0·6000
1·24	1·1071	0·7764	1·2400	4·4800	1·1035	1·0955	0·6200
1·28	1·1228	0·7735	1·2800	4·5600	1·1492	1·1414	0·6400
1·32	1·1379	0·7704	1·3200	4·6400	1·1951	1·1877	0·6600
1·36	1·1525	0·7671	1·3600	4·7200	1·2413	1·2341	0·6800
1·40	1·1667	0·7636	1·4000	4·8000	1·2876	1·2808	0·7000
1·44	1·1803	0·7598	1·4400	4·8800	1·3341	1·3276	0·7200
1·48	1·1935	0·7560	1·4800	4·9600	1·3808	1·3747	0·7400
1·52	1·2063	0·7519	1·5200	5·0400	1·4276	1·4219	0·7600
1·56	1·2187	0·7478	1·5600	5·1200	1·4746	1·4693	0·7800
1·60	1·2308	0·7436	1·6000	5·2000	1·5218	1·5169	0·8000
1·64	1·2424	0·7392	1·6400	5·2800	1·5691	1·5646	0·8200
1·68	1·2537	0·7348	1·6800	5·3600	1·6165	1·6125	0·8400
1·72	1·2647	0·7303	1·7200	5·4400	1·6641	1·6605	0·8600
1·76	1·2754	0·7258	1·7600	5·5200	1·7117	1·7086	0·8800
1·80	1·2857	0·7213	1·8000	5·6000	1·7595	1·7569	0·9000
1·84	1·2958	0·7167	1·8400	5·6800	1·8074	1·8053	0·9200
1·88	1·3056	0·7121	1·8800	5·7600	1·8554	1·8538	0·9400
1·92	1·3151	0·7074	1·9200	5·8400	1·9035	1·9024	0·9600
1·96	1·3243	0·7028	1·9600	5·9200	1·9517	1·9512	0·9800
2·00	1·3333	0·6981	2·0000	6·0000	2·0000	2·0000	1·0000
2·05	1·3443	0·6923	2·0500	6·1000	2·0605	2·0612	1·0250
2·10	1·3548	0·6865	2·1000	6·2000	2·1212	2·1225	1·0500
2·15	1·3651	0·6807	2·1500	6·3000	2·1819	2·1840	1·0750
2·20	1·3750	0·6749	2·2000	6·4000	2·2428	2·2456	1·1000
2·25	1·3846	0·6692	2·2500	6·5000	2·3038	2·3073	1·1250
2·30	1·3939	0·6635	2·3000	6·6000	2·3649	2·3692	1·1500
2·35	1·4030	0·6578	2·3500	6·7000	2·4261	2·4311	1·1750
2·40	1·4118	0·6522	2·4000	6·8000	2·4874	2·4932	1·2000
2·45	1·4203	0·6466	2·4500	6·9000	2·5488	2·5554	1·2250
2·50	1·4286	0·6411	2·5000	7·0000	2·6103	2·6177	1·2500
2·55	1·4366	0·6357	2·5500	7·1000	2·6718	2·6800	1·2750
2·60	1·4444	0·6302	2·6000	7·2000	2·7335	2·7425	1·3000
2·65	1·4521	0·6249	2·6500	7·3000	2·7952	2·8051	1·3250
2·70	1·4595	0·6196	2·7000	7·4000	2·8570	2·8677	1·3500
2·75	1·4667	0·6143	2·7500	7·5000	2·9189	2·9304	1·3750

C16 Arch culvert - radius 0·5 breadth (running full - modified table)

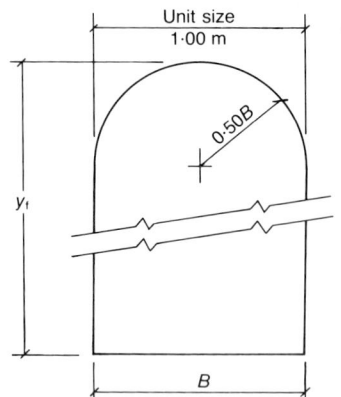

Reference discharges for discharge ratios are for free surface rectangular channel (Table C14)

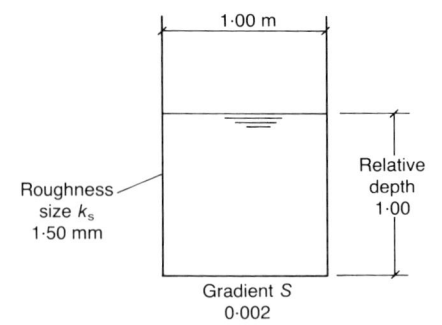

Medial case for discharge ratio for Colebrook–White solutions $\Theta_{ep} = 360$

Rel. dpth Y	U.equ. dia. $D_{ep(u)}$ (m)	Equiv. disch. factor J	Unit sect. area A_u (m²)	Unit wetted perim. P_u (m)	Discharge ratios $Q/Q_{1.00}*$ medial	Mann'g	Unit dist. cntrd $y_{d(u)}$ (m)
0·50	0·6110	0·7467	0·3927	2·5708	0·2397	0·2334	0·2878
0·54	0·6529	0·7738	0·4327	2·6508	0·2755	0·2688	0·3093
0·58	0·6924	0·7965	0·4727	2·7308	0·3125	0·3054	0·3305
0·62	0·7296	0·8155	0·5127	2·8108	0·3504	0·3430	0·3515
0·66	0·7648	0·8311	0·5527	2·8908	0·3893	0·3815	0·3724
0·70	0·7980	0·8439	0·5927	2·9708	0·4289	0·4209	0·3931
0·74	0·8296	0·8542	0·6327	3·0508	0·4692	0·4611	0·4138
0·78	0·8595	0·8624	0·6727	3·1308	0·5102	0·5020	0·4344
0·82	0·8879	0·8687	0·7127	3·2108	0·5518	0·5435	0·4549
0·86	0·9149	0·8734	0·7527	3·2908	0·5939	0·5856	0·4754
0·90	0·9407	0·8767	0·7927	3·3708	0·6366	0·6282	0·4958
0·94	0·9652	0·8787	0·8327	3·4508	0·6797	0·6713	0·5162
0·96	0·9771	0·8793	0·8527	3·4908	0·7014	0·6931	0·5263
0·98	0·9887	0·8797	0·8727	3·5308	0·7232	0·7149	0·5365
1·00	1·0000	0·8798	0·8927	3·5708	0·7451	0·7369	0·5467
1·04	1·0219	0·8794	0·9327	3·6508	0·7892	0·7811	0·5670
1·08	1·0429	0·8782	0·9727	3·7308	0·8337	0·8257	0·5872
1·12	1·0630	0·8763	1·0127	3·8108	0·8784	0·8707	0·6075
1·16	1·0822	0·8738	1·0527	3·8908	0·9235	0·9160	0·6277
1·20	1·1007	0·8709	1·0927	3·9708	0·9689	0·9616	0·6480
1·24	1·1185	0·8674	1·1327	4·0508	1·0145	1·0075	0·6682
1·28	1·1356	0·8636	1·1727	4·1308	1·0604	1·0537	0·6883
1·32	1·1520	0·8595	1·2127	4·2108	1·1065	1·1001	0·7085
1·36	1·1678	0·8550	1·2527	4·2908	1·1528	1·1467	0·7287
1·40	1·1830	0·8503	1·2927	4·3708	1·1994	1·1936	0·7488
1·44	1·1977	0·8454	1·3327	4·4508	1·2461	1·2407	0·7690
1·48	1·2119	0·8403	1·3727	4·5308	1·2930	1·2880	0·7891
1·52	1·2256	0·8350	1·4127	4·6108	1·3401	1·3355	0·8092
1·56	1·2388	0·8296	1·4527	4·6908	1·3873	1·3832	0·8294
1·60	1·2515	0·8241	1·4927	4·7708	1·4347	1·4310	0·8495
1·64	1·2639	0·8185	1·5327	4·8508	1·4822	1·4790	0·8696
1·68	1·2758	0·8129	1·5727	4·9308	1·5299	1·5271	0·8897
1·72	1·2874	0·8071	1·6127	5·0108	1·5777	1·5754	0·9098
1·76	1·2986	0·8013	1·6527	5·0908	1·6256	1·6239	0·9299
1·80	1·3094	0·7955	1·6927	5·1708	1·6737	1·6724	0·9500
1·84	1·3200	0·7897	1·7327	5·2508	1·7218	1·7211	0·9701
1·88	1·3302	0·7839	1·7727	5·3308	1·7701	1·7699	0·9901
1·92	1·3401	0·7780	1·8127	5·4108	1·8185	1·8188	1·0102
1·96	1·3497	0·7722	1·8527	5·4908	1·8669	1·8678	1·0303
2·00	1·3590	0·7664	1·8927	5·5708	1·9155	1·9169	1·0504

Arch culvert - radius 0·75 breadth (running full - modified table) C17

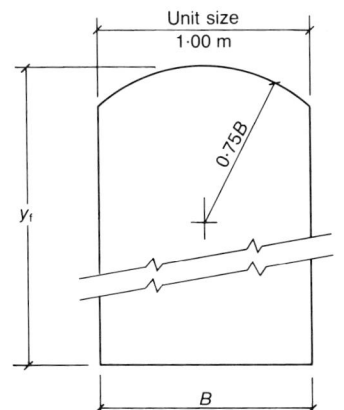

Unit size
1·00 m

0·75B

y_f

B

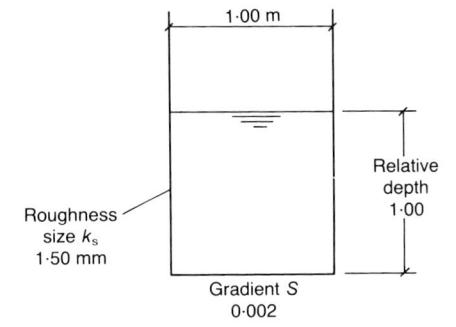

Reference discharges for discharge ratios are for free surface rectangular channel (Table C14)

1·00 m

Relative depth 1·00

Roughness size k_s 1·50 mm

Gradient S 0·002

Medial case for discharge ratio for Colebrook–White solutions $\Theta_{ep} = 360$

Rel. dpth Y	U.equ. dia. $D_{ep(u)}$ (m)	Equiv. disch. factor J	Unit sect. area A_u (m²)	Unit wetted perim. P_u (m)	Discharge ratios $Q/Q_{1·00}*$ medial	Mann'g	Unit dist. cntrd $y_{d(u)}$ (m)
0·50	0·6488	0·7514	0·4400	2·7126	0·2790	0·2722	0·2765
0·54	0·6875	0·7734	0·4800	2·7926	0·3159	0·3086	0·2968
0·58	0·7240	0·7918	0·5200	2·8726	0·3537	0·3461	0·3170
0·62	0·7586	0·8072	0·5600	2·9526	0·3924	0·3845	0·3372
0·66	0·7914	0·8198	0·6000	3·0326	0·4318	0·4237	0·3574
0·70	0·8224	0·8301	0·6400	3·1126	0·4720	0·4637	0·3776
0·74	0·8519	0·8383	0·6800	3·1926	0·5129	0·5044	0·3977
0·78	0·8800	0·8447	0·7200	3·2726	0·5543	0·5458	0·4179
0·82	0·9067	0·8496	0·7600	3·3526	0·5963	0·5877	0·4380
0·86	0·9322	0·8532	0·8000	3·4326	0·6388	0·6302	0·4581
0·90	0·9565	0·8555	0·8400	3·5126	0·6817	0·6731	0·4782
0·94	0·9798	0·8567	0·8800	3·5926	0·7251	0·7166	0·4983
0·96	0·9910	0·8570	0·9000	3·6326	0·7469	0·7385	0·5083
0·98	1·0020	0·8571	0·9200	3·6726	0·7688	0·7604	0·5183
1·00	1·0127	0·8570	0·9400	3·7126	0·7909	0·7825	0·5284
1·04	1·0336	0·8561	0·9800	3·7926	0·8352	0·8270	0·5484
1·08	1·0535	0·8546	1·0200	3·8726	0·8798	0·8718	0·5685
1·12	1·0727	0·8526	1·0600	3·9526	0·9247	0·9169	0·5886
1·16	1·0911	0·8500	1·1000	4·0326	0·9700	0·9623	0·6086
1·20	1·1088	0·8470	1·1400	4·1126	1·0154	1·0081	0·6287
1·24	1·1258	0·8435	1·1800	4·1926	1·0612	1·0541	0·6487
1·28	1·1421	0·8398	1·2200	4·2726	1·1072	1·1004	0·6687
1·32	1·1579	0·8357	1·2600	4·3526	1·1533	1·1469	0·6888
1·36	1·1731	0·8314	1·3000	4·4326	1·1997	1·1936	0·7088
1·40	1·1878	0·8269	1·3400	4·5126	1·2463	1·2406	0·7289
1·44	1·2019	0·8221	1·3800	4·5926	1·2931	1·2877	0·7489
1·48	1·2156	0·8173	1·4200	4·6726	1·3401	1·3351	0·7689
1·52	1·2288	0·8122	1·4600	4·7526	1·3872	1·3826	0·7890
1·56	1·2415	0·8071	1·5000	4·8326	1·4345	1·4303	0·8090
1·60	1·2539	0·8018	1·5400	4·9126	1·4819	1·4782	0·8290
1·64	1·2659	0·7965	1·5800	4·9926	1·5295	1·5262	0·8490
1·68	1·2774	0·7911	1·6200	5·0726	1·5771	1·5744	0·8691
1·72	1·2886	0·7857	1·6600	5·1526	1·6250	1·6227	0·8891
1·76	1·2995	0·7802	1·7000	5·2326	1·6729	1·6711	0·9091
1·80	1·3101	0·7747	1·7400	5·3126	1·7210	1·7197	0·9291
1·84	1·3203	0·7692	1·7800	5·3926	1·7691	1·7684	0·9491
1·88	1·3302	0·7636	1·8200	5·4726	1·8174	1·8172	0·9692
1·92	1·3399	0·7581	1·8600	5·5526	1·8657	1·8661	0·9892
1·96	1·3493	0·7525	1·9000	5·6326	1·9142	1·9151	1·0092
2·00	1·3584	0·7470	1·9400	5·7126	1·9627	1·9642	1·0292

C18   Box culvert (9% splays) (free surface)

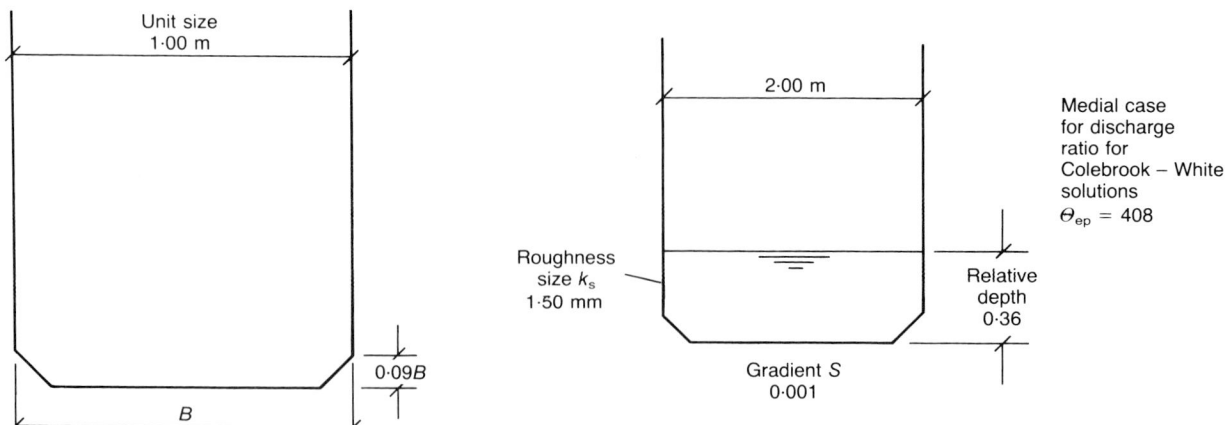

Rel. dpth Y	U.equ. dia. $D_{ep(u)}$ (m)	Equiv. disch. factor J	Unit sect. area A_u (m²)	Unit wetted perim. P_u (m)	Unit surf. brdth $B_{s(u)}$ (m)	Unit mean depth $y_{m(u)}$ (m)	Discharge ratios $Q/Q_{0.36}$ medial	Mann'g	U.crit. disch. $Q_{c(u)}$ (m³s⁻¹)	Unit depth cntrd $y_{d(u)}$ (m)
0·02	0·0767	0·2747	0·0168	0·8766	0·8600	0·0195	0·0099	0·0094	0·0074	0·0099
0·04	0·1475	0·4964	0·0344	0·9331	0·9000	0·0382	0·0314	0·0299	0·0211	0·0197
0·06	0·2134	0·6774	0·0528	0·9897	0·9400	0·0562	0·0614	0·0587	0·0392	0·0293
0·08	0·2753	0·8265	0·0720	1·0463	0·9800	0·0735	0·0987	0·0949	0·0611	0·0388
0·10	0·3358	0·9639	0·0919	1·0946	1·0000	0·0919	0·1431	0·1383	0·0872	0·0482
0·12	0·3945	1·0924	0·1119	1·1346	1·0000	0·1119	0·1931	0·1874	0·1172	0·0578
0·14	0·4492	1·2014	0·1319	1·1746	1·0000	0·1319	0·2472	0·2409	0·1500	0·0675
0·16	0·5003	1·2940	0·1519	1·2146	1·0000	0·1519	0·3047	0·2981	0·1854	0·0773
0·18	0·5481	1·3724	0·1719	1·2546	1·0000	0·1719	0·3653	0·3585	0·2232	0·0872
0·20	0·5929	1·4389	0·1919	1·2946	1·0000	0·1919	0·4285	0·4217	0·2633	0·0970
0·22	0·6351	1·4950	0·2119	1·3346	1·0000	0·2119	0·4940	0·4875	0·3055	0·1069
0·24	0·6748	1·5423	0·2319	1·3746	1·0000	0·2319	0·5616	0·5556	0·3497	0·1169
0·26	0·7123	1·5819	0·2519	1·4146	1·0000	0·2519	0·6311	0·6256	0·3959	0·1268
0·28	0·7477	1·6149	0·2719	1·4546	1·0000	0·2719	0·7021	0·6975	0·4440	0·1367
0·30	0·7812	1·6421	0·2919	1·4946	1·0000	0·2919	0·7747	0·7710	0·4939	0·1467
0·32	0·8130	1·6644	0·3119	1·5346	1·0000	0·3119	0·8486	0·8460	0·5455	0·1566
0·34	0·8432	1·6822	0·3319	1·5746	1·0000	0·3319	0·9238	0·9224	0·5988	0·1666
0·36	0·8718	1·6963	0·3519	1·6146	1·0000	0·3519	1·0000	1·0000	0·6537	0·1765
0·38	0·8991	1·7071	0·3719	1·6546	1·0000	0·3719	1·0773	1·0788	0·7102	0·1865
0·40	0·9251	1·7150	0·3919	1·6946	1·0000	0·3919	1·1555	1·1586	0·7683	0·1965
0·42	0·9499	1·7203	0·4119	1·7346	1·0000	0·4119	1·2346	1·2394	0·8278	0·2065
0·44	0·9735	1·7235	0·4319	1·7746	1·0000	0·4319	1·3145	1·3210	0·8889	0·2164
0·46	0·9962	1·7246	0·4519	1·8146	1·0000	0·4519	1·3951	1·4035	0·9513	0·2264
0·48	1·0178	1·7241	0·4719	1·8546	1·0000	0·4719	1·4764	1·4868	1·0152	0·2364
0·50	1·0386	1·7221	0·4919	1·8946	1·0000	0·4919	1·5583	1·5708	1·0804	0·2464
0·52	1·0584	1·7188	0·5119	1·9346	1·0000	0·5119	1·6408	1·6555	1·1469	0·2564
0·54	1·0775	1·7143	0·5319	1·9746	1·0000	0·5319	1·7239	1·7408	1·2148	0·2663
0·56	1·0958	1·7088	0·5519	2·0146	1·0000	0·5519	1·8075	1·8266	1·2840	0·2763
0·58	1·1134	1·7025	0·5719	2·0546	1·0000	0·5719	1·8915	1·9130	1·3544	0·2863
0·60	1·1304	1·6954	0·5919	2·0946	1·0000	0·5919	1·9760	2·0000	1·4260	0·2963
0·64	1·1624	1·6792	0·6319	2·1746	1·0000	0·6319	2·1463	2·1752	1·5730	0·3163
0·68	1·1921	1·6610	0·6719	2·2546	1·0000	0·6719	2·3180	2·3522	1·7247	0·3363
0·72	1·2198	1·6414	0·7119	2·3346	1·0000	0·7119	2·4911	2·5307	1·8810	0·3562
0·76	1·2456	1·6206	0·7519	2·4146	1·0000	0·7519	2·6653	2·7105	2·0417	0·3762
0·80	1·2698	1·5991	0·7919	2·4946	1·0000	0·7919	2·8406	2·8915	2·2068	0·3962
0·84	1·2925	1·5771	0·8319	2·5746	1·0000	0·8319	3·0168	3·0736	2·3761	0·4162
0·88	1·3138	1·5548	0·8719	2·6546	1·0000	0·8719	3·1939	3·2568	2·5495	0·4362
0·92	1·3339	1·5324	0·9119	2·7346	1·0000	0·9119	3·3718	3·4408	2·7270	0·4562
0·96	1·3528	1·5100	0·9519	2·8146	1·0000	0·9519	3·5504	3·6256	2·9084	0·4762
1·00	1·3707	1·4877	0·9919	2·8946	1·0000	0·9919	3·7296	3·8112	3·0936	0·4962

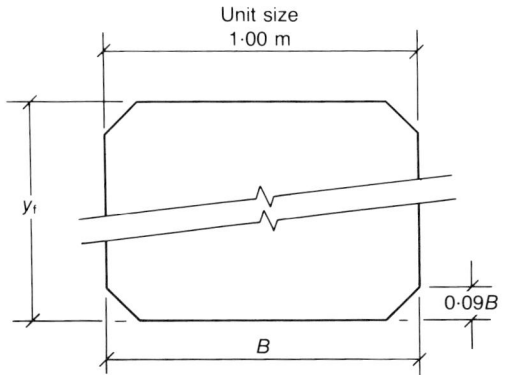

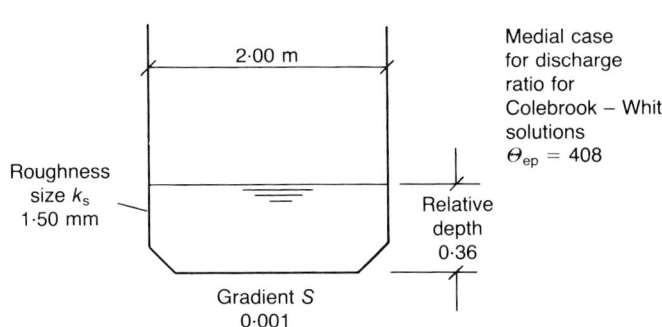

Reference discharges for discharge ratios
are for free surface 9% splay culvert (Table C18)

Medial case
for discharge
ratio for
Colebrook – White
solutions
$\Theta_{ep} = 408$

Rel. dpth Y	U.equ. dia. $D_{ep(u)}$ (m)	Equiv. disch. factor J	Unit sect. area A_u (m²)	Unit wetted perim. P_u (m)	Discharge ratios $Q/Q_{0.36}*$ medial	Mann'g	Unit dist. cntrd $y_{d(u)}$ (m)
0·33	0·5125	0·6574	0·3138	2·4491	0·6392	0·6258	0·1650
0·35	0·5364	0·6770	0·3338	2·4891	0·6998	0·6862	0·1750
0·36	0·5481	0·6862	0·3438	2·5091	0·7306	0·7170	0·1800
0·38	0·5709	0·7035	0·3638	2·5491	0·7932	0·7796	0·1900
0·40	0·5929	0·7195	0·3838	2·5891	0·8570	0·8435	0·2000
0·42	0·6144	0·7341	0·4038	2·6291	0·9220	0·9087	0·2100
0·44	0·6351	0·7475	0·4238	2·6691	0·9881	0·9750	0·2200
0·46	0·6553	0·7599	0·4438	2·7091	1·0552	1·0426	0·2300
0·48	0·6748	0·7712	0·4638	2·7491	1·1232	1·1111	0·2400
0·50	0·6938	0·7815	0·4838	2·7891	1·1922	1·1807	0·2500
0·52	0·7123	0·7910	0·5038	2·8291	1·2621	1·2512	0·2600
0·54	0·7303	0·7996	0·5238	2·8691	1·3328	1·3227	0·2700
0·56	0·7477	0·8075	0·5438	2·9091	1·4043	1·3950	0·2800
0·58	0·7647	0·8146	0·5638	2·9491	1·4765	1·4681	0·2900
0·60	0·7812	0·8211	0·5838	2·9891	1·5494	1·5420	0·3000
0·62	0·7973	0·8269	0·6038	3·0291	1·6230	1·6166	0·3100
0·64	0·8130	0·8322	0·6238	3·0691	1·6972	1·6920	0·3200
0·66	0·8283	0·8369	0·6438	3·1091	1·7721	1·7681	0·3300
0·68	0·8432	0·8411	0·6638	3·1491	1·8475	1·8448	0·3400
0·70	0·8577	0·8449	0·6838	3·1891	1·9235	1·9221	0·3500
0·71	0·8648	0·8466	0·6938	3·2091	1·9617	1·9610	0·3550
0·73	0·8788	0·8497	0·7138	3·2491	2·0385	2·0392	0·3650
0·75	0·8924	0·8524	0·7338	3·2891	2·1158	2·1179	0·3750
0·76	0·8991	0·8536	0·7438	3·3091	2·1546	2·1575	0·3800
0·77	0·9057	0·8547	0·7538	3·3291	2·1936	2·1972	0·3850
0·78	0·9122	0·8557	0·7638	3·3491	2·2326	2·2371	0·3900
0·80	0·9251	0·8575	0·7838	3·3891	2·3111	2·3172	0·4000
0·82	0·9376	0·8590	0·8038	3·4291	2·3899	2·3977	0·4100
0·84	0·9499	0·8602	0·8238	3·4691	2·4692	2·4787	0·4200
0·85	0·9559	0·8607	0·8338	3·4891	2·5090	2·5194	0·4250
0·86	0·9618	0·8611	0·8438	3·5091	2·5489	2·5602	0·4300
0·87	0·9677	0·8614	0·8538	3·5291	2·5889	2·6011	0·4350
0·88	0·9735	0·8617	0·8638	3·5491	2·6290	2·6421	0·4400
0·89	0·9793	0·8620	0·8738	3·5691	2·6691	2·6832	0·4450
0·90	0·9850	0·8621	0·8838	3·5891	2·7094	2·7244	0·4500
0·91	0·9906	0·8623	0·8938	3·6091	2·7498	2·7657	0·4550
0·92	0·9962	0·8623	0·9038	3·6291	2·7902	2·8071	0·4600
0·94	1·0071	0·8623	0·9238	3·6691	2·8713	2·8902	0·4700
0·95	1·0125	0·8622	0·9338	3·6891	2·9120	2·9319	0·4750
1·00	1·0386	0·8611	0·9838	3·7891	3·1166	3·1416	0·5000

C20

Box culvert (16% splays) (free surface)

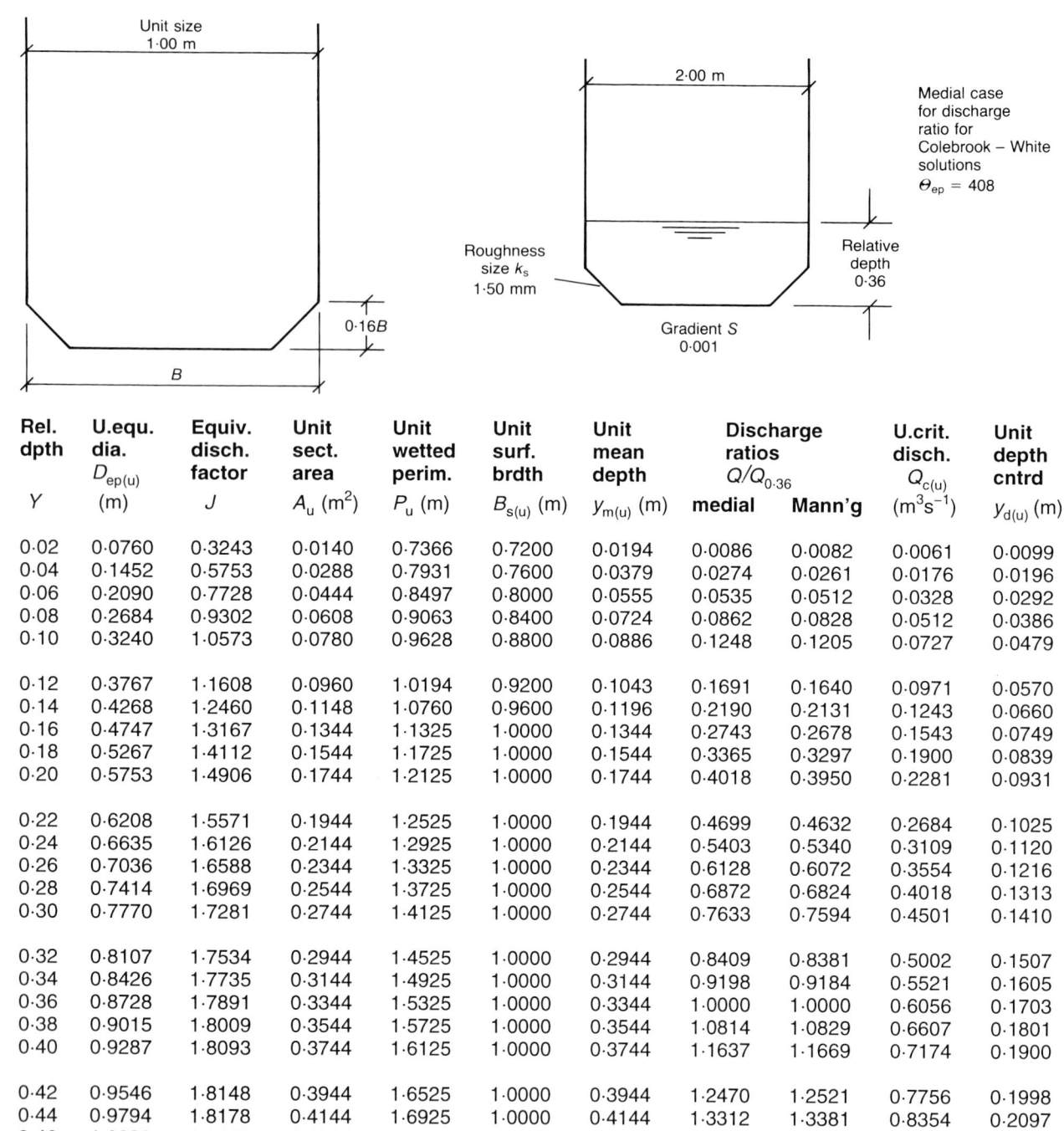

Unit size 1·00 m

0·16B

B

2·00 m

Medial case for discharge ratio for Colebrook – White solutions $\Theta_{ep} = 408$

Roughness size k_s 1·50 mm

Relative depth 0·36

Gradient S 0·001

Rel. dpth Y	U.equ. dia. $D_{ep(u)}$ (m)	Equiv. disch. factor J	Unit sect. area A_u (m²)	Unit wetted perim. P_u (m)	Unit surf. brdth $B_{s(u)}$ (m)	Unit mean depth $y_{m(u)}$ (m)	Discharge ratios $Q/Q_{0.36}$ medial	Mann'g	U.crit. disch. $Q_{c(u)}$ (m³s⁻¹)	Unit depth cntrd $y_{d(u)}$ (m)
0·02	0·0760	0·3243	0·0140	0·7366	0·7200	0·0194	0·0086	0·0082	0·0061	0·0099
0·04	0·1452	0·5753	0·0288	0·7931	0·7600	0·0379	0·0274	0·0261	0·0176	0·0196
0·06	0·2090	0·7728	0·0444	0·8497	0·8000	0·0555	0·0535	0·0512	0·0328	0·0292
0·08	0·2684	0·9302	0·0608	0·9063	0·8400	0·0724	0·0862	0·0828	0·0512	0·0386
0·10	0·3240	1·0573	0·0780	0·9628	0·8800	0·0886	0·1248	0·1205	0·0727	0·0479
0·12	0·3767	1·1608	0·0960	1·0194	0·9200	0·1043	0·1691	0·1640	0·0971	0·0570
0·14	0·4268	1·2460	0·1148	1·0760	0·9600	0·1196	0·2190	0·2131	0·1243	0·0660
0·16	0·4747	1·3167	0·1344	1·1325	1·0000	0·1344	0·2743	0·2678	0·1543	0·0749
0·18	0·5267	1·4112	0·1544	1·1725	1·0000	0·1544	0·3365	0·3297	0·1900	0·0839
0·20	0·5753	1·4906	0·1744	1·2125	1·0000	0·1744	0·4018	0·3950	0·2281	0·0931
0·22	0·6208	1·5571	0·1944	1·2525	1·0000	0·1944	0·4699	0·4632	0·2684	0·1025
0·24	0·6635	1·6126	0·2144	1·2925	1·0000	0·2144	0·5403	0·5340	0·3109	0·1120
0·26	0·7036	1·6588	0·2344	1·3325	1·0000	0·2344	0·6128	0·6072	0·3554	0·1216
0·28	0·7414	1·6969	0·2544	1·3725	1·0000	0·2544	0·6872	0·6824	0·4018	0·1313
0·30	0·7770	1·7281	0·2744	1·4125	1·0000	0·2744	0·7633	0·7594	0·4501	0·1410
0·32	0·8107	1·7534	0·2944	1·4525	1·0000	0·2944	0·8409	0·8381	0·5002	0·1507
0·34	0·8426	1·7735	0·3144	1·4925	1·0000	0·3144	0·9198	0·9184	0·5521	0·1605
0·36	0·8728	1·7891	0·3344	1·5325	1·0000	0·3344	1·0000	1·0000	0·6056	0·1703
0·38	0·9015	1·8009	0·3544	1·5725	1·0000	0·3544	1·0814	1·0829	0·6607	0·1801
0·40	0·9287	1·8093	0·3744	1·6125	1·0000	0·3744	1·1637	1·1669	0·7174	0·1900
0·42	0·9546	1·8148	0·3944	1·6525	1·0000	0·3944	1·2470	1·2521	0·7756	0·1998
0·44	0·9794	1·8178	0·4144	1·6925	1·0000	0·4144	1·3312	1·3381	0·8354	0·2097
0·46	1·0029	1·8185	0·4344	1·7325	1·0000	0·4344	1·4162	1·4251	0·8966	0·2196
0·48	1·0254	1·8174	0·4544	1·7725	1·0000	0·4544	1·5019	1·5130	0·9592	0·2295
0·50	1·0469	1·8145	0·4744	1·8125	1·0000	0·4744	1·5883	1·6016	1·0232	0·2394
0·52	1·0675	1·8103	0·4944	1·8525	1·0000	0·4944	1·6753	1·6909	1·0886	0·2493
0·54	1·0872	1·8047	0·5144	1·8925	1·0000	0·5144	1·7630	1·7809	1·1553	0·2592
0·56	1·1061	1·7981	0·5344	1·9325	1·0000	0·5344	1·8511	1·8715	1·2234	0·2691
0·58	1·1242	1·7905	0·5544	1·9725	1·0000	0·5544	1·9398	1·9627	1·2927	0·2791
0·60	1·1416	1·7821	0·5744	2·0125	1·0000	0·5744	2·0290	2·0544	1·3633	0·2890
0·64	1·1745	1·7632	0·6144	2·0925	1·0000	0·6144	2·2086	2·2394	1·5081	0·3089
0·68	1·2049	1·7422	0·6544	2·1725	1·0000	0·6544	2·3898	2·4262	1·6578	0·3288
0·72	1·2331	1·7197	0·6944	2·2525	1·0000	0·6944	2·5724	2·6146	1·8121	0·3487
0·76	1·2594	1·6962	0·7344	2·3325	1·0000	0·7344	2·7562	2·8044	1·9709	0·3686
0·80	1·2840	1·6719	0·7744	2·4125	1·0000	0·7744	2·9412	2·9954	2·1341	0·3885
0·84	1·3069	1·6472	0·8144	2·4925	1·0000	0·8144	3·1271	3·1876	2·3015	0·4085
0·88	1·3285	1·6223	0·8544	2·5725	1·0000	0·8544	3·3139	3·3808	2·4732	0·4284
0·92	1·3487	1·5974	0·8944	2·6525	1·0000	0·8944	3·5015	3·5750	2·6489	0·4484
0·96	1·3678	1·5725	0·9344	2·7325	1·0000	0·9344	3·6899	3·7700	2·8285	0·4683
1·00	1·3858	1·5479	0·9744	2·8125	1·0000	0·9744	3·8789	3·9658	3·0121	0·4883

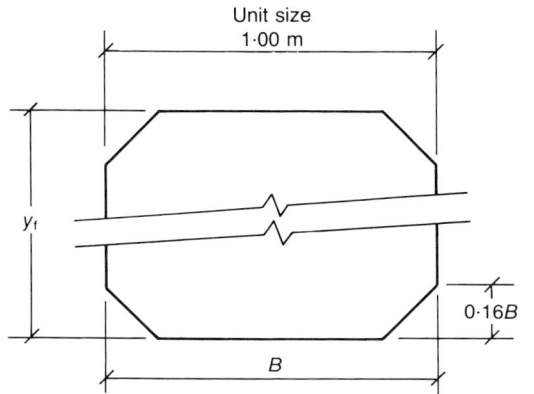

Unit size
1·00 m

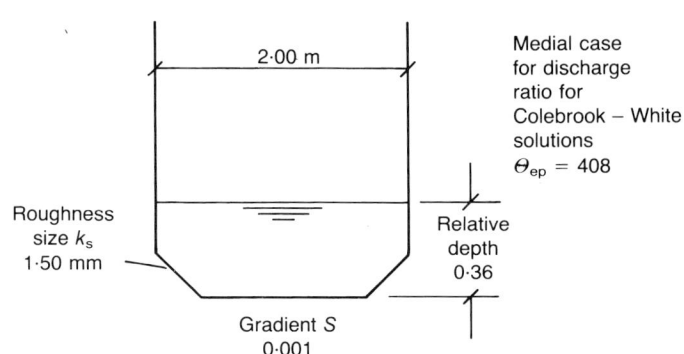

Reference discharges for discharge ratios
are for free surface 16% splay culvert (Table C20)

Medial case
for discharge
ratio for
Colebrook − White
solutions
$\Theta_{ep} = 408$

Roughness
size k_s
1·50 mm

Relative
depth
0·36

Gradient S
0·001

Rel. dpth	U.equ. dia. $D_{ep(u)}$	Equiv. disch. factor	Unit sect. area	Unit wetted perim.	Discharge ratios $Q/Q_{0\cdot36}*$		Unit dist. cntrd
Y	(m)	J	A_u (m²)	P_u (m)	medial	Mann'g	$y_{d(u)}$ (m)
0·33	0·4880	0·6709	0·2788	2·2851	0·5790	0·5659	0·1650
0·35	0·5140	0·6945	0·2988	2·3251	0·6413	0·6278	0·1750
0·36	0·5267	0·7056	0·3088	2·3451	0·6730	0·6595	0·1800
0·38	0·5514	0·7263	0·3288	2·3851	0·7376	0·7240	0·1900
0·40	0·5753	0·7453	0·3488	2·4251	0·8036	0·7900	0·2000
0·42	0·5984	0·7626	0·3688	2·4651	0·8710	0·8576	0·2100
0·44	0·6208	0·7785	0·3888	2·5051	0·9397	0·9265	0·2200
0·46	0·6425	0·7931	0·4088	2·5451	1·0096	0·9967	0·2300
0·48	0·6635	0·8063	0·4288	2·5851	1·0805	1·0681	0·2400
0·50	0·6839	0·8184	0·4488	2·6251	1·1526	1·1407	0·2500
0·52	0·7036	0·8294	0·4688	2·6651	1·2256	1·2143	0·2600
0·54	0·7228	0·8394	0·4888	2·7051	1·2995	1·2890	0·2700
0·56	0·7414	0·8485	0·5088	2·7451	1·3744	1·3647	0·2800
0·58	0·7595	0·8567	0·5288	2·7851	1·4500	1·4413	0·2900
0·60	0·7770	0·8641	0·5488	2·8251	1·5265	1·5188	0·3000
0·62	0·7941	0·8707	0·5688	2·8651	1·6038	1·5971	0·3100
0·64	0·8107	0·8767	0·5888	2·9051	1·6817	1·6762	0·3200
0·66	0·8269	0·8820	0·6088	2·9451	1·7604	1·7561	0·3300
0·68	0·8426	0·8867	0·6288	2·9851	1·8396	1·8367	0·3400
0·70	0·8579	0·8909	0·6488	3·0251	1·9195	1·9180	0·3500
0·71	0·8654	0·8928	0·6588	3·0451	1·9597	1·9589	0·3550
0·73	0·8801	0·8962	0·6788	3·0851	2·0405	2·0412	0·3650
0·75	0·8944	0·8991	0·6988	3·1251	2·1218	2·1241	0·3750
0·76	0·9015	0·9004	0·7088	3·1451	2·1627	2·1658	0·3800
0·77	0·9084	0·9016	0·7188	3·1651	2·2037	2·2076	0·3850
0·78	0·9153	0·9027	0·7288	3·1851	2·2448	2·2496	0·3900
0·80	0·9287	0·9046	0·7488	3·2251	2·3274	2·3339	0·4000
0·82	0·9418	0·9062	0·7688	3·2651	2·4105	2·4187	0·4100
0·84	0·9546	0·9074	0·7888	3·3051	2·4941	2·5041	0·4200
0·85	0·9609	0·9079	0·7988	3·3251	2·5360	2·5470	0·4250
0·86	0·9671	0·9083	0·8088	3·3451	2·5780	2·5900	0·4300
0·87	0·9733	0·9086	0·8188	3·3651	2·6202	2·6331	0·4350
0·88	0·9794	0·9089	0·8288	3·3851	2·6624	2·6763	0·4400
0·89	0·9853	0·9091	0·8388	3·4051	2·7048	2·7196	0·4450
0·90	0·9913	0·9092	0·8488	3·4251	2·7472	2·7631	0·4500
0·91	0·9971	0·9093	0·8588	3·4451	2·7897	2·8066	0·4550
0·92	1·0029	0·9093	0·8688	3·4651	2·8324	2·8503	0·4600
0·94	1·0143	0·9091	0·8888	3·5051	2·9179	2·9379	0·4700
0·95	1·0199	0·9089	0·8988	3·5251	2·9608	2·9819	0·4750
1·00	1·0469	0·9073	0·9488	3·6251	3·1766	3·2031	0·5000

C22 Box culvert upright (23% splays) (free surface)

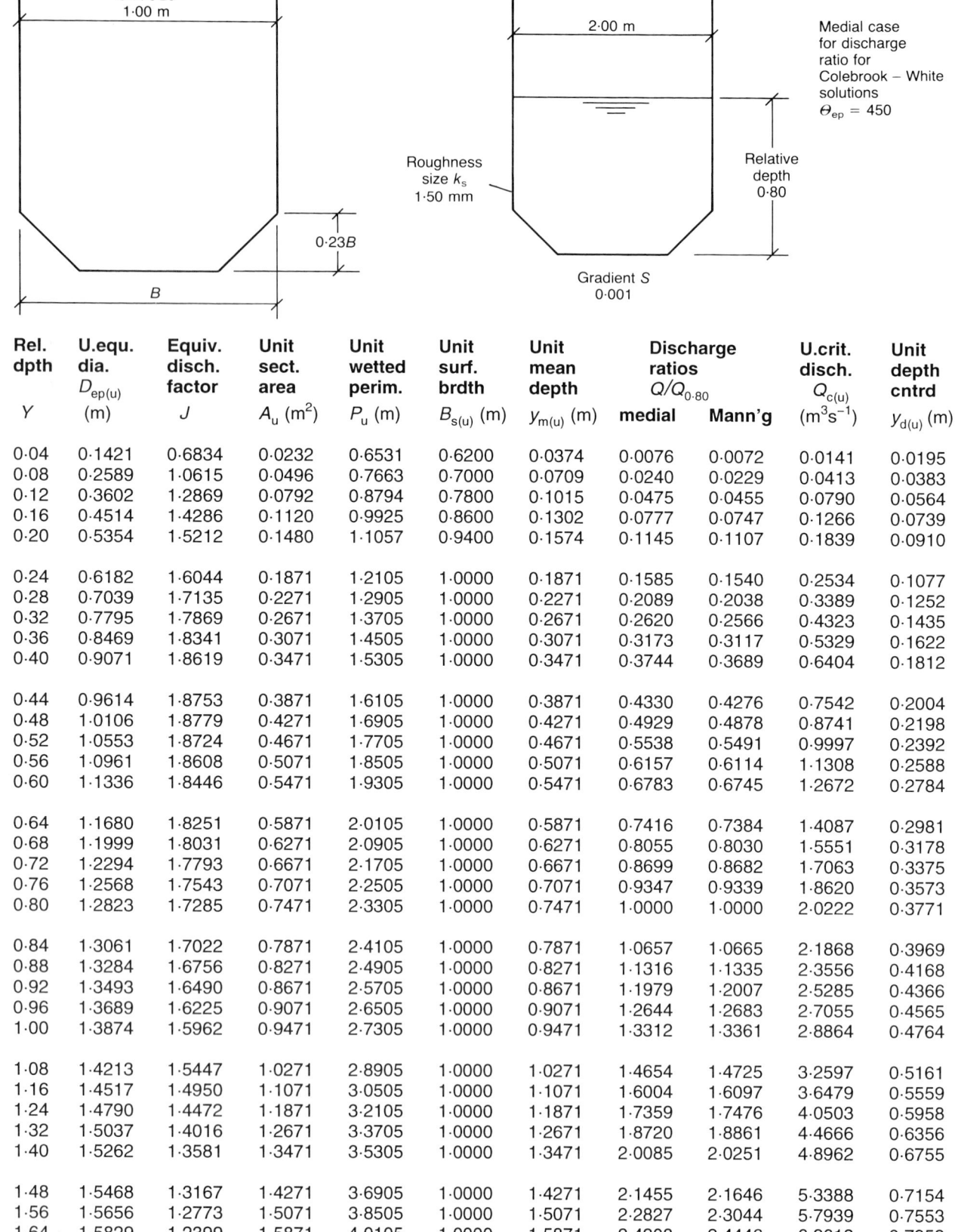

Unit size 1·00 m

0·23B

B

2·00 m

Medial case for discharge ratio for Colebrook – White solutions $\Theta_{ep} = 450$

Roughness size k_s 1·50 mm

Relative depth 0·80

Gradient S 0·001

Rel. dpth Y	U.equ. dia. $D_{ep(u)}$ (m)	Equiv. disch. factor J	Unit sect. area A_u (m²)	Unit wetted perim. P_u (m)	Unit surf. brdth $B_{s(u)}$ (m)	Unit mean depth $y_{m(u)}$ (m)	Discharge ratios $Q/Q_{0.80}$ medial	Mann'g	U.crit. disch. $Q_{c(u)}$ (m³s⁻¹)	Unit depth cntrd $y_{d(u)}$ (m)
0·04	0·1421	0·6834	0·0232	0·6531	0·6200	0·0374	0·0076	0·0072	0·0141	0·0195
0·08	0·2589	1·0615	0·0496	0·7663	0·7000	0·0709	0·0240	0·0229	0·0413	0·0383
0·12	0·3602	1·2869	0·0792	0·8794	0·7800	0·1015	0·0475	0·0455	0·0790	0·0564
0·16	0·4514	1·4286	0·1120	0·9925	0·8600	0·1302	0·0777	0·0747	0·1266	0·0739
0·20	0·5354	1·5212	0·1480	1·1057	0·9400	0·1574	0·1145	0·1107	0·1839	0·0910
0·24	0·6182	1·6044	0·1871	1·2105	1·0000	0·1871	0·1585	0·1540	0·2534	0·1077
0·28	0·7039	1·7135	0·2271	1·2905	1·0000	0·2271	0·2089	0·2038	0·3389	0·1252
0·32	0·7795	1·7869	0·2671	1·3705	1·0000	0·2671	0·2620	0·2566	0·4323	0·1435
0·36	0·8469	1·8341	0·3071	1·4505	1·0000	0·3071	0·3173	0·3117	0·5329	0·1622
0·40	0·9071	1·8619	0·3471	1·5305	1·0000	0·3471	0·3744	0·3689	0·6404	0·1812
0·44	0·9614	1·8753	0·3871	1·6105	1·0000	0·3871	0·4330	0·4276	0·7542	0·2004
0·48	1·0106	1·8779	0·4271	1·6905	1·0000	0·4271	0·4929	0·4878	0·8741	0·2198
0·52	1·0553	1·8724	0·4671	1·7705	1·0000	0·4671	0·5538	0·5491	0·9997	0·2392
0·56	1·0961	1·8608	0·5071	1·8505	1·0000	0·5071	0·6157	0·6114	1·1308	0·2588
0·60	1·1336	1·8446	0·5471	1·9305	1·0000	0·5471	0·6783	0·6745	1·2672	0·2784
0·64	1·1680	1·8251	0·5871	2·0105	1·0000	0·5871	0·7416	0·7384	1·4087	0·2981
0·68	1·1999	1·8031	0·6271	2·0905	1·0000	0·6271	0·8055	0·8030	1·5551	0·3178
0·72	1·2294	1·7793	0·6671	2·1705	1·0000	0·6671	0·8699	0·8682	1·7063	0·3375
0·76	1·2568	1·7543	0·7071	2·2505	1·0000	0·7071	0·9347	0·9339	1·8620	0·3573
0·80	1·2823	1·7285	0·7471	2·3305	1·0000	0·7471	1·0000	1·0000	2·0222	0·3771
0·84	1·3061	1·7022	0·7871	2·4105	1·0000	0·7871	1·0657	1·0665	2·1868	0·3969
0·88	1·3284	1·6756	0·8271	2·4905	1·0000	0·8271	1·1316	1·1335	2·3556	0·4168
0·92	1·3493	1·6490	0·8671	2·5705	1·0000	0·8671	1·1979	1·2007	2·5285	0·4366
0·96	1·3689	1·6225	0·9071	2·6505	1·0000	0·9071	1·2644	1·2683	2·7055	0·4565
1·00	1·3874	1·5962	0·9471	2·7305	1·0000	0·9471	1·3312	1·3361	2·8864	0·4764
1·08	1·4213	1·5447	1·0271	2·8905	1·0000	1·0271	1·4654	1·4725	3·2597	0·5161
1·16	1·4517	1·4950	1·1071	3·0505	1·0000	1·1071	1·6004	1·6097	3·6479	0·5559
1·24	1·4790	1·4472	1·1871	3·2105	1·0000	1·1871	1·7359	1·7476	4·0503	0·5958
1·32	1·5037	1·4016	1·2671	3·3705	1·0000	1·2671	1·8720	1·8861	4·4666	0·6356
1·40	1·5262	1·3581	1·3471	3·5305	1·0000	1·3471	2·0085	2·0251	4·8962	0·6755
1·48	1·5468	1·3167	1·4271	3·6905	1·0000	1·4271	2·1455	2·1646	5·3388	0·7154
1·56	1·5656	1·2773	1·5071	3·8505	1·0000	1·5071	2·2827	2·3044	5·7939	0·7553
1·64	1·5829	1·2399	1·5871	4·0105	1·0000	1·5871	2·4203	2·4446	6·2613	0·7952
1·72	1·5989	1·2044	1·6671	4·1705	1·0000	1·6671	2·5581	2·5851	6·7407	0·8351
1·80	1·6137	1·1707	1·7471	4·3305	1·0000	1·7471	2·6962	2·7259	7·2316	0·8751
1·88	1·6275	1·1386	1·8271	4·4905	1·0000	1·8271	2·8345	2·8669	7·7340	0·9150
1·96	1·6403	1·1081	1·9071	4·6505	1·0000	1·9071	2·9730	3·0081	8·2475	0·9549
2·04	1·6523	1·0790	1·9871	4·8105	1·0000	1·9871	3·1116	3·1495	8·7718	0·9949
2·12	1·6635	1·0514	2·0671	4·9705	1·0000	2·0671	3·2504	3·2911	9·3068	1·0348
2·20	1·6740	1·0250	2·1471	5·1305	1·0000	2·1471	3·3893	3·4328	9·8523	1·0748

Box culvert upright (30% splays) (free surface)

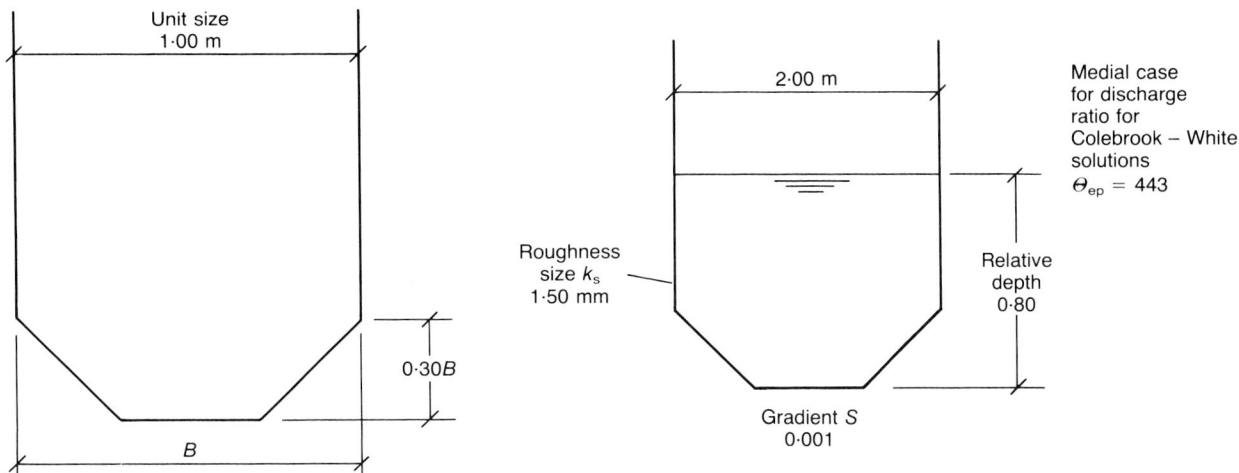

Rel. dpth Y	U.equ. dia. $D_{ep(u)}$ (m)	Equiv. disch. factor J	Unit sect. area A_u (m²)	Unit wetted perim. P_u (m)	Unit surf. brdth $B_{s(u)}$ (m)	Unit mean depth $y_{m(u)}$ (m)	Discharge ratios $Q/Q_{0.80}$ medial	Mann'g	U.crit. disch. $Q_{c(u)}$ (m³s⁻¹)	Unit depth cntrd $y_{d(u)}$ (m)
0.04	0.1372	0.8399	0.0176	0.5131	0.4800	0.0367	0.0059	0.0056	0.0106	0.0194
0.08	0.2453	1.2303	0.0384	0.6263	0.5600	0.0686	0.0191	0.0181	0.0315	0.0378
0.12	0.3376	1.4342	0.0624	0.7394	0.6400	0.0975	0.0381	0.0365	0.0610	0.0554
0.16	0.4204	1.5491	0.0896	0.8525	0.7200	0.1244	0.0631	0.0606	0.0990	0.0724
0.20	0.4971	1.6170	0.1200	0.9657	0.8000	0.1500	0.0940	0.0908	0.1455	0.0889
0.24	0.5695	1.6584	0.1536	1.0788	0.8800	0.1745	0.1312	0.1272	0.2010	0.1050
0.28	0.6389	1.6840	0.1904	1.1920	0.9600	0.1983	0.1750	0.1703	0.2655	0.1208
0.32	0.7140	1.7408	0.2300	1.2885	1.0000	0.2300	0.2267	0.2215	0.3454	0.1365
0.36	0.7892	1.8116	0.2700	1.3685	1.0000	0.2700	0.2835	0.2779	0.4393	0.1533
0.40	0.8560	1.8566	0.3100	1.4485	1.0000	0.3100	0.3425	0.3369	0.5405	0.1710
0.44	0.9159	1.8824	0.3500	1.5285	1.0000	0.3500	0.4035	0.3979	0.6484	0.1891
0.48	0.9698	1.8941	0.3900	1.6085	1.0000	0.3900	0.4659	0.4606	0.7627	0.2077
0.52	1.0186	1.8952	0.4300	1.6885	1.0000	0.4300	0.5297	0.5247	0.8830	0.2265
0.56	1.0630	1.8883	0.4700	1.7685	1.0000	0.4700	0.5946	0.5901	1.0090	0.2455
0.60	1.1036	1.8755	0.5100	1.8485	1.0000	0.5100	0.6605	0.6565	1.1406	0.2647
0.64	1.1408	1.8583	0.5500	1.9285	1.0000	0.5500	0.7271	0.7238	1.2773	0.2840
0.68	1.1750	1.8378	0.5900	2.0085	1.0000	0.5900	0.7945	0.7919	1.4192	0.3034
0.72	1.2066	1.8149	0.6300	2.0885	1.0000	0.6300	0.8625	0.8607	1.5659	0.3229
0.76	1.2359	1.7904	0.6700	2.1685	1.0000	0.6700	0.9310	0.9301	1.7174	0.3424
0.80	1.2630	1.7647	0.7100	2.2485	1.0000	0.7100	1.0000	1.0000	1.8735	0.3620
0.84	1.2884	1.7382	0.7500	2.3285	1.0000	0.7500	1.0695	1.0704	2.0340	0.3816
0.88	1.3120	1.7113	0.7900	2.4085	1.0000	0.7900	1.1393	1.1412	2.1989	0.4013
0.92	1.3341	1.6842	0.8300	2.4885	1.0000	0.8300	1.2094	1.2125	2.3680	0.4210
0.96	1.3549	1.6571	0.8700	2.5685	1.0000	0.8700	1.2799	1.2840	2.5412	0.4407
1.00	1.3743	1.6302	0.9100	2.6485	1.0000	0.9100	1.3507	1.3559	2.7185	0.4604
1.08	1.4100	1.5772	0.9900	2.8085	1.0000	0.9900	1.4929	1.5005	3.0847	0.5000
1.16	1.4418	1.5258	1.0700	2.9685	1.0000	1.0700	1.6360	1.6461	3.4661	0.5396
1.24	1.4703	1.4764	1.1500	3.1285	1.0000	1.1500	1.7797	1.7924	3.8620	0.5793
1.32	1.4961	1.4292	1.2300	3.2885	1.0000	1.2300	1.9241	1.9394	4.2719	0.6190
1.40	1.5195	1.3842	1.3100	3.4485	1.0000	1.3100	2.0690	2.0870	4.6953	0.6588
1.48	1.5408	1.3414	1.3900	3.6085	1.0000	1.3900	2.2143	2.2351	5.1320	0.6986
1.56	1.5603	1.3007	1.4700	3.7685	1.0000	1.4700	2.3600	2.3837	5.5813	0.7384
1.64	1.5782	1.2620	1.5500	3.9285	1.0000	1.5500	2.5061	2.5326	6.0431	0.7782
1.72	1.5947	1.2253	1.6300	4.0885	1.0000	1.6300	2.6524	2.6819	6.5169	0.8180
1.80	1.6100	1.1905	1.7100	4.2485	1.0000	1.7100	2.7990	2.8314	7.0025	0.8579
1.88	1.6241	1.1574	1.7900	4.4085	1.0000	1.7900	2.9458	2.9812	7.4996	0.8978
1.96	1.6373	1.1259	1.8700	4.5685	1.0000	1.8700	3.0928	3.1313	8.0080	0.9376
2.04	1.6496	1.0959	1.9500	4.7285	1.0000	1.9500	3.2400	3.2815	8.5273	0.9775
2.12	1.6610	1.0674	2.0300	4.8885	1.0000	2.0300	3.3874	3.4320	9.0574	1.0174
2.20	1.6718	1.0403	2.1100	5.0485	1.0000	2.1100	3.5349	3.5826	9.5981	1.0573

Standard Horseshoe

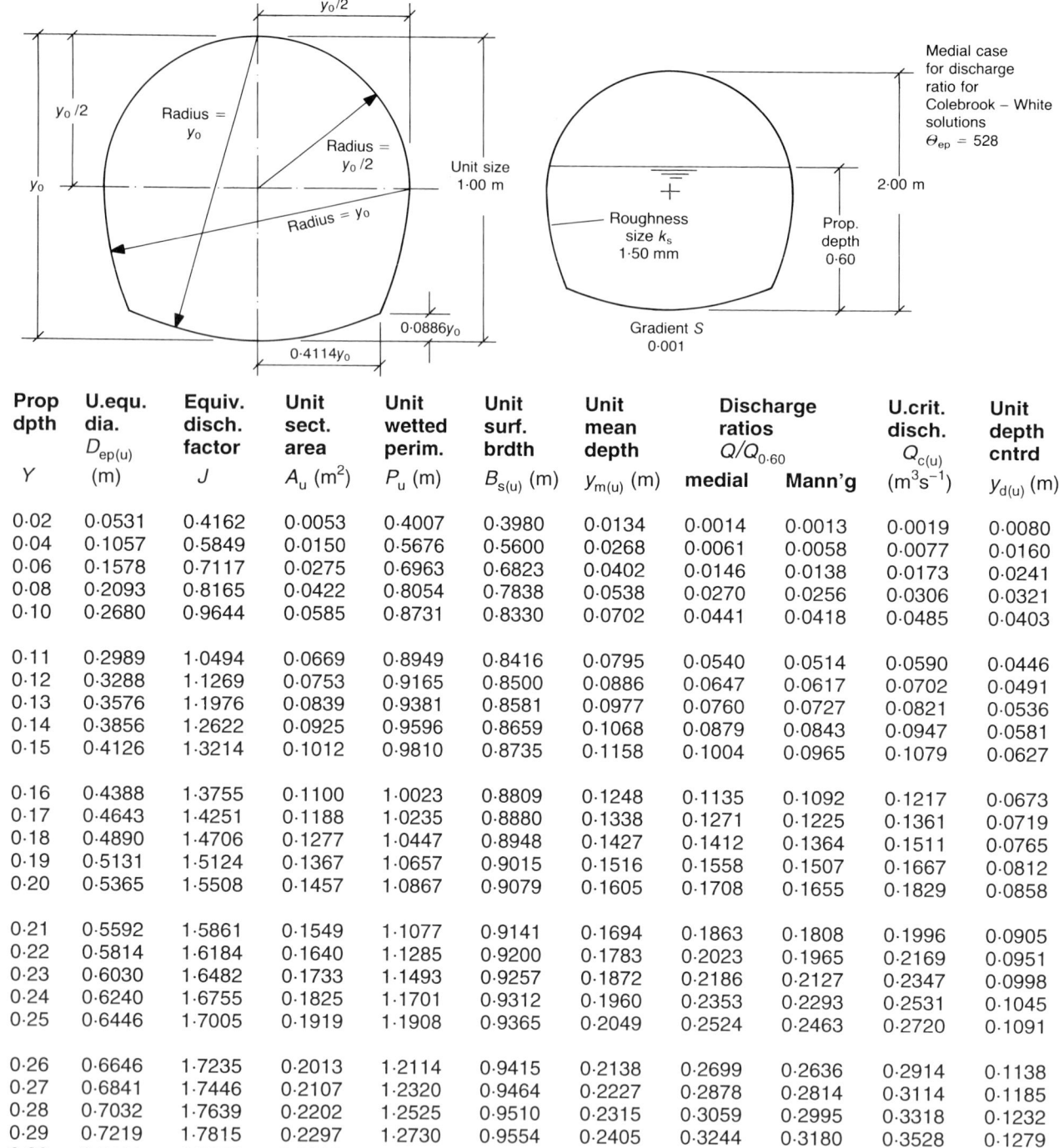

Prop dpth	U.equ. dia. $D_{ep(u)}$	Equiv. disch. factor	Unit sect. area	Unit wetted perim.	Unit surf. brdth	Unit mean depth	Discharge ratios $Q/Q_{0.60}$		U.crit. disch. $Q_{c(u)}$	Unit depth cntrd
Y	(m)	J	A_u (m²)	P_u (m)	$B_{s(u)}$ (m)	$y_{m(u)}$ (m)	medial	Mann'g	(m³s⁻¹)	$y_{d(u)}$ (m)
0·02	0·0531	0·4162	0·0053	0·4007	0·3980	0·0134	0·0014	0·0013	0·0019	0·0080
0·04	0·1057	0·5849	0·0150	0·5676	0·5600	0·0268	0·0061	0·0058	0·0077	0·0160
0·06	0·1578	0·7117	0·0275	0·6963	0·6823	0·0402	0·0146	0·0138	0·0173	0·0241
0·08	0·2093	0·8165	0·0422	0·8054	0·7838	0·0538	0·0270	0·0256	0·0306	0·0321
0·10	0·2680	0·9644	0·0585	0·8731	0·8330	0·0702	0·0441	0·0418	0·0485	0·0403
0·11	0·2989	1·0494	0·0669	0·8949	0·8416	0·0795	0·0540	0·0514	0·0590	0·0446
0·12	0·3288	1·1269	0·0753	0·9165	0·8500	0·0886	0·0647	0·0617	0·0702	0·0491
0·13	0·3576	1·1976	0·0839	0·9381	0·8581	0·0977	0·0760	0·0727	0·0821	0·0536
0·14	0·3856	1·2622	0·0925	0·9596	0·8659	0·1068	0·0879	0·0843	0·0947	0·0581
0·15	0·4126	1·3214	0·1012	0·9810	0·8735	0·1158	0·1004	0·0965	0·1079	0·0627
0·16	0·4388	1·3755	0·1100	1·0023	0·8809	0·1248	0·1135	0·1092	0·1217	0·0673
0·17	0·4643	1·4251	0·1188	1·0235	0·8880	0·1338	0·1271	0·1225	0·1361	0·0719
0·18	0·4890	1·4706	0·1277	1·0447	0·8948	0·1427	0·1412	0·1364	0·1511	0·0765
0·19	0·5131	1·5124	0·1367	1·0657	0·9015	0·1516	0·1558	0·1507	0·1667	0·0812
0·20	0·5365	1·5508	0·1457	1·0867	0·9079	0·1605	0·1708	0·1655	0·1829	0·0858
0·21	0·5592	1·5861	0·1549	1·1077	0·9141	0·1694	0·1863	0·1808	0·1996	0·0905
0·22	0·5814	1·6184	0·1640	1·1285	0·9200	0·1783	0·2023	0·1965	0·2169	0·0951
0·23	0·6030	1·6482	0·1733	1·1493	0·9257	0·1872	0·2186	0·2127	0·2347	0·0998
0·24	0·6240	1·6755	0·1825	1·1701	0·9312	0·1960	0·2353	0·2293	0·2531	0·1045
0·25	0·6446	1·7005	0·1919	1·1908	0·9365	0·2049	0·2524	0·2463	0·2720	0·1091
0·26	0·6646	1·7235	0·2013	1·2114	0·9415	0·2138	0·2699	0·2636	0·2914	0·1138
0·27	0·6841	1·7446	0·2107	1·2320	0·9464	0·2227	0·2878	0·2814	0·3114	0·1185
0·28	0·7032	1·7639	0·2202	1·2525	0·9510	0·2315	0·3059	0·2995	0·3318	0·1232
0·29	0·7219	1·7815	0·2297	1·2730	0·9554	0·2405	0·3244	0·3180	0·3528	0·1279
0·30	0·7401	1·7976	0·2393	1·2934	0·9596	0·2494	0·3433	0·3368	0·3742	0·1325
0·31	0·7579	1·8122	0·2489	1·3138	0·9636	0·2583	0·3624	0·3559	0·3962	0·1372
0·32	0·7753	1·8255	0·2586	1·3342	0·9673	0·2673	0·3818	0·3753	0·4187	0·1419
0·33	0·7923	1·8375	0·2683	1·3545	0·9709	0·2763	0·4015	0·3951	0·4416	0·1466
0·34	0·8089	1·8484	0·2780	1·3747	0·9742	0·2853	0·4215	0·4151	0·4650	0·1513
0·35	0·8251	1·8581	0·2878	1·3950	0·9774	0·2944	0·4418	0·4354	0·4889	0·1560
0·36	0·8410	1·8669	0·2975	1·4152	0·9803	0·3035	0·4622	0·4560	0·5133	0·1607
0·37	0·8565	1·8746	0·3074	1·4354	0·9830	0·3127	0·4830	0·4768	0·5382	0·1654
0·38	0·8717	1·8814	0·3172	1·4555	0·9855	0·3219	0·5039	0·4979	0·5635	0·1701
0·39	0·8866	1·8874	0·3271	1·4757	0·9879	0·3311	0·5251	0·5192	0·5893	0·1748
0·40	0·9011	1·8925	0·3370	1·4958	0·9900	0·3404	0·5464	0·5407	0·6156	0·1796
0·41	0·9153	1·8969	0·3469	1·5159	0·9919	0·3497	0·5680	0·5624	0·6424	0·1843
0·42	0·9292	1·9005	0·3568	1·5360	0·9936	0·3591	0·5898	0·5844	0·6696	0·1890
0·43	0·9428	1·9034	0·3667	1·5560	0·9951	0·3685	0·6117	0·6065	0·6972	0·1938
0·44	0·9561	1·9057	0·3767	1·5761	0·9964	0·3781	0·6338	0·6288	0·7253	0·1985
0·45	0·9690	1·9073	0·3867	1·5961	0·9975	0·3876	0·6560	0·6513	0·7539	0·2033

Prop dpth	U.equ. dia. $D_{ep(u)}$	Equiv. disch. factor	Unit sect. area	Unit wetted perim.	Unit surf. brdth	Unit mean depth	Discharge ratios $Q/Q_{0.60}$		U.crit. disch. $Q_{c(u)}$	Unit depth cntrd
Y	(m)	J	A_u (m²)	P_u (m)	$B_{s(u)}$ (m)	$y_{m(u)}$ (m)	medial	Mann'g	(m³s⁻¹)	$y_{d(u)}$ (m)
0·46	0·9817	1·9084	0·3966	1·6161	0·9984	0·3973	0·6784	0·6739	0·7829	0·2080
0·47	0·9941	1·9089	0·4066	1·6361	0·9991	0·4070	0·7009	0·6967	0·8124	0·2128
0·48	1·0063	1·9088	0·4166	1·6561	0·9996	0·4168	0·7236	0·7196	0·8423	0·2176
0·49	1·0181	1·9082	0·4266	1·6761	0·9999	0·4267	0·7463	0·7426	0·8727	0·2224
0·50	1·0297	1·9072	0·4366	1·6961	1·0000	0·4366	0·7692	0·7658	0·9035	0·2271
0·51	1·0410	1·9057	0·4466	1·7161	0·9998	0·4467	0·7922	0·7891	0·9348	0·2319
0·52	1·0520	1·9036	0·4566	1·7361	0·9992	0·4570	0·8152	0·8124	0·9666	0·2368
0·53	1·0628	1·9012	0·4666	1·7562	0·9982	0·4674	0·8383	0·8358	0·9990	0·2416
0·54	1·0733	1·8982	0·4766	1·7762	0·9968	0·4781	0·8614	0·8593	1·0320	0·2464
0·55	1·0834	1·8948	0·4865	1·7963	0·9950	0·4890	0·8846	0·8828	1·0654	0·2513
0·56	1·0933	1·8909	0·4965	1·8164	0·9928	0·5001	0·9077	0·9063	1·0995	0·2561
0·57	1·1029	1·8865	0·5064	1·8366	0·9902	0·5114	0·9309	0·9298	1·1341	0·2610
0·58	1·1122	1·8817	0·5163	1·8568	0·9871	0·5230	0·9540	0·9532	1·1692	0·2659
0·59	1·1212	1·8764	0·5261	1·8771	0·9837	0·5349	0·9770	0·9767	1·2050	0·2709
0·60	1·1298	1·8706	0·5360	1·8975	0·9798	0·5470	1·0000	1·0000	1·2413	0·2758
0·61	1·1382	1·8643	0·5457	1·9179	0·9755	0·5594	1·0229	1·0232	1·2782	0·2808
0·62	1·1462	1·8575	0·5555	1·9385	0·9708	0·5722	1·0457	1·0464	1·3158	0·2858
0·63	1·1539	1·8502	0·5651	1·9591	0·9656	0·5853	1·0683	1·0694	1·3539	0·2908
0·64	1·1612	1·8425	0·5748	1·9799	0·9600	0·5987	1·0908	1·0922	1·3927	0·2958
0·65	1·1682	1·8342	0·5843	2·0008	0·9539	0·6126	1·1131	1·1148	1·4322	0·3009
0·66	1·1749	1·8255	0·5939	2·0219	0·9474	0·6268	1·1351	1·1373	1·4723	0·3060
0·67	1·1812	1·8162	0·6033	2·0430	0·9404	0·6415	1·1570	1·1595	1·5132	0·3111
0·68	1·1871	1·8065	0·6127	2·0644	0·9330	0·6567	1·1786	1·1814	1·5547	0·3163
0·69	1·1927	1·7962	0·6219	2·0859	0·9250	0·6724	1·1999	1·2031	1·5971	0·3215
0·70	1·1978	1·7854	0·6312	2·1076	0·9165	0·6886	1·2210	1·2244	1·6402	0·3268
0·71	1·2026	1·7741	0·6403	2·1296	0·9075	0·7055	1·2417	1·2454	1·6842	0·3320
0·72	1·2070	1·7623	0·6493	2·1517	0·8980	0·7231	1·2620	1·2661	1·7290	0·3373
0·73	1·2110	1·7499	0·6582	2·1741	0·8879	0·7413	1·2820	1·2863	1·7748	0·3427
0·74	1·2146	1·7370	0·6671	2·1968	0·8773	0·7604	1·3016	1·3061	1·8216	0·3481
0·75	1·2178	1·7235	0·6758	2·2197	0·8660	0·7803	1·3207	1·3255	1·8694	0·3535
0·76	1·2205	1·7094	0·6844	2·2430	0·8542	0·8012	1·3393	1·3444	1·9184	0·3590
0·77	1·2227	1·6948	0·6929	2·2666	0·8417	0·8232	1·3575	1·3627	1·9686	0·3646
0·78	1·2245	1·6795	0·7012	2·2905	0·8285	0·8464	1·3751	1·3805	2·0202	0·3702
0·79	1·2259	1·6636	0·7094	2·3149	0·8146	0·8709	1·3921	1·3977	2·0732	0·3758
0·80	1·2267	1·6471	0·7175	2·3396	0·8000	0·8969	1·4085	1·4142	2·1279	0·3815
0·81	1·2270	1·6300	0·7254	2·3649	0·7846	0·9246	1·4243	1·4301	2·1844	0·3873
0·82	1·2268	1·6121	0·7332	2·3906	0·7684	0·9542	1·4394	1·4452	2·2428	0·3932
0·83	1·2260	1·5935	0·7408	2·4169	0·7513	0·9861	1·4538	1·4595	2·3036	0·3991
0·84	1·2246	1·5742	0·7482	2·4439	0·7332	1·0205	1·4673	1·4731	2·3669	0·4051
0·85	1·2226	1·5541	0·7554	2·4715	0·7141	1·0578	1·4800	1·4857	2·4332	0·4111
0·86	1·2200	1·5331	0·7625	2·4999	0·6940	1·0987	1·4918	1·4974	2·5029	0·4173
0·87	1·2167	1·5113	0·7693	2·5292	0·6726	1·1438	1·5027	1·5081	2·5766	0·4236
0·88	1·2127	1·4885	0·7759	2·5594	0·6499	1·1939	1·5125	1·5177	2·6550	0·4299
0·89	1·2078	1·4646	0·7823	2·5908	0·6258	1·2501	1·5212	1·5261	2·7392	0·4364
0·90	1·2022	1·4396	0·7884	2·6234	0·6000	1·3141	1·5287	1·5333	2·8304	0·4429
0·91	1·1956	1·4133	0·7943	2·6575	0·5724	1·3878	1·5348	1·5390	2·9303	0·4496
0·92	1·1879	1·3856	0·7999	2·6934	0·5426	1·4742	1·5395	1·5432	3·0414	0·4564
0·93	1·1791	1·3562	0·8052	2·7314	0·5103	1·5778	1·5425	1·5457	3·1672	0·4634
0·94	1·1690	1·3248	0·8101	2·7720	0·4750	1·7055	1·5437	1·5462	3·3130	0·4706
0·95	1·1572	1·2910	0·8146	2·8159	0·4359	1·8689	1·5427	1·5444	3·4876	0·4779
0·96	1·1435	1·2542	0·8188	2·8642	0·3919	2·0892	1·5391	1·5400	3·7061	0·4855
0·97	1·1271	1·2132	0·8225	2·9188	0·3412	2·4107	1·5323	1·5321	3·9989	0·4933
0·98	1·1070	1·1658	0·8256	2·9831	0·2800	2·9485	1·5211	1·5196	4·4393	0·5014
0·99	1·0800	1·1064	0·8280	3·0666	0·1990	4·1608	1·5024	1·4992	5·2890	0·5099
1·00	1·0154	0·9764	0·8293	3·2669	0·0000	-	1·4485	1·4411	-	0·5191

C25

Metcalf and Eddy (M and E) Horseshoe

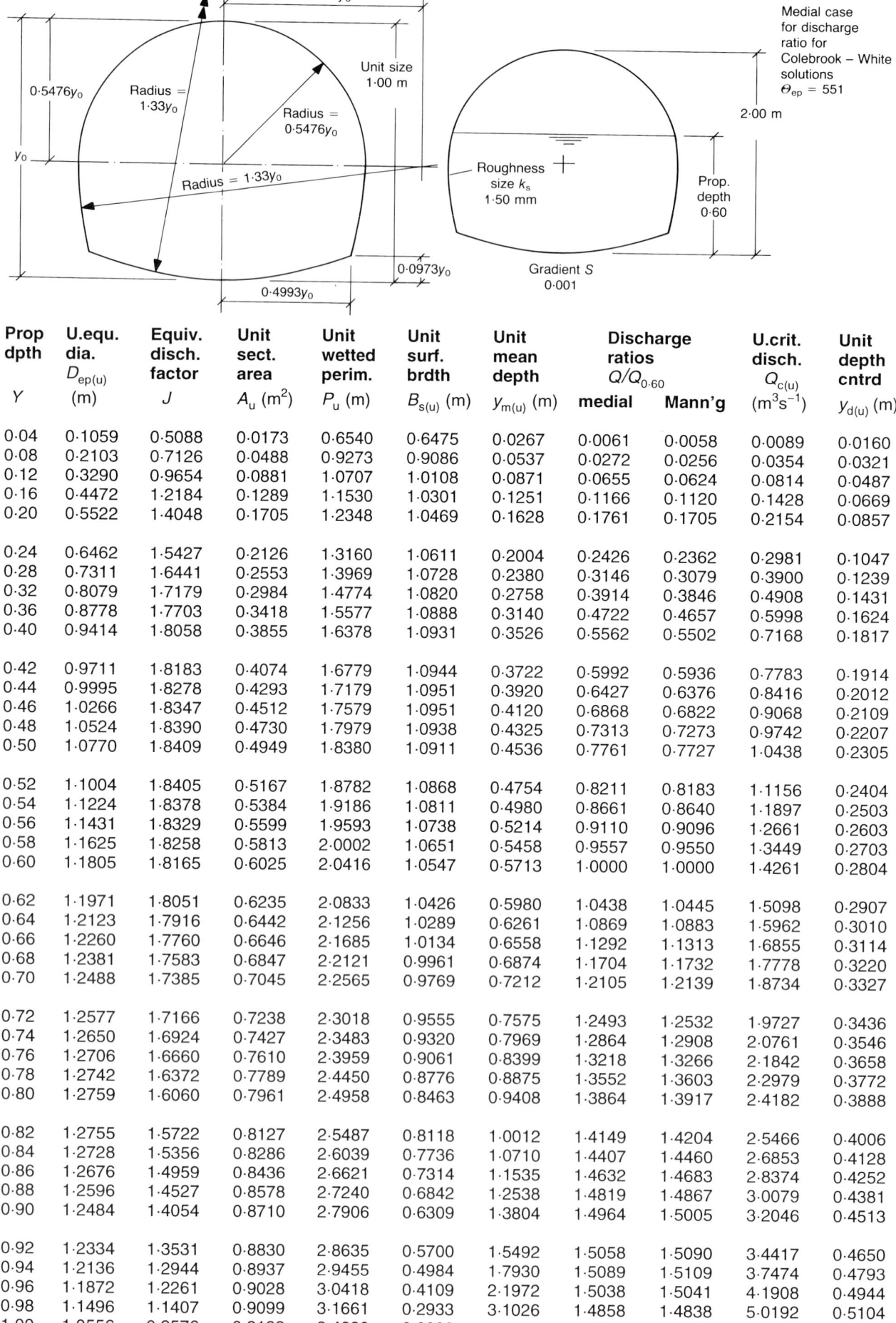

Prop dpth Y	U.equ. dia. $D_{ep(u)}$ (m)	Equiv. disch. factor J	Unit sect. area A_u (m²)	Unit wetted perim. P_u (m)	Unit surf. brdth $B_{s(u)}$ (m)	Unit mean depth $y_{m(u)}$ (m)	Discharge ratios $Q/Q_{0.60}$ medial	Mann'g	U.crit. disch. $Q_{c(u)}$ (m³s⁻¹)	Unit depth cntrd $y_{d(u)}$ (m)
0.04	0.1059	0.5088	0.0173	0.6540	0.6475	0.0267	0.0061	0.0058	0.0089	0.0160
0.08	0.2103	0.7126	0.0488	0.9273	0.9086	0.0537	0.0272	0.0256	0.0354	0.0321
0.12	0.3290	0.9654	0.0881	1.0707	1.0108	0.0871	0.0655	0.0624	0.0814	0.0487
0.16	0.4472	1.2184	0.1289	1.1530	1.0301	0.1251	0.1166	0.1120	0.1428	0.0669
0.20	0.5522	1.4048	0.1705	1.2348	1.0469	0.1628	0.1761	0.1705	0.2154	0.0857
0.24	0.6462	1.5427	0.2126	1.3160	1.0611	0.2004	0.2426	0.2362	0.2981	0.1047
0.28	0.7311	1.6441	0.2553	1.3969	1.0728	0.2380	0.3146	0.3079	0.3900	0.1239
0.32	0.8079	1.7179	0.2984	1.4774	1.0820	0.2758	0.3914	0.3846	0.4908	0.1431
0.36	0.8778	1.7703	0.3418	1.5577	1.0888	0.3140	0.4722	0.4657	0.5998	0.1624
0.40	0.9414	1.8058	0.3855	1.6378	1.0931	0.3526	0.5562	0.5502	0.7168	0.1817
0.42	0.9711	1.8183	0.4074	1.6779	1.0944	0.3722	0.5992	0.5936	0.7783	0.1914
0.44	0.9995	1.8278	0.4293	1.7179	1.0951	0.3920	0.6427	0.6376	0.8416	0.2012
0.46	1.0266	1.8347	0.4512	1.7579	1.0951	0.4120	0.6868	0.6822	0.9068	0.2109
0.48	1.0524	1.8390	0.4730	1.7979	1.0938	0.4325	0.7313	0.7273	0.9742	0.2207
0.50	1.0770	1.8409	0.4949	1.8380	1.0911	0.4536	0.7761	0.7727	1.0438	0.2305
0.52	1.1004	1.8405	0.5167	1.8782	1.0868	0.4754	0.8211	0.8183	1.1156	0.2404
0.54	1.1224	1.8378	0.5384	1.9186	1.0811	0.4980	0.8661	0.8640	1.1897	0.2503
0.56	1.1431	1.8329	0.5599	1.9593	1.0738	0.5214	0.9110	0.9096	1.2661	0.2603
0.58	1.1625	1.8258	0.5813	2.0002	1.0651	0.5458	0.9557	0.9550	1.3449	0.2703
0.60	1.1805	1.8165	0.6025	2.0416	1.0547	0.5713	1.0000	1.0000	1.4261	0.2804
0.62	1.1971	1.8051	0.6235	2.0833	1.0426	0.5980	1.0438	1.0445	1.5098	0.2907
0.64	1.2123	1.7916	0.6442	2.1256	1.0289	0.6261	1.0869	1.0883	1.5962	0.3010
0.66	1.2260	1.7760	0.6646	2.1685	1.0134	0.6558	1.1292	1.1313	1.6855	0.3114
0.68	1.2381	1.7583	0.6847	2.2121	0.9961	0.6874	1.1704	1.1732	1.7778	0.3220
0.70	1.2488	1.7385	0.7045	2.2565	0.9769	0.7212	1.2105	1.2139	1.8734	0.3327
0.72	1.2577	1.7166	0.7238	2.3018	0.9555	0.7575	1.2493	1.2532	1.9727	0.3436
0.74	1.2650	1.6924	0.7427	2.3483	0.9320	0.7969	1.2864	1.2908	2.0761	0.3546
0.76	1.2706	1.6660	0.7610	2.3959	0.9061	0.8399	1.3218	1.3266	2.1842	0.3658
0.78	1.2742	1.6372	0.7789	2.4450	0.8776	0.8875	1.3552	1.3603	2.2979	0.3772
0.80	1.2759	1.6060	0.7961	2.4958	0.8463	0.9408	1.3864	1.3917	2.4182	0.3888
0.82	1.2755	1.5722	0.8127	2.5487	0.8118	1.0012	1.4149	1.4204	2.5466	0.4006
0.84	1.2728	1.5356	0.8286	2.6039	0.7736	1.0710	1.4407	1.4460	2.6853	0.4128
0.86	1.2676	1.4959	0.8436	2.6621	0.7314	1.1535	1.4632	1.4683	2.8374	0.4252
0.88	1.2596	1.4527	0.8578	2.7240	0.6842	1.2538	1.4819	1.4867	3.0079	0.4381
0.90	1.2484	1.4054	0.8710	2.7906	0.6309	1.3804	1.4964	1.5005	3.2046	0.4513
0.92	1.2334	1.3531	0.8830	2.8635	0.5700	1.5492	1.5058	1.5090	3.4417	0.4650
0.94	1.2136	1.2944	0.8937	2.9455	0.4984	1.7930	1.5089	1.5109	3.7474	0.4793
0.96	1.1872	1.2261	0.9028	3.0418	0.4109	2.1972	1.5038	1.5041	4.1908	0.4944
0.98	1.1496	1.1407	0.9099	3.1661	0.2933	3.1026	1.4858	1.4838	5.0192	0.5104
1.00	1.0556	0.9576	0.9139	3.4630	0.0000	-	1.4154	1.4078	-	0.5282

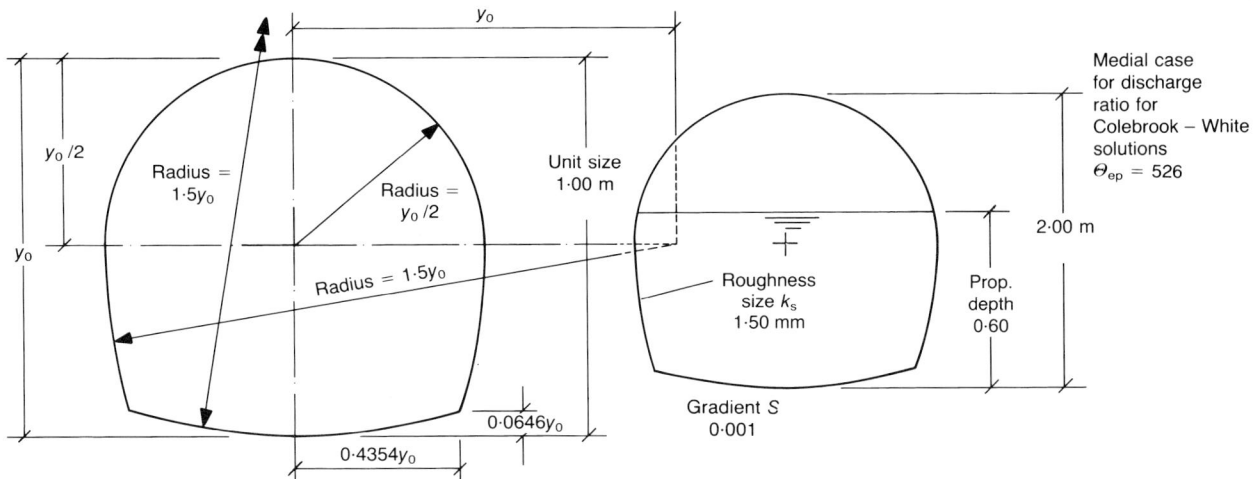

| Prop dpth | U.equ. dia. $D_{ep(u)}$ | Equiv. disch. factor | Unit sect. area | Unit wetted perim. | Unit surf. brdth | Unit mean depth | Discharge ratios $Q/Q_{0.60}$ | | U.crit. disch. $Q_{c(u)}$ | Unit depth cntrd |
Y	(m)	J	A_u (m²)	P_u (m)	$B_{s(u)}$ (m)	$y_{m(u)}$ (m)	medial	Mann'g	(m³s⁻¹)	$y_{d(u)}$ (m)
0.04	0.1060	0.4796	0.0184	0.6944	0.6882	0.0267	0.0073	0.0069	0.0094	0.0160
0.08	0.2235	0.7666	0.0512	0.9157	0.8800	0.0581	0.0332	0.0314	0.0386	0.0324
0.12	0.3477	1.0936	0.0868	0.9987	0.9021	0.0962	0.0749	0.0716	0.0843	0.0509
0.16	0.4562	1.3255	0.1233	1.0811	0.9219	0.1337	0.1265	0.1219	0.1412	0.0699
0.20	0.5521	1.4913	0.1605	1.1630	0.9394	0.1709	0.1859	0.1803	0.2078	0.0890
0.24	0.6378	1.6100	0.1984	1.2445	0.9546	0.2079	0.2515	0.2453	0.2833	0.1082
0.28	0.7148	1.6941	0.2369	1.3255	0.9676	0.2448	0.3225	0.3160	0.3670	0.1274
0.32	0.7845	1.7525	0.2758	1.4062	0.9783	0.2819	0.3980	0.3915	0.4585	0.1466
0.36	0.8478	1.7915	0.3151	1.4867	0.9869	0.3193	0.4773	0.4711	0.5576	0.1658
0.40	0.9055	1.8154	0.3547	1.5669	0.9933	0.3571	0.5597	0.5541	0.6638	0.1850
0.42	0.9324	1.8228	0.3746	1.6070	0.9957	0.3762	0.6020	0.5967	0.7195	0.1947
0.44	0.9582	1.8276	0.3945	1.6471	0.9976	0.3955	0.6448	0.6399	0.7770	0.2043
0.46	0.9828	1.8300	0.4145	1.6871	0.9989	0.4149	0.6882	0.6838	0.8362	0.2140
0.48	1.0063	1.8304	0.4345	1.7271	0.9997	0.4346	0.7321	0.7282	0.8970	0.2237
0.50	1.0288	1.8290	0.4545	1.7671	1.0000	0.4545	0.7763	0.7730	0.9595	0.2334
0.52	1.0503	1.8258	0.4745	1.8071	0.9992	0.4749	0.8209	0.8182	1.0239	0.2432
0.54	1.0707	1.8210	0.4944	1.8472	0.9968	0.4960	0.8657	0.8636	1.0905	0.2529
0.56	1.0901	1.8144	0.5143	1.8874	0.9928	0.5181	0.9106	0.9092	1.1594	0.2628
0.58	1.1083	1.8061	0.5341	1.9278	0.9871	0.5411	0.9554	0.9547	1.2305	0.2727
0.60	1.1254	1.7961	0.5538	1.9685	0.9798	0.5652	1.0000	1.0000	1.3039	0.2826
0.62	1.1413	1.7842	0.5733	2.0095	0.9708	0.5906	1.0442	1.0449	1.3798	0.2927
0.64	1.1559	1.7705	0.5926	2.0509	0.9600	0.6173	1.0879	1.0893	1.4582	0.3028
0.66	1.1692	1.7550	0.6117	2.0928	0.9474	0.6457	1.1309	1.1330	1.5393	0.3130
0.68	1.1811	1.7376	0.6305	2.1354	0.9330	0.6758	1.1730	1.1758	1.6232	0.3234
0.70	1.1916	1.7183	0.6490	2.1786	0.9165	0.7081	1.2141	1.2174	1.7103	0.3339
0.72	1.2007	1.6970	0.6672	2.2227	0.8980	0.7430	1.2539	1.2578	1.8009	0.3446
0.74	1.2081	1.6736	0.6849	2.2677	0.8773	0.7807	1.2922	1.2966	1.8952	0.3554
0.76	1.2139	1.6481	0.7022	2.3139	0.8542	0.8221	1.3288	1.3337	1.9940	0.3664
0.78	1.2180	1.6203	0.7191	2.3615	0.8285	0.8679	1.3634	1.3687	2.0979	0.3775
0.80	1.2202	1.5902	0.7354	2.4106	0.8000	0.9192	1.3959	1.4014	2.2079	0.3890
0.82	1.2204	1.5575	0.7511	2.4616	0.7684	0.9775	1.4258	1.4315	2.3253	0.4006
0.84	1.2185	1.5221	0.7661	2.5149	0.7332	1.0448	1.4529	1.4585	2.4522	0.4126
0.86	1.2141	1.4836	0.7804	2.5709	0.6940	1.1245	1.4767	1.4822	2.5914	0.4248
0.88	1.2071	1.4417	0.7938	2.6304	0.6499	1.2214	1.4968	1.5019	2.7473	0.4375
0.90	1.1970	1.3957	0.8063	2.6944	0.6000	1.3439	1.5125	1.5171	2.9271	0.4505
0.92	1.1833	1.3447	0.8178	2.7644	0.5426	1.5071	1.5231	1.5268	3.1438	0.4641
0.94	1.1649	1.2873	0.8280	2.8430	0.4750	1.7432	1.5273	1.5298	3.4232	0.4783
0.96	1.1402	1.2203	0.8367	2.9352	0.3919	2.1348	1.5230	1.5239	3.8281	0.4932
0.98	1.1047	1.1363	0.8434	3.0541	0.2800	3.0123	1.5056	1.5042	4.5842	0.5091
1.00	1.0152	0.9555	0.8472	3.3379	0.0000	-	1.4352	1.4282	-	0.5269

M and E Semi-elliptical

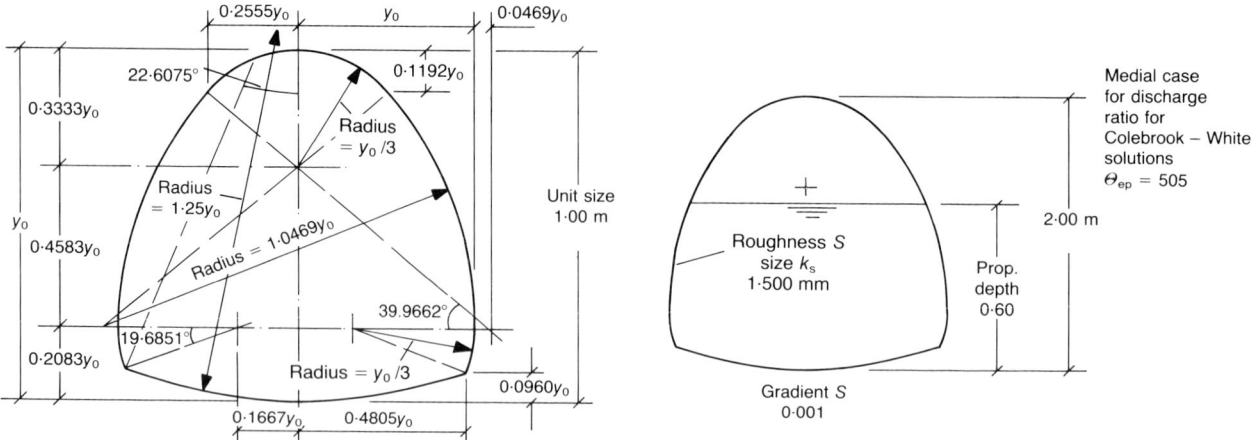

Prop dpth Y	U.equ. dia. $D_{ep(u)}$ (m)	Equiv. disch. factor J	Unit sect. area A_u (m²)	Unit wetted perim. P_u (m)	Unit surf. brdth $B_{s(u)}$ (m)	Unit mean depth $y_{m(u)}$ (m)	Discharge ratios $Q/Q_{0.60}$ medial	Mann'g	U.crit. disch. $Q_{c(u)}$ (m³s⁻¹)	Unit depth cntrd $y_{d(u)}$ (m)
0·04	0·1059	0·5245	0·0168	0·6342	0·6274	0·0268	0·0069	0·0065	0·0086	0·0160
0·08	0·2101	0·7341	0·0472	0·8993	0·8800	0·0537	0·0307	0·0291	0·0343	0·0321
0·12	0·3289	0·9966	0·0852	1·0367	0·9762	0·0873	0·0740	0·0708	0·0789	0·0487
0·16	0·4458	1·2521	0·1247	1·1185	0·9930	0·1255	0·1314	0·1268	0·1383	0·0670
0·20	0·5490	1·4387	0·1645	1·1988	0·9998	0·1646	0·1979	0·1923	0·2090	0·0859
0·24	0·6397	1·5715	0·2045	1·2788	0·9990	0·2047	0·2708	0·2647	0·2898	0·1052
0·28	0·7194	1·6632	0·2444	1·3589	0·9951	0·2456	0·3483	0·3421	0·3793	0·1248
0·32	0·7896	1·7234	0·2841	1·4392	0·9881	0·2875	0·4291	0·4230	0·4770	0·1445
0·36	0·8512	1·7593	0·3234	1·5199	0·9779	0·3307	0·5119	0·5063	0·5825	0·1645
0·40	0·9051	1·7761	0·3623	1·6010	0·9646	0·3756	0·5958	0·5909	0·6953	0·1847
0·42	0·9295	1·7786	0·3815	1·6418	0·9568	0·3987	0·6378	0·6333	0·7544	0·1949
0·44	0·9522	1·7777	0·4005	1·6827	0·9481	0·4225	0·6798	0·6757	0·8153	0·2052
0·46	0·9732	1·7737	0·4194	1·7238	0·9386	0·4469	0·7215	0·7180	0·8780	0·2155
0·48	0·9928	1·7669	0·4381	1·7651	0·9283	0·4719	0·7630	0·7599	0·9425	0·2259
0·50	1·0108	1·7577	0·4565	1·8066	0·9171	0·4978	0·8040	0·8015	1·0087	0·2364
0·52	1·0274	1·7461	0·4748	1·8484	0·9051	0·5246	0·8446	0·8426	1·0768	0·2469
0·54	1·0426	1·7326	0·4927	1·8904	0·8921	0·5523	0·8846	0·8831	1·1467	0·2575
0·56	1·0564	1·7171	0·5104	1·9328	0·8783	0·5812	0·9239	0·9229	1·2186	0·2683
0·58	1·0689	1·6999	0·5279	1·9754	0·8636	0·6112	0·9624	0·9619	1·2924	0·2791
0·60	1·0801	1·6811	0·5450	2·0183	0·8479	0·6427	1·0000	1·0000	1·3682	0·2900
0·62	1·0899	1·6608	0·5618	2·0617	0·8313	0·6758	1·0367	1·0371	1·4462	0·3010
0·64	1·0986	1·6392	0·5782	2·1054	0·8137	0·7106	1·0723	1·0731	1·5264	0·3122
0·66	1·1060	1·6164	0·5943	2·1495	0·7951	0·7475	1·1067	1·1079	1·6091	0·3235
0·68	1·1121	1·5924	0·6100	2·1941	0·7755	0·7867	1·1398	1·1414	1·6944	0·3349
0·70	1·1171	1·5673	0·6253	2·2391	0·7547	0·8286	1·1717	1·1735	1·7825	0·3464
0·72	1·1209	1·5412	0·6402	2·2847	0·7329	0·8736	1·2020	1·2041	1·8738	0·3582
0·74	1·1234	1·5142	0·6546	2·3308	0·7099	0·9222	1·2309	1·2332	1·9686	0·3700
0·76	1·1248	1·4863	0·6686	2·3776	0·6857	0·9751	1·2581	1·2605	2·0675	0·3821
0·78	1·1251	1·4575	0·6821	2·4250	0·6603	1·0330	1·2836	1·2860	2·1709	0·3944
0·80	1·1241	1·4279	0·6950	2·4731	0·6335	1·0970	1·3072	1·3097	2·2796	0·4068
0·82	1·1220	1·3976	0·7074	2·5220	0·6054	1·1684	1·3290	1·3314	2·3945	0·4195
0·84	1·1187	1·3665	0·7192	2·5717	0·5759	1·2488	1·3487	1·3509	2·5169	0·4325
0·86	1·1142	1·3348	0·7304	2·6223	0·5449	1·3405	1·3663	1·3683	2·6483	0·4457
0·88	1·1085	1·3023	0·7410	2·6739	0·5123	1·4465	1·3817	1·3834	2·7909	0·4592
0·90	1·1011	1·2680	0·7509	2·7279	0·4761	1·5772	1·3944	1·3956	2·9531	0·4730
0·92	1·0910	1·2300	0·7600	2·7865	0·4333	1·7540	1·4033	1·4039	3·1520	0·4872
0·94	1·0774	1·1868	0·7682	2·8519	0·3816	2·0131	1·4074	1·4072	3·4131	0·5019
0·96	1·0589	1·1361	0·7752	2·9282	0·3166	2·4480	1·4051	1·4037	3·7981	0·5173
0·98	1·0319	1·0713	0·7807	3·0260	0·2274	3·4323	1·3926	1·3896	4·5291	0·5336
1·00	0·9622	0·9277	0·7837	3·2581	0·0000	-	1·3387	1·3314	-	0·5515

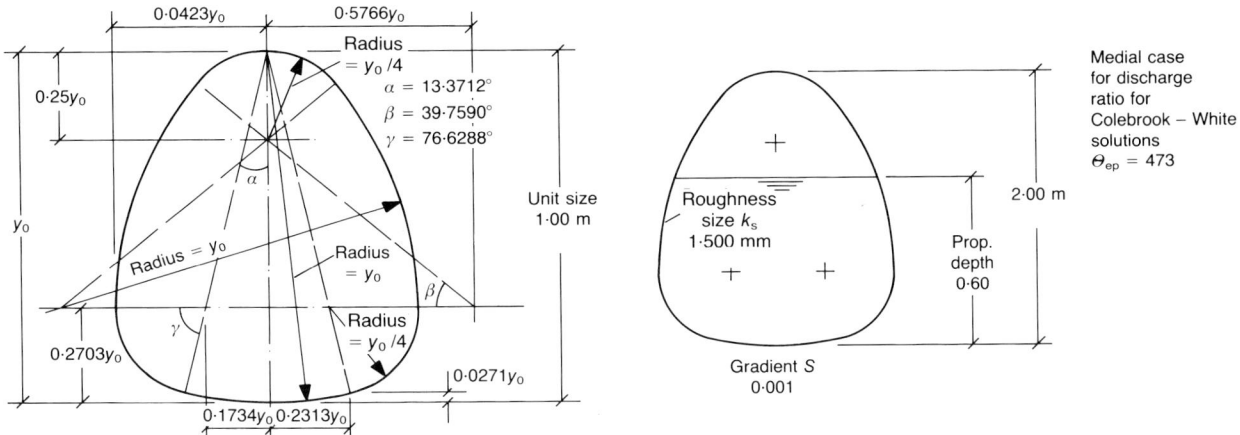

Prop dpth	U.equ. dia. $D_{ep(u)}$	Equiv. disch. factor	Unit sect. area	Unit wetted perim.	Unit surf. brdth	Unit mean depth	Discharge ratios $Q/Q_{0.60}$		U.crit. disch. $Q_{c(u)}$	Unit depth cntrd
Y	(m)	J	A_u (m²)	P_u (m)	$B_{s(u)}$ (m)	$y_{m(u)}$ (m)	medial	Mann'g	(m³s⁻¹)	$y_{d(u)}$ (m)
0·04	0·1084	0·6196	0·0149	0·5497	0·5413	0·0275	0·0078	0·0073	0·0077	0·0161
0·08	0·2245	1·0036	0·0394	0·7028	0·6711	0·0588	0·0332	0·0316	0·0299	0·0332
0·12	0·3341	1·2914	0·0679	0·8129	0·7464	0·0910	0·0739	0·0709	0·0641	0·0508
0·16	0·4358	1·5097	0·0988	0·9069	0·7956	0·1242	0·1273	0·1231	0·1091	0·0686
0·20	0·5290	1·6739	0·1313	0·9929	0·8267	0·1588	0·1913	0·1862	0·1639	0·0866
0·24	0·6132	1·7927	0·1648	1·0746	0·8432	0·1954	0·2634	0·2578	0·2281	0·1050
0·28	0·6879	1·8713	0·1986	1·1548	0·8468	0·2345	0·3412	0·3355	0·3012	0·1237
0·32	0·7529	1·9154	0·2324	1·2348	0·8444	0·2753	0·4226	0·4171	0·3819	0·1428
0·36	0·8094	1·9336	0·2661	1·3150	0·8388	0·3172	0·5062	0·5011	0·4693	0·1622
0·40	0·8584	1·9324	0·2995	1·3955	0·8300	0·3608	0·5909	0·5865	0·5634	0·1819
0·42	0·8804	1·9261	0·3160	1·4359	0·8244	0·3834	0·6334	0·6294	0·6128	0·1918
0·44	0·9007	1·9165	0·3325	1·4764	0·8179	0·4065	0·6759	0·6723	0·6638	0·2018
0·46	0·9195	1·9040	0·3487	1·5171	0·8106	0·4302	0·7182	0·7150	0·7163	0·2119
0·48	0·9368	1·8891	0·3649	1·5579	0·8024	0·4547	0·7602	0·7574	0·7705	0·2221
0·50	0·9527	1·8719	0·3808	1·5989	0·7934	0·4800	0·8017	0·7995	0·8263	0·2324
0·52	0·9673	1·8527	0·3966	1·6401	0·7835	0·5062	0·8428	0·8410	0·8836	0·2428
0·54	0·9805	1·8317	0·4122	1·6816	0·7728	0·5334	0·8833	0·8820	0·9426	0·2532
0·56	0·9924	1·8091	0·4275	1·7232	0·7611	0·5617	0·9231	0·9222	1·0034	0·2638
0·58	1·0030	1·7851	0·4426	1·7651	0·7486	0·5913	0·9620	0·9616	1·0658	0·2744
0·60	1·0124	1·7598	0·4575	1·8074	0·7351	0·6223	1·0000	1·0000	1·1301	0·2852
0·62	1·0206	1·7332	0·4720	1·8499	0·7206	0·6550	1·0370	1·0374	1·1963	0·2961
0·64	1·0276	1·7056	0·4863	1·8928	0·7052	0·6895	1·0729	1·0736	1·2645	0·3071
0·66	1·0335	1·6770	0·5002	1·9360	0·6888	0·7262	1·1076	1·1086	1·3349	0·3183
0·68	1·0382	1·6475	0·5138	1·9796	0·6714	0·7653	1·1409	1·1422	1·4076	0·3296
0·70	1·0418	1·6172	0·5271	2·0237	0·6529	0·8073	1·1728	1·1743	1·4830	0·3411
0·72	1·0442	1·5861	0·5399	2·0682	0·6333	0·8526	1·2031	1·2049	1·5612	0·3527
0·74	1·0455	1·5543	0·5524	2·1133	0·6126	0·9017	1·2319	1·2337	1·6426	0·3645
0·76	1·0458	1·5218	0·5644	2·1589	0·5907	0·9555	1·2589	1·2608	1·7277	0·3765
0·78	1·0449	1·4886	0·5760	2·2051	0·5676	1·0148	1·2841	1·2859	1·8170	0·3888
0·80	1·0429	1·4549	0·5871	2·2519	0·5433	1·0807	1·3073	1·3091	1·9113	0·4012
0·82	1·0398	1·4206	0·5977	2·2994	0·5177	1·1547	1·3285	1·3301	2·0114	0·4139
0·84	1·0356	1·3858	0·6078	2·3477	0·4906	1·2388	1·3475	1·3489	2·1185	0·4269
0·86	1·0303	1·3504	0·6173	2·3968	0·4622	1·3357	1·3643	1·3654	2·2343	0·4402
0·88	1·0239	1·3146	0·6263	2·4467	0·4322	1·4491	1·3787	1·3794	2·3609	0·4537
0·90	1·0163	1·2783	0·6346	2·4977	0·4006	1·5841	1·3906	1·3909	2·5013	0·4676
0·92	1·0074	1·2410	0·6423	2·5502	0·3666	1·7520	1·3998	1·3995	2·6624	0·4819
0·94	0·9958	1·1995	0·6492	2·6080	0·3250	1·9979	1·4047	1·4036	2·8737	0·4967
0·96	0·9798	1·1506	0·6552	2·6750	0·2713	2·4152	1·4035	1·4014	3·1887	0·5121
0·98	0·9563	1·0883	0·6599	2·7604	0·1960	3·3678	1·3925	1·3888	3·7926	0·5283
1·00	0·8948	0·9492	0·6626	2·9617	0·0000	-	1·3416	1·3340	-	0·5462

C29

M and E Basket-handle

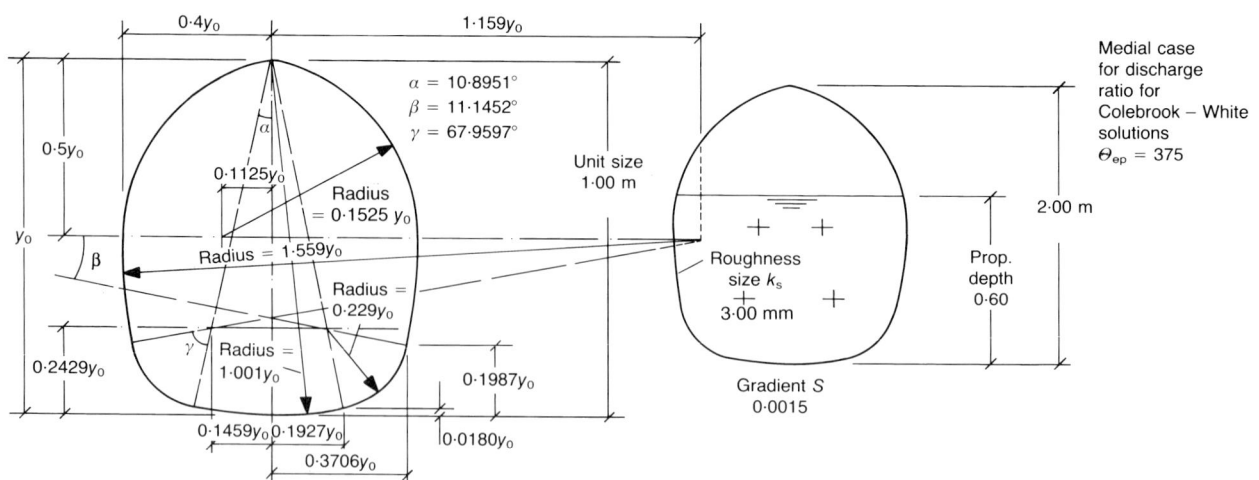

Prop dpth	U.equ. dia. $D_{ep(u)}$	Equiv. disch. factor	Unit sect. area	Unit wetted perim.	Unit surf. brdth	Unit mean depth	Discharge ratios $Q/Q_{0.60}$		U.crit. disch. $Q_{c(u)}$	Unit depth cntrd
Y	(m)	J	A_u (m²)	P_u (m)	$B_{s(u)}$ (m)	$y_{m(u)}$ (m)	medial	Mann'g	(m³s⁻¹)	$y_{d(u)}$ (m)
0·04	0·1125	0·6872	0·0145	0·5143	0·5041	0·0287	0·0080	0·0078	0·0077	0·0164
0·08	0·2278	1·1003	0·0371	0·6505	0·6137	0·0604	0·0329	0·0318	0·0285	0·0338
0·12	0·3344	1·3940	0·0630	0·7535	0·6783	0·0929	0·0720	0·0699	0·0601	0·0515
0·16	0·4316	1·6078	0·0910	0·8433	0·7188	0·1266	0·1227	0·1196	0·1014	0·0694
0·20	0·5191	1·7597	0·1203	0·9267	0·7417	0·1621	0·1826	0·1788	0·1516	0·0877
0·24	0·5961	1·8579	0·1502	1·0080	0·7563	0·1986	0·2492	0·2450	0·2097	0·1062
0·28	0·6639	1·9152	0·1807	1·0890	0·7688	0·2351	0·3211	0·3166	0·2744	0·1249
0·32	0·7240	1·9445	0·2117	1·1696	0·7791	0·2717	0·3974	0·3930	0·3456	0·1437
0·36	0·7777	1·9544	0·2430	1·2501	0·7874	0·3087	0·4774	0·4732	0·4228	0·1626
0·40	0·8259	1·9503	0·2747	1·3303	0·7936	0·3461	0·5605	0·5566	0·5060	0·1815
0·42	0·8481	1·9443	0·2906	1·3704	0·7959	0·3651	0·6030	0·5993	0·5498	0·1911
0·44	0·8692	1·9361	0·3065	1·4104	0·7977	0·3842	0·6460	0·6427	0·5950	0·2006
0·46	0·8893	1·9261	0·3225	1·4504	0·7990	0·4036	0·6895	0·6865	0·6415	0·2102
0·48	0·9083	1·9145	0·3385	1·4904	0·7997	0·4232	0·7334	0·7308	0·6895	0·2198
0·50	0·9264	1·9016	0·3545	1·5305	0·8000	0·4431	0·7777	0·7755	0·7388	0·2294
0·52	0·9435	1·8874	0·3704	1·5705	0·7992	0·4635	0·8223	0·8204	0·7898	0·2391
0·54	0·9597	1·8720	0·3864	1·6105	0·7969	0·4849	0·8669	0·8655	0·8426	0·2488
0·56	0·9749	1·8553	0·4023	1·6507	0·7930	0·5074	0·9116	0·9106	0·8974	0·2585
0·58	0·9890	1·8372	0·4181	1·6911	0·7874	0·5310	0·9560	0·9555	0·9541	0·2684
0·60	1·0020	1·8177	0·4338	1·7317	0·7803	0·5559	1·0000	1·0000	1·0129	0·2783
0·62	1·0139	1·7967	0·4493	1·7727	0·7715	0·5824	1·0435	1·0439	1·0738	0·2884
0·64	1·0245	1·7743	0·4646	1·8141	0·7610	0·6106	1·0862	1·0871	1·1370	0·2985
0·66	1·0340	1·7503	0·4797	1·8559	0·7488	0·6407	1·1280	1·1294	1·2026	0·3088
0·68	1·0422	1·7247	0·4946	1·8983	0·7347	0·6732	1·1687	1·1704	1·2708	0·3193
0·70	1·0490	1·6975	0·5091	1·9414	0·7187	0·7084	1·2080	1·2101	1·3419	0·3299
0·72	1·0544	1·6686	0·5233	1·9852	0·7008	0·7468	1·2457	1·2481	1·4162	0·3406
0·74	1·0584	1·6379	0·5371	2·0300	0·6807	0·7891	1·2816	1·2843	1·4943	0·3516
0·76	1·0608	1·6055	0·5505	2·0758	0·6583	0·8363	1·3155	1·3183	1·5766	0·3628
0·78	1·0617	1·5711	0·5635	2·1229	0·6335	0·8894	1·3470	1·3500	1·6641	0·3743
0·80	1·0608	1·5347	0·5759	2·1714	0·6060	0·9502	1·3760	1·3789	1·7578	⁻0·3860
0·82	1·0581	1·4962	0·5877	2·2217	0·5756	1·0209	1·4020	1·4048	1·8595	0·3980
0·84	1·0534	1·4553	0·5989	2·2740	0·5420	1·1050	1·4247	1·4273	1·9713	0·4104
0·86	1·0466	1·4119	0·6093	2·3287	0·5045	1·2077	1·4437	1·4461	2·0970	0·4232
0·88	1·0375	1·3657	0·6190	2·3866	0·4628	1·3376	1·4586	1·4605	2·2420	0·4364
0·90	1·0257	1·3161	0·6278	2·4483	0·4158	1·5099	1·4687	1·4700	2·4158	0·4502
0·92	1·0109	1·2627	0·6356	2·5150	0·3624	1·7539	1·4734	1·4739	2·6360	0·4645
0·94	0·9924	1·2043	0·6422	2·5887	0·3006	2·1368	1·4716	1·4711	2·9399	0·4796
0·96	0·9692	1·1393	0·6475	2·6725	0·2269	2·8537	1·4618	1·4600	3·4256	0·4956
0·98	0·9392	1·0638	0·6512	2·7735	0·1342	4·8524	1·4412	1·4377	4·4921	0·5128
1·00	0·8960	0·9661	0·6527	2·9137	0·0000	-	1·4022	1·3964	-	0·5316

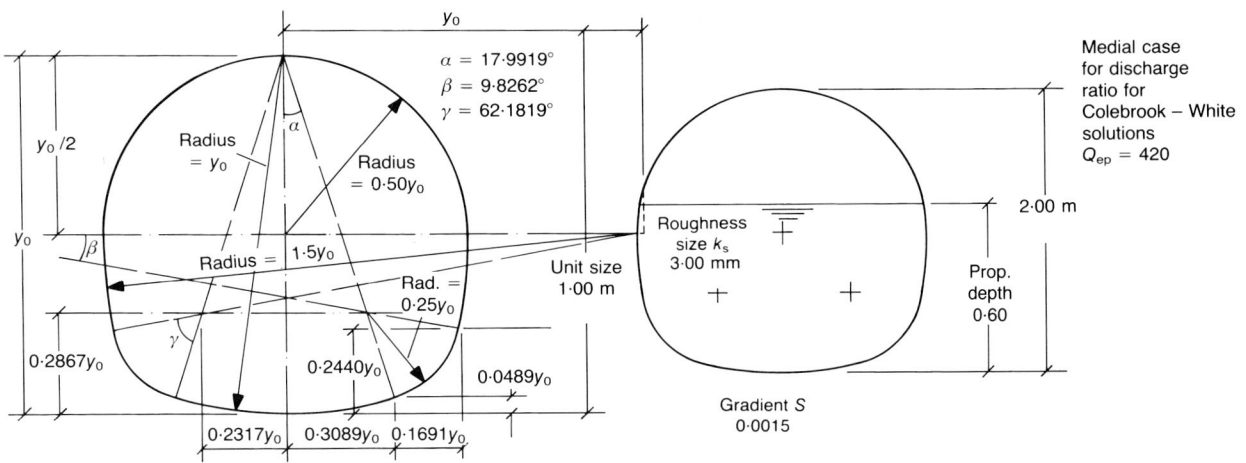

Prop dpth Y	U.equ. dia. $D_{ep(u)}$ (m)	Equiv. disch. factor J	Unit sect. area A_u (m²)	Unit wetted perim. P_u (m)	Unit surf. brdth $B_{s(u)}$ (m)	Unit mean depth $y_{m(u)}$ (m)	Discharge ratios $Q/Q_{0.60}$ medial	Mann'g	U.crit. disch. $Q_{c(u)}$ (m³s⁻¹)	Unit depth cntrd $y_{d(u)}$ (m)
0·04	0·1057	0·5849	0·0150	0·5676	0·5600	0·0268	0·0059	0·0057	0·0077	0·0160
0·08	0·2165	0·8834	0·0417	0·7698	0·7446	0·0559	0·0266	0·0256	0·0309	0·0324
0·12	0·3294	1·1608	0·0734	0·8915	0·8360	0·0878	0·0618	0·0596	0·0681	0·0496
0·16	0·4365	1·3840	0·1081	0·9907	0·8944	0·1209	0·1092	0·1059	0·1177	0·0672
0·20	0·5362	1·5607	0·1447	1·0794	0·9323	0·1552	0·1668	0·1627	0·1785	0·0851
0·24	0·6279	1·6968	0·1825	1·1625	0·9545	0·1912	0·2326	0·2279	0·2499	0·1033
0·28	0·7107	1·7953	0·2209	1·2436	0·9676	0·2283	0·3048	0·2997	0·3306	0·1218
0·32	0·7849	1·8620	0·2599	1·3243	0·9783	0·2656	0·3818	0·3766	0·4194	0·1406
0·36	0·8519	1·9051	0·2992	1·4048	0·9869	0·3031	0·4629	0·4579	0·5158	0·1595
0·40	0·9125	1·9305	0·3388	1·4850	0·9933	0·3411	0·5474	0·5428	0·6196	0·1785
0·42	0·9407	1·9378	0·3587	1·5251	0·9957	0·3602	0·5908	0·5865	0·6741	0·1880
0·44	0·9676	1·9422	0·3786	1·5651	0·9976	0·3795	0·6348	0·6308	0·7304	0·1976
0·46	0·9932	1·9439	0·3986	1·6052	0·9989	0·3990	0·6794	0·6758	0·7884	0·2072
0·48	1·0177	1·9433	0·4186	1·6452	0·9997	0·4187	0·7245	0·7212	0·8481	0·2168
0·50	1·0410	1·9406	0·4386	1·6852	1·0000	0·4386	0·7699	0·7672	0·9095	0·2265
0·52	1·0632	1·9361	0·4586	1·7252	0·9992	0·4589	0·8158	0·8135	0·9728	0·2362
0·54	1·0843	1·9297	0·4785	1·7653	0·9968	0·4801	0·8619	0·8602	1·0383	0·2459
0·56	1·1043	1·9214	0·4984	1·8055	0·9928	0·5020	0·9081	0·9069	1·1059	0·2557
0·58	1·1230	1·9112	0·5182	1·8459	0·9871	0·5250	0·9542	0·9535	1·1758	0·2656
0·60	1·1405	1·8992	0·5379	1·8865	0·9798	0·5490	1·0000	1·0000	1·2481	0·2755
0·62	1·1567	1·8852	0·5574	1·9275	0·9708	0·5742	1·0455	1·0461	1·3227	0·2855
0·64	1·1716	1·8693	0·5767	1·9690	0·9600	0·6007	1·0904	1·0916	1·3998	0·2956
0·66	1·1851	1·8514	0·5958	2·0109	0·9474	0·6289	1·1346	1·1363	1·4796	0·3058
0·68	1·1972	1·8316	0·6146	2·0534	0·9330	0·6588	1·1779	1·1802	1·5621	0·3161
0·70	1·2078	1·8097	0·6331	2·0967	0·9165	0·6908	1·2201	1·2229	1·6478	0·3266
0·72	1·2168	1·7857	0·6512	2·1408	0·8980	0·7252	1·2609	1·2642	1·7368	0·3372
0·74	1·2243	1·7595	0·6690	2·1858	0·8773	0·7626	1·3002	1·3039	1·8295	0·3480
0·76	1·2300	1·7311	0·6863	2·2320	0·8542	0·8035	1·3378	1·3418	1·9265	0·3590
0·78	1·2338	1·7004	0·7032	2·2796	0·8285	0·8487	1·3733	1·3776	2·0286	0·3702
0·80	1·2358	1·6672	0·7194	2·3287	0·8000	0·8993	1·4065	1·4110	2·1365	0·3815
0·82	1·2357	1·6313	0·7351	2·3797	0·7684	0·9567	1·4371	1·4417	2·2518	0·3932
0·84	1·2333	1·5925	0·7502	2·4329	0·7332	1·0231	1·4647	1·4693	2·3761	0·4051
0·86	1·2285	1·5506	0·7644	2·4890	0·6940	1·1015	1·4889	1·4934	2·5124	0·4174
0·88	1·2209	1·5050	0·7779	2·5485	0·6499	1·1969	1·5093	1·5134	2·6650	0·4300
0·90	1·2102	1·4553	0·7904	2·6125	0·6000	1·3173	1·5251	1·5287	2·8408	0·4430
0·92	1·1957	1·4003	0·8018	2·6825	0·5426	1·4778	1·5355	1·5384	3·0525	0·4566
0·94	1·1764	1·3385	0·8120	2·7610	0·4750	1·7096	1·5393	1·5412	3·3249	0·4707
0·96	1·1506	1·2668	0·8207	2·8533	0·3919	2·0941	1·5343	1·5348	3·7193	0·4856
0·98	1·1137	1·1771	0·8275	2·9722	0·2800	2·9554	1·5157	1·5142	4·4550	0·5015
1·00	1·0212	0·9853	0·8313	3·2560	0·0000	-	1·4419	1·4357	-	0·5193

C31

B and B Barrel

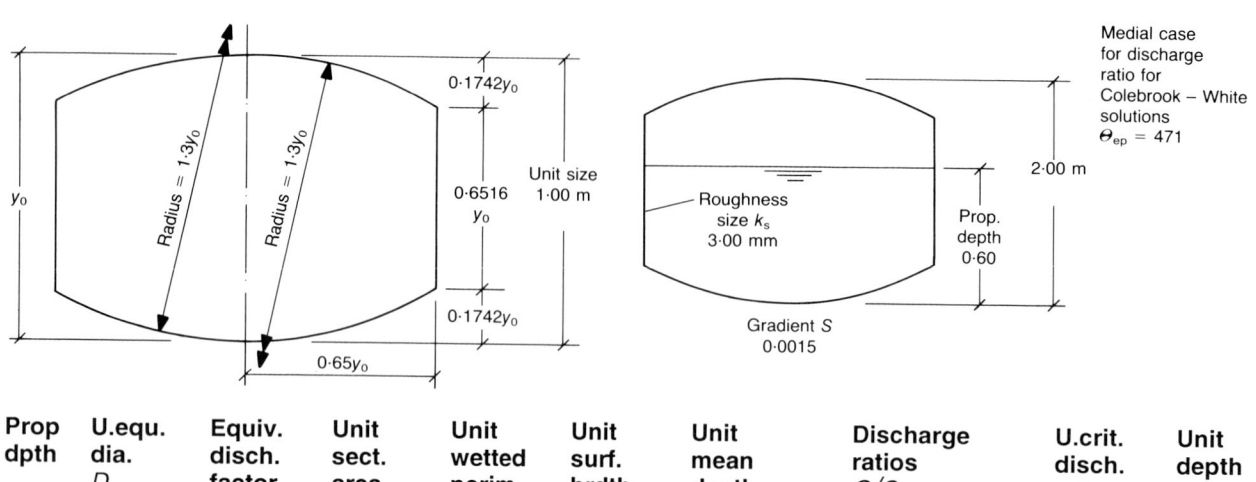

Prop dpth Y	U.equ. dia. $D_{ep(u)}$ (m)	Equiv. disch. factor J	Unit sect. area A_u (m²)	Unit wetted perim. P_u (m)	Unit surf. brdth $B_{s(u)}$ (m)	Unit mean depth $y_{m(u)}$ (m)	Discharge ratios $Q/Q_{0.60}$ medial	Mann'g	U.crit. disch. $Q_{c(u)}$ (m³s⁻¹)	Unit depth cntrd $y_{d(u)}$ (m)
0·04	0·1059	0·5145	0·0171	0·6466	0·6400	0·0267	0·0048	0·0046	0·0088	0·0160
0·08	0·2103	0·7204	0·0482	0·9169	0·8980	0·0537	0·0214	0·0205	0·0350	0·0321
0·12	0·3131	0·8735	0·0881	1·1259	1·0911	0·0808	0·0508	0·0488	0·0784	0·0482
0·16	0·4143	0·9985	0·1350	1·3036	1·2496	0·1081	0·0935	0·0902	0·1390	0·0643
0·20	0·5284	1·1749	0·1867	1·4130	1·3000	0·1436	0·1511	0·1467	0·2215	0·0810
0·24	0·6394	1·3455	0·2387	1·4930	1·3000	0·1836	0·2183	0·2129	0·3203	0·0990
0·28	0·7391	1·4762	0·2907	1·5730	1·3000	0·2236	0·2915	0·2856	0·4304	0·1177
0·32	0·8292	1·5759	0·3427	1·6530	1·3000	0·2636	0·3696	0·3636	0·5509	0·1368
0·36	0·9109	1·6513	0·3947	1·7330	1·3000	0·3036	0·4517	0·4458	0·6810	0·1561
0·40	0·9855	1·7076	0·4467	1·8130	1·3000	0·3436	0·5372	0·5317	0·8199	0·1756
0·42	1·0203	1·7298	0·4727	1·8530	1·3000	0·3636	0·5810	0·5758	0·8925	0·1854
0·44	1·0537	1·7487	0·4987	1·8930	1·3000	0·3836	0·6255	0·6207	0·9672	0·1952
0·46	1·0857	1·7645	0·5247	1·9330	1·3000	0·4036	0·6706	0·6662	1·0438	0·2051
0·48	1·1164	1·7776	0·5507	1·9730	1·3000	0·4236	0·7163	0·7124	1·1224	0·2149
0·50	1·1459	1·7883	0·5767	2·0130	1·3000	0·4436	0·7624	0·7591	1·2028	0·2248
0·52	1·1742	1·7968	0·6027	2·0530	1·3000	0·4636	0·8091	0·8063	1·2850	0·2346
0·54	1·2015	1·8033	0·6287	2·0930	1·3000	0·4836	0·8562	0·8541	1·3691	0·2445
0·56	1·2277	1·8081	0·6547	2·1330	1·3000	0·5036	0·9038	0·9023	1·4549	0·2544
0·58	1·2530	1·8114	0·6807	2·1730	1·3000	0·5236	0·9517	0·9509	1·5424	0·2643
0·60	1·2773	1·8132	0·7067	2·2130	1·3000	0·5436	1·0000	1·0000	1·6316	0·2742
0·62	1·3008	1·8138	0·7327	2·2530	1·3000	0·5636	1·0487	1·0495	1·7225	0·2841
0·64	1·3234	1·8132	0·7587	2·2930	1·3000	0·5836	1·0977	1·0993	1·8150	0·2941
0·66	1·3453	1·8116	0·7847	2·3330	1·3000	0·6036	1·1470	1·1495	1·9091	0·3040
0·68	1·3665	1·8090	0·8107	2·3730	1·3000	0·6236	1·1966	1·2000	2·0047	0·3139
0·70	1·3869	1·8056	0·8367	2·4130	1·3000	0·6436	1·2465	1·2508	2·1020	0·3238
0·72	1·4067	1·8015	0·8627	2·4530	1·3000	0·6636	1·2966	1·3019	2·2007	0·3338
0·74	1·4259	1·7968	0·8887	2·4930	1·3000	0·6836	1·3470	1·3533	2·3009	0·3437
0·76	1·4444	1·7914	0·9147	2·5330	1·3000	0·7036	1·3976	1·4049	2·4026	0·3537
0·78	1·4624	1·7855	0·9407	2·5730	1·3000	0·7236	1·4484	1·4568	2·5058	0·3636
0·80	1·4798	1·7791	0·9667	2·6130	1·3000	0·7436	1·4995	1·5089	2·6104	0·3736
0·82	1·4967	1·7722	0·9927	2·6530	1·3000	0·7636	1·5507	1·5613	2·7164	0·3835
0·84	1·4962	1·7265	1·0183	2·7225	1·2496	0·8149	1·5905	1·6012	2·8787	0·3936
0·86	1·4850	1·6612	1·0426	2·8083	1·1737	0·8883	1·6207	1·6312	3·0770	0·4042
0·88	1·4692	1·5915	1·0652	2·9001	1·0911	0·9763	1·6450	1·6548	3·2961	0·4154
0·90	1·4484	1·5169	1·0861	2·9996	1·0000	1·0861	1·6625	1·6714	3·5448	0·4272
0·92	1·4218	1·4366	1·1052	3·1092	0·8980	1·2307	1·6721	1·6797	3·8393	0·4397
0·94	1·3881	1·3488	1·1220	3·2330	0·7808	1·4370	1·6724	1·6782	4·2118	0·4530
0·96	1·3449	1·2502	1·1362	3·3794	0·6400	1·7754	1·6605	1·6641	4·7410	0·4672
0·98	1·2857	1·1316	1·1473	3·5694	0·4543	2·5253	1·6302	1·6306	5·7094	0·4826
1·00	1·1459	0·8941	1·1533	4·0260	0·0000	-	1·5249	1·5181	-	0·5000

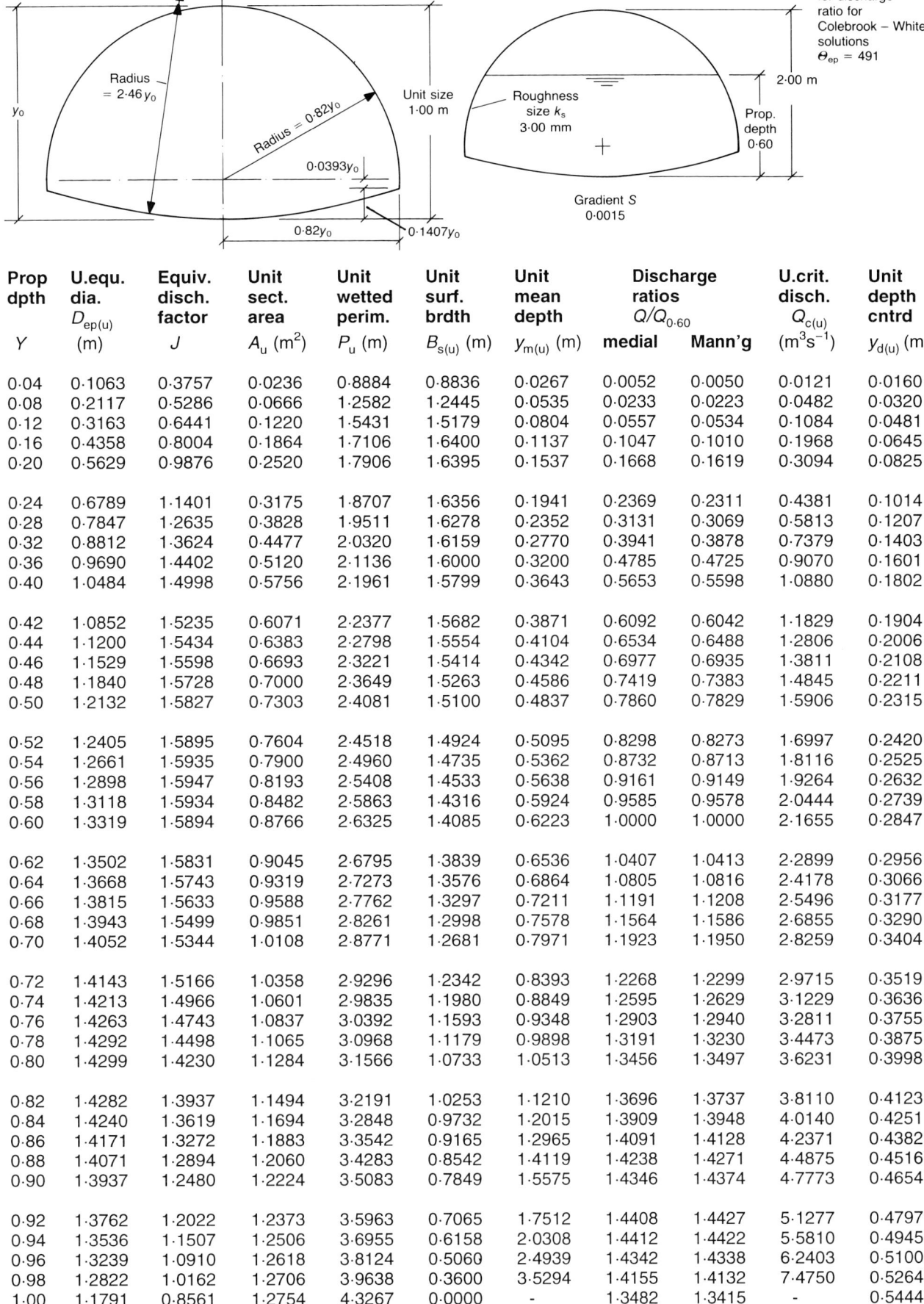

Prop dpth Y	U.equ. dia. $D_{ep(u)}$ (m)	Equiv. disch. factor J	Unit sect. area A_u (m²)	Unit wetted perim. P_u (m)	Unit surf. brdth $B_{s(u)}$ (m)	Unit mean depth $y_{m(u)}$ (m)	Discharge ratios $Q/Q_{0.60}$ medial	Mann'g	U.crit. disch. $Q_{c(u)}$ (m³s⁻¹)	Unit depth cntrd $y_{d(u)}$ (m)
0·04	0·1063	0·3757	0·0236	0·8884	0·8836	0·0267	0·0052	0·0050	0·0121	0·0160
0·08	0·2117	0·5286	0·0666	1·2582	1·2445	0·0535	0·0233	0·0223	0·0482	0·0320
0·12	0·3163	0·6441	0·1220	1·5431	1·5179	0·0804	0·0557	0·0534	0·1084	0·0481
0·16	0·4358	0·8004	0·1864	1·7106	1·6400	0·1137	0·1047	0·1010	0·1968	0·0645
0·20	0·5629	0·9876	0·2520	1·7906	1·6395	0·1537	0·1668	0·1619	0·3094	0·0825
0·24	0·6789	1·1401	0·3175	1·8707	1·6356	0·1941	0·2369	0·2311	0·4381	0·1014
0·28	0·7847	1·2635	0·3828	1·9511	1·6278	0·2352	0·3131	0·3069	0·5813	0·1207
0·32	0·8812	1·3624	0·4477	2·0320	1·6159	0·2770	0·3941	0·3878	0·7379	0·1403
0·36	0·9690	1·4402	0·5120	2·1136	1·6000	0·3200	0·4785	0·4725	0·9070	0·1601
0·40	1·0484	1·4998	0·5756	2·1961	1·5799	0·3643	0·5653	0·5598	1·0880	0·1802
0·42	1·0852	1·5235	0·6071	2·2377	1·5682	0·3871	0·6092	0·6042	1·1829	0·1904
0·44	1·1200	1·5434	0·6383	2·2798	1·5554	0·4104	0·6534	0·6488	1·2806	0·2006
0·46	1·1529	1·5598	0·6693	2·3221	1·5414	0·4342	0·6977	0·6935	1·3811	0·2108
0·48	1·1840	1·5728	0·7000	2·3649	1·5263	0·4586	0·7419	0·7383	1·4845	0·2211
0·50	1·2132	1·5827	0·7303	2·4081	1·5100	0·4837	0·7860	0·7829	1·5906	0·2315
0·52	1·2405	1·5895	0·7604	2·4518	1·4924	0·5095	0·8298	0·8273	1·6997	0·2420
0·54	1·2661	1·5935	0·7900	2·4960	1·4735	0·5362	0·8732	0·8713	1·8116	0·2525
0·56	1·2898	1·5947	0·8193	2·5408	1·4533	0·5638	0·9161	0·9149	1·9264	0·2632
0·58	1·3118	1·5934	0·8482	2·5863	1·4316	0·5924	0·9585	0·9578	2·0444	0·2739
0·60	1·3319	1·5894	0·8766	2·6325	1·4085	0·6223	1·0000	1·0000	2·1655	0·2847
0·62	1·3502	1·5831	0·9045	2·6795	1·3839	0·6536	1·0407	1·0413	2·2899	0·2956
0·64	1·3668	1·5743	0·9319	2·7273	1·3576	0·6864	1·0805	1·0816	2·4178	0·3066
0·66	1·3815	1·5633	0·9588	2·7762	1·3297	0·7211	1·1191	1·1208	2·5496	0·3177
0·68	1·3943	1·5499	0·9851	2·8261	1·2998	0·7578	1·1564	1·1586	2·6855	0·3290
0·70	1·4052	1·5344	1·0108	2·8771	1·2681	0·7971	1·1923	1·1950	2·8259	0·3404
0·72	1·4143	1·5166	1·0358	2·9296	1·2342	0·8393	1·2268	1·2299	2·9715	0·3519
0·74	1·4213	1·4966	1·0601	2·9835	1·1980	0·8849	1·2595	1·2629	3·1229	0·3636
0·76	1·4263	1·4743	1·0837	3·0392	1·1593	0·9348	1·2903	1·2940	3·2811	0·3755
0·78	1·4292	1·4498	1·1065	3·0968	1·1179	0·9898	1·3191	1·3230	3·4473	0·3875
0·80	1·4299	1·4230	1·1284	3·1566	1·0733	1·0513	1·3456	1·3497	3·6231	0·3998
0·82	1·4282	1·3937	1·1494	3·2191	1·0253	1·1210	1·3696	1·3737	3·8110	0·4123
0·84	1·4240	1·3619	1·1694	3·2848	0·9732	1·2015	1·3909	1·3948	4·0140	0·4251
0·86	1·4171	1·3272	1·1883	3·3542	0·9165	1·2965	1·4091	1·4128	4·2371	0·4382
0·88	1·4071	1·2894	1·2060	3·4283	0·8542	1·4119	1·4238	1·4271	4·4875	0·4516
0·90	1·3937	1·2480	1·2224	3·5083	0·7849	1·5575	1·4346	1·4374	4·7773	0·4654
0·92	1·3762	1·2022	1·2373	3·5963	0·7065	1·7512	1·4408	1·4427	5·1277	0·4797
0·94	1·3536	1·1507	1·2506	3·6955	0·6158	2·0308	1·4412	1·4422	5·5810	0·4945
0·96	1·3239	1·0910	1·2618	3·8124	0·5060	2·4939	1·4342	1·4338	6·2403	0·5100
0·98	1·2822	1·0162	1·2706	3·9638	0·3600	3·5294	1·4155	1·4132	7·4750	0·5264
1·00	1·1791	0·8561	1·2754	4·3267	0·0000	-	1·3482	1·3415	-	0·5444

C33

7% deposit in circular pipe

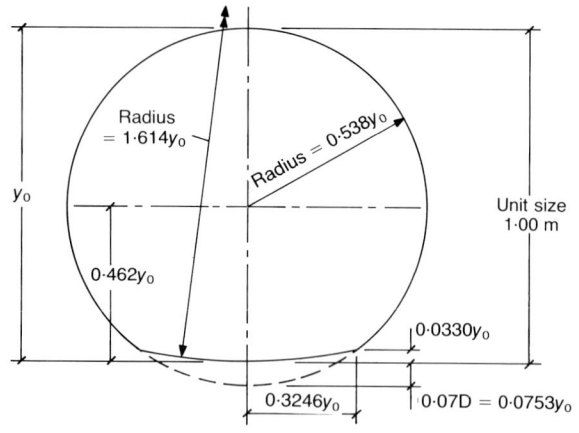

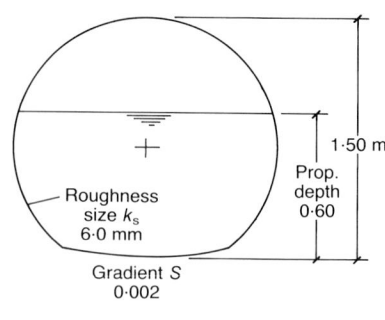

Medial case
for discharge
ratio for
Colebrook-White
solutions
$\Theta_{ep} = 216$

Prop dpth	U.equ. dia. $D_{ep(u)}$	Equiv. disch. factor	Unit sect. area	Unit wetted perim.	Unit surf. brdth	Unit mean depth	Discharge ratios $Q/Q_{0.60}$		U.crit. disch. $Q_{c(u)}$	Unit depth cntrd
Y	(m)	J	A_u (m²)	P_u (m)	$B_{s(u)}$ (m)	$y_{m(u)}$ (m)	medial	Mann'g	(m³s⁻¹)	$y_{d(u)}$ (m)
0·04	0·1119	0·5194	0·0189	0·6767	0·6674	0·0284	0·0068	0·0069	0·0100	0·0161
0·08	0·2383	0·9388	0·0475	0·7973	0·7577	0·0627	0·0290	0·0288	0·0372	0·0341
0·12	0·3503	1·2151	0·0793	0·9056	0·8306	0·0955	0·0629	0·0622	0·0767	0·0523
0·16	0·4525	1·4138	0·1138	1·0056	0·8905	0·1278	0·1071	0·1059	0·1273	0·0703
0·20	0·5471	1·5631	0·1504	1·0996	0·9398	0·1600	0·1606	0·1589	0·1884	0·0883
0·24	0·6352	1·6778	0·1888	1·1892	0·9801	0·1927	0·2225	0·2203	0·2596	0·1062
0·28	0·7172	1·7664	0·2287	1·2756	1·0126	0·2259	0·2919	0·2893	0·3404	0·1242
0·32	0·7937	1·8340	0·2697	1·3595	1·0378	0·2599	0·3678	0·3651	0·4306	0·1422
0·36	0·8647	1·8843	0·3117	1·4417	1·0565	0·2950	0·4494	0·4466	0·5301	0·1604
0·40	0·9304	1·9197	0·3542	1·5226	1·0688	0·3314	0·5356	0·5330	0·6385	0·1788
0·42	0·9613	1·9324	0·3756	1·5628	1·0727	0·3501	0·5801	0·5776	0·6960	0·1880
0·44	0·9909	1·9421	0·3971	1·6029	1·0751	0·3693	0·6254	0·6231	0·7557	0·1973
0·46	1·0192	1·9488	0·4186	1·6429	1·0760	0·3890	0·6714	0·6693	0·8176	0·2066
0·48	1·0461	1·9527	0·4401	1·6829	1·0754	0·4093	0·7179	0·7161	0·8817	0·2160
0·50	1·0716	1·9539	0·4616	1·7230	1·0733	0·4301	0·7648	0·7632	0·9480	0·2255
0·52	1·0958	1·9526	0·4830	1·7631	1·0697	0·4515	0·8120	0·8106	1·0164	0·2351
0·54	1·1187	1·9487	0·5044	1·8035	1·0646	0·4738	0·8592	0·8582	1·0872	0·2447
0·56	1·1401	1·9424	0·5256	1·8440	1·0580	0·4968	0·9064	0·9057	1·1601	0·2544
0·58	1·1602	1·9337	0·5467	1·8848	1·0498	0·5208	0·9534	0·9530	1·2354	0·2642
0·60	1·1788	1·9227	0·5676	1·9260	1·0400	0·5458	1·0000	1·0000	1·3131	0·2741
0·62	1·1959	1·9094	0·5883	1·9676	1·0286	0·5719	1·0461	1·0465	1·3932	0·2841
0·64	1·2115	1·8938	0·6087	2·0097	1·0154	0·5995	1·0916	1·0922	1·4759	0·2942
0·66	1·2256	1·8760	0·6289	2·0524	1·0005	0·6286	1·1361	1·1372	1·5614	0·3045
0·68	1·2381	1·8559	0·6487	2·0958	0·9837	0·6595	1·1797	1·1810	1·6497	0·3149
0·70	1·2490	1·8336	0·6682	2·1400	0·9650	0·6925	1·2220	1·2236	1·7413	0·3254
0·72	1·2582	1·8089	0·6873	2·1851	0·9442	0·7279	1·2628	1·2647	1·8364	0·3361
0·74	1·2656	1·7820	0·7060	2·2312	0·9212	0·7663	1·3020	1·3042	1·9353	0·3469
0·76	1·2712	1·7527	0·7241	2·2786	0·8959	0·8083	1·3393	1·3417	2·0388	0·3580
0·78	1·2749	1·7209	0·7418	2·3274	0·8679	0·8547	1·3745	1·3770	2·1475	0·3692
0·80	1·2765	1·6865	0·7588	2·3778	0·8371	0·9065	1·4072	1·4099	2·2625	0·3807
0·82	1·2760	1·6494	0·7752	2·4303	0·8032	0·9652	1·4373	1·4400	2·3851	0·3924
0·84	1·2731	1·6093	0·7909	2·4852	0·7657	1·0330	1·4642	1·4669	2·5174	0·4045
0·86	1·2676	1·5660	0·8058	2·5429	0·7240	1·1131	1·4877	1·4902	2·6624	0·4168
0·88	1·2592	1·5190	0·8199	2·6043	0·6774	1·2103	1·5072	1·5095	2·8246	0·4295
0·90	1·2476	1·4677	0·8329	2·6704	0·6248	1·3330	1·5220	1·5240	3·0115	0·4426
0·92	1·2321	1·4112	0·8448	2·7428	0·5646	1·4964	1·5314	1·5330	3·2363	0·4562
0·94	1·2116	1·3478	0·8554	2·8241	0·4938	1·7323	1·5340	1·5350	3·5257	0·4705
0·96	1·1844	1·2744	0·8645	2·9195	0·4071	2·1233	1·5277	1·5279	3·9446	0·4854
0·98	1·1457	1·1829	0·8715	3·0428	0·2907	2·9984	1·5076	1·5066	4·7259	0·5014
1·00	1·0493	0·9878	0·8754	3·3371	0·0000	-	1·4309	1·4272	-	0·5192

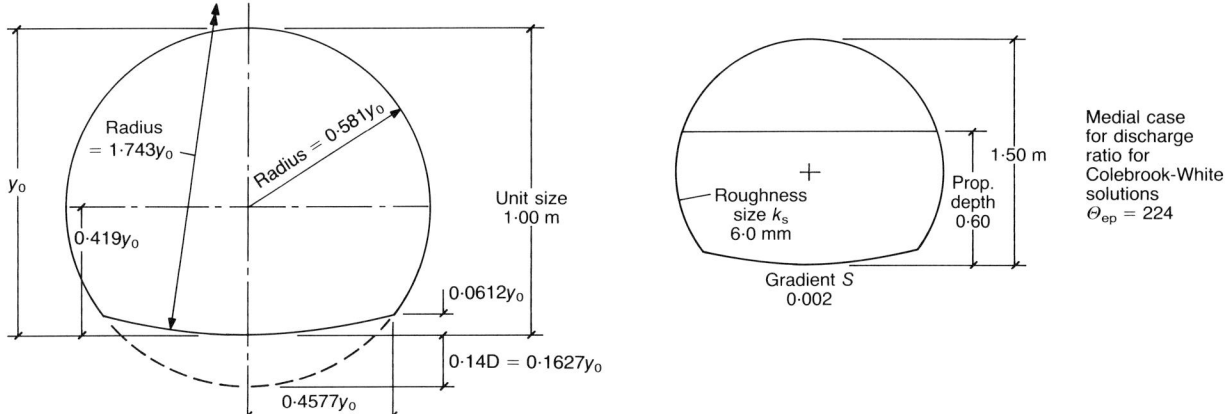

Prop dpth Y	U.equ. dia. $D_{ep(u)}$ (m)	Equiv. disch. factor J	Unit sect. area A_u (m²)	Unit wetted perim. P_u (m)	Unit surf. brdth $B_{s(u)}$ (m)	Unit mean depth $y_{m(u)}$ (m)	Discharge ratios $Q/Q_{0.60}$ medial	Mann'g	U.crit. disch. $Q_{c(u)}$ (m³s⁻¹)	Unit depth cntrd $y_{d(u)}$ (m)
0·04	0·1061	0·4454	0·0198	0·7483	0·7425	0·0267	0·0060	0·0061	0·0102	0·0160
0·08	0·2259	0·7291	0·0550	0·9734	0·9437	0·0583	0·0283	0·0282	0·0415	0·0325
0·12	0·3509	1·0312	0·0938	1·0692	0·9963	0·0942	0·0653	0·0645	0·0901	0·0507
0·16	0·4638	1·2557	0·1346	1·1604	1·0402	0·1294	0·1129	0·1115	0·1516	0·0692
0·20	0·5669	1·4268	0·1769	1·2482	1·0763	0·1644	0·1695	0·1675	0·2246	0·0879
0·24	0·6617	1·5589	0·2206	1·3334	1·1055	0·1995	0·2340	0·2315	0·3085	0·1065
0·28	0·7490	1·6611	0·2653	1·4166	1·1283	0·2351	0·3052	0·3025	0·4028	0·1252
0·32	0·8296	1·7393	0·3108	1·4983	1·1450	0·2714	0·3822	0·3793	0·5070	0·1439
0·36	0·9038	1·7980	0·3568	1·5791	1·1560	0·3086	0·4640	0·4611	0·6207	0·1628
0·40	0·9719	1·8400	0·4032	1·6593	1·1614	0·3471	0·5496	0·5469	0·7438	0·1817
0·42	1·0037	1·8555	0·4264	1·6993	1·1620	0·3669	0·5935	0·5909	0·8089	0·1913
0·44	1·0340	1·8676	0·4496	1·7393	1·1612	0·3872	0·6380	0·6356	0·8761	0·2009
0·46	1·0629	1·8766	0·4728	1·7794	1·1591	0·4079	0·6830	0·6808	0·9457	0·2105
0·48	1·0903	1·8825	0·4960	1·8195	1·1556	0·4292	0·7283	0·7264	1·0176	0·2202
0·50	1·1163	1·8856	0·5190	1·8598	1·1507	0·4511	0·7738	0·7722	1·0917	0·2300
0·52	1·1408	1·8859	0·5420	1·9004	1·1443	0·4736	0·8195	0·8181	1·1681	0·2398
0·54	1·1639	1·8837	0·5648	1·9411	1·1365	0·4970	0·8650	0·8640	1·2469	0·2498
0·56	1·1855	1·8788	0·5874	1·9822	1·1273	0·5211	0·9104	0·9097	1·3280	0·2597
0·58	1·2056	1·8716	0·6099	2·0236	1·1165	0·5463	0·9554	0·9551	1·4116	0·2698
0·60	1·2241	1·8619	0·6321	2·0654	1·1042	0·5725	1·0000	1·0000	1·4977	0·2800
0·62	1·2412	1·8499	0·6540	2·1078	1·0902	0·5999	1·0440	1·0443	1·5864	0·2903
0·64	1·2567	1·8356	0·6757	2·1507	1·0747	0·6288	1·0872	1·0878	1·6779	0·3006
0·66	1·2706	1·8190	0·6970	2·1943	1·0573	0·6592	1·1294	1·1304	1·7722	0·3111
0·68	1·2829	1·8002	0·7180	2·2387	1·0382	0·6916	1·1706	1·1719	1·8698	0·3218
0·70	1·2934	1·7791	0·7385	2·2839	1·0171	0·7261	1·2105	1·2121	1·9708	0·3325
0·72	1·3023	1·7558	0·7586	2·3301	0·9939	0·7633	1·2489	1·2508	2·0756	0·3435
0·74	1·3094	1·7302	0·7783	2·3775	0·9685	0·8035	1·2857	1·2878	2·1847	0·3545
0·76	1·3146	1·7022	0·7974	2·4262	0·9408	0·8475	1·3206	1·3229	2·2988	0·3658
0·78	1·3179	1·6718	0·8159	2·4764	0·9105	0·8961	1·3534	1·3558	2·4187	0·3773
0·80	1·3191	1·6389	0·8338	2·5284	0·8773	0·9504	1·3839	1·3864	2·5455	0·3890
0·82	1·3181	1·6034	0·8510	2·5825	0·8409	1·0120	1·4117	1·4142	2·6808	0·4009
0·84	1·3147	1·5650	0·8674	2·6391	0·8008	1·0831	1·4366	1·4391	2·8269	0·4131
0·86	1·3087	1·5234	0·8830	2·6988	0·7565	1·1671	1·4581	1·4605	2·9872	0·4257
0·88	1·2998	1·4783	0·8976	2·7623	0·7072	1·2692	1·4759	1·4780	3·1667	0·4386
0·90	1·2876	1·4291	0·9112	2·8306	0·6518	1·3981	1·4892	1·4910	3·3740	0·4519
0·92	1·2715	1·3748	0·9236	2·9056	0·5884	1·5697	1·4973	1·4987	3·6238	0·4657
0·94	1·2505	1·3139	0·9347	2·9898	0·5143	1·8175	1·4991	1·4998	3·9460	0·4800
0·96	1·2226	1·2434	0·9441	3·0889	0·4237	2·2282	1·4924	1·4923	4·4132	0·4952
0·98	1·1831	1·1554	0·9514	3·2168	0·3023	3·1477	1·4725	1·4713	5·2861	0·5113
1·00	1·0850	0·9676	0·9555	3·5226	0·0000	-	1·3986	1·3947	-	0·5291

C35

21% deposit in circular pipe

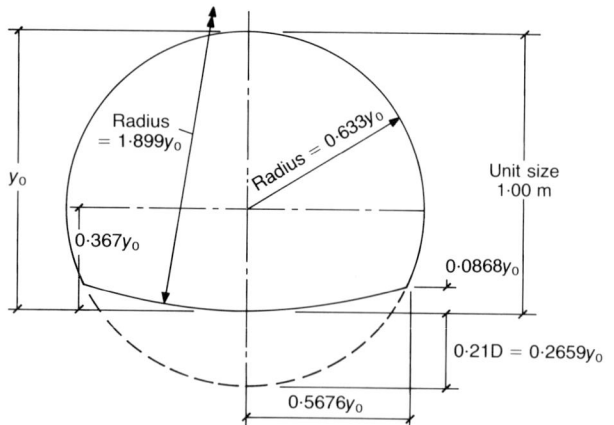

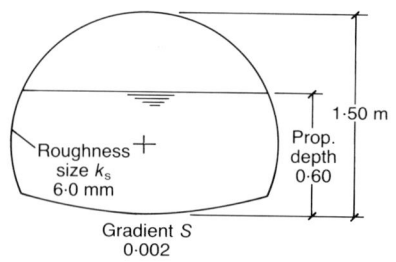

Medial case for discharge ratio for Colebrook-White solutions $\Theta_{ep} = 231$

Prop dpth Y	U.equ. dia. $D_{ep(u)}$ (m)	Equiv. disch. factor J	Unit sect. area A_u (m²)	Unit wetted perim. P_u (m)	Unit surf. brdth $B_{s(u)}$ (m)	Unit mean depth $y_{m(u)}$ (m)	Discharge ratios $Q/Q_{0.60}$ medial	Mann'g	U.crit. disch. $Q_{c(u)}$ (m³s⁻¹)	Unit depth cntrd $y_{d(u)}$ (m)
0·04	0·1061	0·4270	0·0207	0·7809	0·7754	0·0267	0·0056	0·0057	0·0106	0·0160
0·08	0·2112	0·5998	0·0584	1·1063	1·0908	0·0536	0·0256	0·0255	0·0423	0·0321
0·12	0·3400	0·8713	0·1042	1·2259	1·1656	0·0894	0·0631	0·0624	0·0976	0·0491
0·16	0·4619	1·1064	0·1515	1·3116	1·1964	0·1266	0·1127	0·1113	0·1688	0·0675
0·20	0·5729	1·2897	0·1998	1·3953	1·2211	0·1636	0·1715	0·1694	0·2531	0·0863
0·24	0·6743	1·4336	0·2491	1·4776	1·2403	0·2008	0·2380	0·2354	0·3495	0·1053
0·28	0·7672	1·5462	0·2990	1·5588	1·2540	0·2384	0·3109	0·3080	0·4572	0·1244
0·32	0·8524	1·6336	0·3493	1·6393	1·2625	0·2767	0·3891	0·3860	0·5754	0·1436
0·36	0·9304	1·7000	0·3999	1·7193	1·2659	0·3159	0·4715	0·4685	0·7039	0·1629
0·40	1·0015	1·7486	0·4505	1·7994	1·2643	0·3564	0·5572	0·5544	0·8422	0·1823
0·42	1·0346	1·7670	0·4758	1·8395	1·2616	0·3771	0·6010	0·5983	0·9150	0·1921
0·44	1·0661	1·7818	0·5010	1·8797	1·2576	0·3984	0·6451	0·6427	0·9902	0·2020
0·46	1·0960	1·7933	0·5261	1·9200	1·2523	0·4201	0·6897	0·6874	1·0678	0·2118
0·48	1·1243	1·8015	0·5511	1·9606	1·2457	0·4424	0·7344	0·7324	1·1478	0·2218
0·50	1·1510	1·8068	0·5759	2·0013	1·2377	0·4653	0·7792	0·7775	1·2302	0·2318
0·52	1·1762	1·8092	0·6006	2·0424	1·2285	0·4889	0·8240	0·8226	1·3150	0·2419
0·54	1·1998	1·8088	0·6250	2·0838	1·2178	0·5132	0·8686	0·8675	1·4023	0·2520
0·56	1·2218	1·8058	0·6493	2·1256	1·2057	0·5385	0·9129	0·9121	1·4920	0·2520
0·58	1·2423	1·8002	0·6733	2·1678	1·1922	0·5647	0·9567	0·9563	1·5844	0·2622
0·60	1·2611	1·7922	0·6969	2·2106	1·1771	0·5921	1·0000	1·0000	1·6794	0·2829
0·62	1·2784	1·7818	0·7203	2·2539	1·1605	0·6207	1·0426	1·0429	1·7772	0·2934
0·64	1·2940	1·7691	0·7434	2·2979	1·1422	0·6508	1·0844	1·0850	1·8780	0·3040
0·66	1·3080	1·7540	0·7660	2·3426	1·1222	0·6826	1·1251	1·1261	1·9818	0·3147
0·68	1·3202	1·7367	0·7882	2·3882	1·1004	0·7163	1·1647	1·1661	2·0891	0·3256
0·70	1·3308	1·7171	0·8100	2·4347	1·0767	0·7523	1·2030	1·2046	2·2002	0·3366
0·72	1·3396	1·6953	0·8313	2·4823	1·0509	0·7910	1·2399	1·2417	2·3153	0·3477
0·74	1·3465	1·6712	0·8520	2·5311	1·0229	0·8330	1·2750	1·2771	2·4352	0·3590
0·76	1·3515	1·6448	0·8722	2·5814	0·9925	0·8788	1·3083	1·3105	2·5605	0·3705
0·78	1·3545	1·6160	0·8917	2·6332	0·9594	0·9294	1·3395	1·3419	2·6921	0·3821
0·80	1·3555	1·5847	0·9105	2·6870	0·9235	0·9860	1·3684	1·3708	2·8314	0·3940
0·82	1·3542	1·5509	0·9286	2·7430	0·8843	1·0502	1·3947	1·3972	2·9801	0·4061
0·84	1·3504	1·5142	0·9459	2·8017	0·8413	1·1243	1·4181	1·4205	3·1408	0·4186
0·86	1·3441	1·4745	0·9622	2·8636	0·7941	1·2118	1·4383	1·4406	3·3171	0·4313
0·88	1·3348	1·4314	0·9776	2·9296	0·7417	1·3181	1·4548	1·4568	3·5149	0·4443
0·90	1·3222	1·3843	0·9919	3·0006	0·6829	1·4524	1·4671	1·4688	3·7433	0·4578
0·92	1·3056	1·3323	1·0049	3·0786	0·6161	1·6312	1·4743	1·4756	4·0190	0·4718
0·94	1·2841	1·2740	1·0164	3·1663	0·5380	1·8893	1·4754	1·4761	4·3752	0·4863
0·96	1·2556	1·2064	1·0263	3·2695	0·4429	2·3172	1·4684	1·4682	4·8923	0·5015
0·98	1·2154	1·1220	1·0340	3·4029	0·3157	3·2749	1·4488	1·4474	5·8595	0·5177
1·00	1·1157	0·9417	1·0382	3·7220	0·0000	-	1·3767	1·3728	-	0·5356

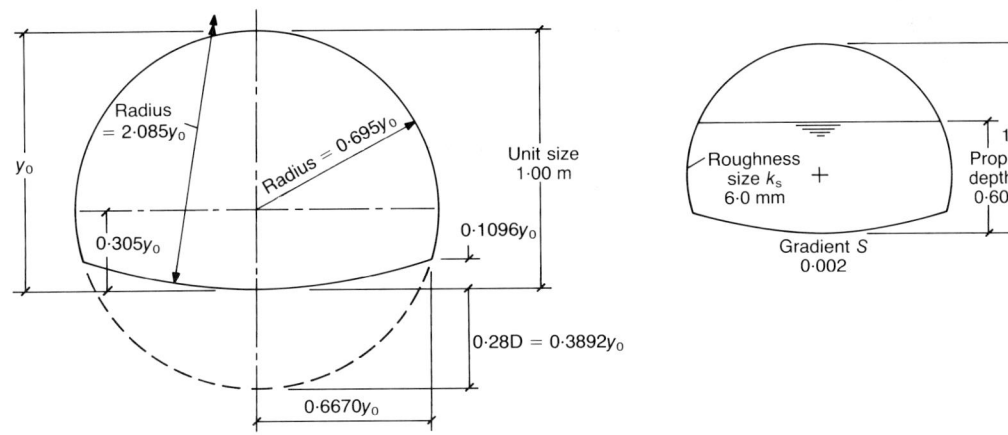

Prop dpth Y	U.equ. dia. $D_{ep(u)}$ (m)	Equiv. disch. factor J	Unit sect. area A_u (m²)	Unit wetted perim. P_u (m)	Unit surf. brdth $B_{s(u)}$ (m)	Unit mean depth $y_{m(u)}$ (m)	Discharge ratios $Q/Q_{0.60}$ medial	Mann'g	U.crit. disch. $Q_{c(u)}$ (m³s⁻¹)	Unit depth cntrd $y_{d(u)}$ (m)
0.04	0.1062	0.4077	0.0217	0.8181	0.8129	0.0267	0.0053	0.0054	0.0111	0.0160
0.08	0.2114	0.5731	0.0613	1.1589	1.1440	0.0535	0.0241	0.0240	0.0444	0.0321
0.12	0.3245	0.7390	0.1119	1.3795	1.3399	0.0835	0.0591	0.0584	0.1013	0.0482
0.16	0.4540	0.9756	0.1659	1.4619	1.3594	0.1220	0.1097	0.1082	0.1815	0.0660
0.20	0.5718	1.1640	0.2206	1.5432	1.3740	0.1605	0.1700	0.1678	0.2768	0.0847
0.24	0.6793	1.3143	0.2758	1.6238	1.3839	0.1993	0.2381	0.2353	0.3855	0.1037
0.28	0.7776	1.4336	0.3313	1.7040	1.3891	0.2385	0.3124	0.3093	0.5066	0.1230
0.32	0.8674	1.5273	0.3868	1.7840	1.3897	0.2784	0.3918	0.3885	0.6392	0.1425
0.36	0.9492	1.5997	0.4424	1.8641	1.3856	0.3192	0.4750	0.4718	0.7827	0.1621
0.40	1.0236	1.6536	0.4976	1.9446	1.3770	0.3614	0.5611	0.5581	0.9368	0.1818
0.42	1.0581	1.6745	0.5251	1.9851	1.3708	0.3831	0.6049	0.6021	1.0178	0.1918
0.44	1.0909	1.6917	0.5525	2.0257	1.3635	0.4052	0.6491	0.6465	1.1012	0.2018
0.46	1.1219	1.7054	0.5796	2.0666	1.3550	0.4278	0.6934	0.6911	1.1872	0.2119
0.48	1.1512	1.7158	0.6066	2.1078	1.3452	0.4510	0.7379	0.7359	1.2758	0.2220
0.50	1.1789	1.7231	0.6334	2.1493	1.3342	0.4748	0.7824	0.7806	1.3668	0.2322
0.52	1.2048	1.7274	0.6600	2.1912	1.3218	0.4993	0.8267	0.8252	1.4605	0.2424
0.54	1.2291	1.7289	0.6863	2.2335	1.3081	0.5246	0.8707	0.8696	1.5567	0.2528
0.56	1.2518	1.7276	0.7123	2.2762	1.2931	0.5509	0.9144	0.9136	1.6556	0.2632
0.58	1.2727	1.7238	0.7380	2.3195	1.2766	0.5781	0.9575	0.9571	1.7573	0.2737
0.60	1.2920	1.7175	0.7634	2.3633	1.2586	0.6065	1.0000	1.0000	1.8618	0.2842
0.62	1.3096	1.7087	0.7884	2.4079	1.2390	0.6363	1.0418	1.0421	1.9692	0.2949
0.64	1.3255	1.6975	0.8129	2.4531	1.2179	0.6675	1.0826	1.0832	2.0799	0.3057
0.66	1.3397	1.6840	0.8371	2.4992	1.1950	0.7005	1.1223	1.1233	2.1939	0.3166
0.68	1.3521	1.6683	0.8607	2.5462	1.1703	0.7355	1.1609	1.1622	2.3115	0.3276
0.70	1.3628	1.6503	0.8839	2.5943	1.1437	0.7728	1.1981	1.1997	2.4332	0.3388
0.72	1.3716	1.6300	0.9064	2.6435	1.1150	0.8130	1.2338	1.2357	2.5594	0.3501
0.74	1.3785	1.6075	0.9284	2.6940	1.0841	0.8564	1.2678	1.2699	2.6907	0.3616
0.76	1.3835	1.5827	0.9498	2.7461	1.0507	0.9039	1.3000	1.3022	2.8279	0.3732
0.78	1.3864	1.5555	0.9705	2.8000	1.0147	0.9564	1.3300	1.3324	2.9720	0.3851
0.80	1.3871	1.5259	0.9904	2.8558	0.9757	1.0150	1.3578	1.3603	3.1246	0.3971
0.82	1.3856	1.4938	1.0095	2.9141	0.9334	1.0815	1.3830	1.3855	3.2875	0.4094
0.84	1.3817	1.4590	1.0277	2.9751	0.8872	1.1583	1.4054	1.4078	3.4635	0.4220
0.86	1.3751	1.4212	1.0449	3.0396	0.8367	1.2489	1.4246	1.4268	3.6569	0.4349
0.88	1.3655	1.3800	1.0611	3.1084	0.7808	1.3590	1.4402	1.4422	3.8738	0.4481
0.90	1.3525	1.3351	1.0761	3.1825	0.7183	1.4981	1.4517	1.4533	4.1246	0.4617
0.92	1.3355	1.2855	1.0898	3.2639	0.6475	1.6832	1.4583	1.4595	4.4275	0.4758
0.94	1.3135	1.2297	1.1019	3.3556	0.5650	1.9504	1.4589	1.4595	4.8192	0.4904
0.96	1.2845	1.1651	1.1123	3.4635	0.4648	2.3932	1.4516	1.4514	5.3884	0.5058
0.98	1.2437	1.0844	1.1203	3.6031	0.3311	3.3840	1.4321	1.4307	6.4537	0.5221
1.00	1.1426	0.9116	1.1247	3.9374	0.0000	-	1.3614	1.3575	-	0.5400

1·30 Pipe arch (corrugated sheet metal)

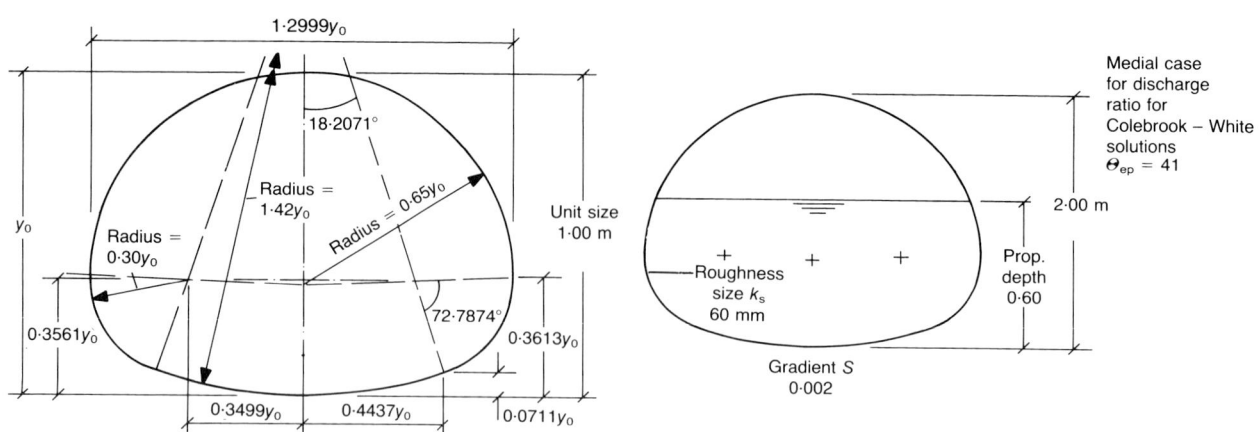

Prop dpth Y	U.equ. dia. $D_{ep(u)}$ (m)	Equiv. disch. factor J	Unit sect. area A_u (m²)	Unit wetted perim. P_u (m)	Unit surf. brdth $B_{s(u)}$ (m)	Unit mean depth $y_{m(u)}$ (m)	Discharge ratios $Q/Q_{0.60}$ medial	Mann'g	U.crit. disch. $Q_{c(u)}$ (m³s⁻¹)	Unit depth cntrd $y_{d(u)}$ (m)
0·04	0·1060	0·4927	0·0179	0·6757	0·6693	0·0267	0·0038	0·0049	0·0092	0·0160
0·08	0·2115	0·6972	0·0504	0·9531	0·9347	0·0539	0·0191	0·0220	0·0366	0·0321
0·12	0·3268	0·9243	0·0908	1·1108	1·0701	0·0848	0·0485	0·0529	0·0828	0·0487
0·16	0·4413	1·1299	0·1354	1·2269	1·1540	0·1173	0·0909	0·0964	0·1452	0·0660
0·20	0·5513	1·3060	0·1827	1·3260	1·2123	0·1507	0·1448	0·1510	0·2222	0·0837
0·24	0·6557	1·4548	0·2321	1·4159	1·2531	0·1852	0·2088	0·2152	0·3128	0·1016
0·28	0·7539	1·5785	0·2828	1·5005	1·2803	0·2209	0·2815	0·2879	0·4163	0·1198
0·32	0·8454	1·6789	0·3344	1·5820	1·2955	0·2581	0·3614	0·3673	0·5320	0·1382
0·36	0·9297	1·7571	0·3863	1·6621	1·2998	0·2972	0·4467	0·4521	0·6595	0·1569
0·40	1·0062	1·8143	0·4382	1·7422	1·2961	0·3381	0·5360	0·5407	0·7980	0·1760
0·42	1·0416	1·8358	0·4641	1·7824	1·2924	0·3591	0·5818	0·5860	0·8710	0·1856
0·44	1·0752	1·8531	0·4899	1·8227	1·2875	0·3805	0·6280	0·6318	0·9465	0·1953
0·46	1·1070	1·8664	0·5156	1·8632	1·2812	0·4024	0·6746	0·6780	1·0243	0·2051
0·48	1·1370	1·8761	0·5412	1·9039	1·2737	0·4249	0·7215	0·7244	1·1047	0·2149
0·50	1·1653	1·8822	0·5666	1·9449	1·2649	0·4479	0·7685	0·7709	1·1874	0·2249
0·52	1·1918	1·8851	0·5918	1·9861	1·2548	0·4716	0·8154	0·8173	1·2726	0·2349
0·54	1·2166	1·8848	0·6167	2·0278	1·2432	0·4961	0·8622	0·8636	1·3603	0·2449
0·56	1·2397	1·8816	0·6415	2·0698	1·2303	0·5214	0·9087	0·9096	1·4506	0·2551
0·58	1·2611	1·8755	0·6659	2·1123	1·2159	0·5477	0·9547	0·9551	1·5434	0·2654
0·60	1·2807	1·8667	0·6901	2·1554	1·2000	0·5751	1·0000	1·0000	1·6389	0·2757
0·62	1·2986	1·8553	0·7139	2·1990	1·1825	0·6037	1·0446	1·0442	1·7372	0·2862
0·64	1·3148	1·8412	0·7374	2·2433	1·1634	0·6338	1·0883	1·0874	1·8384	0·2968
0·66	1·3292	1·8247	0·7605	2·2884	1·1426	0·6655	1·1309	1·1296	1·9428	0·3075
0·68	1·3418	1·8058	0·7831	2·3344	1·1200	0·6992	1·1722	1·1706	2·0506	0·3183
0·70	1·3526	1·7844	0·8053	2·3813	1·0954	0·7351	1·2121	1·2101	2·1620	0·3293
0·72	1·3615	1·7606	0·8269	2·4294	1·0688	0·7736	1·2504	1·2481	2·2776	0·3404
0·74	1·3684	1·7344	0·8480	2·4787	1·0400	0·8154	1·2868	1·2843	2·3979	0·3517
0·76	1·3734	1·7057	0·8685	2·5294	1·0088	0·8609	1·3212	1·3185	2·5235	0·3631
0·78	1·3762	1·6746	0·8883	2·5819	0·9749	0·9112	1·3534	1·3505	2·6555	0·3748
0·80	1·3769	1·6408	0·9075	2·6362	0·9381	0·9674	1·3830	1·3800	2·7950	0·3867
0·82	1·3752	1·6044	0·9258	2·6929	0·8980	1·0310	1·4098	1·4068	2·9439	0·3988
0·84	1·3710	1·5650	0·9434	2·7522	0·8542	1·1044	1·4334	1·4305	3·1046	0·4112
0·86	1·3641	1·5225	0·9600	2·8148	0·8060	1·1911	1·4535	1·4508	3·2808	0·4239
0·88	1·3542	1·4764	0·9756	2·8816	0·7526	1·2963	1·4697	1·4672	3·4782	0·4370
0·90	1·3408	1·4262	0·9900	2·9535	0·6928	1·4290	1·4812	1·4791	3·7061	0·4505
0·92	1·3233	1·3710	1·0032	3·0324	0·6248	1·6056	1·4873	1·4858	3·9808	0·4644
0·94	1·3007	1·3092	1·0149	3·1212	0·5455	1·8605	1·4867	1·4859	4·3352	0·4789
0·96	1·2709	1·2378	1·0249	3·2257	0·4490	2·2827	1·4773	1·4776	4·8493	0·4942
0·98	1·2291	1·1489	1·0327	3·3609	0·3200	3·2272	1·4540	1·4559	5·8095	0·5104
1·00	1·1259	0·9600	1·0370	3·6842	0·0000	-	1·3730	1·3789	-	0·5283

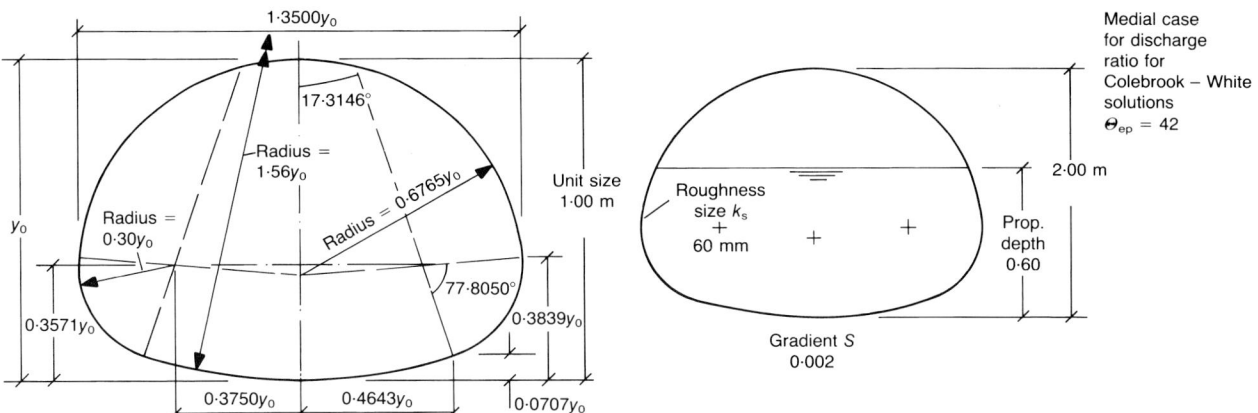

Prop dpth Y	U.equ. dia. $D_{ep(u)}$ (m)	Equiv. disch. factor J	Unit sect. area A_u (m²)	Unit wetted perim. P_u (m)	Unit surf. brdth $B_{s(u)}$ (m)	Unit mean depth $y_{m(u)}$ (m)	Discharge ratios $Q/Q_{0.60}$ medial	Mann'g	U.crit. disch. $Q_{c(u)}$ (m³s⁻¹)	Unit depth cntrd $y_{d(u)}$ (m)
0·04	0·1060	0·4704	0·0188	0·7081	0·7020	0·0267	0·0038	0·0049	0·0096	0·0160
0·08	0·2120	0·6675	0·0529	0·9975	0·9799	0·0539	0·0192	0·0220	0·0384	0·0321
0·12	0·3287	0·8923	0·0951	1·1572	1·1176	0·0851	0·0487	0·0531	0·0869	0·0488
0·16	0·4446	1·0964	0·1416	1·2739	1·2023	0·1178	0·0912	0·0967	0·1522	0·0661
0·20	0·5561	1·2722	0·1909	1·3733	1·2612	0·1514	0·1452	0·1514	0·2327	0·0838
0·24	0·6622	1·4215	0·2423	1·4634	1·3024	0·1860	0·2094	0·2158	0·3272	0·1018
0·28	0·7621	1·5465	0·2949	1·5481	1·3298	0·2218	0·2822	0·2885	0·4350	0·1200
0·32	0·8554	1·6490	0·3485	1·6296	1·3454	0·2590	0·3622	0·3682	0·5554	0·1385
0·36	0·9415	1·7298	0·4024	1·7098	1·3500	0·2981	0·4479	0·4532	0·6881	0·1573
0·40	1·0197	1·7896	0·4563	1·7901	1·3443	0·3395	0·5375	0·5420	0·8326	0·1749
0·42	1·0559	1·8123	0·4832	1·8304	1·3392	0·3608	0·5832	0·5874	0·9089	0·1846
0·44	1·0902	1·8306	0·5099	1·8709	1·3328	0·3826	0·6295	0·6332	0·9877	0·1944
0·46	1·1226	1·8448	0·5365	1·9116	1·3252	0·4048	0·6761	0·6794	1·0690	0·2043
0·48	1·1531	1·8553	0·5629	1·9526	1·3163	0·4276	0·7229	0·7257	1·1527	0·2143
0·50	1·1819	1·8622	0·5891	1·9939	1·3061	0·4510	0·7698	0·7721	1·2390	0·2243
0·52	1·2088	1·8657	0·6151	2·0355	1·2947	0·4751	0·8166	0·8184	1·3278	0·2344
0·54	1·2340	1·8660	0·6409	2·0775	1·2818	0·5000	0·8631	0·8645	1·4192	0·2445
0·56	1·2574	1·8633	0·6664	2·1199	1·2676	0·5257	0·9093	0·9102	1·5131	0·2548
0·58	1·2790	1·8577	0·6916	2·1629	1·2520	0·5524	0·9550	0·9554	1·6097	0·2652
0·60	1·2989	1·8494	0·7165	2·2064	1·2348	0·5802	1·0000	1·0000	1·7091	0·2756
0·62	1·3170	1·8383	0·7410	2·2506	1·2161	0·6093	1·0442	1·0438	1·8113	0·2862
0·64	1·3333	1·8247	0·7651	2·2954	1·1958	0·6398	1·0875	1·0866	1·9165	0·2968
0·66	1·3477	1·8085	0·7888	2·3411	1·1737	0·6720	1·1297	1·1284	2·0250	0·3076
0·68	1·3604	1·7899	0·8120	2·3877	1·1499	0·7062	1·1705	1·1689	2·1370	0·3185
0·70	1·3712	1·7688	0·8348	2·4353	1·1241	0·7426	1·2099	1·2080	2·2528	0·3296
0·72	1·3800	1·7453	0·8570	2·4840	1·0962	0·7818	1·2477	1·2454	2·3729	0·3408
0·74	1·3869	1·7193	0·8786	2·5341	1·0662	0·8241	1·2836	1·2811	2·4978	0·3521
0·76	1·3917	1·6909	0·8996	2·5856	1·0337	0·8703	1·3174	1·3148	2·6282	0·3637
0·78	1·3945	1·6601	0·9200	2·6389	0·9985	0·9213	1·3490	1·3462	2·7652	0·3754
0·80	1·3950	1·6266	0·9395	2·6941	0·9604	0·9783	1·3781	1·3752	2·9101	0·3874
0·82	1·3931	1·5904	0·9583	2·7517	0·9190	1·0428	1·4044	1·4015	3·0647	0·3996
0·84	1·3887	1·5514	0·9763	2·8121	0·8738	1·1173	1·4275	1·4247	3·2316	0·4121
0·86	1·3816	1·5092	0·9933	2·8758	0·8242	1·2052	1·4472	1·4445	3·4147	0·4249
0·88	1·3713	1·4635	1·0092	2·9437	0·7693	1·3118	1·4629	1·4605	3·6198	0·4380
0·90	1·3576	1·4137	1·0240	3·0170	0·7080	1·4464	1·4740	1·4720	3·8566	0·4515
0·92	1·3398	1·3589	1·0375	3·0974	0·6382	1·6255	1·4797	1·4783	4·1422	0·4655
0·94	1·3168	1·2977	1·0495	3·1879	0·5571	1·8839	1·4788	1·4782	4·5108	0·4801
0·96	1·2866	1·2269	1·0596	3·2944	0·4583	2·3119	1·4692	1·4696	5·0455	0·4954
0·98	1·2442	1·1388	1·0676	3·4322	0·3266	3·2692	1·4460	1·4479	6·0447	0·5116
1·00	1·1398	0·9518	1·0719	3·7620	0·0000	-	1·3654	1·3713	-	0·5295

1·40 Pipe arch (corrugated sheet metal)

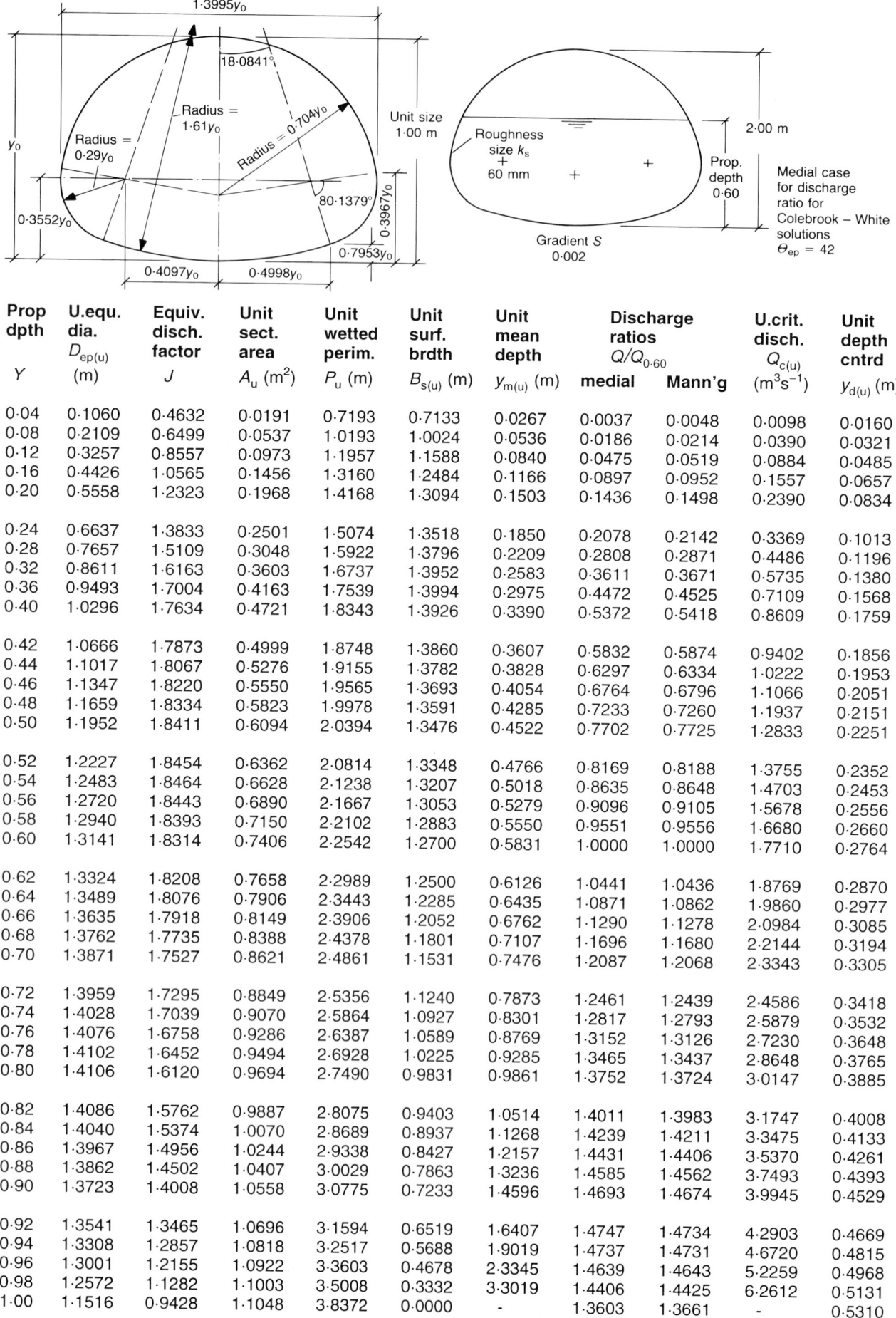

Prop dpth Y	U.equ. dia. $D_{ep(u)}$ (m)	Equiv. disch. factor J	Unit sect. area A_u (m²)	Unit wetted perim. P_u (m)	Unit surf. brdth $B_{s(u)}$ (m)	Unit mean depth $y_{m(u)}$ (m)	Discharge ratios $Q/Q_{0.60}$ medial	Mann'g	U.crit. disch. $Q_{c(u)}$ (m³s⁻¹)	Unit depth cntrd $y_{d(u)}$ (m)
0·04	0·1060	0·4632	0·0191	0·7193	0·7133	0·0267	0·0037	0·0048	0·0098	0·0160
0·08	0·2109	0·6499	0·0537	1·0193	1·0024	0·0536	0·0186	0·0214	0·0390	0·0321
0·12	0·3257	0·8557	0·0973	1·1957	1·1588	0·0840	0·0475	0·0519	0·0884	0·0485
0·16	0·4426	1·0565	0·1456	1·3160	1·2484	0·1166	0·0897	0·0952	0·1557	0·0657
0·20	0·5558	1·2323	0·1968	1·4168	1·3094	0·1503	0·1436	0·1498	0·2390	0·0834
0·24	0·6637	1·3833	0·2501	1·5074	1·3518	0·1850	0·2078	0·2142	0·3369	0·1013
0·28	0·7657	1·5109	0·3048	1·5922	1·3796	0·2209	0·2808	0·2871	0·4486	0·1196
0·32	0·8611	1·6163	0·3603	1·6737	1·3952	0·2583	0·3611	0·3671	0·5735	0·1380
0·36	0·9493	1·7004	0·4163	1·7539	1·3994	0·2975	0·4472	0·4525	0·7109	0·1568
0·40	1·0296	1·7634	0·4721	1·8343	1·3926	0·3390	0·5372	0·5418	0·8609	0·1759
0·42	1·0666	1·7873	0·4999	1·8748	1·3860	0·3607	0·5832	0·5874	0·9402	0·1856
0·44	1·1017	1·8067	0·5276	1·9155	1·3782	0·3828	0·6297	0·6334	1·0222	0·1953
0·46	1·1347	1·8220	0·5550	1·9565	1·3693	0·4054	0·6764	0·6796	1·1066	0·2051
0·48	1·1659	1·8334	0·5823	1·9978	1·3591	0·4285	0·7233	0·7260	1·1937	0·2151
0·50	1·1952	1·8411	0·6094	2·0394	1·3476	0·4522	0·7702	0·7725	1·2833	0·2251
0·52	1·2227	1·8454	0·6362	2·0814	1·3348	0·4766	0·8169	0·8188	1·3755	0·2352
0·54	1·2483	1·8464	0·6628	2·1238	1·3207	0·5018	0·8635	0·8648	1·4703	0·2453
0·56	1·2720	1·8443	0·6890	2·1667	1·3053	0·5279	0·9096	0·9105	1·5678	0·2556
0·58	1·2940	1·8393	0·7150	2·2102	1·3053	0·5279	0·9096	0·9105	1·5678	0·2556
0·60	1·3141	1·8314	0·7406	2·2542	1·2700	0·5831	1·0000	1·0000	1·7710	0·2764
0·62	1·3324	1·8208	0·7658	2·2989	1·2500	0·6126	1·0441	1·0436	1·8769	0·2870
0·64	1·3489	1·8076	0·7906	2·3443	1·2285	0·6435	1·0871	1·0862	1·9860	0·2977
0·66	1·3635	1·7918	0·8149	2·3906	1·2052	0·6762	1·1290	1·1278	2·0984	0·3085
0·68	1·3762	1·7735	0·8388	2·4378	1·1801	0·7107	1·1696	1·1680	2·2144	0·3194
0·70	1·3871	1·7527	0·8621	2·4861	1·1531	0·7476	1·2087	1·2068	2·3343	0·3305
0·72	1·3959	1·7295	0·8849	2·5356	1·1240	0·7873	1·2461	1·2439	2·4586	0·3418
0·74	1·4028	1·7039	0·9070	2·5864	1·0927	0·8301	1·2817	1·2793	2·5879	0·3532
0·76	1·4076	1·6758	0·9286	2·6387	1·0589	0·8769	1·3152	1·3126	2·7230	0·3648
0·78	1·4102	1·6452	0·9494	2·6928	1·0225	0·9285	1·3465	1·3437	2·8648	0·3765
0·80	1·4106	1·6120	0·9694	2·7490	0·9831	0·9861	1·3752	1·3724	3·0147	0·3885
0·82	1·4086	1·5762	0·9887	2·8075	0·9403	1·0514	1·4011	1·3983	3·1747	0·4008
0·84	1·4040	1·5374	1·0070	2·8689	0·8937	1·1268	1·4239	1·4211	3·3475	0·4133
0·86	1·3967	1·4956	1·0244	2·9338	0·8427	1·2157	1·4431	1·4406	3·5370	0·4261
0·88	1·3862	1·4502	1·0407	3·0029	0·7863	1·3236	1·4585	1·4562	3·7493	0·4393
0·90	1·3723	1·4008	1·0558	3·0775	0·7233	1·4596	1·4693	1·4674	3·9945	0·4529
0·92	1·3541	1·3465	1·0696	3·1594	0·6519	1·6407	1·4747	1·4734	4·2903	0·4669
0·94	1·3308	1·2857	1·0818	3·2517	0·5688	1·9019	1·4737	1·4731	4·6720	0·4815
0·96	1·3001	1·2155	1·0922	3·3603	0·4678	2·3345	1·4639	1·4643	5·2259	0·4968
0·98	1·2572	1·1282	1·1003	3·5008	0·3332	3·3019	1·4406	1·4425	6·2612	0·5131
1·00	1·1516	0·9428	1·1048	3·8372	0·0000	-	1·3603	1·3661	-	0·5310

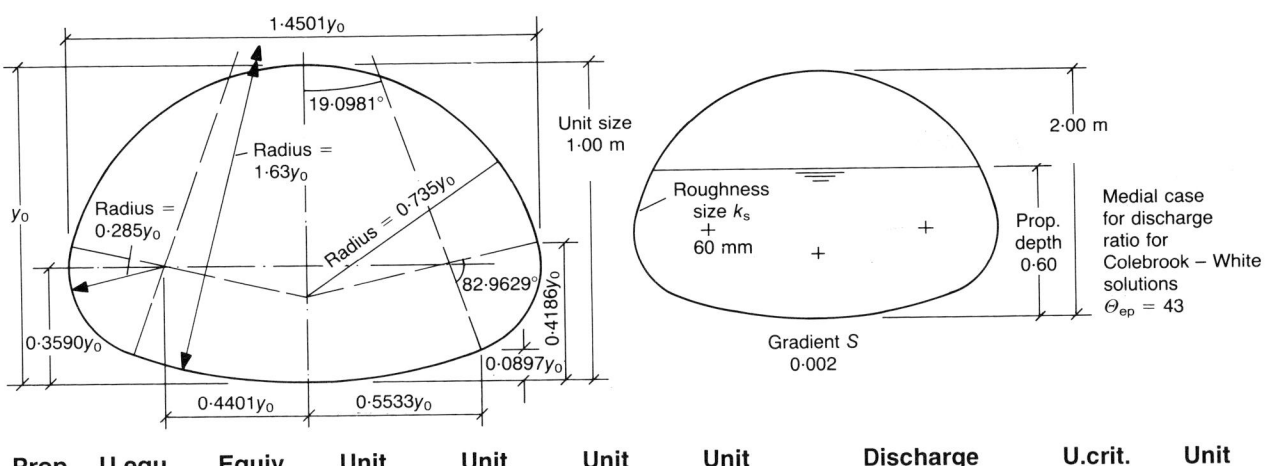

Prop dpth Y	U.equ. dia. $D_{ep(u)}$ (m)	Equiv. disch. factor J	Unit sect. area A_u (m²)	Unit wetted perim. P_u (m)	Unit surf. brdth $B_{s(u)}$ (m)	Unit mean depth $y_{m(u)}$ (m)	Discharge ratios $Q/Q_{0.60}$ medial	Mann'g	U.crit. disch. $Q_{c(u)}$ (m³s⁻¹)	Unit depth cntrd $y_{d(u)}$ (m)
0·04	0·1061	0·4604	0·0192	0·7237	0·7178	0·0267	0·0036	0·0047	0·0098	0·0160
0·08	0·2109	0·6460	0·0541	1·0256	1·0088	0·0536	0·0180	0·0208	0·0392	0·0321
0·12	0·3217	0·8252	0·0985	1·2249	1·1906	0·0828	0·0458	0·0501	0·0888	0·0483
0·16	0·4388	1·0201	0·1482	1·3513	1·2881	0·1151	0·0873	0·0927	0·1575	0·0653
0·20	0·5532	1·1947	0·2012	1·4546	1·3531	0·1487	0·1407	0·1468	0·2429	0·0828
0·24	0·6628	1·3464	0·2562	1·5464	1·3980	0·1833	0·2046	0·2110	0·3435	0·1007
0·28	0·7667	1·4761	0·3128	1·6318	1·4278	0·2191	0·2775	0·2839	0·4585	0·1189
0·32	0·8643	1·5844	0·3703	1·7137	1·4448	0·2563	0·3580	0·3640	0·5870	0·1373
0·36	0·9548	1·6721	0·4282	1·7939	1·4501	0·2953	0·4445	0·4498	0·7287	0·1560
0·40	1·0375	1·7391	0·4861	1·8742	1·4442	0·3366	0·5352	0·5397	0·8833	0·1750
0·42	1·0757	1·7648	0·5150	1·9149	1·4369	0·3584	0·5815	0·5857	0·9654	0·1847
0·44	1·1117	1·7856	0·5436	1·9559	1·4277	0·3808	0·6283	0·6320	1·0504	0·1944
0·46	1·1457	1·8021	0·5721	1·9973	1·4173	0·4036	0·6753	0·6785	1·1381	0·2043
0·48	1·1777	1·8145	0·6003	2·0389	1·4057	0·4270	0·7224	0·7252	1·2285	0·2142
0·50	1·2077	1·8232	0·6283	2·0809	1·3928	0·4511	0·7696	0·7719	1·3214	0·2242
0·52	1·2358	1·8283	0·6560	2·1234	1·3787	0·4758	0·8165	0·8184	1·4170	0·2343
0·54	1·2619	1·8301	0·6834	2·1663	1·3632	0·5013	0·8632	0·8646	1·5153	0·2445
0·56	1·2862	1·8286	0·7105	2·2097	1·3464	0·5277	0·9094	0·9103	1·6164	0·2548
0·58	1·3086	1·8242	0·7373	2·2536	1·3282	0·5551	0·9551	0·9555	1·7202	0·2652
0·60	1·3291	1·8168	0·7636	2·2982	1·3084	0·5836	1·0000	1·0000	1·8269	0·2757
0·62	1·3477	1·8066	0·7896	2·3435	1·2872	0·6134	1·0441	1·0436	1·9366	0·2863
0·64	1·3644	1·7938	0·8151	2·3896	1·2643	0·6447	1·0871	1·0862	2·0496	0·2970
0·66	1·3792	1·7783	0·8402	2·4366	1·2397	0·6777	1·1289	1·1277	2·1659	0·3079
0·68	1·3921	1·7603	0·8647	2·4845	1·2133	0·7127	1·1694	1·1679	2·2860	0·3189
0·70	1·4030	1·7397	0·8887	2·5335	1·1849	0·7500	1·2084	1·2065	2·4101	0·3300
0·72	1·4120	1·7167	0·9121	2·5838	1·1545	0·7900	1·2457	1·2435	2·5387	0·3413
0·74	1·4189	1·6913	0·9348	2·6355	1·1218	0·8333	1·2811	1·2787	2·6724	0·3527
0·76	1·4236	1·6634	0·9569	2·6887	1·0866	0·8806	1·3144	1·3118	2·8121	0·3643
0·78	1·4262	1·6329	0·9783	2·7438	1·0488	0·9328	1·3454	1·3427	2·9588	0·3762
0·80	1·4265	1·5999	0·9989	2·8009	1·0080	0·9910	1·3739	1·3712	3·1138	0·3882
0·82	1·4243	1·5642	1·0186	2·8606	0·9637	1·0569	1·3996	1·3968	3·2792	0·4005
0·84	1·4195	1·5256	1·0374	2·9231	0·9156	1·1330	1·4221	1·4194	3·4578	0·4131
0·86	1·4120	1·4839	1·0552	2·9893	0·8630	1·2227	1·4411	1·4386	3·6537	0·4259
0·88	1·4012	1·4387	1·0719	3·0597	0·8050	1·3315	1·4562	1·4540	3·8733	0·4391
0·90	1·3870	1·3895	1·0873	3·1358	0·7403	1·4688	1·4668	1·4649	4·1268	0·4528
0·92	1·3685	1·3354	1·1014	3·2194	0·6669	1·6515	1·4720	1·4707	4·4325	0·4668
0·94	1·3447	1·2749	1·1139	3·3135	0·5817	1·9149	1·4707	1·4701	4·8271	0·4815
0·96	1·3136	1·2051	1·1246	3·4244	0·4783	2·3510	1·4607	1·4612	5·3998	0·4968
0·98	1·2700	1·1183	1·1328	3·5679	0·3406	3·3261	1·4373	1·4392	6·4699	0·5131
1·00	1·1631	0·9341	1·1374	3·9116	0·0000	-	1·3569	1·3627	-	0·5311

C41

1·50 Pipe arch (corrugated sheet metal)

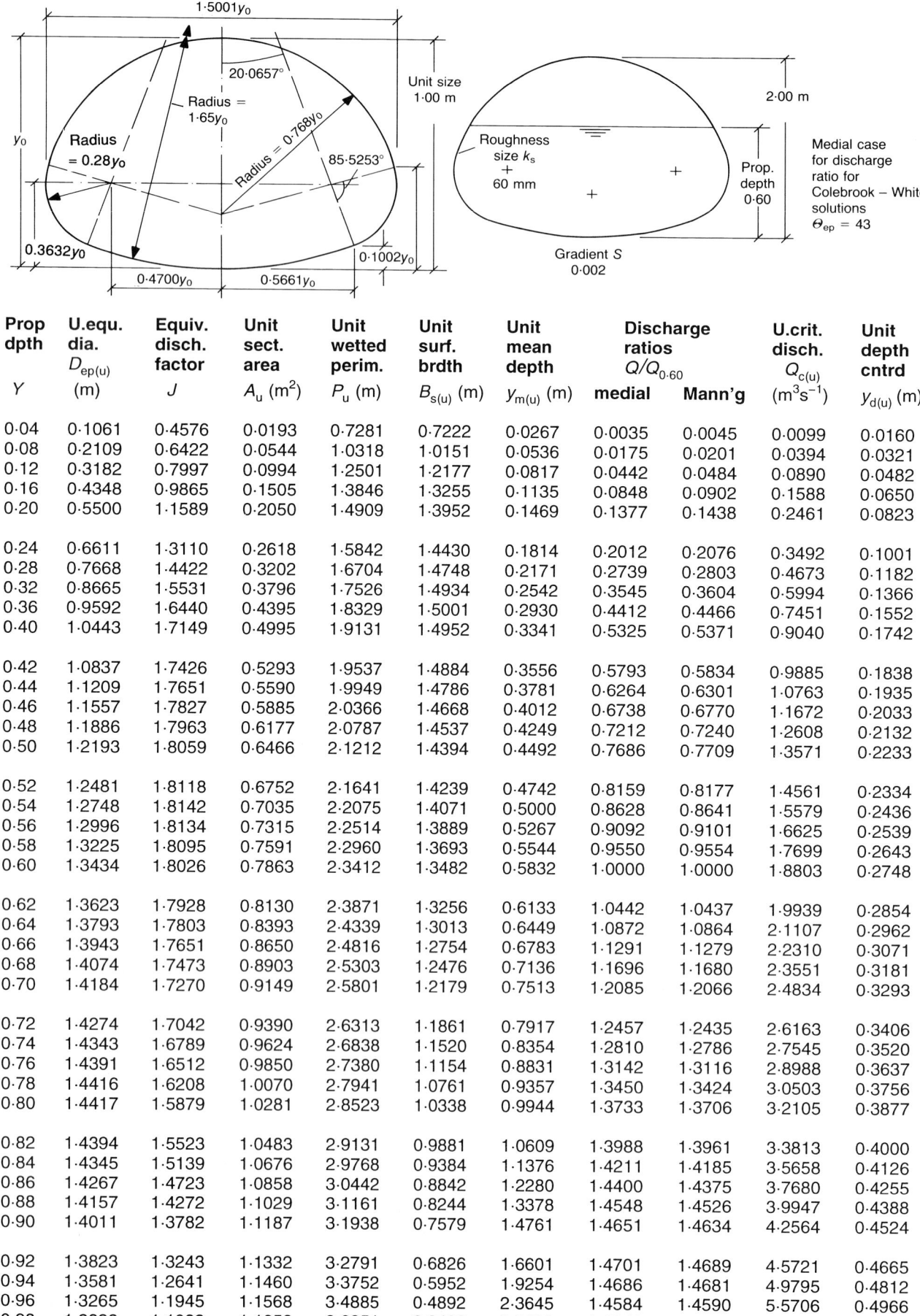

Prop dpth Y	U.equ. dia. $D_{ep(u)}$ (m)	Equiv. disch. factor J	Unit sect. area A_u (m²)	Unit wetted perim. P_u (m)	Unit surf. brdth $B_{s(u)}$ (m)	Unit mean depth $y_{m(u)}$ (m)	Discharge ratios $Q/Q_{0.60}$ medial	Mann'g	U.crit. disch. $Q_{c(u)}$ (m³s⁻¹)	Unit depth cntrd $y_{d(u)}$ (m)
0·04	0·1061	0·4576	0·0193	0·7281	0·7222	0·0267	0·0035	0·0045	0·0099	0·0160
0·08	0·2109	0·6422	0·0544	1·0318	1·0151	0·0536	0·0175	0·0201	0·0394	0·0321
0·12	0·3182	0·7997	0·0994	1·2501	1·2177	0·0817	0·0442	0·0484	0·0890	0·0482
0·16	0·4348	0·9865	0·1505	1·3846	1·3255	0·1135	0·0848	0·0902	0·1588	0·0650
0·20	0·5500	1·1589	0·2050	1·4909	1·3952	0·1469	0·1377	0·1438	0·2461	0·0823
0·24	0·6611	1·3110	0·2618	1·5842	1·4430	0·1814	0·2012	0·2076	0·3492	0·1001
0·28	0·7668	1·4422	0·3202	1·6704	1·4748	0·2171	0·2739	0·2803	0·4673	0·1182
0·32	0·8665	1·5531	0·3796	1·7526	1·4934	0·2542	0·3545	0·3604	0·5994	0·1366
0·36	0·9592	1·6440	0·4395	1·8329	1·5001	0·2930	0·4412	0·4466	0·7451	0·1552
0·40	1·0443	1·7149	0·4995	1·9131	1·4952	0·3341	0·5325	0·5371	0·9040	0·1742
0·42	1·0837	1·7426	0·5293	1·9537	1·4884	0·3556	0·5793	0·5834	0·9885	0·1838
0·44	1·1209	1·7651	0·5590	1·9949	1·4786	0·3781	0·6264	0·6301	1·0763	0·1935
0·46	1·1557	1·7827	0·5885	2·0366	1·4668	0·4012	0·6738	0·6770	1·1672	0·2033
0·48	1·1886	1·7963	0·6177	2·0787	1·4537	0·4249	0·7212	0·7240	1·2608	0·2132
0·50	1·2193	1·8059	0·6466	2·1212	1·4394	0·4492	0·7686	0·7709	1·3571	0·2233
0·52	1·2481	1·8118	0·6752	2·1641	1·4239	0·4742	0·8159	0·8177	1·4561	0·2334
0·54	1·2748	1·8142	0·7035	2·2075	1·4071	0·5000	0·8628	0·8641	1·5579	0·2436
0·56	1·2996	1·8134	0·7315	2·2514	1·3889	0·5267	0·9092	0·9101	1·6625	0·2539
0·58	1·3225	1·8095	0·7591	2·2960	1·3693	0·5544	0·9550	0·9554	1·7699	0·2643
0·60	1·3434	1·8026	0·7863	2·3412	1·3482	0·5832	1·0000	1·0000	1·8803	0·2748
0·62	1·3623	1·7928	0·8130	2·3871	1·3256	0·6133	1·0442	1·0437	1·9939	0·2854
0·64	1·3793	1·7803	0·8393	2·4339	1·3013	0·6449	1·0872	1·0864	2·1107	0·2962
0·66	1·3943	1·7651	0·8650	2·4816	1·2754	0·6783	1·1291	1·1279	2·2310	0·3071
0·68	1·4074	1·7473	0·8903	2·5303	1·2476	0·7136	1·1696	1·1680	2·3551	0·3181
0·70	1·4184	1·7270	0·9149	2·5801	1·2179	0·7513	1·2085	1·2066	2·4834	0·3293
0·72	1·4274	1·7042	0·9390	2·6313	1·1861	0·7917	1·2457	1·2435	2·6163	0·3406
0·74	1·4343	1·6789	0·9624	2·6838	1·1520	0·8354	1·2810	1·2786	2·7545	0·3520
0·76	1·4391	1·6512	0·9850	2·7380	1·1154	0·8831	1·3142	1·3116	2·8988	0·3637
0·78	1·4416	1·6208	1·0070	2·7941	1·0761	0·9357	1·3450	1·3424	3·0503	0·3756
0·80	1·4417	1·5879	1·0281	2·8523	1·0338	0·9944	1·3733	1·3706	3·2105	0·3877
0·82	1·4394	1·5523	1·0483	2·9131	0·9881	1·0609	1·3988	1·3961	3·3813	0·4000
0·84	1·4345	1·5139	1·0676	2·9768	0·9384	1·1376	1·4211	1·4185	3·5658	0·4126
0·86	1·4267	1·4723	1·0858	3·0442	0·8842	1·2280	1·4400	1·4375	3·7680	0·4255
0·88	1·4157	1·4272	1·1029	3·1161	0·8244	1·3378	1·4548	1·4526	3·9947	0·4388
0·90	1·4011	1·3782	1·1187	3·1938	0·7579	1·4761	1·4651	1·4634	4·2564	0·4524
0·92	1·3823	1·3243	1·1332	3·2791	0·6826	1·6601	1·4701	1·4689	4·5721	0·4665
0·94	1·3581	1·2641	1·1460	3·3752	0·5952	1·9254	1·4686	1·4681	4·9795	0·4812
0·96	1·3265	1·1945	1·1568	3·4885	0·4892	2·3645	1·4584	1·4590	5·5706	0·4966
0·98	1·2823	1·1082	1·1653	3·6351	0·3483	3·3461	1·4348	1·4368	6·6753	0·5129
1·00	1·1739	0·9251	1·1700	3·9864	0·0000	-	1·3543	1·3601	-	0·5308

1·55 Pipe arch (corrugated sheet metal)

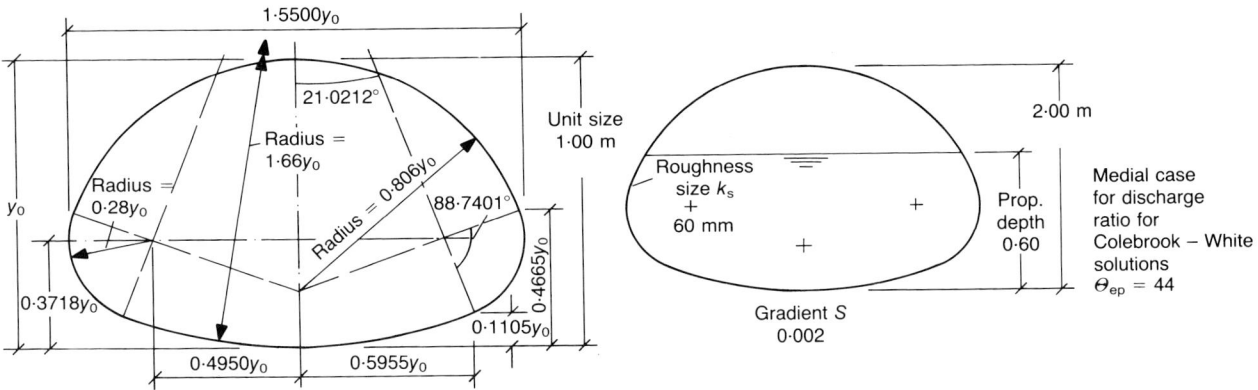

Prop dpth Y	U.equ. dia. $D_{ep(u)}$ (m)	Equiv. disch. factor J	Unit sect. area A_u (m²)	Unit wetted perim. P_u (m)	Unit surf. brdth $B_{s(u)}$ (m)	Unit mean depth $y_{m(u)}$ (m)	Discharge ratios $Q/Q_{0·60}$ medial	Mann'g	U.crit. disch. $Q_{c(u)}$ (m³s⁻¹)	Unit depth cntrd $y_{d(u)}$ (m)
0·04	0·1061	0·4563	0·0194	0·7303	0·7244	0·0267	0·0034	0·0044	0·0099	0·0160
0·08	0·2109	0·6403	0·0546	1·0349	1·0182	0·0536	0·0169	0·0195	0·0396	0·0321
0·12	0·3156	0·7831	0·0999	1·2659	1·2348	0·0809	0·0426	0·0467	0·0889	0·0482
0·16	0·4305	0·9579	0·1519	1·4117	1·3562	0·1120	0·0821	0·0874	0·1592	0·0647
0·20	0·5460	1·1269	0·2078	1·5222	1·4322	0·1451	0·1340	0·1400	0·2479	0·0819
0·24	0·6582	1·2781	0·2662	1·6177	1·4841	0·1794	0·1968	0·2032	0·3530	0·0995
0·28	0·7655	1·4103	0·3263	1·7051	1·5191	0·2148	0·2691	0·2754	0·4736	0·1175
0·32	0·8670	1·5233	0·3875	1·7880	1·5404	0·2516	0·3495	0·3554	0·6087	0·1357
0·36	0·9619	1·6173	0·4494	1·8686	1·5495	0·2900	0·4363	0·4417	0·7578	0·1543
0·40	1·0496	1·6921	0·5113	1·9487	1·5472	0·3305	0·5281	0·5327	0·9205	0·1732
0·42	1·0904	1·7222	0·5422	1·9891	1·5417	0·3517	0·5753	0·5795	1·0070	0·1827
0·44	1·1291	1·7473	0·5730	2·0300	1·5332	0·3737	0·6230	0·6267	1·0969	0·1924
0·46	1·1653	1·7672	0·6035	2·0716	1·5216	0·3967	0·6709	0·6742	1·1903	0·2022
0·48	1·1992	1·7819	0·6338	2·1142	1·5071	0·4206	0·7189	0·7217	1·2872	0·2120
0·50	1·2309	1·7926	0·6638	2·1572	1·4913	0·4451	0·7668	0·7691	1·3869	0·2220
0·52	1·2605	1·7994	0·6935	2·2007	1·4743	0·4704	0·8144	0·8163	1·4894	0·2321
0·54	1·2880	1·8026	0·7228	2·2447	1·4559	0·4964	0·8618	0·8631	1·5948	0·2423
0·56	1·3134	1·8024	0·7517	2·2892	1·4362	0·5234	0·9085	0·9094	1·7030	0·2526
0·58	1·3369	1·7990	0·7802	2·3345	1·4151	0·5513	0·9547	0·9551	1·8142	0·2630
0·60	1·3582	1·7925	0·8083	2·3804	1·3926	0·5804	1·0000	1·0000	1·9284	0·2735
0·62	1·3776	1·7831	0·8359	2·4271	1·3684	0·6108	1·0444	1·0440	2·0459	0·2841
0·64	1·3950	1·7709	0·8630	2·4747	1·3427	0·6427	1·0877	1·0869	2·1667	0·2949
0·66	1·4103	1·7559	0·8896	2·5232	1·3153	0·6764	1·1297	1·1285	2·2911	0·3058
0·68	1·4236	1·7383	0·9156	2·5728	1·2860	0·7120	1·1704	1·1688	2·4194	0·3168
0·70	1·4348	1·7181	0·9410	2·6235	1·2548	0·7500	1·2094	1·2075	2·5520	0·3280
0·72	1·4439	1·6953	0·9658	2·6756	1·2214	0·7907	1·2467	1·2446	2·6894	0·3393
0·74	1·4508	1·6700	0·9899	2·7292	1·1858	0·8348	1·2820	1·2797	2·8322	0·3508
0·76	1·4555	1·6422	1·0132	2·7844	1·1477	0·8828	1·3152	1·3127	2·9813	0·3625
0·78	1·4580	1·6119	1·0358	2·8416	1·1068	0·9358	1·3460	1·3434	3·1377	0·3744
0·80	1·4580	1·5789	1·0575	2·9010	1·0628	0·9950	1·3743	1·3716	3·3031	0·3865
0·82	1·4556	1·5432	1·0782	2·9631	1·0154	1·0619	1·3997	1·3970	3·4795	0·3989
0·84	1·4504	1·5047	1·0980	3·0282	0·9640	1·1391	1·4219	1·4193	3·6699	0·4115
0·86	1·4423	1·4630	1·1168	3·0971	0·9079	1·2300	1·4405	1·4381	3·8787	0·4244
0·88	1·4311	1·4179	1·1343	3·1706	0·8463	1·3404	1·4552	1·4531	4·1126	0·4377
0·90	1·4161	1·3688	1·1506	3·2500	0·7777	1·4795	1·4653	1·4636	4·3826	0·4514
0·92	1·3968	1·3149	1·1654	3·3373	0·7002	1·6644	1·4701	1·4689	4·7082	0·4655
0·94	1·3721	1·2546	1·1785	3·4356	0·6103	1·9310	1·4684	1·4679	5·1284	0·4802
0·96	1·3399	1·1852	1·1897	3·5516	0·5015	2·3721	1·4580	1·4585	5·7379	0·4956
0·98	1·2949	1·0990	1·1983	3·7017	0·3569	3·3579	1·4341	1·4361	6·8765	0·5120
1·00	1·1849	0·9165	1·2031	4·0615	0·0000	-	1·3532	1·3589	-	0·5299

1·60 Pipe arch (corrugated sheet metal)

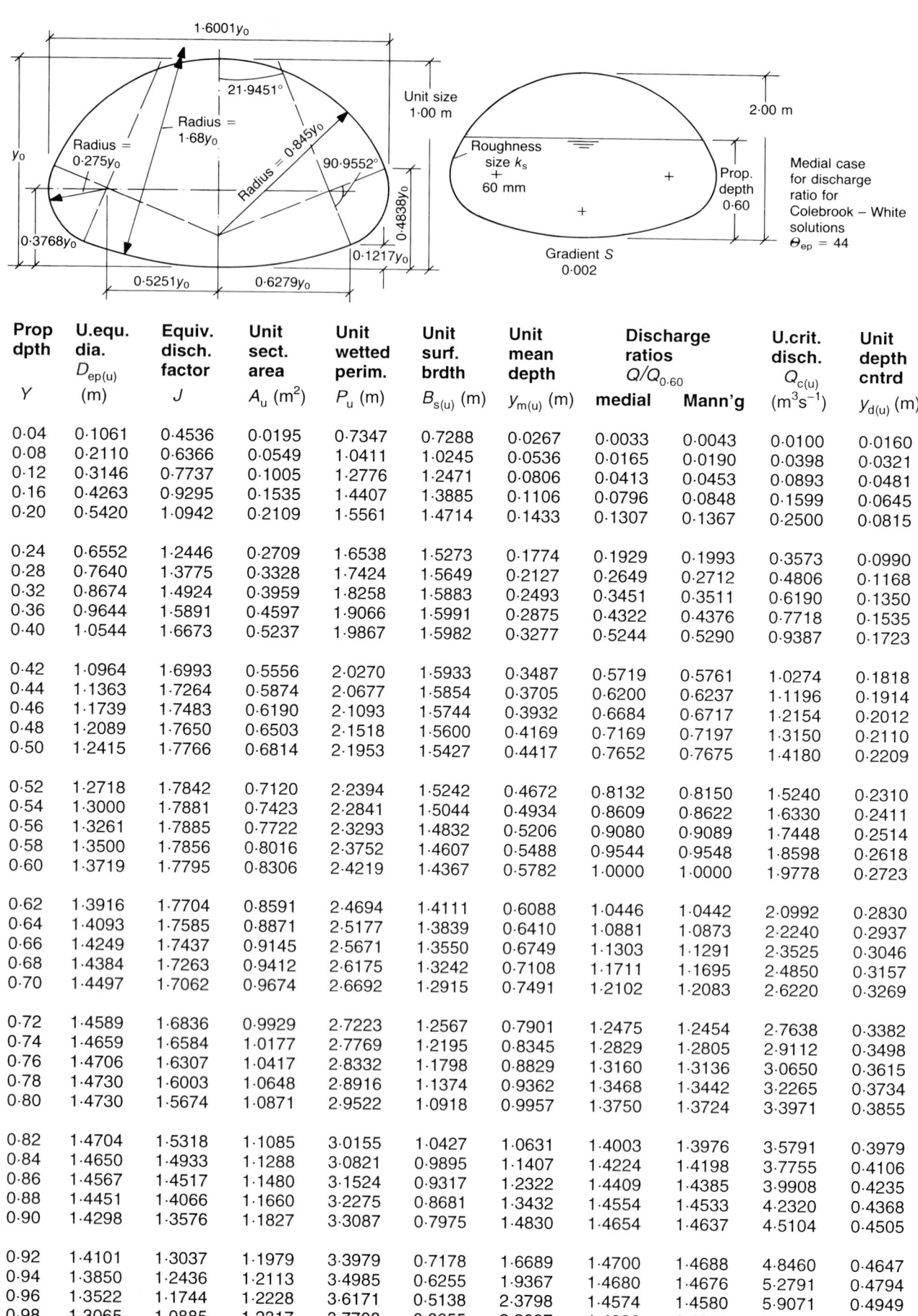

Prop dpth Y	U.equ. dia. $D_{ep(u)}$ (m)	Equiv. disch. factor J	Unit sect. area A_u (m²)	Unit wetted perim. P_u (m)	Unit surf. brdth $B_{s(u)}$ (m)	Unit mean depth $y_{m(u)}$ (m)	Discharge ratios $Q/Q_{0.60}$ medial	Mann'g	U.crit. disch. $Q_{c(u)}$ (m³s⁻¹)	Unit depth cntrd $y_{d(u)}$ (m)
0·04	0·1061	0·4536	0·0195	0·7347	0·7288	0·0267	0·0033	0·0043	0·0100	0·0160
0·08	0·2110	0·6366	0·0549	1·0411	1·0245	0·0536	0·0165	0·0190	0·0398	0·0321
0·12	0·3146	0·7737	0·1005	1·2776	1·2471	0·0806	0·0413	0·0453	0·0893	0·0481
0·16	0·4263	0·9295	0·1535	1·4407	1·3885	0·1106	0·0796	0·0848	0·1599	0·0645
0·20	0·5420	1·0942	0·2109	1·5561	1·4714	0·1433	0·1307	0·1367	0·2500	0·0815
0·24	0·6552	1·2446	0·2709	1·6538	1·5273	0·1774	0·1929	0·1993	0·3573	0·0990
0·28	0·7640	1·3775	0·3328	1·7424	1·5649	0·2127	0·2649	0·2712	0·4806	0·1168
0·32	0·8674	1·4924	0·3959	1·8258	1·5883	0·2493	0·3451	0·3511	0·6190	0·1350
0·36	0·9644	1·5891	0·4597	1·9066	1·5991	0·2875	0·4322	0·4376	0·7718	0·1535
0·40	1·0544	1·6673	0·5237	1·9867	1·5982	0·3277	0·5244	0·5290	0·9387	0·1723
0·42	1·0964	1·6993	0·5556	2·0270	1·5933	0·3487	0·5719	0·5761	1·0274	0·1818
0·44	1·1363	1·7264	0·5874	2·0677	1·5854	0·3705	0·6200	0·6237	1·1196	0·1914
0·46	1·1739	1·7483	0·6190	2·1093	1·5744	0·3932	0·6684	0·6717	1·2154	0·2012
0·48	1·2089	1·7650	0·6503	2·1518	1·5600	0·4169	0·7169	0·7197	1·3150	0·2110
0·50	1·2415	1·7766	0·6814	2·1953	1·5427	0·4417	0·7652	0·7675	1·4180	0·2209
0·52	1·2718	1·7842	0·7120	2·2394	1·5242	0·4672	0·8132	0·8150	1·5240	0·2310
0·54	1·3000	1·7881	0·7423	2·2841	1·5044	0·4934	0·8609	0·8622	1·6330	0·2411
0·56	1·3261	1·7885	0·7722	2·3293	1·4832	0·5206	0·9080	0·9089	1·7448	0·2514
0·58	1·3500	1·7856	0·8016	2·3752	1·4607	0·5488	0·9544	0·9548	1·8598	0·2618
0·60	1·3719	1·7795	0·8306	2·4219	1·4367	0·5782	1·0000	1·0000	1·9778	0·2723
0·62	1·3916	1·7704	0·8591	2·4694	1·4111	0·6088	1·0446	1·0442	2·0992	0·2830
0·64	1·4093	1·7585	0·8871	2·5177	1·3839	0·6410	1·0881	1·0873	2·2240	0·2937
0·66	1·4249	1·7437	0·9145	2·5671	1·3550	0·6749	1·1303	1·1291	2·3525	0·3046
0·68	1·4384	1·7263	0·9412	2·6175	1·3242	0·7108	1·1711	1·1695	2·4850	0·3157
0·70	1·4497	1·7062	0·9674	2·6692	1·2915	0·7491	1·2102	1·2083	2·6220	0·3269
0·72	1·4589	1·6836	0·9929	2·7223	1·2567	0·7901	1·2475	1·2454	2·7638	0·3382
0·74	1·4659	1·6584	1·0177	2·7769	1·2195	0·8345	1·2829	1·2805	2·9112	0·3498
0·76	1·4706	1·6307	1·0417	2·8332	1·1798	0·8829	1·3160	1·3136	3·0650	0·3615
0·78	1·4730	1·6003	1·0648	2·8916	1·1374	0·9362	1·3468	1·3442	3·2265	0·3734
0·80	1·4730	1·5674	1·0871	2·9522	1·0918	0·9957	1·3750	1·3724	3·3971	0·3855
0·82	1·4704	1·5318	1·1085	3·0155	1·0427	1·0631	1·4003	1·3976	3·5791	0·3979
0·84	1·4650	1·4933	1·1288	3·0821	0·9895	1·1407	1·4224	1·4198	3·7755	0·4106
0·86	1·4567	1·4517	1·1480	3·1524	0·9317	1·2322	1·4409	1·4385	3·9908	0·4235
0·88	1·4451	1·4066	1·1660	3·2275	0·8681	1·3432	1·4554	1·4533	4·2320	0·4368
0·90	1·4298	1·3576	1·1827	3·3087	0·7975	1·4830	1·4654	1·4637	4·5104	0·4505
0·92	1·4101	1·3037	1·1979	3·3979	0·7178	1·6689	1·4700	1·4688	4·8460	0·4647
0·94	1·3850	1·2436	1·2113	3·4985	0·6255	1·9367	1·4680	1·4676	5·2791	0·4794
0·96	1·3522	1·1744	1·2228	3·6171	0·5138	2·3798	1·4574	1·4580	5·9071	0·4949
0·98	1·3065	1·0885	1·2317	3·7708	0·3655	3·3697	1·4333	1·4353	7·0801	0·5112
1·00	1·1949	0·9069	1·2365	4·1392	0·0000	-	1·3520	1·3578	-	0·5292

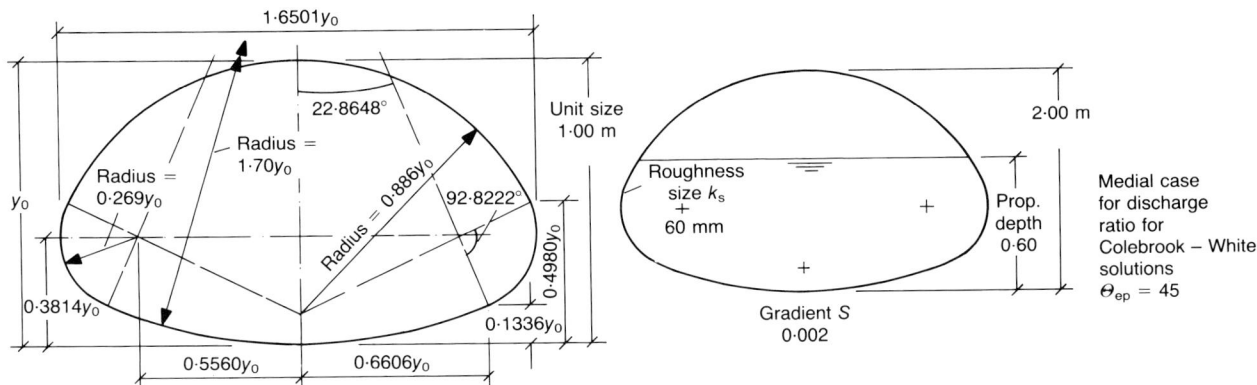

Prop dpth Y	U.equ. dia. $D_{ep(u)}$ (m)	Equiv. disch. factor J	Unit sect. area A_u (m²)	Unit wetted perim. P_u (m)	Unit surf. brdth $B_{s(u)}$ (m)	Unit mean depth $y_{m(u)}$ (m)	Discharge ratios $Q/Q_{0.60}$ medial	Mann'g	U.crit. disch. $Q_{c(u)}$ (m³s⁻¹)	Unit depth cntrd $y_{d(u)}$ (m)
0·04	0·1061	0·4509	0·0196	0·7390	0·7332	0·0267	0·0032	0·0041	0·0100	0·0160
0·08	0·2110	0·6329	0·0552	1·0472	1·0307	0·0536	0·0160	0·0185	0·0400	0·0321
0·12	0·3147	0·7693	0·1011	1·2851	1·2548	0·0806	0·0402	0·0442	0·0899	0·0481
0·16	0·4223	0·9044	0·1549	1·4670	1·4175	0·1093	0·0771	0·0823	0·1603	0·0643
0·20	0·5376	1·0629	0·2136	1·5889	1·5092	0·1415	0·1273	0·1333	0·2516	0·0811
0·24	0·6517	1·2118	0·2752	1·6893	1·5697	0·1753	0·1890	0·1953	0·3609	0·0984
0·28	0·7619	1·3452	0·3389	1·7792	1·6103	0·2104	0·2606	0·2669	0·4868	0·1162
0·32	0·8670	1·4618	0·4038	1·8632	1·6358	0·2469	0·3407	0·3467	0·6284	0·1343
0·36	0·9660	1·5609	0·4696	1·9443	1·6483	0·2849	0·4278	0·4333	0·7848	0·1527
0·40	1·0582	1·6422	0·5355	2·0244	1·6488	0·3248	0·5204	0·5251	0·9558	0·1714
0·42	1·1014	1·6759	0·5685	2·0646	1·6445	0·3457	0·5682	0·5724	1·0467	0·1809
0·44	1·1425	1·7048	0·6013	2·1053	1·6371	0·3673	0·6167	0·6205	1·1412	0·1905
0·46	1·1813	1·7287	0·6340	2·1467	1·6266	0·3897	0·6656	0·6689	1·2394	0·2001
0·48	1·2176	1·7474	0·6663	2·1890	1·6126	0·4132	0·7146	0·7174	1·3414	0·2099
0·50	1·2512	1·7605	0·6984	2·2328	1·5950	0·4379	0·7634	0·7657	1·4473	0·2198
0·52	1·2824	1·7689	0·7301	2·2775	1·5750	0·4636	0·8119	0·8137	1·5568	0·2298
0·54	1·3112	1·7734	0·7614	2·3228	1·5537	0·4901	0·8599	0·8612	1·6692	0·2400
0·56	1·3379	1·7744	0·7923	2·3687	1·5311	0·5174	0·9074	0·9082	1·7847	0·2503
0·58	1·3624	1·7719	0·8227	2·4154	1·5071	0·5459	0·9541	0·9545	1·9033	0·2606
0·60	1·3847	1·7663	0·8525	2·4628	1·4816	0·5754	1·0000	1·0000	2·0252	0·2712
0·62	1·4048	1·7575	0·8819	2·5111	1·4546	0·6063	1·0449	1·0445	2·1504	0·2818
0·64	1·4228	1·7458	0·9107	2·5603	1·4259	0·6387	1·0886	1·0878	2·2792	0·2926
0·66	1·4387	1·7313	0·9389	2·6105	1·3955	0·6728	1·1310	1·1298	2·4118	0·3035
0·68	1·4524	1·7140	0·9665	2·6619	1·3633	0·7090	1·1719	1·1704	2·5485	0·3145
0·70	1·4639	1·6941	0·9935	2·7146	1·3291	0·7475	1·2111	1·2093	2·6898	0·3257
0·72	1·4732	1·6716	1·0197	2·7686	1·2927	0·7888	1·2486	1·2465	2·8360	0·3371
0·74	1·4802	1·6465	1·0451	2·8243	1·2540	0·8335	1·2840	1·2817	2·9880	0·3486
0·76	1·4850	1·6188	1·0698	2·8818	1·2127	0·8822	1·3172	1·3147	3·1466	0·3604
0·78	1·4873	1·5885	1·0936	2·9413	1·1687	0·9358	1·3480	1·3454	3·3131	0·3723
0·80	1·4871	1·5557	1·1165	3·0032	1·1214	0·9956	1·3761	1·3735	3·4889	0·3845
0·82	1·4844	1·5200	1·1385	3·0678	1·0706	1·0634	1·4014	1·3987	3·6764	0·3969
0·84	1·4789	1·4816	1·1593	3·1358	1·0157	1·1414	1·4234	1·4208	3·8788	0·4095
0·86	1·4703	1·4400	1·1791	3·2077	0·9560	1·2333	1·4418	1·4394	4·1006	0·4225
0·88	1·4584	1·3950	1·1975	3·2844	0·8905	1·3448	1·4562	1·4541	4·3490	0·4359
0·90	1·4428	1·3460	1·2146	3·3674	0·8178	1·4853	1·4660	1·4643	4·6356	0·4496
0·92	1·4228	1·2923	1·2302	3·4586	0·7358	1·6719	1·4704	1·4693	4·9812	0·4638
0·94	1·3971	1·2324	1·2440	3·5616	0·6410	1·9407	1·4683	1·4679	5·4270	0·4785
0·96	1·3638	1·1633	1·2557	3·6829	0·5264	2·3854	1·4575	1·4581	6·0733	0·4940
0·98	1·3174	1·0777	1·2648	3·8402	0·3744	3·3784	1·4331	1·4351	7·2801	0·5103
1·00	1·2043	0·8971	1·2698	4·2174	0·0000	-	1·3514	1·3571	-	0·5283

1·70 Pipe arch (corrugated sheet metal)

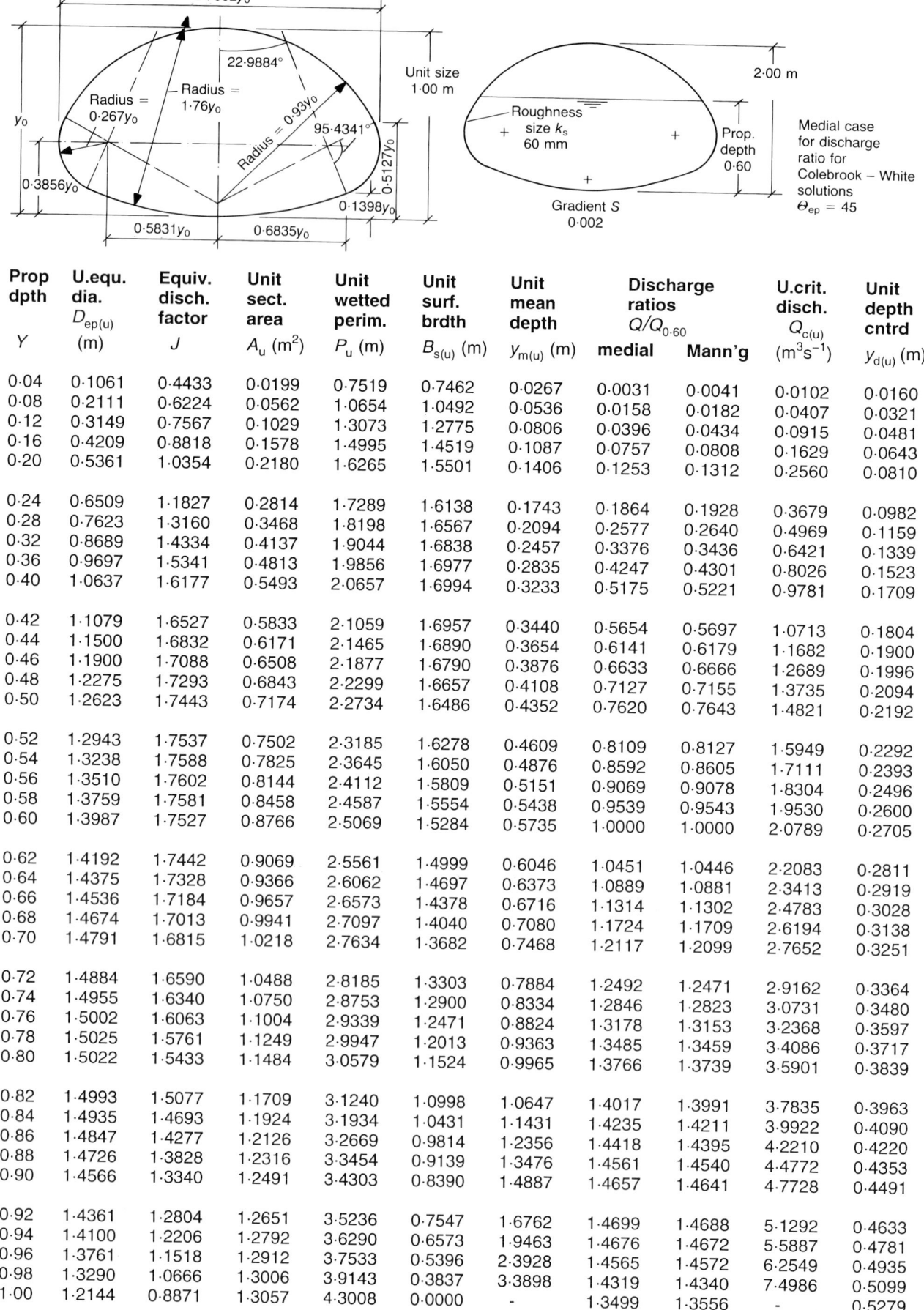

Prop dpth Y	U.equ. dia. $D_{ep(u)}$ (m)	Equiv. disch. factor J	Unit sect. area A_u (m²)	Unit wetted perim. P_u (m)	Unit surf. brdth $B_{s(u)}$ (m)	Unit mean depth $y_{m(u)}$ (m)	Discharge ratios $Q/Q_{0.60}$ medial	Mann'g	U.crit. disch. $Q_{c(u)}$ (m³s⁻¹)	Unit depth cntrd $y_{d(u)}$ (m)
0·04	0·1061	0·4433	0·0199	0·7519	0·7462	0·0267	0·0031	0·0041	0·0102	0·0160
0·08	0·2111	0·6224	0·0562	1·0654	1·0492	0·0536	0·0158	0·0182	0·0407	0·0321
0·12	0·3149	0·7567	0·1029	1·3073	1·2775	0·0806	0·0396	0·0434	0·0915	0·0481
0·16	0·4209	0·8818	0·1578	1·4995	1·4519	0·1087	0·0757	0·0808	0·1629	0·0643
0·20	0·5361	1·0354	0·2180	1·6265	1·5501	0·1406	0·1253	0·1312	0·2560	0·0810
0·24	0·6509	1·1827	0·2814	1·7289	1·6138	0·1743	0·1864	0·1928	0·3679	0·0982
0·28	0·7623	1·3160	0·3468	1·8198	1·6567	0·2094	0·2577	0·2640	0·4969	0·1159
0·32	0·8689	1·4334	0·4137	1·9044	1·6838	0·2457	0·3376	0·3436	0·6421	0·1339
0·36	0·9697	1·5341	0·4813	1·9856	1·6977	0·2835	0·4247	0·4301	0·8026	0·1523
0·40	1·0637	1·6177	0·5493	2·0657	1·6994	0·3233	0·5175	0·5221	0·9781	0·1709
0·42	1·1079	1·6527	0·5833	2·1059	1·6957	0·3440	0·5654	0·5697	1·0713	0·1804
0·44	1·1500	1·6832	0·6171	2·1465	1·6890	0·3654	0·6141	0·6179	1·1682	0·1900
0·46	1·1900	1·7088	0·6508	2·1877	1·6790	0·3876	0·6633	0·6666	1·2689	0·1996
0·48	1·2275	1·7293	0·6843	2·2299	1·6657	0·4108	0·7127	0·7155	1·3735	0·2094
0·50	1·2623	1·7443	0·7174	2·2734	1·6486	0·4352	0·7620	0·7643	1·4821	0·2192
0·52	1·2943	1·7537	0·7502	2·3185	1·6278	0·4609	0·8109	0·8127	1·5949	0·2292
0·54	1·3238	1·7588	0·7825	2·3645	1·6050	0·4876	0·8592	0·8605	1·7111	0·2393
0·56	1·3510	1·7602	0·8144	2·4112	1·5809	0·5151	0·9069	0·9078	1·8304	0·2496
0·58	1·3759	1·7581	0·8458	2·4587	1·5554	0·5438	0·9539	0·9543	1·9530	0·2600
0·60	1·3987	1·7527	0·8766	2·5069	1·5284	0·5735	1·0000	1·0000	2·0789	0·2705
0·62	1·4192	1·7442	0·9069	2·5561	1·4999	0·6046	1·0451	1·0446	2·2083	0·2811
0·64	1·4375	1·7328	0·9366	2·6062	1·4697	0·6373	1·0889	1·0881	2·3413	0·2919
0·66	1·4536	1·7184	0·9657	2·6573	1·4378	0·6716	1·1314	1·1302	2·4783	0·3028
0·68	1·4674	1·7013	0·9941	2·7097	1·4040	0·7080	1·1724	1·1709	2·6194	0·3138
0·70	1·4791	1·6815	1·0218	2·7634	1·3682	0·7468	1·2117	1·2099	2·7652	0·3251
0·72	1·4884	1·6590	1·0488	2·8185	1·3303	0·7884	1·2492	1·2471	2·9162	0·3364
0·74	1·4955	1·6340	1·0750	2·8753	1·2900	0·8334	1·2846	1·2823	3·0731	0·3480
0·76	1·5002	1·6063	1·1004	2·9339	1·2471	0·8824	1·3178	1·3153	3·2368	0·3597
0·78	1·5025	1·5761	1·1249	2·9947	1·2013	0·9363	1·3485	1·3459	3·4086	0·3717
0·80	1·5022	1·5433	1·1484	3·0579	1·1524	0·9965	1·3766	1·3739	3·5901	0·3839
0·82	1·4993	1·5077	1·1709	3·1240	1·0998	1·0647	1·4017	1·3991	3·7835	0·3963
0·84	1·4935	1·4693	1·1924	3·1934	1·0431	1·1431	1·4235	1·4211	3·9922	0·4090
0·86	1·4847	1·4277	1·2126	3·2669	0·9814	1·2356	1·4418	1·4395	4·2210	0·4220
0·88	1·4726	1·3828	1·2316	3·3454	0·9139	1·3476	1·4561	1·4540	4·4772	0·4353
0·90	1·4566	1·3340	1·2491	3·4303	0·8390	1·4887	1·4657	1·4641	4·7728	0·4491
0·92	1·4361	1·2804	1·2651	3·5236	0·7547	1·6762	1·4699	1·4688	5·1292	0·4633
0·94	1·4100	1·2206	1·2792	3·6290	0·6573	1·9463	1·4676	1·4672	5·5887	0·4781
0·96	1·3761	1·1518	1·2912	3·7533	0·5396	2·3928	1·4565	1·4572	6·2549	0·4935
0·98	1·3290	1·0666	1·3006	3·9143	0·3837	3·3898	1·4319	1·4340	7·4986	0·5099
1·00	1·2144	0·8871	1·3057	4·3008	0·0000	-	1·3499	1·3556	-	0·5279

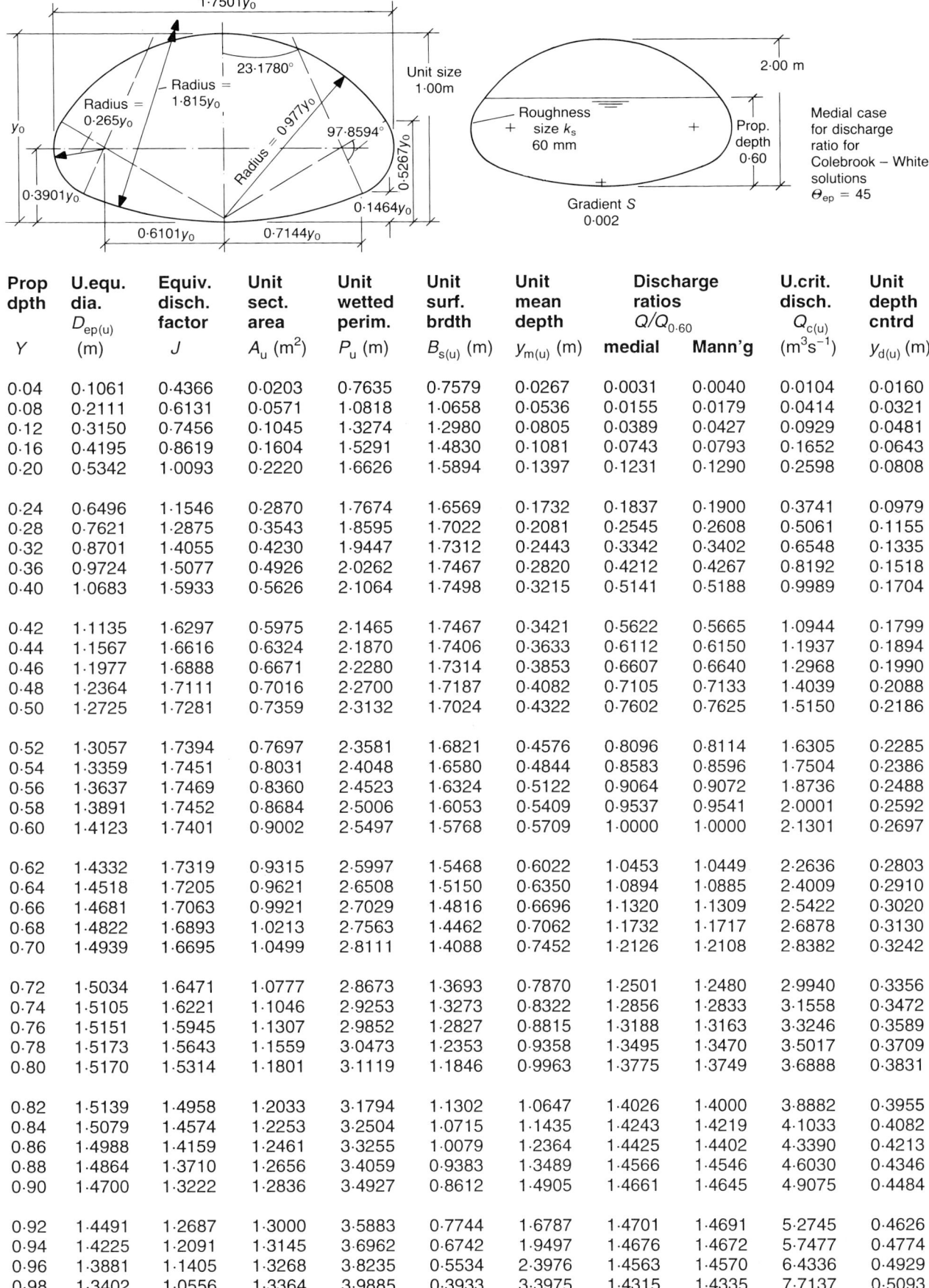

Prop dpth	U.equ. dia. $D_{ep(u)}$	Equiv. disch. factor	Unit sect. area	Unit wetted perim.	Unit surf. brdth	Unit mean depth	Discharge ratios $Q/Q_{0.60}$		U.crit. disch. $Q_{c(u)}$	Unit depth cntrd
Y	(m)	J	A_u (m²)	P_u (m)	$B_{s(u)}$ (m)	$y_{m(u)}$ (m)	medial	Mann'g	(m³s⁻¹)	$y_{d(u)}$ (m)
0·04	0·1061	0·4366	0·0203	0·7635	0·7579	0·0267	0·0031	0·0040	0·0104	0·0160
0·08	0·2111	0·6131	0·0571	1·0818	1·0658	0·0536	0·0155	0·0179	0·0414	0·0321
0·12	0·3150	0·7456	0·1045	1·3274	1·2980	0·0805	0·0389	0·0427	0·0929	0·0481
0·16	0·4195	0·8619	0·1604	1·5291	1·4830	0·1081	0·0743	0·0793	0·1652	0·0643
0·20	0·5342	1·0093	0·2220	1·6626	1·5894	0·1397	0·1231	0·1290	0·2598	0·0808
0·24	0·6496	1·1546	0·2870	1·7674	1·6569	0·1732	0·1837	0·1900	0·3741	0·0979
0·28	0·7621	1·2875	0·3543	1·8595	1·7022	0·2081	0·2545	0·2608	0·5061	0·1155
0·32	0·8701	1·4055	0·4230	1·9447	1·7312	0·2443	0·3342	0·3402	0·6548	0·1335
0·36	0·9724	1·5077	0·4926	2·0262	1·7467	0·2820	0·4212	0·4267	0·8192	0·1518
0·40	1·0683	1·5933	0·5626	2·1064	1·7498	0·3215	0·5141	0·5188	0·9989	0·1704
0·42	1·1135	1·6297	0·5975	2·1465	1·7467	0·3421	0·5622	0·5665	1·0944	0·1799
0·44	1·1567	1·6616	0·6324	2·1870	1·7406	0·3633	0·6112	0·6150	1·1937	0·1894
0·46	1·1977	1·6888	0·6671	2·2280	1·7314	0·3853	0·6607	0·6640	1·2968	0·1990
0·48	1·2364	1·7111	0·7016	2·2700	1·7187	0·4082	0·7105	0·7133	1·4039	0·2088
0·50	1·2725	1·7281	0·7359	2·3132	1·7024	0·4322	0·7602	0·7625	1·5150	0·2186
0·52	1·3057	1·7394	0·7697	2·3581	1·6821	0·4576	0·8096	0·8114	1·6305	0·2285
0·54	1·3359	1·7451	0·8031	2·4048	1·6580	0·4844	0·8583	0·8596	1·7504	0·2386
0·56	1·3637	1·7469	0·8360	2·4523	1·6324	0·5122	0·9064	0·9072	1·8736	0·2488
0·58	1·3891	1·7452	0·8684	2·5006	1·6053	0·5409	0·9537	0·9541	2·0001	0·2592
0·60	1·4123	1·7401	0·9002	2·5497	1·5768	0·5709	1·0000	1·0000	2·1301	0·2697
0·62	1·4332	1·7319	0·9315	2·5997	1·5468	0·6022	1·0453	1·0449	2·2636	0·2803
0·64	1·4518	1·7205	0·9621	2·6508	1·5150	0·6350	1·0894	1·0885	2·4009	0·2910
0·66	1·4681	1·7063	0·9921	2·7029	1·4816	0·6696	1·1320	1·1309	2·5422	0·3020
0·68	1·4822	1·6893	1·0213	2·7563	1·4462	0·7062	1·1732	1·1717	2·6878	0·3130
0·70	1·4939	1·6695	1·0499	2·8111	1·4088	0·7452	1·2126	1·2108	2·8382	0·3242
0·72	1·5034	1·6471	1·0777	2·8673	1·3693	0·7870	1·2501	1·2480	2·9940	0·3356
0·74	1·5105	1·6221	1·1046	2·9253	1·3273	0·8322	1·2856	1·2833	3·1558	0·3472
0·76	1·5151	1·5945	1·1307	2·9852	1·2827	0·8815	1·3188	1·3163	3·3246	0·3589
0·78	1·5173	1·5643	1·1559	3·0473	1·2353	0·9358	1·3495	1·3470	3·5017	0·3709
0·80	1·5170	1·5314	1·1801	3·1119	1·1846	0·9963	1·3775	1·3749	3·6888	0·3831
0·82	1·5139	1·4958	1·2033	3·1794	1·1302	1·0647	1·4026	1·4000	3·8882	0·3955
0·84	1·5079	1·4574	1·2253	3·2504	1·0715	1·1435	1·4243	1·4219	4·1033	0·4082
0·86	1·4988	1·4159	1·2461	3·3255	1·0079	1·2364	1·4425	1·4402	4·3390	0·4213
0·88	1·4864	1·3710	1·2656	3·4059	0·9383	1·3489	1·4566	1·4546	4·6030	0·4346
0·90	1·4700	1·3222	1·2836	3·4927	0·8612	1·4905	1·4661	1·4645	4·9075	0·4484
0·92	1·4491	1·2687	1·3000	3·5883	0·7744	1·6787	1·4701	1·4691	5·2745	0·4626
0·94	1·4225	1·2091	1·3145	3·6962	0·6742	1·9497	1·4676	1·4672	5·7477	0·4774
0·96	1·3881	1·1405	1·3268	3·8235	0·5534	2·3976	1·4563	1·4570	6·4336	0·4929
0·98	1·3402	1·0556	1·3364	3·9885	0·3933	3·3975	1·4315	1·4335	7·7137	0·5093
1·00	1·2240	0·8770	1·3416	4·3845	0·0000	-	1·3489	1·3547	-	0·5273

1·35 Horizontal ellipse (corrugated sheet metal)

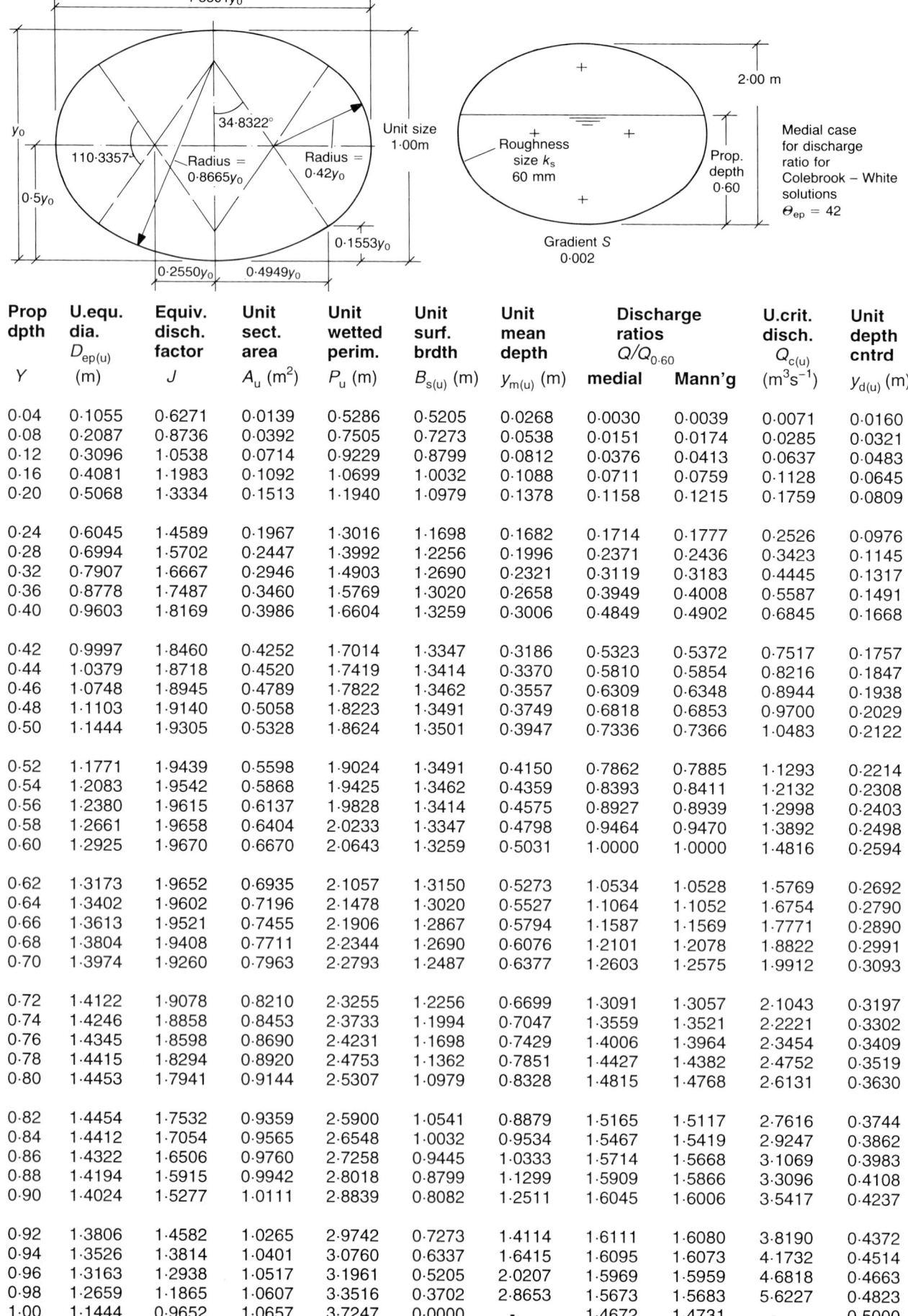

Prop dpth Y	U.equ. dia. $D_{ep(u)}$ (m)	Equiv. disch. factor J	Unit sect. area A_u (m²)	Unit wetted perim. P_u (m)	Unit surf. brdth $B_{s(u)}$ (m)	Unit mean depth $y_{m(u)}$ (m)	Discharge ratios $Q/Q_{0.60}$ medial	Mann'g	U.crit. disch. $Q_{c(u)}$ (m³s⁻¹)	Unit depth cntrd $y_{d(u)}$ (m)
0·04	0·1055	0·6271	0·0139	0·5286	0·5205	0·0268	0·0030	0·0039	0·0071	0·0160
0·08	0·2087	0·8736	0·0392	0·7505	0·7273	0·0538	0·0151	0·0174	0·0285	0·0321
0·12	0·3096	1·0538	0·0714	0·9229	0·8799	0·0812	0·0376	0·0413	0·0637	0·0483
0·16	0·4081	1·1983	0·1092	1·0699	1·0032	0·1088	0·0711	0·0759	0·1128	0·0645
0·20	0·5068	1·3334	0·1513	1·1940	1·0979	0·1378	0·1158	0·1215	0·1759	0·0809
0·24	0·6045	1·4589	0·1967	1·3016	1·1698	0·1682	0·1714	0·1777	0·2526	0·0976
0·28	0·6994	1·5702	0·2447	1·3992	1·2256	0·1996	0·2371	0·2436	0·3423	0·1145
0·32	0·7907	1·6667	0·2946	1·4903	1·2690	0·2321	0·3119	0·3183	0·4445	0·1317
0·36	0·8778	1·7487	0·3460	1·5769	1·3020	0·2658	0·3949	0·4008	0·5587	0·1491
0·40	0·9603	1·8169	0·3986	1·6604	1·3259	0·3006	0·4849	0·4902	0·6845	0·1668
0·42	0·9997	1·8460	0·4252	1·7014	1·3347	0·3186	0·5323	0·5372	0·7517	0·1757
0·44	1·0379	1·8718	0·4520	1·7419	1·3414	0·3370	0·5810	0·5854	0·8216	0·1847
0·46	1·0748	1·8945	0·4789	1·7822	1·3462	0·3557	0·6309	0·6348	0·8944	0·1938
0·48	1·1103	1·9140	0·5058	1·8223	1·3491	0·3749	0·6818	0·6853	0·9700	0·2029
0·50	1·1444	1·9305	0·5328	1·8624	1·3501	0·3947	0·7336	0·7366	1·0483	0·2122
0·52	1·1771	1·9439	0·5598	1·9024	1·3491	0·4150	0·7862	0·7885	1·1293	0·2214
0·54	1·2083	1·9542	0·5868	1·9425	1·3462	0·4359	0·8393	0·8411	1·2132	0·2308
0·56	1·2380	1·9615	0·6137	1·9828	1·3414	0·4575	0·8927	0·8939	1·2998	0·2403
0·58	1·2661	1·9658	0·6404	2·0233	1·3347	0·4798	0·9464	0·9470	1·3892	0·2498
0·60	1·2925	1·9670	0·6670	2·0643	1·3259	0·5031	1·0000	1·0000	1·4816	0·2594
0·62	1·3173	1·9652	0·6935	2·1057	1·3150	0·5273	1·0534	1·0528	1·5769	0·2692
0·64	1·3402	1·9602	0·7196	2·1478	1·3020	0·5527	1·1064	1·1052	1·6754	0·2790
0·66	1·3613	1·9521	0·7455	2·1906	1·2867	0·5794	1·1587	1·1569	1·7771	0·2890
0·68	1·3804	1·9408	0·7711	2·2344	1·2690	0·6076	1·2101	1·2078	1·8822	0·2991
0·70	1·3974	1·9260	0·7963	2·2793	1·2487	0·6377	1·2603	1·2575	1·9912	0·3093
0·72	1·4122	1·9078	0·8210	2·3255	1·2256	0·6699	1·3091	1·3057	2·1043	0·3197
0·74	1·4246	1·8858	0·8453	2·3733	1·1994	0·7047	1·3559	1·3521	2·2221	0·3302
0·76	1·4345	1·8598	0·8690	2·4231	1·1698	0·7429	1·4006	1·3964	2·3454	0·3409
0·78	1·4415	1·8294	0·8920	2·4753	1·1362	0·7851	1·4427	1·4382	2·4752	0·3519
0·80	1·4453	1·7941	0·9144	2·5307	1·0979	0·8328	1·4815	1·4768	2·6131	0·3630
0·82	1·4454	1·7532	0·9359	2·5900	1·0541	0·8879	1·5165	1·5117	2·7616	0·3744
0·84	1·4412	1·7054	0·9565	2·6548	1·0032	0·9534	1·5467	1·5419	2·9247	0·3862
0·86	1·4322	1·6506	0·9760	2·7258	0·9445	1·0333	1·5714	1·5668	3·1069	0·3983
0·88	1·4194	1·5915	0·9942	2·8018	0·8799	1·1299	1·5909	1·5866	3·3096	0·4108
0·90	1·4024	1·5277	1·0111	2·8839	0·8082	1·2511	1·6045	1·6006	3·5417	0·4237
0·92	1·3806	1·4582	1·0265	2·9742	0·7273	1·4114	1·6111	1·6080	3·8190	0·4372
0·94	1·3526	1·3814	1·0401	3·0760	0·6337	1·6415	1·6095	1·6073	4·1732	0·4514
0·96	1·3163	1·2938	1·0517	3·1961	0·5205	2·0207	1·5969	1·5959	4·6818	0·4663
0·98	1·2659	1·1865	1·0607	3·3516	0·3702	2·8653	1·5673	1·5683	5·6227	0·4823
1·00	1·1444	0·9652	1·0657	3·7247	0·0000	-	1·4672	1·4731	-	0·5000

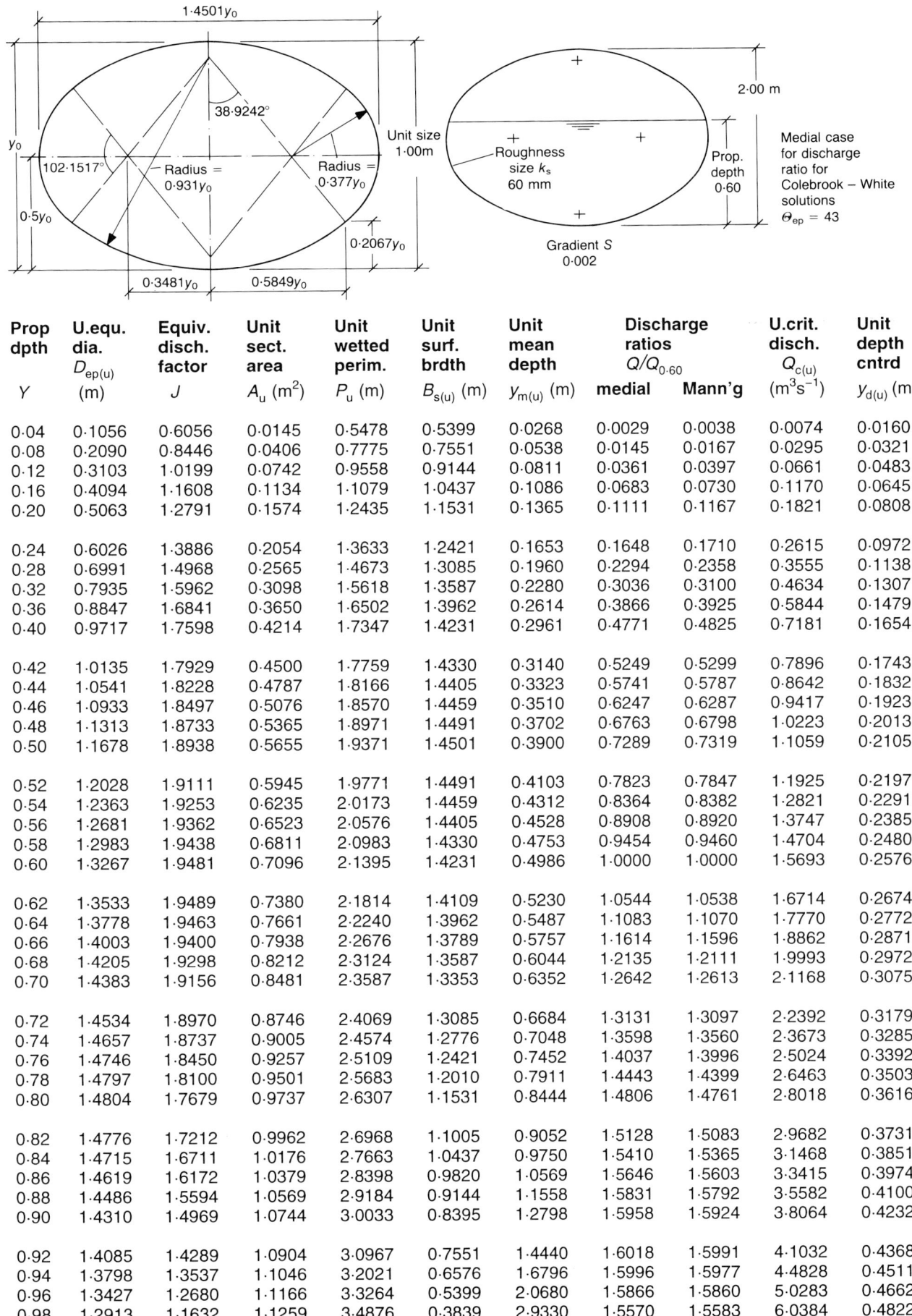

Prop dpth	U.equ. dia. $D_{ep(u)}$	Equiv. disch. factor	Unit sect. area	Unit wetted perim.	Unit surf. brdth	Unit mean depth	Discharge ratios $Q/Q_{0.60}$		U.crit. disch. $Q_{c(u)}$	Unit depth cntrd
Y	(m)	J	A_u (m²)	P_u (m)	$B_{s(u)}$ (m)	$y_{m(u)}$ (m)	medial	Mann'g	(m³s⁻¹)	$y_{d(u)}$ (m)
0·04	0·1056	0·6056	0·0145	0·5478	0·5399	0·0268	0·0029	0·0038	0·0074	0·0160
0·08	0·2090	0·8446	0·0406	0·7775	0·7551	0·0538	0·0145	0·0167	0·0295	0·0321
0·12	0·3103	1·0199	0·0742	0·9558	0·9144	0·0811	0·0361	0·0397	0·0661	0·0483
0·16	0·4094	1·1608	0·1134	1·1079	1·0437	0·1086	0·0683	0·0730	0·1170	0·0645
0·20	0·5063	1·2791	0·1574	1·2435	1·1531	0·1365	0·1111	0·1167	0·1821	0·0808
0·24	0·6026	1·3886	0·2054	1·3633	1·2421	0·1653	0·1648	0·1710	0·2615	0·0972
0·28	0·6991	1·4968	0·2565	1·4673	1·3085	0·1960	0·2294	0·2358	0·3555	0·1138
0·32	0·7935	1·5962	0·3098	1·5618	1·3587	0·2280	0·3036	0·3100	0·4634	0·1307
0·36	0·8847	1·6841	0·3650	1·6502	1·3962	0·2614	0·3866	0·3925	0·5844	0·1479
0·40	0·9717	1·7598	0·4214	1·7347	1·4231	0·2961	0·4771	0·4825	0·7181	0·1654
0·42	1·0135	1·7929	0·4500	1·7759	1·4330	0·3140	0·5249	0·5299	0·7896	0·1743
0·44	1·0541	1·8228	0·4787	1·8166	1·4405	0·3323	0·5741	0·5787	0·8642	0·1832
0·46	1·0933	1·8497	0·5076	1·8570	1·4459	0·3510	0·6247	0·6287	0·9417	0·1923
0·48	1·1313	1·8733	0·5365	1·8971	1·4491	0·3702	0·6763	0·6798	1·0223	0·2013
0·50	1·1678	1·8938	0·5655	1·9371	1·4501	0·3900	0·7289	0·7319	1·1059	0·2105
0·52	1·2028	1·9111	0·5945	1·9771	1·4491	0·4103	0·7823	0·7847	1·1925	0·2197
0·54	1·2363	1·9253	0·6235	2·0173	1·4459	0·4312	0·8364	0·8382	1·2821	0·2291
0·56	1·2681	1·9362	0·6523	2·0576	1·4405	0·4528	0·8908	0·8920	1·3747	0·2385
0·58	1·2983	1·9438	0·6811	2·0983	1·4330	0·4753	0·9454	0·9460	1·4704	0·2480
0·60	1·3267	1·9481	0·7096	2·1395	1·4231	0·4986	1·0000	1·0000	1·5693	0·2576
0·62	1·3533	1·9489	0·7380	2·1814	1·4109	0·5230	1·0544	1·0538	1·6714	0·2674
0·64	1·3778	1·9463	0·7661	2·2240	1·3962	0·5487	1·1083	1·1070	1·7770	0·2772
0·66	1·4003	1·9400	0·7938	2·2676	1·3789	0·5757	1·1614	1·1596	1·8862	0·2871
0·68	1·4205	1·9298	0·8212	2·3124	1·3587	0·6044	1·2135	1·2111	1·9993	0·2972
0·70	1·4383	1·9156	0·8481	2·3587	1·3353	0·6352	1·2642	1·2613	2·1168	0·3075
0·72	1·4534	1·8970	0·8746	2·4069	1·3085	0·6684	1·3131	1·3097	2·2392	0·3179
0·74	1·4657	1·8737	0·9005	2·4574	1·2776	0·7048	1·3598	1·3560	2·3673	0·3285
0·76	1·4746	1·8450	0·9257	2·5109	1·2421	0·7452	1·4037	1·3996	2·5024	0·3392
0·78	1·4797	1·8100	0·9501	2·5683	1·2010	0·7911	1·4443	1·4399	2·6463	0·3503
0·80	1·4804	1·7679	0·9737	2·6307	1·1531	0·8444	1·4806	1·4761	2·8018	0·3616
0·82	1·4776	1·7212	0·9962	2·6968	1·1005	0·9052	1·5128	1·5083	2·9682	0·3731
0·84	1·4715	1·6711	1·0176	2·7663	1·0437	0·9750	1·5410	1·5365	3·1468	0·3851
0·86	1·4619	1·6172	1·0379	2·8398	0·9820	1·0569	1·5646	1·5603	3·3415	0·3974
0·88	1·4486	1·5594	1·0569	2·9184	0·9144	1·1558	1·5831	1·5792	3·5582	0·4100
0·90	1·4310	1·4969	1·0744	3·0033	0·8395	1·2798	1·5958	1·5924	3·8064	0·4232
0·92	1·4085	1·4289	1·0904	3·0967	0·7551	1·4440	1·6018	1·5991	4·1032	0·4368
0·94	1·3798	1·3537	1·1046	3·2021	0·6576	1·6796	1·5996	1·5977	4·4828	0·4511
0·96	1·3427	1·2680	1·1166	3·3264	0·5399	2·0680	1·5866	1·5860	5·0283	0·4662
0·98	1·2913	1·1632	1·1259	3·4876	0·3839	2·9330	1·5570	1·5583	6·0384	0·4822
1·00	1·1678	0·9469	1·1310	3·8742	0·0000	-	1·4579	1·4638	-	0·5000

1·55 Horizontal ellipse (corrugated sheet metal)

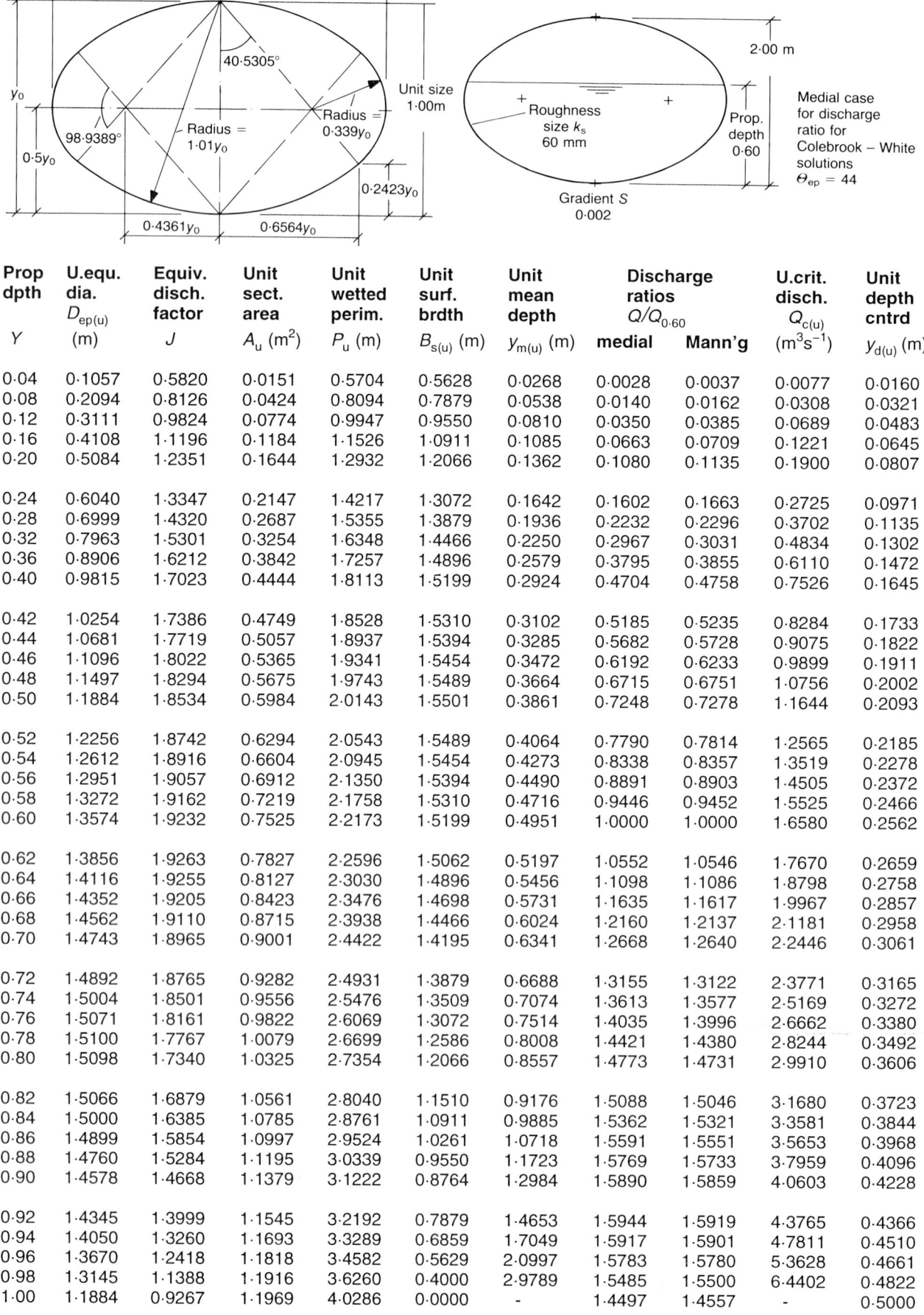

Prop dpth Y	U.equ. dia. $D_{ep(u)}$ (m)	Equiv. disch. factor J	Unit sect. area A_u (m²)	Unit wetted perim. P_u (m)	Unit surf. brdth $B_{s(u)}$ (m)	Unit mean depth $y_{m(u)}$ (m)	Discharge ratios $Q/Q_{0.60}$ medial	Mann'g	U.crit. disch. $Q_{c(u)}$ (m³s⁻¹)	Unit depth cntrd $y_{d(u)}$ (m)
0·04	0·1057	0·5820	0·0151	0·5704	0·5628	0·0268	0·0028	0·0037	0·0077	0·0160
0·08	0·2094	0·8126	0·0424	0·8094	0·7879	0·0538	0·0140	0·0162	0·0308	0·0321
0·12	0·3111	0·9824	0·0774	0·9947	0·9550	0·0810	0·0350	0·0385	0·0689	0·0483
0·16	0·4108	1·1196	0·1184	1·1526	1·0911	0·1085	0·0663	0·0709	0·1221	0·0645
0·20	0·5084	1·2351	0·1644	1·2932	1·2066	0·1362	0·1080	0·1135	0·1900	0·0807
0·24	0·6040	1·3347	0·2147	1·4217	1·3072	0·1642	0·1602	0·1663	0·2725	0·0971
0·28	0·6999	1·4320	0·2687	1·5355	1·3879	0·1936	0·2232	0·2296	0·3702	0·1135
0·32	0·7963	1·5301	0·3254	1·6348	1·4466	0·2250	0·2967	0·3031	0·4834	0·1302
0·36	0·8906	1·6212	0·3842	1·7257	1·4896	0·2579	0·3795	0·3855	0·6110	0·1472
0·40	0·9815	1·7023	0·4444	1·8113	1·5199	0·2924	0·4704	0·4758	0·7526	0·1645
0·42	1·0254	1·7386	0·4749	1·8528	1·5310	0·3102	0·5185	0·5235	0·8284	0·1733
0·44	1·0681	1·7719	0·5057	1·8937	1·5394	0·3285	0·5682	0·5728	0·9075	0·1822
0·46	1·1096	1·8022	0·5365	1·9341	1·5454	0·3472	0·6192	0·6233	0·9899	0·1911
0·48	1·1497	1·8294	0·5675	1·9743	1·5489	0·3664	0·6715	0·6751	1·0756	0·2002
0·50	1·1884	1·8534	0·5984	2·0143	1·5501	0·3861	0·7248	0·7278	1·1644	0·2093
0·52	1·2256	1·8742	0·6294	2·0543	1·5489	0·4064	0·7790	0·7814	1·2565	0·2185
0·54	1·2612	1·8916	0·6604	2·0945	1·5454	0·4273	0·8338	0·8357	1·3519	0·2278
0·56	1·2951	1·9057	0·6912	2·1350	1·5394	0·4490	0·8891	0·8903	1·4505	0·2372
0·58	1·3272	1·9162	0·7219	2·1758	1·5310	0·4716	0·9446	0·9452	1·5525	0·2466
0·60	1·3574	1·9232	0·7525	2·2173	1·5199	0·4951	1·0000	1·0000	1·6580	0·2562
0·62	1·3856	1·9263	0·7827	2·2596	1·5062	0·5197	1·0552	1·0546	1·7670	0·2659
0·64	1·4116	1·9255	0·8127	2·3030	1·4896	0·5456	1·1098	1·1086	1·8798	0·2758
0·66	1·4352	1·9205	0·8423	2·3476	1·4698	0·5731	1·1635	1·1617	1·9967	0·2857
0·68	1·4562	1·9110	0·8715	2·3938	1·4466	0·6024	1·2160	1·2137	2·1181	0·2958
0·70	1·4743	1·8965	0·9001	2·4422	1·4195	0·6341	1·2668	1·2640	2·2446	0·3061
0·72	1·4892	1·8765	0·9282	2·4931	1·3879	0·6688	1·3155	1·3122	2·3771	0·3165
0·74	1·5004	1·8501	0·9556	2·5476	1·3509	0·7074	1·3613	1·3577	2·5169	0·3272
0·76	1·5071	1·8161	0·9822	2·6069	1·3072	0·7514	1·4035	1·3996	2·6662	0·3380
0·78	1·5100	1·7767	1·0079	2·6699	1·2586	0·8008	1·4421	1·4380	2·8244	0·3492
0·80	1·5098	1·7340	1·0325	2·7354	1·2066	0·8557	1·4773	1·4731	2·9910	0·3606
0·82	1·5066	1·6879	1·0561	2·8040	1·1510	0·9176	1·5088	1·5046	3·1680	0·3723
0·84	1·5000	1·6385	1·0785	2·8761	1·0911	0·9885	1·5362	1·5321	3·3581	0·3844
0·86	1·4899	1·5854	1·0997	2·9524	1·0261	1·0718	1·5591	1·5551	3·5653	0·3968
0·88	1·4760	1·5284	1·1195	3·0339	0·9550	1·1723	1·5769	1·5733	3·7959	0·4096
0·90	1·4578	1·4668	1·1379	3·1222	0·8764	1·2984	1·5890	1·5859	4·0603	0·4228
0·92	1·4345	1·3999	1·1545	3·2192	0·7879	1·4653	1·5944	1·5919	4·3765	0·4366
0·94	1·4050	1·3260	1·1693	3·3289	0·6859	1·7049	1·5917	1·5901	4·7811	0·4510
0·96	1·3670	1·2418	1·1818	3·4582	0·5629	2·0997	1·5783	1·5780	5·3628	0·4661
0·98	1·3145	1·1388	1·1916	3·6260	0·4000	2·9789	1·5485	1·5500	6·4402	0·4822
1·00	1·1884	0·9267	1·1969	4·0286	0·0000	-	1·4497	1·4557	-	0·5000

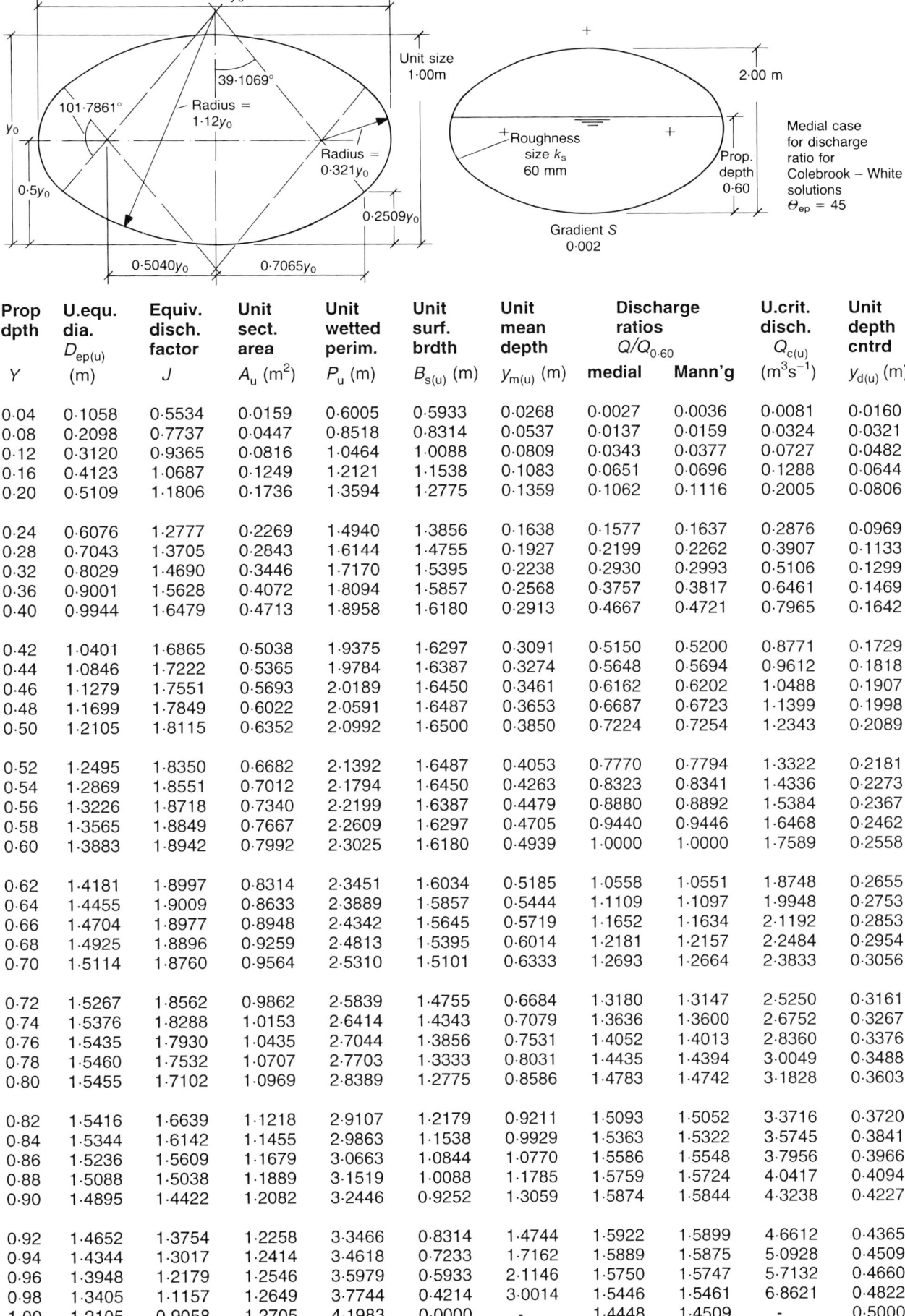

Prop dpth Y	U.equ. dia. $D_{ep(u)}$ (m)	Equiv. disch. factor J	Unit sect. area A_u (m²)	Unit wetted perim. P_u (m)	Unit surf. brdth $B_{s(u)}$ (m)	Unit mean depth $y_{m(u)}$ (m)	Discharge ratios $Q/Q_{0·60}$ medial	Mann'g	U.crit. disch. $Q_{c(u)}$ (m³s⁻¹)	Unit depth cntrd $y_{d(u)}$ (m)
0·04	0·1058	0·5534	0·0159	0·6005	0·5933	0·0268	0·0027	0·0036	0·0081	0·0160
0·08	0·2098	0·7737	0·0447	0·8518	0·8314	0·0537	0·0137	0·0159	0·0324	0·0321
0·12	0·3120	0·9365	0·0816	1·0464	1·0088	0·0809	0·0343	0·0377	0·0727	0·0482
0·16	0·4123	1·0687	0·1249	1·2121	1·1538	0·1083	0·0651	0·0696	0·1288	0·0644
0·20	0·5109	1·1806	0·1736	1·3594	1·2775	0·1359	0·1062	0·1116	0·2005	0·0806
0·24	0·6076	1·2777	0·2269	1·4940	1·3856	0·1638	0·1577	0·1637	0·2876	0·0969
0·28	0·7043	1·3705	0·2843	1·6144	1·4755	0·1927	0·2199	0·2262	0·3907	0·1133
0·32	0·8029	1·4690	0·3446	1·7170	1·5395	0·2238	0·2930	0·2993	0·5106	0·1299
0·36	0·9001	1·5628	0·4072	1·8094	1·5857	0·2568	0·3757	0·3817	0·6461	0·1469
0·40	0·9944	1·6479	0·4713	1·8958	1·6180	0·2913	0·4667	0·4721	0·7965	0·1642
0·42	1·0401	1·6865	0·5038	1·9375	1·6297	0·3091	0·5150	0·5200	0·8771	0·1729
0·44	1·0846	1·7222	0·5365	1·9784	1·6387	0·3274	0·5648	0·5694	0·9612	0·1818
0·46	1·1279	1·7551	0·5693	2·0189	1·6450	0·3461	0·6162	0·6202	1·0488	0·1907
0·48	1·1699	1·7849	0·6022	2·0591	1·6487	0·3653	0·6687	0·6723	1·1399	0·1998
0·50	1·2105	1·8115	0·6352	2·0992	1·6500	0·3850	0·7224	0·7254	1·2343	0·2089
0·52	1·2495	1·8350	0·6682	2·1392	1·6487	0·4053	0·7770	0·7794	1·3322	0·2181
0·54	1·2869	1·8551	0·7012	2·1794	1·6450	0·4263	0·8323	0·8341	1·4336	0·2273
0·56	1·3226	1·8718	0·7340	2·2199	1·6387	0·4479	0·8880	0·8892	1·5384	0·2367
0·58	1·3565	1·8849	0·7667	2·2609	1·6297	0·4705	0·9440	0·9446	1·6468	0·2462
0·60	1·3883	1·8942	0·7992	2·3025	1·6180	0·4939	1·0000	1·0000	1·7589	0·2558
0·62	1·4181	1·8997	0·8314	2·3451	1·6034	0·5185	1·0558	1·0551	1·8748	0·2655
0·64	1·4455	1·9009	0·8633	2·3889	1·5857	0·5444	1·1109	1·1097	1·9948	0·2753
0·66	1·4704	1·8977	0·8948	2·4342	1·5645	0·5719	1·1652	1·1634	2·1192	0·2853
0·68	1·4925	1·8896	0·9259	2·4813	1·5395	0·6014	1·2181	1·2157	2·2484	0·2954
0·70	1·5114	1·8760	0·9564	2·5310	1·5101	0·6333	1·2693	1·2664	2·3833	0·3056
0·72	1·5267	1·8562	0·9862	2·5839	1·4755	0·6684	1·3180	1·3147	2·5250	0·3161
0·74	1·5376	1·8288	1·0153	2·6414	1·4343	0·7079	1·3636	1·3600	2·6752	0·3267
0·76	1·5435	1·7930	1·0435	2·7044	1·3856	0·7531	1·4052	1·4013	2·8360	0·3376
0·78	1·5460	1·7532	1·0707	2·7703	1·3333	0·8031	1·4435	1·4394	3·0049	0·3488
0·80	1·5455	1·7102	1·0969	2·8389	1·2775	0·8586	1·4783	1·4742	3·1828	0·3603
0·82	1·5416	1·6639	1·1218	2·9107	1·2179	0·9211	1·5093	1·5052	3·3716	0·3720
0·84	1·5344	1·6142	1·1455	2·9863	1·1538	0·9929	1·5363	1·5322	3·5745	0·3841
0·86	1·5236	1·5609	1·1679	3·0663	1·0844	1·0770	1·5586	1·5548	3·7956	0·3966
0·88	1·5088	1·5038	1·1889	3·1519	1·0088	1·1785	1·5759	1·5724	4·0417	0·4094
0·90	1·4895	1·4422	1·2082	3·2446	0·9252	1·3059	1·5874	1·5844	4·3238	0·4227
0·92	1·4652	1·3754	1·2258	3·3466	0·8314	1·4744	1·5922	1·5899	4·6612	0·4365
0·94	1·4344	1·3017	1·2414	3·4618	0·7233	1·7162	1·5889	1·5875	5·0928	0·4509
0·96	1·3948	1·2179	1·2546	3·5979	0·5933	2·1146	1·5750	1·5747	5·7132	0·4660
0·98	1·3405	1·1157	1·2649	3·7744	0·4214	3·0014	1·5446	1·5461	6·8621	0·4822
1·00	1·2105	0·9058	1·2705	4·1983	0·0000	-	1·4448	1·4509	-	0·5000

C51 Trapezoidal channel - 0·125 to 1 side-slope

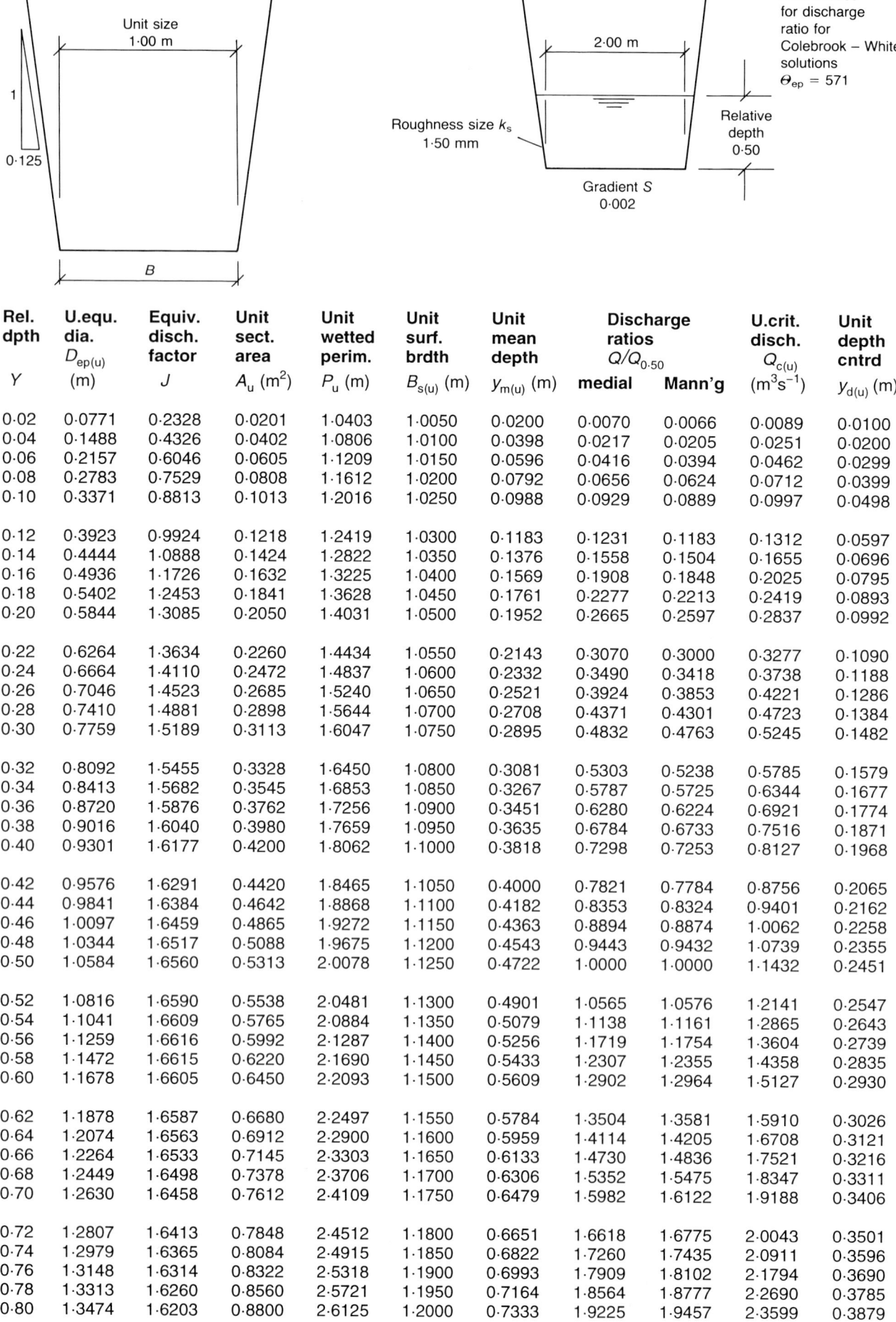

Rel. dpth Y	U.equ. dia. $D_{ep(u)}$ (m)	Equiv. disch. factor J	Unit sect. area A_u (m²)	Unit wetted perim. P_u (m)	Unit surf. brdth $B_{s(u)}$ (m)	Unit mean depth $y_{m(u)}$ (m)	Discharge ratios $Q/Q_{0.50}$ medial	Mann'g	U.crit. disch. $Q_{c(u)}$ (m³s⁻¹)	Unit depth cntrd $y_{d(u)}$ (m)
0·02	0·0771	0·2328	0·0201	1·0403	1·0050	0·0200	0·0070	0·0066	0·0089	0·0100
0·04	0·1488	0·4326	0·0402	1·0806	1·0100	0·0398	0·0217	0·0205	0·0251	0·0200
0·06	0·2157	0·6046	0·0605	1·1209	1·0150	0·0596	0·0416	0·0394	0·0462	0·0299
0·08	0·2783	0·7529	0·0808	1·1612	1·0200	0·0792	0·0656	0·0624	0·0712	0·0399
0·10	0·3371	0·8813	0·1013	1·2016	1·0250	0·0988	0·0929	0·0889	0·0997	0·0498
0·12	0·3923	0·9924	0·1218	1·2419	1·0300	0·1183	0·1231	0·1183	0·1312	0·0597
0·14	0·4444	1·0888	0·1424	1·2822	1·0350	0·1376	0·1558	0·1504	0·1655	0·0696
0·16	0·4936	1·1726	0·1632	1·3225	1·0400	0·1569	0·1908	0·1848	0·2025	0·0795
0·18	0·5402	1·2453	0·1841	1·3628	1·0450	0·1761	0·2277	0·2213	0·2419	0·0893
0·20	0·5844	1·3085	0·2050	1·4031	1·0500	0·1952	0·2665	0·2597	0·2837	0·0992
0·22	0·6264	1·3634	0·2260	1·4434	1·0550	0·2143	0·3070	0·3000	0·3277	0·1090
0·24	0·6664	1·4110	0·2472	1·4837	1·0600	0·2332	0·3490	0·3418	0·3738	0·1188
0·26	0·7046	1·4523	0·2685	1·5240	1·0650	0·2521	0·3924	0·3853	0·4221	0·1286
0·28	0·7410	1·4881	0·2898	1·5644	1·0700	0·2708	0·4371	0·4301	0·4723	0·1384
0·30	0·7759	1·5189	0·3113	1·6047	1·0750	0·2895	0·4832	0·4763	0·5245	0·1482
0·32	0·8092	1·5455	0·3328	1·6450	1·0800	0·3081	0·5303	0·5238	0·5785	0·1579
0·34	0·8413	1·5682	0·3545	1·6853	1·0850	0·3267	0·5787	0·5725	0·6344	0·1677
0·36	0·8720	1·5876	0·3762	1·7256	1·0900	0·3451	0·6280	0·6224	0·6921	0·1774
0·38	0·9016	1·6040	0·3980	1·7659	1·0950	0·3635	0·6784	0·6733	0·7516	0·1871
0·40	0·9301	1·6177	0·4200	1·8062	1·1000	0·3818	0·7298	0·7253	0·8127	0·1968
0·42	0·9576	1·6291	0·4420	1·8465	1·1050	0·4000	0·7821	0·7784	0·8756	0·2065
0·44	0·9841	1·6384	0·4642	1·8868	1·1100	0·4182	0·8353	0·8324	0·9401	0·2162
0·46	1·0097	1·6459	0·4865	1·9272	1·1150	0·4363	0·8894	0·8874	1·0062	0·2258
0·48	1·0344	1·6517	0·5088	1·9675	1·1200	0·4543	0·9443	0·9432	1·0739	0·2355
0·50	1·0584	1·6560	0·5313	2·0078	1·1250	0·4722	1·0000	1·0000	1·1432	0·2451
0·52	1·0816	1·6590	0·5538	2·0481	1·1300	0·4901	1·0565	1·0576	1·2141	0·2547
0·54	1·1041	1·6609	0·5765	2·0884	1·1350	0·5079	1·1138	1·1161	1·2865	0·2643
0·56	1·1259	1·6616	0·5992	2·1287	1·1400	0·5256	1·1719	1·1754	1·3604	0·2739
0·58	1·1472	1·6615	0·6220	2·1690	1·1450	0·5433	1·2307	1·2355	1·4358	0·2835
0·60	1·1678	1·6605	0·6450	2·2093	1·1500	0·5609	1·2902	1·2964	1·5127	0·2930
0·62	1·1878	1·6587	0·6680	2·2497	1·1550	0·5784	1·3504	1·3581	1·5910	0·3026
0·64	1·2074	1·6563	0·6912	2·2900	1·1600	0·5959	1·4114	1·4205	1·6708	0·3121
0·66	1·2264	1·6533	0·7145	2·3303	1·1650	0·6133	1·4730	1·4836	1·7521	0·3216
0·68	1·2449	1·6498	0·7378	2·3706	1·1700	0·6306	1·5352	1·5475	1·8347	0·3311
0·70	1·2630	1·6458	0·7612	2·4109	1·1750	0·6479	1·5982	1·6122	1·9188	0·3406
0·72	1·2807	1·6413	0·7848	2·4512	1·1800	0·6651	1·6618	1·6775	2·0043	0·3501
0·74	1·2979	1·6365	0·8084	2·4915	1·1850	0·6822	1·7260	1·7435	2·0911	0·3596
0·76	1·3148	1·6314	0·8322	2·5318	1·1900	0·6993	1·7909	1·8102	2·1794	0·3690
0·78	1·3313	1·6260	0·8560	2·5721	1·1950	0·7164	1·8564	1·8777	2·2690	0·3785
0·80	1·3474	1·6203	0·8800	2·6125	1·2000	0·7333	1·9225	1·9457	2·3599	0·3879

Rel. dpth	U.equ. dia. $D_{ep(u)}$	Equiv. disch. factor	Unit sect. area	Unit wetted perim.	Unit surf. brdth	Unit mean depth	Discharge ratios $Q/Q_{0.50}$		U.crit. disch. $Q_{c(u)}$	Unit depth cntrd
Y	(m)	J	A_u (m^2)	P_u (m)	$B_{s(u)}$ (m)	$y_{m(u)}$ (m)	medial	Mann'g	(m^{3}s^{-1})	$y_{d(u)}$ (m)
0·82	1·3632	1·6143	0·9041	2·6528	1·2050	0·7502	1·9892	2·0145	2·4522	0·3973
0·84	1·3786	1·6082	0·9282	2·6931	1·2100	0·7671	2·0566	2·0839	2·5458	0·4067
0·86	1·3938	1·6019	0·9525	2·7334	1·2150	0·7839	2·1246	2·1540	2·6408	0·4161
0·88	1·4087	1·5955	0·9768	2·7737	1·2200	0·8007	2·1931	2·2248	2·7371	0·4255
0·90	1·4232	1·5889	1·0012	2·8140	1·2250	0·8173	2·2623	2·2961	2·8347	0·4348
0·92	1·4375	1·5822	1·0258	2·8543	1·2300	0·8340	2·3320	2·3682	2·9336	0·4442
0·94	1·4516	1·5754	1·0504	2·8946	1·2350	0·8506	2·4024	2·4409	3·0338	0·4535
0·96	1·4654	1·5685	1·0752	2·9349	1·2400	0·8671	2·4733	2·5142	3·1353	0·4629
0·98	1·4789	1·5616	1·1000	2·9753	1·2450	0·8836	2·5448	2·5881	3·2381	0·4722
1·00	1·4923	1·5546	1·1250	3·0156	1·2500	0·9000	2·6169	2·6627	3·3422	0·4815
1·02	1·5054	1·5476	1·1500	3·0559	1·2550	0·9164	2·6896	2·7379	3·4476	0·4908
1·04	1·5183	1·5405	1·1752	3·0962	1·2600	0·9327	2·7628	2·8137	3·5542	0·5001
1·06	1·5309	1·5334	1·2004	3·1365	1·2650	0·9490	2·8366	2·8902	3·6621	0·5093
1·08	1·5434	1·5263	1·2258	3·1768	1·2700	0·9652	2·9110	2·9672	3·7713	0·5186
1·10	1·5557	1·5192	1·2512	3·2171	1·2750	0·9814	2·9859	3·0449	3·8817	0·5278
1·12	1·5679	1·5121	1·2768	3·2574	1·2800	0·9975	3·0614	3·1232	3·9934	0·5371
1·14	1·5798	1·5050	1·3024	3·2977	1·2850	1·0136	3·1375	3·2021	4·1063	0·5463
1·16	1·5916	1·4979	1·3282	3·3381	1·2900	1·0296	3·2141	3·2816	4·2205	0·5555
1·18	1·6032	1·4908	1·3540	3·3784	1·2950	1·0456	3·2913	3·3618	4·3359	0·5647
1·20	1·6147	1·4838	1·3800	3·4187	1·3000	1·0615	3·3690	3·4425	4·4525	0·5739
1·24	1·6371	1·4697	1·4322	3·4993	1·3100	1·0933	3·5262	3·6058	4·6895	0·5923
1·28	1·6590	1·4559	1·4848	3·5799	1·3200	1·1248	3·6856	3·7715	4·9315	0·6106
1·32	1·6804	1·4421	1·5378	3·6605	1·3300	1·1562	3·8472	3·9396	5·1783	0·6288
1·36	1·7013	1·4286	1·5912	3·7412	1·3400	1·1875	4·0110	4·1101	5·4299	0·6471
1·40	1·7217	1·4152	1·6450	3·8218	1·3500	1·2185	4·1770	4·2830	5·6865	0·6652
1·44	1·7417	1·4021	1·6992	3·9024	1·3600	1·2494	4·3452	4·4583	5·9478	0·6834
1·48	1·7613	1·3892	1·7538	3·9830	1·3700	1·2801	4·5155	4·6359	6·2140	0·7015
1·52	1·7805	1·3764	1·8088	4·0637	1·3800	1·3107	4·6881	4·8160	6·4849	0·7196
1·56	1·7993	1·3639	1·8642	4·1443	1·3900	1·3412	4·8629	4·9984	6·7607	0·7376
1·60	1·8178	1·3517	1·9200	4·2249	1·4000	1·3714	5·0398	5·1833	7·0412	0·7556
1·64	1·8360	1·3396	1·9762	4·3055	1·4100	1·4016	5·2190	5·3705	7·3265	0·7735
1·68	1·8538	1·3278	2·0328	4·3861	1·4200	1·4315	5·4003	5·5601	7·6165	0·7914
1·72	1·8714	1·3162	2·0898	4·4668	1·4300	1·4614	5·5839	5·7521	7·9113	0·8093
1·76	1·8887	1·3048	2·1472	4·5474	1·4400	1·4911	5·7696	5·9465	8·2108	0·8271
1·80	1·9058	1·2937	2·2050	4·6280	1·4500	1·5207	5·9575	6·1432	8·5151	0·8449
1·84	1·9226	1·2827	2·2632	4·7086	1·4600	1·5501	6·1476	6·3424	8·8241	0·8627
1·88	1·9392	1·2720	2·3218	4·7893	1·4700	1·5795	6·3399	6·5440	9·1377	0·8804
1·92	1·9555	1·2615	2·3808	4·8699	1·4800	1·6086	6·5344	6·7480	9·4561	0·8981
1·96	1·9717	1·2512	2·4402	4·9505	1·4900	1·6377	6·7312	6·9543	9·7792	0·9157
2·00	1·9876	1·2411	2·5000	5·0311	1·5000	1·6667	6·9301	7·1631	10·107	0·9333
2·05	2·0073	1·2288	2·5753	5·1319	1·5125	1·7027	7·1819	7·4275	10·523	0·9553
2·10	2·0267	1·2167	2·6512	5·2327	1·5250	1·7385	7·4371	7·6957	10·947	0·9772
2·15	2·0458	1·2050	2·7278	5·3335	1·5375	1·7742	7·6958	7·9677	11·378	0·9991
2·20	2·0647	1·1936	2·8050	5·4342	1·5500	1·8097	7·9580	8·2434	11·817	1·0209
2·25	2·0833	1·1824	2·8828	5·5350	1·5625	1·8450	8·2237	8·5230	12·262	1·0427
2·30	2·1017	1·1716	2·9613	5·6358	1·5750	1·8802	8·4929	8·8065	12·715	1·0644
2·35	2·1199	1·1609	3·0403	5·7366	1·5875	1·9152	8·7655	9·0937	13·176	1·0861
2·40	2·1380	1·1506	3·1200	5·8374	1·6000	1·9500	9·0418	9·3848	13·644	1·1077
2·45	2·1558	1·1405	3·2003	5·9381	1·6125	1·9847	9·3215	9·6798	14·119	1·1293
2·50	2·1734	1·1306	3·2813	6·0389	1·6250	2·0192	9·6048	9·9787	14·601	1·1508
2·55	2·1909	1·1210	3·3628	6·1397	1·6375	2·0536	9·8916	10·281	15·091	1·1723
2·60	2·2082	1·1116	3·4450	6·2405	1·6500	2·0879	10·182	10·588	15·588	1·1937
2·65	2·2253	1·1024	3·5278	6·3412	1·6625	2·1220	10·476	10·899	16·093	1·2151
2·70	2·2423	1·0935	3·6113	6·4420	1·6750	2·1560	10·773	11·213	16·605	1·2364
2·75	2·2592	1·0847	3·6953	6·5428	1·6875	2·1898	11·075	11·532	17·124	1·2578

C52

Trapezoidal channel - 0·25 to 1 side-slope

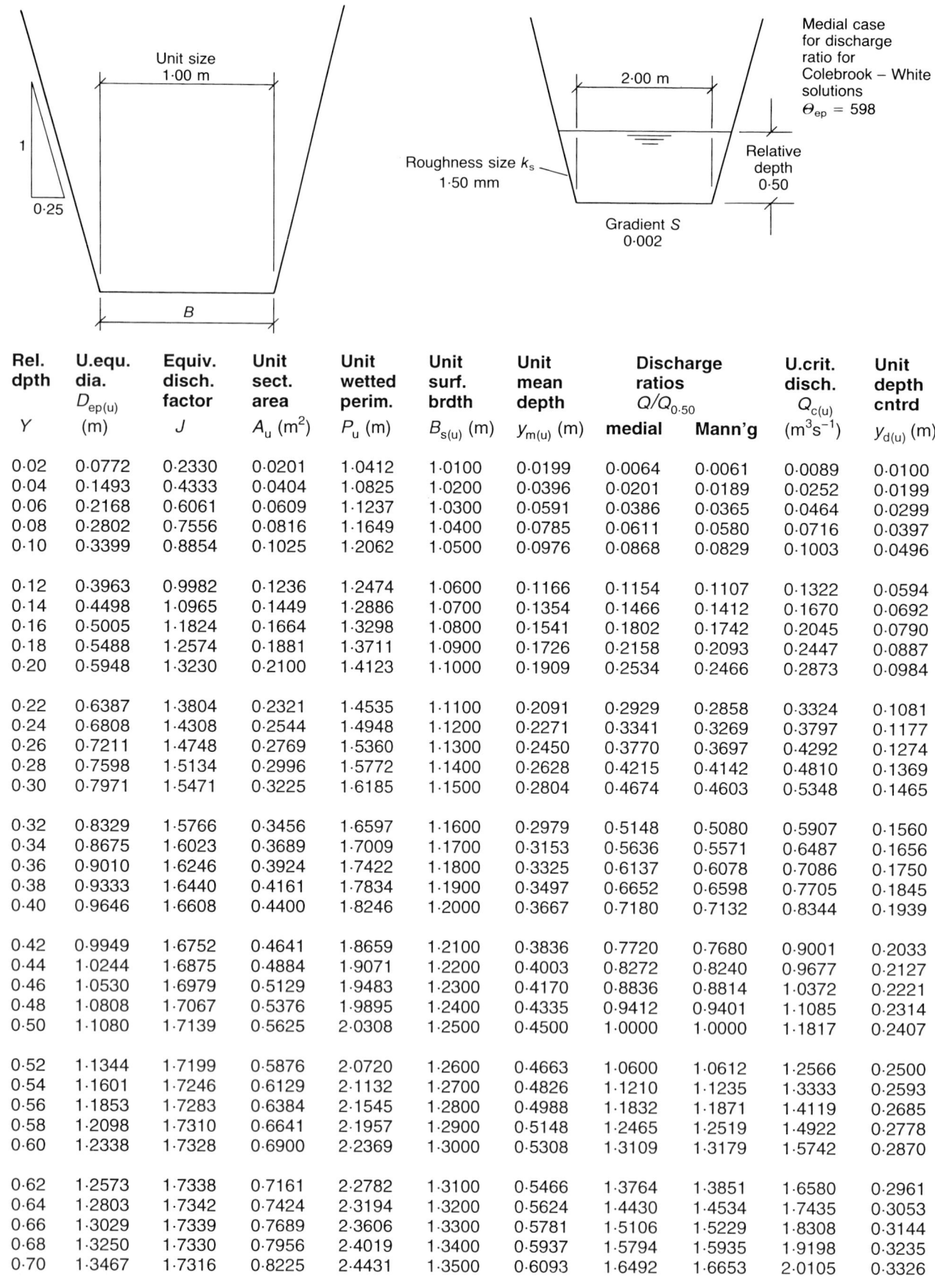

Rel. dpth Y	U.equ. dia. $D_{ep(u)}$ (m)	Equiv. disch. factor J	Unit sect. area A_u (m²)	Unit wetted perim. P_u (m)	Unit surf. brdth $B_{s(u)}$ (m)	Unit mean depth $y_{m(u)}$ (m)	Discharge ratios $Q/Q_{0.50}$ medial	Discharge ratios $Q/Q_{0.50}$ Mann'g	U.crit. disch. $Q_{c(u)}$ (m³s⁻¹)	Unit depth cntrd $y_{d(u)}$ (m)
0·02	0·0772	0·2330	0·0201	1·0412	1·0100	0·0199	0·0064	0·0061	0·0089	0·0100
0·04	0·1493	0·4333	0·0404	1·0825	1·0200	0·0396	0·0201	0·0189	0·0252	0·0199
0·06	0·2168	0·6061	0·0609	1·1237	1·0300	0·0591	0·0386	0·0365	0·0464	0·0299
0·08	0·2802	0·7556	0·0816	1·1649	1·0400	0·0785	0·0611	0·0580	0·0716	0·0397
0·10	0·3399	0·8854	0·1025	1·2062	1·0500	0·0976	0·0868	0·0829	0·1003	0·0496
0·12	0·3963	0·9982	0·1236	1·2474	1·0600	0·1166	0·1154	0·1107	0·1322	0·0594
0·14	0·4498	1·0965	0·1449	1·2886	1·0700	0·1354	0·1466	0·1412	0·1670	0·0692
0·16	0·5005	1·1824	0·1664	1·3298	1·0800	0·1541	0·1802	0·1742	0·2045	0·0790
0·18	0·5488	1·2574	0·1881	1·3711	1·0900	0·1726	0·2158	0·2093	0·2447	0·0887
0·20	0·5948	1·3230	0·2100	1·4123	1·1000	0·1909	0·2534	0·2466	0·2873	0·0984
0·22	0·6387	1·3804	0·2321	1·4535	1·1100	0·2091	0·2929	0·2858	0·3324	0·1081
0·24	0·6808	1·4308	0·2544	1·4948	1·1200	0·2271	0·3341	0·3269	0·3797	0·1177
0·26	0·7211	1·4748	0·2769	1·5360	1·1300	0·2450	0·3770	0·3697	0·4292	0·1274
0·28	0·7598	1·5134	0·2996	1·5772	1·1400	0·2628	0·4215	0·4142	0·4810	0·1369
0·30	0·7971	1·5471	0·3225	1·6185	1·1500	0·2804	0·4674	0·4603	0·5348	0·1465
0·32	0·8329	1·5766	0·3456	1·6597	1·1600	0·2979	0·5148	0·5080	0·5907	0·1560
0·34	0·8675	1·6023	0·3689	1·7009	1·1700	0·3153	0·5636	0·5571	0·6487	0·1656
0·36	0·9010	1·6246	0·3924	1·7422	1·1800	0·3325	0·6137	0·6078	0·7086	0·1750
0·38	0·9333	1·6440	0·4161	1·7834	1·1900	0·3497	0·6652	0·6598	0·7705	0·1845
0·40	0·9646	1·6608	0·4400	1·8246	1·2000	0·3667	0·7180	0·7132	0·8344	0·1939
0·42	0·9949	1·6752	0·4641	1·8659	1·2100	0·3836	0·7720	0·7680	0·9001	0·2033
0·44	1·0244	1·6875	0·4884	1·9071	1·2200	0·4003	0·8272	0·8240	0·9677	0·2127
0·46	1·0530	1·6979	0·5129	1·9483	1·2300	0·4170	0·8836	0·8814	1·0372	0·2221
0·48	1·0808	1·7067	0·5376	1·9895	1·2400	0·4335	0·9412	0·9401	1·1085	0·2314
0·50	1·1080	1·7139	0·5625	2·0308	1·2500	0·4500	1·0000	1·0000	1·1817	0·2407
0·52	1·1344	1·7199	0·5876	2·0720	1·2600	0·4663	1·0600	1·0612	1·2566	0·2500
0·54	1·1601	1·7246	0·6129	2·1132	1·2700	0·4826	1·1210	1·1235	1·3333	0·2593
0·56	1·1853	1·7283	0·6384	2·1545	1·2800	0·4988	1·1832	1·1871	1·4119	0·2685
0·58	1·2098	1·7310	0·6641	2·1957	1·2900	0·5148	1·2465	1·2519	1·4922	0·2778
0·60	1·2338	1·7328	0·6900	2·2369	1·3000	0·5308	1·3109	1·3179	1·5742	0·2870
0·62	1·2573	1·7338	0·7161	2·2782	1·3100	0·5466	1·3764	1·3851	1·6580	0·2961
0·64	1·2803	1·7342	0·7424	2·3194	1·3200	0·5624	1·4430	1·4534	1·7435	0·3053
0·66	1·3029	1·7339	0·7689	2·3606	1·3300	0·5781	1·5106	1·5229	1·8308	0·3144
0·68	1·3250	1·7330	0·7956	2·4019	1·3400	0·5937	1·5794	1·5935	1·9198	0·3235
0·70	1·3467	1·7316	0·8225	2·4431	1·3500	0·6093	1·6492	1·6653	2·0105	0·3326
0·72	1·3679	1·7298	0·8496	2·4843	1·3600	0·6247	1·7200	1·7383	2·1029	0·3417
0·74	1·3888	1·7276	0·8769	2·5255	1·3700	0·6401	1·7919	1·8124	2·1970	0·3507
0·76	1·4094	1·7250	0·9044	2·5668	1·3800	0·6554	1·8649	1·8876	2·2928	0·3598
0·78	1·4296	1·7220	0·9321	2·6080	1·3900	0·6706	1·9388	1·9640	2·3903	0·3688
0·80	1·4495	1·7188	0·9600	2·6492	1·4000	0·6857	2·0139	2·0415	2·4894	0·3778

Trapezoidal channel - 0·25 to 1 side-slope

Rel. dpth Y	U.equ. dia. $D_{ep(u)}$ (m)	Equiv. disch. factor J	Unit sect. area A_u (m²)	Unit wetted perim. P_u (m)	Unit surf. brdth $B_{s(u)}$ (m)	Unit mean depth $y_{m(u)}$ (m)	Discharge ratios $Q/Q_{0.50}$ medial	Mann'g	Crit. disch. $Q_{c(u)}$ (m³s⁻¹)	Unit depth cntrd $y_{d(u)}$ (m)
0·82	1·4690	1·7153	0·9881	2·6905	1·4100	0·7008	2·0900	2·1201	2·5903	0·3867
0·84	1·4883	1·7116	1·0164	2·7317	1·4200	0·7158	2·1671	2·1998	2·6929	0·3957
0·86	1·5073	1·7076	1·0449	2·7729	1·4300	0·7307	2·2453	2·2807	2·7971	0·4046
0·88	1·5260	1·7035	1·0736	2·8142	1·4400	0·7456	2·3244	2·3627	2·9030	0·4136
0·90	1·5444	1·6992	1·1025	2·8554	1·4500	0·7603	2·4047	2·4458	3·0105	0·4224
0·92	1·5626	1·6948	1·1316	2·8966	1·4600	0·7751	2·4859	2·5301	3·1198	0·4313
0·94	1·5806	1·6902	1·1609	2·9379	1·4700	0·7897	2·5682	2·6154	3·2307	0·4402
0·96	1·5983	1·6855	1·1904	2·9791	1·4800	0·8043	2·6516	2·7019	3·3433	0·4490
0·98	1·6159	1·6807	1·2201	3·0203	1·4900	0·8189	2·7359	2·7895	3·4575	0·4579
1·00	1·6332	1·6758	1·2500	3·0616	1·5000	0·8333	2·8213	2·8782	3·5734	0·4667
1·02	1·6503	1·6709	1·2801	3·1028	1·5100	0·8477	2·9078	2·9681	3·6909	0·4755
1·04	1·6672	1·6658	1·3104	3·1440	1·5200	0·8621	2·9952	3·0591	3·8102	0·4842
1·06	1·6839	1·6608	1·3409	3·1852	1·5300	0·8764	3·0837	3·1512	3·9311	0·4930
1·08	1·7004	1·6557	1·3716	3·2265	1·5400	0·8906	3·1733	3·2444	4·0536	0·5017
1·10	1·7168	1·6505	1·4025	3·2677	1·5500	0·9048	3·2638	3·3387	4·1778	0·5105
1·12	1·7330	1·6453	1·4336	3·3089	1·5600	0·9190	3·3554	3·4342	4·3037	0·5192
1·14	1·7490	1·6401	1·4649	3·3502	1·5700	0·9331	3·4481	3·5308	4·4312	0·5279
1·16	1·7649	1·6349	1·4964	3·3914	1·5800	0·9471	3·5417	3·6285	4·5604	0·5365
1·18	1·7807	1·6297	1·5281	3·4326	1·5900	0·9611	3·6365	3·7274	4·6913	0·5452
1·20	1·7963	1·6244	1·5600	3·4739	1·6000	0·9750	3·7322	3·8274	4·8238	0·5538
1·24	1·8271	1·6139	1·6244	3·5563	1·6200	1·0027	3·9269	4·0308	5·0938	0·5711
1·28	1·8573	1·6035	1·6896	3·6388	1·6400	1·0302	4·1257	4·2388	5·3705	0·5883
1·32	1·8871	1·5931	1·7556	3·7212	1·6600	1·0576	4·3287	4·4513	5·6539	0·6054
1·36	1·9164	1·5828	1·8224	3·8037	1·6800	1·0848	4·5359	4·6684	5·9439	0·6225
1·40	1·9454	1·5726	1·8900	3·8862	1·7000	1·1118	4·7474	4·8902	6·2406	0·6395
1·44	1·9739	1·5625	1·9584	3·9686	1·7200	1·1386	4·9630	5·1166	6·5441	0·6565
1·48	2·0020	1·5525	2·0276	4·0511	1·7400	1·1653	5·1830	5·3476	6·8542	0·6734
1·52	2·0298	1·5427	2·0976	4·1336	1·7600	1·1918	5·4072	5·5833	7·1711	0·6902
1·56	2·0573	1·5330	2·1684	4·2160	1·7800	1·2182	5·6357	5·8237	7·4948	0·7071
1·60	2·0845	1·5234	2·2400	4·2985	1·8000	1·2444	5·8685	6·0689	7·8252	0·7238
1·64	2·1113	1·5140	2·3124	4·3809	1·8200	1·2705	6·1057	6·3187	8·1624	0·7405
1·68	2·1379	1·5047	2·3856	4·4634	1·8400	1·2965	6·3472	6·5734	8·5064	0·7572
1·72	2·1642	1·4957	2·4596	4·5459	1·8600	1·3224	6·5930	6·8328	8·8573	0·7738
1·76	2·1903	1·4867	2·5344	4·6283	1·8800	1·3481	6·8432	7·0971	9·2150	0·7904
1·80	2·2162	1·4779	2·6100	4·7108	1·9000	1·3737	7·0979	7·3662	9·5795	0·8069
1·84	2·2418	1·4693	2·6864	4·7933	1·9200	1·3992	7·3570	7·6401	9·9510	0·8234
1·88	2·2672	1·4608	2·7636	4·8757	1·9400	1·4245	7·6205	7·9190	10·329	0·8398
1·92	2·2925	1·4525	2·8416	4·9582	1·9600	1·4498	7·8884	8·2028	10·715	0·8562
1·96	2·3175	1·4443	2·9204	5·0406	1·9800	1·4749	8·1609	8·4915	11·107	0·8726
2·00	2·3423	1·4363	3·0000	5·1231	2·0000	1·5000	8·4379	8·7852	11·506	0·8889
2·05	2·3731	1·4265	3·1006	5·2262	2·0250	1·5312	8·7905	9·1593	12·015	0·9092
2·10	2·4037	1·4170	3·2025	5·3293	2·0500	1·5622	9·1502	9·5413	12·535	0·9295
2·15	2·4340	1·4076	3·3056	5·4323	2·0750	1·5931	9·5171	9·9312	13·066	0·9497
2·20	2·4641	1·3985	3·4100	5·5354	2·1000	1·6238	9·8911	10·329	13·608	0·9699
2·25	2·4940	1·3896	3·5156	5·6385	2·1250	1·6544	10·272	10·735	14·161	0·9900
2·30	2·5237	1·3809	3·6225	5·7416	2·1500	1·6849	10·661	11·149	14·725	1·0101
2·35	2·5532	1·3723	3·7306	5·8446	2·1750	1·7152	11·057	11·571	15·300	1·0301
2·40	2·5825	1·3640	3·8400	5·9477	2·2000	1·7455	11·460	12·001	15·887	1·0500
2·45	2·6116	1·3559	3·9506	6·0508	2·2250	1·7756	11·871	12·440	16·485	1·0699
2·50	2·6406	1·3480	4·0625	6·1539	2·2500	1·8056	12·289	12·886	17·095	1·0897
2·55	2·6694	1·3403	4·1756	6·2570	2·2750	1·8354	12·715	13·341	17·715	1·1095
2·60	2·6981	1·3327	4·2900	6·3600	2·3000	1·8652	13·148	13·805	18·348	1·1293
2·65	2·7266	1·3253	4·4056	6·4631	2·3250	1·8949	13·589	14·277	18·992	1·1490
2·70	2·7550	1·3181	4·5225	6·5662	2·3500	1·9245	14·037	14·757	19·647	1·1687
2·75	2·7833	1·3111	4·6406	6·6693	2·3750	1·9539	14·493	15·246	20·314	1·1883

C53 Trapezoidal channel - 0·375 to 1 side-slope

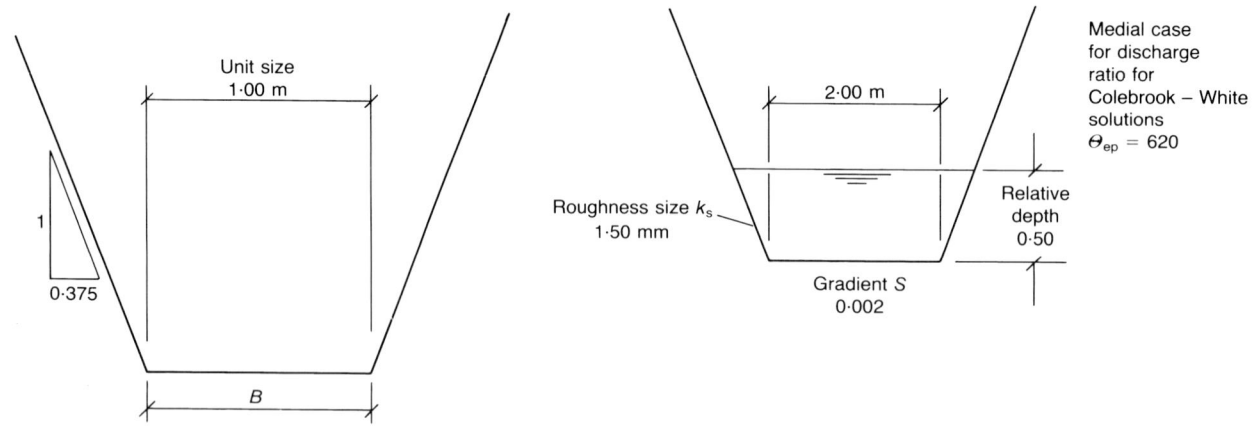

Rel. dpth Y	U.equ. dia. $D_{ep(u)}$ (m)	Equiv. disch. factor J	Unit sect. area A_u (m²)	Unit wetted perim. P_u (m)	Unit surf. brdth $B_{s(u)}$ (m)	Unit mean depth $y_{m(u)}$ (m)	Discharge ratios $Q/Q_{0.50}$ medial	Mann'g	U.crit. disch. $Q_{c(u)}$ (m³s⁻¹)	Unit depth cntrd $y_{d(u)}$ (m)
0·02	0·0773	0·2329	0·0202	1·0427	1·0150	0·0199	0·0060	0·0056	0·0089	0·0100
0·04	0·1496	0·4330	0·0406	1·0854	1·0300	0·0394	0·0187	0·0176	0·0252	0·0199
0·06	0·2175	0·6057	0·0614	1·1282	1·0450	0·0587	0·0361	0·0341	0·0466	0·0298
0·08	0·2815	0·7553	0·0824	1·1709	1·0600	0·0777	0·0573	0·0544	0·0719	0·0396
0·10	0·3420	0·8852	0·1037	1·2136	1·0750	0·0965	0·0817	0·0779	0·1009	0·0494
0·12	0·3993	0·9984	0·1254	1·2563	1·0900	0·1150	0·1090	0·1044	0·1332	0·0591
0·14	0·4537	1·0972	0·1473	1·2990	1·1050	0·1333	0·1389	0·1336	0·1685	0·0688
0·16	0·5056	1·1838	0·1696	1·3418	1·1200	0·1514	0·1712	0·1653	0·2067	0·0785
0·18	0·5552	1·2597	0·1922	1·3845	1·1350	0·1693	0·2058	0·1993	0·2476	0·0881
0·20	0·6026	1·3264	0·2150	1·4272	1·1500	0·1870	0·2424	0·2356	0·2911	0·0977
0·22	0·6481	1·3850	0·2381	1·4699	1·1650	0·2044	0·2810	0·2739	0·3372	0·1072
0·24	0·6918	1·4367	0·2616	1·5126	1·1800	0·2217	0·3216	0·3142	0·3857	0·1167
0·26	0·7338	1·4822	0·2854	1·5554	1·1950	0·2388	0·3640	0·3565	0·4367	0·1262
0·28	0·7744	1·5224	0·3094	1·5981	1·2100	0·2557	0·4081	0·4007	0·4899	0·1356
0·30	0·8136	1·5578	0·3338	1·6408	1·2250	0·2724	0·4540	0·4467	0·5455	0·1449
0·32	0·8515	1·5890	0·3584	1·6835	1·2400	0·2890	0·5015	0·4945	0·6034	0·1543
0·34	0·8883	1·6166	0·3834	1·7262	1·2550	0·3055	0·5507	0·5440	0·6635	0·1636
0·36	0·9239	1·6408	0·4086	1·7690	1·2700	0·3217	0·6015	0·5953	0·7258	0·1729
0·38	0·9586	1·6622	0·4341	1·8117	1·2850	0·3379	0·6539	0·6482	0·7903	0·1821
0·40	0·9922	1·6809	0·4600	1·8544	1·3000	0·3538	0·7078	0·7028	0·8569	0·1913
0·42	1·0250	1·6974	0·4862	1·8971	1·3150	0·3697	0·7632	0·7590	0·9257	0·2005
0·44	1·0570	1·7118	0·5126	1·9398	1·3300	0·3854	0·8202	0·8169	0·9966	0·2096
0·46	1·0882	1·7243	0·5394	1·9826	1·3450	0·4010	0·8787	0·8763	1·0696	0·2187
0·48	1·1187	1·7352	0·5664	2·0253	1·3600	0·4165	0·9386	0·9374	1·1447	0·2278
0·50	1·1485	1·7446	0·5938	2·0680	1·3750	0·4318	1·0000	1·0000	1·2218	0·2368
0·52	1·1776	1·7527	0·6214	2·1107	1·3900	0·4471	1·0629	1·0642	1·3011	0·2459
0·54	1·2062	1·7596	0·6494	2·1534	1·4050	0·4622	1·1273	1·1300	1·3824	0·2548
0·56	1·2342	1·7654	0·6776	2·1962	1·4200	0·4772	1·1931	1·1973	1·4658	0·2638
0·58	1·2616	1·7703	0·7062	2·2389	1·4350	0·4921	1·2603	1·2662	1·5512	0·2727
0·60	1·2886	1·7742	0·7350	2·2816	1·4500	0·5069	1·3290	1·3366	1·6387	0·2816
0·62	1·3151	1·7774	0·7642	2·3243	1·4650	0·5216	1·3991	1·4086	1·7283	0·2905
0·64	1·3411	1·7799	0·7936	2·3670	1·4800	0·5362	1·4706	1·4822	1·8198	0·2994
0·66	1·3667	1·7817	0·8234	2·4098	1·4950	0·5507	1·5436	1·5572	1·9134	0·3082
0·68	1·3919	1·7830	0·8534	2·4525	1·5100	0·5652	1·6180	1·6338	2·0091	0·3170
0·70	1·4167	1·7837	0·8837	2·4952	1·5250	0·5795	1·6939	1·7120	2·1068	0·3257
0·72	1·4412	1·7839	0·9144	2·5379	1·5400	0·5938	1·7712	1·7917	2·2065	0·3345
0·74	1·4653	1·7838	0·9453	2·5806	1·5550	0·6079	1·8499	1·8730	2·3083	0·3432
0·76	1·4891	1·7832	0·9766	2·6234	1·5700	0·6220	1·9300	1·9558	2·4120	0·3519
0·78	1·5126	1·7823	1·0081	2·6661	1·5850	0·6361	2·0116	2·0401	2·5179	0·3606
0·80	1·5357	1·7811	1·0400	2·7088	1·6000	0·6500	2·0946	2·1260	2·6257	0·3692

Trapezoidal channel - 0·375 to 1 side-slope

Rel. dpth Y	U.equ. dia. $D_{ep(u)}$ (m)	Equiv. disch. factor J	Unit sect. area A_u (m²)	Unit wetted perim. P_u (m)	Unit surf. brdth $B_{s(u)}$ (m)	Unit mean depth $y_{m(u)}$ (m)	Discharge ratios $Q/Q_{0.50}$ medial	Mann'g	U.crit. disch. $Q_{c(u)}$ (m³s⁻¹)	Unit depth cntrd $y_{d(u)}$ (m)
0·82	1·5586	1·7795	1·0721	2·7515	1·6150	0·6639	2·1790	2·2135	2·7356	0·3779
0·84	1·5813	1·7778	1·1046	2·7942	1·6300	0·6777	2·2649	2·3025	2·8476	0·3865
0·86	1·6036	1·7758	1·1374	2·8370	1·6450	0·6914	2·3522	2·3930	2·9615	0·3950
0·88	1·6257	1·7736	1·1704	2·8797	1·6600	0·7051	2·4409	2·4852	3·0776	0·4036
0·90	1·6476	1·7712	1·2037	2·9224	1·6750	0·7187	2·5311	2·5789	3·1956	0·4121
0·92	1·6693	1·7686	1·2374	2·9651	1·6900	0·7322	2·6228	2·6741	3·3158	0·4207
0·94	1·6907	1·7659	1·2714	3·0078	1·7050	0·7457	2·7159	2·7710	3·4379	0·4292
0·96	1·7119	1·7630	1·3056	3·0506	1·7200	0·7591	2·8104	2·8694	3·5621	0·4376
0·98	1·7330	1·7600	1·3402	3·0933	1·7350	0·7724	2·9064	2·9694	3·6884	0·4461
1·00	1·7538	1·7569	1·3750	3·1360	1·7500	0·7857	3·0039	3·0710	3·8168	0·4545
1·02	1·7745	1·7537	1·4101	3·1787	1·7650	0·7990	3·1028	3·1742	3·9472	0·4630
1·04	1·7950	1·7504	1·4456	3·2214	1·7800	0·8121.	3·2032	3·2790	4·0796	0·4714
1·06	1·8153	1·7471	1·4813	3·2642	1·7950	0·8253	3·3051	3·3854	4·2142	0·4797
1·08	1·8354	1·7437	1·5174	3·3069	1·8100	0·8383	3·4084	3·4934	4·3508	0·4881
1·10	1·8554	1·7402	1·5537	3·3496	1·8250	0·8514	3·5133	3·6030	4·4895	0·4965
1·12	1·8753	1·7366	1·5904	3·3923	1·8400	0·8643	3·6196	3·7143	4·6303	0·5048
1·14	1·8950	1·7331	1·6273	3·4350	1·8550	0·8773	3·7274	3·8271	4·7732	0·5131
1·16	1·9146	1·7295	1·6646	3·4778	1·8700	0·8902	3·8367	3·9416	4·9182	0·5214
1·18	1·9340	1·7258	1·7021	3·5205	1·8850	0·9030	3·9475	4·0578	5·0653	0·5297
1·20	1·9533	1·7221	1·7400	3·5632	1·9000	0·9158	4·0598	4·1756	5·2144	0·5379
1·24	1·9915	1·7147	1·8166	3·6486	1·9300	0·9412	4·2890	4·4161	5·5191	0·5544
1·28	2·0293	1·7073	1·8944	3·7341	1·9600	0·9665	4·5242	4·6633	5·8323	0·5708
1·32	2·0666	1·6998	1·9734	3·8195	1·9900	0·9917	4·7656	4·9172	6·1540	0·5872
1·36	2·1036	1·6923	2·0536	3·9050	2·0200	1·0166	5·0131	5·1778	6·4842	0·6034
1·40	2·1401	1·6849	2·1350	3·9904	2·0500	1·0415	5·2667	5·4452	6·8231	0·6197
1·44	2·1763	1·6774	2·2176	4·0758	2·0800	1·0662	5·5267	5·7195	7·1706	0·6358
1·48	2·2122	1·6701	2·3014	4·1613	2·1100	1·0907	5·7929	6·0006	7·5267	0·6520
1·52	2·2478	1·6628	2·3864	4·2467	2·1400	1·1151	6·0654	6·2887	7·8917	0·6680
1·56	2·2830	1·6556	2·4726	4·3322	2·1700	1·1394	6·3443	6·5839	8·2653	0·6840
1·60	2·3180	1·6484	2·5600	4·4176	2·2000	1·1636	6·6296	6·8860	8·6479	0·7000
1·64	2·3527	1·6414	2·6486	4·5030	2·2300	1·1877	6·9213	7·1953	9·0393	0·7159
1·68	2·3872	1·6344	2·7384	4·5885	2·2600	1·2117	7·2195	7·5118	9·4396	0·7318
1·72	2·4214	1·6275	2·8294	4·6739	2·2900	1·2355	7·5242	7·8355	9·8488	0·7476
1·76	2·4555	1·6208	2·9216	4·7594	2·3200	1·2593	7·8356	8·1664	10·267	0·7634
1·80	2·4893	1·6141	3·0150	4·8448	2·3500	1·2830	8·1535	8·5046	10·694	0·7791
1·84	2·5229	1·6076	3·1096	4·9302	2·3800	1·3066	8·4781	8·8503	11·131	0·7948
1·88	2·5563	1·6011	3·2054	5·0157	2·4100	1·3300	8·8094	9·2033	11·576	0·8104
1·92	2·5895	1·5948	3·3024	5·1011	2·4400	1·3534	9·1474	9·5639	12·031	0·8260
1·96	2·6226	1·5885	3·4006	5·1866	2·4700	1·3768	9·4922	9·9319	12·495	0·8416
2·00	2·6555	1·5824	3·5000	5·2720	2·5000	1·4000	9·8438	10·308	12·969	0·8571
2·05	2·6965	1·5749	3·6259	5·3788	2·5375	1·4289	10·293	10·788	13·573	0·8765
2·10	2·7372	1·5675	3·7537	5·4856	2·5750	1·4578	10·753	11·280	14·193	0·8958
2·15	2·7777	1·5603	3·8834	5·5924	2·6125	1·4865	11·224	11·785	14·827	0·9151
2·20	2·8179	1·5533	4·0150	5·6992	2·6500	1·5151	11·706	12·302	15·476	0·9342
2·25	2·8580	1·5464	4·1484	5·8060	2·6875	1·5436	12·199	12·831	16·140	0·9534
2·30	2·8979	1·5397	4·2838	5·9128	2·7250	1·5720	12·703	13·372	16·820	0·9725
2·35	2·9377	1·5331	4·4209	6·0196	2·7625	1·6003	13·218	13·926	17·514	0·9915
2·40	2·9773	1·5267	4·5600	6·1264	2·8000	1·6286	13·745	14·493	18·223	1·0105
2·45	3·0167	1·5204	4·7009	6·2332	2·8375	1·6567	14·283	15·073	18·948	1·0295
2·50	3·0560	1·5143	4·8438	6·3400	2·8750	1·6848	14·833	15·665	19·689	1·0484
2·55	3·0951	1·5083	4·9884	6·4468	2·9125	1·7128	15·394	16·271	20·444	1·0673
2·60	3·1342	1·5024	5·1350	6·5536	2·9500	1·7407	15·967	16·889	21·216	1·0861
2·65	3·1730	1·4966	5·2834	6·6604	2·9875	1·7685	16·551	17·521	22·003	1·1049
2·70	3·2118	1·4910	5·4338	6·7672	3·0250	1·7963	17·147	18·166	22·806	1·1236
2·75	3·2505	1·4855	5·5859	6·8740	3·0625	1·8240	17·756	18·824	23·625	1·1423

111

Trapezoidal channel - 0·50 to 1 side-slope

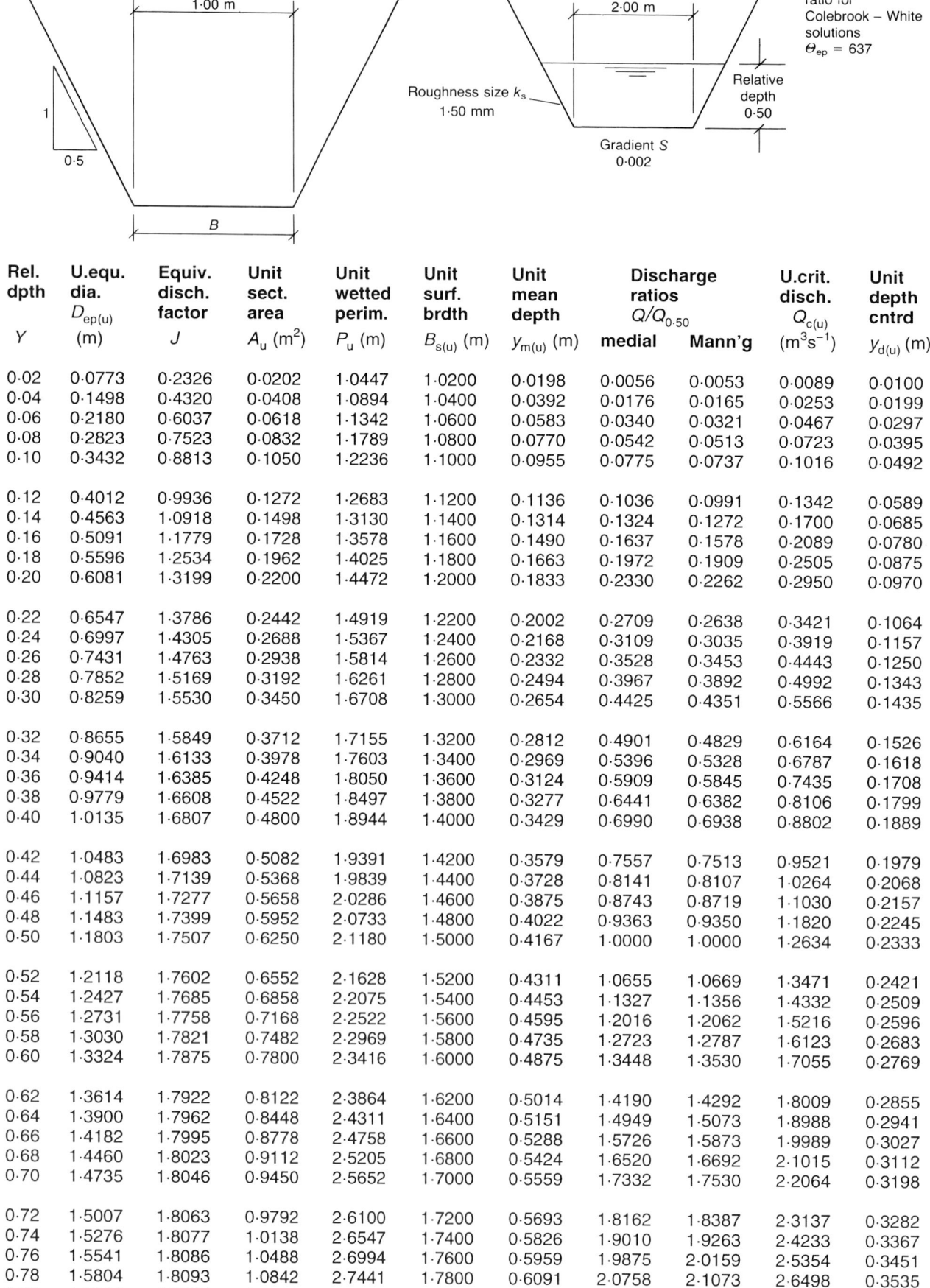

Rel. dpth Y	U.equ. dia. $D_{ep(u)}$ (m)	Equiv. disch. factor J	Unit sect. area A_u (m²)	Unit wetted perim. P_u (m)	Unit surf. brdth $B_{s(u)}$ (m)	Unit mean depth $y_{m(u)}$ (m)	Discharge ratios $Q/Q_{0.50}$ medial	Mann'g	U.crit. disch. $Q_{c(u)}$ (m³s⁻¹)	Unit depth cntrd $y_{d(u)}$ (m)
0·02	0·0773	0·2326	0·0202	1·0447	1·0200	0·0198	0·0056	0·0053	0·0089	0·0100
0·04	0·1498	0·4320	0·0408	1·0894	1·0400	0·0392	0·0176	0·0165	0·0253	0·0199
0·06	0·2180	0·6037	0·0618	1·1342	1·0600	0·0583	0·0340	0·0321	0·0467	0·0297
0·08	0·2823	0·7523	0·0832	1·1789	1·0800	0·0770	0·0542	0·0513	0·0723	0·0395
0·10	0·3432	0·8813	0·1050	1·2236	1·1000	0·0955	0·0775	0·0737	0·1016	0·0492
0·12	0·4012	0·9936	0·1272	1·2683	1·1200	0·1136	0·1036	0·0991	0·1342	0·0589
0·14	0·4563	1·0918	0·1498	1·3130	1·1400	0·1314	0·1324	0·1272	0·1700	0·0685
0·16	0·5091	1·1779	0·1728	1·3578	1·1600	0·1490	0·1637	0·1578	0·2089	0·0780
0·18	0·5596	1·2534	0·1962	1·4025	1·1800	0·1663	0·1972	0·1909	0·2505	0·0875
0·20	0·6081	1·3199	0·2200	1·4472	1·2000	0·1833	0·2330	0·2262	0·2950	0·0970
0·22	0·6547	1·3786	0·2442	1·4919	1·2200	0·2002	0·2709	0·2638	0·3421	0·1064
0·24	0·6997	1·4305	0·2688	1·5367	1·2400	0·2168	0·3109	0·3035	0·3919	0·1157
0·26	0·7431	1·4763	0·2938	1·5814	1·2600	0·2332	0·3528	0·3453	0·4443	0·1250
0·28	0·7852	1·5169	0·3192	1·6261	1·2800	0·2494	0·3967	0·3892	0·4992	0·1343
0·30	0·8259	1·5530	0·3450	1·6708	1·3000	0·2654	0·4425	0·4351	0·5566	0·1435
0·32	0·8655	1·5849	0·3712	1·7155	1·3200	0·2812	0·4901	0·4829	0·6164	0·1526
0·34	0·9040	1·6133	0·3978	1·7603	1·3400	0·2969	0·5396	0·5328	0·6787	0·1618
0·36	0·9414	1·6385	0·4248	1·8050	1·3600	0·3124	0·5909	0·5845	0·7435	0·1708
0·38	0·9779	1·6608	0·4522	1·8497	1·3800	0·3277	0·6441	0·6382	0·8106	0·1799
0·40	1·0135	1·6807	0·4800	1·8944	1·4000	0·3429	0·6990	0·6938	0·8802	0·1889
0·42	1·0483	1·6983	0·5082	1·9391	1·4200	0·3579	0·7557	0·7513	0·9521	0·1979
0·44	1·0823	1·7139	0·5368	1·9839	1·4400	0·3728	0·8141	0·8107	1·0264	0·2068
0·46	1·1157	1·7277	0·5658	2·0286	1·4600	0·3875	0·8743	0·8719	1·1030	0·2157
0·48	1·1483	1·7399	0·5952	2·0733	1·4800	0·4022	0·9363	0·9350	1·1820	0·2245
0·50	1·1803	1·7507	0·6250	2·1180	1·5000	0·4167	1·0000	1·0000	1·2634	0·2333
0·52	1·2118	1·7602	0·6552	2·1628	1·5200	0·4311	1·0655	1·0669	1·3471	0·2421
0·54	1·2427	1·7685	0·6858	2·2075	1·5400	0·4453	1·1327	1·1356	1·4332	0·2509
0·56	1·2731	1·7758	0·7168	2·2522	1·5600	0·4595	1·2016	1·2062	1·5216	0·2596
0·58	1·3030	1·7821	0·7482	2·2969	1·5800	0·4735	1·2723	1·2787	1·6123	0·2683
0·60	1·3324	1·7875	0·7800	2·3416	1·6000	0·4875	1·3448	1·3530	1·7055	0·2769
0·62	1·3614	1·7922	0·8122	2·3864	1·6200	0·5014	1·4190	1·4292	1·8009	0·2855
0·64	1·3900	1·7962	0·8448	2·4311	1·6400	0·5151	1·4949	1·5073	1·8988	0·2941
0·66	1·4182	1·7995	0·8778	2·4758	1·6600	0·5288	1·5726	1·5873	1·9989	0·3027
0·68	1·4460	1·8023	0·9112	2·5205	1·6800	0·5424	1·6520	1·6692	2·1015	0·3112
0·70	1·4735	1·8046	0·9450	2·5652	1·7000	0·5559	1·7332	1·7530	2·2064	0·3198
0·72	1·5007	1·8063	0·9792	2·6100	1·7200	0·5693	1·8162	1·8387	2·3137	0·3282
0·74	1·5276	1·8077	1·0138	2·6547	1·7400	0·5826	1·9010	1·9263	2·4233	0·3367
0·76	1·5541	1·8086	1·0488	2·6994	1·7600	0·5959	1·9875	2·0159	2·5354	0·3451
0·78	1·5804	1·8093	1·0842	2·7441	1·7800	0·6091	2·0758	2·1073	2·6498	0·3535
0·80	1·6064	1·8095	1·1200	2·7889	1·8000	0·6222	2·1659	2·2007	2·7666	0·3619

Rel. dpth Y	U.equ. dia. $D_{ep(u)}$ (m)	Equiv. disch. factor J	Unit sect. area A_u (m²)	Unit wetted perim. P_u (m)	Unit surf. brdth $B_{s(u)}$ (m)	Unit mean depth $y_{m(u)}$ (m)	Discharge ratios $Q/Q_{0.50}$ medial	Mann'g	U.crit. disch. $Q_{c(u)}$ (m³s⁻¹)	Unit depth cntrd $y_{d(u)}$ (m)
0·82	1·6321	1·8095	1·1562	2·8336	1·8200	0·6353	2·2578	2·2961	2·8859	0·3703
0·84	1·6576	1·8092	1·1928	2·8783	1·8400	0·6483	2·3515	2·3934	3·0075	0·3786
0·86	1·6829	1·8087	1·2298	2·9230	1·8600	0·6612	2·4470	2·4926	3·1315	0·3869
0·88	1·7080	1·8080	1·2672	2·9677	1·8800	0·6740	2·5443	2·5939	3·2580	0·3952
0·90	1·7328	1·8070	1·3050	3·0125	1·9000	0·6868	2·6434	2·6971	3·3869	0·4034
0·92	1·7574	1·8059	1·3432	3·0572	1·9200	0·6996	2·7444	2·8023	3·5182	0·4117
0·94	1·7819	1·8046	1·3818	3·1019	1·9400	0·7123	2·8472	2·9095	3·6520	0·4199
0·96	1·8061	1·8032	1·4208	3·1466	1·9600	0·7249	2·9519	3·0187	3·7882	0·4281
0·98	1·8302	1·8016	1·4602	3·1913	1·9800	0·7375	3·0584	3·1299	3·9269	0·4363
1·00	1·8541	1·7999	1·5000	3·2361	2·0000	0·7500	3·1668	3·2431	4·0680	0·4444
1·02	1·8778	1·7981	1·5402	3·2808	2·0200	0·7625	3·2770	3·3584	4·2116	0·4526
1·04	1·9014	1·7962	1·5808	3·3255	2·0400	0·7749	3·3891	3·4757	4·3577	0·4607
1·06	1·9249	1·7942	1·6218	3·3702	2·0600	0·7873	3·5031	3·5951	4·5063	0·4688
1·08	1·9481	1·7922	1·6632	3·4150	2·0800	0·7996	3·6190	3·7165	4·6574	0·4769
1·10	1·9713	1·7900	1·7050	3·4597	2·1000	0·8119	3·7368	3·8401	4·8110	0·4849
1·12	1·9943	1·7878	1·7472	3·5044	2·1200	0·8242	3·8565	3·9657	4·9671	0·4930
1·14	2·0172	1·7855	1·7898	3·5491	2·1400	0·8364	3·9781	4·0934	5·1258	0·5010
1·16	2·0399	1·7832	1·8328	3·5938	2·1600	0·8485	4·1016	4·2232	5·2870	0·5090
1·18	2·0626	1·7808	1·8762	3·6386	2·1800	0·8606	4·2271	4·3551	5·4507	0·5170
1·20	2·0851	1·7784	1·9200	3·6833	2·2000	0·8727	4·3545	4·4892	5·6170	0·5250
1·24	2·1298	1·7735	2·0088	3·7727	2·2400	0·8968	4·6151	4·7637	5·9572	0·5409
1·28	2·1741	1·7684	2·0992	3·8622	2·2800	0·9207	4·8837	5·0469	6·3077	0·5567
1·32	2·2180	1·7633	2·1912	3·9516	2·3200	0·9445	5·1602	5·3388	6·6687	0·5725
1·36	2·2616	1·7582	2·2848	4·0411	2·3600	0·9681	5·4446	5·6395	7·0401	0·5883
1·40	2·3048	1·7530	2·3800	4·1305	2·4000	0·9917	5·7372	5·9491	7·4220	0·6039
1·44	2·3477	1·7477	2·4768	4·2199	2·4400	1·0151	6·0378	6·2676	7·8145	0·6195
1·48	2·3903	1·7425	2·5752	4·3094	2·4800	1·0384	6·3467	6·5952	8·2177	0·6351
1·52	2·4327	1·7373	2·6752	4·3988	2·5200	1·0616	6·6639	6·9320	8·6317	0·6506
1·56	2·4747	1·7322	2·7768	4·4883	2·5600	1·0847	6·9894	7·2780	9·0564	0·6661
1·60	2·5165	1·7270	2·8800	4·5777	2·6000	1·1077	7·3233	7·6333	9·4921	0·6815
1·64	2·5581	1·7219	2·9848	4·6672	2·6400	1·1306	7·6656	7·9980	9·9387	0·6968
1·68	2·5995	1·7169	3·0912	4·7566	2·6800	1·1534	8·0166	8·3721	10·396	0·7122
1·72	2·6407	1·7119	3·1992	4·8460	2·7200	1·1762	8·3761	8·7559	10·865	0·7275
1·76	2·6816	1·7069	3·3088	4·9355	2·7600	1·1988	8·7443	9·1493	11·345	0·7427
1·80	2·7224	1·7020	3·4200	5·0249	2·8000	1·2214	9·1213	9·5524	11·836	0·7579
1·84	2·7630	1·6972	3·5328	5·1144	2·8400	1·2439	9·5071	9·9653	12·339	0·7731
1·88	2·8035	1·6925	3·6472	5·2038	2·8800	1·2664	9·9017	10·388	12·853	0·7882
1·92	2·8438	1·6878	3·7632	5·2933	2·9200	1·2888	10·305	10·821	13·378	0·8033
1·96	2·8839	1·6831	3·8808	5·3827	2·9600	1·3111	10·718	11·264	13·915	0·8183
2·00	2·9239	1·6786	4·0000	5·4721	3·0000	1·3333	11·140	11·717	14·464	0·8333
2·05	2·9737	1·6730	4·1512	5·5839	3·0500	1·3611	11·680	12·298	15·166	0·8521
2·10	3·0233	1·6675	4·3050	5·6957	3·1000	1·3887	12·234	12·895	15·887	0·8707
2·15	3·0727	1·6622	4·4613	5·8075	3·1500	1·4163	12·803	13·508	16·626	0·8894
2·20	3·1220	1·6569	4·6200	5·9193	3·2000	1·4438	13·387	14·138	17·384	0·9079
2·25	3·1710	1·6517	4·7813	6·0312	3·2500	1·4712	13·985	14·784	18·161	0·9265
2·30	3·2199	1·6467	4·9450	6·1430	3·3000	1·4985	14·599	15·447	18·956	0·9450
2·35	3·2687	1·6417	5·1112	6·2548	3·3500	1·5257	15·227	16·127	19·771	0·9634
2·40	3·3173	1·6369	5·2800	6·3666	3·4000	1·5529	15·871	16·824	20·605	0·9818
2·45	3·3658	1·6322	5·4513	6·4784	3·4500	1·5801	16·530	17·539	21·458	1·0002
2·50	3·4142	1·6275	5·6250	6·5902	3·5000	1·6071	17·204	18·271	22·331	1·0185
2·55	3·4624	1·6230	5·8012	6·7020	3·5500	1·6342	17·894	19·021	23·224	1·0368
2·60	3·5105	1·6185	5·9800	6·8138	3·6000	1·6611	18·600	19·788	24·136	1·0551
2·65	3·5585	1·6142	6·1612	6·9256	3·6500	1·6880	19·321	20·573	25·068	1·0733
2·70	3·6065	1·6099	6·3450	7·0374	3·7000	1·7149	20·058	21·376	26·020	1·0915
2·75	3·6543	1·6058	6·5312	7·1492	3·7500	1·7417	20·812	22·198	26·992	1·1096

C55 Trapezoidal channel - 0·625 to 1 side-slope

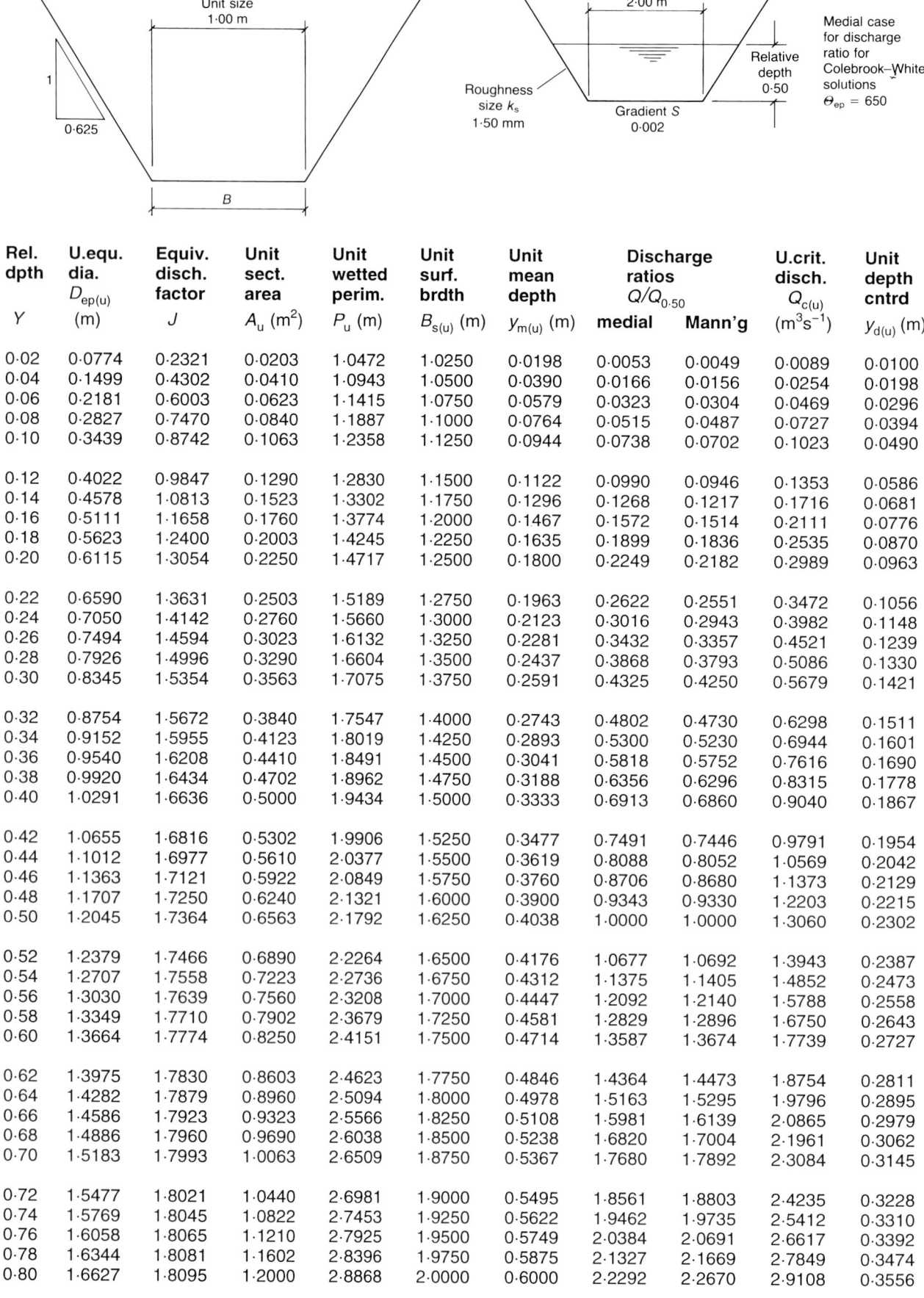

Rel. dpth Y	U.equ. dia. $D_{ep(u)}$ (m)	Equiv. disch. factor J	Unit sect. area A_u (m²)	Unit wetted perim. P_u (m)	Unit surf. brdth $B_{s(u)}$ (m)	Unit mean depth $y_{m(u)}$ (m)	Discharge ratios $Q/Q_{0.50}$ medial	Mann'g	U.crit. disch. $Q_{c(u)}$ (m³s⁻¹)	Unit depth cntrd $y_{d(u)}$ (m)
0·02	0·0774	0·2321	0·0203	1·0472	1·0250	0·0198	0·0053	0·0049	0·0089	0·0100
0·04	0·1499	0·4302	0·0410	1·0943	1·0500	0·0390	0·0166	0·0156	0·0254	0·0198
0·06	0·2181	0·6003	0·0623	1·1415	1·0750	0·0579	0·0323	0·0304	0·0469	0·0296
0·08	0·2827	0·7470	0·0840	1·1887	1·1000	0·0764	0·0515	0·0487	0·0727	0·0394
0·10	0·3439	0·8742	0·1063	1·2358	1·1250	0·0944	0·0738	0·0702	0·1023	0·0490
0·12	0·4022	0·9847	0·1290	1·2830	1·1500	0·1122	0·0990	0·0946	0·1353	0·0586
0·14	0·4578	1·0813	0·1523	1·3302	1·1750	0·1296	0·1268	0·1217	0·1716	0·0681
0·16	0·5111	1·1658	0·1760	1·3774	1·2000	0·1467	0·1572	0·1514	0·2111	0·0776
0·18	0·5623	1·2400	0·2003	1·4245	1·2250	0·1635	0·1899	0·1836	0·2535	0·0870
0·20	0·6115	1·3054	0·2250	1·4717	1·2500	0·1800	0·2249	0·2182	0·2989	0·0963
0·22	0·6590	1·3631	0·2503	1·5189	1·2750	0·1963	0·2622	0·2551	0·3472	0·1056
0·24	0·7050	1·4142	0·2760	1·5660	1·3000	0·2123	0·3016	0·2943	0·3982	0·1148
0·26	0·7494	1·4594	0·3023	1·6132	1·3250	0·2281	0·3432	0·3357	0·4521	0·1239
0·28	0·7926	1·4996	0·3290	1·6604	1·3500	0·2437	0·3868	0·3793	0·5086	0·1330
0·30	0·8345	1·5354	0·3563	1·7075	1·3750	0·2591	0·4325	0·4250	0·5679	0·1421
0·32	0·8754	1·5672	0·3840	1·7547	1·4000	0·2743	0·4802	0·4730	0·6298	0·1511
0·34	0·9152	1·5955	0·4123	1·8019	1·4250	0·2893	0·5300	0·5230	0·6944	0·1601
0·36	0·9540	1·6208	0·4410	1·8491	1·4500	0·3041	0·5818	0·5752	0·7616	0·1690
0·38	0·9920	1·6434	0·4702	1·8962	1·4750	0·3188	0·6356	0·6296	0·8315	0·1778
0·40	1·0291	1·6636	0·5000	1·9434	1·5000	0·3333	0·6913	0·6860	0·9040	0·1867
0·42	1·0655	1·6816	0·5302	1·9906	1·5250	0·3477	0·7491	0·7446	0·9791	0·1954
0·44	1·1012	1·6977	0·5610	2·0377	1·5500	0·3619	0·8088	0·8052	1·0569	0·2042
0·46	1·1363	1·7121	0·5922	2·0849	1·5750	0·3760	0·8706	0·8680	1·1373	0·2129
0·48	1·1707	1·7250	0·6240	2·1321	1·6000	0·3900	0·9343	0·9330	1·2203	0·2215
0·50	1·2045	1·7364	0·6563	2·1792	1·6250	0·4038	1·0000	1·0000	1·3060	0·2302
0·52	1·2379	1·7466	0·6890	2·2264	1·6500	0·4176	1·0677	1·0692	1·3943	0·2387
0·54	1·2707	1·7558	0·7223	2·2736	1·6750	0·4312	1·1375	1·1405	1·4852	0·2473
0·56	1·3030	1·7639	0·7560	2·3208	1·7000	0·4447	1·2092	1·2140	1·5788	0·2558
0·58	1·3349	1·7710	0·7902	2·3679	1·7250	0·4581	1·2829	1·2896	1·6750	0·2643
0·60	1·3664	1·7774	0·8250	2·4151	1·7500	0·4714	1·3587	1·3674	1·7739	0·2727
0·62	1·3975	1·7830	0·8603	2·4623	1·7750	0·4846	1·4364	1·4473	1·8754	0·2811
0·64	1·4282	1·7879	0·8960	2·5094	1·8000	0·4978	1·5163	1·5295	1·9796	0·2895
0·66	1·4586	1·7923	0·9323	2·5566	1·8250	0·5108	1·5981	1·6139	2·0865	0·2979
0·68	1·4886	1·7960	0·9690	2·6038	1·8500	0·5238	1·6820	1·7004	2·1961	0·3062
0·70	1·5183	1·7993	1·0063	2·6509	1·8750	0·5367	1·7680	1·7892	2·3084	0·3145
0·72	1·5477	1·8021	1·0440	2·6981	1·9000	0·5495	1·8561	1·8803	2·4235	0·3228
0·74	1·5769	1·8045	1·0822	2·7453	1·9250	0·5622	1·9462	1·9735	2·5412	0·3310
0·76	1·6058	1·8065	1·1210	2·7925	1·9500	0·5749	2·0384	2·0691	2·6617	0·3392
0·78	1·6344	1·8081	1·1602	2·8396	1·9750	0·5875	2·1327	2·1669	2·7849	0·3474
0·80	1·6627	1·8095	1·2000	2·8868	2·0000	0·6000	2·2292	2·2670	2·9108	0·3556

Rel. dpth Y	U.equ. dia. $D_{ep(u)}$ (m)	Equiv. disch. factor J	Unit sect. area A_u (m²)	Unit wetted perim. P_u (m)	Unit surf. brdth $B_{s(u)}$ (m)	Unit mean depth $y_{m(u)}$ (m)	Discharge ratios $Q/Q_{0.50}$ medial	Mann'g	U.crit. disch. $Q_{c(u)}$ (m³s⁻¹)	Unit depth cntrd $y_{d(u)}$ (m)
0·82	1·6909	1·8105	1·2402	2·9340	2·0250	0·6125	2·3277	2·3694	3·0396	0·3637
0·84	1·7188	1·8113	1·2810	2·9811	2·0500	0·6249	2·4284	2·4741	3·1711	0·3718
0·86	1·7465	1·8118	1·3222	3·0283	2·0750	0·6372	2·5313	2·5811	3·3054	0·3799
0·88	1·7740	1·8121	1·3640	3·0755	2·1000	0·6495	2·6363	2·6905	3·4425	0·3880
0·90	1·8014	1·8122	1·4062	3·1226	2·1250	0·6618	2·7435	2·8023	3·5824	0·3960
0·92	1·8285	1·8122	1·4490	3·1698	2·1500	0·6740	2·8528	2·9164	3·7251	0·4040
0·94	1·8555	1·8119	1·4922	3·2170	2·1750	0·6861	2·9644	3·0329	3·8707	0·4120
0·96	1·8823	1·8115	1·5360	3·2642	2·2000	0·6982	3·0781	3·1518	4·0192	0·4200
0·98	1·9089	1·8110	1·5803	3·3113	2·2250	0·7102	3·1941	3·2731	4·1705	0·4280
1·00	1·9354	1·8104	1·6250	3·3585	2·2500	0·7222	3·3123	3·3969	4·3246	0·4359
1·02	1·9617	1·8096	1·6702	3·4057	2·2750	0·7342	3·4327	3·5231	4·4817	0·4438
1·04	1·9879	1·8087	1·7160	3·4528	2·3000	0·7461	3·5554	3·6517	4·6416	0·4517
1·06	2·0140	1·8077	1·7622	3·5000	2·3250	0·7580	3·6804	3·7829	4·8045	0·4596
1·08	2·0399	1·8066	1·8090	3·5472	2·3500	0·7698	3·8076	3·9165	4·9703	0·4675
1·10	2·0657	1·8055	1·8562	3·5943	2·3750	0·7816	3·9371	4·0526	5·1391	0·4753
1·12	2·0914	1·8043	1·9040	3·6415	2·4000	0·7933	4·0689	4·1913	5·3107	0·4831
1·14	2·1170	1·8030	1·9522	3·6887	2·4250	0·8051	4·2030	4·3325	5·4854	0·4909
1·16	2·1425	1·8016	2·0010	3·7359	2·4500	0·8167	4·3395	4·4762	5·6630	0·4987
1·18	2·1678	1·8002	2·0502	3·7830	2·4750	0·8284	4·4783	4·6225	5·8436	0·5065
1·20	2·1931	1·7988	2·1000	3·8302	2·5000	0·8400	4·6194	4·7713	6·0273	0·5143
1·24	2·2433	1·7957	2·2010	3·9245	2·5500	0·8631	4·9087	5·0769	6·4035	0·5298
1·28	2·2932	1·7926	2·3040	4·0189	2·6000	0·8862	5·2076	5·3929	6·7920	0·5452
1·32	2·3427	1·7893	2·4090	4·1132	2·6500	0·9091	5·5161	5·7196	7·1927	0·5605
1·36	2·3919	1·7859	2·5160	4·2076	2·7000	0·9319	5·8344	6·0569	7·6058	0·5759
1·40	2·4408	1·7824	2·6250	4·3019	2·7500	0·9545	6·1625	6·4052	8·0313	0·5911
1·44	2·4894	1·7789	2·7360	4·3962	2·8000	0·9771	6·5004	6·7644	8·4695	0·6063
1·48	2·5378	1·7754	2·8490	4·4906	2·8500	0·9996	6·8484	7·1347	8·9202	0·6215
1·52	2·5859	1·7718	2·9640	4·5849	2·9000	1·0221	7·2065	7·5162	9·3838	0·6366
1·56	2·6338	1·7682	3·0810	4·6793	2·9500	1·0444	7·5749	7·9090	9·8602	0·6516
1·60	2·6814	1·7647	3·2000	4·7736	3·0000	1·0667	7·9535	8·3133	10·350	0·6667
1·64	2·7289	1·7611	3·3210	4·8679	3·0500	1·0889	8·3425	8·7292	10·852	0·6816
1·68	2·7761	1·7575	3·4440	4·9623	3·1000	1·1110	8·7420	9·1567	11·368	0·6966
1·72	2·8232	1·7540	3·5690	5·0566	3·1500	1·1330	9·1520	9·5961	11·897	0·7115
1·76	2·8701	1·7505	3·6960	5·1510	3·2000	1·1550	9·5728	10·047	12·439	0·7263
1·80	2·9169	1·7470	3·8250	5·2453	3·2500	1·1769	10·004	10·511	12·995	0·7412
1·84	2·9635	1·7435	3·9560	5·3396	3·3000	1·1988	10·447	10·986	13·564	0·7560
1·88	3·0100	1·7401	4·0890	5·4340	3·3500	1·2206	10·900	11·474	14·147	0·7707
1·92	3·0563	1·7368	4·2240	5·5283	3·4000	1·2424	11·364	11·974	14·744	0·7855
1·96	3·1025	1·7334	4·3610	5·6227	3·4500	1·2641	11·840	12·486	15·354	0·8001
2·00	3·1485	1·7301	4·5000	5·7170	3·5000	1·2857	12·327	13·012	15·979	0·8148
2·05	3·2059	1·7261	4·6766	5·8349	3·5625	1·3127	12·951	13·686	16·779	0·8331
2·10	3·2631	1·7221	4·8562	5·9528	3·6250	1·3397	13·593	14·381	17·602	0·8514
2·15	3·3202	1·7182	5·0391	6·0708	3·6875	1·3665	14·253	15·095	18·447	0·8696
2·20	3·3771	1·7143	5·2250	6·1887	3·7500	1·3933	14·932	15·831	19·314	0·8877
2·25	3·4339	1·7105	5·4141	6·3066	3·8125	1·4201	15·628	16·587	20·204	0·9058
2·30	3·4905	1·7068	5·6062	6·4245	3·8750	1·4468	16·344	17·364	21·117	0·9239
2·35	3·5470	1·7032	5·8016	6·5425	3·9375	1·4734	17·078	18·162	22·053	0·9420
2·40	3·6034	1·6996	6·0000	6·6604	4·0000	1·5000	17·831	18·982	23·012	0·9600
2·45	3·6596	1·6961	6·2016	6·7783	4·0625	1·5265	18·602	19·823	23·995	0·9780
2·50	3·7158	1·6927	6·4063	6·8962	4·1250	1·5530	19·394	20·687	25·001	0·9959
2·55	3·7718	1·6893	6·6141	7·0142	4·1875	1·5795	20·204	21·572	26·031	1·0139
2·60	3·8278	1·6860	6·8250	7·1321	4·2500	1·6059	21·034	22·479	27·084	1·0317
2·65	3·8836	1·6828	7·0391	7·2500	4·3125	1·6322	21·884	23·409	28·162	1·0496
2·70	3·9394	1·6797	7·2563	7·3679	4·3750	1·6586	22·754	24·362	29·264	1·0674
2·75	3·9950	1·6766	7·4766	7·4859	4·4375	1·6849	23·643	25·338	30·391	1·0852

C56

Trapezoidal channel - 0·75 to 1 side-slope

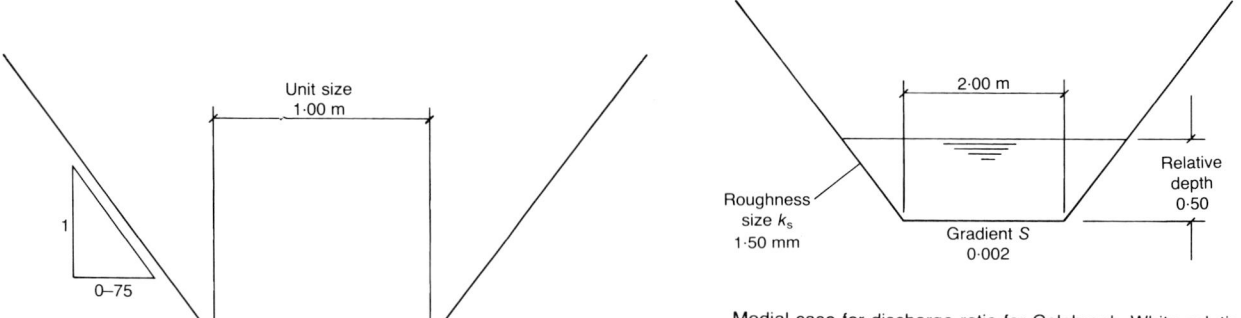

Unit size 1·00 m

2·00 m

Roughness size k_s 1·50 mm

Gradient S 0·002

Relative depth 0·50

B

Medial case for discharge ratio for Colebrook–White solutions
$\Theta_{ep} = 660$

Rel. dpth Y	U.equ. dia. $D_{ep(u)}$ (m)	Equiv. disch. factor J	Unit sect. area A_u (m²)	Unit wetted perim. P_u (m)	Unit surf. brdth $B_{s(u)}$ (m)	Unit mean depth $y_{m(u)}$ (m)	Discharge ratios $Q/Q_{0.50}$ medial	Mann'g	U.crit. disch. $Q_{c(u)}$ (m³s⁻¹)	Unit depth cntrd $y_{d(u)}$ (m)
0·02	0·0773	0·2314	0·0203	1·0500	1·0300	0·0197	0·0050	0·0047	0·0089	0·0100
0·04	0·1498	0·4279	0·0412	1·1000	1·0600	0·0389	0·0158	0·0148	0·0254	0·0198
0·06	0·2181	0·5958	0·0627	1·1500	1·0900	0·0575	0·0307	0·0289	0·0471	0·0296
0·08	0·2827	0·7400	0·0848	1·2000	1·1200	0·0757	0·0492	0·0465	0·0731	0·0392
0·10	0·3440	0·8645	0·1075	1·2500	1·1500	0·0935	0·0707	0·0672	0·1029	0·0488
0·12	0·4025	0·9726	0·1308	1·3000	1·1800	0·1108	0·0950	0·0907	0·1364	0·0583
0·14	0·4584	1·0667	0·1547	1·3500	1·2100	0·1279	0·1220	0·1170	0·1732	0·0678
0·16	0·5120	1·1489	0·1792	1·4000	1·2400	0·1445	0·1516	0·1459	0·2133	0·0771
0·18	0·5636	1·2210	0·2043	1·4500	1·2700	0·1609	0·1836	0·1774	0·2566	0·0864
0·20	0·6133	1·2845	0·2300	1·5000	1·3000	0·1769	0·2179	0·2113	0·3030	0·0957
0·22	0·6614	1·3406	0·2563	1·5500	1·3300	0·1927	0·2546	0·2476	0·3523	0·1048
0·24	0·7080	1·3901	0·2832	1·6000	1·3600	0·2082	0·2936	0·2862	0·4047	0·1139
0·26	0·7532	1·4341	0·3107	1·6500	1·3900	0·2235	0·3348	0·3273	0·4600	0·1229
0·28	0·7972	1·4731	0·3388	1·7000	1·4200	0·2386	0·3782	0·3706	0·5182	0·1319
0·30	0·8400	1·5079	0·3675	1·7500	1·4500	0·2534	0·4238	0·4163	0·5794	0·1408
0·32	0·8818	1·5390	0·3968	1·8000	1·4800	0·2681	0·4716	0·4643	0·6434	0·1497
0·34	0·9226	1·5667	0·4267	1·8500	1·5100	0·2826	0·5216	0·5145	0·7103	0·1585
0·36	0·9625	1·5915	0·4572	1·9000	1·5400	0·2969	0·5737	0·5671	0·7801	0·1672
0·38	1·0016	1·6137	0·4883	1·9500	1·5700	0·3110	0·6281	0·6220	0·8528	0·1760
0·40	1·0400	1·6336	0·5200	2·0000	1·6000	0·3250	0·6846	0·6792	0·9283	0·1846
0·42	1·0777	1·6515	0·5523	2·0500	1·6300	0·3388	0·7433	0·7387	1·0068	0·1932
0·44	1·1147	1·6675	0·5852	2·1000	1·6600	0·3525	0·8042	0·8005	1·0881	0·2018
0·46	1·1511	1·6819	0·6187	2·1500	1·6900	0·3661	0·8673	0·8647	1·1723	0·2103
0·48	1·1869	1·6949	0·6528	2·2000	1·7200	0·3795	0·9325	0·9311	1·2594	0·2188
0·50	1·2222	1·7065	0·6875	2·2500	1·7500	0·3929	1·0000	1·0000	1·3494	0·2273
0·52	1·2570	1·7170	0·7228	2·3000	1·7800	0·4061	1·0697	1·0712	1·4424	0·2357
0·54	1·2914	1·7264	0·7587	2·3500	1·8100	0·4192	1·1417	1·1448	1·5382	0·2441
0·56	1·3253	1·7348	0·7952	2·4000	1·8400	0·4322	1·2158	1·2208	1·6371	0·2524
0·58	1·3589	1·7424	0·8323	2·4500	1·8700	0·4451	1·2923	1·2992	1·7388	0·2607
0·60	1·3920	1·7492	0·8700	2·5000	1·9000	0·4579	1·3710	1·3801	1·8436	0·2690
0·62	1·4248	1·7553	0·9083	2·5500	1·9300	0·4706	1·4519	1·4634	1·9513	0·2772
0·64	1·4572	1·7607	0·9472	2·6000	1·9600	0·4833	1·5352	1·5491	2·0620	0·2854
0·66	1·4894	1·7656	0·9867	2·6500	1·9900	0·4958	1·6208	1·6374	2·1758	0·2936
0·68	1·5212	1·7699	1·0268	2·7000	2·0200	0·5083	1·7087	1·7281	2·2925	0·3017
0·70	1·5527	1·7738	1·0675	2·7500	2·0500	0·5207	1·7989	1·8213	2·4123	0·3098
0·72	1·5840	1·7772	1·1088	2·8000	2·0800	0·5331	1·8915	1·9171	2·5352	0·3179
0·74	1·6150	1·7802	1·1507	2·8500	2·1100	0·5454	1·9864	2·0155	2·6611	0·3260
0·76	1·6458	1·7829	1·1932	2·9000	2·1400	0·5576	2·0838	2·1164	2·7901	0·3340
0·78	1·6763	1·7852	1·2363	2·9500	2·1700	0·5697	2·1835	2·2199	2·9222	0·3420
0·80	1·7067	1·7872	1·2800	3·0000	2·2000	0·5818	2·2856	2·3260	3·0575	0·3500

Rel. dpth Y	U.equ. dia. $D_{ep(u)}$ (m)	Equiv. disch. factor J	Unit sect. area A_u (m²)	Unit wetted perim. P_u (m)	Unit surf. brdth $B_{s(u)}$ (m)	Unit mean depth $y_{m(u)}$ (m)	Discharge ratios $Q/Q_{0.50}$ medial	Mann'g	U.crit. disch. $Q_{c(u)}$ (m³s⁻¹)	Unit depth cntrd $y_{d(u)}$ (m)
0·82	1·7368	1·7889	1·3243	3·0500	2·2300	0·5939	2·3901	2·4347	3·1959	0·3580
0·84	1·7667	1·7904	1·3692	3·1000	2·2600	0·6058	2·4971	2·5461	3·3374	0·3659
0·86	1·7964	1·7916	1·4147	3·1500	2·2900	0·6178	2·6066	2·6601	3·4821	0·3738
0·88	1·8260	1·7926	1·4608	3·2000	2·3200	0·6297	2·7185	2·7768	3·6300	0·3817
0·90	1·8554	1·7935	1·5075	3·2500	2·3500	0·6415	2·8329	2·8963	3·7810	0·3896
0·92	1·8846	1·7941	1·5548	3·3000	2·3800	0·6533	2·9498	3·0184	3·9354	0·3974
0·94	1·9137	1·7946	1·6027	3·3500	2·4100	0·6650	3·0692	3·1433	4·0929	0·4052
0·96	1·9426	1·7949	1·6512	3·4000	2·4400	0·6767	3·1912	3·2710	4·2537	0·4130
0·98	1·9714	1·7951	1·7003	3·4500	2·4700	0·6884	3·3157	3·4014	4·4177	0·4208
1·00	2·0000	1·7952	1·7500	3·5000	2·5000	0·7000	3·4428	3·5347	4·5851	0·4286
1·02	2·0285	1·7951	1·8003	3·5500	2·5300	0·7116	3·5724	3·6708	4·7557	0·4363
1·04	2·0569	1·7949	1·8512	3·6000	2·5600	0·7231	3·7047	3·8097	4·9297	0·4440
1·06	2·0852	1·7947	1·9027	3·6500	2·5900	0·7346	3·8395	3·9514	5·1070	0·4518
1·08	2·1133	1·7943	1·9548	3·7000	2·6200	0·7461	3·9770	4·0961	5·2877	0·4594
1·10	2·1413	1·7939	2·0075	3·7500	2·6500	0·7575	4·1172	4·2436	5·4717	0·4671
1·12	2·1693	1·7934	2·0608	3·8000	2·6800	0·7690	4·2600	4·3941	5·6591	0·4748
1·14	2·1971	1·7928	2·1147	3·8500	2·7100	0·7803	4·4054	4·5475	5·8499	0·4824
1·16	2·2248	1·7921	2·1692	3·9000	2·7400	0·7917	4·5536	4·7039	6·0441	0·4901
1·18	2·2525	1·7914	2·2243	3·9500	2·7700	0·8030	4·7045	4·8632	6·2418	0·4977
1·20	2·2800	1·7907	2·2800	4·0000	2·8000	0·8143	4·8581	5·0256	6·4429	0·5053
1·24	2·3348	1·7890	2·3932	4·1000	2·8600	0·8368	5·1735	5·3593	6·8556	0·5204
1·28	2·3893	1·7872	2·5088	4·2000	2·9200	0·8592	5·5000	5·7053	7·2823	0·5355
1·32	2·4435	1·7852	2·6268	4·3000	2·9800	0·8815	5·8377	6·0636	7·7231	0·5506
1·36	2·4975	1·7831	2·7472	4·4000	3·0400	0·9037	6·1868	6·4345	8·1782	0·5655
1·40	2·5511	1·7810	2·8700	4·5000	3·1000	0·9258	6·5473	6·8181	8·6477	0·5805
1·44	2·6045	1·7787	2·9952	4·6000	3·1600	0·9478	6·9193	7·2145	9·1318	0·5954
1·48	2·6577	1·7764	3·1228	4·7000	3·2200	0·9698	7·3031	7·6239	9·6305	0·6102
1·52	2·7107	1·7741	3·2528	4·8000	3·2800	0·9917	7·6987	8·0464	10·144	0·6250
1·56	2·7634	1·7717	3·3852	4·9000	3·3400	1·0135	8·1062	8·4822	10·672	0·6398
1·60	2·8160	1·7693	3·5200	5·0000	3·4000	1·0353	8·5258	8·9315	11·216	0·6545
1·64	2·8684	1·7669	3·6572	5·1000	3·4600	1·0570	8·9576	9·3944	11·775	0·6692
1·68	2·9206	1·7645	3·7968	5·2000	3·5200	1·0786	9·4016	9·8710	12·349	0·6839
1·72	2·9727	1·7620	3·9388	5·3000	3·5800	1·1002	9·8581	10·362	12·938	0·6985
1·76	3·0246	1·7596	4·0832	5·4000	3·6400	1·1218	10·327	10·866	13·543	0·7131
1·80	3·0764	1·7572	4·2300	5·5000	3·7000	1·1432	10·809	11·385	14·163	0·7277
1·84	3·1280	1·7548	4·3792	5·6000	3·7600	1·1647	11·303	11·918	14·800	0·7422
1·88	3·1795	1·7524	4·5308	5·7000	3·8200	1·1861	11·811	12·465	15·452	0·7567
1·92	3·2309	1·7500	4·6848	5·8000	3·8800	1·2074	12·331	13·028	16·121	0·7711
1·96	3·2822	1·7476	4·8412	5·9000	3·9400	1·2287	12·864	13·605	16·805	0·7856
2·00	3·3333	1·7453	5·0000	6·0000	4·0000	1·2500	13·411	14·197	17·506	0·8000
2·05	3·3971	1·7424	5·2019	6·1250	4·0750	1·2765	14·113	14·958	18·405	0·8180
2·10	3·4608	1·7395	5·4075	6·2500	4·1500	1·3030	14·836	15·743	19·330	0·8359
2·15	3·5243	1·7367	5·6169	6·3750	4·2250	1·3294	15·581	16·552	20·281	0·8538
2·20	3·5877	1·7340	5·8300	6·5000	4·3000	1·3558	16·347	17·385	21·258	0·8717
2·25	3·6509	1·7312	6·0469	6·6250	4·3750	1·3821	17·134	18·243	22·262	0·8895
2·30	3·7141	1·7286	6·2675	6·7500	4·4500	1·4084	17·944	19·126	23·293	0·9073
2·35	3·7771	1·7259	6·4919	6·8750	4·5250	1·4347	18·776	20·034	24·350	0·9251
2·40	3·8400	1·7233	6·7200	7·0000	4·6000	1·4609	19·630	20·968	25·435	0·9429
2·45	3·9028	1·7208	6·9519	7·1250	4·6750	1·4870	20·507	21·927	26·547	0·9606
2·50	3·9655	1·7183	7·1875	7·2500	4·7500	1·5132	21·406	22·912	27·687	0·9783
2·55	4·0281	1·7159	7·4269	7·3750	4·8250	1·5392	22·329	23·924	28·855	0·9959
2·60	4·0907	1·7135	7·6700	7·5000	4·9000	1·5653	23·274	24·962	30·051	1·0136
2·65	4·1531	1·7111	7·9169	7·6250	4·9750	1·5913	24·244	26·027	31·275	1·0312
2·70	4·2155	1·7088	8·1675	7·7500	5·0500	1·6173	25·236	27·119	32·527	1·0488
2·75	4·2778	1·7065	8·4219	7·8750	5·1250	1·6433	26·253	28·239	33·809	1·0663

C57

Trapezoidal channel - 0·875 to 1 side-slope

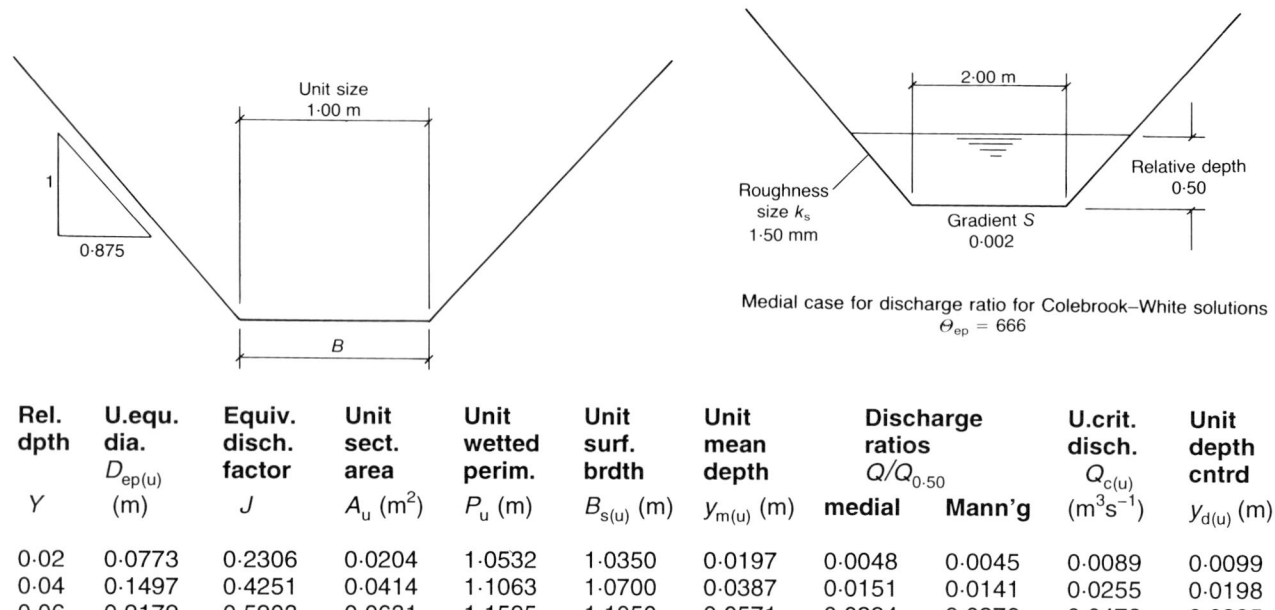

Medial case for discharge ratio for Colebrook–White solutions
$\Theta_{ep} = 666$

Rel. dpth Y	U.equ. dia. $D_{ep(u)}$ (m)	Equiv. disch. factor J	Unit sect. area A_u (m²)	Unit wetted perim. P_u (m)	Unit surf. brdth $B_{s(u)}$ (m)	Unit mean depth $y_{m(u)}$ (m)	Discharge ratios $Q/Q_{0.50}$ medial	Mann'g	U.crit. disch. $Q_{c(u)}$ (m³s⁻¹)	Unit depth cntrd $y_{d(u)}$ (m)
0·02	0·0773	0·2306	0·0204	1·0532	1·0350	0·0197	0·0048	0·0045	0·0089	0·0099
0·04	0·1497	0·4251	0·0414	1·1063	1·0700	0·0387	0·0151	0·0141	0·0255	0·0198
0·06	0·2179	0·5903	0·0631	1·1595	1·1050	0·0571	0·0294	0·0276	0·0473	0·0295
0·08	0·2824	0·7315	0·0856	1·2126	1·1400	0·0751	0·0471	0·0445	0·0735	0·0391
0·10	0·3437	0·8530	0·1088	1·2658	1·1750	0·0926	0·0679	0·0645	0·1036	0·0487
0·12	0·4022	0·9579	0·1326	1·3189	1·2100	0·1096	0·0915	0·0873	0·1375	0·0581
0·14	0·4581	1·0490	0·1571	1·3721	1·2450	0·1262	0·1178	0·1129	0·1748	0·0675
0·16	0·5119	1·1284	0·1824	1·4252	1·2800	0·1425	0·1466	0·1411	0·2156	0·0767
0·18	0·5637	1·1979	0·2083	1·4784	1·3150	0·1584	0·1780	0·1719	0·2597	0·0859
0·20	0·6138	1·2590	0·2350	1·5315	1·3500	0·1741	0·2118	0·2052	0·3070	0·0950
0·22	0·6622	1·3128	0·2624	1·5847	1·3850	0·1894	0·2479	0·2410	0·3576	0·1041
0·24	0·7092	1·3604	0·2904	1·6378	1·4200	0·2045	0·2865	0·2792	0·4113	0·1131
0·26	0·7550	1·4026	0·3192	1·6910	1·4550	0·2193	0·3273	0·3199	0·4681	0·1220
0·28	0·7995	1·4401	0·3486	1·7441	1·4900	0·2340	0·3706	0·3630	0·5280	0·1308
0·30	0·8429	1·4734	0·3788	1·7973	1·5250	0·2484	0·4161	0·4086	0·5911	0·1396
0·32	0·8854	1·5032	0·4096	1·8504	1·5600	0·2626	0·4639	0·4566	0·6573	0·1483
0·34	0·9270	1·5299	0·4412	1·9036	1·5950	0·2766	0·5141	0·5071	0·7265	0·1570
0·36	0·9677	1·5537	0·4734	1·9567	1·6300	0·2904	0·5666	0·5600	0·7989	0·1656
0·38	1·0077	1·5751	0·5063	2·0099	1·6650	0·3041	0·6215	0·6153	0·8744	0·1742
0·40	1·0470	1·5944	0·5400	2·0630	1·7000	0·3176	0·6786	0·6731	0·9531	0·1827
0·42	1·0856	1·6117	0·5744	2·1162	1·7350	0·3310	0·7382	0·7335	1·0348	0·1912
0·44	1·1237	1·6273	0·6094	2·1693	1·7700	0·3443	0·8000	0·7963	1·1198	0·1996
0·46	1·1611	1·6413	0·6452	2·2225	1·8050	0·3574	0·8643	0·8617	1·2078	0·2080
0·48	1·1981	1·6540	0·6816	2·2756	1·8400	0·3704	0·9310	0·9295	1·2991	0·2163
0·50	1·2346	1·6654	0·7188	2·3288	1·8750	0·3833	1·0000	1·0000	1·3936	0·2246
0·52	1·2706	1·6758	0·7566	2·3819	1·9100	0·3961	1·0715	1·0730	1·4912	0·2329
0·54	1·3062	1·6851	0·7952	2·4351	1·9450	0·4088	1·1454	1·1487	1·5921	0·2411
0·56	1·3414	1·6935	0·8344	2·4882	1·9800	0·4214	1·2218	1·2269	1·6962	0·2493
0·58	1·3762	1·7012	0·8743	2·5414	2·0150	0·4339	1·3006	1·3078	1·8036	0·2575
0·60	1·4107	1·7081	0·9150	2·5945	2·0500	0·4463	1·3819	1·3914	1·9143	0·2656
0·62	1·4448	1·7143	0·9564	2·6477	2·0850	0·4587	1·4658	1·4777	2·0283	0·2737
0·64	1·4787	1·7199	0·9984	2·7008	2·1200	0·4709	1·5521	1·5666	2·1456	0·2817
0·66	1·5122	1·7250	1·0412	2·7540	2·1550	0·4831	1·6410	1·6583	2·2662	0·2897
0·68	1·5455	1·7296	1·0846	2·8071	2·1900	0·4953	1·7325	1·7528	2·3902	0·2977
0·70	1·5785	1·7337	1·1287	2·8603	2·2250	0·5073	1·8266	1·8500	2·5176	0·3057
0·72	1·6113	1·7374	1·1736	2·9134	2·2600	0·5193	1·9232	1·9501	2·6484	0·3136
0·74	1·6438	1·7408	1·2192	2·9666	2·2950	0·5312	2·0225	2·0530	2·7826	0·3215
0·76	1·6762	1·7438	1·2654	3·0197	2·3300	0·5431	2·1244	2·1587	2·9203	0·3294
0·78	1·7083	1·7465	1·3123	3·0729	2·3650	0·5549	2·2290	2·2673	3·0614	0·3373
0·80	1·7402	1·7488	1·3600	3·1260	2·4000	0·5667	2·3362	2·3788	3·2060	0·3451

Rel. dpth Y	U.equ. dia. $D_{ep(u)}$ (m)	Equiv. disch. factor J	Unit sect. area A_u (m²)	Unit wetted perim. P_u (m)	Unit surf. brdth $B_{s(u)}$ (m)	Unit mean depth $y_{m(u)}$ (m)	Discharge ratios $Q/Q_{0.50}$ medial	Mann'g	U.crit. disch. $Q_{c(u)}$ (m³s⁻¹)	Unit depth cntrd $y_{d(u)}$ (m)
0·82	1·7720	1·7510	1·4083	3·1792	2·4350	0·5784	2·4461	2·4932	3·3541	0·3529
0·84	1·8035	1·7529	1·4574	3·2323	2·4700	0·5900	2·5588	2·6106	3·5057	0·3607
0·86	1·8349	1·7545	1·5072	3·2855	2·5050	0·6017	2·6742	2·7310	3·6609	0·3685
0·88	1·8662	1·7560	1·5576	3·3386	2·5400	0·6132	2·7924	2·8543	3·8197	0·3762
0·90	1·8972	1·7572	1·6087	3·3918	2·5750	0·6248	2·9133	2·9807	3·9820	0·3839
0·92	1·9282	1·7583	1·6606	3·4449	2·6100	0·6362	3·0370	3·1101	4·1480	0·3916
0·94	1·9590	1·7593	1·7131	3·4981	2·6450	0·6477	3·1636	3·2426	4·3176	0·3993
0·96	1·9896	1·7601	1·7664	3·5512	2·6800	0·6591	3·2930	3·3782	4·4908	0·4070
0·98	2·0202	1·7607	1·8204	3·6044	2·7150	0·6705	3·4252	3·5169	4·6678	0·4146
1·00	2·0506	1·7613	1·8750	3·6575	2·7500	0·6818	3·5603	3·6587	4·8484	0·4222
1·02	2·0809	1·7617	1·9303	3·7107	2·7850	0·6931	3·6983	3·8037	5·0327	0·4298
1·04	2·1110	1·7620	1·9864	3·7638	2·8200	0·7044	3·8393	3·9520	5·2208	0·4374
1·06	2·1411	1·7622	2·0431	3·8170	2·8550	0·7156	3·9831	4·1034	5·4126	0·4450
1·08	2·1711	1·7623	2·1006	3·8701	2·8900	0·7269	4·1299	4·2580	5·6082	0·4525
1·10	2·2010	1·7624	2·1588	3·9233	2·9250	0·7380	4·2797	4·4160	5·8077	0·4601
1·12	2·2307	1·7624	2·2176	3·9764	2·9600	0·7492	4·4325	4·5772	6·0109	0·4676
1·14	2·2604	1·7623	2·2771	4·0296	2·9950	0·7603	4·5883	4·7417	6·2180	0·4751
1·16	2·2900	1·7621	2·3374	4·0827	3·0300	0·7714	4·7471	4·9095	6·4289	0·4826
1·18	2·3195	1·7619	2·3983	4·1359	3·0650	0·7825	4·9090	5·0808	6·6438	0·4901
1·20	2·3490	1·7616	2·4600	4·1890	3·1000	0·7935	5·0739	5·2554	6·8625	0·4976
1·24	2·4076	1·7609	2·5854	4·2953	3·1700	0·8156	5·4131	5·6148	7·3118	0·5125
1·28	2·4660	1·7600	2·7136	4·4016	3·2400	0·8375	5·7648	5·9881	7·7769	0·5273
1·32	2·5241	1·7590	2·8446	4·5079	3·3100	0·8594	6·1291	6·3753	8·2581	0·5421
1·36	2·5819	1·7578	2·9784	4·6142	3·3800	0·8812	6·5062	6·7768	8·7554	0·5568
1·40	2·6395	1·7566	3·1150	4·7206	3·4500	0·9029	6·8963	7·1927	9·2691	0·5715
1·44	2·6969	1·7553	3·2544	4·8269	3·5200	0·9245	7·2996	7·6231	9·7993	0·5862
1·48	2·7541	1·7539	3·3966	4·9332	3·5900	0·9461	7·7160	8·0682	10·346	0·6008
1·52	2·8111	1·7524	3·5416	5·0395	3·6600	0·9677	8·1459	8·5283	10·910	0·6154
1·56	2·8679	1·7509	3·6894	5·1458	3·7300	0·9891	8·5893	9·0036	11·491	0·6299
1·60	2·9246	1·7493	3·8400	5·2521	3·8000	1·0105	9·0465	9·4941	12·088	0·6444
1·64	2·9811	1·7477	3·9934	5·3584	3·8700	1·0319	9·5174	10·000	12·703	0·6589
1·68	3·0374	1·7461	4·1496	5·4647	3·9400	1·0532	10·002	10·522	13·336	0·6734
1·72	3·0936	1·7445	4·3086	5·5710	4·0100	1·0745	10·501	11·059	13·986	0·6878
1·76	3·1497	1·7429	4·4704	5·6773	4·0800	1·0957	11·015	11·613	14·654	0·7022
1·80	3·2056	1·7412	4·6350	5·7836	4·1500	1·1169	11·543	12·183	15·339	0·7165
1·84	3·2615	1·7396	4·8024	5·8899	4·2200	1·1380	12·085	12·769	16·043	0·7308
1·88	3·3172	1·7379	4·9726	5·9962	4·2900	1·1591	12·642	13·371	16·765	0·7451
1·92	3·3728	1·7363	5·1456	6·1025	4·3600	1·1802	13·214	13·991	17·505	0·7594
1·96	3·4283	1·7347	5·3214	6·2088	4·4300	1·2012	13·800	14·627	18·264	0·7737
2·00	3·4837	1·7330	5·5000	6·3151	4·5000	1·2222	14·402	15·281	19·041	0·7879
2·05	3·5529	1·7310	5·7272	6·4479	4·5875	1·2484	15·176	16·122	20·039	0·8056
2·10	3·6219	1·7290	5·9587	6·5808	4·6750	1·2746	15·974	16·990	21·067	0·8233
2·15	3·6908	1·7270	6·1947	6·7137	4·7625	1·3007	16·796	17·886	22·124	0·8410
2·20	3·7595	1·7250	6·4350	6·8466	4·8500	1·3268	17·643	18·810	23·212	0·8587
2·25	3·8282	1·7231	6·6797	6·9795	4·9375	1·3528	18·514	19·762	24·330	0·8763
2·30	3·8968	1·7212	6·9287	7·1123	5·0250	1·3789	19·411	20·743	25·479	0·8939
2·35	3·9652	1·7193	7·1822	7·2452	5·1125	1·4048	20·333	21·753	26·658	0·9115
2·40	4·0336	1·7175	7·4400	7·3781	5·2000	1·4308	21·281	22·792	27·869	0·9290
2·45	4·1018	1·7156	7·7022	7·5110	5·2875	1·4567	22·254	23·861	29·111	0·9466
2·50	4·1700	1·7138	7·9688	7·6438	5·3750	1·4826	23·254	24·959	30·385	0·9641
2·55	4·2381	1·7121	8·2397	7·7767	5·4625	1·5084	24·280	26·088	31·691	0·9815
2·60	4·3062	1·7103	8·5150	7·9096	5·5500	1·5342	25·333	27·247	33·029	0·9990
2·65	4·3741	1·7086	8·7947	8·0425	5·6375	1·5600	26·412	28·438	34·399	1·0164
2·70	4·4420	1·7069	9·0788	8·1753	5·7250	1·5858	27·519	29·659	35·802	1·0338
2·75	4·5098	1·7053	9·3672	8·3082	5·8125	1·6116	28·653	30·912	37·239	1·0512

C58

Trapezoidal channel - 1·0 to 1 side-slope

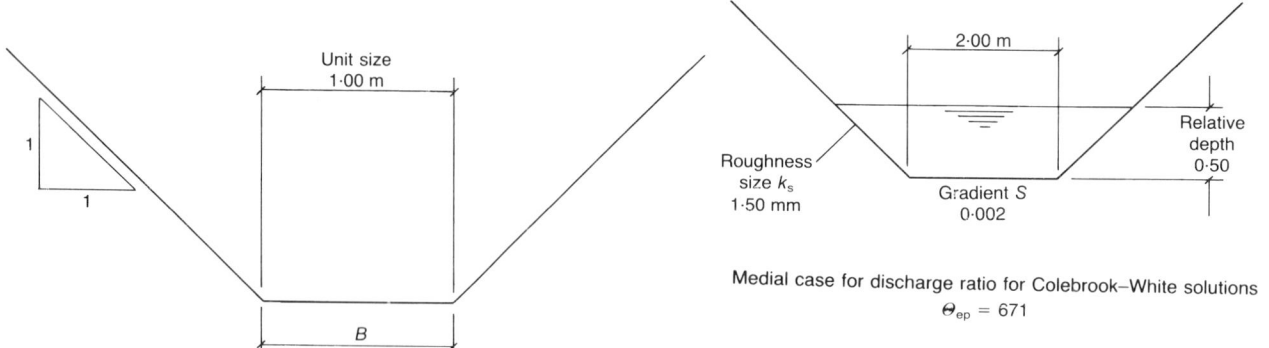

Unit size 1·00 m

B

2·00 m

Roughness size k_s 1·50 mm

Gradient S 0·002

Relative depth 0·50

Medial case for discharge ratio for Colebrook–White solutions
$\Theta_{ep} = 671$

Rel. dpth Y	U.equ. dia. $D_{ep(u)}$ (m)	Equiv. disch. factor J	Unit sect. area A_u (m²)	Unit wetted perim. P_u (m)	Unit surf. brdth $B_{s(u)}$ (m)	Unit mean depth $y_{m(u)}$ (m)	Discharge ratios $Q/Q_{0.50}$ medial	Mann'g	U.crit. disch. $Q_{c(u)}$ (m³s⁻¹)	Unit depth cntrd $y_{d(u)}$ (m)
0·02	0·0772	0·2296	0·0204	1·0566	1·0400	0·0196	0·0046	0·0043	0·0089	0·0099
0·04	0·1495	0·4219	0·0416	1·1131	1·0800	0·0385	0·0145	0·0135	0·0256	0·0197
0·06	0·2175	0·5841	0·0636	1·1697	1·1200	0·0568	0·0282	0·0265	0·0475	0·0294
0·08	0·2818	0·7220	0·0864	1·2263	1·1600	0·0745	0·0454	0·0428	0·0738	0·0390
0·10	0·3430	0·8399	0·1100	1·2828	1·2000	0·0917	0·0655	0·0622	0·1043	0·0485
0·12	0·4014	0·9414	0·1344	1·3394	1·2400	0·1084	0·0884	0·0844	0·1386	0·0579
0·14	0·4573	1·0291	0·1596	1·3960	1·2800	0·1247	0·1141	0·1093	0·1765	0·0671
0·16	0·5111	1·1054	0·1856	1·4525	1·3200	0·1406	0·1423	0·1369	0·2179	0·0763
0·18	0·5630	1·1719	0·2124	1·5091	1·3600	0·1562	0·1730	0·1671	0·2629	0·0854
0·20	0·6132	1·2303	0·2400	1·5657	1·4000	0·1714	0·2063	0·1998	0·3112	0·0944
0·22	0·6618	1·2816	0·2684	1·6223	1·4400	0·1864	0·2420	0·2351	0·3629	0·1034
0·24	0·7091	1·3269	0·2976	1·6788	1·4800	0·2011	0·2802	0·2730	0·4179	0·1123
0·26	0·7551	1·3669	0·3276	1·7354	1·5200	0·2155	0·3208	0·3134	0·4763	0·1211
0·28	0·8000	1·4025	0·3584	1·7920	1·5600	0·2297	0·3638	0·3563	0·5380	0·1298
0·30	0·8439	1·4342	0·3900	1·8485	1·6000	0·2437	0·4092	0·4018	0·6030	0·1385
0·32	0·8869	1·4625	0·4224	1·9051	1·6400	0·2576	0·4571	0·4498	0·6713	0·1471
0·34	0·9290	1·4878	0·4556	1·9617	1·6800	0·2712	0·5075	0·5004	0·7430	0·1556
0·36	0·9704	1·5104	0·4896	2·0182	1·7200	0·2847	0·5603	0·5536	0·8180	0·1641
0·38	1·0110	1·5308	0·5244	2·0748	1·7600	0·2980	0·6156	0·6094	0·8964	0·1726
0·40	1·0510	1·5491	0·5600	2·1314	1·8000	0·3111	0·6733	0·6678	0·9782	0·1810
0·42	1·0903	1·5655	0·5964	2·1879	1·8400	0·3241	0·7336	0·7288	1·0633	0·1893
0·44	1·1292	1·5804	0·6336	2·2445	1·8800	0·3370	0·7964	0·7925	1·1519	0·1976
0·46	1·1675	1·5939	0·6716	2·3011	1·9200	0·3498	0·8617	0·8590	1·2439	0·2058
0·48	1·2053	1·6060	0·7104	2·3576	1·9600	0·3624	0·9296	0·9281	1·3393	0·2141
0·50	1·2426	1·6170	0·7500	2·4142	2·0000	0·3750	1·0000	1·0000	1·4383	0·2222
0·52	1·2796	1·6270	0·7904	2·4708	2·0400	0·3875	1·0731	1·0747	1·5407	0·2304
0·54	1·3162	1·6360	0·8316	2·5274	2·0800	0·3998	1·1488	1·1521	1·6466	0·2384
0·56	1·3524	1·6442	0·8736	2·5839	2·1200	0·4121	1·2271	1·2324	1·7561	0·2465
0·58	1·3882	1·6516	0·9164	2·6405	2·1600	0·4243	1·3081	1·3155	1·8692	0·2545
0·60	1·4238	1·6584	0·9600	2·6971	2·2000	0·4364	1·3918	1·4015	1·9859	0·2625
0·62	1·4590	1·6646	1·0044	2·7536	2·2400	0·4484	1·4782	1·4905	2·1062	0·2705
0·64	1·4940	1·6701	1·0496	2·8102	2·2800	0·4604	1·5674	1·5823	2·2301	0·2784
0·66	1·5287	1·6752	1·0956	2·8668	2·3200	0·4722	1·6593	1·6772	2·3577	0·2863
0·68	1·5631	1·6798	1·1424	2·9233	2·3600	0·4841	1·7540	1·7750	2·4890	0·2941
0·70	1·5974	1·6840	1·1900	2·9799	2·4000	0·4958	1·8515	1·8758	2·6241	0·3020
0·72	1·6314	1·6878	1·2384	3·0365	2·4400	0·5075	1·9518	1·9797	2·7628	0·3098
0·74	1·6652	1·6913	1·2876	3·0930	2·4800	0·5192	2·0550	2·0867	2·9054	0·3175
0·76	1·6988	1·6944	1·3376	3·1496	2·5200	0·5308	2·1610	2·1968	3·0518	0·3253
0·78	1·7322	1·6972	1·3884	3·2062	2·5600	0·5423	2·2700	2·3100	3·2019	0·3330
0·80	1·7654	1·6998	1·4400	3·2627	2·6000	0·5538	2·3819	2·4264	3·3560	0·3407

Rel. dpth Y	U.equ. dia. $D_{ep(u)}$ (m)	Equiv. disch. factor J	Unit sect. area A_u (m²)	Unit wetted perim. P_u (m)	Unit surf. brdth $B_{s(u)}$ (m)	Unit mean depth $y_{m(u)}$ (m)	Discharge ratios $Q/Q_{0.50}$ medial	Mann'g	U.crit. disch. $Q_{c(u)}$ (m³s⁻¹)	Unit depth cntrd $y_{d(u)}$ (m)
0·82	1·7984	1·7021	1·4924	3·3193	2·6400	0·5653	2·4967	2·5460	3·5139	0·3484
0·84	1·8313	1·7042	1·5456	3·3759	2·6800	0·5767	2·6145	2·6688	3·6757	0·3561
0·86	1·8641	1·7061	1·5996	3·4324	2·7200	0·5881	2·7353	2·7949	3·8414	0·3637
0·88	1·8967	1·7078	1·6544	3·4890	2·7600	0·5994	2·8591	2·9242	4·0111	0·3713
0·90	1·9292	1·7093	1·7100	3·5456	2·8000	0·6107	2·9860	3·0569	4·1848	0·3789
0·92	1·9615	1·7107	1·7664	3·6022	2·8400	0·6220	3·1159	3·1929	4·3625	0·3865
0·94	1·9937	1·7119	1·8236	3·6587	2·8800	0·6332	3·2489	3·3323	4·5442	0·3941
0·96	2·0258	1·7129	1·8816	3·7153	2·9200	0·6444	3·3851	3·4751	4·7300	0·4016
0·98	2·0578	1·7139	1·9404	3·7719	2·9600	0·6555	3·5243	3·6213	4·9198	0·4092
1·00	2·0896	1·7147	2·0000	3·8284	3·0000	0·6667	3·6668	3·7710	5·1138	0·4167
1·02	2·1214	1·7154	2·0604	3·8850	3·0400	0·6778	3·8124	3·9241	5·3119	0·4242
1·04	2·1531	1·7160	2·1216	3·9416	3·0800	0·6888	3·9612	4·0808	5·5142	0·4316
1·06	2·1846	1·7166	2·1836	3·9981	3·1200	0·6999	4·1133	4·2410	5·7206	0·4391
1·08	2·2161	1·7170	2·2464	4·0547	3·1600	0·7109	4·2686	4·4047	5·9313	0·4465
1·10	2·2475	1·7174	2·3100	4·1113	3·2000	0·7219	4·4271	4·5721	6·1462	0·4540
1·12	2·2788	1·7176	2·3744	4·1678	3·2400	0·7328	4·5890	4·7431	6·3653	0·4614
1·14	2·3100	1·7179	2·4396	4·2244	3·2800	0·7438	4·7542	4·9178	6·5887	0·4688
1·16	2·3411	1·7180	2·5056	4·2810	3·3200	0·7547	4·9228	5·0961	6·8165	0·4762
1·18	2·3722	1·7181	2·5724	4·3375	3·3600	0·7656	5·0947	5·2782	7·0485	0·4835
1·20	2·4032	1·7182	2·6400	4·3941	3·4000	0·7765	5·2700	5·4640	7·2850	0·4909
1·24	2·4650	1·7181	2·7776	4·5072	3·4800	0·7982	5·6308	5·8469	7·7710	0·5056
1·28	2·5265	1·7179	2·9184	4·6204	3·5600	0·8198	6·0055	6·2451	8·2747	0·5202
1·32	2·5878	1·7175	3·0624	4·7335	3·6400	0·8413	6·3942	6·6588	8·7963	0·5348
1·36	2·6489	1·7170	3·2096	4·8467	3·7200	0·8628	6·7970	7·0882	9·3361	0·5494
1·40	2·7098	1·7164	3·3600	4·9598	3·8000	0·8842	7·2142	7·5336	9·8941	0·5639
1·44	2·7705	1·7157	3·5136	5·0729	3·8800	0·9056	7·6459	7·9952	10·471	0·5784
1·48	2·8310	1·7149	3·6704	5·1861	3·9600	0·9269	8·0924	8·4732	11·066	0·5928
1·52	2·8913	1·7140	3·8304	5·2992	4·0400	0·9481	8·5537	8·9677	11·680	0·6072
1·56	2·9515	1·7131	3·9936	5·4123	4·1200	0·9693	9·0300	9·4791	12·313	0·6216
1·60	3·0115	1·7122	4·1600	5·5255	4·2000	0·9905	9·5216	10·007	12·965	0·6359
1·64	3·0714	1·7112	4·3296	5·6386	4·2800	1·0116	10·029	10·553	13·637	0·6502
1·68	3·1311	1·7102	4·5024	5·7518	4·3600	1·0327	10·551	11·116	14·328	0·6645
1·72	3·1908	1·7091	4·6784	5·8649	4·4400	1·0537	11·089	11·697	15·039	0·6787
1·76	3·2503	1·7081	4·8576	5·9780	4·5200	1·0747	11·643	12·295	15·770	0·6929
1·80	3·3097	1·7070	5·0400	6·0912	4·6000	1·0957	12·213	12·912	16·521	0·7071
1·84	3·3690	1·7059	5·2256	6·2043	4·6800	1·1166	12·800	13·547	17·292	0·7213
1·88	3·4282	1·7048	5·4144	6·3174	4·7600	1·1375	13·402	14·201	18·083	0·7355
1·92	3·4873	1·7037	5·6064	6·4306	4·8400	1·1583	14·021	14·873	18·896	0·7496
1·96	3·5464	1·7025	5·8016	6·5437	4·9200	1·1792	14·657	15·564	19·729	0·7637
2·00	3·6053	1·7014	6·0000	6·6569	5·0000	1·2000	15·310	16·274	20·583	0·7778
2·05	3·6789	1·7000	6·2525	6·7983	5·1000	1·2260	16·150	17·189	21·680	0·7954
2·10	3·7523	1·6986	6·5100	6·9397	5·2000	1·2519	17·016	18·134	22·810	0·8129
2·15	3·8257	1·6972	6·7725	7·0811	5·3000	1·2778	17·910	19·110	23·974	0·8304
2·20	3·8989	1·6959	7·0400	7·2225	5·4000	1·3037	18·831	20·118	25·172	0·8479
2·25	3·9720	1·6945	7·3125	7·3640	5·5000	1·3295	19·780	21·157	26·404	0·8654
2·30	4·0451	1·6932	7·5900	7·5054	5·6000	1·3554	20·758	22·228	27·671	0·8828
2·35	4·1181	1·6918	7·8725	7·6468	5·7000	1·3811	21·763	23·332	28·973	0·9002
2·40	4·1909	1·6905	8·1600	7·7882	5·8000	1·4069	22·797	24·468	30·310	0·9176
2·45	4·2637	1·6892	8·4525	7·9296	5·9000	1·4326	23·860	25·638	31·682	0·9350
2·50	4·3365	1·6879	8·7500	8·0711	6·0000	1·4583	24·952	26·842	33·090	0·9524
2·55	4·4091	1·6866	9·0525	8·2125	6·1000	1·4840	26·074	28·079	34·534	0·9697
2·60	4·4817	1·6854	9·3600	8·3539	6·2000	1·5097	27·225	29·350	36·015	0·9870
2·65	4·5543	1·6841	9·6725	8·4953	6·3000	1·5353	28·407	30·657	37·532	1·0043
2·70	4·6267	1·6829	9·9900	8·6368	6·4000	1·5609	29·619	31·998	39·086	1·0216
2·75	4·6992	1·6817	10·312	8·7782	6·5000	1·5865	30·861	33·375	40·677	1·0389

C59 Trapezoidal channel - 1·25 to 1 side-slope

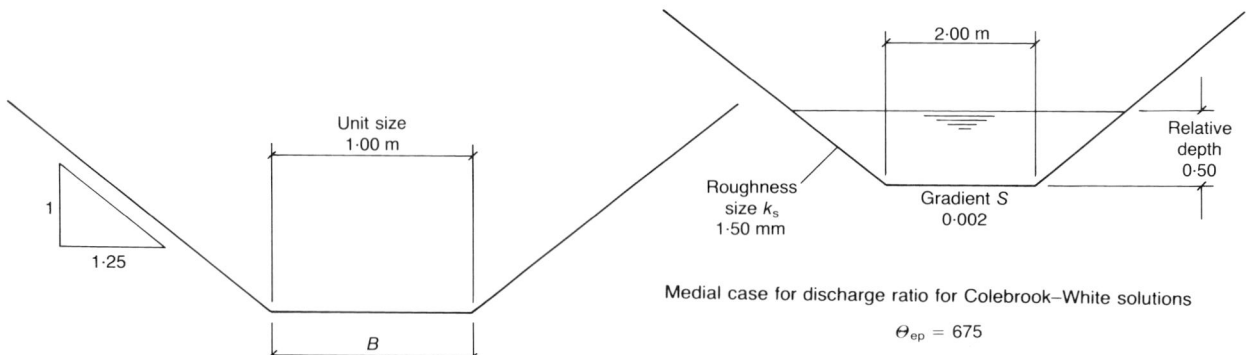

Unit size 1·00 m

2·00 m

Relative depth 0·50

Roughness size k_s 1·50 mm

Gradient S 0·002

Medial case for discharge ratio for Colebrook–White solutions

$\Theta_{ep} = 675$

Rel. dpth Y	U.equ. dia. $D_{ep(u)}$ (m)	Equiv. disch. factor J	Unit sect. area A_u (m²)	Unit wetted perim. P_u (m)	Unit surf. brdth $B_{s(u)}$ (m)	Unit mean depth $y_{m(u)}$ (m)	Discharge ratios $Q/Q_{0.50}$ medial	Mann'g	U.crit. disch. $Q_{c(u)}$ (m³s⁻¹)	Unit depth cntrd $y_{d(u)}$ (m)
0·02	0·0771	0·2275	0·0205	1·0640	1·0500	0·0195	0·0042	0·0039	0·0090	0·0099
0·04	0·1489	0·4147	0·0420	1·1281	1·1000	0·0382	0·0134	0·0125	0·0257	0·0197
0·06	0·2164	0·5703	0·0645	1·1921	1·1500	0·0561	0·0263	0·0247	0·0478	0·0293
0·08	0·2802	0·7008	0·0880	1·2561	1·2000	0·0733	0·0423	0·0400	0·0746	0·0388
0·10	0·3409	0·8112	0·1125	1·3202	1·2500	0·0900	0·0614	0·0582	0·1057	0·0481
0·12	0·3988	0·9051	0·1380	1·3842	1·3000	0·1062	0·0832	0·0793	0·1408	0·0574
0·14	0·4544	0·9856	0·1645	1·4482	1·3500	0·1219	0·1077	0·1031	0·1798	0·0665
0·16	0·5079	1·0550	0·1920	1·5122	1·4000	0·1371	0·1348	0·1297	0·2227	0·0756
0·18	0·5595	1·1152	0·2205	1·5763	1·4500	0·1521	0·1646	0·1588	0·2693	0·0845
0·20	0·6096	1·1676	0·2500	1·6403	1·5000	0·1667	0·1970	0·1907	0·3196	0·0933
0·22	0·6583	1·2134	0·2805	1·7043	1·5500	0·1810	0·2319	0·2252	0·3737	0·1021
0·24	0·7057	1·2537	0·3120	1·7684	1·6000	0·1950	0·2694	0·2624	0·4315	0·1108
0·26	0·7520	1·2893	0·3445	1·8324	1·6500	0·2088	0·3095	0·3022	0·4929	0·1194
0·28	0·7973	1·3207	0·3780	1·8964	1·7000	0·2224	0·3522	0·3448	0·5582	0·1279
0·30	0·8416	1·3487	0·4125	1·9605	1·7500	0·2357	0·3975	0·3901	0·6272	0·1364
0·32	0·8852	1·3735	0·4480	2·0245	1·8000	0·2489	0·4455	0·4381	0·6999	0·1448
0·34	0·9279	1·3958	0·4845	2·0885	1·8500	0·2619	0·4961	0·4890	0·7765	0·1531
0·36	0·9700	1·4157	0·5220	2·1526	1·9000	0·2747	0·5494	0·5426	0·8568	0·1614
0·38	1·0115	1·4335	0·5605	2·2166	1·9500	0·2874	0·6054	0·5991	0·9410	0·1696
0·40	1·0523	1·4496	0·6000	2·2806	2·0000	0·3000	0·6642	0·6585	1·0291	0·1778
0·42	1·0927	1·4641	0·6405	2·3447	2·0500	0·3124	0·7257	0·7208	1·1211	0·1859
0·44	1·1326	1·4771	0·6820	2·4087	2·1000	0·3248	0·7900	0·7861	1·2171	0·1940
0·46	1·1720	1·4890	0·7245	2·4727	2·1500	0·3370	0·8571	0·8544	1·3170	0·2020
0·48	1·2110	1·4997	0·7680	2·5367	2·2000	0·3491	0·9271	0·9257	1·4210	0·2100
0·50	1·2496	1·5094	0·8125	2·6008	2·2500	0·3611	1·0000	1·0000	1·5290	0·2179
0·52	1·2879	1·5183	0·8580	2·6648	2·3000	0·3730	1·0758	1·0775	1·6411	0·2259
0·54	1·3258	1·5263	0·9045	2·7288	2·3500	0·3849	1·1546	1·1580	1·7573	0·2337
0·56	1·3635	1·5337	0·9520	2·7929	2·4000	0·3967	1·2363	1·2418	1·8776	0·2416
0·58	1·4008	1·5404	1·0005	2·8569	2·4500	0·4084	1·3211	1·3288	2·0022	0·2494
0·60	1·4379	1·5465	1·0500	2·9209	2·5000	0·4200	1·4089	1·4190	2·1310	0·2571
0·62	1·4747	1·5521	1·1005	2·9850	2·5500	0·4316	1·4997	1·5126	2·2640	0·2649
0·64	1·5113	1·5572	1·1520	3·0490	2·6000	0·4431	1·5937	1·6095	2·4013	0·2726
0·66	1·5477	1·5619	1·2045	3·1130	2·6500	0·4545	1·6908	1·7097	2·5430	0·2803
0·68	1·5839	1·5661	1·2580	3·1771	2·7000	0·4659	1·7911	1·8133	2·6891	0·2879
0·70	1·6198	1·5701	1·3125	3·2411	2·7500	0·4773	1·8946	1·9204	2·8395	0·2956
0·72	1·6556	1·5737	1·3680	3·3051	2·8000	0·4886	2·0013	2·0310	2·9944	0·3032
0·74	1·6912	1·5770	1·4245	3·3692	2·8500	0·4998	2·1113	2·1451	3·1538	0·3107
0·76	1·7267	1·5800	1·4820	3·4332	2·9000	0·5110	2·2246	2·2628	3·3177	0·3183
0·78	1·7620	1·5828	1·5405	3·4972	2·9500	0·5222	2·3412	2·3841	3·4861	0·3258
0·80	1·7971	1·5853	1·6000	3·5612	3·0000	0·5333	2·4612	2·5090	3·6591	0·3333

Rel. dpth Y	U.equ. dia. $D_{ep(u)}$ (m)	Equiv. disch. factor J	Unit sect. area A_u (m^2)	Unit wetted perim. P_u (m)	Unit surf. brdth $B_{s(u)}$ (m)	Unit mean depth $y_{m(u)}$ (m)	Discharge ratios $Q/Q_{0.50}$ medial	Mann'g	U.crit. disch. $Q_{c(u)}$ (m^{3}s^{-1})	Unit depth cntrd $y_{d(u)}$ (m)
0·82	1·8321	1·5877	1·6605	3·6253	3·0500	0·5444	2·5845	2·6375	3·8368	0·3408
0·84	1·8670	1·5898	1·7220	3·6893	3·1000	0·5555	2·7113	2·7698	4·0191	0·3483
0·86	1·9018	1·5918	1·7845	3·7533	3·1500	0·5665	2·8415	2·9059	4·2061	0·3557
0·88	1·9364	1·5936	1·8480	3·8174	3·2000	0·5775	2·9753	3·0457	4·3978	0·3632
0·90	1·9709	1·5952	1·9125	3·8814	3·2500	0·5885	3·1125	3·1894	4·5943	0·3706
0·92	2·0054	1·5967	1·9780	3·9454	3·3000	0·5994	3·2533	3·3369	4·7956	0·3780
0·94	2·0397	1·5981	2·0445	4·0095	3·3500	0·6103	3·3976	3·4883	5·0017	0·3854
0·96	2·0739	1·5994	2·1120	4·0735	3·4000	0·6212	3·5456	3·6437	5·2127	0·3927
0·98	2·1080	1·6006	2·1805	4·1375	3·4500	0·6320	3·6971	3·8030	5·4286	0·4001
1·00	2·1421	1·6016	2·2500	4·2016	3·5000	0·6429	3·8524	3·9664	5·6494	0·4074
1·02	2·1760	1·6026	2·3205	4·2656	3·5500	0·6537	4·0113	4·1338	5·8751	0·4147
1·04	2·2099	1·6035	2·3920	4·3296	3·6000	0·6644	4·1740	4·3053	6·1059	0·4220
1·06	2·2437	1·6043	2·4645	4·3937	3·6500	0·6752	4·3404	4·4809	6·3417	0·4293
1·08	2·2774	1·6050	2·5380	4·4577	3·7000	0·6859	4·5106	4·6606	6·5826	0·4366
1·10	2·3111	1·6056	2·6125	4·5217	3·7500	0·6967	4·6846	4·8446	6·8286	0·4439
1·12	2·3447	1·6062	2·6880	4·5857	3·8000	0·7074	4·8625	5·0328	7·0797	0·4511
1·14	2·3782	1·6068	2·7645	4·6498	3·8500	0·7181	5·0442	5·2252	7·3359	0·4584
1·16	2·4116	1·6072	2·8420	4·7138	3·9000	0·7287	5·2298	5·4219	7·5974	0·4656
1·18	2·4450	1·6077	2·9205	4·7778	3·9500	0·7394	5·4194	5·6230	7·8641	0·4728
1·20	2·4784	1·6080	3·0000	4·8419	4·0000	0·7500	5·6129	5·8285	8·1360	0·4800
1·24	2·5449	1·6086	3·1620	4·9699	4·1000	0·7712	6·0119	6·2527	8·6958	0·4944
1·28	2·6112	1·6091	3·3280	5·0980	4·2000	0·7924	6·4271	6·6948	9·2771	0·5087
1·32	2·6774	1·6094	3·4980	5·2261	4·3000	0·8135	6·8586	7·1551	9·8800	0·5230
1·36	2·7433	1·6096	3·6720	5·3541	4·4000	0·8345	7·3068	7·6338	10·505	0·5373
1·40	2·8091	1·6097	3·8500	5·4822	4·5000	0·8556	7·7718	8·1313	11·152	0·5515
1·44	2·8747	1·6097	4·0320	5·6102	4·6000	0·8765	8·2539	8·6479	11·821	0·5657
1·48	2·9402	1·6097	4·2180	5·7383	4·7000	0·8974	8·7532	9·1837	12·513	0·5799
1·52	3·0056	1·6095	4·4080	5·8664	4·8000	0·9183	9·2700	9·7391	13·228	0·5940
1·56	3·0708	1·6093	4·6020	5·9944	4·9000	0·9392	9·8045	10·314	13·966	0·6081
1·60	3·1360	1·6091	4·8000	6·1225	5·0000	0·9600	10·357	10·910	14·728	0·6222
1·64	3·2010	1·6088	5·0020	6·2506	5·1000	0·9808	10·928	11·525	15·513	0·6363
1·68	3·2659	1·6085	5·2080	6·3786	5·2000	1·0015	11·516	12·162	16·322	0·6503
1·72	3·3307	1·6081	5·4180	6·5067	5·3000	1·0223	12·124	12·819	17·155	0·6643
1·76	3·3955	1·6077	5·6320	6·6347	5·4000	1·0430	12·750	13·497	18·012	0·6783
1·80	3·4601	1·6073	5·8500	6·7628	5·5000	1·0636	13·395	14·197	18·894	0·6923
1·84	3·5247	1·6069	6·0720	6·8909	5·6000	1·0843	14·059	14·919	19·800	0·7063
1·88	3·5891	1·6064	6·2980	7·0189	5·7000	1·1049	14·743	15·662	20·731	0·7202
1·92	3·6536	1·6060	6·5280	7·1470	5·8000	1·1255	15·446	16·428	21·688	0·7341
1·96	3·7179	1·6055	6·7620	7·2751	5·9000	1·1461	16·169	17·216	22·670	0·7480
2·00	3·7822	1·6050	7·0000	7·4031	6·0000	1·1667	16·912	18·027	23·677	0·7619
2·05	3·8625	1·6043	7·3031	7·5632	6·1250	1·1923	17·869	19·073	24·973	0·7792
2·10	3·9426	1·6037	7·6125	7·7233	6·2500	1·2180	18·858	20·155	26·309	0·7966
2·15	4·0227	1·6031	7·9281	7·8834	6·3750	1·2436	19·879	21·274	27·687	0·8138
2·20	4·1027	1·6024	8·2500	8·0434	6·5000	1·2692	20·933	22·430	29·106	0·8311
2·25	4·1827	1·6017	8·5781	8·2035	6·6250	1·2948	22·019	23·624	30·567	0·8484
2·30	4·2625	1·6011	8·9125	8·3636	6·7500	1·3204	23·139	24·856	32·071	0·8656
2·35	4·3423	1·6004	9·2531	8·5237	6·8750	1·3459	24·293	26·127	33·617	0·8828
2·40	4·4221	1·5998	9·6000	8·6837	7·0000	1·3714	25·481	27·437	35·206	0·9000
2·45	4·5017	1·5991	9·9531	8·8438	7·1250	1·3969	26·703	28·787	36·839	0·9172
2·50	4·5813	1·5985	10·312	9·0039	7·2500	1·4224	27·960	30·177	38·516	0·9343
2·55	4·6609	1·5978	10·678	9·1640	7·3750	1·4479	29·252	31·608	40·237	0·9515
2·60	4·7404	1·5972	11·050	9·3241	7·5000	1·4733	30·579	33·080	42·002	0·9686
2·65	4·8199	1·5965	11·428	9·4841	7·6250	1·4988	31·943	34·593	43·813	0·9857
2·70	4·8993	1·5959	11·812	9·6442	7·7500	1·5242	33·343	36·148	45·669	1·0029
2·75	4·9787	1·5953	12·203	9·8043	7·8750	1·5496	34·779	37·746	47·571	1·0200

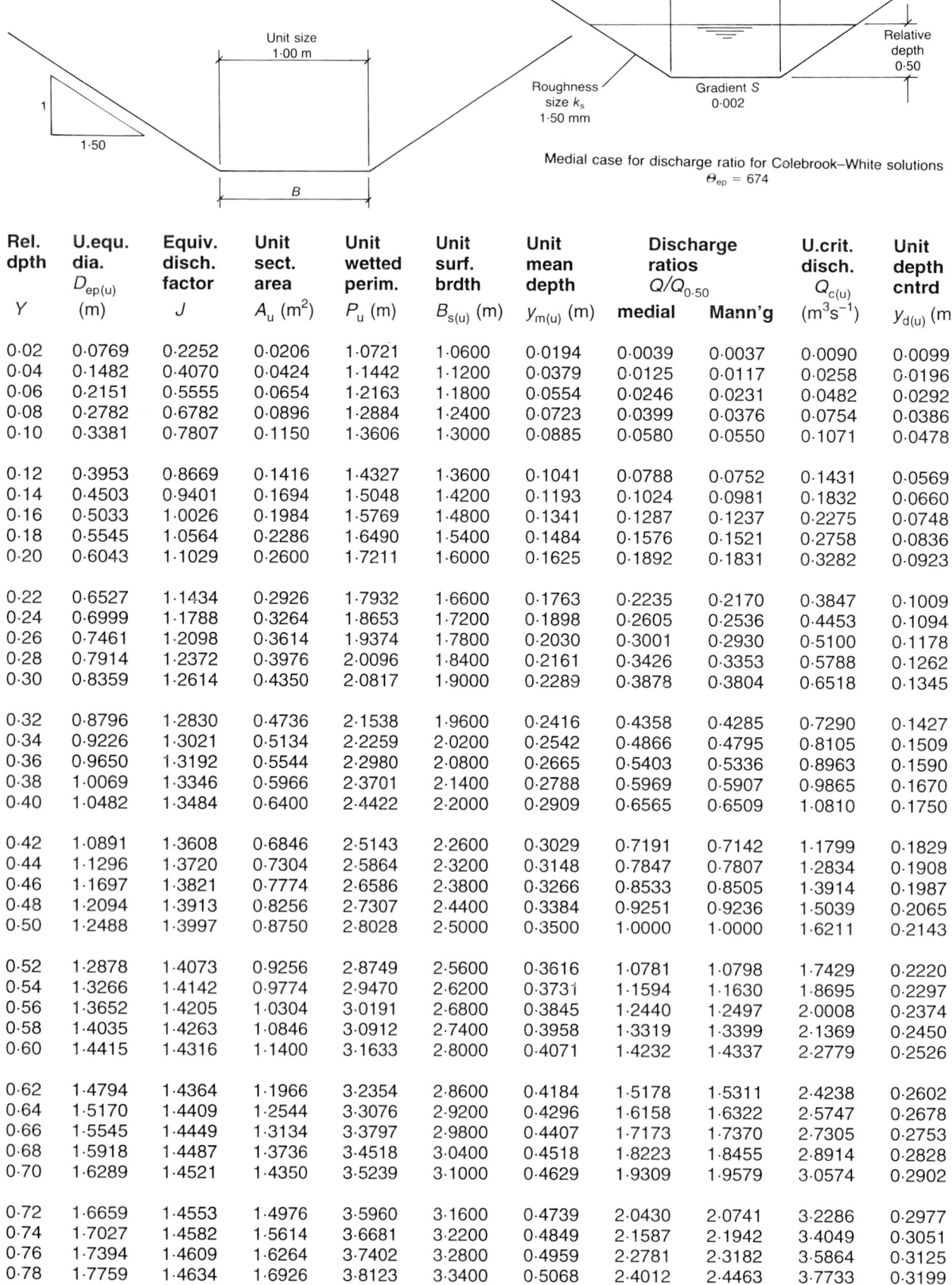

Medial case for discharge ratio for Colebrook–White solutions
$\Theta_{ep} = 674$

Rel. dpth Y	U.equ. dia. $D_{ep(u)}$ (m)	Equiv. disch. factor J	Unit sect. area A_u (m²)	Unit wetted perim. P_u (m)	Unit surf. brdth $B_{s(u)}$ (m)	Unit mean depth $y_{m(u)}$ (m)	Discharge ratios $Q/Q_{0·50}$ medial	Mann'g	U.crit. disch. $Q_{c(u)}$ (m³s⁻¹)	Unit depth cntrd $y_{d(u)}$ (m)
0·02	0·0769	0·2252	0·0206	1·0721	1·0600	0·0194	0·0039	0·0037	0·0090	0·0099
0·04	0·1482	0·4070	0·0424	1·1442	1·1200	0·0379	0·0125	0·0117	0·0258	0·0196
0·06	0·2151	0·5555	0·0654	1·2163	1·1800	0·0554	0·0246	0·0231	0·0482	0·0292
0·08	0·2782	0·6782	0·0896	1·2884	1·2400	0·0723	0·0399	0·0376	0·0754	0·0386
0·10	0·3381	0·7807	0·1150	1·3606	1·3000	0·0885	0·0580	0·0550	0·1071	0·0478
0·12	0·3953	0·8669	0·1416	1·4327	1·3600	0·1041	0·0788	0·0752	0·1431	0·0569
0·14	0·4503	0·9401	0·1694	1·5048	1·4200	0·1193	0·1024	0·0981	0·1832	0·0660
0·16	0·5033	1·0026	0·1984	1·5769	1·4800	0·1341	0·1287	0·1237	0·2275	0·0748
0·18	0·5545	1·0564	0·2286	1·6490	1·5400	0·1484	0·1576	0·1521	0·2758	0·0836
0·20	0·6043	1·1029	0·2600	1·7211	1·6000	0·1625	0·1892	0·1831	0·3282	0·0923
0·22	0·6527	1·1434	0·2926	1·7932	1·6600	0·1763	0·2235	0·2170	0·3847	0·1009
0·24	0·6999	1·1788	0·3264	1·8653	1·7200	0·1898	0·2605	0·2536	0·4453	0·1094
0·26	0·7461	1·2098	0·3614	1·9374	1·7800	0·2030	0·3001	0·2930	0·5100	0·1178
0·28	0·7914	1·2372	0·3976	2·0096	1·8400	0·2161	0·3426	0·3353	0·5788	0·1262
0·30	0·8359	1·2614	0·4350	2·0817	1·9000	0·2289	0·3878	0·3804	0·6518	0·1345
0·32	0·8796	1·2830	0·4736	2·1538	1·9600	0·2416	0·4358	0·4285	0·7290	0·1427
0·34	0·9226	1·3021	0·5134	2·2259	2·0200	0·2542	0·4866	0·4795	0·8105	0·1509
0·36	0·9650	1·3192	0·5544	2·2980	2·0800	0·2665	0·5403	0·5336	0·8963	0·1590
0·38	1·0069	1·3346	0·5966	2·3701	2·1400	0·2788	0·5969	0·5907	0·9865	0·1670
0·40	1·0482	1·3484	0·6400	2·4422	2·2000	0·2909	0·6565	0·6509	1·0810	0·1750
0·42	1·0891	1·3608	0·6846	2·5143	2·2600	0·3029	0·7191	0·7142	1·1799	0·1829
0·44	1·1296	1·3720	0·7304	2·5864	2·3200	0·3148	0·7847	0·7807	1·2834	0·1908
0·46	1·1697	1·3821	0·7774	2·6586	2·3800	0·3266	0·8533	0·8505	1·3914	0·1987
0·48	1·2094	1·3913	0·8256	2·7307	2·4400	0·3384	0·9251	0·9236	1·5039	0·2065
0·50	1·2488	1·3997	0·8750	2·8028	2·5000	0·3500	1·0000	1·0000	1·6211	0·2143
0·52	1·2878	1·4073	0·9256	2·8749	2·5600	0·3616	1·0781	1·0798	1·7429	0·2220
0·54	1·3266	1·4142	0·9774	2·9470	2·6200	0·3731	1·1594	1·1630	1·8695	0·2297
0·56	1·3652	1·4205	1·0304	3·0191	2·6800	0·3845	1·2440	1·2497	2·0008	0·2374
0·58	1·4035	1·4263	1·0846	3·0912	2·7400	0·3958	1·3319	1·3399	2·1369	0·2450
0·60	1·4415	1·4316	1·1400	3·1633	2·8000	0·4071	1·4232	1·4337	2·2779	0·2526
0·62	1·4794	1·4364	1·1966	3·2354	2·8600	0·4184	1·5178	1·5311	2·4238	0·2602
0·64	1·5170	1·4409	1·2544	3·3076	2·9200	0·4296	1·6158	1·6322	2·5747	0·2678
0·66	1·5545	1·4449	1·3134	3·3797	2·9800	0·4407	1·7173	1·7370	2·7305	0·2753
0·68	1·5918	1·4487	1·3736	3·4518	3·0400	0·4518	1·8223	1·8455	2·8914	0·2828
0·70	1·6289	1·4521	1·4350	3·5239	3·1000	0·4629	1·9309	1·9579	3·0574	0·2902
0·72	1·6659	1·4553	1·4976	3·5960	3·1600	0·4739	2·0430	2·0741	3·2286	0·2977
0·74	1·7027	1·4582	1·5614	3·6681	3·2200	0·4849	2·1587	2·1942	3·4049	0·3051
0·76	1·7394	1·4609	1·6264	3·7402	3·2800	0·4959	2·2781	2·3182	3·5864	0·3125
0·78	1·7759	1·4634	1·6926	3·8123	3·3400	0·5068	2·4012	2·4463	3·7733	0·3199
0·80	1·8124	1·4657	1·7600	3·8844	3·4000	0·5176	2·5280	2·5784	3·9654	0·3273

Rel. dpth	U.equ. dia. $D_{ep(u)}$	Equiv. disch. factor	Unit sect. area	Unit wetted perim.	Unit surf. brdth	Unit mean depth	Discharge ratios $Q/Q_{0.50}$		U.crit. disch. $Q_{c(u)}$	Unit depth cntrd
Y	(m)	J	A_u (m²)	P_u (m)	$B_{s(u)}$ (m)	$y_{m(u)}$ (m)	medial	Mann'g	(m³s⁻¹)	$y_{d(u)}$ (m)
0·82	1·8487	1·4679	1·8286	3·9566	3·4600	0·5285	2·6586	2·7146	4·1629	0·3346
0·84	1·8849	1·4698	1·8984	4·0287	3·5200	0·5393	2·7929	2·8549	4·3659	0·3419
0·86	1·9210	1·4716	1·9694	4·1008	3·5800	0·5501	2·9312	2·9993	4·5742	0·3493
0·88	1·9570	1·4733	2·0416	4·1729	3·6400	0·5609	3·0733	3·1480	4·7881	0·3566
0·90	1·9929	1·4749	2·1150	4·2450	3·7000	0·5716	3·2193	3·3010	5·0075	0·3638
0·92	2·0288	1·4763	2·1896	4·3171	3·7600	0·5823	3·3692	3·4583	5·2326	0·3711
0·94	2·0645	1·4776	2·2654	4·3892	3·8200	0·5930	3·5232	3·6199	5·4632	0·3783
0·96	2·1002	1·4789	2·3424	4·4613	3·8800	0·6037	3·6812	3·7859	5·6995	0·3856
0·98	2·1358	1·4800	2·4206	4·5334	3·9400	0·6144	3·8432	3·9564	5·9415	0·3928
1·00	2·1713	1·4811	2·5000	4·6056	4·0000	0·6250	4·0093	4·1313	6·1893	0·4000
1·02	2·2067	1·4820	2·5806	4·6777	4·0600	0·6356	4·1796	4·3108	6·4429	0·4072
1·04	2·2421	1·4830	2·6624	4·7498	4·1200	0·6462	4·3540	4·4949	6·7023	0·4144
1·06	2·2774	1·4838	2·7454	4·8219	4·1800	0·6568	4·5327	4·6836	6·9676	0·4215
1·08	2·3127	1·4846	2·8296	4·8940	4·2400	0·6674	4·7155	4·8769	7·2388	0·4287
1·10	2·3479	1·4853	2·9150	4·9661	4·3000	0·6779	4·9027	5·0750	7·5160	0·4358
1·12	2·3831	1·4859	3·0016	5·0382	4·3600	0·6884	5·0942	5·2778	7·7991	0·4430
1·14	2·4182	1·4865	3·0894	5·1103	4·4200	0·6990	5·2899	5·4853	8·0884	0·4501
1·16	2·4532	1·4871	3·1784	5·1824	4·4800	0·7095	5·4901	5·6978	8·3837	0·4572
1·18	2·4882	1·4876	3·2686	5·2546	4·5400	0·7200	5·6947	5·9150	8·6851	0·4643
1·20	2·5232	1·4881	3·3600	5·3267	4·6000	0·7304	5·9037	6·1373	8·9927	0·4714
1·24	2·5929	1·4889	3·5464	5·4709	4·7200	0·7514	6·3353	6·5966	9·6266	0·4856
1·28	2·6625	1·4896	3·7376	5·6151	4·8400	0·7722	6·7850	7·0761	10·286	0·4997
1·32	2·7320	1·4902	3·9336	5·7593	4·9600	0·7931	7·2531	7·5761	10·970	0·5138
1·36	2·8013	1·4907	4·1344	5·9035	5·0800	0·8139	7·7400	8·0970	11·680	0·5279
1·40	2·8705	1·4911	4·3400	6·0478	5·2000	0·8346	8·2459	8·6390	12·416	0·5419
1·44	2·9395	1·4914	4·5504	6·1920	5·3200	0·8553	8·7710	9·2025	13·179	0·5559
1·48	3·0085	1·4916	4·7656	6·3362	5·4400	0·8760	9·3155	9·7879	13·968	0·5699
1·52	3·0773	1·4918	4·9856	6·4804	5·5600	0·8967	9·8799	10·395	14·784	0·5839
1·56	3·1461	1·4919	5·2104	6·6247	5·6800	0·9173	10·464	11·025	15·628	0·5978
1·60	3·2147	1·4920	5·4400	6·7689	5·8000	0·9379	11·069	11·678	16·499	0·6118
1·64	3·2833	1·4920	5·6744	6·9131	5·9200	0·9585	11·694	12·354	17·397	0·6257
1·68	3·3518	1·4920	5·9136	7·0573	6·0400	0·9791	12·340	13·053	18·324	0·6395
1·72	3·4202	1·4920	6·1576	7·2015	6·1600	0·9996	13·006	13·776	19·279	0·6534
1·76	3·4885	1·4919	6·4064	7·3458	6·2800	1·0201	13·695	14·523	20·263	0·6673
1·80	3·5567	1·4918	6·6600	7·4900	6·4000	1·0406	14·404	15·294	21·276	0·6811
1·84	3·6249	1·4917	6·9184	7·6342	6·5200	1·0611	15·135	16·090	22·317	0·6949
1·88	3·6931	1·4915	7·1816	7·7784	6·6400	1·0816	15·888	16·910	23·389	0·7087
1·92	3·7612	1·4914	7·4496	7·9227	6·7600	1·1020	16·664	17·756	24·490	0·7225
1·96	3·8292	1·4912	7·7224	8·0669	6·8800	1·1224	17·462	18·628	25·621	0·7362
2·00	3·8972	1·4910	8·0000	8·2111	7·0000	1·1429	18·282	19·525	26·782	0·7500
2·05	3·9821	1·4908	8·3537	8·3914	7·1500	1·1684	19·340	20·684	28·277	0·7672
2·10	4·0669	1·4905	8·7150	8·5717	7·3000	1·1938	20·434	21·883	29·819	0·7843
2·15	4·1517	1·4902	9·0838	8·7519	7·4500	1·2193	21·564	23·125	31·411	0·8015
2·20	4·2364	1·4900	9·4600	8·9322	7·6000	1·2447	22·732	24·409	33·051	0·8186
2·25	4·3210	1·4897	9·8437	9·1125	7·7500	1·2702	23·937	25·737	34·742	0·8357
2·30	4·4056	1·4894	10·235	9·2928	7·9000	1·2956	25·180	27·108	36·482	0·8528
2·35	4·4901	1·4890	10·634	9·4730	8·0500	1·3210	26·461	28·523	38·273	0·8699
2·40	4·5746	1·4887	11·040	9·6533	8·2000	1·3463	27·781	29·983	40·115	0·8870
2·45	4·6590	1·4884	11·454	9·8336	8·3500	1·3717	29·141	31·488	42·009	0·9040
2·50	4·7434	1·4881	11·875	10·014	8·5000	1·3971	30·540	33·039	43·954	0·9211
2·55	4·8278	1·4878	12·304	10·194	8·6500	1·4224	31·979	34·637	45·952	0·9381
2·60	4·9121	1·4874	12·740	10·374	8·8000	1·4477	33·458	36·281	48·004	0·9551
2·65	4·9963	1·4871	13·184	10·555	8·9500	1·4730	34·979	37·973	50·108	0·9721
2·70	5·0806	1·4868	13·635	10·735	9·1000	1·4984	36·540	39·713	52·266	0·9891
2·75	5·1648	1·4865	14·094	10·915	9·2500	1·5236	38·144	41·502	54·479	1·0061

C61

Trapezoidal channel - 1·75 to 1 side-slope

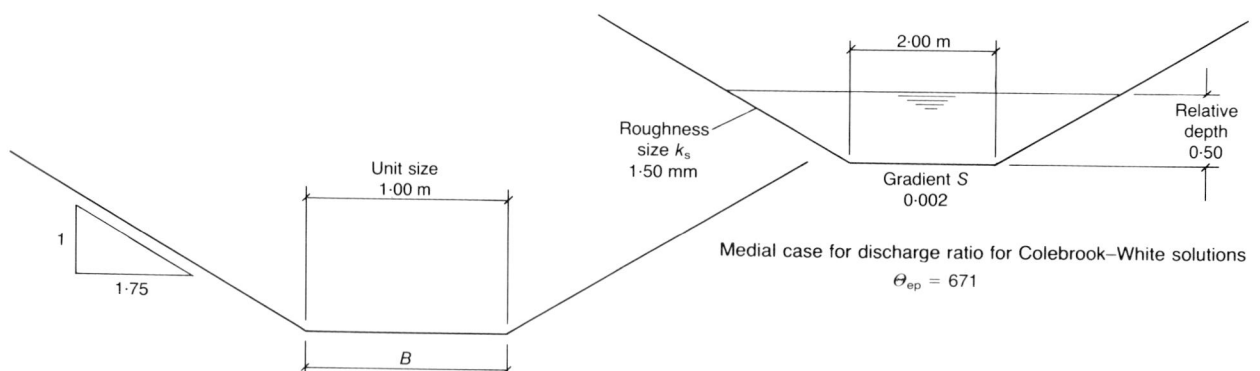

Unit size 1·00 m

Roughness size k_s 1·50 mm

2·00 m

Relative depth 0·50

Gradient S 0·002

Medial case for discharge ratio for Colebrook–White solutions
$\Theta_{ep} = 671$

Rel. dpth Y	U.equ. dia. $D_{ep(u)}$ (m)	Equiv. disch. factor J	Unit sect. area A_u (m²)	Unit wetted perim. P_u (m)	Unit surf. brdth $B_{s(u)}$ (m)	Unit mean depth $y_{m(u)}$ (m)	Discharge ratios $Q/Q_{0.50}$ medial	Mann'g	U.crit. disch. $Q_{c(u)}$ (m³s⁻¹)	Unit depth cntrd $y_{d(u)}$ (m)
0·02	0·0766	0·2228	0·0207	1·0806	1·0700	0·0193	0·0037	0·0034	0·0090	0·0099
0·04	0·1474	0·3988	0·0428	1·1612	1·1400	0·0375	0·0118	0·0110	0·0260	0·0196
0·06	0·2135	0·5402	0·0663	1·2419	1·2100	0·0548	0·0233	0·0218	0·0486	0·0290
0·08	0·2758	0·6553	0·0912	1·3225	1·2800	0·0712	0·0378	0·0356	0·0762	0·0384
0·10	0·3350	0·7500	0·1175	1·4031	1·3500	0·0870	0·0551	0·0523	0·1086	0·0475
0·12	0·3914	0·8288	0·1452	1·4837	1·4200	0·1023	0·0752	0·0717	0·1454	0·0565
0·14	0·4457	0·8950	0·1743	1·5644	1·4900	0·1170	0·0980	0·0938	0·1867	0·0654
0·16	0·4980	0·9511	0·2048	1·6450	1·5600	0·1313	0·1235	0·1187	0·2324	0·0742
0·18	0·5487	0·9989	0·2367	1·7256	1·6300	0·1452	0·1517	0·1463	0·2825	0·0828
0·20	0·5979	1·0400	0·2700	1·8062	1·7000	0·1588	0·1827	0·1768	0·3370	0·0914
0·22	0·6459	1·0755	0·3047	1·8868	1·7700	0·1721	0·2164	0·2100	0·3959	0·0998
0·24	0·6929	1·1063	0·3408	1·9675	1·8400	0·1852	0·2529	0·2461	0·4593	0·1082
0·26	0·7388	1·1333	0·3783	2·0481	1·9100	0·1981	0·2922	0·2852	0·5272	0·1164
0·28	0·7839	1·1569	0·4172	2·1287	1·9800	0·2107	0·3344	0·3272	0·5997	0·1247
0·30	0·8283	1·1778	0·4575	2·2093	2·0500	0·2232	0·3795	0·3722	0·6768	0·1328
0·32	0·8720	1·1962	0·4992	2·2900	2·1200	0·2355	0·4275	0·4203	0·7586	0·1409
0·34	0·9150	1·2126	0·5423	2·3706	2·1900	0·2476	0·4785	0·4715	0·8451	0·1489
0·36	0·9576	1·2272	0·5868	2·4512	2·2600	0·2596	0·5326	0·5258	0·9364	0·1568
0·38	0·9996	1·2403	0·6327	2·5318	2·3300	0·2715	0·5897	0·5834	1·0325	0·1647
0·40	1·0412	1·2520	0·6800	2·6125	2·4000	0·2833	0·6500	0·6443	1·1335	0·1725
0·42	1·0823	1·2626	0·7287	2·6931	2·4700	0·2950	0·7135	0·7086	1·2395	0·1803
0·44	1·1231	1·2721	0·7788	2·7737	2·5400	0·3066	0·7801	0·7762	1·3505	0·1881
0·46	1·1636	1·2806	0·8303	2·8543	2·6100	0·3181	0·8501	0·8473	1·4665	0·1958
0·48	1·2037	1·2884	0·8832	2·9349	2·6800	0·3296	0·9234	0·9218	1·5877	0·2035
0·50	1·2435	1·2955	0·9375	3·0156	2·7500	0·3409	1·0000	1·0000	1·7142	0·2111
0·52	1·2831	1·3019	0·9932	3·0962	2·8200	0·3522	1·0801	1·0818	1·8458	0·2187
0·54	1·3225	1·3078	1·0503	3·1768	2·8900	0·3634	1·1636	1·1672	1·9828	0·2263
0·56	1·3616	1·3131	1·1088	3·2574	2·9600	0·3746	1·2506	1·2564	2·1252	0·2338
0·58	1·4005	1·3180	1·1687	3·3381	3·0300	0·3857	1·3412	1·3494	2·2730	0·2413
0·60	1·4392	1·3225	1·2300	3·4187	3·1000	0·3968	1·4354	1·4462	2·4263	0·2488
0·62	1·4777	1·3266	1·2927	3·4993	3·1700	0·4078	1·5332	1·5469	2·5851	0·2562
0·64	1·5160	1·3304	1·3568	3·5799	3·2400	0·4188	1·6348	1·6516	2·7495	0·2636
0·66	1·5542	1·3338	1·4223	3·6605	3·3100	0·4297	1·7400	1·7603	2·9197	0·2710
0·68	1·5922	1·3370	1·4892	3·7412	3·3800	0·4406	1·8491	1·8730	3·0955	0·2784
0·70	1·6301	1·3400	1·5575	3·8218	3·4500	0·4514	1·9620	1·9899	3·2771	0·2858
0·72	1·6679	1·3427	1·6272	3·9024	3·5200	0·4623	2·0787	2·1109	3·4646	0·2931
0·74	1·7055	1·3452	1·6983	3·9830	3·5900	0·4731	2·1994	2·2362	3·6579	0·3004
0·76	1·7431	1·3475	1·7708	4·0637	3·6600	0·4838	2·3240	2·3657	3·8572	0·3077
0·78	1·7805	1·3497	1·8447	4·1443	3·7300	0·4946	2·4526	2·4996	4·0625	0·3150
0·80	1·8178	1·3517	1·9200	4·2249	3·8000	0·5053	2·5853	2·6379	4·2739	0·3222

Rel. dpth Y	U.equ. dia. $D_{ep(u)}$ (m)	Equiv. disch. factor J	Unit sect. area A_u (m²)	Unit wetted perim. P_u (m)	Unit surf. brdth $B_{s(u)}$ (m)	Unit mean depth $y_{m(u)}$ (m)	Discharge ratios $Q/Q_{0.50}$ medial	Mann'g	U.crit. disch. $Q_{c(u)}$ (m³s⁻¹)	Unit depth cntrd $y_{d(u)}$ (m)
0·82	1·8550	1·3535	1·9967	4·3055	3·8700	0·5159	2·7221	2·7806	4·4913	0·3295
0·84	1·8921	1·3552	2·0748	4·3861	3·9400	0·5266	2·8630	2·9277	4·7149	0·3367
0·86	1·9292	1·3568	2·1543	4·4668	4·0100	0·5372	3·0081	3·0795	4·9448	0·3439
0·88	1·9661	1·3583	2·2352	4·5474	4·0800	0·5478	3·1574	3·2358	5·1809	0·3511
0·90	2·0030	1·3597	2·3175	4·6280	4·1500	0·5584	3·3110	3·3967	5·4233	0·3583
0·92	2·0398	1·3609	2·4012	4·7086	4·2200	0·5690	3·4689	3·5624	5·6721	0·3654
0·94	2·0766	1·3621	2·4863	4·7893	4·2900	0·5796	3·6311	3·7328	5·9274	0·3726
0·96	2·1132	1·3632	2·5728	4·8699	4·3600	0·5901	3·7977	3·9080	6·1891	0·3797
0·98	2·1498	1·3643	2·6607	4·9505	4·4300	0·6006	3·9688	4·0881	6·4573	0·3868
1·00	2·1864	1·3652	2·7500	5·0311	4·5000	0·6111	4·1443	4·2730	6·7321	0·3939
1·02	2·2229	1·3661	2·8407	5·1118	4·5700	0·6216	4·3244	4·4630	7·0136	0·4010
1·04	2·2593	1·3669	2·9328	5·1924	4·6400	0·6321	4·5090	4·6579	7·3017	0·4081
1·06	2·2957	1·3677	3·0263	5·2730	4·7100	0·6425	4·6981	4·8578	7·5966	0·4152
1·08	2·3320	1·3684	3·1212	5·3536	4·7800	0·6530	4·8920	5·0629	7·8982	0·4223
1·10	2·3683	1·3691	3·2175	5·4342	4·8500	0·6634	5·0904	5·2731	8·2067	0·4293
1·12	2·4046	1·3697	3·3152	5·5149	4·9200	0·6738	5·2936	5·4885	8·5220	0·4364
1·14	2·4408	1·3703	3·4143	5·5955	4·9900	0·6842	5·5016	5·7091	8·8443	0·4434
1·16	2·4769	1·3709	3·5148	5·6761	5·0600	0·6946	5·7143	5·9351	9·1735	0·4505
1·18	2·5130	1·3714	3·6167	5·7567	5·1300	0·7050	5·9319	6·1664	9·5098	0·4575
1·20	2·5491	1·3719	3·7200	5·8374	5·2000	0·7154	6·1544	6·4031	9·8531	0·4645
1·24	2·6211	1·3727	3·9308	5·9986	5·3400	0·7361	6·6140	6·8928	10·561	0·4785
1·28	2·6931	1·3735	4·1472	6·1598	5·4800	0·7568	7·0936	7·4047	11·298	0·4925
1·32	2·7648	1·3741	4·3692	6·3211	5·6200	0·7774	7·5935	7·9391	12·064	0·5065
1·36	2·8365	1·3747	4·5968	6·4823	5·7600	0·7981	8·1139	8·4963	12·860	0·5204
1·40	2·9081	1·3751	4·8300	6·6436	5·9000	0·8186	8·6551	9·0769	13·685	0·5343
1·44	2·9795	1·3755	5·0688	6·8048	6·0400	0·8392	9·2175	9·6811	14·541	0·5482
1·48	3·0509	1·3759	5·3132	6·9661	6·1800	0·8597	9·8012	10·309	15·428	0·5620
1·52	3·1222	1·3762	5·5632	7·1273	6·3200	0·8803	10·407	10·962	16·345	0·5759
1·56	3·1934	1·3764	5·8188	7·2886	6·4600	0·9007	11·034	11·639	17·294	0·5897
1·60	3·2645	1·3766	6·0800	7·4498	6·6000	0·9212	11·684	12·342	18·274	0·6035
1·64	3·3356	1·3768	6·3468	7·6111	6·7400	0·9417	12·356	13·069	19·287	0·6173
1·68	3·4066	1·3769	6·6192	7·7723	6·8800	0·9621	13·052	13·823	20·332	0·6311
1·72	3·4775	1·3770	6·8972	7·9335	7·0200	0·9825	13·770	14·603	21·409	0·6448
1·76	3·5484	1·3771	7·1808	8·0948	7·1600	1·0029	14·512	15·409	22·520	0·6586
1·80	3·6192	1·3771	7·4700	8·2560	7·3000	1·0233	15·277	16·242	23·664	0·6723
1·84	3·6899	1·3772	7·7648	8·4173	7·4400	1·0437	16·067	17·103	24·841	0·6860
1·88	3·7606	1·3772	8·0652	8·5785	7·5800	1·0640	16·880	17·991	26·052	0·6997
1·92	3·8313	1·3772	8·3712	8·7398	7·7200	1·0844	17·718	18·906	27·298	0·7134
1·96	3·9019	1·3771	8·6828	8·9010	7·8600	1·1047	18·581	19·850	28·578	0·7271
2·00	3·9725	1·3771	9·0000	9·0623	8·0000	1·1250	19·469	20·823	29·894	0·7407
2·05	4·0607	1·3771	9·4044	9·2638	8·1750	1·1504	20·614	22·079	31·587	0·7578
2·10	4·1488	1·3770	9·8175	9·4654	8·3500	1·1757	21·799	23·381	33·336	0·7749
2·15	4·2369	1·3769	10·239	9·6669	8·5250	1·2011	23·025	24·730	35·142	0·7919
2·20	4·3249	1·3768	10·670	9·8685	8·7000	1·2264	24·292	26·126	37·004	0·8089
2·25	4·4128	1·3767	11·109	10·070	8·8750	1·2518	25·600	27·569	38·923	0·8259
2·30	4·5008	1·3765	11·557	10·272	9·0500	1·2771	26·950	29·061	40·901	0·8430
2·35	4·5886	1·3764	12·014	10·473	9·2250	1·3024	28·343	30·602	42·937	0·8599
2·40	4·6765	1·3763	12·480	10·675	9·4000	1·3277	29·778	32·192	45·032	0·8769
2·45	4·7643	1·3761	12·954	10·876	9·5750	1·3529	31·257	33·833	47·186	0·8939
2·50	4·8520	1·3760	13·437	11·078	9·7500	1·3782	32·780	35·524	49·401	0·9109
2·55	4·9398	1·3758	13·929	11·279	9·9250	1·4035	34·347	37·267	51·676	0·9278
2·60	5·0275	1·3757	14·430	11·481	10·100	1·4287	35·959	39·062	54·013	0·9447
2·65	5·1151	1·3755	14·939	11·682	10·275	1·4540	37·616	40·909	56·412	0·9617
2·70	5·2028	1·3753	15·457	11·884	10·450	1·4792	39·319	42·810	58·872	0·9786
2·75	5·2904	1·3752	15·984	12·086	10·625	1·5044	41·068	44·765	61·396	0·9955

Trapezoidal channel - 2·0 to 1 side-slope

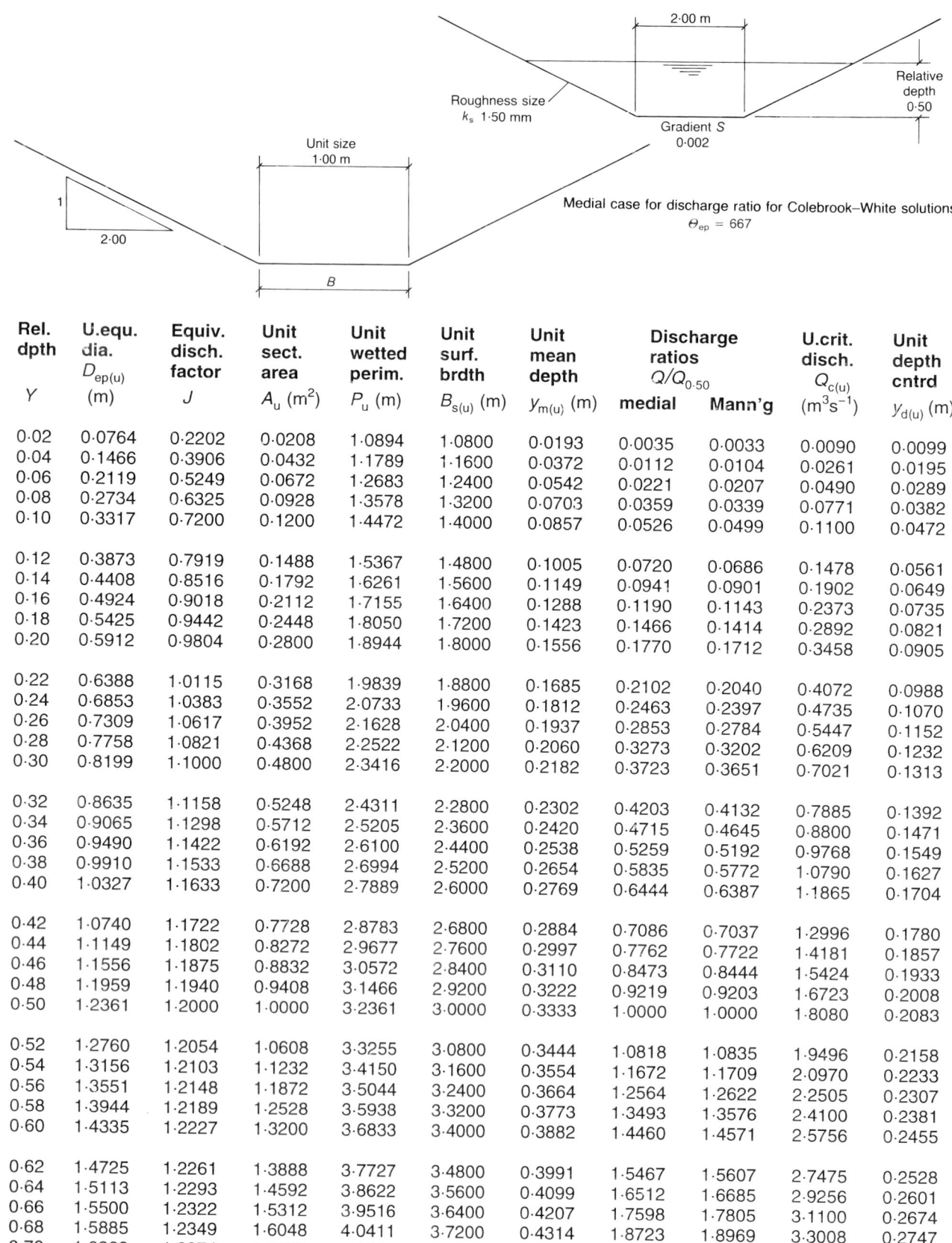

Rel. dpth Y	U.equ. dia. $D_{ep(u)}$ (m)	Equiv. disch. factor J	Unit sect. area A_u (m²)	Unit wetted perim. P_u (m)	Unit surf. brdth $B_{s(u)}$ (m)	Unit mean depth $y_{m(u)}$ (m)	Discharge ratios $Q/Q_{0.50}$ medial	Discharge ratios $Q/Q_{0.50}$ Mann'g	U.crit. disch. $Q_{c(u)}$ (m³s⁻¹)	Unit depth cntrd $y_{d(u)}$ (m)
0·02	0·0764	0·2202	0·0208	1·0894	1·0800	0·0193	0·0035	0·0033	0·0090	0·0099
0·04	0·1466	0·3906	0·0432	1·1789	1·1600	0·0372	0·0112	0·0104	0·0261	0·0195
0·06	0·2119	0·5249	0·0672	1·2683	1·2400	0·0542	0·0221	0·0207	0·0490	0·0289
0·08	0·2734	0·6325	0·0928	1·3578	1·3200	0·0703	0·0359	0·0339	0·0771	0·0382
0·10	0·3317	0·7200	0·1200	1·4472	1·4000	0·0857	0·0526	0·0499	0·1100	0·0472
0·12	0·3873	0·7919	0·1488	1·5367	1·4800	0·1005	0·0720	0·0686	0·1478	0·0561
0·14	0·4408	0·8516	0·1792	1·6261	1·5600	0·1149	0·0941	0·0901	0·1902	0·0649
0·16	0·4924	0·9018	0·2112	1·7155	1·6400	0·1288	0·1190	0·1143	0·2373	0·0735
0·18	0·5425	0·9442	0·2448	1·8050	1·7200	0·1423	0·1466	0·1414	0·2892	0·0821
0·20	0·5912	0·9804	0·2800	1·8944	1·8000	0·1556	0·1770	0·1712	0·3458	0·0905
0·22	0·6388	1·0115	0·3168	1·9839	1·8800	0·1685	0·2102	0·2040	0·4072	0·0988
0·24	0·6853	1·0383	0·3552	2·0733	1·9600	0·1812	0·2463	0·2397	0·4735	0·1070
0·26	0·7309	1·0617	0·3952	2·1628	2·0400	0·1937	0·2853	0·2784	0·5447	0·1152
0·28	0·7758	1·0821	0·4368	2·2522	2·1200	0·2060	0·3273	0·3202	0·6209	0·1232
0·30	0·8199	1·1000	0·4800	2·3416	2·2000	0·2182	0·3723	0·3651	0·7021	0·1313
0·32	0·8635	1·1158	0·5248	2·4311	2·2800	0·2302	0·4203	0·4132	0·7885	0·1392
0·34	0·9065	1·1298	0·5712	2·5205	2·3600	0·2420	0·4715	0·4645	0·8800	0·1471
0·36	0·9490	1·1422	0·6192	2·6100	2·4400	0·2538	0·5259	0·5192	0·9768	0·1549
0·38	0·9910	1·1533	0·6688	2·6994	2·5200	0·2654	0·5835	0·5772	1·0790	0·1627
0·40	1·0327	1·1633	0·7200	2·7889	2·6000	0·2769	0·6444	0·6387	1·1865	0·1704
0·42	1·0740	1·1722	0·7728	2·8783	2·6800	0·2884	0·7086	0·7037	1·2996	0·1780
0·44	1·1149	1·1802	0·8272	2·9677	2·7600	0·2997	0·7762	0·7722	1·4181	0·1857
0·46	1·1556	1·1875	0·8832	3·0572	2·8400	0·3110	0·8473	0·8444	1·5424	0·1933
0·48	1·1959	1·1940	0·9408	3·1466	2·9200	0·3222	0·9219	0·9203	1·6723	0·2008
0·50	1·2361	1·2000	1·0000	3·2361	3·0000	0·3333	1·0000	1·0000	1·8080	0·2083
0·52	1·2760	1·2054	1·0608	3·3255	3·0800	0·3444	1·0818	1·0835	1·9496	0·2158
0·54	1·3156	1·2103	1·1232	3·4150	3·1600	0·3554	1·1672	1·1709	2·0970	0·2233
0·56	1·3551	1·2148	1·1872	3·5044	3·2400	0·3664	1·2564	1·2622	2·2505	0·2307
0·58	1·3944	1·2189	1·2528	3·5938	3·3200	0·3773	1·3493	1·3576	2·4100	0·2381
0·60	1·4335	1·2227	1·3200	3·6833	3·4000	0·3882	1·4460	1·4571	2·5756	0·2455
0·62	1·4725	1·2261	1·3888	3·7727	3·4800	0·3991	1·5467	1·5607	2·7475	0·2528
0·64	1·5113	1·2293	1·4592	3·8622	3·5600	0·4099	1·6512	1·6685	2·9256	0·2601
0·66	1·5500	1·2322	1·5312	3·9516	3·6400	0·4207	1·7598	1·7805	3·1100	0·2674
0·68	1·5885	1·2349	1·6048	4·0411	3·7200	0·4314	1·8723	1·8969	3·3008	0·2747
0·70	1·6269	1·2374	1·6800	4·1305	3·8000	0·4421	1·9890	2·0177	3·4981	0·2819
0·72	1·6652	1·2397	1·7568	4·2199	3·8800	0·4528	2·1098	2·1429	3·7019	0·2892
0·74	1·7034	1·2418	1·8352	4·3094	3·9600	0·4634	2·2348	2·2727	3·9124	0·2964
0·76	1·7416	1·2438	1·9152	4·3988	4·0400	0·4741	2·3640	2·4070	4·1294	0·3036
0·78	1·7796	1·2456	1·9968	4·4883	4·1200	0·4847	2·4974	2·5460	4·3532	0·3108
0·80	1·8175	1·2473	2·0800	4·5777	4·2000	0·4952	2·6352	2·6896	4·5839	0·3179

Rel. dpth Y	U.equ. dia. $D_{ep(u)}$ (m)	Equiv. disch. factor J	Unit sect. area A_u (m²)	Unit wetted perim. P_u (m)	Unit surf. brdth $B_{s(u)}$ (m)	Unit mean depth $y_{m(u)}$ (m)	Discharge ratios $Q/Q_{0.50}$ medial	Mann'g	U.crit. disch. $Q_{c(u)}$ (m³s⁻¹)	Unit depth cntrd $y_{d(u)}$ (m)
0·82	1·8553	1·2489	2·1648	4·6672	4·2800	0·5058	2·7774	2·8380	4·8213	0·3251
0·84	1·8931	1·2503	2·2512	4·7566	4·3600	0·5163	2·9241	2·9911	5·0657	0·3322
0·86	1·9308	1·2517	2·3392	4·8460	4·4400	0·5268	3·0752	3·1492	5·3170	0·3394
0·88	1·9684	1·2529	2·4288	4·9355	4·5200	0·5373	3·2308	3·3122	5·5754	0·3465
0·90	2·0060	1·2541	2·5200	5·0249	4·6000	0·5478	3·3910	3·4801	5·8409	0·3536
0·92	2·0435	1·2552	2·6128	5·1144	4·6800	0·5583	3·5558	3·6531	6·1136	0·3607
0·94	2·0809	1·2563	2·7072	5·2038	4·7600	0·5687	3·7252	3·8312	6·3935	0·3677
0·96	2·1183	1·2572	2·8032	5·2933	4·8400	0·5792	3·8994	4·0144	6·6807	0·3748
0·98	2·1556	1·2581	2·9008	5·3827	4·9200	0·5896	4·0783	4·2028	6·9752	0·3818
1·00	2·1929	1·2589	3·0000	5·4721	5·0000	0·6000	4·2620	4·3965	7·2771	0·3889
1·02	2·2302	1·2597	3·1008	5·5616	5·0800	0·6104	4·4506	4·5955	7·5865	0·3959
1·04	2·2673	1·2605	3·2032	5·6510	5·1600	0·6208	4·6441	4·7999	7·9034	0·4029
1·06	2·3045	1·2611	3·3072	5·7405	5·2400	0·6311	4·8425	5·0097	8·2278	0·4100
1·08	2·3416	1·2618	3·4128	5·8299	5·3200	0·6415	5·0459	5·2250	8·5599	0·4170
1·10	2·3786	1·2624	3·5200	5·9193	5·4000	0·6519	5·2543	5·4459	8·8997	0·4240
1·12	2·4157	1·2630	3·6288	6·0088	5·4800	0·6622	5·4678	5·6723	9·2473	0·4309
1·14	2·4526	1·2635	3·7392	6·0982	5·5600	0·6725	5·6864	5·9044	9·6026	0·4379
1·16	2·4896	1·2640	3·8512	6·1877	5·6400	0·6828	5·9101	6·1421	9·9659	0·4449
1·18	2·5265	1·2644	3·9648	6·2771	5·7200	0·6931	6·1391	6·3857	10·337	0·4519
1·20	2·5634	1·2649	4·0800	6·3666	5·8000	0·7034	6·3732	6·6350	10·716	0·4588
1·24	2·6371	1·2657	4·3152	6·5454	5·9600	0·7240	6·8575	7·1513	11·498	0·4727
1·28	2·7106	1·2664	4·5568	6·7243	6·1200	0·7446	7·3633	7·6915	12·313	0·4866
1·32	2·7841	1·2670	4·8048	6·9032	6·2800	0·7651	7·8909	8·2560	13·161	0·5004
1·36	2·8575	1·2675	5·0592	7·0821	6·4400	0·7856	8·4406	8·8452	14·042	0·5143
1·40	2·9307	1·2680	5·3200	7·2610	6·6000	0·8061	9·0129	9·4595	14·957	0·5281
1·44	3·0039	1·2684	5·5872	7·4399	6·7600	0·8265	9·6079	10·099	15·907	0·5419
1·48	3·0770	1·2688	5·8608	7·6188	6·9200	0·8469	10·226	10·765	16·891	0·5556
1·52	3·1501	1·2691	6·1408	7·7976	7·0800	0·8673	10·868	11·457	17·909	0·5694
1·56	3·2231	1·2694	6·4272	7·9765	7·2400	0·8877	11·533	12·176	18·964	0·5831
1·60	3·2960	1·2696	6·7200	8·1554	7·4000	0·9081	12·223	12·922	20·054	0·5968
1·64	3·3688	1·2698	7·0192	8·3343	7·5600	0·9285	12·936	13·696	21·180	0·6105
1·68	3·4416	1·2700	7·3248	8·5132	7·7200	0·9488	13·675	14·497	22·343	0·6242
1·72	3·5144	1·2702	7·6368	8·6921	7·8800	0·9691	14·439	15·327	23·543	0·6379
1·76	3·5871	1·2703	7·9552	8·8710	8·0400	0·9895	15·227	16·185	24·780	0·6516
1·80	3·6597	1·2704	8·2800	9·0498	8·2000	1·0098	16·042	17·073	26·055	0·6652
1·84	3·7323	1·2705	8·6112	9·2287	8·3600	1·0300	16·882	17·990	27·369	0·6789
1·88	3·8049	1·2706	8·9488	9·4076	8·5200	1·0503	17·749	18·937	28·720	0·6925
1·92	3·8775	1·2707	9·2928	9·5865	8·6800	1·0706	18·642	19·914	30·111	0·7061
1·96	3·9500	1·2707	9·6432	9·7654	8·8400	1·0909	19·562	20·921	31·540	0·7197
2·00	4·0224	1·2707	10·000	9·9443	9·0000	1·1111	20·509	21·960	33·010	0·7333
2·05	4·1130	1·2708	10·455	10·168	9·2000	1·1364	21·731	23·302	34·902	0·7503
2·10	4·2034	1·2708	10·920	10·391	9·4000	1·1617	22·997	24·694	36·858	0·7673
2·15	4·2939	1·2708	11·395	10·615	9·6000	1·1870	24·306	26·137	38·877	0·7843
2·20	4·3843	1·2708	11·880	10·839	9·8000	1·2122	25·660	27·630	40·961	0·8012
2·25	4·4747	1·2707	12·375	11·062	10·000	1·2375	27·059	29·176	43·110	0·8182
2·30	4·5650	1·2707	12·880	11·286	10·200	1·2627	28·503	30·774	45·325	0·8351
2·35	4·6553	1·2707	13·395	11·510	10·400	1·2880	29·994	32·425	47·606	0·8520
2·40	4·7455	1·2706	13·920	11·733	10·600	1·3132	31·530	34·130	49·954	0·8690
2·45	4·8358	1·2706	14·455	11·957	10·800	1·3384	33·114	35·889	52·369	0·8859
2·50	4·9260	1·2705	15·000	12·180	11·000	1·3636	34·746	37·704	54·853	0·9028
2·55	5·0161	1·2704	15·555	12·404	11·200	1·3888	36·426	39·575	57·406	0·9197
2·60	5·1063	1·2704	16·120	12·628	11·400	1·4140	38·154	41·502	60·028	0·9366
2·65	5·1964	1·2703	16·695	12·851	11·600	1·4392	39·931	43·487	62·721	0·9534
2·70	5·2865	1·2702	17·280	13·075	11·800	1·4644	41·758	45·530	65·484	0·9703
2·75	5·3766	1·2701	17·875	13·298	12·000	1·4896	43·636	47·631	68·319	0·9872

C63 Trapezoidal channel - 2·5 to 1 side-slope

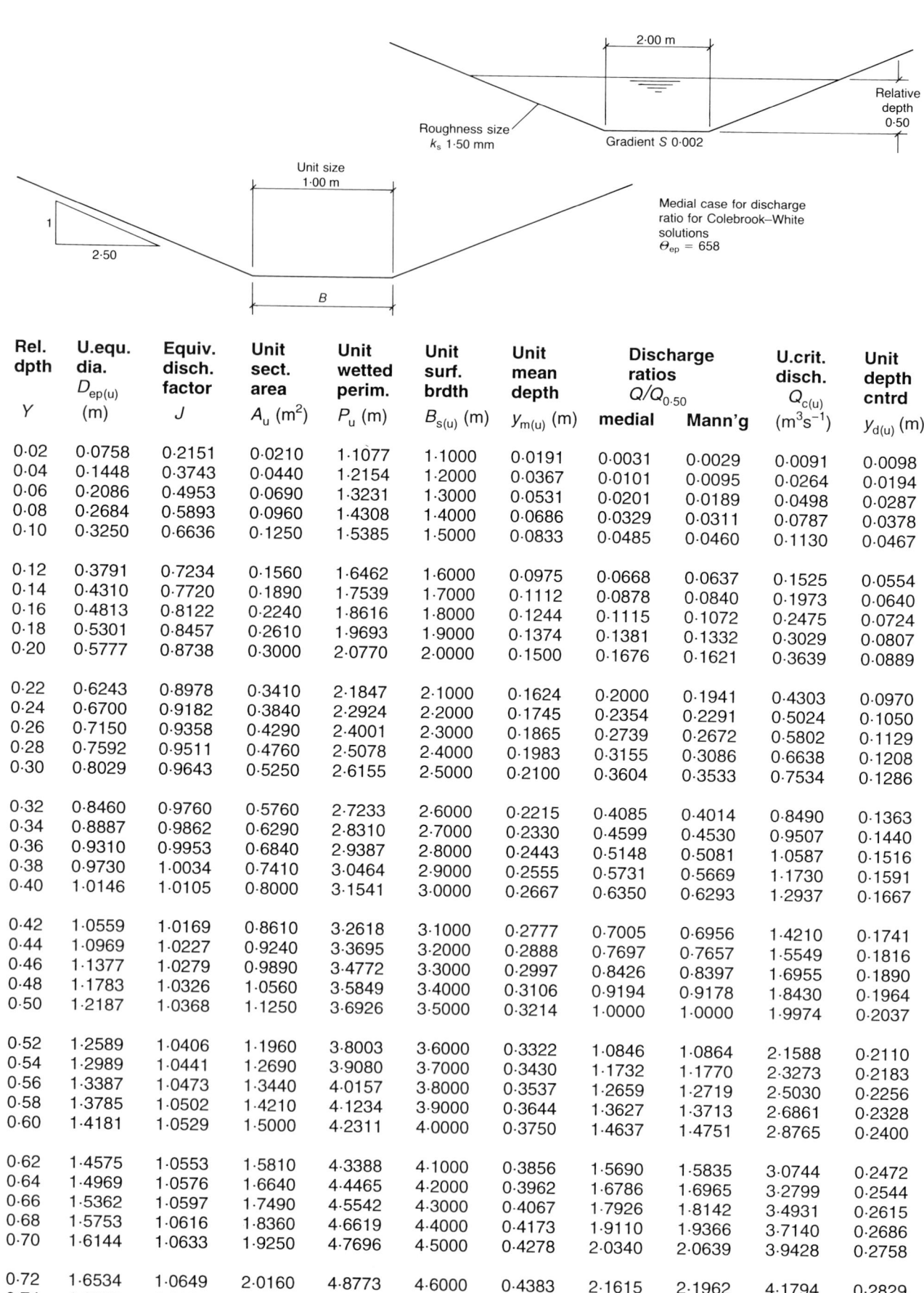

Rel. dpth Y	U.equ. dia. $D_{ep(u)}$ (m)	Equiv. disch. factor J	Unit sect. area A_u (m²)	Unit wetted perim. P_u (m)	Unit surf. brdth $B_{s(u)}$ (m)	Unit mean depth $y_{m(u)}$ (m)	Discharge ratios $Q/Q_{0.50}$ medial	Discharge ratios $Q/Q_{0.50}$ Mann'g	U.crit. disch. $Q_{c(u)}$ (m³s⁻¹)	Unit depth cntrd $y_{d(u)}$ (m)
0·02	0·0758	0·2151	0·0210	1·1077	1·1000	0·0191	0·0031	0·0029	0·0091	0·0098
0·04	0·1448	0·3743	0·0440	1·2154	1·2000	0·0367	0·0101	0·0095	0·0264	0·0194
0·06	0·2086	0·4953	0·0690	1·3231	1·3000	0·0531	0·0201	0·0189	0·0498	0·0287
0·08	0·2684	0·5893	0·0960	1·4308	1·4000	0·0686	0·0329	0·0311	0·0787	0·0378
0·10	0·3250	0·6636	0·1250	1·5385	1·5000	0·0833	0·0485	0·0460	0·1130	0·0467
0·12	0·3791	0·7234	0·1560	1·6462	1·6000	0·0975	0·0668	0·0637	0·1525	0·0554
0·14	0·4310	0·7720	0·1890	1·7539	1·7000	0·1112	0·0878	0·0840	0·1973	0·0640
0·16	0·4813	0·8122	0·2240	1·8616	1·8000	0·1244	0·1115	0·1072	0·2475	0·0724
0·18	0·5301	0·8457	0·2610	1·9693	1·9000	0·1374	0·1381	0·1332	0·3029	0·0807
0·20	0·5777	0·8738	0·3000	2·0770	2·0000	0·1500	0·1676	0·1621	0·3639	0·0889
0·22	0·6243	0·8978	0·3410	2·1847	2·1000	0·1624	0·2000	0·1941	0·4303	0·0970
0·24	0·6700	0·9182	0·3840	2·2924	2·2000	0·1745	0·2354	0·2291	0·5024	0·1050
0·26	0·7150	0·9358	0·4290	2·4001	2·3000	0·1865	0·2739	0·2672	0·5802	0·1129
0·28	0·7592	0·9511	0·4760	2·5078	2·4000	0·1983	0·3155	0·3086	0·6638	0·1208
0·30	0·8029	0·9643	0·5250	2·6155	2·5000	0·2100	0·3604	0·3533	0·7534	0·1286
0·32	0·8460	0·9760	0·5760	2·7233	2·6000	0·2215	0·4085	0·4014	0·8490	0·1363
0·34	0·8887	0·9862	0·6290	2·8310	2·7000	0·2330	0·4599	0·4530	0·9507	0·1440
0·36	0·9310	0·9953	0·6840	2·9387	2·8000	0·2443	0·5148	0·5081	1·0587	0·1516
0·38	0·9730	1·0034	0·7410	3·0464	2·9000	0·2555	0·5731	0·5669	1·1730	0·1591
0·40	1·0146	1·0105	0·8000	3·1541	3·0000	0·2667	0·6350	0·6293	1·2937	0·1667
0·42	1·0559	1·0169	0·8610	3·2618	3·1000	0·2777	0·7005	0·6956	1·4210	0·1741
0·44	1·0969	1·0227	0·9240	3·3695	3·2000	0·2888	0·7697	0·7657	1·5549	0·1816
0·46	1·1377	1·0279	0·9890	3·4772	3·3000	0·2997	0·8426	0·8397	1·6955	0·1890
0·48	1·1783	1·0326	1·0560	3·5849	3·4000	0·3106	0·9194	0·9178	1·8430	0·1964
0·50	1·2187	1·0368	1·1250	3·6926	3·5000	0·3214	1·0000	1·0000	1·9974	0·2037
0·52	1·2589	1·0406	1·1960	3·8003	3·6000	0·3322	1·0846	1·0864	2·1588	0·2110
0·54	1·2989	1·0441	1·2690	3·9080	3·7000	0·3430	1·1732	1·1770	2·3273	0·2183
0·56	1·3387	1·0473	1·3440	4·0157	3·8000	0·3537	1·2659	1·2719	2·5030	0·2256
0·58	1·3785	1·0502	1·4210	4·1234	3·9000	0·3644	1·3627	1·3713	2·6861	0·2328
0·60	1·4181	1·0529	1·5000	4·2311	4·0000	0·3750	1·4637	1·4751	2·8765	0·2400
0·62	1·4575	1·0553	1·5810	4·3388	4·1000	0·3856	1·5690	1·5835	3·0744	0·2472
0·64	1·4969	1·0576	1·6640	4·4465	4·2000	0·3962	1·6786	1·6965	3·2799	0·2544
0·66	1·5362	1·0597	1·7490	4·5542	4·3000	0·4067	1·7926	1·8142	3·4931	0·2615
0·68	1·5753	1·0616	1·8360	4·6619	4·4000	0·4173	1·9110	1·9366	3·7140	0·2686
0·70	1·6144	1·0633	1·9250	4·7696	4·5000	0·4278	2·0340	2·0639	3·9428	0·2758
0·72	1·6534	1·0649	2·0160	4·8773	4·6000	0·4383	2·1615	2·1962	4·1794	0·2829
0·74	1·6923	1·0665	2·1090	4·9850	4·7000	0·4487	2·2936	2·3334	4·4241	0·2899
0·76	1·7311	1·0679	2·2040	5·0927	4·8000	0·4592	2·4304	2·4756	4·6769	0·2970
0·78	1·7699	1·0691	2·3010	5·2004	4·9000	0·4696	2·5720	2·6230	4·9378	0·3041
0·80	1·8085	1·0704	2·4000	5·3081	5·0000	0·4800	2·7183	2·7756	5·2071	0·3111

Rel. dpth Y	U.equ. dia. $D_{ep(u)}$ (m)	Equiv. disch. factor J	Unit sect. area A_u (m²)	Unit wetted perim. P_u (m)	Unit surf. brdth $B_{s(u)}$ (m)	Unit mean depth $y_{m(u)}$ (m)	Discharge ratios $Q/Q_{0.50}$ medial	Mann'g	U.crit. disch. $Q_{c(u)}$ (m³s⁻¹)	Unit depth cntrd $y_{d(u)}$ (m)
0·82	1·8472	1·0715	2·5010	5·4158	5·1000	0·4904	2·8695	2·9334	5·4846	0·3181
0·84	1·8857	1·0725	2·6040	5·5235	5·2000	0·5008	3·0257	3·0966	5·7706	0·3252
0·86	1·9243	1·0735	2·7090	5·6312	5·3000	0·5111	3·1868	3·2652	6·0651	0·3322
0·88	1·9627	1·0744	2·8160	5·7389	5·4000	0·5215	3·3529	3·4393	6·3681	0·3392
0·90	2·0011	1·0753	2·9250	5·8466	5·5000	0·5318	3·5241	3·6189	6·6799	0·3462
0·92	2·0395	1·0761	3·0360	5·9544	5·6000	0·5421	3·7005	3·8040	7·0003	0·3531
0·94	2·0778	1·0768	3·1490	6·0621	5·7000	0·5525	3·8820	3·9949	7·3296	0·3601
0·96	2·1161	1·0775	3·2640	6·1698	5·8000	0·5628	4·0688	4·1915	7·6678	0·3671
0·98	2·1544	1·0781	3·3810	6·2775	5·9000	0·5731	4·2609	4·3939	8·0150	0·3740
1·00	2·1926	1·0788	3·5000	6·3852	6·0000	0·5833	4·4583	4·6022	8·3712	0·3810
1·02	2·2308	1·0793	3·6210	6·4929	6·1000	0·5936	4·6611	4·8164	8·7365	0·3879
1·04	2·2689	1·0799	3·7440	6·6006	6·2000	0·6039	4·8694	5·0366	9·1110	0·3948
1·06	2·3070	1·0804	3·8690	6·7083	6·3000	0·6141	5·0832	5·2629	9·4948	0·4017
1·08	2·3451	1·0809	3·9960	6·8160	6·4000	0·6244	5·3026	5·4953	9·8880	0·4086
1·10	2·3831	1·0813	4·1250	6·9237	6·5000	0·6346	5·5276	5·7339	10·291	0·4156
1·12	2·4211	1·0817	4·2560	7·0314	6·6000	0·6448	5·7583	5·9787	10·703	0·4225
1·14	2·4591	1·0821	4·3890	7·1391	6·7000	0·6551	5·9946	6·2299	11·124	0·4294
1·16	2·4971	1·0825	4·5240	7·2468	6·8000	0·6653	6·2368	6·4874	11·556	0·4362
1·18	2·5350	1·0829	4·6610	7·3545	6·9000	0·6755	6·4848	6·7514	11·996	0·4431
1·20	2·5730	1·0832	4·8000	7·4622	7·0000	0·6857	6·7386	7·0219	12·447	0·4500
1·24	2·6487	1·0838	5·0840	7·6776	7·2000	0·7061	7·2641	7·5827	13·378	0·4637
1·28	2·7244	1·0844	5·3760	7·8930	7·4000	0·7265	7·8136	8·1703	14·349	0·4775
1·32	2·8001	1·0849	5·6760	8·1084	7·6000	0·7468	8·3877	8·7851	15·361	0·4912
1·36	2·8756	1·0853	5·9840	8·3238	7·8000	0·7672	8·9866	9·4276	16·413	0·5048
1·40	2·9511	1·0857	6·3000	8·5392	8·0000	0·7875	9·6107	10·098	17·508	0·5185
1·44	3·0265	1·0860	6·6240	8·7546	8·2000	0·8078	10·260	10·798	18·644	0·5322
1·48	3·1019	1·0864	6·9560	8·9700	8·4000	0·8281	10·936	11·527	19·823	0·5458
1·52	3·1772	1·0866	7·2960	9·1855	8·6000	0·8484	11·638	12·285	21·044	0·5594
1·56	3·2525	1·0869	7·6440	9·4009	8·8000	0·8686	12·367	13·073	22·310	0·5731
1·60	3·3277	1·0871	8·0000	9·6163	9·0000	0·8889	13·123	13·892	23·620	0·5867
1·64	3·4029	1·0873	8·3640	9·8317	9·2000	0·9091	13·907	14·743	24·974	0·6003
1·68	3·4780	1·0875	8·7360	10·047	9·4000	0·9294	14·718	15·624	26·373	0·6138
1·72	3·5531	1·0877	9·1160	10·262	9·6000	0·9496	15·557	16·538	27·818	0·6274
1·76	3·6282	1·0878	9·5040	10·478	9·8000	0·9698	16·425	17·483	29·309	0·6410
1·80	3·7033	1·0880	9·9000	10·693	10·000	0·9900	17·322	18·462	30·847	0·6545
1·84	3·7783	1·0881	10·304	10·909	10·200	1·0102	18·248	19·474	32·432	0·6681
1·88	3·8533	1·0882	10·716	11·124	10·400	1·0304	19·204	20·520	34·064	0·6816
1·92	3·9282	1·0883	11·136	11·340	10·600	1·0506	20·189	21·600	35·744	0·6952
1·96	4·0031	1·0884	11·564	11·555	10·800	1·0707	21·205	22·714	37·472	0·7087
2·00	4·0781	1·0884	12·000	11·770	11·000	1·0909	22·252	23·864	39·250	0·7222
2·05	4·1717	1·0885	12·556	12·040	11·250	1·1161	23·603	25·351	41·541	0·7391
2·10	4·2652	1·0886	13·125	12·309	11·500	1·1413	25·003	26·894	43·910	0·7560
2·15	4·3588	1·0886	13·706	12·578	11·750	1·1665	26·453	28·494	46·357	0·7729
2·20	4·4523	1·0887	14·300	12·847	12·000	1·1917	27·954	30·152	48·885	0·7897
2·25	4·5458	1·0887	14·906	13·117	12·250	1·2168	29·505	31·869	51·493	0·8066
2·30	4·6392	1·0888	15·525	13·386	12·500	1·2420	31·107	33·645	54·182	0·8235
2·35	4·7327	1·0888	16·156	13·655	12·750	1·2672	32·762	35·482	56·953	0·8403
2·40	4·8261	1·0888	16·800	13·924	13·000	1·2923	34·469	37·379	59·807	0·8571
2·45	4·9195	1·0888	17·456	14·194	13·250	1·3175	36·230	39·339	62·745	0·8740
2·50	5·0128	1·0888	18·125	14·463	13·500	1·3426	38·044	41·361	65·767	0·8908
2·55	5·1062	1·0889	18·806	14·732	13·750	1·3677	39·913	43·447	68·875	0·9076
2·60	5·1995	1·0889	19·500	15·001	14·000	1·3929	41·837	45·597	72·069	0·9244
2·65	5·2928	1·0889	20·206	15·271	14·250	1·4180	43·817	47·812	75·350	0·9413
2·70	5·3861	1·0888	20·925	15·540	14·500	1·4431	45·853	50·093	78·718	0·9581
2·75	5·4794	1·0888	21·656	15·809	14·750	1·4682	47·945	52·440	82·175	0·9749

C64

Trapezoidal channel - 3·0 to 1 side-slope

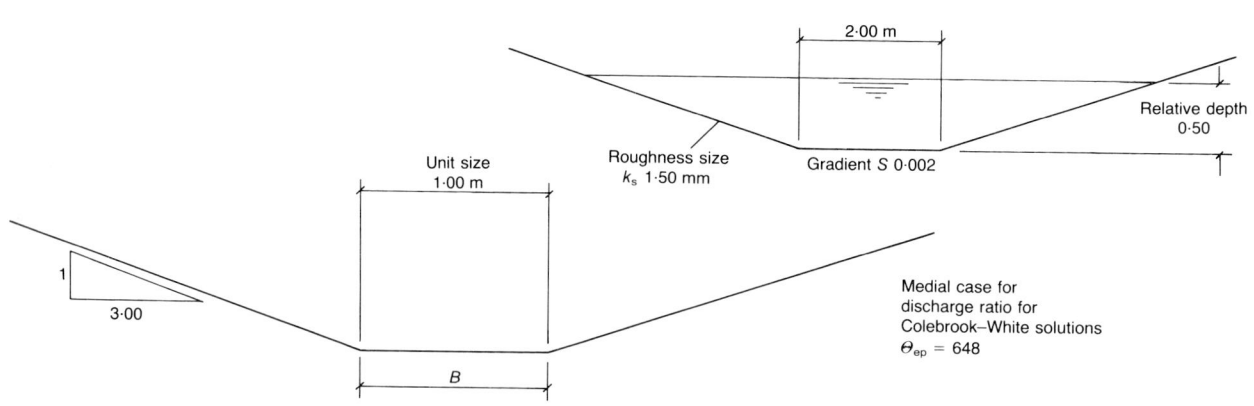

Rel. dpth Y	U.equ. dia. $D_{ep(u)}$ (m)	Equiv. disch. factor J	Unit sect. area A_u (m²)	Unit wetted perim. P_u (m)	Unit surf. brdth $B_{s(u)}$ (m)	Unit mean depth $y_{m(u)}$ (m)	Discharge ratios $Q/Q_{0.50}$ medial	Mann'g	U.crit. disch. $Q_{c(u)}$ (m³s⁻¹)	Unit depth cntrd $y_{d(u)}$ (m)
0·02	0·0753	0·2099	0·0212	1·1265	1·1200	0·0189	0·0029	0·0027	0·0091	0·0098
0·04	0·1430	0·3586	0·0448	1·2530	1·2400	0·0361	0·0093	0·0087	0·0267	0·0193
0·06	0·2053	0·4675	0·0708	1·3795	1·3600	0·0521	0·0186	0·0174	0·0506	0·0285
0·08	0·2635	0·5496	0·0992	1·5060	1·4800	0·0670	0·0306	0·0289	0·0804	0·0374
0·10	0·3185	0·6130	0·1300	1·6325	1·6000	0·0812	0·0452	0·0429	0·1160	0·0462
0·12	0·3711	0·6628	0·1632	1·7589	1·7200	0·0949	0·0626	0·0597	0·1574	0·0547
0·14	0·4218	0·7027	0·1988	1·8854	1·8400	0·1080	0·0827	0·0792	0·2046	0·0631
0·16	0·4708	0·7351	0·2368	2·0119	1·9600	0·1208	0·1056	0·1015	0·2578	0·0714
0·18	0·5185	0·7617	0·2772	2·1384	2·0800	0·1333	0·1314	0·1267	0·3169	0·0795
0·20	0·5651	0·7839	0·3200	2·2649	2·2000	0·1455	0·1601	0·1549	0·3822	0·0875
0·22	0·6109	0·8025	0·3652	2·3914	2·3200	0·1574	0·1919	0·1861	0·4537	0·0954
0·24	0·6558	0·8182	0·4128	2·5179	2·4400	0·1692	0·2267	0·2206	0·5317	0·1033
0·26	0·7000	0·8317	0·4628	2·6444	2·5600	0·1808	0·2648	0·2583	0·6162	0·1110
0·28	0·7437	0·8432	0·5152	2·7709	2·6800	0·1922	0·3061	0·2994	0·7074	0·1187
0·30	0·7869	0·8532	0·5700	2·8974	2·8000	0·2036	0·3508	0·3439	0·8054	0·1263
0·32	0·8297	0·8620	0·6272	3·0239	2·9200	0·2148	0·3990	0·3920	0·9103	0·1339
0·34	0·8720	0·8696	0·6868	3·1503	3·0400	0·2259	0·4506	0·4438	1·0223	0·1414
0·36	0·9141	0·8763	0·7488	3·2768	3·1600	0·2370	0·5059	0·4993	1·1415	0·1488
0·38	0·9558	0·8822	0·8132	3·4033	3·2800	0·2479	0·5648	0·5586	1·2680	0·1563
0·40	0·9972	0·8875	0·8800	3·5298	3·4000	0·2588	0·6275	0·6218	1·4020	0·1636
0·42	1·0384	0·8922	0·9492	3·6563	3·5200	0·2697	0·6940	0·6891	1·5436	0·1710
0·44	1·0794	0·8964	1·0208	3·7828	3·6400	0·2804	0·7645	0·7604	1·6929	0·1783
0·46	1·1202	0·9002	1·0948	3·9093	3·7600	0·2912	0·8389	0·8360	1·8500	0·1855
0·48	1·1608	0·9036	1·1712	4·0358	3·8800	0·3019	0·9174	0·9158	2·0151	0·1928
0·50	1·2013	0·9067	1·2500	4·1623	4·0000	0·3125	1·0000	1·0000	2·1882	0·2000
0·52	1·2416	0·9094	1·3312	4·2888	4·1200	0·3231	1·0869	1·0886	2·3696	0·2072
0·54	1·2817	0·9120	1·4148	4·4153	4·2400	0·3337	1·1780	1·1818	2·5593	0·2144
0·56	1·3218	0·9143	1·5008	4·5418	4·3600	0·3442	1·2735	1·2797	2·7574	0·2215
0·58	1·3617	0·9164	1·5892	4·6682	4·4800	0·3547	1·3734	1·3822	2·9641	0·2286
0·60	1·4015	0·9183	1·6800	4·7947	4·6000	0·3652	1·4779	1·4895	3·1794	0·2357
0·62	1·4413	0·9200	1·7732	4·9212	4·7200	0·3757	1·5869	1·6017	3·4035	0·2428
0·64	1·4809	0·9217	1·8688	5·0477	4·8400	0·3861	1·7005	1·7189	3·6365	0·2499
0·66	1·5205	0·9232	1·9668	5·1742	4·9600	0·3965	1·8189	1·8411	3·8785	0·2569
0·68	1·5599	0·9245	2·0672	5·3007	5·0800	0·4069	1·9420	1·9684	4·1295	0·2639
0·70	1·5994	0·9258	2·1700	5·4272	5·2000	0·4173	2·0700	2·1010	4·3898	0·2710
0·72	1·6387	0·9270	2·2752	5·5537	5·3200	0·4277	2·2029	2·2388	4·6594	0·2780
0·74	1·6780	0·9280	2·3828	5·6802	5·4400	0·4380	2·3408	2·3820	4·9385	0·2850
0·76	1·7172	0·9290	2·4928	5·8067	5·5600	0·4483	2·4837	2·5306	5·2270	0·2920
0·78	1·7564	0·9300	2·6052	5·9332	5·6800	0·4587	2·6318	2·6848	5·5252	0·2989
0·80	1·7955	0·9308	2·7200	6·0596	5·8000	0·4690	2·7850	2·8446	5·8331	0·3059

Rel. dpth	U.equ. dia. $D_{ep(u)}$	Equiv. disch. factor	Unit sect. area	Unit wetted perim.	Unit surf. brdth	Unit mean depth	Discharge ratios $Q/Q_{0.50}$		U.crit. disch. $Q_{c(u)}$	Unit depth cntrd
Y	(m)	J	A_u (m²)	P_u (m)	$B_{s(u)}$ (m)	$y_{m(u)}$ (m)	medial	Mann'g	(m³s⁻¹)	$y_{d(u)}$ (m)
0·82	1·8346	0·9316	2·8372	6·1861	5·9200	0·4793	2·9435	3·0101	6·1508	0·3128
0·84	1·8736	0·9324	2·9568	6·3126	6·0400	0·4895	3·1073	3·1813	6·4785	0·3198
0·86	1·9126	0·9331	3·0788	6·4391	6·1600	0·4998	3·2764	3·3583	6·8162	0·3267
0·88	1·9515	0·9338	3·2032	6·5656	6·2800	0·5101	3·4510	3·5413	7·1640	0·3336
0·90	1·9904	0·9344	3·3300	6·6921	6·4000	0·5203	3·6311	3·7302	7·5221	0·3405
0·92	2·0293	0·9349	3·4592	6·8186	6·5200	0·5306	3·8167	3·9252	7·8904	0·3474
0·94	2·0681	0·9355	3·5908	6·9451	6·6400	0·5408	4·0080	4·1264	8·2692	0·3543
0·96	2·1069	0·9360	3·7248	7·0716	6·7600	0·5510	4·2049	4·3338	8·6585	0·3612
0·98	2·1457	0·9365	3·8612	7·1981	6·8800	0·5612	4·4075	4·5474	9·0584	0·3681
1·00	2·1844	0·9369	4·0000	7·3246	7·0000	0·5714	4·6160	4·7674	9·4689	0·3750
1·02	2·2232	0·9373	4·1412	7·4510	7·1200	0·5816	4·8303	4·9939	9·8903	0·3819
1·04	2·2618	0·9377	4·2848	7·5775	7·2400	0·5918	5·0506	5·2268	10·323	0·3887
1·06	2·3005	0·9381	4·4308	7·7040	7·3600	0·6020	5·2768	5·4663	10·766	0·3956
1·08	2·3392	0·9384	4·5792	7·8305	7·4800	0·6122	5·5091	5·7125	11·220	0·4025
1·10	2·3778	0·9388	4·7300	7·9570	7·6000	0·6224	5·7474	5·9654	11·685	0·4093
1·12	2·4164	0·9391	4·8832	8·0835	7·7200	0·6325	5·9919	6·2251	12·162	0·4161
1·14	2·4550	0·9394	5·0388	8·2100	7·8400	0·6427	6·2426	6·4916	12·650	0·4230
1·16	2·4935	0·9397	5·1968	8·3365	7·9600	0·6529	6·4996	6·7651	13·149	0·4298
1·18	2·5321	0·9399	5·3572	8·4630	8·0800	0·6630	6·7629	7·0456	13·660	0·4367
1·20	2·5706	0·9402	5·5200	8·5895	8·2000	0·6732	7·0326	7·3332	14·183	0·4435
1·24	2·6476	0·9406	5·8528	8·8424	8·4400	0·6935	7·5913	7·9298	15·263	0·4571
1·28	2·7245	0·9410	6·1952	9·0954	8·6800	0·7137	8·1762	8·5555	16·390	0·4707
1·32	2·8014	0·9414	6·5472	9·3484	8·9200	0·7340	8·7877	9·2110	17·566	0·4844
1·36	2·8782	0·9417	6·9088	9·6014	9·1600	0·7542	9·4263	9·8966	18·790	0·4980
1·40	2·9550	0·9420	7·2800	9·8544	9·4000	0·7745	10·092	10·613	20·063	0·5115
1·44	3·0318	0·9423	7·6608	10·107	9·6400	0·7947	10·786	11·361	21·386	0·5251
1·48	3·1085	0·9426	8·0512	10·360	9·8800	0·8149	11·508	12·140	22·760	0·5387
1·52	3·1851	0·9428	8·4512	10·613	10·120	0·8351	12·259	12·952	24·185	0·5522
1·56	3·2618	0·9430	8·8608	10·866	10·360	0·8553	13·040	13·797	25·662	0·5658
1·60	3·3383	0·9432	9·2800	11·119	10·600	0·8755	13·849	14·675	27·191	0·5793
1·64	3·4149	0·9433	9·7088	11·372	10·840	0·8956	14·689	15·586	28·774	0·5928
1·68	3·4914	0·9435	10·147	11·625	11·080	0·9158	15·559	16·533	30·410	0·6064
1·72	3·5679	0·9436	10·595	11·878	11·320	0·9360	16·460	17·514	32·100	0·6199
1·76	3·6444	0·9438	11·053	12·131	11·560	0·9561	17·392	18·531	33·845	0·6334
1·80	3·7209	0·9439	11·520	12·384	11·800	0·9763	18·355	19·583	35·645	0·6469
1·84	3·7973	0·9440	11·997	12·637	12·040	0·9964	19·350	20·672	37·501	0·6604
1·88	3·8737	0·9441	12·483	12·890	12·280	1·0165	20·378	21·798	39·414	0·6739
1·92	3·9501	0·9442	12·979	13·143	12·520	1·0367	21·438	22·961	41·384	0·6873
1·96	4·0265	0·9442	13·485	13·396	12·760	1·0568	22·531	24·162	43·411	0·7008
2·00	4·1028	0·9443	14·000	13·649	13·000	1·0769	23·658	25·401	45·497	0·7143
2·05	4·1983	0·9444	14·657	13·965	13·300	1·1021	25·114	27·005	48·186	0·7311
2·10	4·2936	0·9445	15·330	14·282	13·600	1·1272	26·624	28·670	50·969	0·7479
2·15	4·3890	0·9445	16·018	14·598	13·900	1·1523	28·187	30·397	53·845	0·7648
2·20	4·4844	0·9446	16·720	14·914	14·200	1·1775	29·806	32·189	56·816	0·7816
2·25	4·5797	0·9446	17·438	15·230	14·500	1·2026	31·480	34·044	59·883	0·7984
2·30	4·6750	0·9447	18·170	15·546	14·800	1·2277	33·211	35·965	63·047	0·8152
2·35	4·7703	0·9447	18·917	15·863	15·100	1·2528	34·998	37·951	66·308	0·8320
2·40	4·8656	0·9448	19·680	16·179	15·400	1·2779	36·844	40·005	69·669	0·8488
2·45	4·9608	0·9448	20·458	16·495	15·700	1·3030	38·747	42·126	73·129	0·8656
2·50	5·0561	0·9448	21·250	16·811	16·000	1·3281	40·710	44·317	76·690	0·8824
2·55	5·1513	0·9448	22·057	17·128	16·300	1·3532	42·732	46·576	80·353	0·8991
2·60	5·2465	0·9449	22·880	17·444	16·600	1·3783	44·814	48·907	84·118	0·9159
2·65	5·3418	0·9449	23·718	17·760	16·900	1·4034	46·958	51·308	87·987	0·9327
2·70	5·4370	0·9449	24·570	18·076	17·200	1·4285	49·163	53·782	91·961	0·9495
2·75	5·5321	0·9449	25·437	18·393	17·500	1·4536	51·430	56·329	96·040	0·9662

C65

Regime trapezoidal - 1·5 to 1 side-slope

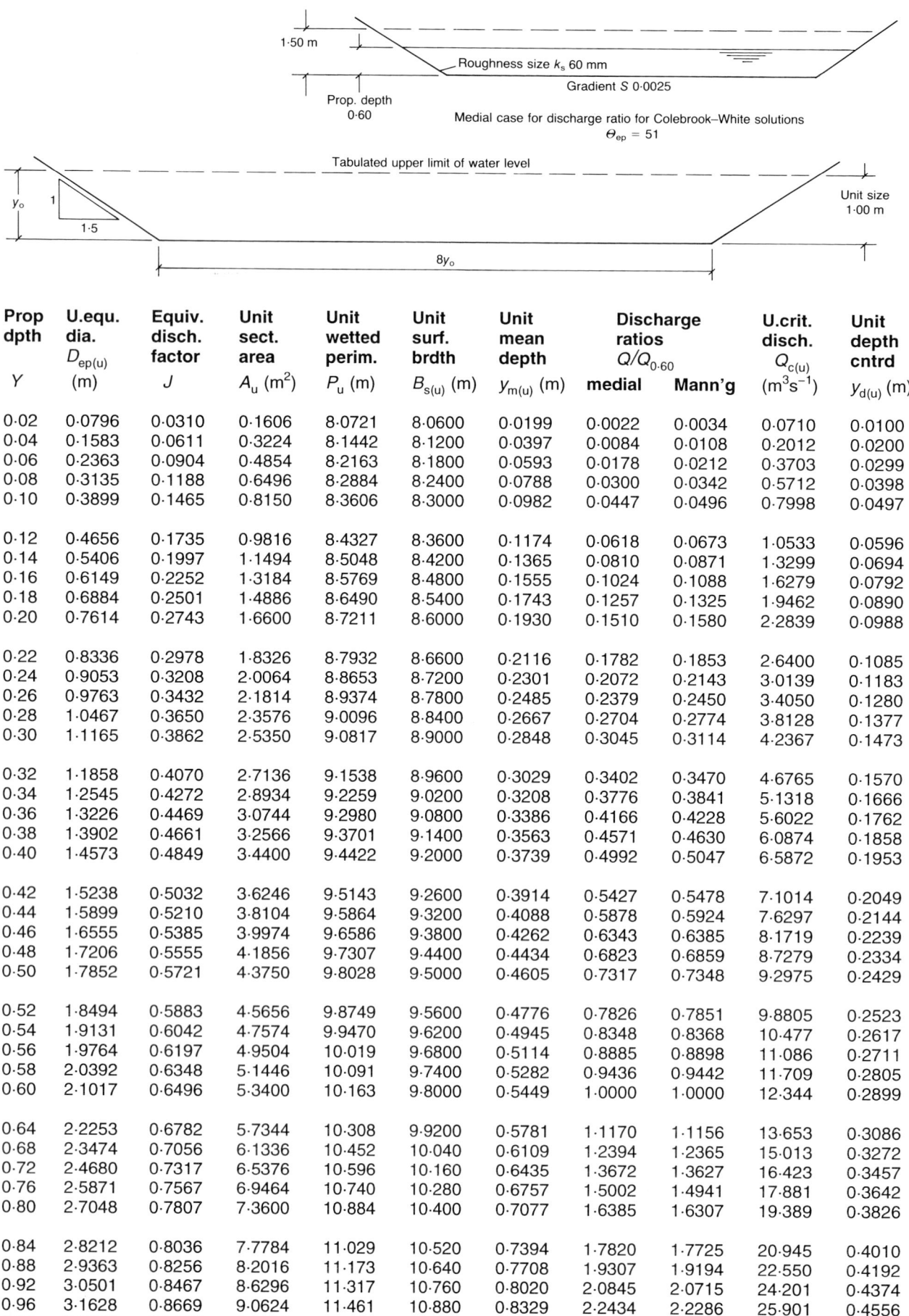

1·50 m

Roughness size k_s 60 mm

Gradient S 0·0025

Prop. depth 0·60

Medial case for discharge ratio for Colebrook–White solutions
$\Theta_{ep} = 51$

Tabulated upper limit of water level

y_o

1

1·5

Unit size 1·00 m

$8y_o$

Prop dpth Y	U.equ. dia. $D_{ep(u)}$ (m)	Equiv. disch. factor J	Unit sect. area A_u (m²)	Unit wetted perim. P_u (m)	Unit surf. brdth $B_{s(u)}$ (m)	Unit mean depth $y_{m(u)}$ (m)	Discharge ratios $Q/Q_{0.60}$ medial	Mann'g	U.crit. disch. $Q_{c(u)}$ (m³s⁻¹)	Unit depth cntrd $y_{d(u)}$ (m)
0·02	0·0796	0·0310	0·1606	8·0721	8·0600	0·0199	0·0022	0·0034	0·0710	0·0100
0·04	0·1583	0·0611	0·3224	8·1442	8·1200	0·0397	0·0084	0·0108	0·2012	0·0200
0·06	0·2363	0·0904	0·4854	8·2163	8·1800	0·0593	0·0178	0·0212	0·3703	0·0299
0·08	0·3135	0·1188	0·6496	8·2884	8·2400	0·0788	0·0300	0·0342	0·5712	0·0398
0·10	0·3899	0·1465	0·8150	8·3606	8·3000	0·0982	0·0447	0·0496	0·7998	0·0497
0·12	0·4656	0·1735	0·9816	8·4327	8·3600	0·1174	0·0618	0·0673	1·0533	0·0596
0·14	0·5406	0·1997	1·1494	8·5048	8·4200	0·1365	0·0810	0·0871	1·3299	0·0694
0·16	0·6149	0·2252	1·3184	8·5769	8·4800	0·1555	0·1024	0·1088	1·6279	0·0792
0·18	0·6884	0·2501	1·4886	8·6490	8·5400	0·1743	0·1257	0·1325	1·9462	0·0890
0·20	0·7614	0·2743	1·6600	8·7211	8·6000	0·1930	0·1510	0·1580	2·2839	0·0988
0·22	0·8336	0·2978	1·8326	8·7932	8·6600	0·2116	0·1782	0·1853	2·6400	0·1085
0·24	0·9053	0·3208	2·0064	8·8653	8·7200	0·2301	0·2072	0·2143	3·0139	0·1183
0·26	0·9763	0·3432	2·1814	8·9374	8·7800	0·2485	0·2379	0·2450	3·4050	0·1280
0·28	1·0467	0·3650	2·3576	9·0096	8·8400	0·2667	0·2704	0·2774	3·8128	0·1377
0·30	1·1165	0·3862	2·5350	9·0817	8·9000	0·2848	0·3045	0·3114	4·2367	0·1473
0·32	1·1858	0·4070	2·7136	9·1538	8·9600	0·3029	0·3402	0·3470	4·6765	0·1570
0·34	1·2545	0·4272	2·8934	9·2259	9·0200	0·3208	0·3776	0·3841	5·1318	0·1666
0·36	1·3226	0·4469	3·0744	9·2980	9·0800	0·3386	0·4166	0·4228	5·6022	0·1762
0·38	1·3902	0·4661	3·2566	9·3701	9·1400	0·3563	0·4571	0·4630	6·0874	0·1858
0·40	1·4573	0·4849	3·4400	9·4422	9·2000	0·3739	0·4992	0·5047	6·5872	0·1953
0·42	1·5238	0·5032	3·6246	9·5143	9·2600	0·3914	0·5427	0·5478	7·1014	0·2049
0·44	1·5899	0·5210	3·8104	9·5864	9·3200	0·4088	0·5878	0·5924	7·6297	0·2144
0·46	1·6555	0·5385	3·9974	9·6586	9·3800	0·4262	0·6343	0·6385	8·1719	0·2239
0·48	1·7206	0·5555	4·1856	9·7307	9·4400	0·4434	0·6823	0·6859	8·7279	0·2334
0·50	1·7852	0·5721	4·3750	9·8028	9·5000	0·4605	0·7317	0·7348	9·2975	0·2429
0·52	1·8494	0·5883	4·5656	9·8749	9·5600	0·4776	0·7826	0·7851	9·8805	0·2523
0·54	1·9131	0·6042	4·7574	9·9470	9·6200	0·4945	0·8348	0·8368	10·477	0·2617
0·56	1·9764	0·6197	4·9504	10·019	9·6800	0·5114	0·8885	0·8898	11·086	0·2711
0·58	2·0392	0·6348	5·1446	10·091	9·7400	0·5282	0·9436	0·9442	11·709	0·2805
0·60	2·1017	0·6496	5·3400	10·163	9·8000	0·5449	1·0000	1·0000	12·344	0·2899
0·64	2·2253	0·6782	5·7344	10·308	9·9200	0·5781	1·1170	1·1156	13·653	0·3086
0·68	2·3474	0·7056	6·1336	10·452	10·040	0·6109	1·2394	1·2365	15·013	0·3272
0·72	2·4680	0·7317	6·5376	10·596	10·160	0·6435	1·3672	1·3627	16·423	0·3457
0·76	2·5871	0·7567	6·9464	10·740	10·280	0·6757	1·5002	1·4941	17·881	0·3642
0·80	2·7048	0·7807	7·3600	10·884	10·400	0·7077	1·6385	1·6307	19·389	0·3826
0·84	2·8212	0·8036	7·7784	11·029	10·520	0·7394	1·7820	1·7725	20·945	0·4010
0·88	2·9363	0·8256	8·2016	11·173	10·640	0·7708	1·9307	1·9194	22·550	0·4192
0·92	3·0501	0·8467	8·6296	11·317	10·760	0·8020	2·0845	2·0715	24·201	0·4374
0·96	3·1628	0·8669	9·0624	11·461	10·880	0·8329	2·2434	2·2286	25·901	0·4556
1·00	3·2743	0·8863	9·5000	11·606	11·000	0·8636	2·4075	2·3908	27·647	0·4737

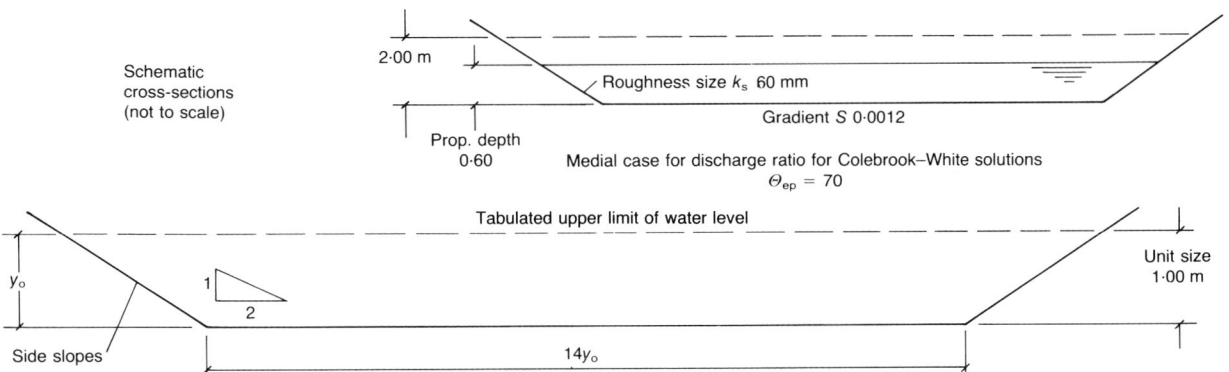

Schematic cross-sections (not to scale)

2·00 m

Roughness size k_s 60 mm

Gradient S 0·0012

Prop. depth 0·60

Medial case for discharge ratio for Colebrook–White solutions $\Theta_{ep} = 70$

Tabulated upper limit of water level

y_o

1 / 2

Side slopes

$14y_o$

Unit size 1·00 m

Prop dpth Y	U.equ. dia. $D_{ep(u)}$ (m)	Equiv. disch. factor J	Unit sect. area A_u (m²)	Unit wetted perim. P_u (m)	Unit surf. brdth $B_{s(u)}$ (m)	Unit mean depth $y_{m(u)}$ (m)	Discharge ratios $Q/Q_{0.60}$ medial	Mann'g	U.crit. disch. $Q_{c(u)}$ (m³s⁻¹)	Unit depth cntrd $y_{d(u)}$ (m)
0·02	0·0797	0·0178	0·2808	14·089	14·080	0·0199	0·0024	0·0034	0·1242	0·0100
0·04	0·1589	0·0352	0·5632	14·179	14·160	0·0398	0·0088	0·0108	0·3517	0·0200
0·06	0·2375	0·0523	0·8472	14·268	14·240	0·0595	0·0185	0·0211	0·6471	0·0299
0·08	0·3156	0·0691	1·1328	14·358	14·320	0·0791	0·0309	0·0342	0·9977	0·0398
0·10	0·3932	0·0855	1·4200	14·447	14·400	0·0986	0·0458	0·0496	1·3964	0·0498
0·12	0·4702	0·1016	1·7088	14·537	14·480	0·1180	0·0630	0·0673	1·8383	0·0597
0·14	0·5467	0·1174	1·9992	14·626	14·560	0·1373	0·0825	0·0870	2·3199	0·0695
0·16	0·6228	0·1330	2·2912	14·716	14·640	0·1565	0·1040	0·1088	2·8385	0·0794
0·18	0·6984	0·1482	2·5848	14·805	14·720	0·1756	0·1275	0·1324	3·3919	0·0892
0·20	0·7734	0·1631	2·8800	14·894	14·800	0·1946	0·1529	0·1579	3·9785	0·0991
0·22	0·8481	0·1778	3·1768	14·984	14·880	0·2135	0·1802	0·1853	4·5967	0·1089
0·24	0·9222	0·1922	3·4752	15·073	14·960	0·2323	0·2093	0·2143	5·2452	0·1187
0·26	0·9959	0·2063	3·7752	15·163	15·040	0·2510	0·2401	0·2450	5·9231	0·1284
0·28	1·0692	0·2202	4·0768	15·252	15·120	0·2696	0·2726	0·2774	6·6292	0·1382
0·30	1·1420	0·2338	4·3800	15·342	15·200	0·2882	0·3068	0·3115	7·3629	0·1479
0·32	1·2144	0·2472	4·6848	15·431	15·280	0·3066	0·3426	0·3471	8·1234	0·1577
0·34	1·2863	0·2604	4·9912	15·521	15·360	0·3249	0·3800	0·3842	8·9099	0·1674
0·36	1·3579	0·2733	5·2992	15·610	15·440	0·3432	0·4189	0·4229	9·7219	0·1771
0·38	1·4290	0·2860	5·6088	15·699	15·520	0·3614	0·4594	0·4632	10·559	0·1867
0·40	1·4998	0·2984	5·9200	15·789	15·600	0·3795	0·5014	0·5049	11·420	0·1964
0·42	1·5701	0·3107	6·2328	15·878	15·680	0·3975	0·5449	0·5480	12·306	0·2060
0·44	1·6401	0·3227	6·5472	15·968	15·760	0·4154	0·5898	0·5926	13·215	0·2157
0·46	1·7097	0·3345	6·8632	16·057	15·840	0·4333	0·6362	0·6387	14·147	0·2253
0·48	1·7789	0·3461	7·1808	16·147	15·920	0·4511	0·6840	0·6862	15·102	0·2349
0·50	1·8477	0·3575	7·5000	16·236	16·000	0·4688	0·7332	0·7350	16·080	0·2444
0·52	1·9162	0·3687	7·8208	16·326	16·080	0·4864	0·7839	0·7853	17·080	0·2540
0·54	1·9843	0·3798	8·1432	16·415	16·160	0·5039	0·8359	0·8369	18·102	0·2636
0·56	2·0521	0·3906	8·4672	16·504	16·240	0·5214	0·8892	0·8899	19·146	0·2731
0·58	2·1195	0·4013	8·7928	16·594	16·320	0·5388	0·9440	0·9443	20·211	0·2826
0·60	2·1866	0·4117	9·1200	16·683	16·400	0·5561	1·0000	1·0000	21·298	0·2921
0·64	2·3198	0·4322	9·7792	16·862	16·560	0·5905	1·1161	1·1154	23·533	0·3111
0·68	2·4517	0·4520	10·445	17·041	16·720	0·6247	1·2375	1·2360	25·852	0·3300
0·72	2·5823	0·4711	11·117	17·220	16·880	0·6586	1·3641	1·3619	28·252	0·3488
0·76	2·7117	0·4896	11·795	17·399	17·040	0·6922	1·4957	1·4929	30·731	0·3676
0·80	2·8400	0·5076	12·480	17·578	17·200	0·7256	1·6324	1·6290	33·290	0·3863
0·84	2·9671	0·5249	13·171	17·757	17·360	0·7587	1·7741	1·7701	35·927	0·4050
0·88	3·0930	0·5418	13·869	17·935	17·520	0·7916	1·9208	1·9163	38·641	0·4236
0·92	3·2180	0·5581	14·573	18·114	17·680	0·8243	2·0723	2·0674	41·432	0·4422
0·96	3·3418	0·5739	15·283	18·293	17·840	0·8567	2·2288	2·2234	44·298	0·4607
1·00	3·4647	0·5892	16·000	18·472	18·000	0·8889	2·3901	2·3844	47·239	0·4792

Regime trapezoidal - 2·5 to 1 side-slope

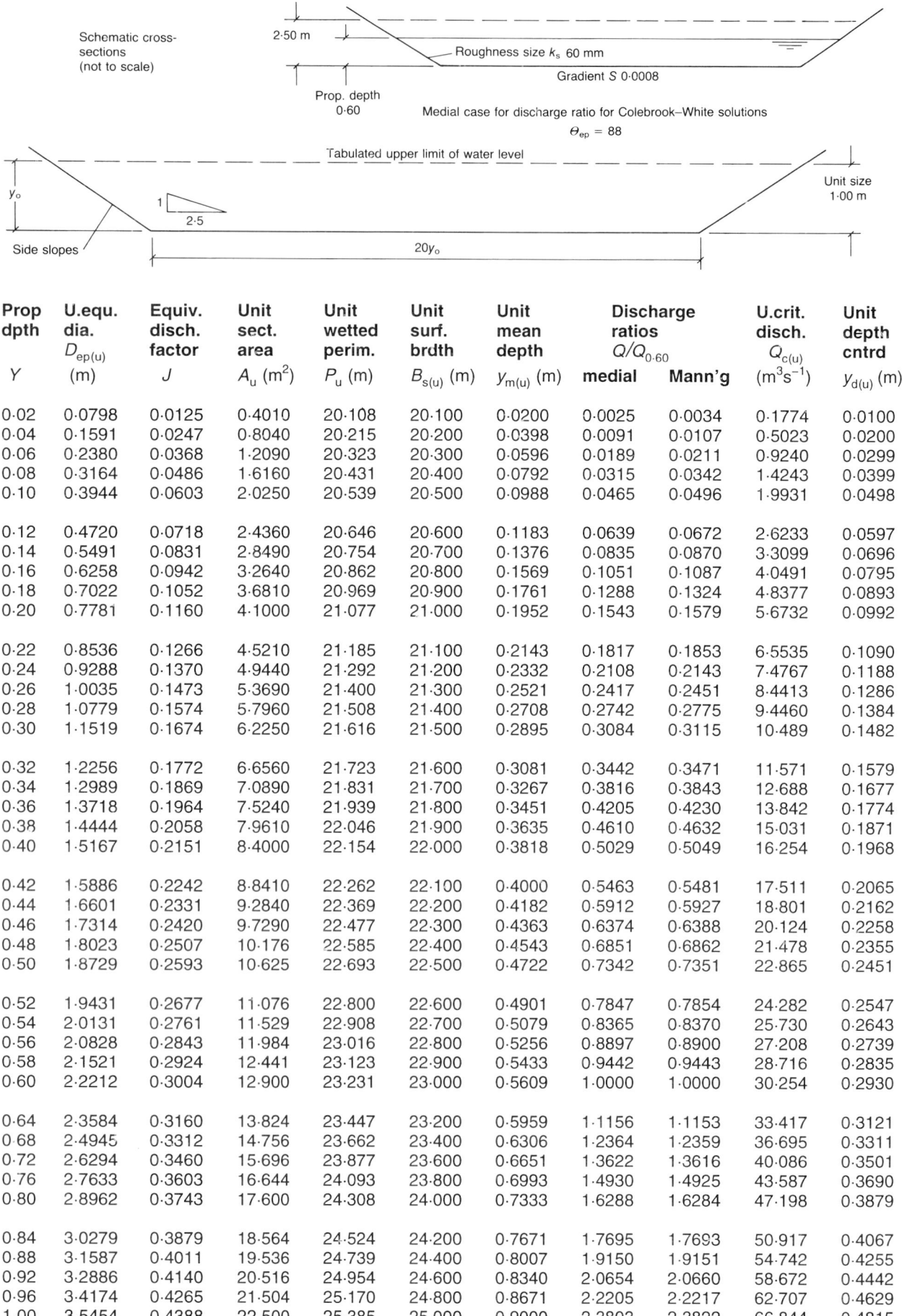

Schematic cross-sections (not to scale)

2·50 m

Roughness size k_s 60 mm

Gradient S 0·0008

Prop. depth 0·60

Medial case for discharge ratio for Colebrook–White solutions
$\Theta_{ep} = 88$

Tabulated upper limit of water level

y_o

1
2·5

Side slopes

$20y_o$

Unit size 1·00 m

Prop dpth Y	U.equ. dia. $D_{ep(u)}$ (m)	Equiv. disch. factor J	Unit sect. area A_u (m²)	Unit wetted perim. P_u (m)	Unit surf. brdth $B_{s(u)}$ (m)	Unit mean depth $y_{m(u)}$ (m)	Discharge ratios $Q/Q_{0.60}$ medial	Mann'g	U.crit. disch. $Q_{c(u)}$ (m³s⁻¹)	Unit depth cntrd $y_{d(u)}$ (m)
0·02	0·0798	0·0125	0·4010	20·108	20·100	0·0200	0·0025	0·0034	0·1774	0·0100
0·04	0·1591	0·0247	0·8040	20·215	20·200	0·0398	0·0091	0·0107	0·5023	0·0200
0·06	0·2380	0·0368	1·2090	20·323	20·300	0·0596	0·0189	0·0211	0·9240	0·0299
0·08	0·3164	0·0486	1·6160	20·431	20·400	0·0792	0·0315	0·0342	1·4243	0·0399
0·10	0·3944	0·0603	2·0250	20·539	20·500	0·0988	0·0465	0·0496	1·9931	0·0498
0·12	0·4720	0·0718	2·4360	20·646	20·600	0·1183	0·0639	0·0672	2·6233	0·0597
0·14	0·5491	0·0831	2·8490	20·754	20·700	0·1376	0·0835	0·0870	3·3099	0·0696
0·16	0·6258	0·0942	3·2640	20·862	20·800	0·1569	0·1051	0·1087	4·0491	0·0795
0·18	0·7022	0·1052	3·6810	20·969	20·900	0·1761	0·1288	0·1324	4·8377	0·0893
0·20	0·7781	0·1160	4·1000	21·077	21·000	0·1952	0·1543	0·1579	5·6732	0·0992
0·22	0·8536	0·1266	4·5210	21·185	21·100	0·2143	0·1817	0·1853	6·5535	0·1090
0·24	0·9288	0·1370	4·9440	21·292	21·200	0·2332	0·2108	0·2143	7·4767	0·1188
0·26	1·0035	0·1473	5·3690	21·400	21·300	0·2521	0·2417	0·2451	8·4413	0·1286
0·28	1·0779	0·1574	5·7960	21·508	21·400	0·2708	0·2742	0·2775	9·4460	0·1384
0·30	1·1519	0·1674	6·2250	21·616	21·500	0·2895	0·3084	0·3115	10·489	0·1482
0·32	1·2256	0·1772	6·6560	21·723	21·600	0·3081	0·3442	0·3471	11·571	0·1579
0·34	1·2989	0·1869	7·0890	21·831	21·700	0·3267	0·3816	0·3843	12·688	0·1677
0·36	1·3718	0·1964	7·5240	21·939	21·800	0·3451	0·4205	0·4230	13·842	0·1774
0·38	1·4444	0·2058	7·9610	22·046	21·900	0·3635	0·4610	0·4632	15·031	0·1871
0·40	1·5167	0·2151	8·4000	22·154	22·000	0·3818	0·5029	0·5049	16·254	0·1968
0·42	1·5886	0·2242	8·8410	22·262	22·100	0·4000	0·5463	0·5481	17·511	0·2065
0·44	1·6601	0·2331	9·2840	22·369	22·200	0·4182	0·5912	0·5927	18·801	0·2162
0·46	1·7314	0·2420	9·7290	22·477	22·300	0·4363	0·6374	0·6388	20·124	0·2258
0·48	1·8023	0·2507	10·176	22·585	22·400	0·4543	0·6851	0·6862	21·478	0·2355
0·50	1·8729	0·2593	10·625	22·693	22·500	0·4722	0·7342	0·7351	22·865	0·2451
0·52	1·9431	0·2677	11·076	22·800	22·600	0·4901	0·7847	0·7854	24·282	0·2547
0·54	2·0131	0·2761	11·529	22·908	22·700	0·5079	0·8365	0·8370	25·730	0·2643
0·56	2·0828	0·2843	11·984	23·016	22·800	0·5256	0·8897	0·8900	27·208	0·2739
0·58	2·1521	0·2924	12·441	23·123	22·900	0·5433	0·9442	0·9443	28·716	0·2835
0·60	2·2212	0·3004	12·900	23·231	23·000	0·5609	1·0000	1·0000	30·254	0·2930
0·64	2·3584	0·3160	13·824	23·447	23·200	0·5959	1·1156	1·1153	33·417	0·3121
0·68	2·4945	0·3312	14·756	23·662	23·400	0·6306	1·2364	1·2359	36·695	0·3311
0·72	2·6294	0·3460	15·696	23·877	23·600	0·6651	1·3622	1·3616	40·086	0·3501
0·76	2·7633	0·3603	16·644	24·093	23·800	0·6993	1·4930	1·4925	43·587	0·3690
0·80	2·8962	0·3743	17·600	24·308	24·000	0·7333	1·6288	1·6284	47·198	0·3879
0·84	3·0279	0·3879	18·564	24·524	24·200	0·7671	1·7695	1·7693	50·917	0·4067
0·88	3·1587	0·4011	19·536	24·739	24·400	0·8007	1·9150	1·9151	54·742	0·4255
0·92	3·2886	0·4140	20·516	24·954	24·600	0·8340	2·0654	2·0660	58·672	0·4442
0·96	3·4174	0·4265	21·504	25·170	24·800	0·8671	2·2205	2·2217	62·707	0·4629
1·00	3·5454	0·4388	22·500	25·385	25·000	0·9000	2·3803	2·3822	66·844	0·4815

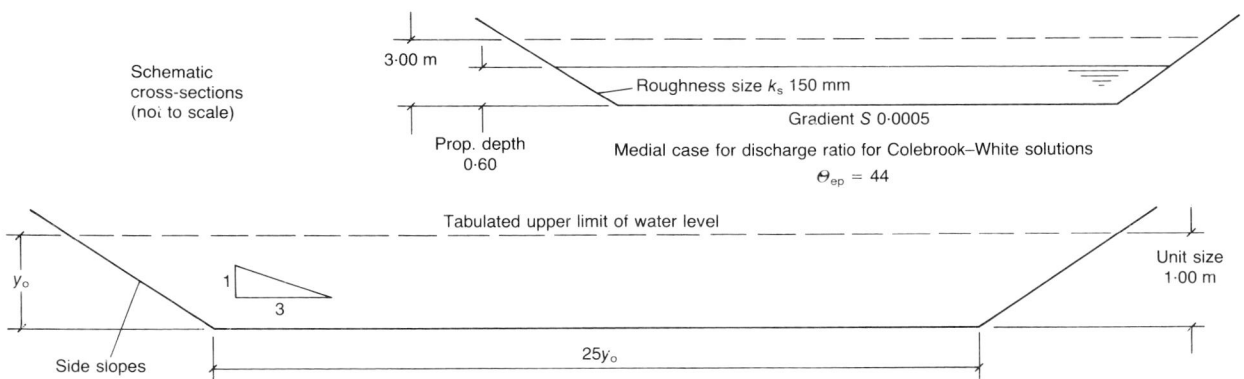

Schematic cross-sections (not to scale)

3·00 m

Roughness size k_s 150 mm

Gradient S 0·0005

Prop. depth 0·60

Medial case for discharge ratio for Colebrook–White solutions
$\Theta_{ep} = 44$

Tabulated upper limit of water level

Unit size 1·00 m

y_o

1

3

Side slopes

$25y_o$

Prop dpth Y	U.equ. dia. $D_{ep(u)}$ (m)	Equiv. disch. factor J	Unit sect. area A_u (m²)	Unit wetted perim. P_u (m)	Unit surf. brdth $B_{s(u)}$ (m)	Unit mean depth $y_{m(u)}$ (m)	Discharge ratios $Q/Q_{0.60}$ medial	Mann'g	U.crit. disch. $Q_{c(u)}$ (m³s⁻¹)	Unit depth cntrd $y_{d(u)}$ (m)
0·02	0·0798	0·0100	0·5012	25·126	25·120	0·0200	0·0020	0·0034	0·2217	0·0100
0·04	0·1592	0·0198	1·0048	25·253	25·240	0·0398	0·0081	0·0107	0·6278	0·0200
0·06	0·2381	0·0295	1·5108	25·379	25·360	0·0596	0·0172	0·0211	1·1548	0·0299
0·08	0·3167	0·0390	2·0192	25·506	25·480	0·0792	0·0292	0·0341	1·7800	0·0399
0·10	0·3948	0·0484	2·5300	25·632	25·600	0·0988	0·0437	0·0496	2·4907	0·0498
0·12	0·4726	0·0576	3·0432	25·759	25·720	0·1183	0·0606	0·0672	3·2781	0·0597
0·14	0·5499	0·0667	3·5588	25·885	25·840	0·1377	0·0797	0·0869	4·1359	0·0696
0·16	0·6269	0·0757	4·0768	26·012	25·960	0·1570	0·1009	0·1087	5·0593	0·0795
0·18	0·7035	0·0846	4·5972	26·138	26·080	0·1763	0·1241	0·1323	6·0443	0·0894
0·20	0·7797	0·0933	5·1200	26·265	26·200	0·1954	0·1494	0·1579	7·0879	0·0992
0·22	0·8556	0·1018	5·6452	26·391	26·320	0·2145	0·1765	0·1852	8·1872	0·1091
0·24	0·9311	0·1103	6·1728	26·518	26·440	0·2335	0·2054	0·2142	9·3401	0·1189
0·26	1·0063	0·1186	6·7028	26·644	26·560	0·2524	0·2361	0·2450	10·545	0·1287
0·28	1·0811	0·1269	7·2352	26·771	26·680	0·2712	0·2685	0·2774	11·799	0·1385
0·30	1·1555	0·1350	7·7700	26·897	26·800	0·2899	0·3027	0·3114	13·102	0·1483
0·32	1·2296	0·1429	8·3072	27·024	26·920	0·3086	0·3385	0·3470	14·451	0·1580
0·34	1·3034	0·1508	8·8468	27·150	27·040	0·3272	0·3759	0·3842	15·847	0·1678
0·36	1·3768	0·1586	9·3888	27·277	27·160	0·3457	0·4150	0·4229	17·287	0·1775
0·38	1·4499	0·1662	9·9332	27·403	27·280	0·3641	0·4556	0·4631	18·770	0·1872
0·40	1·5227	0·1738	10·480	27·530	27·400	0·3825	0·4977	0·5048	20·297	0·1969
0·42	1·5952	0·1812	11·029	27·656	27·520	0·4008	0·5414	0·5480	21·865	0·2066
0·44	1·6673	0·1885	11·581	27·783	27·640	0·4190	0·5866	0·5926	23·475	0·2163
0·46	1·7392	0·1958	12·135	27·909	27·760	0·4371	0·6333	0·6387	25·125	0·2260
0·48	1·8107	0·2029	12·691	28·036	27·880	0·4552	0·6814	0·6862	26·814	0·2356
0·50	1·8820	0·2099	13·250	28·162	28·000	0·4732	0·7310	0·7350	28·543	0·2453
0·52	1·9529	0·2169	13·811	28·289	28·120	0·4912	0·7820	0·7853	30·311	0·2549
0·54	2·0235	0·2237	14·375	28·415	28·240	0·5090	0·8344	0·8370	32·117	0·2645
0·56	2·0939	0·2305	14·941	28·542	28·360	0·5268	0·8882	0·8900	33·960	0·2741
0·58	2·1640	0·2371	15·509	28·668	28·480	0·5446	0·9434	0·9443	35·841	0·2837
0·60	2·2337	0·2437	16·080	28·795	28·600	0·5622	1·0000	1·0000	37·758	0·2933
0·64	2·3725	0·2566	17·229	29·048	28·840	0·5974	1·1173	1·1154	41·701	0·3124
0·68	2·5101	0·2691	18·387	29·301	29·080	0·6323	1·2399	1·2360	45·786	0·3314
0·72	2·6467	0·2813	19·555	29·554	29·320	0·6670	1·3678	1·3617	50·012	0·3505
0·76	2·7823	0·2932	20·733	29·807	29·560	0·7014	1·5009	1·4926	54·374	0·3694
0·80	2·9169	0·3048	21·920	30·060	29·800	0·7356	1·6392	1·6286	58·873	0·3883
0·84	3·0505	0·3161	23·117	30·313	30·040	0·7695	1·7825	1·7695	63·504	0·4072
0·88	3·1831	0·3272	24·323	30·566	30·280	0·8033	1·9310	1·9155	68·267	0·4260
0·92	3·3148	0·3379	25·539	30·819	30·520	0·8368	2·0844	2·0663	73·161	0·4448
0·96	3·4456	0·3484	26·765	31·072	30·760	0·8701	2·2427	2·2221	78·183	0·4635
1·00	3·5755	0·3586	28·000	31·325	31·000	0·9032	2·4060	2·3827	83·333	0·4821

C69

Regime trapezoidal - 4·0 to 1 side-slope

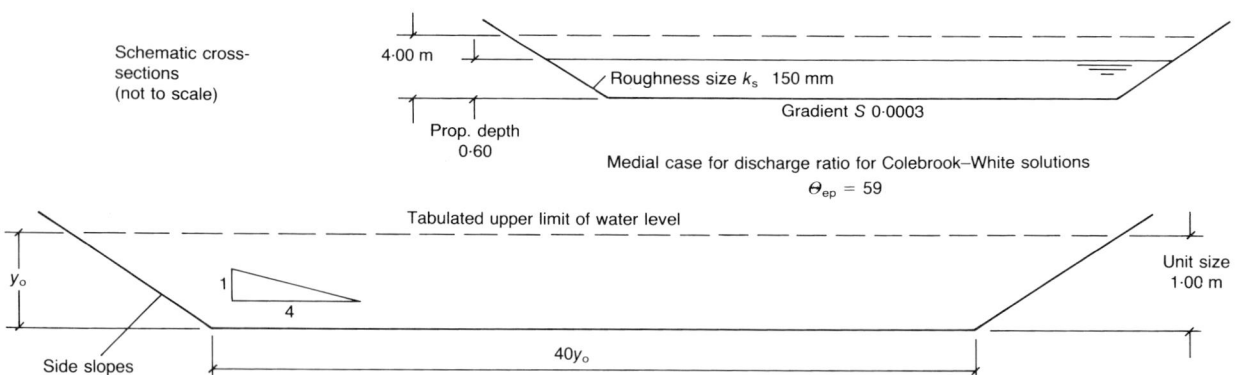

Schematic cross-sections (not to scale)

4·00 m

Roughness size k_s 150 mm

Gradient S 0·0003

Prop. depth 0·60

Medial case for discharge ratio for Colebrook–White solutions
$\Theta_{ep} = 59$

Tabulated upper limit of water level

y_o

Side slopes

Unit size 1·00 m

$40y_o$

Prop dpth Y	U.equ. dia. $D_{ep(u)}$ (m)	Equiv. disch. factor J	Unit sect. area A_u (m²)	Unit wetted perim. P_u (m)	Unit surf. brdth $B_{s(u)}$ (m)	Unit mean depth $y_{m(u)}$ (m)	Discharge ratios $Q/Q_{0.60}$ medial	Mann'g	U.crit. disch. $Q_{c(u)}$ (m³s⁻¹)	Unit depth cntrd $y_{d(u)}$ (m)
0·02	0·0798	0·0062	0·8016	40·165	40·160	0·0200	0·0023	0·0034	0·3547	0·0100
0·04	0·1593	0·0124	1·6064	40·330	40·320	0·0398	0·0085	0·0108	1·0041	0·0200
0·06	0·2385	0·0185	2·4144	40·495	40·480	0·0596	0·0180	0·0212	1·8465	0·0299
0·08	0·3173	0·0245	3·2256	40·660	40·640	0·0794	0·0302	0·0342	2·8458	0·0399
0·10	0·3958	0·0305	4·0400	40·825	40·800	0·0990	0·0450	0·0497	3·9811	0·0498
0·12	0·4740	0·0363	4·8576	40·990	40·960	0·1186	0·0621	0·0673	5·2386	0·0598
0·14	0·5519	0·0421	5·6784	41·154	41·120	0·1381	0·0814	0·0871	6·6080	0·0697
0·16	0·6295	0·0479	6·5024	41·319	41·280	0·1575	0·1029	0·1089	8·0817	0·0796
0·18	0·7067	0·0535	7·3296	41·484	41·440	0·1769	0·1263	0·1326	9·6532	0·0895
0·20	0·7837	0·0591	8·1600	41·649	41·600	0·1962	0·1517	0·1581	11·317	0·0993
0·22	0·8603	0·0646	8·9936	41·814	41·760	0·2154	0·1789	0·1855	13·070	0·1092
0·24	0·9367	0·0701	9·8304	41·979	41·920	0·2345	0·2080	0·2146	14·908	0·1191
0·26	1·0128	0·0755	10·670	42·144	42·080	0·2536	0·2388	0·2453	16·826	0·1289
0·28	1·0885	0·0808	11·514	42·309	42·240	0·2726	0·2713	0·2778	18·824	0·1387
0·30	1·1640	0·0861	12·360	42·474	42·400	0·2915	0·3055	0·3118	20·898	0·1485
0·32	1·2392	0·0913	13·210	42·639	42·560	0·3104	0·3414	0·3474	23·046	0·1583
0·34	1·3141	0·0964	14·062	42·804	42·720	0·3292	0·3788	0·3846	25·266	0·1681
0·36	1·3888	0·1015	14·918	42·969	42·880	0·3479	0·4178	0·4234	27·556	0·1779
0·38	1·4631	0·1066	15·778	43·134	43·040	0·3666	0·4583	0·4636	29·915	0·1877
0·40	1·5372	0·1115	16·640	43·298	43·200	0·3852	0·5004	0·5053	32·341	0·1974
0·42	1·6111	0·1164	17·506	43·463	43·360	0·4037	0·5440	0·5485	34·832	0·2072
0·44	1·6846	0·1213	18·374	43·628	43·520	0·4222	0·5890	0·5931	37·388	0·2169
0·46	1·7579	0·1261	19·246	43·793	43·680	0·4406	0·6355	0·6391	40·008	0·2266
0·48	1·8310	0·1309	20·122	43·958	43·840	0·4590	0·6834	0·6866	42·689	0·2363
0·50	1·9038	0·1355	21·000	44·123	44·000	0·4773	0·7327	0·7354	45·432	0·2460
0·52	1·9763	0·1402	21·882	44·288	44·160	0·4955	0·7834	0·7856	48·235	0·2557
0·54	2·0486	0·1448	22·766	44·453	44·320	0·5137	0·8356	0·8372	51·098	0·2654
0·56	2·1206	0·1493	23·654	44·618	44·480	0·5318	0·8890	0·8901	54·019	0·2751
0·58	2·1924	0·1538	24·546	44·783	44·640	0·5499	0·9439	0·9444	56·998	0·2847
0·60	2·2640	0·1582	25·440	44·948	44·800	0·5679	1·0000	1·0000	60·034	0·2943
0·64	2·4063	0·1670	27·238	45·278	45·120	0·6037	1·1163	1·1151	66·275	0·3136
0·68	2·5478	0·1755	29·050	45·607	45·440	0·6393	1·2378	1·2354	72·736	0·3328
0·72	2·6883	0·1838	30·874	45·937	45·760	0·6747	1·3645	1·3609	79·414	0·3519
0·76	2·8280	0·1920	32·710	46·267	46·080	0·7099	1·4962	1·4913	86·304	0·3711
0·80	2·9667	0·2000	34·560	46·597	46·400	0·7448	1·6329	1·6268	93·403	0·3901
0·84	3·1046	0·2078	36·422	46·927	46·720	0·7796	1·7745	1·7672	100·71	0·4092
0·88	3·2417	0·2155	38·298	47·257	47·040	0·8141	1·9210	1·9124	108·21	0·4281
0·92	3·3779	0·2230	40·186	47·587	47·360	0·8485	2·0723	2·0626	115·92	0·4471
0·96	3·5133	0·2303	42·086	47·916	47·680	0·8827	2·2284	2·2175	123·82	0·4660
1·00	3·6480	0·2375	44·000	48·246	48·000	0·9167	2·3892	2·3771	131·92	0·4848

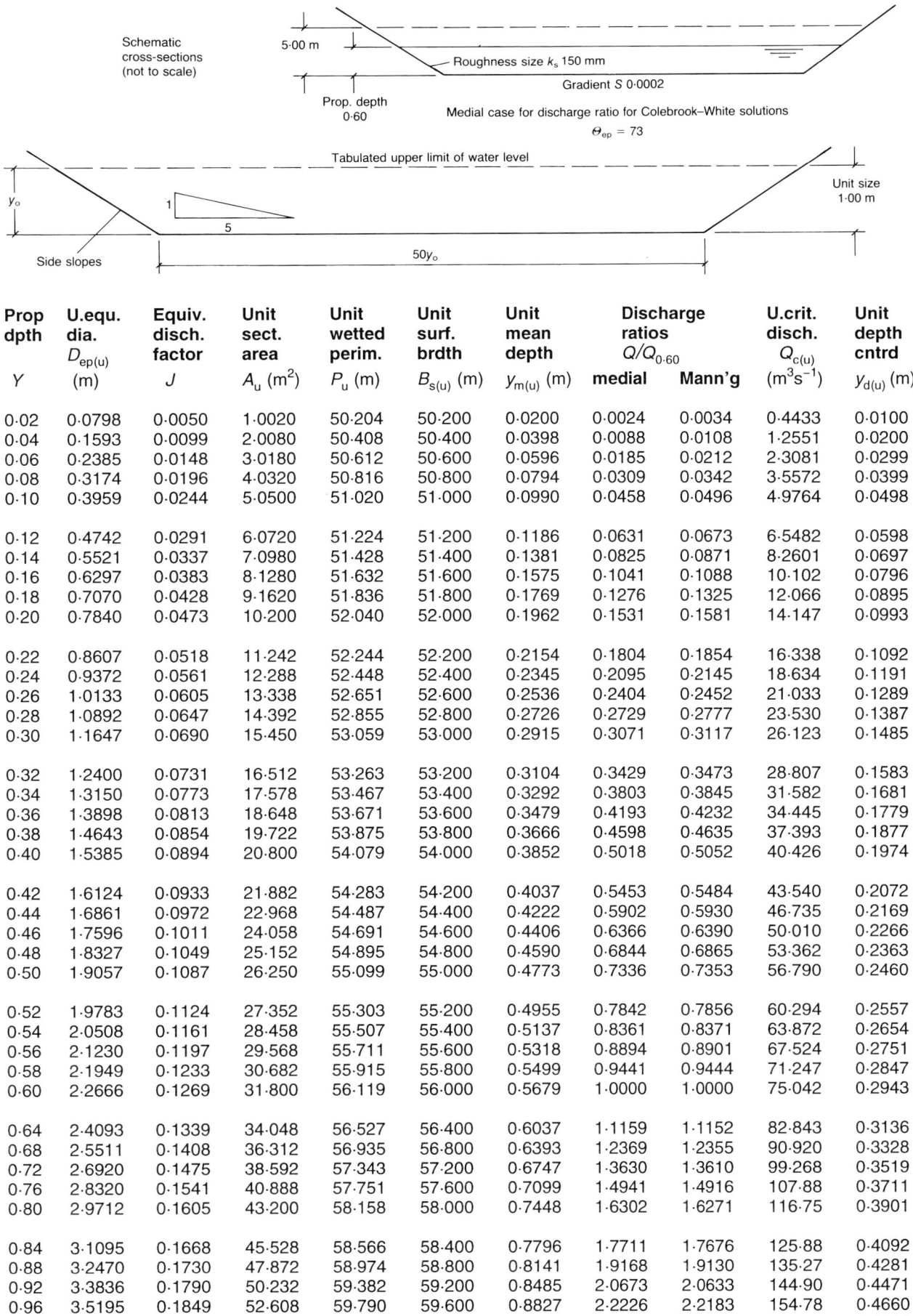

Schematic cross-sections (not to scale)

5·00 m

Roughness size k_s 150 mm

Gradient S 0·0002

Prop. depth 0·60

Medial case for discharge ratio for Colebrook–White solutions

$\Theta_{ep} = 73$

Tabulated upper limit of water level

y_o

Unit size 1·00 m

1

5

Side slopes

$50y_o$

Prop dpth Y	U.equ. dia. $D_{ep(u)}$ (m)	Equiv. disch. factor J	Unit sect. area A_u (m²)	Unit wetted perim. P_u (m)	Unit surf. brdth $B_{s(u)}$ (m)	Unit mean depth $y_{m(u)}$ (m)	Discharge ratios $Q/Q_{0·60}$ medial	Mann'g	U.crit. disch. $Q_{c(u)}$ (m³s⁻¹)	Unit depth cntrd $y_{d(u)}$ (m)
0·02	0·0798	0·0050	1·0020	50·204	50·200	0·0200	0·0024	0·0034	0·4433	0·0100
0·04	0·1593	0·0099	2·0080	50·408	50·400	0·0398	0·0088	0·0108	1·2551	0·0200
0·06	0·2385	0·0148	3·0180	50·612	50·600	0·0596	0·0185	0·0212	2·3081	0·0299
0·08	0·3174	0·0196	4·0320	50·816	50·800	0·0794	0·0309	0·0342	3·5572	0·0399
0·10	0·3959	0·0244	5·0500	51·020	51·000	0·0990	0·0458	0·0496	4·9764	0·0498
0·12	0·4742	0·0291	6·0720	51·224	51·200	0·1186	0·0631	0·0673	6·5482	0·0598
0·14	0·5521	0·0337	7·0980	51·428	51·400	0·1381	0·0825	0·0871	8·2601	0·0697
0·16	0·6297	0·0383	8·1280	51·632	51·600	0·1575	0·1041	0·1088	10·102	0·0796
0·18	0·7070	0·0428	9·1620	51·836	51·800	0·1769	0·1276	0·1325	12·066	0·0895
0·20	0·7840	0·0473	10·200	52·040	52·000	0·1962	0·1531	0·1581	14·147	0·0993
0·22	0·8607	0·0518	11·242	52·244	52·200	0·2154	0·1804	0·1854	16·338	0·1092
0·24	0·9372	0·0561	12·288	52·448	52·400	0·2345	0·2095	0·2145	18·634	0·1191
0·26	1·0133	0·0605	13·338	52·651	52·600	0·2536	0·2404	0·2452	21·033	0·1289
0·28	1·0892	0·0647	14·392	52·855	52·800	0·2726	0·2729	0·2777	23·530	0·1387
0·30	1·1647	0·0690	15·450	53·059	53·000	0·2915	0·3071	0·3117	26·123	0·1485
0·32	1·2400	0·0731	16·512	53·263	53·200	0·3104	0·3429	0·3473	28·807	0·1583
0·34	1·3150	0·0773	17·578	53·467	53·400	0·3292	0·3803	0·3845	31·582	0·1681
0·36	1·3898	0·0813	18·648	53·671	53·600	0·3479	0·4193	0·4232	34·445	0·1779
0·38	1·4643	0·0854	19·722	53·875	53·800	0·3666	0·4598	0·4635	37·393	0·1877
0·40	1·5385	0·0894	20·800	54·079	54·000	0·3852	0·5018	0·5052	40·426	0·1974
0·42	1·6124	0·0933	21·882	54·283	54·200	0·4037	0·5453	0·5484	43·540	0·2072
0·44	1·6861	0·0972	22·968	54·487	54·400	0·4222	0·5902	0·5930	46·735	0·2169
0·46	1·7596	0·1011	24·058	54·691	54·600	0·4406	0·6366	0·6390	50·010	0·2266
0·48	1·8327	0·1049	25·152	54·895	54·800	0·4590	0·6844	0·6865	53·362	0·2363
0·50	1·9057	0·1087	26·250	55·099	55·000	0·4773	0·7336	0·7353	56·790	0·2460
0·52	1·9783	0·1124	27·352	55·303	55·200	0·4955	0·7842	0·7856	60·294	0·2557
0·54	2·0508	0·1161	28·458	55·507	55·400	0·5137	0·8361	0·8371	63·872	0·2654
0·56	2·1230	0·1197	29·568	55·711	55·600	0·5318	0·8894	0·8901	67·524	0·2751
0·58	2·1949	0·1233	30·682	55·915	55·800	0·5499	0·9441	0·9444	71·247	0·2847
0·60	2·2666	0·1269	31·800	56·119	56·000	0·5679	1·0000	1·0000	75·042	0·2943
0·64	2·4093	0·1339	34·048	56·527	56·400	0·6037	1·1159	1·1152	82·843	0·3136
0·68	2·5511	0·1408	36·312	56·935	56·800	0·6393	1·2369	1·2355	90·920	0·3328
0·72	2·6920	0·1475	38·592	57·343	57·200	0·6747	1·3630	1·3610	99·268	0·3519
0·76	2·8320	0·1541	40·888	57·751	57·600	0·7099	1·4941	1·4916	107·88	0·3711
0·80	2·9712	0·1605	43·200	58·158	58·000	0·7448	1·6302	1·6271	116·75	0·3901
0·84	3·1095	0·1668	45·528	58·566	58·400	0·7796	1·7711	1·7676	125·88	0·4092
0·88	3·2470	0·1730	47·872	58·974	58·800	0·8141	1·9168	1·9130	135·27	0·4281
0·92	3·3836	0·1790	50·232	59·382	59·200	0·8485	2·0673	2·0633	144·90	0·4471
0·96	3·5195	0·1849	52·608	59·790	59·600	0·8827	2·2226	2·2183	154·78	0·4660
1·00	3·6546	0·1907	55·000	60·198	60·000	0·9167	2·3825	2·3782	164·90	0·4848

C71　Wide rectangular channel (free surface)

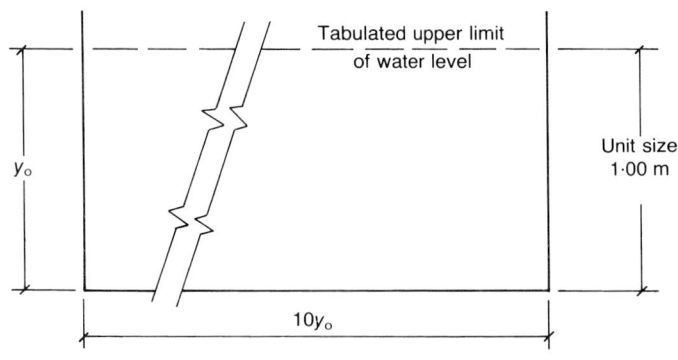

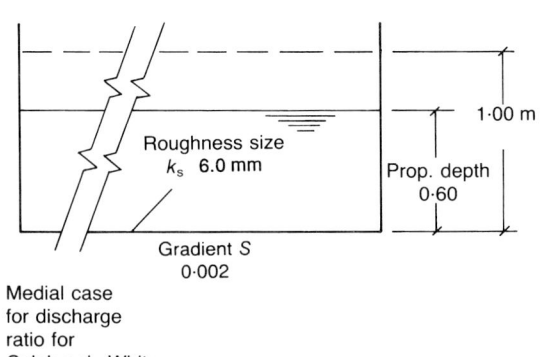

Medial case
for discharge
ratio for
Colebrook–White
solutions
$\Theta_{ep} = 297$

Prop dpth	U.equ. dia. $D_{ep(u)}$	Equiv. disch. factor	Unit sect. area	Unit wetted perim.	Unit surf. brdth	Unit mean depth	Discharge ratios $Q/Q_{0.60}$		U.crit. disch. $Q_{c(u)}$	Unit depth cntrd
Y	(m)	J	A_u (m²)	P_u (m)	$B_{s(u)}$ (m)	$y_{m(u)}$ (m)	medial	Mann'g	(m³s⁻¹)	$y_{d(u)}$ (m)
0·02	0·0797	0·0249	0·2000	10·040	10·000	0·0200	0·0035	0·0037	0·0886	0·0100
0·04	0·1587	0·0495	0·4000	10·080	10·000	0·0400	0·0115	0·0118	0·2505	0·0200
0·06	0·2372	0·0736	0·6000	10·120	10·000	0·0600	0·0230	0·0231	0·4602	0·0300
0·08	0·3150	0·0974	0·8000	10·160	10·000	0·0800	0·0374	0·0371	0·7086	0·0400
0·10	0·3922	0·1208	1·0000	10·200	10·000	0·1000	0·0544	0·0537	0·9903	0·0500
0·12	0·4687	0·1438	1·2000	10·240	10·000	0·1200	0·0737	0·0726	1·3018	0·0600
0·14	0·5447	0·1665	1·4000	10·280	10·000	0·1400	0·0951	0·0936	1·6404	0·0700
0·16	0·6202	0·1888	1·6000	10·320	10·000	0·1600	0·1186	0·1167	2·0042	0·0800
0·18	0·6950	0·2107	1·8000	10·360	10·000	0·1800	0·1440	0·1416	2·3915	0·0900
0·20	0·7692	0·2324	2·0000	10·400	10·000	0·2000	0·1711	0·1684	2·8009	0·1000
0·22	0·8429	0·2536	2·2000	10·440	10·000	0·2200	0·2000	0·1968	3·2314	0·1100
0·24	0·9160	0·2746	2·4000	10·480	10·000	0·2400	0·2305	0·2270	3·6819	0·1200
0·26	0·9886	0·2952	2·6000	10·520	10·000	0·2600	0·2625	0·2587	4·1516	0·1300
0·28	1·0606	0·3155	2·8000	10·560	10·000	0·2800	0·2960	0·2920	4·6398	0·1400
0·30	1·1321	0·3355	3·0000	10·600	10·000	0·3000	0·3310	0·3268	5·1457	0·1500
0·32	1·2030	0·3552	3·2000	10·640	10·000	0·3200	0·3674	0·3630	5·6687	0·1600
0·34	1·2734	0·3746	3·4000	10·680	10·000	0·3400	0·4051	0·4005	6·2084	0·1700
0·36	1·3433	0·3937	3·6000	10·720	10·000	0·3600	0·4441	0·4395	6·7642	0·1800
0·38	1·4126	0·4124	3·8000	10·760	10·000	0·3800	0·4843	0·4797	7·3356	0·1900
0·40	1·4815	0·4309	4·0000	10·800	10·000	0·4000	0·5257	0·5212	7·9223	0·2000
0·42	1·5498	0·4491	4·2000	10·840	10·000	0·4200	0·5684	0·5640	8·5238	0·2100
0·44	1·6176	0·4671	4·4000	10·880	10·000	0·4400	0·6121	0·6080	9·1399	0·2200
0·46	1·6850	0·4847	4·6000	10·920	10·000	0·4600	0·6570	0·6531	9·7701	0·2300
0·48	1·7518	0·5021	4·8000	10·960	10·000	0·4800	0·7030	0·6994	10·414	0·2400
0·50	1·8182	0·5193	5·0000	11·000	10·000	0·5000	0·7500	0·7469	11·072	0·2500
0·52	1·8841	0·5361	5·2000	11·040	10·000	0·5200	0·7981	0·7954	11·743	0·2600
0·54	1·9495	0·5527	5·4000	11·080	10·000	0·5400	0·8471	0·8450	12·427	0·2700
0·56	2·0144	0·5691	5·6000	11·120	10·000	0·5600	0·8971	0·8956	13·123	0·2800
0·58	2·0789	0·5852	5·8000	11·160	10·000	0·5800	0·9481	0·9473	13·833	0·2900
0·60	2·1429	0·6011	6·0000	11·200	10·000	0·6000	1·0000	1·0000	14·554	0·3000
0·64	2·2695	0·6321	6·4000	11·280	10·000	0·6400	1·1066	1·1083	16·034	0·3200
0·68	2·3944	0·6621	6·8000	11·360	10·000	0·6800	1·2166	1·2204	17·560	0·3400
0·72	2·5175	0·6913	7·2000	11·440	10·000	0·7200	1·3300	1·3361	19·132	0·3600
0·76	2·6389	0·7196	7·6000	11·520	10·000	0·7600	1·4466	1·4553	20·748	0·3800
0·80	2·7586	0·7471	8·0000	11·600	10·000	0·8000	1·5662	1·5779	22·408	0·4000
0·84	2·8767	0·7737	8·4000	11·680	10·000	0·8400	1·6889	1·7037	24·109	0·4200
0·88	2·9932	0·7996	8·8000	11·760	10·000	0·8800	1·8144	1·8327	25·851	0·4400
0·92	3·1081	0·8247	9·2000	11·840	10·000	0·9200	1·9426	1·9647	27·634	0·4600
0·96	3·2215	0·8490	9·6000	11·920	10·000	0·9600	2·0735	2·0997	29·456	0·4800
1·00	3·3333	0·8726	10·000	12·000	10·000	1·0000	2·2070	2·2375	31·316	0·5000

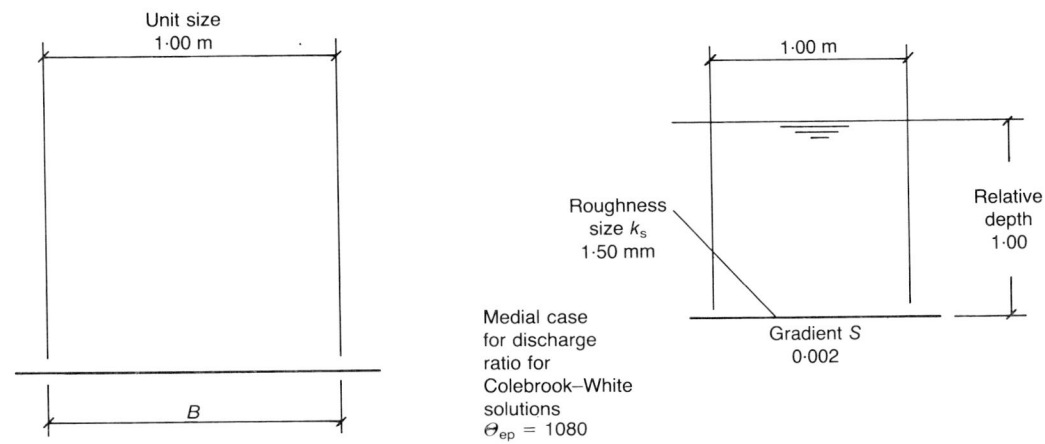

Unit size 1·00 m

B

1·00 m

Roughness size k_s 1·50 mm

Medial case for discharge ratio for Colebrook–White solutions $\Theta_{ep} = 1080$

Gradient S 0·002

Relative depth 1·00

Prop dpth Y	U.equ. dia. $D_{ep(u)}$ (m)	Equiv. disch. factor J	Unit sect. area A_u (m²)	Unit wetted perim. P_u (m)	Unit surf. brdth $B_{s(u)}$ (m)	Unit mean depth $y_{m(u)}$ (m)	Discharge ratios $Q/Q_{0.60}$ medial	Mann'g	U.crit. disch. $Q_{c(u)}$ (m³s⁻¹)	Unit depth cntrd $y_{d(u)}$ (m)
0·04	0·1600	0·5026	0·0400	1·0000	1·0000	0·0400	0·0051	0·0047	0·0251	0·0200
0·08	0·3200	1·0053	0·0800	1·0000	1·0000	0·0800	0·0163	0·0149	0·0709	0·0400
0·12	0·4800	1·5079	0·1200	1·0000	1·0000	0·1200	0·0318	0·0292	0·1302	0·0600
0·16	0·6400	2·0106	0·1600	1·0000	1·0000	0·1600	0·0510	0·0472	0·2004	0·0800
0·20	0·8000	2·5132	0·2000	1·0000	1·0000	0·2000	0·0736	0·0684	0·2801	0·1000
0·24	0·9600	3·0159	0·2400	1·0000	1·0000	0·2400	0·0991	0·0927	0·3682	0·1200
0·28	1·1200	3·5185	0·2800	1·0000	1·0000	0·2800	0·1274	0·1198	0·4640	0·1400
0·32	1·2800	4·0211	0·3200	1·0000	1·0000	0·3200	0·1583	0·1497	0·5669	0·1600
0·36	1·4400	4·5238	0·3600	1·0000	1·0000	0·3600	0·1917	0·1822	0·6764	0·1800
0·40	1·6000	5·0264	0·4000	1·0000	1·0000	0·4000	0·2275	0·2172	0·7922	0·2000
0·44	1·7600	5·5291	0·4400	1·0000	1·0000	0·4400	0·2655	0·2545	0·9140	0·2200
0·48	1·9200	6·0317	0·4800	1·0000	1·0000	0·4800	0·3057	0·2943	1·0414	0·2400
0·52	2·0800	6·5344	0·5200	1·0000	1·0000	0·5200	0·3480	0·3363	1·1743	0·2600
0·56	2·2400	7·0370	0·5600	1·0000	1·0000	0·5600	0·3923	0·3805	1·3123	0·2800
0·60	2·4000	7·5396	0·6000	1·0000	1·0000	0·6000	0·4386	0·4268	1·4554	0·3000
0·64	2·5600	8·0423	0·6400	1·0000	1·0000	0·6400	0·4869	0·4753	1·6034	0·3200
0·68	2·7200	8·5449	0·6800	1·0000	1·0000	0·6800	0·5370	0·5258	1·7560	0·3400
0·72	2·8800	9·0476	0·7200	1·0000	1·0000	0·7200	0·5889	0·5784	1·9132	0·3600
0·76	3·0400	9·5502	0·7600	1·0000	1·0000	0·7600	0·6426	0·6329	2·0748	0·3800
0·80	3·2000	10·053	0·8000	1·0000	1·0000	0·8000	0·6980	0·6894	2·2408	0·4000
0·84	3·3600	10·556	0·8400	1·0000	1·0000	0·8400	0·7551	0·7478	2·4109	0·4200
0·88	3·5200	11·058	0·8800	1·0000	1·0000	0·8800	0·8139	0·8081	2·5851	0·4400
0·92	3·6800	11·561	0·9200	1·0000	1·0000	0·9200	0·8744	0·8703	2·7634	0·4600
0·96	3·8400	12·063	0·9600	1·0000	1·0000	0·9600	0·9364	0·9342	2·9456	0·4800
1·00	4·0000	12·566	1·0000	1·0000	1·0000	1·0000	1·0000	1·0000	3·1316	0·5000
1·10	4·4000	13·823	1·1000	1·0000	1·0000	1·1000	1·1658	1·1722	3·6128	0·5500
1·20	4·8000	15·079	1·2000	1·0000	1·0000	1·2000	1·3409	1·3551	4·1165	0·6000
1·30	5·2000	16·336	1·3000	1·0000	1·0000	1·3000	1·5250	1·5485	4·6417	0·6500
1·40	5·6000	17·593	1·4000	1·0000	1·0000	1·4000	1·7178	1·7521	5·1874	0·7000
1·50	6·0000	18·849	1·5000	1·0000	1·0000	1·5000	1·9190	1·9656	5·7530	0·7500
1·60	6·4000	20·106	1·6000	1·0000	1·0000	1·6000	2·1283	2·1888	6·3378	0·8000
1·70	6·8000	21·362	1·7000	1·0000	1·0000	1·7000	2·3457	2·4215	6·9412	0·8500
1·80	7·2000	22·619	1·8000	1·0000	1·0000	1·8000	2·5708	2·6635	7·5626	0·9000
1·90	7·6000	23·876	1·9000	1·0000	1·0000	1·9000	2·8035	2·9147	8·2015	0·9500
2·00	8·0000	25·132	2·0000	1·0000	1·0000	2·0000	3·0437	3·1748	8·8574	1·0000
2·10	8·4000	26·389	2·1000	1·0000	1·0000	2·1000	3·2910	3·4438	9·5299	1·0500
2·20	8·8000	27·645	2·2000	1·0000	1·0000	2·2000	3·5455	3·7214	10·219	1·1000
2·30	9·2000	28·902	2·3000	1·0000	1·0000	2·3000	3·8070	4·0076	10·923	1·1500
2·40	9·6000	30·159	2·4000	1·0000	1·0000	2·4000	4·0753	4·3021	11·643	1·2000
2·50	10·000	31·415	2·5000	1·0000	1·0000	2·5000	4·3503	4·6050	12·379	1·2500

C73

Arc invert (free surface)

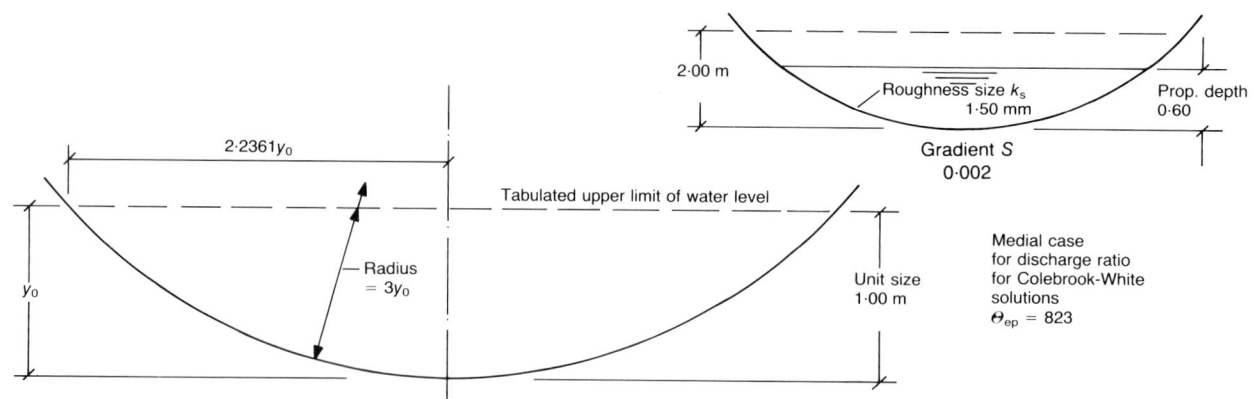

2·00 m

Roughness size k_s
1·50 mm

Prop. depth
0·60

Gradient S
0·002

2·2361y_0

Tabulated upper limit of water level

Radius
= 3y_0

y_0

Unit size
1·00 m

Medial case
for discharge ratio
for Colebrook-White
solutions
$\Theta_{ep} = 823$

Prop dpth Y	U.equ. dia. $D_{ep(u)}$ (m)	Equiv. disch. factor J	Unit sect. area A_u (m²)	Unit wetted perim. P_u (m)	Unit surf. brdth $B_{s(u)}$ (m)	Unit mean depth $y_{m(u)}$ (m)	Discharge ratios $Q/Q_{0.60}$ medial	Mann'g	U.crit. disch. $Q_{c(u)}$ (m³s⁻¹)	Unit depth cntrd $y_{d(u)}$ (m)
0·02	0·0533	0·2413	0·0092	0·6932	0·6917	0·0133	0·0007	0·0007	0·0033	0·0080
0·04	0·1063	0·3406	0·0261	0·9809	0·9765	0·0267	0·0033	0·0030	0·0133	0·0160
0·06	0·1593	0·4162	0·0479	1·2020	1·1940	0·0401	0·0078	0·0072	0·0300	0·0240
0·08	0·2120	0·4796	0·0736	1·3887	1·3764	0·0535	0·0145	0·0134	0·0533	0·0320
0·10	0·2646	0·5350	0·1028	1·5535	1·5362	0·0669	0·0233	0·0217	0·0832	0·0401
0·12	0·3170	0·5849	0·1349	1·7028	1·6800	0·0803	0·0343	0·0322	0·1198	0·0481
0·14	0·3693	0·6304	0·1699	1·8402	1·8115	0·0938	0·0476	0·0449	0·1629	0·0561
0·16	0·4213	0·6724	0·2073	1·9684	1·9333	0·1072	0·0632	0·0598	0·2126	0·0641
0·18	0·4733	0·7117	0·2472	2·0890	2·0470	0·1207	0·0811	0·0770	0·2689	0·0722
0·20	0·5250	0·7486	0·2892	2·2032	2·1541	0·1342	0·1013	0·0966	0·3318	0·0802
0·22	0·5766	0·7834	0·3333	2·3121	2·2553	0·1478	0·1239	0·1185	0·4012	0·0883
0·24	0·6280	0·8165	0·3794	2·4163	2·3515	0·1613	0·1488	0·1427	0·4772	0·0963
0·26	0·6792	0·8480	0·4273	2·5164	2·4433	0·1749	0·1760	0·1694	0·5596	0·1044
0·28	0·7303	0·8781	0·4771	2·6129	2·5311	0·1885	0·2056	0·1985	0·6486	0·1125
0·30	0·7812	0·9069	0·5285	2·7062	2·6153	0·2021	0·2376	0·2300	0·7441	0·1205
0·32	0·8320	0·9346	0·5817	2·7965	2·6964	0·2157	0·2719	0·2640	0·8460	0·1286
0·34	0·8825	0·9613	0·6364	2·8843	2·7745	0·2294	0·3086	0·3004	0·9544	0·1367
0·36	0·9329	0·9869	0·6926	2·9696	2·8498	0·2430	0·3477	0·3393	1·0693	0·1448
0·38	0·9832	1·0118	0·7503	3·0528	2·9227	0·2567	0·3891	0·3806	1·1906	0·1529
0·40	1·0332	1·0357	0·8095	3·1339	2·9933	0·2704	0·4329	0·4245	1·3183	0·1610
0·42	1·0831	1·0590	0·8701	3·2132	3·0618	0·2842	0·4791	0·4708	1·4524	0·1691
0·44	1·1328	1·0815	0·9320	3·2907	3·1282	0·2979	0·5276	0·5196	1·5930	0·1772
0·46	1·1824	1·1033	0·9952	3·3666	3·1927	0·3117	0·5784	0·5709	1·7399	0·1853
0·48	1·2318	1·1246	1·0597	3·4411	3·2555	0·3255	0·6317	0·6247	1·8932	0·1934
0·50	1·2810	1·1452	1·1254	3·5141	3·3166	0·3393	0·6872	0·6810	2·0529	0·2015
0·52	1·3300	1·1652	1·1923	3·5858	3·3762	0·3532	0·7451	0·7398	2·2189	0·2096
0·54	1·3789	1·1848	1·2604	3·6563	3·4342	0·3670	0·8054	0·8011	2·3912	0·2178
0·56	1·4276	1·2038	1·3297	3·7256	3·4908	0·3809	0·8679	0·8649	2·5699	0·2259
0·58	1·4761	1·2223	1·4000	3·7938	3·5460	0·3948	0·9328	0·9312	2·7549	0·2340
0·60	1·5245	1·2404	1·4715	3·8610	3·6000	0·4088	1·0000	1·0000	2·9461	0·2422
0·64	1·6207	1·2753	1·6176	3·9924	3·7043	0·4367	1·1413	1·1451	3·3475	0·2585
0·68	1·7162	1·3085	1·7678	4·1203	3·8040	0·4647	1·2918	1·3001	3·7739	0·2748
0·72	1·8110	1·3403	1·9219	4·2449	3·8995	0·4928	1·4513	1·4650	4·2251	0·2912
0·76	1·9051	1·3706	2·0797	4·3666	3·9912	0·5211	1·6199	1·6397	4·7012	0·3076
0·80	1·9985	1·3997	2·2411	4·4855	4·0792	0·5494	1·7974	1·8243	5·2020	0·3240
0·84	2·0913	1·4276	2·4060	4·6020	4·1638	0·5778	1·9837	2·0186	5·7274	0·3404
0·88	2·1833	1·4544	2·5742	4·7161	4·2453	0·6064	2·1788	2·2227	6·2772	0·3568
0·92	2·2746	1·4800	2·7456	4·8282	4·3237	0·6350	2·3825	2·4363	6·8515	0·3733
0·96	2·3653	1·5047	2·9200	4·9382	4·3993	0·6638	2·5948	2·6595	7·4500	0·3898
1·00	2·4552	1·5284	3·0975	5·0464	4·4721	0·6926	2·8156	2·8922	8·0726	0·4063

10% concave bed river

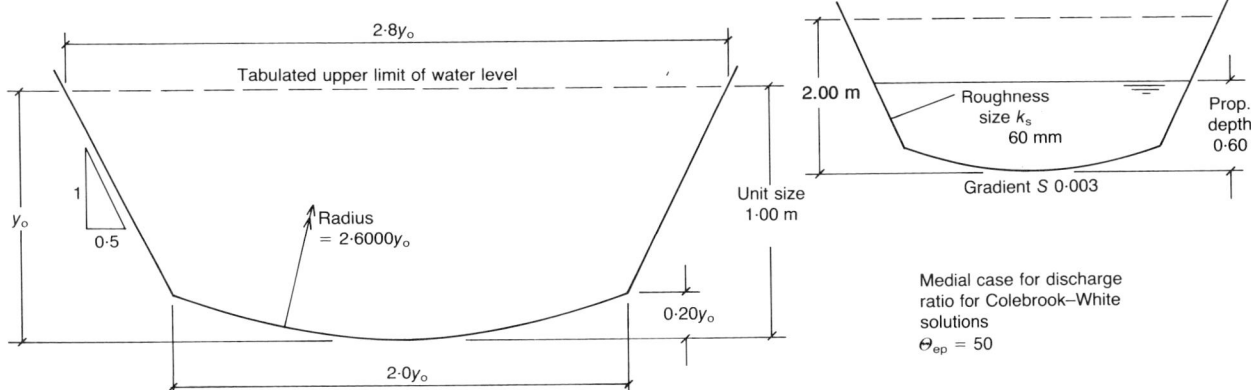

Prop dpth	U.equ. dia. $D_{ep(u)}$	Equiv. disch. factor	Unit sect. area	Unit wetted perim.	Unit surf. brdth	Unit mean depth	Discharge ratios $Q/Q_{0.60}$		U.crit. disch. $Q_{c(u)}$	Unit depth cntrd
Y	(m)	J	A_u (m²)	P_u (m)	$B_{s(u)}$ (m)	$y_{m(u)}$ (m)	medial	Mann'g	(m³s⁻¹)	$y_{d(u)}$ (m)
0·02	0·0532	0·2591	0·0086	0·6454	0·6437	0·0133	0·0005	0·0008	0·0031	0·0080
0·04	0·1063	0·3656	0·0243	0·9133	0·9086	0·0267	0·0027	0·0035	0·0124	0·0160
0·06	0·1591	0·4466	0·0445	1·1193	1·1107	0·0401	0·0070	0·0085	0·0279	0·0240
0·08	0·2118	0·5145	0·0685	1·2933	1·2800	0·0535	0·0136	0·0158	0·0496	0·0320
0·10	0·2643	0·5738	0·0956	1·4469	1·4283	0·0669	0·0227	0·0255	0·0774	0·0401
0·12	0·3165	0·6270	0·1255	1·5860	1·5615	0·0804	0·0343	0·0377	0·1114	0·0481
0·14	0·3686	0·6756	0·1580	1·7142	1·6833	0·0938	0·0485	0·0526	0·1516	0·0561
0·16	0·4205	0·7204	0·1928	1·8338	1·7960	0·1073	0·0654	0·0701	0·1978	0·0642
0·18	0·4722	0·7622	0·2298	1·9463	1·9012	0·1209	0·0851	0·0902	0·2501	0·0722
0·20	0·5237	0·8014	0·2688	2·0529	2·0000	0·1344	0·1075	0·1131	0·3086	0·0803
0·22	0·5892	0·8824	0·3090	2·0976	2·0200	0·1530	0·1347	0·1406	0·3784	0·0885
0·24	0·6527	0·9571	0·3496	2·1424	2·0400	0·1714	0·1643	0·1703	0·4532	0·0971
0·26	0·7144	1·0261	0·3906	2·1871	2·0600	0·1896	0·1960	0·2021	0·5326	0·1058
0·28	0·7742	1·0898	0·4320	2·2318	2·0800	0·2077	0·2297	0·2358	0·6165	0·1147
0·30	0·8325	1·1488	0·4738	2·2765	2·1000	0·2256	0·2654	0·2714	0·7047	0·1237
0·32	0·8892	1·2034	0·5160	2·3212	2·1200	0·2434	0·3030	0·3089	0·7972	0·1328
0·34	0·9444	1·2539	0·5586	2·3660	2·1400	0·2610	0·3424	0·3481	0·8937	0·1419
0·36	0·9982	1·3008	0·6016	2·4107	2·1600	0·2785	0·3835	0·3890	0·9942	0·1510
0·38	1·0507	1·3443	0·6450	2·4554	2·1800	0·2959	0·4263	0·4316	1·0987	0·1602
0·40	1·1020	1·3847	0·6888	2·5001	2·2000	0·3131	0·4708	0·4758	1·2069	0·1694
0·42	1·1521	1·4222	0·7330	2·5448	2·2200	0·3302	0·5169	0·5215	1·3190	0·1786
0·44	1·2011	1·4571	0·7776	2·5896	2·2400	0·3471	0·5646	0·5688	1·4347	0·1877
0·46	1·2490	1·4895	0·8226	2·6343	2·2600	0·3640	0·6139	0·6177	1·5541	0·1969
0·48	1·2960	1·5197	0·8680	2·6790	2·2800	0·3807	0·6647	0·6680	1·6771	0·2061
0·50	1·3420	1·5478	0·9138	2·7237	2·3000	0·3973	0·7169	0·7198	1·8037	0·2153
0·52	1·3870	1·5739	0·9600	2·7685	2·3200	0·4138	0·7707	0·7730	1·9338	0·2244
0·54	1·4312	1·5983	1·0066	2·8132	2·3400	0·4302	0·8259	0·8276	2·0674	0·2336
0·56	1·4746	1·6210	1·0536	2·8579	2·3600	0·4464	0·8825	0·8837	2·2045	0·2427
0·58	1·5172	1·6421	1·1010	2·9026	2·3800	0·4626	0·9406	0·9412	2·3450	0·2518
0·60	1·5591	1·6618	1·1488	2·9473	2·4000	0·4787	1·0000	1·0000	2·4889	0·2609
0·64	1·6407	1·6973	1·2456	3·0368	2·4400	0·5105	1·1231	1·1218	2·7869	0·2791
0·68	1·7196	1·7280	1·3440	3·1262	2·4800	0·5419	1·2516	1·2489	3·0983	0·2972
0·72	1·7962	1·7548	1·4440	3·2157	2·5200	0·5730	1·3855	1·3814	3·4230	0·3152
0·76	1·8705	1·7780	1·5456	3·3051	2·5600	0·6037	1·5248	1·5191	3·7608	0·3332
0·80	1·9429	1·7980	1·6488	3·3946	2·6000	0·6341	1·6693	1·6620	4·1117	0·3511
0·84	2·0133	1·8154	1·7536	3·4840	2·6400	0·6642	1·8190	1·8101	4·4756	0·3689
0·88	2·0820	1·8304	1·8600	3·5734	2·6800	0·6940	1·9740	1·9634	4·8524	0·3866
0·92	2·1491	1·8432	1·9680	3·6629	2·7200	0·7235	2·1341	2·1218	5·2422	0·4043
0·96	2·2147	1·8542	2·0776	3·7523	2·7600	0·7527	2·2994	2·2853	5·6448	0·4219
1·00	2·2789	1·8636	2·1888	3·8418	2·8000	0·7817	2·4699	2·4540	6·0602	0·4395

C75

5% concave bed river

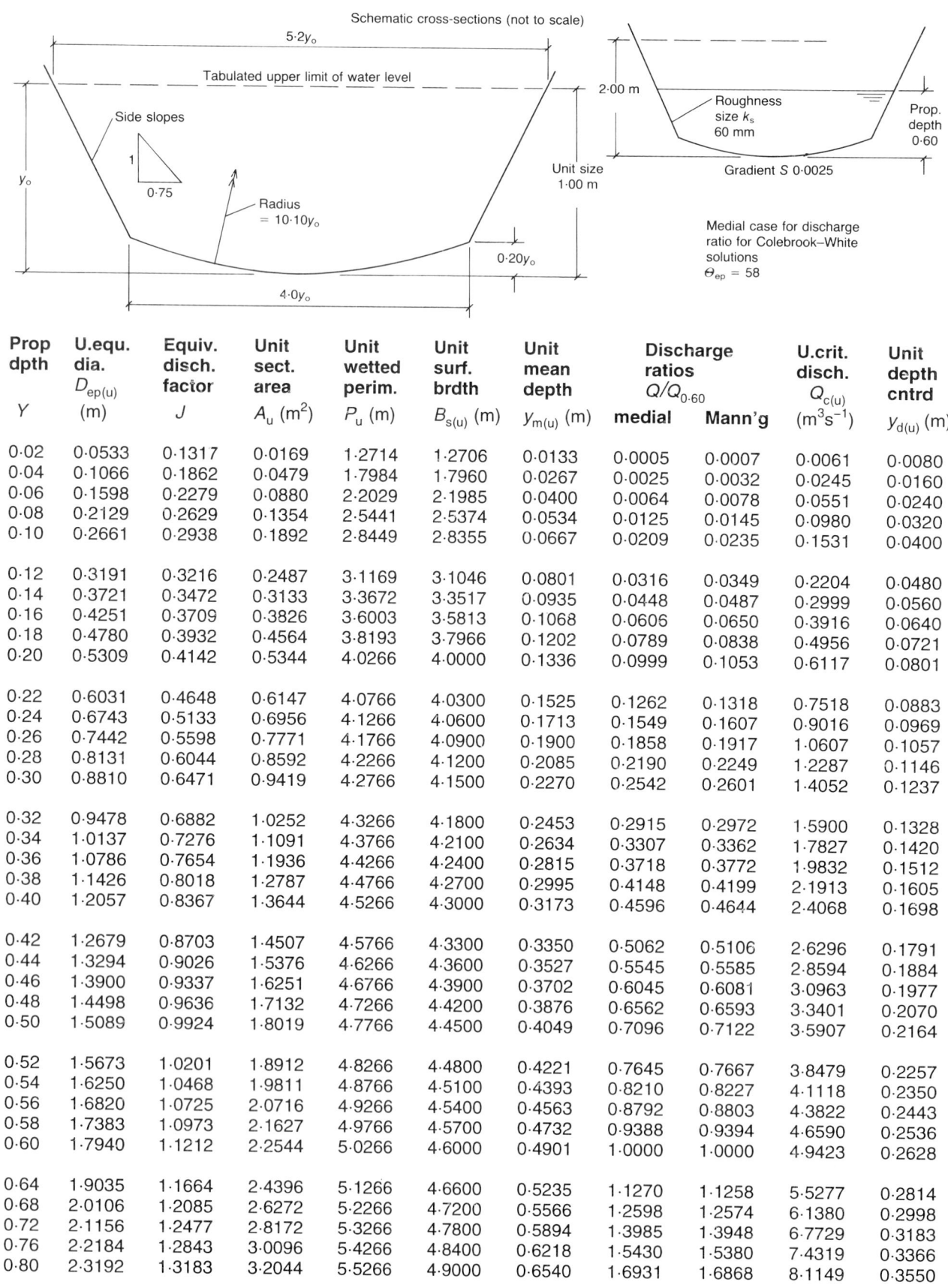

Schematic cross-sections (not to scale)

5·2y_o

Tabulated upper limit of water level

Side slopes

1 / 0.75

y_o

Radius = 10·10y_o

0·20y_o

4·0y_o

2·00 m

Roughness size k_s 60 mm

Prop. depth 0·60

Unit size 1·00 m

Gradient S 0·0025

Medial case for discharge ratio for Colebrook–White solutions $\Theta_{ep} = 58$

Prop dpth Y	U.equ. dia. $D_{ep(u)}$ (m)	Equiv. disch. factor J	Unit sect. area A_u (m²)	Unit wetted perim. P_u (m)	Unit surf. brdth $B_{s(u)}$ (m)	Unit mean depth $y_{m(u)}$ (m)	Discharge ratios $Q/Q_{0.60}$ medial	Discharge ratios Mann'g	U.crit. disch. $Q_{c(u)}$ (m³s⁻¹)	Unit depth cntrd $y_{d(u)}$ (m)
0·02	0·0533	0·1317	0·0169	1·2714	1·2706	0·0133	0·0005	0·0007	0·0061	0·0080
0·04	0·1066	0·1862	0·0479	1·7984	1·7960	0·0267	0·0025	0·0032	0·0245	0·0160
0·06	0·1598	0·2279	0·0880	2·2029	2·1985	0·0400	0·0064	0·0078	0·0551	0·0240
0·08	0·2129	0·2629	0·1354	2·5441	2·5374	0·0534	0·0125	0·0145	0·0980	0·0320
0·10	0·2661	0·2938	0·1892	2·8449	2·8355	0·0667	0·0209	0·0235	0·1531	0·0400
0·12	0·3191	0·3216	0·2487	3·1169	3·1046	0·0801	0·0316	0·0349	0·2204	0·0480
0·14	0·3721	0·3472	0·3133	3·3672	3·3517	0·0935	0·0448	0·0487	0·2999	0·0560
0·16	0·4251	0·3709	0·3826	3·6003	3·5813	0·1068	0·0606	0·0650	0·3916	0·0640
0·18	0·4780	0·3932	0·4564	3·8193	3·7966	0·1202	0·0789	0·0838	0·4956	0·0721
0·20	0·5309	0·4142	0·5344	4·0266	4·0000	0·1336	0·0999	0·1053	0·6117	0·0801
0·22	0·6031	0·4648	0·6147	4·0766	4·0300	0·1525	0·1262	0·1318	0·7518	0·0883
0·24	0·6743	0·5133	0·6956	4·1266	4·0600	0·1713	0·1549	0·1607	0·9016	0·0969
0·26	0·7442	0·5598	0·7771	4·1766	4·0900	0·1900	0·1858	0·1917	1·0607	0·1057
0·28	0·8131	0·6044	0·8592	4·2266	4·1200	0·2085	0·2190	0·2249	1·2287	0·1146
0·30	0·8810	0·6471	0·9419	4·2766	4·1500	0·2270	0·2542	0·2601	1·4052	0·1237
0·32	0·9478	0·6882	1·0252	4·3266	4·1800	0·2453	0·2915	0·2972	1·5900	0·1328
0·34	1·0137	0·7276	1·1091	4·3766	4·2100	0·2634	0·3307	0·3362	1·7827	0·1420
0·36	1·0786	0·7654	1·1936	4·4266	4·2400	0·2815	0·3718	0·3772	1·9832	0·1512
0·38	1·1426	0·8018	1·2787	4·4766	4·2700	0·2995	0·4148	0·4199	2·1913	0·1605
0·40	1·2057	0·8367	1·3644	4·5266	4·3000	0·3173	0·4596	0·4644	2·4068	0·1698
0·42	1·2679	0·8703	1·4507	4·5766	4·3300	0·3350	0·5062	0·5106	2·6296	0·1791
0·44	1·3294	0·9026	1·5376	4·6266	4·3600	0·3527	0·5545	0·5585	2·8594	0·1884
0·46	1·3900	0·9337	1·6251	4·6766	4·3900	0·3702	0·6045	0·6081	3·0963	0·1977
0·48	1·4498	0·9636	1·7132	4·7266	4·4200	0·3876	0·6562	0·6593	3·3401	0·2070
0·50	1·5089	0·9924	1·8019	4·7766	4·4500	0·4049	0·7096	0·7122	3·5907	0·2164
0·52	1·5673	1·0201	1·8912	4·8266	4·4800	0·4221	0·7645	0·7667	3·8479	0·2257
0·54	1·6250	1·0468	1·9811	4·8766	4·5100	0·4393	0·8210	0·8227	4·1118	0·2350
0·56	1·6820	1·0725	2·0716	4·9266	4·5400	0·4563	0·8792	0·8803	4·3822	0·2443
0·58	1·7383	1·0973	2·1627	4·9766	4·5700	0·4732	0·9388	0·9394	4·6590	0·2536
0·60	1·7940	1·1212	2·2544	5·0266	4·6000	0·4901	1·0000	1·0000	4·9423	0·2628
0·64	1·9035	1·1664	2·4396	5·1266	4·6600	0·5235	1·1270	1·1258	5·5277	0·2814
0·68	2·0106	1·2085	2·6272	5·2266	4·7200	0·5566	1·2598	1·2574	6·1380	0·2998
0·72	2·1156	1·2477	2·8172	5·3266	4·7800	0·5894	1·3985	1·3948	6·7729	0·3183
0·76	2·2184	1·2843	3·0096	5·4266	4·8400	0·6218	1·5430	1·5380	7·4319	0·3366
0·80	2·3192	1·3183	3·2044	5·5266	4·9000	0·6540	1·6931	1·6868	8·1149	0·3550
0·84	2·4182	1·3502	3·4016	5·6266	4·9600	0·6858	1·8488	1·8412	8·8215	0·3732
0·88	2·5154	1·3799	3·6012	5·7266	5·0200	0·7174	2·0100	2·0011	9·5517	0·3914
0·92	2·6109	1·4077	3·8032	5·8266	5·0800	0·7487	2·1767	2·1666	10·305	0·4096
0·96	2·7048	1·4337	4·0076	5·9266	5·1400	0·7797	2·3489	2·3374	11·082	0·4277
1·00	2·7972	1·4581	4·2144	6·0266	5·2000	0·8105	2·5265	2·5137	11·881	0·4457

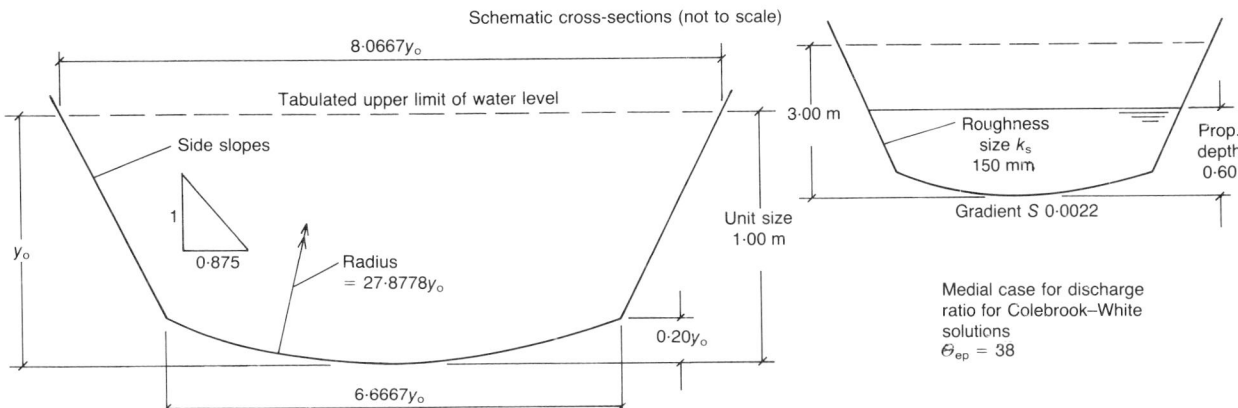

Schematic cross-sections (not to scale)

Prop dpth	U.equ. dia. $D_{ep(u)}$	Equiv. disch. factor	Unit sect. area	Unit wetted perim.	Unit surf. brdth	Unit mean depth	Discharge ratios $Q/Q_{0.60}$		U.crit. disch. $Q_{c(u)}$	Unit depth cntrd
Y	(m)	J	A_u (m²)	P_u (m)	$B_{s(u)}$ (m)	$y_{m(u)}$ (m)	medial	Mann'g	(m³s⁻¹)	$y_{d(u)}$ (m)
0.02	0.0533	0.0793	0.0282	2.1121	2.1116	0.0133	0.0004	0.0007	0.0102	0.0066
0.04	0.1066	0.1121	0.0796	2.9871	2.9857	0.0267	0.0021	0.0031	0.0407	0.0159
0.06	0.1599	0.1373	0.1463	3.6587	3.6561	0.0400	0.0057	0.0076	0.0916	0.0239
0.08	0.2132	0.1585	0.2252	4.2250	4.2209	0.0533	0.0113	0.0141	0.1629	0.0321
0.10	0.2664	0.1772	0.3147	4.7239	4.7183	0.0667	0.0192	0.0229	0.2545	0.0402
0.12	0.3197	0.1941	0.4136	5.1751	5.1677	0.0800	0.0293	0.0340	0.3664	0.0479
0.14	0.3729	0.2096	0.5211	5.5901	5.5807	0.0934	0.0418	0.0475	0.4987	0.0560
0.16	0.4261	0.2240	0.6366	5.9764	5.9650	0.1067	0.0567	0.0634	0.6513	0.0641
0.18	0.4793	0.2375	0.7596	6.3393	6.3257	0.1201	0.0742	0.0818	0.8242	0.0722
0.20	0.5324	0.2503	0.8895	6.6827	6.6667	0.1334	0.0943	0.1027	1.0175	0.0800
0.22	0.6076	0.2834	1.0232	6.7358	6.7017	0.1527	0.1201	0.1291	1.2520	0.0883
0.24	0.6820	0.3156	1.1576	6.7890	6.7367	0.1718	0.1483	0.1577	1.5027	0.0969
0.26	0.7557	0.3470	1.2927	6.8421	6.7717	0.1909	0.1789	0.1886	1.7687	0.1057
0.28	0.8287	0.3775	1.4285	6.8953	6.8067	0.2099	0.2117	0.2216	2.0493	0.1147
0.30	0.9009	0.4073	1.5649	6.9484	6.8417	0.2287	0.2468	0.2567	2.3438	0.1238
0.32	0.9724	0.4363	1.7021	7.0016	6.8767	0.2475	0.2839	0.2937	2.6519	0.1330
0.34	1.0433	0.4646	1.8400	7.0547	6.9117	0.2662	0.3231	0.3328	2.9730	0.1423
0.36	1.1135	0.4921	1.9786	7.1079	6.9467	0.2848	0.3643	0.3737	3.3068	0.1516
0.38	1.1830	0.5190	2.1179	7.1610	6.9817	0.3033	0.4075	0.4165	3.6529	0.1610
0.40	1.2519	0.5452	2.2579	7.2142	7.0167	0.3218	0.4526	0.4611	4.0109	0.1704
0.42	1.3202	0.5707	2.3985	7.2673	7.0517	0.3401	0.4995	0.5075	4.3806	0.1798
0.44	1.3879	0.5956	2.5399	7.3205	7.0867	0.3584	0.5482	0.5556	4.7618	0.1893
0.46	1.4549	0.6199	2.6820	7.3736	7.1217	0.3766	0.5988	0.6055	5.1542	0.1987
0.48	1.5214	0.6436	2.8248	7.4268	7.1567	0.3947	0.6511	0.6570	5.5576	0.2082
0.50	1.5873	0.6667	2.9683	7.4799	7.1917	0.4127	0.7051	0.7102	5.9718	0.2176
0.52	1.6527	0.6892	3.1125	7.5331	7.2267	0.4307	0.7608	0.7650	6.3966	0.2271
0.54	1.7175	0.7112	3.2573	7.5862	7.2617	0.4486	0.8181	0.8214	6.8318	0.2365
0.56	1.7818	0.7327	3.4029	7.6394	7.2967	0.4664	0.8772	0.8794	7.2774	0.2460
0.58	1.8455	0.7537	3.5492	7.6925	7.3317	0.4841	0.9378	0.9389	7.7332	0.2554
0.60	1.9088	0.7742	3.6962	7.7457	7.3667	0.5017	1.0000	1.0000	8.1989	0.2649
0.64	2.0338	0.8137	3.9923	7.8520	7.4367	0.5368	1.1292	1.1267	9.1601	0.2837
0.68	2.1568	0.8514	4.2911	7.9583	7.5067	0.5716	1.2646	1.2595	10.160	0.3026
0.72	2.2780	0.8874	4.5928	8.0646	7.5767	0.6062	1.4060	1.3980	11.198	0.3214
0.76	2.3974	0.9218	4.8973	8.1709	7.6467	0.6404	1.5533	1.5424	12.273	0.3402
0.80	2.5151	0.9546	5.2045	8.2772	7.7167	0.6745	1.7064	1.6924	13.385	0.3589
0.84	2.6312	0.9860	5.5146	8.3835	7.7867	0.7082	1.8653	1.8479	14.533	0.3776
0.88	2.7456	1.0160	5.8275	8.4898	7.8567	0.7417	2.0298	2.0090	15.717	0.3962
0.92	2.8586	1.0447	6.1431	8.5961	7.9267	0.7750	2.1999	2.1755	16.936	0.4149
0.96	2.9700	1.0722	6.4616	8.7024	7.9967	0.8080	2.3755	2.3474	18.189	0.4334
1.00	3.0801	1.0985	6.7829	8.8087	8.0667	0.8409	2.5565	2.5246	19.477	0.4519

2% concave bed river

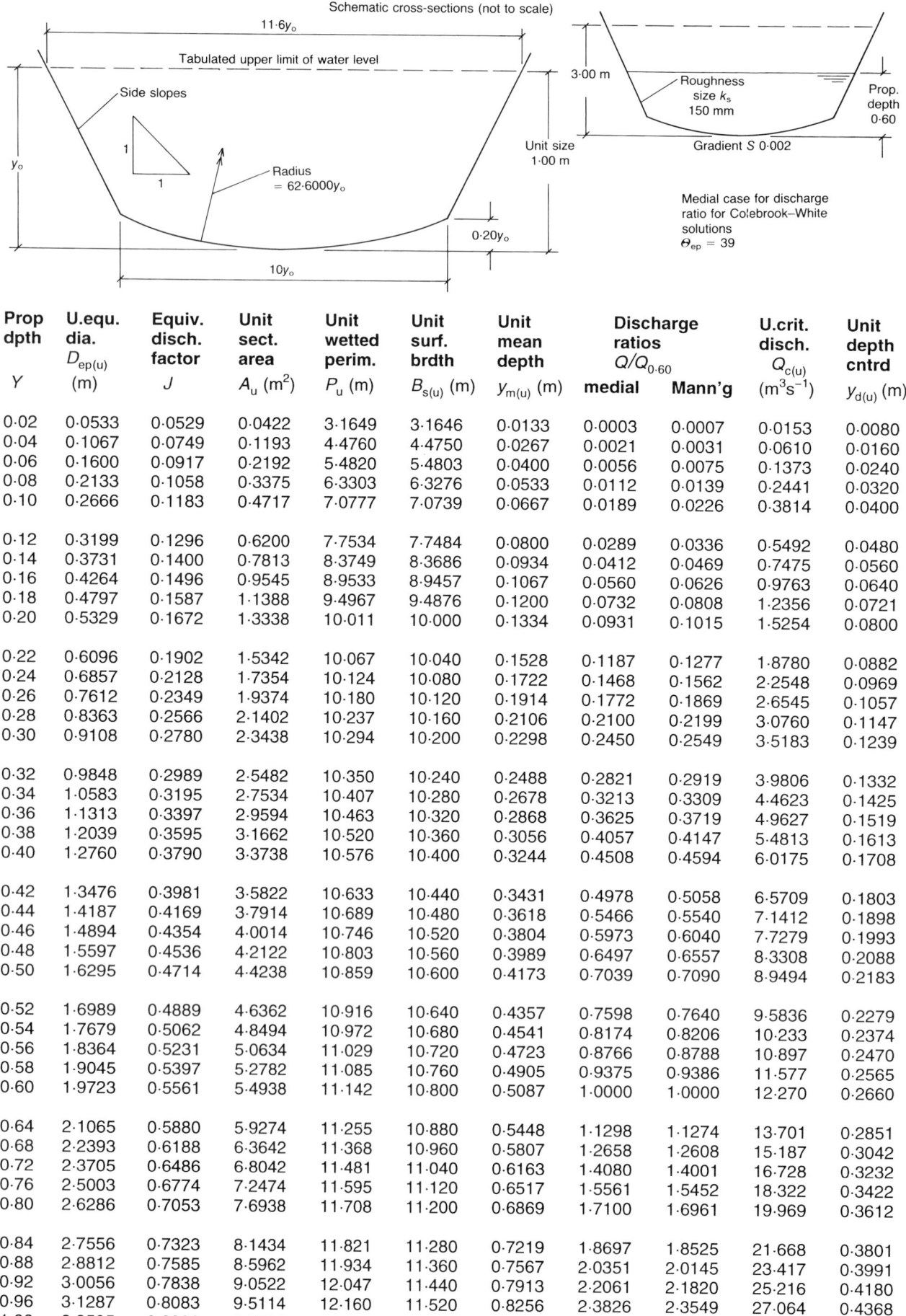

Prop dpth Y	U.equ. dia. $D_{ep(u)}$ (m)	Equiv. disch. factor J	Unit sect. area A_u (m²)	Unit wetted perim. P_u (m)	Unit surf. brdth $B_{s(u)}$ (m)	Unit mean depth $y_{m(u)}$ (m)	Discharge ratios $Q/Q_{0.60}$ medial	Mann'g	U.crit. disch. $Q_{c(u)}$ (m³s⁻¹)	Unit depth cntrd $y_{d(u)}$ (m)
0·02	0·0533	0·0529	0·0422	3·1649	3·1646	0·0133	0·0003	0·0007	0·0153	0·0080
0·04	0·1067	0·0749	0·1193	4·4760	4·4750	0·0267	0·0021	0·0031	0·0610	0·0160
0·06	0·1600	0·0917	0·2192	5·4820	5·4803	0·0400	0·0056	0·0075	0·1373	0·0240
0·08	0·2133	0·1058	0·3375	6·3303	6·3276	0·0533	0·0112	0·0139	0·2441	0·0320
0·10	0·2666	0·1183	0·4717	7·0777	7·0739	0·0667	0·0189	0·0226	0·3814	0·0400
0·12	0·3199	0·1296	0·6200	7·7534	7·7484	0·0800	0·0289	0·0336	0·5492	0·0480
0·14	0·3731	0·1400	0·7813	8·3749	8·3686	0·0934	0·0412	0·0469	0·7475	0·0560
0·16	0·4264	0·1496	0·9545	8·9533	8·9457	0·1067	0·0560	0·0626	0·9763	0·0640
0·18	0·4797	0·1587	1·1388	9·4967	9·4876	0·1200	0·0732	0·0808	1·2356	0·0721
0·20	0·5329	0·1672	1·3338	10·011	10·000	0·1334	0·0931	0·1015	1·5254	0·0800
0·22	0·6096	0·1902	1·5342	10·067	10·040	0·1528	0·1187	0·1277	1·8780	0·0882
0·24	0·6857	0·2128	1·7354	10·124	10·080	0·1722	0·1468	0·1562	2·2548	0·0969
0·26	0·7612	0·2349	1·9374	10·180	10·120	0·1914	0·1772	0·1869	2·6545	0·1057
0·28	0·8363	0·2566	2·1402	10·237	10·160	0·2106	0·2100	0·2199	3·0760	0·1147
0·30	0·9108	0·2780	2·3438	10·294	10·200	0·2298	0·2450	0·2549	3·5183	0·1239
0·32	0·9848	0·2989	2·5482	10·350	10·240	0·2488	0·2821	0·2919	3·9806	0·1332
0·34	1·0583	0·3195	2·7534	10·407	10·280	0·2678	0·3213	0·3309	4·4623	0·1425
0·36	1·1313	0·3397	2·9594	10·463	10·320	0·2868	0·3625	0·3719	4·9627	0·1519
0·38	1·2039	0·3595	3·1662	10·520	10·360	0·3056	0·4057	0·4147	5·4813	0·1613
0·40	1·2760	0·3790	3·3738	10·576	10·400	0·3244	0·4508	0·4594	6·0175	0·1708
0·42	1·3476	0·3981	3·5822	10·633	10·440	0·3431	0·4978	0·5058	6·5709	0·1803
0·44	1·4187	0·4169	3·7914	10·689	10·480	0·3618	0·5466	0·5540	7·1412	0·1898
0·46	1·4894	0·4354	4·0014	10·746	10·520	0·3804	0·5973	0·6040	7·7279	0·1993
0·48	1·5597	0·4536	4·2122	10·803	10·560	0·3989	0·6497	0·6557	8·3308	0·2088
0·50	1·6295	0·4714	4·4238	10·859	10·600	0·4173	0·7039	0·7090	8·9494	0·2183
0·52	1·6989	0·4889	4·6362	10·916	10·640	0·4357	0·7598	0·7640	9·5836	0·2279
0·54	1·7679	0·5062	4·8494	10·972	10·680	0·4541	0·8174	0·8206	10·233	0·2374
0·56	1·8364	0·5231	5·0634	11·029	10·720	0·4723	0·8766	0·8788	10·897	0·2470
0·58	1·9045	0·5397	5·2782	11·085	10·760	0·4905	0·9375	0·9386	11·577	0·2565
0·60	1·9723	0·5561	5·4938	11·142	10·800	0·5087	1·0000	1·0000	12·270	0·2660
0·64	2·1065	0·5880	5·9274	11·255	10·880	0·5448	1·1298	1·1274	13·701	0·2851
0·68	2·2393	0·6188	6·3642	11·368	10·960	0·5807	1·2658	1·2608	15·187	0·3042
0·72	2·3705	0·6486	6·8042	11·481	11·040	0·6163	1·4080	1·4001	16·728	0·3232
0·76	2·5003	0·6774	7·2474	11·595	11·120	0·6517	1·5561	1·5452	18·322	0·3422
0·80	2·6286	0·7053	7·6938	11·708	11·200	0·6869	1·7100	1·6961	19·969	0·3612
0·84	2·7556	0·7323	8·1434	11·821	11·280	0·7219	1·8697	1·8525	21·668	0·3801
0·88	2·8812	0·7585	8·5962	11·934	11·360	0·7567	2·0351	2·0145	23·417	0·3991
0·92	3·0056	0·7838	9·0522	12·047	11·440	0·7913	2·2061	2·1820	25·216	0·4180
0·96	3·1287	0·8083	9·5114	12·160	11·520	0·8256	2·3826	2·3549	27·064	0·4368
1·00	3·2505	0·8320	9·9738	12·273	11·600	0·8598	2·5644	2·5331	28·961	0·4556

1·5% concave bed river   C78

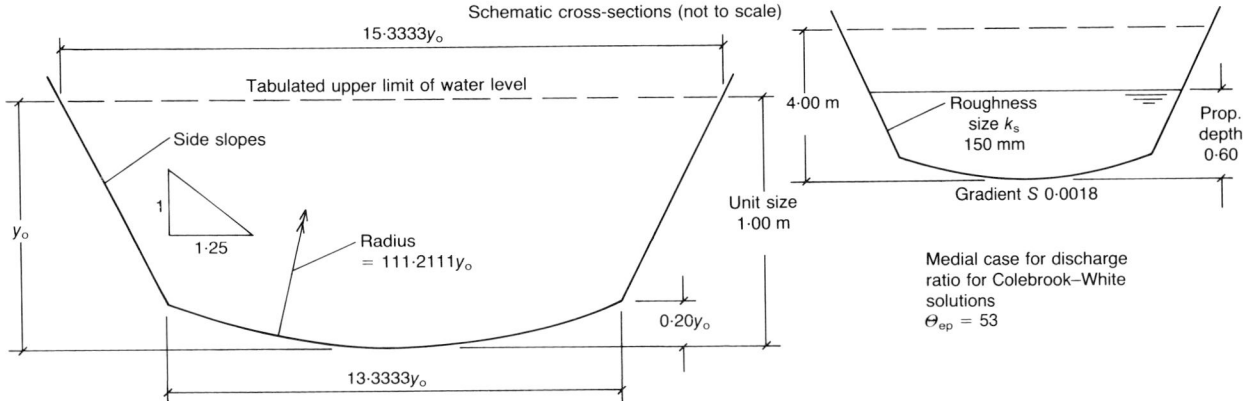

Prop dpth Y	U.equ. dia. $D_{ep(u)}$ (m)	Equiv. disch. factor J	Unit sect. area A_u (m²)	Unit wetted perim. P_u (m)	Unit surf. brdth $B_{s(u)}$ (m)	Unit mean depth $y_{m(u)}$ (m)	Discharge ratios $Q/Q_{0.60}$ medial	Mann'g	U.crit. disch. $Q_{c(u)}$ (m³s⁻¹)	Unit depth cntrd $y_{d(u)}$ (m)
0·02	0·0533	0·0397	0·0562	4·2183	4·2181	0·0133	0·0004	0·0007	0·0203	0·0080
0·04	0·1067	0·0562	0·1591	5·9657	5·9650	0·0267	0·0022	0·0031	0·0813	0·0160
0·06	0·1600	0·0688	0·2922	7·3066	7·3053	0·0400	0·0059	0·0074	0·1830	0·0240
0·08	0·2133	0·0794	0·4499	8·4370	8·4350	0·0533	0·0116	0·0138	0·3254	0·0320
0·10	0·2666	0·0888	0·6287	9·4330	9·4302	0·0667	0·0194	0·0224	0·5084	0·0400
0·12	0·3199	0·0973	0·8264	10·334	10·330	0·0800	0·0295	0·0333	0·7320	0·0480
0·14	0·3732	0·1050	1·0415	11·162	11·157	0·0933	0·0420	0·0465	0·9965	0·0560
0·16	0·4265	0·1123	1·2724	11·932	11·927	0·1067	0·0569	0·0621	1·3014	0·0640
0·18	0·4798	0·1191	1·5182	12·657	12·650	0·1200	0·0742	0·0802	1·6471	0·0721
0·20	0·5331	0·1255	1·7781	13·341	13·333	0·1334	0·0941	0·1007	2·0334	0·0800
0·22	0·6103	0·1430	2·0453	13·405	13·383	0·1528	0·1198	0·1268	2·5038	0·0882
0·24	0·6870	0·1602	2·3134	13·469	13·433	0·1722	0·1479	0·1552	3·0064	0·0969
0·26	0·7633	0·1772	2·5826	13·533	13·483	0·1915	0·1784	0·1858	3·5395	0·1057
0·28	0·8392	0·1939	2·8528	13·597	13·533	0·2108	0·2112	0·2187	4·1016	0·1148
0·30	0·9147	0·2103	3·1239	13·661	13·583	0·2300	0·2462	0·2536	4·6915	0·1239
0·32	0·9897	0·2265	3·3961	13·726	13·633	0·2491	0·2832	0·2906	5·3080	0·1332
0·34	1·0644	0·2425	3·6693	13·790	13·683	0·2682	0·3224	0·3296	5·9502	0·1425
0·36	1·1386	0·2582	3·9434	13·854	13·733	0·2871	0·3636	0·3705	6·6173	0·1519
0·38	1·2124	0·2737	4·2186	13·918	13·783	0·3061	0·4067	0·4133	7·3086	0·1614
0·40	1·2859	0·2889	4·4948	13·982	13·833	0·3249	0·4517	0·4579	8·0234	0·1708
0·42	1·3590	0·3040	4·7719	14·046	13·883	0·3437	0·4986	0·5044	8·7610	0·1803
0·44	1·4317	0·3188	5·0501	14·110	13·933	0·3624	0·5474	0·5527	9·5210	0·1898
0·46	1·5040	0·3333	5·3293	14·174	13·983	0·3811	0·5980	0·6027	10·303	0·1994
0·48	1·5759	0·3477	5·6094	14·238	14·033	0·3997	0·6503	0·6545	11·106	0·2089
0·50	1·6475	0·3619	5·8906	14·302	14·083	0·4183	0·7044	0·7080	11·930	0·2185
0·52	1·7187	0·3759	6·1728	14·366	14·133	0·4368	0·7602	0·7631	12·775	0·2280
0·54	1·7896	0·3896	6·4559	14·430	14·183	0·4552	0·8177	0·8199	13·640	0·2376
0·56	1·8601	0·4032	6·7401	14·494	14·233	0·4735	0·8768	0·8783	14·525	0·2471
0·58	1·9303	0·4165	7·0253	14·558	14·283	0·4919	0·9376	0·9384	15·429	0·2567
0·60	2·0001	0·4297	7·3114	14·622	14·333	0·5101	1·0000	1·0000	16·353	0·2663
0·64	2·1388	0·4555	7·8868	14·750	14·433	0·5464	1·1296	1·1280	18·257	0·2854
0·68	2·2761	0·4806	8·4661	14·878	14·533	0·5825	1·2655	1·2621	20·235	0·3045
0·72	2·4122	0·5050	9·0494	15·006	14·633	0·6184	1·4075	1·4023	22·285	0·3236
0·76	2·5470	0·5287	9·6368	15·134	14·733	0·6541	1·5554	1·5485	24·407	0·3426
0·80	2·6806	0·5518	10·228	15·262	14·833	0·6895	1·7093	1·7005	26·597	0·3617
0·84	2·8130	0·5742	10·823	15·390	14·933	0·7248	1·8689	1·8583	28·856	0·3807
0·88	2·9443	0·5960	11·423	15·518	15·033	0·7598	2·0343	2·0217	31·181	0·3996
0·92	3·0745	0·6173	12·026	15·646	15·133	0·7947	2·2052	2·1908	33·572	0·4186
0·96	3·2035	0·6380	12·633	15·775	15·233	0·8293	2·3817	2·3654	36·028	0·4375
1·00	3·3315	0·6581	13·245	15·903	15·333	0·8638	2·5637	2·5454	38·549	0·4564

C79

1·25% concave bed river

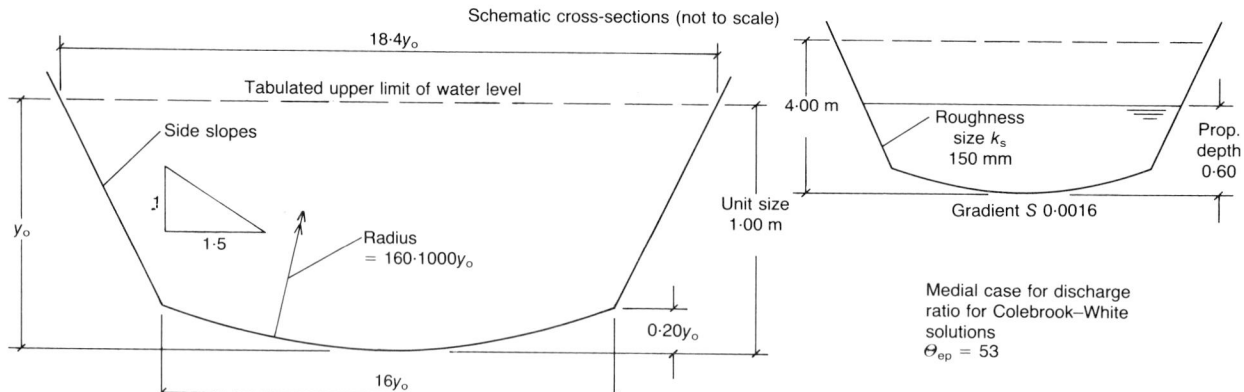

Schematic cross-sections (not to scale)

18·4y_o

Tabulated upper limit of water level

Side slopes

1 : 1·5

Radius = 160·1000y_o

16y_o

0·20y_o

4·00 m

Roughness size k_s 150 mm

Unit size 1·00 m

Gradient S 0·0016

Prop. depth 0·60

Medial case for discharge ratio for Colebrook–White solutions $\Theta_{ep} = 53$

Prop dpth Y	U.equ. dia. $D_{ep(u)}$ (m)	Equiv. disch. factor J	Unit sect. area A_u (m²)	Unit wetted perim. P_u (m)	Unit surf. brdth $B_{s(u)}$ (m)	Unit mean depth $y_{m(u)}$ (m)	Discharge ratios $Q/Q_{0.60}$ medial	Mann'g	U.crit. disch. $Q_{c(u)}$ (m³s⁻¹)	Unit depth cntrd $y_{d(u)}$ (m)
0·02	0·0533	0·0331	0·0675	5·0613	5·0611	0·0133	0·0004	0·0007	0·0244	0·0080
0·04	0·1067	0·0468	0·1909	7·1578	7·1572	0·0267	0·0022	0·0031	0·0976	0·0160
0·06	0·1600	0·0573	0·3506	8·7666	8·7655	0·0400	0·0059	0·0074	0·2196	0·0240
0·08	0·2133	0·0662	0·5398	10·123	10·121	0·0533	0·0115	0·0138	0·3904	0·0320
0·10	0·2666	0·0740	0·7544	11·318	11·315	0·0667	0·0193	0·0224	0·6100	0·0400
0·12	0·3200	0·0811	0·9917	12·398	12·395	0·0800	0·0294	0·0332	0·8785	0·0480
0·14	0·3733	0·0876	1·2497	13·392	13·388	0·0933	0·0418	0·0463	1·1957	0·0560
0·16	0·4266	0·0936	1·5268	14·316	14·312	0·1067	0·0567	0·0619	1·5617	0·0640
0·18	0·4799	0·0993	1·8217	15·185	15·179	0·1200	0·0739	0·0799	1·9762	0·0721
0·20	0·5332	0·1046	2·1336	16·007	16·000	0·1334	0·0938	0·1004	2·4399	0·0800
0·22	0·6105	0·1193	2·4542	16·079	16·060	0·1528	0·1194	0·1263	3·0044	0·0882
0·24	0·6875	0·1337	2·7760	16·151	16·120	0·1722	0·1475	0·1547	3·6075	0·0969
0·26	0·7641	0·1480	3·0990	16·223	16·180	0·1915	0·1779	0·1853	4·2472	0·1057
0·28	0·8403	0·1620	3·4232	16·295	16·240	0·2108	0·2106	0·2181	4·9217	0·1148
0·30	0·9161	0·1758	3·7486	16·367	16·300	0·2300	0·2455	0·2529	5·6295	0·1239
0·32	0·9916	0·1895	4·0752	16·439	16·360	0·2491	0·2825	0·2899	6·3693	0·1332
0·34	1·0667	0·2029	4·4030	16·511	16·420	0·2681	0·3217	0·3288	7·1400	0·1425
0·36	1·1414	0·2162	4·7320	16·584	16·480	0·2871	0·3628	0·3697	7·9405	0·1519
0·38	1·2157	0·2293	5·0622	16·656	16·540	0·3061	0·4059	0·4125	8·7700	0·1614
0·40	1·2897	0·2422	5·3936	16·728	16·600	0·3249	0·4509	0·4571	9·6277	0·1708
0·42	1·3634	0·2549	5·7262	16·800	16·660	0·3437	0·4979	0·5036	10·513	0·1803
0·44	1·4367	0·2675	6·0600	16·872	16·720	0·3624	0·5467	0·5519	11·425	0·1898
0·46	1·5097	0·2799	6·3950	16·944	16·780	0·3811	0·5973	0·6020	12·363	0·1994
0·48	1·5823	0·2921	6·7312	17·016	16·840	0·3997	0·6497	0·6538	13·327	0·2089
0·50	1·6546	0·3042	7·0686	17·088	16·900	0·4183	0·7038	0·7074	14·316	0·2185
0·52	1·7266	0·3161	7·4072	17·160	16·960	0·4367	0·7597	0·7626	15·330	0·2280
0·54	1·7982	0·3278	7·7470	17·233	17·020	0·4552	0·8173	0·8195	16·367	0·2376
0·56	1·8696	0·3394	8·0880	17·305	17·080	0·4735	0·8765	0·8780	17·429	0·2471
0·58	1·9406	0·3508	8·4302	17·377	17·140	0·4918	0·9375	0·9382	18·514	0·2567
0·60	2·0113	0·3621	8·7736	17·449	17·200	0·5101	1·0000	1·0000	19·623	0·2663
0·64	2·1518	0·3842	9·4640	17·593	17·320	0·5464	1·1300	1·1284	21·908	0·2854
0·68	2·2910	0·4058	10·159	17·737	17·440	0·5825	1·2663	1·2630	24·282	0·3045
0·72	2·4291	0·4268	10·859	17·882	17·560	0·6184	1·4088	1·4037	26·742	0·3236
0·76	2·5661	0·4472	11·564	18·026	17·680	0·6541	1·5574	1·5505	29·287	0·3426
0·80	2·7019	0·4672	12·274	18·170	17·800	0·6895	1·7120	1·7032	31·916	0·3617
0·84	2·8367	0·4866	12·988	18·314	17·920	0·7248	1·8724	1·8618	34·626	0·3807
0·88	2·9704	0·5055	13·707	18·458	18·040	0·7598	2·0386	2·0261	37·417	0·3996
0·92	3·1030	0·5240	14·431	18·603	18·160	0·7947	2·2106	2·1962	40·286	0·4186
0·96	3·2347	0·5421	15·160	18·747	18·280	0·8293	2·3882	2·3719	43·234	0·4375
1·00	3·3653	0·5596	15·894	18·891	18·400	0·8638	2·5713	2·5532	46·258	0·4564

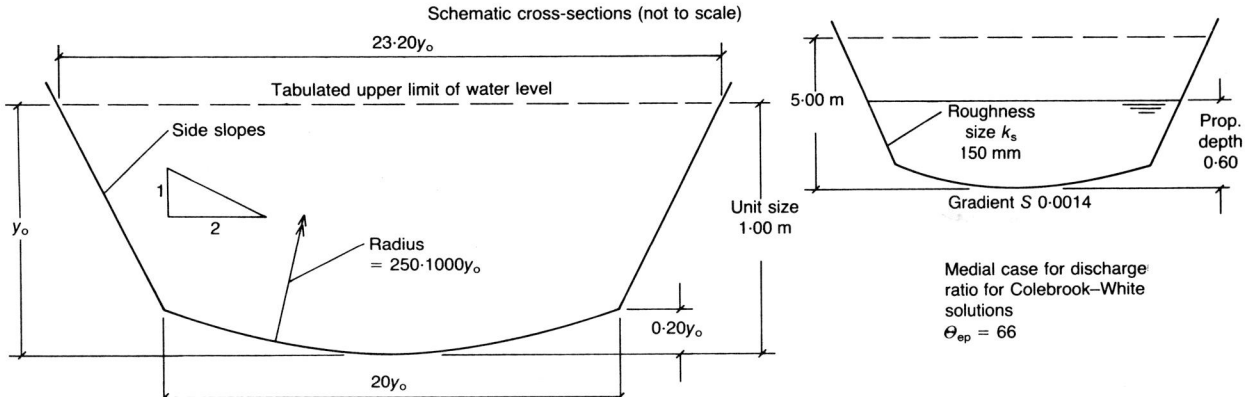

Schematic cross-sections (not to scale)

23·20y_o

Tabulated upper limit of water level

Side slopes

y_o

Radius = 250·1000y_o

0·20y_o

20y_o

5·00 m

Roughness size k_s 150 mm

Prop. depth 0·60

Unit size 1·00 m

Gradient S 0·0014

Medial case for discharge ratio for Colebrook–White solutions $\Theta_{ep} = 66$

Prop dpth Y	U.equ. dia. $D_{ep(u)}$ (m)	Equiv. disch. factor J	Unit sect. area A_u (m²)	Unit wetted perim. P_u (m)	Unit surf. brdth $B_{s(u)}$ (m)	Unit mean depth $y_{m(u)}$ (m)	Discharge ratios $Q/Q_{0.60}$ medial	Mann'g	U.crit. disch. $Q_{c(u)}$ (m³s⁻¹)	Unit depth cntrd $y_{d(u)}$ (m)
0·02	0·0533	0·0265	0·0844	6·3259	6·3257	0·0133	0·0004	0·0007	0·0305	–·0577
0·04	0·1067	0·0375	0·2385	8·9462	8·9457	0·0267	0·0023	0·0031	0·1220	0·0300
0·06	0·1600	0·0459	0·4383	10·957	10·956	0·0400	0·0061	0·0074	0·2745	0·0087
0·08	0·2133	0·0530	0·6747	12·652	12·651	0·0533	0·0118	0·0137	0·4879	0·0569
0·10	0·2666	0·0592	0·9428	14·145	14·144	0·0667	0·0198	0·0223	0·7623	0·0696
0·12	0·3200	0·0649	1·2395	15·496	15·493	0·0800	0·0299	0·0331	1·0979	0·0428
0·14	0·3733	0·0701	1·5621	16·737	16·734	0·0933	0·0425	0·0462	1·4946	0·0346
0·16	0·4266	0·0749	1·9084	17·893	17·889	0·1067	0·0574	0·0617	1·9520	0·0513
0·18	0·4799	0·0794	2·2770	18·979	18·974	0·1200	0·0748	0·0796	2·4702	0·0778
0·20	0·5332	0·0837	2·6669	20·005	20·000	0·1333	0·0947	0·1000	3·0497	0·0801
0·22	0·6106	0·0955	3·0677	20·095	20·080	0·1528	0·1204	0·1259	3·7549	0·0883
0·24	0·6877	0·1070	3·4701	20·184	20·160	0·1721	0·1485	0·1542	4·5084	0·0969
0·26	0·7644	0·1184	3·8741	20·274	20·240	0·1914	0·1789	0·1847	5·3077	0·1058
0·28	0·8407	0·1297	4·2797	20·363	20·320	0·2106	0·2116	0·2174	6·1506	0·1148
0·30	0·9166	0·1408	4·6869	20·453	20·400	0·2297	0·2465	0·2522	7·0351	0·1240
0·32	0·9922	0·1517	5·0957	20·542	20·480	0·2488	0·2835	0·2891	7·9597	0·1332
0·34	1·0675	0·1625	5·5061	20·631	20·560	0·2678	0·3226	0·3280	8·9230	0·1425
0·36	1·1424	0·1732	5·9181	20·721	20·640	0·2867	0·3637	0·3688	9·9238	0·1519
0·38	1·2170	0·1837	6·3317	20·810	20·720	0·3056	0·4067	0·4116	10·961	0·1613
0·40	1·2913	0·1941	6·7469	20·900	20·800	0·3244	0·4517	0·4562	12·033	0·1708
0·42	1·3652	0·2043	7·1637	20·989	20·880	0·3431	0·4986	0·5027	13·140	0·1803
0·44	1·4388	0·2144	7·5821	21·079	20·960	0·3617	0·5473	0·5510	14·281	0·1898
0·46	1·5121	0·2244	8·0021	21·168	21·040	0·3803	0·5978	0·6012	15·454	0·1993
0·48	1·5851	0·2342	8·4237	21·258	21·120	0·3988	0·6501	0·6530	16·660	0·2088
0·50	1·6577	0·2440	8·8469	21·347	21·200	0·4173	0·7042	0·7066	17·897	0·2183
0·52	1·7301	0·2535	9·2717	21·436	21·280	0·4357	0·7600	0·7620	19·165	0·2279
0·54	1·8021	0·2630	9·6981	21·526	21·360	0·4540	0·8175	0·8190	20·464	0·2374
0·56	1·8739	0·2723	10·126	21·615	21·440	0·4723	0·8767	0·8777	21·793	0·2470
0·58	1·9453	0·2816	10·556	21·705	21·520	0·4905	0·9375	0·9380	23·151	0·2565
0·60	2·0165	0·2907	10·987	21·794	21·600	0·5087	1·0000	1·0000	24·538	0·2660
0·64	2·1579	0·3085	11·854	21·973	21·760	0·5448	1·1299	1·1288	27·399	0·2851
0·68	2·2982	0·3259	12·728	22·152	21·920	0·5806	1·2661	1·2640	30·371	0·3042
0·72	2·4375	0·3429	13·608	22·331	22·080	0·6163	1·4085	1·4054	33·453	0·3232
0·76	2·5756	0·3595	14·494	22·510	22·240	0·6517	1·5571	1·5530	36·642	0·3422
0·80	2·7127	0·3756	15·387	22·689	22·400	0·6869	1·7117	1·7067	39·936	0·3612
0·84	2·8488	0·3914	16·286	22·867	22·560	0·7219	1·8723	1·8663	43·333	0·3802
0·88	2·9838	0·4067	17·192	23·046	22·720	0·7567	2·0387	2·0319	46·831	0·3991
0·92	3·1179	0·4217	18·104	23·225	22·880	0·7912	2·2109	2·2033	50·429	0·4180
0·96	3·2511	0·4364	19·022	23·404	23·040	0·8256	2·3888	2·3805	54·126	0·4368
1·00	3·3833	0·4507	19·947	23·583	23·200	0·8598	2·5723	2·5635	57·920	0·4556

0·75% concave bed river

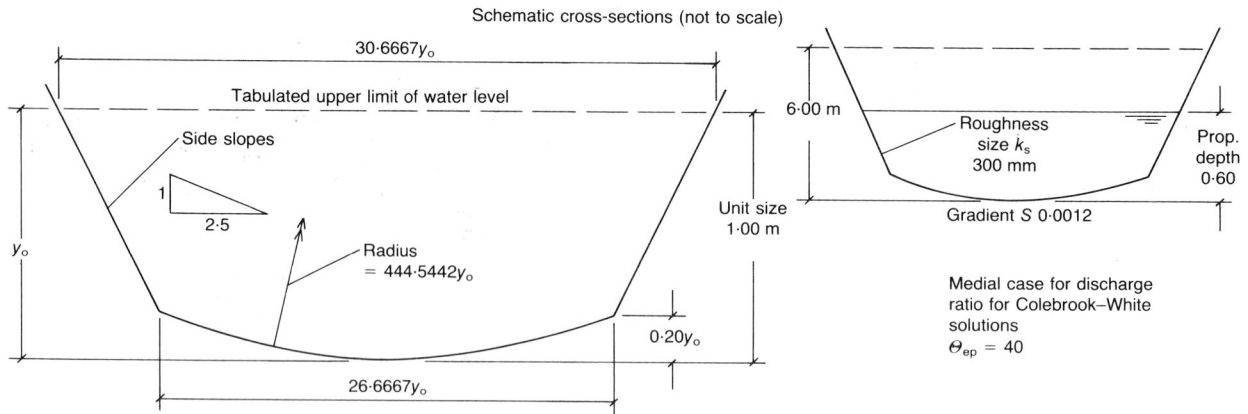

Schematic cross-sections (not to scale)

Prop dpth	U.equ. dia. $D_{ep(u)}$	Equiv. disch. factor	Unit sect. area	Unit wetted perim.	Unit surf. brdth	Unit mean depth	Discharge ratios $Q/Q_{0.60}$		U.crit. disch. $Q_{c(u)}$	Unit depth cntrd
Y	(m)	J	A_u (m^2)	P_u (m)	$B_{s(u)}$ (m)	$y_{m(u)}$ (m)	medial	Mann'g	(m^{3}s^{-1})	$y_{d(u)}$ (m)
0·02	0·0533	0·0198	0·1123	8·4337	8·4336	0·0133	0·0003	0·0007	0·0406	0·0080
0·04	0·1067	0·0281	0·3180	11·927	11·927	0·0267	0·0021	0·0031	0·1626	0·0160
0·06	0·1600	0·0344	0·5843	14·608	14·607	0·0400	0·0055	0·0073	0·3659	0·0240
0·08	0·2133	0·0397	0·8994	16·868	16·867	0·0533	0·0110	0·0137	0·6503	0·0320
0·10	0·2667	0·0444	1·2574	18·859	18·857	0·0667	0·0186	0·0222	1·0168	0·0400
0·12	0·3199	0·0487	1·6523	20·659	20·657	0·0800	0·0283	0·0330	1·4634	0·0480
0·14	0·3733	0·0526	2·0824	22·314	22·312	0·0933	0·0404	0·0461	1·9923	0·0560
0·16	0·4266	0·0562	2·5441	23·855	23·852	0·1067	0·0549	0·0615	2·6019	0·0640
0·18	0·4800	0·0596	3·0362	25·302	25·298	0·1200	0·0719	0·0794	3·2938	0·0721
0·20	0·5333	0·0628	3·5557	26·671	26·667	0·1333	0·0914	0·0998	4·0660	0·0800
0·22	0·6109	0·0717	4·0900	26·778	26·767	0·1528	0·1167	0·1257	5·0068	0·0884
0·24	0·6883	0·0804	4·6264	26·886	26·867	0·1722	0·1445	0·1539	6·0120	0·0970
0·26	0·7653	0·0891	5·1647	26·994	26·967	0·1915	0·1747	0·1844	7·0781	0·1058
0·28	0·8420	0·0976	5·7050	27·101	27·067	0·2108	0·2072	0·2171	8·2022	0·1149
0·30	0·9184	0·1060	6·2474	27·209	27·167	0·2300	0·2420	0·2519	9·3819	0·1240
0·32	0·9945	0·1144	6·7917	27·317	27·267	0·2491	0·2789	0·2887	10·615	0·1333
0·34	1·0703	0·1226	7·3380	27·425	27·367	0·2681	0·3180	0·3276	11·899	0·1426
0·36	1·1458	0·1307	7·8864	27·532	27·467	0·2871	0·3591	0·3685	13·233	0·1520
0·38	1·2209	0·1388	8·4367	27·640	27·567	0·3060	0·4022	0·4112	14·616	0·1614
0·40	1·2958	0·1467	8·9890	27·748	27·667	0·3249	0·4473	0·4559	16·045	0·1709
0·42	1·3704	0·1546	9·5434	27·855	27·767	0·3437	0·4944	0·5024	17·521	0·1804
0·44	1·4447	0·1623	10·100	27·963	27·867	0·3624	0·5433	0·5507	19·041	0·1899
0·46	1·5187	0·1700	10·658	28·071	27·967	0·3811	0·5941	0·6009	20·604	0·1994
0·48	1·5925	0·1775	11·218	28·179	28·067	0·3997	0·6468	0·6528	22·211	0·2090
0·50	1·6659	0·1850	11·781	28·286	28·167	0·4183	0·7013	0·7064	23·859	0·2185
0·52	1·7391	0·1924	12·345	28·394	28·267	0·4367	0·7576	0·7618	25·548	0·2281
0·54	1·8120	0·1997	12·911	28·502	28·367	0·4552	0·8156	0·8188	27·278	0·2376
0·56	1·8847	0·2070	13·480	28·609	28·467	0·4735	0·8754	0·8776	29·048	0·2472
0·58	1·9570	0·2141	14·050	28·717	28·567	0·4918	0·9368	0·9380	30·857	0·2568
0·60	2·0291	0·2212	14·622	28·825	28·667	0·5101	1·0000	1·0000	32·704	0·2663
0·64	2·1726	0·2350	15·773	29·040	28·867	0·5464	1·1314	1·1289	36·512	0·2854
0·68	2·3150	0·2486	16·932	29·256	29·067	0·5825	1·2694	1·2643	40·468	0·3045
0·72	2·4564	0·2618	18·098	29·471	29·267	0·6184	1·4138	1·4059	44·569	0·3236
0·76	2·5969	0·2748	19·273	29·686	29·467	0·6541	1·5645	1·5537	48·811	0·3427
0·80	2·7364	0·2875	20·456	29·902	29·667	0·6895	1·7215	1·7075	53·192	0·3617
0·84	2·8750	0·2999	21·646	30·117	29·867	0·7248	1·8846	1·8674	57·709	0·3807
0·88	3·0126	0·3120	22·845	30·333	30·067	0·7598	2·0538	2·0333	62·360	0·3997
0·92	3·1494	0·3239	24·052	30·548	30·267	0·7947	2·2290	2·2050	67·142	0·4186
0·96	3·2853	0·3355	25·266	30·763	30·467	0·8293	2·4101	2·3825	72·055	0·4375
1·00	3·4203	0·3468	26·489	30·979	30·667	0·8638	2·5970	2·5657	77·095	0·4564

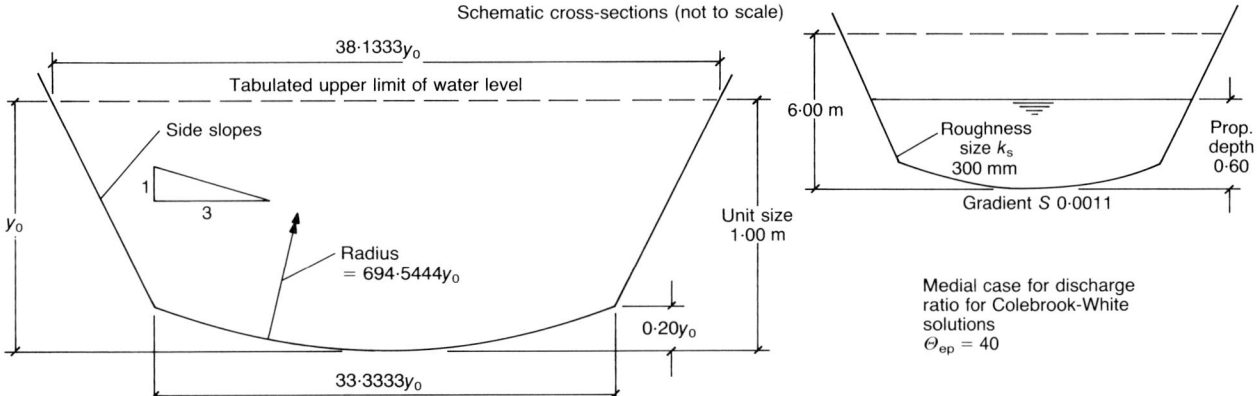

Prop dpth Y	U.equ. dia. $D_{ep(u)}$ (m)	Equiv. disch. factor J	Unit sect. area A_u (m²)	Unit wetted perim. P_u (m)	Unit surf. brdth $B_{s(u)}$ (m)	Unit mean depth $y_{m(u)}$ (m)	Discharge ratios $Q/Q_{0.60}$ medial	Mann'g	U.crit. disch. $Q_{c(u)}$ (m³s⁻¹)	Unit depth cntrd $y_{d(u)}$ (m)
0·02	0·0533	0·0159	0·1405	10·542	10·542	0·0133	0·0003	0·0007	0·0508	0·0080
0·04	0·1067	0·0225	0·3975	14·908	14·908	0·0267	0·0021	0·0030	0·2033	0·0160
0·06	0·1599	0·0275	0·7299	18·259	18·258	0·0400	0·0055	0·0073	0·4570	0·0240
0·08	0·2133	0·0318	1·1242	21·084	21·083	0·0533	0·0110	0·0137	0·8129	0·0320
0·10	0·2666	0·0355	1·5713	23·572	23·571	0·0667	0·0185	0·0222	1·2705	0·0400
0·12	0·3200	0·0389	2·0659	25·822	25·821	0·0800	0·0283	0·0330	1·8299	0·0480
0·14	0·3734	0·0421	2·6034	27·891	27·889	0·0933	0·0404	0·0460	2·4908	0·0560
0·16	0·4266	0·0449	3·1800	29·817	29·815	0·1067	0·0549	0·0614	3·2523	0·0640
0·18	0·4800	0·0477	3·7952	31·626	31·623	0·1200	0·0718	0·0793	4·1174	0·0721
0·20	0·5333	0·0503	4·4446	33·337	33·333	0·1333	0·0913	0·0996	5·0824	0·0800
0·22	0·6111	0·0574	5·1124	33·463	33·453	0·1528	0·1166	0·1255	6·2587	0·0882
0·24	0·6886	0·0644	5·7827	33·590	33·573	0·1722	0·1444	0·1537	7·5155	0·0969
0·26	0·7659	0·0714	6·4554	33·716	33·693	0·1916	0·1746	0·1842	8·8485	0·1057
0·28	0·8428	0·0782	7·1304	33·842	33·813	0·2109	0·2071	0·2169	10·254	0·1148
0·30	0·9194	0·0850	7·8079	33·969	33·933	0·2301	0·2418	0·2517	11·729	0·1239
0·32	0·9958	0·0917	8·4878	34·095	34·053	0·2492	0·2787	0·2886	13·270	0·1332
0·34	1·0718	0·0984	9·1700	34·222	34·173	0·2683	0·3178	0·3274	14·876	0·1426
0·36	1·1476	0·1050	9·8547	34·348	34·293	0·2874	0·3589	0·3683	16·543	0·1520
0·38	1·2231	0·1115	10·542	34·475	34·413	0·3063	0·4020	0·4110	18·271	0·1614
0·40	1·2984	0·1179	11·231	34·601	34·533	0·3252	0·4471	0·4557	20·058	0·1709
0·42	1·3733	0·1242	11·923	34·728	34·653	0·3441	0·4942	0·5022	21·901	0·1804
0·44	1·4480	0·1305	12·617	34·854	34·773	0·3628	0·5432	0·5506	23·801	0·1899
0·46	1·5224	0·1367	13·314	34·981	34·893	0·3816	0·5940	0·6007	25·755	0·1994
0·48	1·5966	0·1429	14·013	35·107	35·013	0·4002	0·6467	0·6526	27·762	0·2090
0·50	1·6705	0·1489	14·715	35·234	35·133	0·4188	0·7012	0·7063	29·821	0·2186
0·52	1·7441	0·1550	15·418	35·360	35·253	0·4374	0·7575	0·7617	31·932	0·2281
0·54	1·8175	0·1609	16·125	35·487	35·373	0·4558	0·8155	0·8187	34·093	0·2377
0·56	1·8907	0·1668	16·833	35·613	35·493	0·4743	0·8753	0·8775	36·303	0·2473
0·58	1·9636	0·1726	17·544	35·740	35·613	0·4926	0·9368	0·9379	38·562	0·2568
0·60	2·0362	0·1784	18·258	35·866	35·733	0·5109	1·0000	1·0000	40·870	0·2664
0·64	2·1808	0·1897	19·692	36·119	35·973	0·5474	1·1315	1·1290	45·625	0·2855
0·68	2·3244	0·2008	21·136	36·372	36·213	0·5836	1·2695	1·2644	50·565	0·3047
0·72	2·4670	0·2116	22·589	36·625	36·453	0·6197	1·4140	1·4061	55·685	0·3238
0·76	2·6088	0·2222	24·052	36·878	36·693	0·6555	1·5648	1·5540	60·981	0·3429
0·80	2·7497	0·2326	25·525	37·131	36·933	0·6911	1·7219	1·7080	66·449	0·3619
0·84	2·8896	0·2428	27·007	37·384	37·173	0·7265	1·8851	1·8680	72·086	0·3810
0·88	3·0287	0·2528	28·498	37·637	37·413	0·7617	2·0544	2·0339	77·889	0·4000
0·92	3·1670	0·2626	30·000	37·890	37·653	0·7967	2·2297	2·2057	83·856	0·4190
0·96	3·3045	0·2722	31·511	38·143	37·893	0·8316	2·4109	2·3834	89·984	0·4379
1·00	3·4411	0·2815	33·031	38·396	38·133	0·8662	2·5980	2·5668	96·271	0·4568

C83

0·50% concave bed river

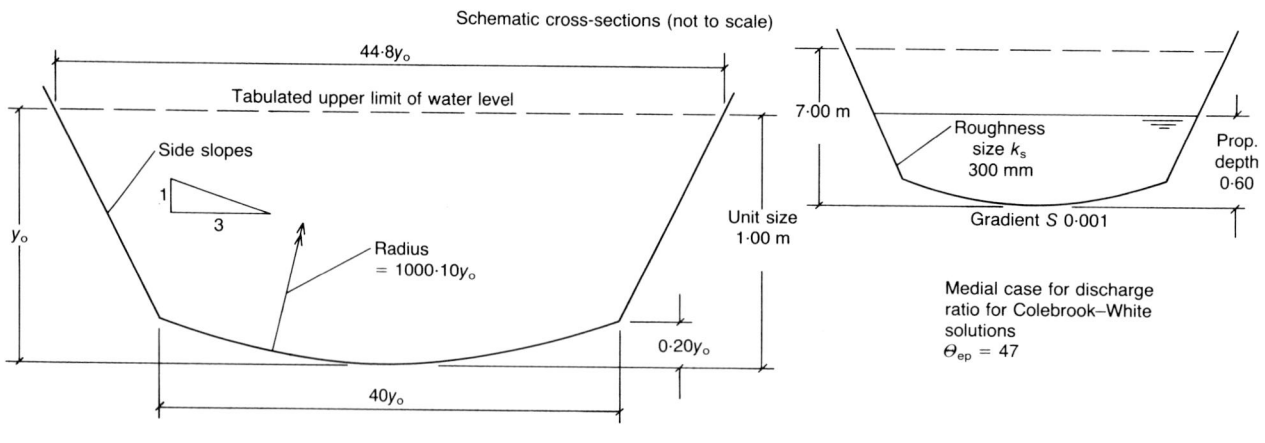

Schematic cross-sections (not to scale)

44·8y₀

Tabulated upper limit of water level

Side slopes
1
3

Radius = 1000·10y₀

y₀

40y₀

0·20y₀

Unit size 1·00 m

7·00 m

Roughness size k_s 300 mm

Prop. depth 0·60

Gradient S 0·001

Medial case for discharge ratio for Colebrook–White solutions $\Theta_{ep} = 47$

Prop dpth Y	U.equ. dia. $D_{ep(u)}$ (m)	Equiv. disch. factor J	Unit sect. area A_u (m²)	Unit wetted perim. P_u (m)	Unit surf. brdth $B_{s(u)}$ (m)	Unit mean depth $y_{m(u)}$ (m)	Discharge ratios $Q/Q_{0.60}$ medial	Mann'g	U.crit. disch. $Q_{c(u)}$ (m³s⁻¹)	Unit depth cntrd $y_{d(u)}$ (m)
0·02	0·0533	0·0132	0·1685	12·650	12·650	0·0133	0·0004	0·0007	0·0609	0·0080
0·04	0·1067	0·0187	0·4770	17·889	17·889	0·0267	0·0021	0·0030	0·2439	0·0160
0·06	0·1599	0·0229	0·8757	21·910	21·910	0·0400	0·0057	0·0073	0·5482	0·0240
0·08	0·2134	0·0265	1·3496	25·300	25·299	0·0533	0·0112	0·0137	0·9761	0·0320
0·10	0·2666	0·0296	1·8854	28·286	28·285	0·0667	0·0189	0·0222	1·5244	0·0400
0·12	0·3200	0·0324	2·4790	30·986	30·984	0·0800	0·0288	0·0329	2·1959	0·0480
0·14	0·3732	0·0350	3·1229	33·468	33·467	0·0933	0·0410	0·0460	2·9873	0·0560
0·16	0·4266	0·0375	3·8160	35·779	35·777	0·1067	0·0556	0·0614	3·9027	0·0640
0·18	0·4800	0·0397	4·5541	37·950	37·948	0·1200	0·0727	0·0793	4·9405	0·0721
0·20	0·5333	0·0419	5·3334	40·003	40·000	0·1333	0·0922	0·0996	6·0988	0·0800
0·22	0·6115	0·0479	6·1346	40·129	40·120	0·1529	0·1176	0·1255	7·5121	0·0882
0·24	0·6894	0·0538	6·9382	40·256	40·240	0·1724	0·1455	0·1537	9·0221	0·0969
0·26	0·7671	0·0597	7·7442	40·382	40·360	0·1919	0·1758	0·1843	10·623	0·1057
0·28	0·8445	0·0655	8·5526	40·509	40·480	0·2113	0·2084	0·2170	12·311	0·1148
0·30	0·9217	0·0713	9·3634	40·635	40·600	0·2306	0·2433	0·2518	14·082	0·1240
0·32	0·9986	0·0770	10·177	40·762	40·720	0·2499	0·2802	0·2887	15·932	0·1333
0·34	1·0753	0·0826	10·992	40·888	40·840	0·2692	0·3193	0·3276	17·859	0·1427
0·36	1·1518	0·0882	11·810	41·015	40·960	0·2883	0·3605	0·3685	19·860	0·1521
0·38	1·2280	0·0938	12·631	41·141	41·080	0·3075	0·4036	0·4113	21·932	0·1616
0·40	1·3040	0·0993	13·453	41·268	41·200	0·3265	0·4487	0·4560	24·075	0·1711
0·42	1·3798	0·1047	14·279	41·394	41·320	0·3456	0·4957	0·5025	26·285	0·1806
0·44	1·4553	0·1101	15·106	41·521	41·440	0·3645	0·5446	0·5508	28·562	0·1902
0·46	1·5306	0·1155	15·936	41·647	41·560	0·3835	0·5954	0·6010	30·903	0·1997
0·48	1·6057	0·1208	16·769	41·774	41·680	0·4023	0·6479	0·6529	33·308	0·2093
0·50	1·6805	0·1260	17·603	41·900	41·800	0·4211	0·7023	0·7065	35·774	0·2189
0·52	1·7551	0·1312	18·441	42·027	41·920	0·4399	0·7584	0·7619	38·301	0·2285
0·54	1·8295	0·1363	19·280	42·153	42·040	0·4586	0·8163	0·8189	40·888	0·2381
0·56	1·9037	0·1415	20·122	42·280	42·160	0·4773	0·8758	0·8776	43·534	0·2478
0·58	1·9777	0·1465	20·967	42·406	42·280	0·4959	0·9371	0·9380	46·237	0·2574
0·60	2·0515	0·1515	21·813	42·532	42·400	0·5145	1·0000	1·0000	48·996	0·2670
0·64	2·1983	0·1614	23·514	42·785	42·640	0·5515	1·1308	1·1288	54·682	0·2862
0·68	2·3444	0·1711	25·225	43·038	42·880	0·5883	1·2681	1·2640	60·586	0·3055
0·72	2·4896	0·1807	26·945	43·291	43·120	0·6249	1·4117	1·4054	66·701	0·3247
0·76	2·6340	0·1900	28·674	43·544	43·360	0·6613	1·5615	1·5529	73·022	0·3439
0·80	2·7776	0·1992	30·413	43·797	43·600	0·6976	1·7174	1·7064	79·545	0·3631
0·84	2·9205	0·2083	32·162	44·050	43·840	0·7336	1·8793	1·8659	86·267	0·3823
0·88	3·0626	0·2172	33·921	44·303	44·080	0·7695	2·0471	2·0312	93·183	0·4014
0·92	3·2039	0·2259	35·689	44·556	44·320	0·8052	2·2208	2·2023	100·29	0·4205
0·96	3·3445	0·2345	37·466	44·809	44·560	0·8408	2·4002	2·3792	107·58	0·4396
1·00	3·4844	0·2429	39·253	45·062	44·800	0·8762	2·5853	2·5617	115·06	0·4587

0·40% concave bed river

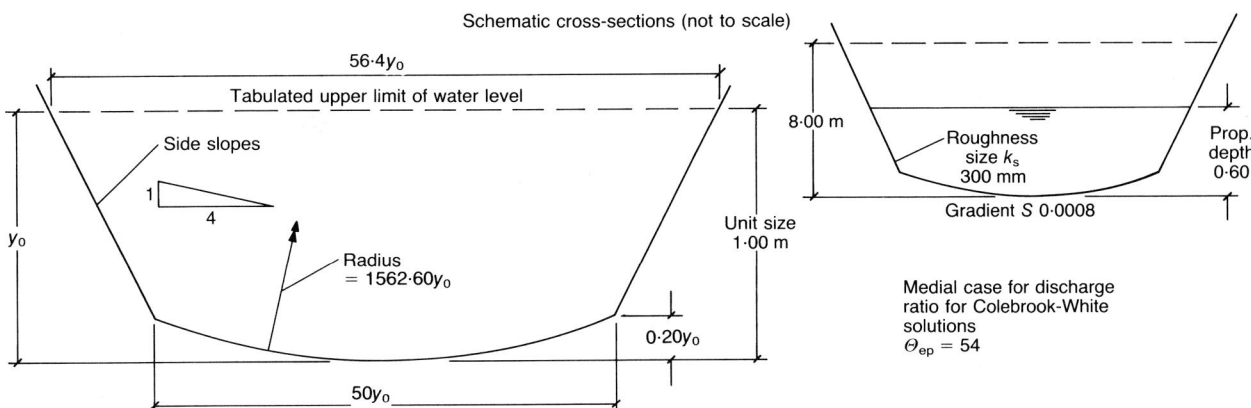

Schematic cross-sections (not to scale)

56·4y₀

Tabulated upper limit of water level

Side slopes

1 / 4

Radius = 1562·60y₀

0·20y₀

50y₀

y₀

Unit size 1·00 m

8·00 m

Roughness size k_s 300 mm

Prop. depth 0·60

Gradient S 0·0008

Medial case for discharge ratio for Colebrook-White solutions $\Theta_{ep} = 54$

Prop dpth Y	U.equ. dia. $D_{ep(u)}$ (m)	Equiv. disch. factor J	Unit sect. area A_u (m²)	Unit wetted perim. P_u (m)	Unit surf. brdth $B_{s(u)}$ (m)	Unit mean depth $y_{m(u)}$ (m)	Discharge ratios $Q/Q_{0.60}$ medial	Mann'g	U.crit. disch. $Q_{c(u)}$ (m³s⁻¹)	Unit depth cntrd $y_{d(u)}$ (m)
0·02	0·0534	0·0106	0·2110	15·812	15·812	0·0133	0·0004	0·0007	0·0763	0·0080
0·04	0·1067	0·0150	0·5962	22·361	22·361	0·0267	0·0022	0·0030	0·3049	0·0160
0·06	0·1600	0·0184	1·0958	27·387	27·387	0·0400	0·0058	0·0073	0·6864	0·0240
0·08	0·2133	0·0212	1·6867	31·624	31·623	0·0533	0·0114	0·0137	1·2198	0·0320
0·10	0·2668	0·0237	2·3579	35·357	35·356	0·0667	0·0192	0·0222	1·9069	0·0400
0·12	0·3197	0·0259	3·0957	38·731	38·730	0·0799	0·0291	0·0329	2·7408	0·0480
0·14	0·3735	0·0280	3·9059	41·835	41·833	0·0934	0·0415	0·0460	3·7375	0·0560
0·16	0·4266	0·0300	4·7699	44·723	44·722	0·1067	0·0561	0·0614	4·8783	0·0640
0·18	0·4799	0·0318	5·6908	47·436	47·434	0·1200	0·0733	0·0792	6·1726	0·0721
0·20	0·5333	0·0335	6·6668	50·002	50·000	0·1333	0·0930	0·0995	7·6234	0·0800
0·22	0·6114	0·0383	7·6684	50·167	50·160	0·1529	0·1185	0·1254	9·3893	0·0882
0·24	0·6893	0·0430	8·6732	50·332	50·320	0·1724	0·1464	0·1536	11·276	0·0969
0·26	0·7669	0·0477	9·6812	50·497	50·480	0·1918	0·1767	0·1841	13·277	0·1057
0·28	0·8442	0·0523	10·692	50·662	50·640	0·2111	0·2093	0·2168	15·386	0·1148
0·30	0·9213	0·0569	11·707	50·827	50·800	0·2304	0·2442	0·2516	17·599	0·1240
0·32	0·9982	0·0615	12·724	50·992	50·960	0·2497	0·2812	0·2885	19·911	0·1333
0·34	1·0748	0·0660	13·745	51·157	51·120	0·2689	0·3202	0·3274	22·320	0·1426
0·36	1·1511	0·0705	14·769	51·322	51·280	0·2880	0·3613	0·3682	24·821	0·1520
0·38	1·2272	0·0749	15·796	51·486	51·440	0·3071	0·4044	0·4110	27·412	0·1615
0·40	1·3031	0·0793	16·827	51·651	51·600	0·3261	0·4495	0·4557	30·091	0·1710
0·42	1·3787	0·0836	17·860	51·816	51·760	0·3451	0·4965	0·5022	32·855	0·1805
0·44	1·4542	0·0879	18·897	51·981	51·920	0·3640	0·5453	0·5506	35·702	0·1901
0·46	1·5293	0·0921	19·937	52·146	52·080	0·3828	0·5960	0·6007	38·630	0·1996
0·48	1·6043	0·0963	20·980	52·311	52·240	0·4016	0·6485	0·6526	41·637	0·2092
0·50	1·6790	0·1005	22·027	52·476	52·400	0·4204	0·7027	0·7063	44·722	0·2188
0·52	1·7535	0·1046	23·076	52·641	52·560	0·4390	0·7588	0·7617	47·883	0·2284
0·54	1·8278	0·1087	24·129	52·806	52·720	0·4577	0·8166	0·8188	51·119	0·2380
0·56	1·9018	0·1128	25·185	52·971	52·880	0·4763	0·8760	0·8775	54·429	0·2476
0·58	1·9756	0·1168	26·244	53·136	53·040	0·4948	0·9372	0·9379	57·811	0·2572
0·60	2·0493	0·1208	27·307	53·301	53·200	0·5133	1·0000	1·0000	61·265	0·2668
0·64	2·1959	0·1286	29·441	53·630	53·520	0·5501	1·1306	1·1290	68·381	0·2860
0·68	2·3416	0·1363	31·588	53·960	53·840	0·5867	1·2676	1·2643	75·770	0·3052
0·72	2·4865	0·1439	33·748	54·290	54·160	0·6231	1·4110	1·4060	83·426	0·3244
0·76	2·6306	0·1513	35·921	54·620	54·480	0·6593	1·5606	1·5538	91·341	0·3436
0·80	2·7739	0·1586	38·107	54·950	54·800	0·6954	1·7162	1·7076	99·511	0·3627
0·84	2·9165	0·1657	40·305	55·280	55·120	0·7312	1·8779	1·8675	107·93	0·3818
0·88	3·0582	0·1728	42·516	55·610	55·440	0·7669	2·0455	2·0333	116·60	0·4009
0·92	3·1992	0·1797	44·740	55·939	55·760	0·8024	2·2190	2·2049	125·50	0·4200
0·96	3·3395	0·1864	46·977	56·269	56·080	0·8377	2·3982	2·3823	134·64	0·4391
1·00	3·4790	0·1931	49·227	56·599	56·400	0·8728	2·5831	2·5655	144·02	0·4581

C85

0·30% concave bed river

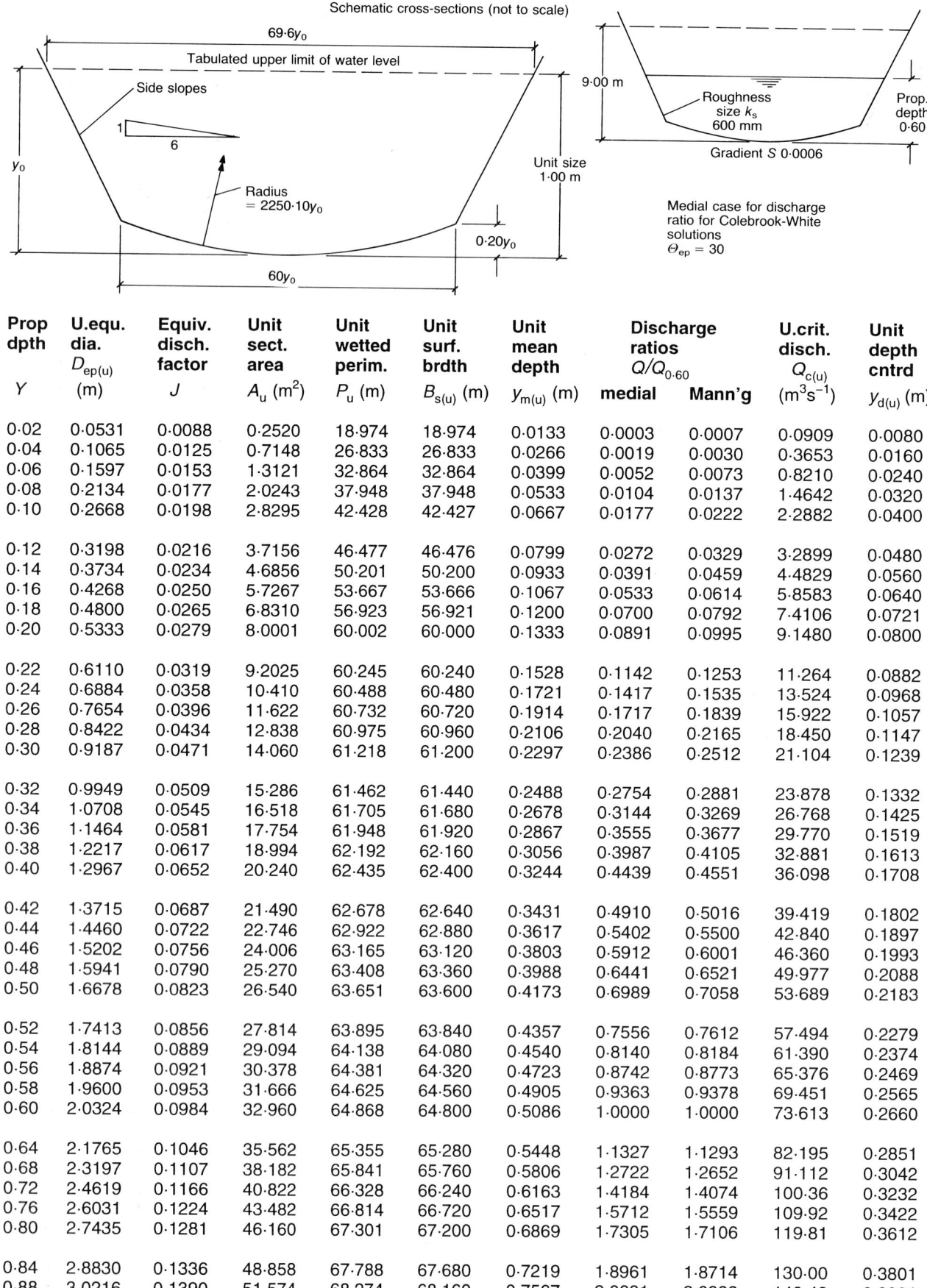

Schematic cross-sections (not to scale)

Prop dpth Y	U.equ. dia. $D_{ep(u)}$ (m)	Equiv. disch. factor J	Unit sect. area A_u (m²)	Unit wetted perim. P_u (m)	Unit surf. brdth $B_{s(u)}$ (m)	Unit mean depth $y_{m(u)}$ (m)	Discharge ratios $Q/Q_{0.60}$ medial	Discharge ratios $Q/Q_{0.60}$ Mann'g	U.crit. disch. $Q_{c(u)}$ (m³s⁻¹)	Unit depth cntrd $y_{d(u)}$ (m)
0·02	0·0531	0·0088	0·2520	18·974	18·974	0·0133	0·0003	0·0007	0·0909	0·0080
0·04	0·1065	0·0125	0·7148	26·833	26·833	0·0266	0·0019	0·0030	0·3653	0·0160
0·06	0·1597	0·0153	1·3121	32·864	32·864	0·0399	0·0052	0·0073	0·8210	0·0240
0·08	0·2134	0·0177	2·0243	37·948	37·948	0·0533	0·0104	0·0137	1·4642	0·0320
0·10	0·2668	0·0198	2·8295	42·428	42·427	0·0667	0·0177	0·0222	2·2882	0·0400
0·12	0·3198	0·0216	3·7156	46·477	46·476	0·0799	0·0272	0·0329	3·2899	0·0480
0·14	0·3734	0·0234	4·6856	50·201	50·200	0·0933	0·0391	0·0459	4·4829	0·0560
0·16	0·4268	0·0250	5·7267	53·667	53·666	0·1067	0·0533	0·0614	5·8583	0·0640
0·18	0·4800	0·0265	6·8310	56·923	56·921	0·1200	0·0700	0·0792	7·4106	0·0721
0·20	0·5333	0·0279	8·0001	60·002	60·000	0·1333	0·0891	0·0995	9·1480	0·0800
0·22	0·6110	0·0319	9·2025	60·245	60·240	0·1528	0·1142	0·1253	11·264	0·0882
0·24	0·6884	0·0358	10·410	60·488	60·480	0·1721	0·1417	0·1535	13·524	0·0968
0·26	0·7654	0·0396	11·622	60·732	60·720	0·1914	0·1717	0·1839	15·922	0·1057
0·28	0·8422	0·0434	12·838	60·975	60·960	0·2106	0·2040	0·2165	18·450	0·1147
0·30	0·9187	0·0471	14·060	61·218	61·200	0·2297	0·2386	0·2512	21·104	0·1239
0·32	0·9949	0·0509	15·286	61·462	61·440	0·2488	0·2754	0·2881	23·878	0·1332
0·34	1·0708	0·0545	16·518	61·705	61·680	0·2678	0·3144	0·3269	26·768	0·1425
0·36	1·1464	0·0581	17·754	61·948	61·920	0·2867	0·3555	0·3677	29·770	0·1519
0·38	1·2217	0·0617	18·994	62·192	62·160	0·3056	0·3987	0·4105	32·881	0·1613
0·40	1·2967	0·0652	20·240	62·435	62·400	0·3244	0·4439	0·4551	36·098	0·1708
0·42	1·3715	0·0687	21·490	62·678	62·640	0·3431	0·4910	0·5016	39·419	0·1802
0·44	1·4460	0·0722	22·746	62·922	62·880	0·3617	0·5402	0·5500	42·840	0·1897
0·46	1·5202	0·0756	24·006	63·165	63·120	0·3803	0·5912	0·6001	46·360	0·1993
0·48	1·5941	0·0790	25·270	63·408	63·360	0·3988	0·6441	0·6521	49·977	0·2088
0·50	1·6678	0·0823	26·540	63·651	63·600	0·4173	0·6989	0·7058	53·689	0·2183
0·52	1·7413	0·0856	27·814	63·895	63·840	0·4357	0·7556	0·7612	57·494	0·2279
0·54	1·8144	0·0889	29·094	64·138	64·080	0·4540	0·8140	0·8184	61·390	0·2374
0·56	1·8874	0·0921	30·378	64·381	64·320	0·4723	0·8742	0·8773	65·376	0·2469
0·58	1·9600	0·0953	31·666	64·625	64·560	0·4905	0·9363	0·9378	69·451	0·2565
0·60	2·0324	0·0984	32·960	64·868	64·800	0·5086	1·0000	1·0000	73·613	0·2660
0·64	2·1765	0·1046	35·562	65·355	65·280	0·5448	1·1327	1·1293	82·195	0·2851
0·68	2·3197	0·1107	38·182	65·841	65·760	0·5806	1·2722	1·2652	91·112	0·3042
0·72	2·4619	0·1166	40·822	66·328	66·240	0·6163	1·4184	1·4074	100·36	0·3232
0·76	2·6031	0·1224	43·482	66·814	66·720	0·6517	1·5712	1·5559	109·92	0·3422
0·80	2·7435	0·1281	46·160	67·301	67·200	0·6869	1·7305	1·7106	119·81	0·3612
0·84	2·8830	0·1336	48·858	67·788	67·680	0·7219	1·8961	1·8714	130·00	0·3801
0·88	3·0216	0·1390	51·574	68·274	68·160	0·7567	2·0681	2·0383	140·49	0·3991
0·92	3·1594	0·1443	54·310	68·761	68·640	0·7912	2·2462	2·2111	151·29	0·4179
0·96	3·2963	0·1495	57·066	69·248	69·120	0·8256	2·4306	2·3900	162·38	0·4368
1·00	3·4325	0·1546	59·840	69·734	69·600	0·8598	2·6211	2·5747	173·76	0·4556

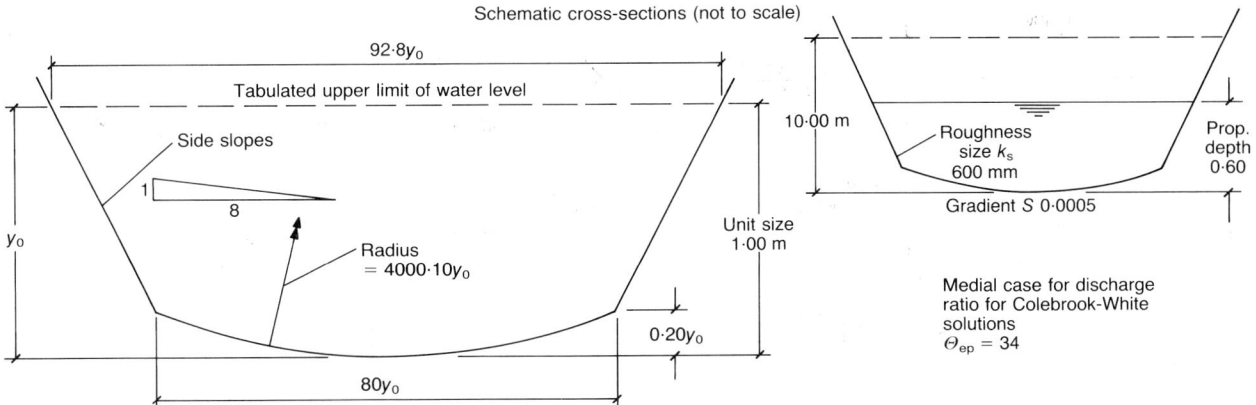

Schematic cross-sections (not to scale)

Prop dpth	U.equ. dia. $D_{ep(u)}$	Equiv. disch. factor	Unit sect. area	Unit wetted perim.	Unit surf. brdth	Unit mean depth	Discharge ratios $Q/Q_{0.60}$		U.crit. disch. $Q_{c(u)}$	Unit depth cntrd
Y	(m)	J	A_u (m²)	P_u (m)	$B_{s(u)}$ (m)	$y_{m(u)}$ (m)	medial	Mann'g	(m³s⁻¹)	$y_{d(u)}$ (m)
0·02	0·0533	0·0066	0·3369	25·299	25·299	0·0133	0·0003	0·0007	0·1218	0·0080
0·04	0·1064	0·0093	0·9521	35·778	35·777	0·0266	0·0019	0·0030	0·4864	0·0160
0·06	0·1597	0·0114	1·7494	43·818	43·818	0·0399	0·0053	0·0073	1·0947	0·0240
0·08	0·2137	0·0133	2·7029	50·597	50·597	0·0534	0·0106	0·0137	1·9563	0·0320
0·10	0·2669	0·0148	3·7745	56·569	56·569	0·0667	0·0180	0·0222	3·0533	0·0400
0·12	0·3203	0·0162	4·9617	61·969	61·968	0·0801	0·0277	0·0329	4·3967	0·0480
0·14	0·3730	0·0175	6·2417	66·934	66·933	0·0933	0·0395	0·0459	5·9690	0·0560
0·16	0·4266	0·0187	7·6317	71·555	71·554	0·1067	0·0538	0·0613	7·8051	0·0640
0·18	0·4798	0·0199	9·1041	75·896	75·895	0·1200	0·0706	0·0791	9·8745	0·0721
0·20	0·5333	0·0209	10·667	80·001	80·000	0·1333	0·0899	0·0995	12·197	0·0800
0·22	0·6110	0·0239	12·270	80·324	80·320	0·1528	0·1150	0·1253	15·018	0·0882
0·24	0·6884	0·0268	13·880	80·646	80·640	0·1721	0·1426	0·1534	18·032	0·0968
0·26	0·7655	0·0297	15·496	80·969	80·960	0·1914	0·1726	0·1838	21·229	0·1057
0·28	0·8423	0·0326	17·118	81·291	81·280	0·2106	0·2050	0·2165	24·601	0·1147
0·30	0·9188	0·0354	18·747	81·614	81·600	0·2297	0·2396	0·2512	28·139	0·1239
0·32	0·9950	0·0381	20·382	81·936	81·920	0·2488	0·2764	0·2880	31·837	0·1332
0·34	1·0709	0·0409	22·024	82·259	82·240	0·2678	0·3154	0·3268	35·690	0·1425
0·36	1·1466	0·0436	23·672	82·581	82·560	0·2867	0·3565	0·3676	39·693	0·1519
0·38	1·2219	0·0463	25·326	82·904	82·880	0·3056	0·3997	0·4104	43·841	0·1613
0·40	1·2970	0·0490	26·987	83·226	83·200	0·3244	0·4448	0·4550	48·131	0·1708
0·42	1·3718	0·0516	28·654	83·549	83·520	0·3431	0·4919	0·5016	52·559	0·1802
0·44	1·4464	0·0542	30·328	83·871	83·840	0·3617	0·5410	0·5499	57·120	0·1897
0·46	1·5207	0·0567	32·008	84·194	84·160	0·3803	0·5920	0·6001	61·814	0·1993
0·48	1·5947	0·0593	33·694	84·516	84·480	0·3988	0·6448	0·6520	66·637	0·2088
0·50	1·6684	0·0618	35·387	84·839	84·800	0·4173	0·6995	0·7057	71·585	0·2183
0·52	1·7419	0·0643	37·086	85·161	85·120	0·4357	0·7561	0·7612	76·658	0·2279
0·54	1·8152	0·0667	38·792	85·484	85·440	0·4540	0·8144	0·8184	81·853	0·2374
0·56	1·8881	0·0691	40·504	85·806	85·760	0·4723	0·8745	0·8772	87·168	0·2469
0·58	1·9609	0·0715	42·222	86·129	86·080	0·4905	0·9364	0·9378	92·601	0·2565
0·60	2·0334	0·0739	43·947	86·451	86·400	0·5086	1·0000	1·0000	98·151	0·2660
0·64	2·1776	0·0785	47·416	87·096	87·040	0·5448	1·1324	1·1294	109·59	0·2851
0·68	2·3209	0·0831	50·910	87·741	87·680	0·5806	1·2716	1·2652	121·48	0·3042
0·72	2·4633	0·0876	54·430	88·386	88·320	0·6163	1·4174	1·4075	133·81	0·3232
0·76	2·6047	0·0919	57·976	89·031	88·960	0·6517	1·5697	1·5560	146·57	0·3422
0·80	2·7453	0·0962	61·547	89·676	89·600	0·6869	1·7284	1·7108	159·74	0·3612
0·84	2·8850	0·1003	65·144	90·321	90·240	0·7219	1·8936	1·8717	173·33	0·3801
0·88	3·0238	0·1044	68·766	90·966	90·880	0·7567	2·0649	2·0386	187·32	0·3991
0·92	3·1618	0·1084	72·414	91·611	91·520	0·7912	2·2425	2·2116	201·71	0·4179
0·96	3·2990	0·1123	76·088	92·256	92·160	0·8256	2·4262	2·3905	216·50	0·4368
1·00	3·4353	0·1162	79·787	92·901	92·800	0·8598	2·6160	2·5754	231·68	0·4556

1·0 to 1 tangent river

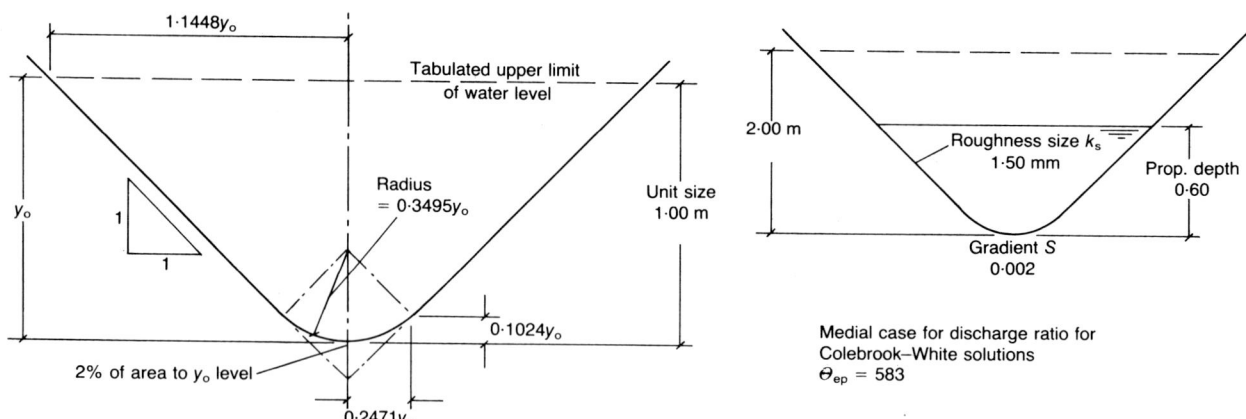

Medial case for discharge ratio for Colebrook–White solutions $\Theta_{ep} = 583$

Prop dpth Y	U.equ. dia. $D_{ep(u)}$ (m)	Equiv. disch. factor J	Unit sect. area A_u (m²)	Unit wetted perim. P_u (m)	Unit surf. brdth $B_{s(u)}$ (m)	Unit mean depth $y_{m(u)}$ (m)	Discharge ratios $Q/Q_{0.60}$ medial	Mann'g	U.crit. disch. $Q_{c(u)}$ (m³s⁻¹)	Unit depth cntrd $y_{d(u)}$ (m)
0·02	0·0526	0·6957	0·0031	0·2376	0·2331	0·0134	0·0008	0·0008	0·0011	0·0080
0·04	0·1038	0·9657	0·0088	0·3377	0·3247	0·0270	0·0037	0·0035	0·0045	0·0161
0·06	0·1535	1·1603	0·0160	0·4157	0·3916	0·0407	0·0087	0·0082	0·0101	0·0242
0·08	0·2018	1·3140	0·0243	0·4824	0·4450	0·0547	0·0159	0·0150	0·0178	0·0323
0·10	0·2486	1·4401	0·0337	0·5422	0·4895	0·0688	0·0252	0·0239	0·0277	0·0405
0·12	0·2931	1·5378	0·0439	0·5988	0·5295	0·0829	0·0366	0·0348	0·0396	0·0488
0·14	0·3349	1·6053	0·0549	0·6554	0·5695	0·0964	0·0498	0·0476	0·0533	0·0570
0·16	0·3745	1·6526	0·0667	0·7120	0·6095	0·1094	0·0649	0·0623	0·0690	0·0651
0·18	0·4125	1·6861	0·0793	0·7686	0·6495	0·1220	0·0821	0·0789	0·0867	0·0732
0·20	0·4491	1·7100	0·0926	0·8251	0·6895	0·1344	0·1013	0·0977	0·1064	0·0811
0·22	0·4847	1·7270	0·1068	0·8817	0·7295	0·1465	0·1225	0·1185	0·1280	0·0890
0·24	0·5194	1·7390	0·1218	0·9383	0·7695	0·1583	0·1460	0·1415	0·1518	0·0968
0·26	0·5533	1·7473	0·1376	0·9948	0·8095	0·1700	0·1716	0·1667	0·1777	0·1046
0·28	0·5867	1·7530	0·1542	1·0514	0·8495	0·1815	0·1995	0·1942	0·2057	0·1122
0·30	0·6195	1·7566	0·1716	1·1080	0·8895	0·1929	0·2297	0·2241	0·2360	0·1198
0·32	0·6519	1·7586	0·1898	1·1645	0·9295	0·2042	0·2623	0·2564	0·2686	0·1274
0·34	0·6839	1·7595	0·2088	1·2211	0·9695	0·2153	0·2973	0·2913	0·3034	0·1349
0·36	0·7156	1·7595	0·2286	1·2777	1·0095	0·2264	0·3348	0·3286	0·3406	0·1423
0·38	0·7470	1·7588	0·2492	1·3342	1·0495	0·2374	0·3749	0·3687	0·3802	0·1497
0·40	0·7781	1·7576	0·2705	1·3908	1·0895	0·2483	0·4176	0·4113	0·4222	0·1571
0·42	0·8090	1·7560	0·2927	1·4474	1·1295	0·2592	0·4630	0·4568	0·4667	0·1644
0·44	0·8397	1·7541	0·3157	1·5039	1·1695	0·2700	0·5110	0·5051	0·5137	0·1717
0·46	0·8703	1·7520	0·3395	1·5605	1·2095	0·2807	0·5619	0·5562	0·5633	0·1790
0·48	0·9007	1·7497	0·3641	1·6171	1·2495	0·2914	0·6156	0·6103	0·6155	0·1862
0·50	0·9309	1·7473	0·3895	1·6736	1·2895	0·3021	0·6721	0·6674	0·6704	0·1934
0·52	0·9610	1·7449	0·4157	1·7302	1·3295	0·3127	0·7316	0·7275	0·7279	0·2006
0·54	0·9910	1·7424	0·4427	1·7868	1·3695	0·3232	0·7941	0·7908	0·7882	0·2078
0·56	1·0209	1·7399	0·4705	1·8434	1·4095	0·3338	0·8596	0·8573	0·8512	0·2149
0·58	1·0507	1·7373	0·4991	1·8999	1·4495	0·3443	0·9283	0·9270	0·9170	0·2220
0·60	1·0804	1·7348	0·5285	1·9565	1·4895	0·3548	1·0000	1·0000	0·9857	0·2291
0·64	1·1396	1·7298	0·5896	2·0696	1·5695	0·3757	1·1532	1·1562	1·1317	0·2432
0·68	1·1985	1·7249	0·6540	2·1828	1·6495	0·3965	1·3195	1·3262	1·2896	0·2573
0·72	1·2572	1·7202	0·7216	2·2959	1·7295	0·4172	1·4994	1·5106	1·4596	0·2713
0·76	1·3157	1·7157	0·7924	2·4090	1·8095	0·4379	1·6932	1·7099	1·6420	0·2853
0·80	1·3740	1·7114	0·8664	2·5222	1·8895	0·4585	1·9013	1·9244	1·8371	0·2992
0·84	1·4321	1·7072	0·9435	2·6353	1·9695	0·4791	2·1242	2·1545	2·0451	0·3131
0·88	1·4902	1·7033	1·0239	2·7485	2·0495	0·4996	2·3620	2·4008	2·2664	0·3269
0·92	1·5481	1·6995	1·1075	2·8616	2·1295	0·5201	2·6153	2·6636	2·5011	0·3407
0·96	1·6059	1·6959	1·1943	2·9747	2·2095	0·5405	2·8843	2·9434	2·7496	0·3545
1·00	1·6636	1·6925	1·2843	3·0879	2·2895	0·5609	3·1695	3·2406	3·0121	0·3683

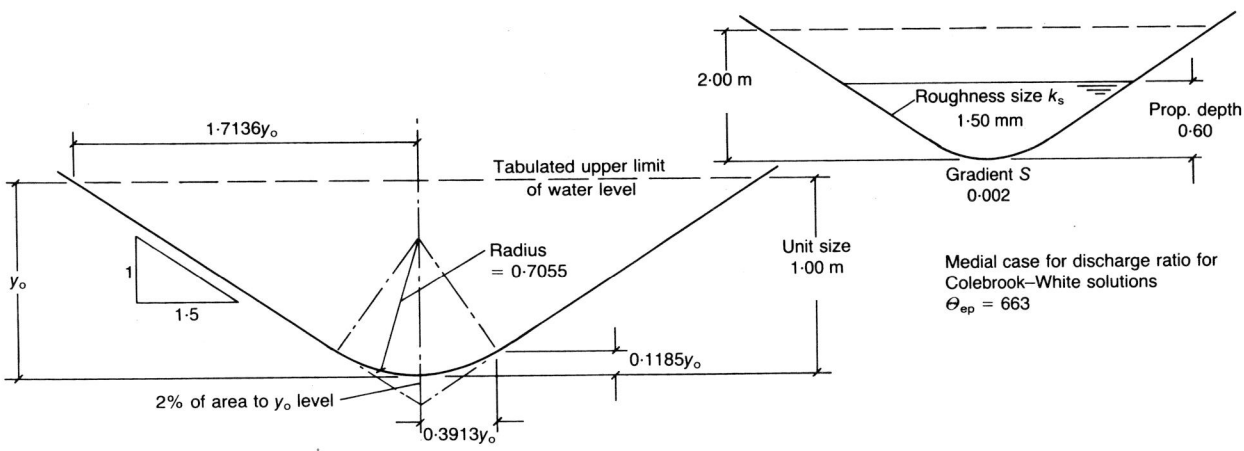

Prop dpth	U.equ. dia. $D_{ep(u)}$	Equiv. disch. factor	Unit sect. area	Unit wetted perim.	Unit surf. brdth	Unit mean depth	Discharge ratios $Q/Q_{0.60}$		U.crit. disch. $Q_{c(u)}$	Unit depth cntrd
Y	(m)	J	A_u (m²)	P_u (m)	$B_{s(u)}$ (m)	$y_{m(u)}$ (m)	medial	Mann'g	(m³s⁻¹)	$y_{d(u)}$ (m)
0·02	0·0530	0·4942	0·0045	0·3368	0·3336	0·0134	0·0007	0·0007	0·0016	0·0080
0·04	0·1053	0·6926	0·0126	0·4774	0·4684	0·0268	0·0033	0·0031	0·0064	0·0160
0·06	0·1568	0·8405	0·0230	0·5861	0·5694	0·0404	0·0079	0·0074	0·0145	0·0241
0·08	0·2077	0·9615	0·0352	0·6785	0·6526	0·0540	0·0146	0·0137	0·0256	0·0322
0·10	0·2578	1·0649	0·0490	0·7604	0·7242	0·0677	0·0233	0·0220	0·0399	0·0403
0·12	0·3072	1·1555	0·0641	0·8351	0·7872	0·0815	0·0341	0·0323	0·0573	0·0484
0·14	0·3548	1·2287	0·0805	0·9072	0·8472	0·0950	0·0470	0·0447	0·0777	0·0565
0·16	0·4003	1·2842	0·0980	0·9793	0·9072	0·1080	0·0618	0·0590	0·1009	0·0646
0·18	0·4442	1·3272	0·1168	1·0514	0·9672	0·1207	0·0786	0·0753	0·1270	0·0726
0·20	0·4867	1·3608	0·1367	1·1236	1·0272	0·1331	0·0975	0·0937	0·1562	0·0805
0·22	0·5281	1·3875	0·1579	1·1957	1·0872	0·1452	0·1185	0·1142	0·1884	0·0884
0·24	0·5685	1·4088	0·1802	1·2678	1·1472	0·1571	0·1417	0·1369	0·2236	0·0962
0·26	0·6082	1·4261	0·2037	1·3399	1·2072	0·1688	0·1672	0·1620	0·2621	0·1039
0·28	0·6473	1·4401	0·2285	1·4120	1·2672	0·1803	0·1949	0·1893	0·3038	0·1115
0·30	0·6857	1·4515	0·2544	1·4841	1·3272	0·1917	0·2250	0·2191	0·3489	0·1191
0·32	0·7237	1·4610	0·2816	1·5562	1·3872	0·2030	0·2576	0·2513	0·3973	0·1267
0·34	0·7613	1·4688	0·3099	1·6283	1·4472	0·2141	0·2926	0·2861	0·4491	0·1342
0·36	0·7985	1·4752	0·3395	1·7004	1·5072	0·2252	0·3302	0·3235	0·5045	0·1416
0·38	0·8354	1·4806	0·3702	1·7726	1·5672	0·2362	0·3703	0·3636	0·5635	0·1490
0·40	0·8720	1·4851	0·4022	1·8447	1·6272	0·2471	0·4132	0·4065	0·6261	0·1564
0·42	0·9084	1·4888	0·4353	1·9168	1·6872	0·2580	0·4587	0·4521	0·6924	0·1637
0·44	0·9445	1·4919	0·4696	1·9889	1·7472	0·2688	0·5070	0·5006	0·7625	0·1710
0·46	0·9805	1·4945	0·5052	2·0610	1·8072	0·2795	0·5582	0·5521	0·8364	0·1783
0·48	1·0162	1·4966	0·5419	2·1331	1·8672	0·2902	0·6122	0·6066	0·9143	0·1855
0·50	1·0518	1·4984	0·5799	2·2052	1·9272	0·3009	0·6692	0·6641	0·9961	0·1927
0·52	1·0873	1·4998	0·6190	2·2773	1·9872	0·3115	0·7291	0·7248	1·0819	0·1999
0·54	1·1226	1·5010	0·6594	2·3494	2·0472	0·3221	0·7922	0·7886	1·1718	0·2070
0·56	1·1578	1·5020	0·7009	2·4216	2·1072	0·3326	0·8583	0·8557	1·2659	0·2142
0·58	1·1929	1·5028	0·7436	2·4937	2·1672	0·3431	0·9275	0·9262	1·3641	0·2213
0·60	1·2278	1·5034	0·7876	2·5658	2·2272	0·3536	1·0000	1·0000	1·4667	0·2284
0·64	1·2975	1·5041	0·8791	2·7100	2·3472	0·3745	1·1548	1·1580	1·6847	0·2425
0·68	1·3669	1·5045	0·9754	2·8542	2·4672	0·3953	1·3230	1·3303	1·9205	0·2566
0·72	1·4360	1·5045	1·0765	2·9984	2·5872	0·4161	1·5051	1·5172	2·1744	0·2706
0·76	1·5049	1·5043	1·1823	3·1427	2·7072	0·4367	1·7014	1·7193	2·4469	0·2846
0·80	1·5736	1·5040	1·2930	3·2869	2·8272	0·4574	1·9123	1·9370	2·7384	0·2985
0·84	1·6421	1·5035	1·4085	3·4311	2·9472	0·4779	2·1381	2·1708	3·0493	0·3123
0·88	1·7104	1·5029	1·5288	3·5753	3·0672	0·4984	2·3794	2·4212	3·3800	0·3262
0·92	1·7786	1·5022	1·6539	3·7196	3·1872	0·5189	2·6364	2·6884	3·7309	0·3400
0·96	1·8467	1·5015	1·7838	3·8638	3·3072	0·5394	2·9095	2·9731	4·1025	0·3538
1·00	1·9146	1·5007	1·9185	4·0080	3·4272	0·5598	3·1990	3·2756	4·4950	0·3675

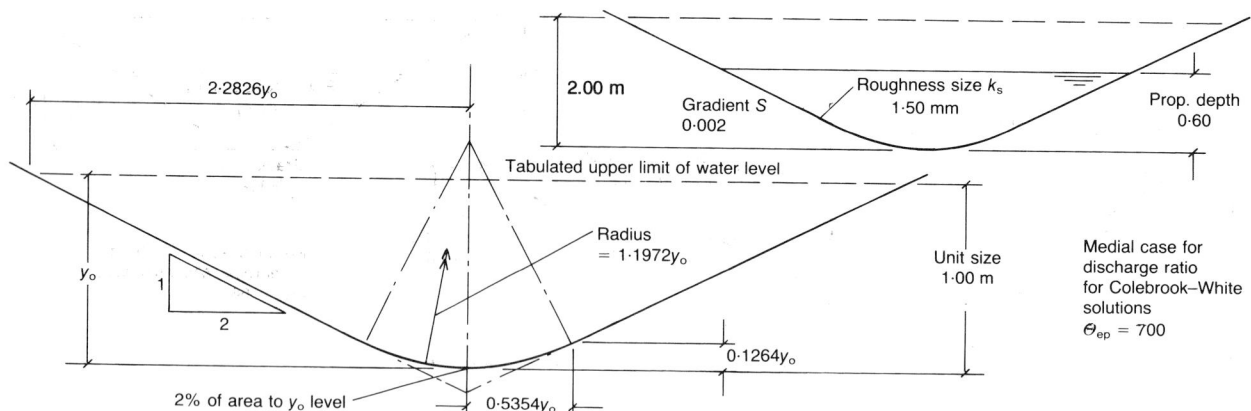

Prop dpth	U.equ. dia.	Equiv. disch. factor	Unit sect. area	Unit wetted perim.	Unit surf. brdth	Unit mean depth	Discharge ratios $Q/Q_{0.60}$		U.crit. disch.	Unit depth cntrd
Y	$D_{ep(u)}$ (m)	J	A_u (m²)	P_u (m)	$B_{s(u)}$ (m)	$y_{m(u)}$ (m)	medial	Mann'g	$Q_{c(u)}$ (m³s⁻¹)	$y_{d(u)}$ (m)
0·02	0·0531	0·3808	0·0058	0·4383	0·4358	0·0134	0·0007	0·0007	0·0021	0·0080
0·04	0·1058	0·5357	0·0164	0·6207	0·6138	0·0268	0·0032	0·0030	0·0084	0·0160
0·06	0·1581	0·6525	0·0301	0·7613	0·7485	0·0402	0·0076	0·0071	0·0189	0·0241
0·08	0·2100	0·7494	0·0462	0·8803	0·8606	0·0537	0·0140	0·0131	0·0335	0·0321
0·10	0·2614	0·8333	0·0644	0·9856	0·9580	0·0672	0·0225	0·0212	0·0523	0·0401
0·12	0·3125	0·9079	0·0845	1·0812	1·0449	0·0808	0·0330	0·0312	0·0752	0·0482
0·14	0·3627	0·9729	0·1062	1·1710	1·1252	0·0944	0·0457	0·0434	0·1021	0·0563
0·16	0·4109	1·0240	0·1295	1·2605	1·2052	0·1074	0·0603	0·0575	0·1329	0·0643
0·18	0·4574	1·0646	0·1544	1·3499	1·2852	0·1201	0·0770	0·0736	0·1675	0·0723
0·20	0·5027	1·0971	0·1809	1·4393	1·3652	0·1325	0·0957	0·0918	0·2062	0·0802
0·22	0·5468	1·1236	0·2090	1·5288	1·4452	0·1446	0·1166	0·1122	0·2489	0·0881
0·24	0·5900	1·1454	0·2387	1·6182	1·5252	0·1565	0·1397	0·1348	0·2957	0·0959
0·26	0·6324	1·1635	0·2700	1·7077	1·6052	0·1682	0·1651	0·1597	0·3468	0·1036
0·28	0·6742	1·1786	0·3029	1·7971	1·6852	0·1797	0·1928	0·1870	0·4021	0·1112
0·30	0·7154	1·1913	0·3374	1·8866	1·7652	0·1911	0·2228	0·2167	0·4619	0·1188
0·32	0·7561	1·2021	0·3735	1·9760	1·8452	0·2024	0·2553	0·2489	0·5262	0·1264
0·34	0·7964	1·2113	0·4112	2·0654	1·9252	0·2136	0·2904	0·2837	0·5951	0·1339
0·36	0·8363	1·2192	0·4505	2·1549	2·0052	0·2247	0·3279	0·3211	0·6687	0·1413
0·38	0·8759	1·2260	0·4914	2·2443	2·0852	0·2357	0·3681	0·3612	0·7471	0·1487
0·40	0·9151	1·2319	0·5339	2·3338	2·1652	0·2466	0·4110	0·4041	0·8303	0·1561
0·42	0·9542	1·2370	0·5780	2·4232	2·2452	0·2574	0·4567	0·4499	0·9185	0·1634
0·44	0·9930	1·2415	0·6237	2·5127	2·3252	0·2682	0·5051	0·4985	1·0117	0·1707
0·46	1·0315	1·2454	0·6710	2·6021	2·4052	0·2790	0·5564	0·5501	1·1100	0·1779
0·48	1·0699	1·2488	0·7199	2·6915	2·4852	0·2897	0·6106	0·6048	1·2135	0·1852
0·50	1·1082	1·2518	0·7705	2·7810	2·5652	0·3003	0·6678	0·6625	1·3223	0·1924
0·52	1·1463	1·2545	0·8226	2·8704	2·6452	0·3110	0·7279	0·7234	1·4364	0·1996
0·54	1·1842	1·2569	0·8763	2·9599	2·7252	0·3215	0·7912	0·7876	1·5560	0·2067
0·56	1·2220	1·2590	0·9316	3·0493	2·8052	0·3321	0·8576	0·8550	1·6811	0·2138
0·58	1·2597	1·2608	0·9885	3·1388	2·8852	0·3426	0·9272	0·9258	1·8118	0·2209
0·60	1·2973	1·2625	1·0470	3·2282	2·9652	0·3531	1·0000	1·0000	1·9482	0·2280
0·64	1·3722	1·2652	1·1688	3·4071	3·1252	0·3740	1·1556	1·1589	2·2383	0·2422
0·68	1·4467	1·2674	1·2970	3·5860	3·2852	0·3948	1·3247	1·3322	2·5520	0·2562
0·72	1·5210	1·2692	1·4316	3·7649	3·4452	0·4155	1·5079	1·5204	2·8899	0·2703
0·76	1·5951	1·2706	1·5726	3·9437	3·6052	0·4362	1·7053	1·7239	3·2526	0·2842
0·80	1·6689	1·2717	1·7200	4·1226	3·7652	0·4568	1·9176	1·9432	3·6405	0·2981
0·84	1·7425	1·2726	1·8738	4·3015	3·9252	0·4774	2·1450	2·1788	4·0544	0·3120
0·88	1·8160	1·2733	2·0340	4·4804	4·0852	0·4979	2·3879	2·4311	4·4946	0·3258
0·92	1·8893	1·2738	2·2007	4·6593	4·2452	0·5184	2·6468	2·7005	4·9618	0·3397
0·96	1·9625	1·2743	2·3737	4·8382	4·4052	0·5388	2·9218	2·9876	5·4564	0·3534
1·00	2·0355	1·2746	2·5531	5·0171	4·5652	0·5592	3·2135	3·2927	5·9789	0·3672

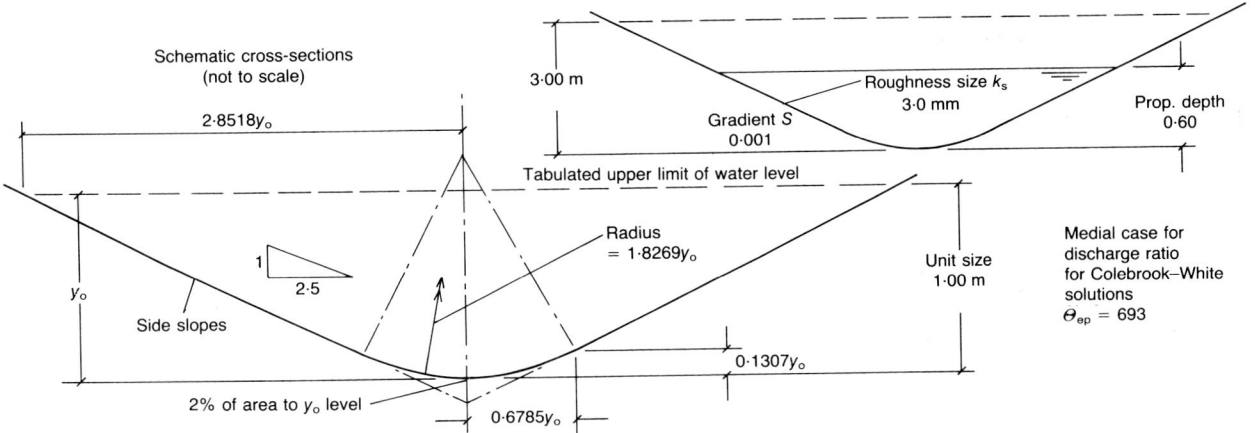

Schematic cross-sections (not to scale)

2·8518y_o

3·00 m

Roughness size k_s 3·0 mm

Gradient S 0·001

Prop. depth 0·60

Tabulated upper limit of water level

1 : 2·5

Radius = 1·8269y_o

y_o

Side slopes

Unit size 1·00 m

Medial case for discharge ratio for Colebrook–White solutions $\Theta_{ep} = 693$

0·1307y_o

2% of area to y_o level

0·6785y_o

Prop dpth Y	U.equ. dia. $D_{ep(u)}$ (m)	Equiv. disch. factor J	Unit sect. area A_u (m²)	Unit wetted perim. P_u (m)	Unit surf. brdth $B_{s(u)}$ (m)	Unit mean depth $y_{m(u)}$ (m)	Discharge ratios $Q/Q_{0.60}$ medial	Mann'g	U.crit. disch. $Q_{c(u)}$ (m³s⁻¹)	Unit depth cntrd $y_{d(u)}$ (m)
0·02	0·0532	0·3088	0·0072	0·5412	0·5392	0·0133	0·0007	0·0006	0·0026	0·0080
0·04	0·1061	0·4352	0·0203	0·7660	0·7604	0·0267	0·0031	0·0029	0·0104	0·0160
0·06	0·1588	0·5312	0·0373	0·9390	0·9287	0·0401	0·0073	0·0069	0·0234	0·0240
0·08	0·2111	0·6112	0·0573	1·0853	1·0694	0·0536	0·0136	0·0128	0·0415	0·0321
0·10	0·2633	0·6809	0·0799	1·2145	1·1923	0·0670	0·0219	0·0207	0·0648	0·0401
0·12	0·3151	0·7433	0·1049	1·3317	1·3024	0·0805	0·0323	0·0307	0·0932	0·0481
0·14	0·3664	0·7991	0·1320	1·4406	1·4037	0·0940	0·0448	0·0427	0·1267	0·0562
0·16	0·4160	0·8442	0·1610	1·5483	1·5037	0·1071	0·0593	0·0567	0·1650	0·0642
0·18	0·4640	0·8803	0·1921	1·6560	1·6037	0·1198	0·0758	0·0727	0·2082	0·0722
0·20	0·5107	0·9097	0·2252	1·7637	1·7037	0·1322	0·0945	0·0909	0·2564	0·0801
0·22	0·5563	0·9339	0·2603	1·8714	1·8037	0·1443	0·1153	0·1112	0·3096	0·0880
0·24	0·6009	0·9539	0·2973	1·9791	1·9037	0·1562	0·1383	0·1337	0·3680	0·0957
0·26	0·6448	0·9707	0·3364	2·0868	2·0037	0·1679	0·1636	0·1586	0·4317	0·1034
0·28	0·6880	0·9850	0·3775	2·1945	2·1037	0·1794	0·1912	0·1858	0·5007	0·1111
0·30	0·7307	0·9971	0·4206	2·3022	2·2037	0·1908	0·2212	0·2155	0·5753	0·1187
0·32	0·7729	1·0075	0·4656	2·4099	2·3037	0·2021	0·2537	0·2476	0·6556	0·1262
0·34	0·8146	1·0165	0·5127	2·5176	2·4037	0·2133	0·2887	0·2824	0·7415	0·1337
0·36	0·8559	1·0242	0·5618	2·6253	2·5037	0·2244	0·3263	0·3198	0·8333	0·1411
0·38	0·8970	1·0310	0·6128	2·7330	2·6037	0·2354	0·3665	0·3600	0·9311	0·1485
0·40	0·9377	1·0370	0·6659	2·8407	2·7037	0·2463	0·4094	0·4029	1·0349	0·1559
0·42	0·9781	1·0422	0·7210	2·9484	2·8037	0·2572	0·4551	0·4487	1·1450	0·1632
0·44	1·0184	1·0468	0·7781	3·0561	2·9037	0·2680	0·5036	0·4974	1·2613	0·1705
0·46	1·0584	1·0509	0·8371	3·1638	3·0037	0·2787	0·5550	0·5491	1·3840	0·1778
0·48	1·0982	1·0546	0·8982	3·2715	3·1037	0·2894	0·6093	0·6038	1·5132	0·1850
0·50	1·1379	1·0578	0·9613	3·3792	3·2037	0·3001	0·6666	0·6617	1·6490	0·1922
0·52	1·1774	1·0607	1·0264	3·4869	3·3037	0·3107	0·7270	0·7227	1·7915	0·1994
0·54	1·2167	1·0634	1·0934	3·5946	3·4037	0·3213	0·7905	0·7870	1·9408	0·2065
0·56	1·2560	1·0657	1·1625	3·7023	3·5037	0·3318	0·8571	0·8546	2·0970	0·2137
0·58	1·2951	1·0678	1·2336	3·8100	3·6037	0·3423	0·9269	0·9256	2·2602	0·2208
0·60	1·3341	1·0698	1·3067	3·9177	3·7037	0·3528	1·0000	1·0000	2·4304	0·2279
0·64	1·4118	1·0731	1·4588	4·1332	3·9037	0·3737	1·1562	1·1594	2·7927	0·2420
0·68	1·4892	1·0758	1·6189	4·3486	4·1037	0·3945	1·3261	1·3333	3·1844	0·2561
0·72	1·5663	1·0781	1·7871	4·5640	4·3037	0·4152	1·5101	1·5221	3·6063	0·2701
0·76	1·6431	1·0800	1·9632	4·7794	4·5037	0·4359	1·7086	1·7264	4·0592	0·2840
0·80	1·7197	1·0816	2·1474	4·9948	4·7037	0·4565	1·9220	1·9465	4·5437	0·2979
0·84	1·7961	1·0830	2·3395	5·2102	4·9037	0·4771	2·1507	2·1831	5·0605	0·3118
0·88	1·8724	1·0841	2·5397	5·4256	5·1037	0·4976	2·3951	2·4364	5·6103	0·3257
0·92	1·9485	1·0851	2·7478	5·6410	5·3037	0·5181	2·6555	2·7071	6·1938	0·3395
0·96	2·0244	1·0860	2·9640	5·8564	5·5037	0·5385	2·9324	2·9954	6·8116	0·3532
1·00	2·1003	1·0867	3·1881	6·0718	5·7037	0·5590	3·2260	3·3019	7·4642	0·3670

C91

3.0 to 1 tangent river

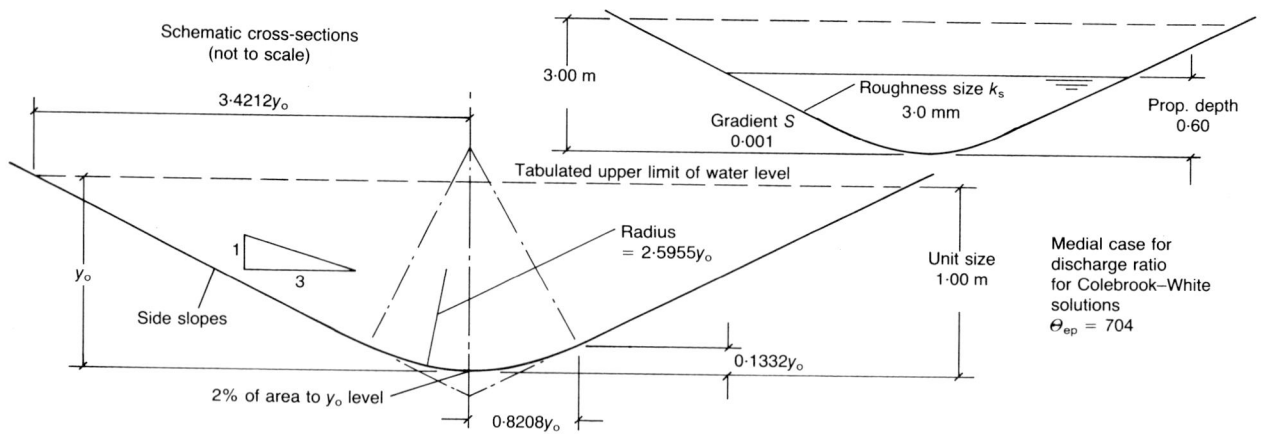

Prop dpth Y	U.equ. dia. $D_{ep(u)}$ (m)	Equiv. disch. factor J	Unit sect. area A_u (m²)	Unit wetted perim. P_u (m)	Unit surf. brdth $B_{s(u)}$ (m)	Unit mean depth $y_{m(u)}$ (m)	Discharge ratios $Q/Q_{0.60}$ medial	Mann'g	U.crit. disch. $Q_{c(u)}$ (m³s⁻¹)	Unit depth cntrd $y_{d(u)}$ (m)
0·02	0·0532	0·2594	0·0086	0·6448	0·6432	0·0133	0·0007	0·0006	0·0031	0·0080
0·04	0·1063	0·3659	0·0242	0·9125	0·9078	0·0267	0·0030	0·0028	0·0124	0·0160
0·06	0·1591	0·4470	0·0445	1·1183	1·1097	0·0401	0·0073	0·0068	0·0279	0·0240
0·08	0·2118	0·5149	0·0684	1·2922	1·2789	0·0535	0·0135	0·0127	0·0496	0·0320
0·10	0·2643	0·5743	0·0955	1·4456	1·4270	0·0669	0·0217	0·0205	0·0774	0·0401
0·12	0·3165	0·6275	0·1254	1·5847	1·5602	0·0804	0·0320	0·0304	0·1113	0·0481
0·14	0·3685	0·6757	0·1578	1·7133	1·6824	0·0938	0·0444	0·0423	0·1514	0·0561
0·16	0·4189	0·7154	0·1927	1·8398	1·8024	0·1069	0·0589	0·0562	0·1973	0·0642
0·18	0·4678	0·7474	0·2299	1·9663	1·9224	0·1196	0·0754	0·0722	0·2490	0·0721
0·20	0·5153	0·7735	0·2696	2·0927	2·0424	0·1320	0·0939	0·0903	0·3067	0·0800
0·22	0·5617	0·7951	0·3116	2·2192	2·1624	0·1441	0·1147	0·1106	0·3705	0·0879
0·24	0·6072	0·8132	0·3561	2·3457	2·2824	0·1560	0·1377	0·1331	0·4404	0·0956
0·26	0·6519	0·8284	0·4029	2·4722	2·4024	0·1677	0·1629	0·1579	0·5167	0·1033
0·28	0·6960	0·8414	0·4522	2·5987	2·5224	0·1793	0·1905	0·1851	0·5995	0·1110
0·30	0·7395	0·8525	0·5038	2·7252	2·6424	0·1907	0·2205	0·2147	0·6889	0·1186
0·32	0·7825	0·8620	0·5579	2·8517	2·7624	0·2020	0·2530	0·2469	0·7851	0·1261
0·34	0·8251	0·8703	0·6143	2·9782	2·8824	0·2131	0·2880	0·2816	0·8881	0·1336
0·36	0·8673	0·8776	0·6732	3·1047	3·0024	0·2242	0·3256	0·3191	0·9982	0·1410
0·38	0·9092	0·8839	0·7344	3·2312	3·1224	0·2352	0·3658	0·3592	1·1154	0·1484
0·40	0·9507	0·8895	0·7981	3·3577	3·2424	0·2461	0·4087	0·4022	1·2399	0·1558
0·42	0·9920	0·8945	0·8641	3·4841	3·3624	0·2570	0·4545	0·4480	1·3718	0·1631
0·44	1·0331	0·8989	0·9326	3·6106	3·4824	0·2678	0·5030	0·4967	1·5112	0·1704
0·46	1·0740	0·9028	1·0034	3·7371	3·6024	0·2785	0·5544	0·5484	1·6584	0·1777
0·48	1·1147	0·9063	1·0767	3·8636	3·7224	0·2892	0·6088	0·6032	1·8133	0·1849
0·50	1·1552	0·9095	1·1523	3·9901	3·8424	0·2999	0·6662	0·6612	1·9761	0·1921
0·52	1·1955	0·9123	1·2303	4·1166	3·9624	0·3105	0·7266	0·7223	2·1470	0·1993
0·54	1·2357	0·9149	1·3108	4·2431	4·0824	0·3211	0·7902	0·7867	2·3260	0·2064
0·56	1·2758	0·9172	1·3936	4·3696	4·2024	0·3316	0·8569	0·8544	2·5133	0·2136
0·58	1·3157	0·9193	1·4789	4·4961	4·3224	0·3421	0·9268	0·9255	2·7090	0·2207
0·60	1·3556	0·9212	1·5665	4·6226	4·4424	0·3526	1·0000	1·0000	2·9132	0·2278
0·64	1·4349	0·9246	1·7490	4·8755	4·6824	0·3735	1·1565	1·1597	3·3475	0·2419
0·68	1·5140	0·9274	1·9411	5·1285	4·9224	0·3943	1·3267	1·3339	3·8173	0·2560
0·72	1·5927	0·9298	2·1428	5·3815	5·1624	0·4151	1·5110	1·5231	4·3233	0·2700
0·76	1·6712	0·9318	2·3541	5·6345	5·4024	0·4358	1·7099	1·7278	4·8664	0·2839
0·80	1·7495	0·9335	2·5750	5·8875	5·6424	0·4564	1·9238	1·9485	5·4475	0·2978
0·84	1·8276	0·9350	2·8055	6·1405	5·8824	0·4769	2·1529	2·1856	6·0674	0·3117
0·88	1·9055	0·9363	3·0456	6·3934	6·1224	0·4975	2·3978	2·4396	6·7268	0·3256
0·92	1·9832	0·9374	3·2953	6·6464	6·3624	0·5179	2·6589	2·7109	7·4267	0·3394
0·96	2·0608	0·9384	3·5546	6·8994	6·6024	0·5384	2·9363	3·0001	8·1676	0·3531
1·00	2·1383	0·9392	3·8235	7·1524	6·8424	0·5588	3·2307	3·3074	8·9505	0·3669

160

4·0 to 1 tangent river

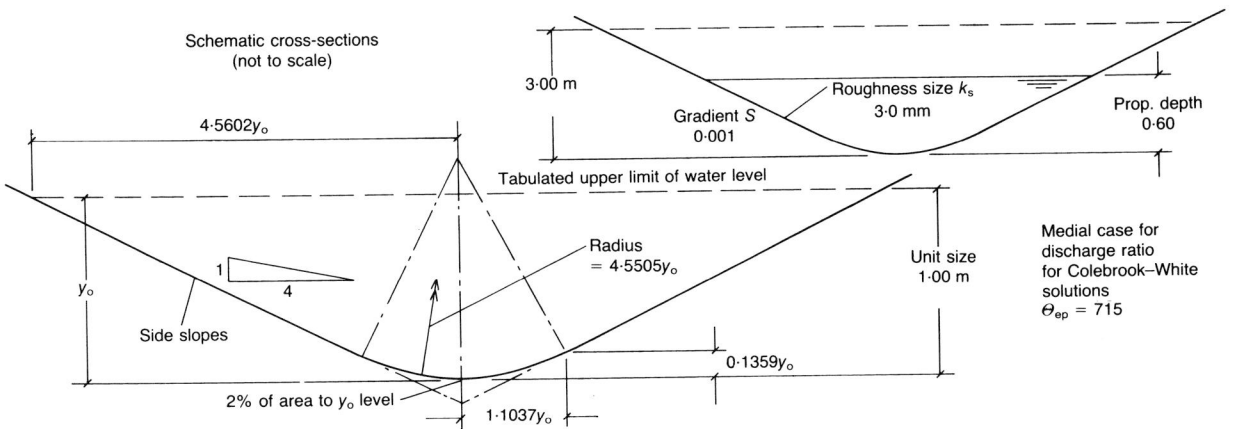

Schematic cross-sections
(not to scale)

3·00 m

Gradient S
0·001

Roughness size k_s
3·0 mm

Prop. depth
0·60

4·5602y_o

Tabulated upper limit of water level

Radius
= 4·5505y_o

Unit size
1·00 m

Medial case for
discharge ratio
for Colebrook–White
solutions
Θ_{ep} = 715

y_o

1

4

Side slopes

2% of area to y_o level

1·1037y_o

0·1359y_o

Prop dpth Y	U.equ. dia. $D_{ep(u)}$ (m)	Equiv. disch. factor J	Unit sect. area A_u (m²)	Unit wetted perim. P_u (m)	Unit surf. brdth $B_{s(u)}$ (m)	Unit mean depth $y_{m(u)}$ (m)	Discharge ratios $Q/Q_{0·60}$ medial	Mann'g	U.crit. disch. $Q_{c(u)}$ (m³s⁻¹)	Unit depth cntrd $y_{d(u)}$ (m)
0·02	0·0533	0·1961	0·0114	0·8536	0·8523	0·0133	0·0007	0·0006	0·0041	0·0080
0·04	0·1064	0·2769	0·0321	1·2076	1·2041	0·0267	0·0030	0·0028	0·0164	0·0160
0·06	0·1595	0·3387	0·0590	1·4795	1·4730	0·0401	0·0072	0·0067	0·0370	0·0240
0·08	0·2125	0·3905	0·0908	1·7091	1·6990	0·0534	0·0133	0·0125	0·0657	0·0320
0·10	0·2653	0·4360	0·1268	1·9115	1·8975	0·0668	0·0215	0·0203	0·1026	0·0400
0·12	0·3180	0·4770	0·1665	2·0947	2·0763	0·0802	0·0317	0·0300	0·1477	0·0481
0·14	0·3706	0·5143	0·2097	2·2636	2·2404	0·0936	0·0440	0·0419	0·2010	0·0561
0·16	0·4219	0·5457	0·2561	2·4286	2·4004	0·1067	0·0584	0·0558	0·2620	0·0641
0·18	0·4716	0·5712	0·3057	2·5935	2·5604	0·1194	0·0748	0·0717	0·3309	0·0720
0·20	0·5199	0·5922	0·3586	2·7584	2·7204	0·1318	0·0934	0·0897	0·4076	0·0799
0·22	0·5672	0·6096	0·4146	2·9233	2·8804	0·1439	0·1141	0·1099	0·4925	0·0878
0·24	0·6136	0·6242	0·4738	3·0883	3·0404	0·1558	0·1370	0·1324	0·5857	0·0956
0·26	0·6593	0·6366	0·5362	3·2532	3·2004	0·1675	0·1622	0·1572	0·6873	0·1033
0·28	0·7042	0·6472	0·6018	3·4181	3·3604	0·1791	0·1898	0·1843	0·7975	0·1109
0·30	0·7486	0·6564	0·6706	3·5830	3·5204	0·1905	0·2198	0·2140	0·9165	0·1185
0·32	0·7925	0·6643	0·7426	3·7480	3·6804	0·2018	0·2522	0·2461	1·0446	0·1260
0·34	0·8360	0·6712	0·8178	3·9129	3·8404	0·2129	0·2872	0·2809	1·1818	0·1335
0·36	0·8791	0·6773	0·8962	4·0778	4·0004	0·2240	0·3248	0·3183	1·3284	0·1409
0·38	0·9219	0·6826	0·9778	4·2427	4·1604	0·2350	0·3651	0·3584	1·4845	0·1483
0·40	0·9644	0·6873	1·0626	4·4076	4·3204	0·2460	0·4080	0·4014	1·6503	0·1557
0·42	1·0066	0·6915	1·1506	4·5726	4·4804	0·2568	0·4538	0·4472	1·8260	0·1630
0·44	1·0485	0·6953	1·2418	4·7375	4·6404	0·2676	0·5024	0·4960	2·0118	0·1703
0·46	1·0903	0·6987	1·3363	4·9024	4·8004	0·2784	0·5538	0·5478	2·2078	0·1776
0·48	1·1318	0·7017	1·4339	5·0673	4·9604	0·2891	0·6083	0·6026	2·4141	0·1848
0·50	1·1732	0·7044	1·5347	5·2323	5·1204	0·2997	0·6657	0·6606	2·6311	0·1920
0·52	1·2145	0·7069	1·6387	5·3972	5·2804	0·3103	0·7262	0·7218	2·8587	0·1992
0·54	1·2556	0·7091	1·7459	5·5621	5·4404	0·3209	0·7899	0·7863	3·0972	0·2063
0·56	1·2965	0·7112	1·8563	5·7270	5·6004	0·3315	0·8567	0·8541	3·3467	0·2135
0·58	1·3373	0·7131	1·9699	5·8920	5·7604	0·3420	0·9267	0·9253	3·6075	0·2206
0·60	1·3781	0·7148	2·0867	6·0569	5·9204	0·3525	1·0000	1·0000	3·8795	0·2277
0·64	1·4592	0·7178	2·3299	6·3867	6·2404	0·3734	1·1567	1·1600	4·4583	0·2418
0·68	1·5400	0·7203	2·5859	6·7166	6·5604	0·3942	1·3273	1·3345	5·0842	0·2559
0·72	1·6205	0·7225	2·8548	7·0464	6·8804	0·4149	1·5120	1·5242	5·7585	0·2699
0·76	1·7008	0·7244	3·1364	7·3763	7·2004	0·4356	1·7113	1·7294	6·4822	0·2838
0·80	1·7808	0·7260	3·4308	7·7061	7·5204	0·4562	1·9256	1·9506	7·2566	0·2977
0·84	1·8606	0·7274	3·7380	8·0360	7·8404	0·4768	2·1553	2·1883	8·0826	0·3116
0·88	1·9403	0·7286	4·0580	8·3658	8·1604	0·4973	2·4007	2·4429	8·9614	0·3254
0·92	2·0198	0·7297	4·3908	8·6957	8·4804	0·5178	2·6624	2·7150	9·8940	0·3393
0·96	2·0991	0·7306	4·7364	9·0255	8·8004	0·5382	2·9405	3·0049	10·881	0·3530
1·00	2·1784	0·7315	5·0949	9·3554	9·1204	0·5586	3·2356	3·3132	11·925	0·3668

5 to 1 tangent river

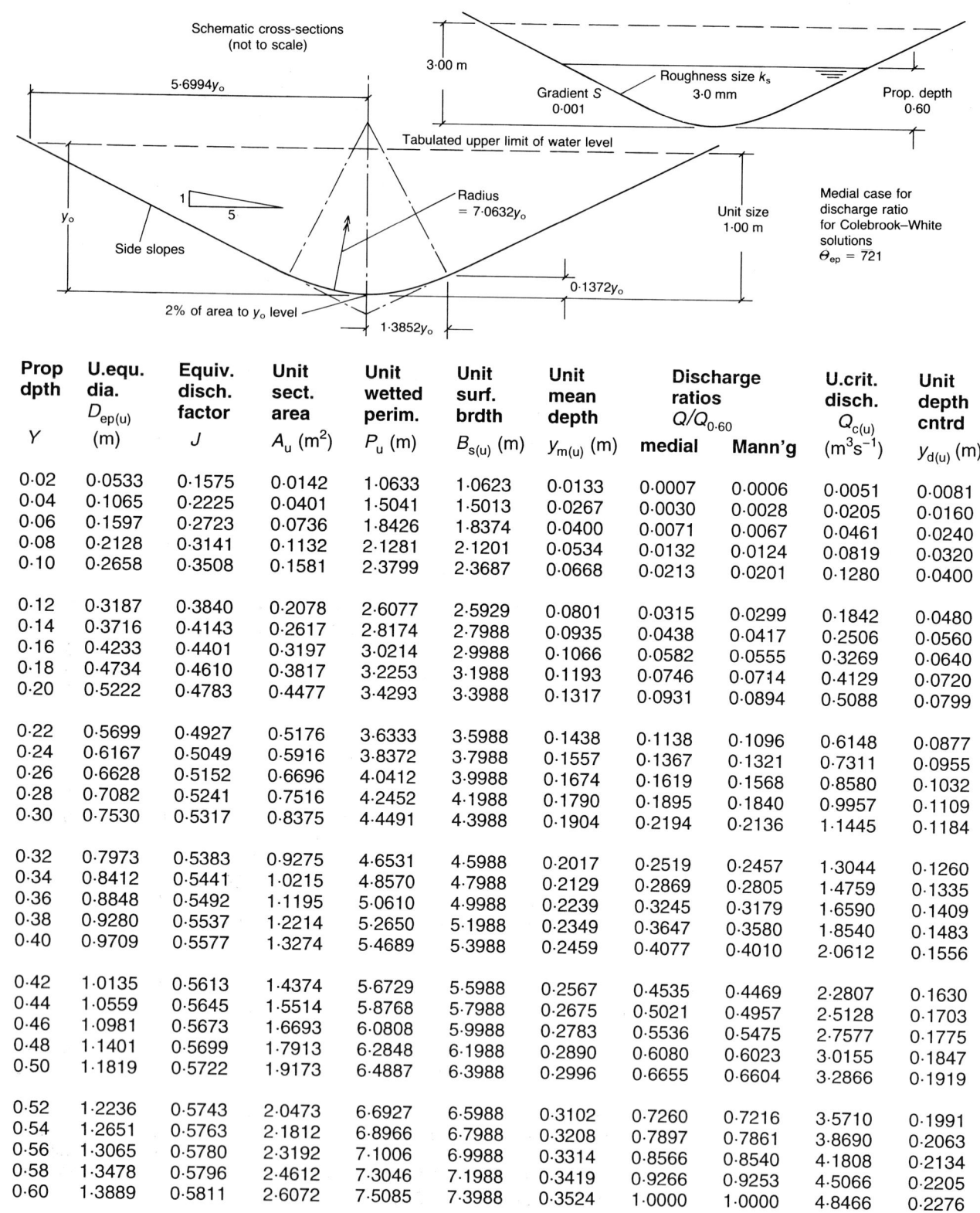

Prop dpth Y	U.equ. dia. $D_{ep(u)}$ (m)	Equiv. disch. factor J	Unit sect. area A_u (m²)	Unit wetted perim. P_u (m)	Unit surf. brdth $B_{s(u)}$ (m)	Unit mean depth $y_{m(u)}$ (m)	Discharge ratios $Q/Q_{0.60}$ medial	Mann'g	U.crit. disch. $Q_{c(u)}$ (m³s⁻¹)	Unit depth cntrd $y_{d(u)}$ (m)
0·02	0·0533	0·1575	0·0142	1·0633	1·0623	0·0133	0·0007	0·0006	0·0051	0·0081
0·04	0·1065	0·2225	0·0401	1·5041	1·5013	0·0267	0·0030	0·0028	0·0205	0·0160
0·06	0·1597	0·2723	0·0736	1·8426	1·8374	0·0400	0·0071	0·0067	0·0461	0·0240
0·08	0·2128	0·3141	0·1132	2·1281	2·1201	0·0534	0·0132	0·0124	0·0819	0·0320
0·10	0·2658	0·3508	0·1581	2·3799	2·3687	0·0668	0·0213	0·0201	0·1280	0·0400
0·12	0·3187	0·3840	0·2078	2·6077	2·5929	0·0801	0·0315	0·0299	0·1842	0·0480
0·14	0·3716	0·4143	0·2617	2·8174	2·7988	0·0935	0·0438	0·0417	0·2506	0·0560
0·16	0·4233	0·4401	0·3197	3·0214	2·9988	0·1066	0·0582	0·0555	0·3269	0·0640
0·18	0·4734	0·4610	0·3817	3·2253	3·1988	0·1193	0·0746	0·0714	0·4129	0·0720
0·20	0·5222	0·4783	0·4477	3·4293	3·3988	0·1317	0·0931	0·0894	0·5088	0·0799
0·22	0·5699	0·4927	0·5176	3·6333	3·5988	0·1438	0·1138	0·1096	0·6148	0·0877
0·24	0·6167	0·5049	0·5916	3·8372	3·7988	0·1557	0·1367	0·1321	0·7311	0·0955
0·26	0·6628	0·5152	0·6696	4·0412	3·9988	0·1674	0·1619	0·1568	0·8580	0·1032
0·28	0·7082	0·5241	0·7516	4·2452	4·1988	0·1790	0·1895	0·1840	0·9957	0·1109
0·30	0·7530	0·5317	0·8375	4·4491	4·3988	0·1904	0·2194	0·2136	1·1445	0·1184
0·32	0·7973	0·5383	0·9275	4·6531	4·5988	0·2017	0·2519	0·2457	1·3044	0·1260
0·34	0·8412	0·5441	1·0215	4·8570	4·7988	0·2129	0·2869	0·2805	1·4759	0·1335
0·36	0·8848	0·5492	1·1195	5·0610	4·9988	0·2239	0·3245	0·3179	1·6590	0·1409
0·38	0·9280	0·5537	1·2214	5·2650	5·1988	0·2349	0·3647	0·3580	1·6590	0·1483
0·40	0·9709	0·5577	1·3274	5·4689	5·3988	0·2459	0·4077	0·4010	2·0612	0·1556
0·42	1·0135	0·5613	1·4374	5·6729	5·5988	0·2567	0·4535	0·4469	2·2807	0·1630
0·44	1·0559	0·5645	1·5514	5·8768	5·7988	0·2675	0·5021	0·4957	2·5128	0·1703
0·46	1·0981	0·5673	1·6693	6·0808	5·9988	0·2783	0·5536	0·5475	2·7577	0·1775
0·48	1·1401	0·5699	1·7913	6·2848	6·1988	0·2890	0·6080	0·6023	3·0155	0·1847
0·50	1·1819	0·5722	1·9173	6·4887	6·3988	0·2996	0·6655	0·6604	3·2866	0·1919
0·52	1·2236	0·5743	2·0473	6·6927	6·5988	0·3102	0·7260	0·7216	3·5710	0·1991
0·54	1·2651	0·5763	2·1812	6·8966	6·7988	0·3208	0·7897	0·7861	3·8690	0·2063
0·56	1·3065	0·5780	2·3192	7·1006	6·9988	0·3314	0·8566	0·8540	4·1808	0·2134
0·58	1·3478	0·5796	2·4612	7·3046	7·1988	0·3419	0·9266	0·9253	4·5066	0·2205
0·60	1·3889	0·5811	2·6072	7·5085	7·3988	0·3524	1·0000	1·0000	4·8466	0·2276
0·64	1·4709	0·5837	2·9111	7·9164	7·7988	0·3733	1·1569	1·1601	5·5698	0·2417
0·68	1·5526	0·5859	3·2311	8·3244	8·1988	0·3941	1·3275	1·3348	6·3519	0·2558
0·72	1·6339	0·5878	3·5670	8·7323	8·5988	0·4148	1·5124	1·5247	7·1945	0·2698
0·76	1·7150	0·5895	3·9190	9·1402	8·9988	0·4355	1·7119	1·7301	8·0989	0·2838
0·80	1·7959	0·5909	4·2869	9·5481	9·3988	0·4561	1·9264	1·9516	9·0666	0·2977
0·84	1·8766	0·5921	4·6709	9·9561	9·7988	0·4767	2·1564	2·1896	10·099	0·3116
0·88	1·9571	0·5932	5·0708	10·364	10·199	0·4972	2·4021	2·4446	11·197	0·3254
0·92	2·0374	0·5942	5·4868	10·772	10·599	0·5177	2·6641	2·7170	12·363	0·3392
0·96	2·1177	0·5951	5·9187	11·180	10·999	0·5381	2·9426	3·0073	13·597	0·3392
1·00	2·1977	0·5958	6·3667	11·588	11·399	0·5585	3·2380	3·3160	14·901	0·3667

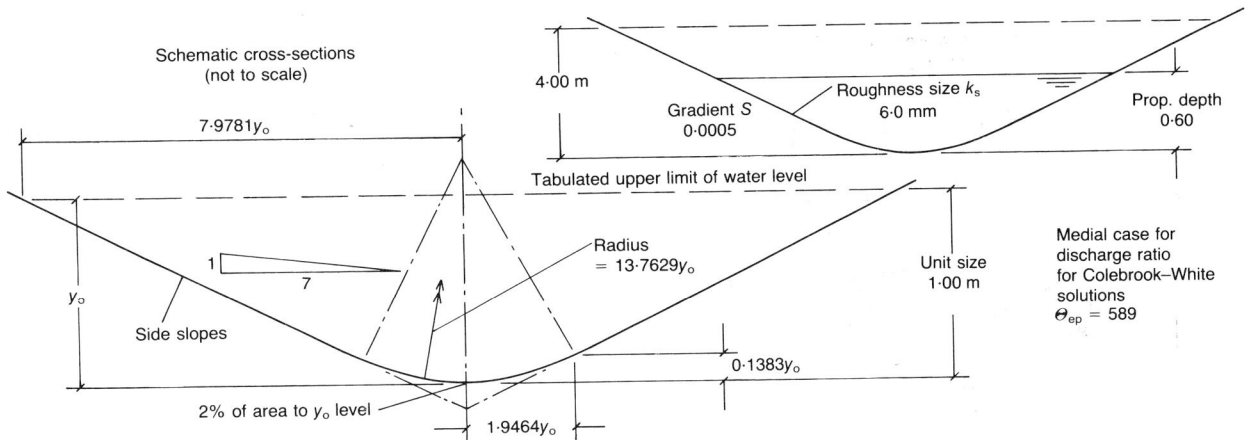

Schematic cross-sections (not to scale)

7·9781y₀

4·00 m

Gradient S 0·0005

Roughness size k_s 6·0 mm

Prop. depth 0·60

Tabulated upper limit of water level

Radius = 13·7629y₀

1

7

Side slopes

y_0

Unit size 1·00 m

Medial case for discharge ratio for Colebrook–White solutions $\Theta_{ep} = 589$

0·1383y₀

2% of area to y₀ level

1·9464y₀

Prop dpth Y	U.equ. dia. $D_{ep(u)}$ (m)	Equiv. disch. factor J	Unit sect. area A_u (m²)	Unit wetted perim. P_u (m)	Unit surf. brdth $B_{s(u)}$ (m)	Unit mean depth $y_{m(u)}$ (m)	Discharge ratios $Q/Q_{0.60}$ medial	Mann'g	U.crit. disch. $Q_{c(u)}$ (m³s⁻¹)	Unit depth cntrd $y_{d(u)}$ (m)
0·02	0·0533	0·1129	0·0198	1·4841	1·4834	0·0133	0·0006	0·0006	0·0072	0·0081
0·04	0·1066	0·1595	0·0559	2·0991	2·0971	0·0267	0·0029	0·0028	0·0286	0·0161
0·06	0·1598	0·1953	0·1027	2·5712	2·5674	0·0400	0·0070	0·0066	0·0644	0·0240
0·08	0·2130	0·2254	0·1581	2·9693	2·9636	0·0534	0·0130	0·0124	0·1144	0·0320
0·10	0·2662	0·2519	0·2210	3·3202	3·3121	0·0667	0·0210	0·0200	0·1787	0·0401
0·12	0·3193	0·2758	0·2904	3·6375	3·6270	0·0801	0·0311	0·0297	0·2573	0·0480
0·14	0·3724	0·2978	0·3659	3·9295	3·9162	0·0934	0·0433	0·0415	0·3502	0·0560
0·16	0·4245	0·3166	0·4470	4·2124	4·1962	0·1065	0·0576	0·0553	0·4569	0·0640
0·18	0·4749	0·3319	0·5337	4·4952	4·4762	0·1192	0·0739	0·0712	0·5771	0·0720
0·20	0·5241	0·3446	0·6261	4·7781	4·7562	0·1316	0·0923	0·0892	0·7113	0·0799
0·22	0·5722	0·3552	0·7240	5·0609	5·0362	0·1438	0·1129	0·1094	0·8596	0·0877
0·24	0·6194	0·3641	0·8275	5·3437	5·3162	0·1557	0·1358	0·1318	1·0224	0·0955
0·26	0·6659	0·3718	0·9366	5·6266	5·5962	0·1674	0·1609	0·1565	1·1999	0·1032
0·28	0·7116	0·3783	1·0513	5·9094	5·8762	0·1789	0·1884	0·1837	1·3926	0·1108
0·30	0·7569	0·3840	1·1717	6·1923	6·1562	0·1903	0·2183	0·2133	1·6007	0·1184
0·32	0·8016	0·3889	1·2976	6·4751	6·4362	0·2016	0·2508	0·2454	1·8245	0·1259
0·34	0·8459	0·3932	1·4291	6·7580	6·7162	0·2128	0·2857	0·2801	2·0644	0·1334
0·36	0·8898	0·3970	1·5662	7·0408	6·9962	0·2239	0·3233	0·3176	2·3207	0·1409
0·38	0·9334	0·4004	1·7090	7·3236	7·2762	0·2349	0·3636	0·3577	2·5936	0·1482
0·40	0·9767	0·4034	1·8573	7·6065	7·5562	0·2458	0·4066	0·4007	2·8836	0·1556
0·42	1·0197	0·4060	2·0112	7·8893	7·8362	0·2567	0·4524	0·4465	3·1908	0·1629
0·44	1·0625	0·4084	2·1707	8·1722	8·1162	0·2675	0·5010	0·4954	3·5156	0·1702
0·46	1·1051	0·4106	2·3359	8·4550	8·3962	0·2782	0·5526	0·5472	3·8582	0·1775
0·48	1·1475	0·4125	2·5066	8·7379	8·6762	0·2889	0·6071	0·6021	4·2191	0·1847
0·50	1·1897	0·4143	2·6829	9·0207	8·9562	0·2996	0·6647	0·6602	4·5984	0·1919
0·52	1·2317	0·4159	2·8648	9·3035	9·2362	0·3102	0·7253	0·7214	4·9965	0·1991
0·54	1·2736	0·4174	3·0524	9·5864	9·5162	0·3208	0·7892	0·7860	5·4135	0·2062
0·56	1·3154	0·4187	3·2455	9·8692	9·7962	0·3313	0·8562	0·8539	5·8499	0·2134
0·58	1·3570	0·4199	3·4442	10·152	10·076	0·3418	0·9264	0·9252	6·3059	0·2205
0·60	1·3986	0·4211	3·6485	10·435	10·356	0·3523	1·0000	1·0000	6·7817	0·2276
0·64	1·4814	0·4230	4·0740	11·001	10·916	0·3732	1·1574	1·1602	7·7939	0·2417
0·68	1·5638	0·4247	4·5218	11·566	11·476	0·3940	1·3286	1·3351	8·8886	0·2558
0·72	1·6459	0·4262	4·9921	12·132	12·036	0·4148	1·5141	1·5251	10·068	0·2698
0·76	1·7278	0·4275	5·4847	12·698	12·596	0·4354	1·7144	1·7308	11·334	0·2837
0·80	1·8094	0·4286	5·9998	13·263	13·156	0·4560	1·9299	1·9525	12·688	0·2976
0·84	1·8909	0·4295	6·5372	13·829	13·716	0·4766	2·1609	2·1907	14·133	0·3115
0·88	1·9721	0·4304	7·0971	14·395	14·276	0·4971	2·4078	2·4460	15·670	0·3253
0·92	2·0532	0·4312	7·6793	14·960	14·836	0·5176	2·6710	2·7188	17·301	0·3392
0·96	2·1342	0·4318	8·2840	15·526	15·396	0·5381	2·9510	3·0094	19·029	0·3529
1·00	2·2150	0·4324	8·9110	16·092	15·956	0·5585	3·2480	3·3185	20·854	0·3667

C95

10 to 1 tangent river

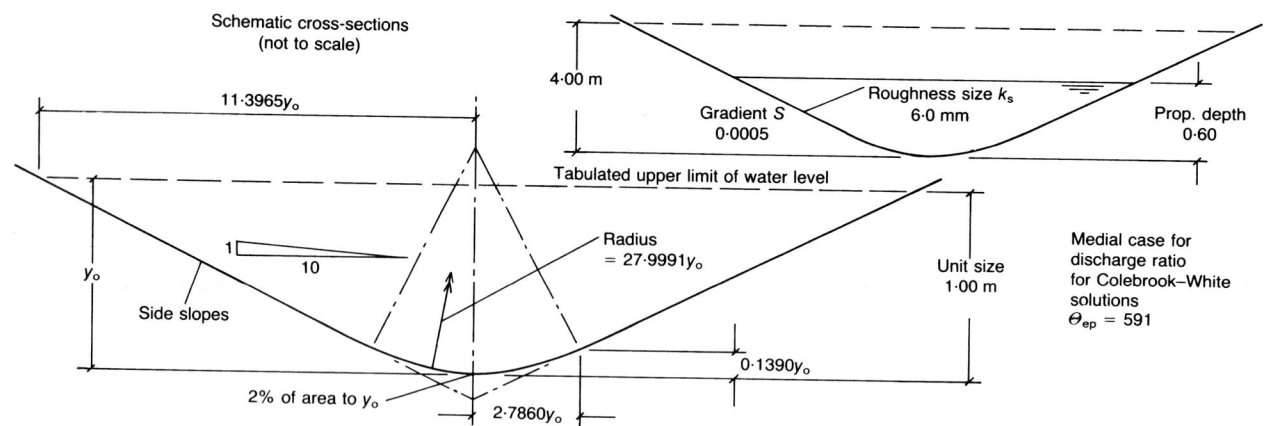

Schematic cross-sections (not to scale)

11·3965y₀

4·00 m

Gradient S 0·0005

Roughness size k_s 6·0 mm

Prop. depth 0·60

Tabulated upper limit of water level

Radius = 27·9991y₀

y_0

Side slopes

1 / 10

Unit size 1·00 m

Medial case for discharge ratio for Colebrook–White solutions Θ_{ep} = 591

2% of area to y₀

2·7860y₀

0·1390y₀

Prop dpth Y	U.equ. dia. $D_{ep(u)}$ (m)	Equiv. disch. factor J	Unit sect. area A_u (m²)	Unit wetted perim. P_u (m)	Unit surf. brdth $B_{s(u)}$ (m)	Unit mean depth $y_{m(u)}$ (m)	Discharge ratios $Q/Q_{0.60}$ medial	Mann'g	U.crit. disch. $Q_{c(u)}$ (m³s⁻¹)	Unit depth cntrd $y_{d(u)}$ (m)
0·02	0·0533	0·0791	0·0282	2·1167	2·1162	0·0133	0·0006	0·0006	0·0102	0·0080
0·04	0·1066	0·1119	0·0798	2·9936	2·9922	0·0267	0·0029	0·0027	0·0408	0·0159
0·06	0·1599	0·1370	0·1466	3·6667	3·6640	0·0400	0·0069	0·0066	0·0918	0·0240
0·08	0·2132	0·1582	0·2257	4·2341	4·2301	0·0533	0·0129	0·0123	0·1632	0·0317
0·10	0·2664	0·1768	0·3154	4·7342	4·7286	0·0667	0·0210	0·0200	0·2550	0·0398
0·12	0·3197	0·1936	0·4145	5·1864	5·1789	0·0800	0·0310	0·0297	0·3672	0·0480
0·14	0·3729	0·2091	0·5223	5·6023	5·5929	0·0934	0·0432	0·0414	0·4998	0·0560
0·16	0·4251	0·2224	0·6381	6·0043	5·9929	0·1065	0·0575	0·0552	0·6521	0·0640
0·18	0·4758	0·2333	0·7620	6·4063	6·3929	0·1192	0·0738	0·0711	0·8238	0·0720
0·20	0·5252	0·2423	0·8938	6·8083	6·7929	0·1316	0·0922	0·0891	1·0154	0·0799
0·22	0·5735	0·2499	1·0337	7·2103	7·1929	0·1437	0·1128	0·1092	1·2272	0·0877
0·24	0·6209	0·2562	1·1816	7·6123	7·5929	0·1556	0·1356	0·1316	1·4596	0·0954
0·26	0·6675	0·2617	1·3374	8·0143	7·9929	0·1673	0·1608	0·1564	1·7132	0·1032
0·28	0·7135	0·2663	1·5013	8·4162	8·3929	0·1789	0·1883	0·1835	1·9884	0·1108
0·30	0·7589	0·2704	1·6731	8·8182	8·7929	0·1903	0·2182	0·2131	2·2855	0·1184
0·32	0·8039	0·2739	1·8530	9·2202	9·1929	0·2016	0·2506	0·2452	2·6052	0·1259
0·34	0·8484	0·2770	2·0409	9·6222	9·5929	0·2127	0·2856	0·2799	2·9478	0·1334
0·36	0·8925	0·2797	2·2367	10·024	9·9929	0·2238	0·3231	0·3174	3·3138	0·1408
0·38	0·9363	0·2821	2·4406	10·426	10·393	0·2348	0·3634	0·3575	3·7036	0·1482
0·40	0·9798	0·2843	2·6524	10·828	10·793	0·2458	0·4064	0·4005	4·1177	0·1556
0·42	1·0231	0·2862	2·8723	11·230	11·193	0·2566	0·4522	0·4464	4·5565	0·1629
0·44	1·0661	0·2879	3·1001	11·632	11·593	0·2674	0·5009	0·4952	5·0204	0·1702
0·46	1·1088	0·2895	3·3360	12·034	11·993	0·2782	0·5524	0·5470	5·5098	0·1774
0·48	1·1514	0·2909	3·5799	12·436	12·393	0·2889	0·6070	0·6020	6·0252	0·1847
0·50	1·1939	0·2921	3·8317	12·838	12·793	0·2995	0·6646	0·6600	6·5670	0·1919
0·52	1·2361	0·2933	4·0916	13·240	13·193	0·3101	0·7252	0·7213	7·1355	0·1991
0·54	1·2782	0·2943	4·3594	13·642	13·593	0·3207	0·7891	0·7859	7·7313	0·2062
0·56	1·3202	0·2953	4·6353	14·044	13·993	0·3313	0·8561	0·8538	8·3545	0·2133
0·58	1·3621	0·2962	4·9192	14·446	14·393	0·3418	0·9264	0·9252	9·0058	0·2204
0·60	1·4038	0·2970	5·2110	14·848	14·793	0·3523	1·0000	1·0000	9·6854	0·2275
0·64	1·4870	0·2985	5·8187	15·652	15·593	0·3732	1·1574	1·1603	11·131	0·2417
0·68	1·5699	0·2997	6·4585	16·456	16·393	0·3940	1·3287	1·3353	12·695	0·2557
0·72	1·6524	0·3008	7·1302	17·260	17·193	0·4147	1·5144	1·5254	14·379	0·2697
0·76	1·7347	0·3017	7·8339	18·064	17·993	0·4354	1·7148	1·7311	16·187	0·2837
0·80	1·8167	0·3025	8·5696	18·868	18·793	0·4560	1·9303	1·9530	18·122	0·2976
0·84	1·8986	0·3032	9·3373	19·672	19·593	0·4766	2·1614	2·1914	20·186	0·3115
0·88	1·9803	0·3038	10·137	20·476	20·393	0·4971	2·4085	2·4468	22·381	0·3253
0·92	2·0618	0·3044	10·969	21·280	21·193	0·5176	2·6718	2·7197	24·712	0·3391
0·96	2·1432	0·3049	11·832	22·084	21·993	0·5380	2·9519	3·0106	27·179	0·3529
1·00	2·2244	0·3053	12·728	22·888	22·793	0·5584	3·2491	3·3198	29·786	0·3667

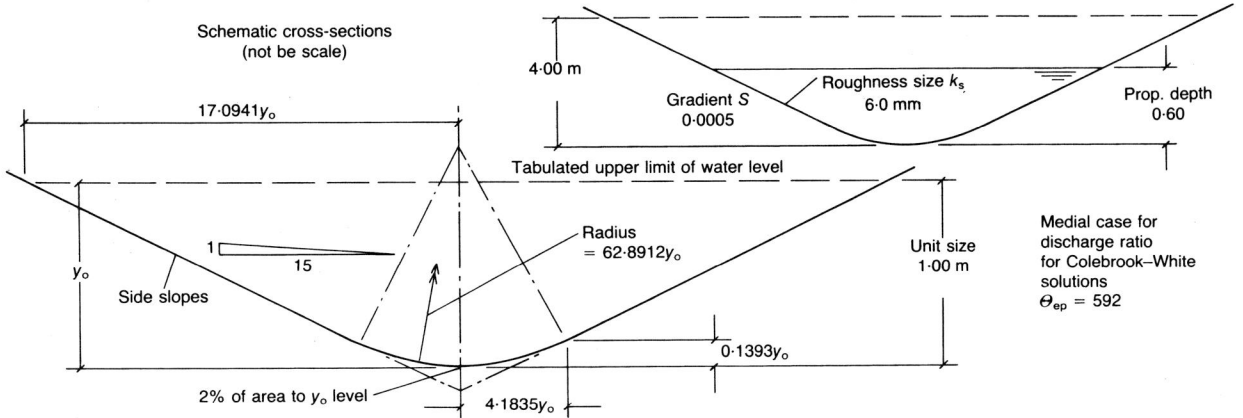

Schematic cross-sections (not be scale)

4·00 m

Gradient S 0·0005

Roughness size k_s 6·0 mm

Prop. depth 0·60

17·0941y_o

Tabulated upper limit of water level

Radius = 62·8912y_o

Unit size 1·00 m

Medial case for discharge ratio for Colebrook–White solutions $\Theta_{ep} = 592$

y_o

1 : 15

Side slopes

2% of area to y_o level

0·1393y_o

4·1835y_o

Prop dpth Y	U.equ. dia. $D_{ep(u)}$ (m)	Equiv. disch. factor J	Unit sect. area A_u (m²)	Unit wetted perim. P_u (m)	Unit surf. brdth $B_{s(u)}$ (m)	Unit mean depth $y_{m(u)}$ (m)	Discharge ratios $Q/Q_{0.60}$ medial	Mann'g	U.crit. disch. $Q_{c(u)}$ (m³s⁻¹)	Unit depth cntrd $y_{d(u)}$ (m)
0·02	0·0533	0·0528	0·0423	3·1722	3·1719	0·0133	0·0006	0·0006	0·0153	0·0017
0·04	0·1067	0·0747	0·1196	4·4863	4·4854	0·0267	0·0029	0·0027	0·0612	0·0141
0·06	0·1600	0·0915	0·2197	5·4948	5·4930	0·0400	0·0069	0·0066	0·1376	0·0231
0·08	0·2133	0·1056	0·3383	6·3450	6·3423	0·0533	0·0129	0·0123	0·2447	0·0308
0·10	0·2666	0·1180	0·4728	7·0941	7·0903	0·0667	0·0209	0·0200	0·3823	0·0402
0·12	0·3199	0·1293	0·6214	7·7714	7·7665	0·0800	0·0310	0·0296	0·5505	0·0481
0·14	0·3731	0·1396	0·7831	8·3943	8·3881	0·0934	0·0432	0·0414	0·7492	0·0560
0·16	0·4255	0·1486	0·9568	8·9957	8·9881	0·1065	0·0574	0·0552	0·9776	0·0640
0·18	0·4762	0·1559	1·1426	9·5970	9·5881	0·1192	0·0737	0·0710	1·2352	0·0720
0·20	0·5257	0·1619	1·3403	10·198	10·188	0·1316	0·0921	0·0890	1·5224	0·0798
0·22	0·5741	0·1670	1·5501	10·800	10·788	0·1437	0·1127	0·1091	1·8401	0·0877
0·24	0·6217	0·1713	1·7719	11·401	11·388	0·1556	0·1355	0·1315	2·1887	0·0954
0·26	0·6684	0·1750	2·0056	12·002	11·988	0·1673	0·1607	0·1563	2·5690	0·1031
0·28	0·7145	0·1781	2·2514	12·604	12·588	0·1789	0·1882	0·1834	2·9817	0·1108
0·30	0·7601	0·1808	2·5092	13·205	13·188	0·1903	0·2181	0·2130	3·4274	0·1184
0·32	0·8051	0·1832	2·7789	13·806	13·788	0·2015	0·2505	0·2451	3·9068	0·1259
0·34	0·8497	0·1853	3·0607	14·408	14·388	0·2127	0·2855	0·2799	4·4206	0·1334
0·36	0·8940	0·1871	3·3544	15·009	14·988	0·2238	0·3230	0·3173	4·9696	0·1408
0·38	0·9379	0·1887	3·6602	15·610	15·588	0·2348	0·3633	0·3574	5·5542	0·1482
0·40	0·9815	0·1902	3·9780	16·212	16·188	0·2457	0·4063	0·4004	6·1752	0·1556
0·42	1·0249	0·1915	4·3077	16·813	16·788	0·2566	0·4521	0·4463	6·8333	0·1629
0·44	1·0680	0·1927	4·6495	17·414	17·388	0·2674	0·5008	0·4951	7·5291	0·1702
0·46	1·1109	0·1937	5·0033	18·016	17·988	0·2781	0·5524	0·5469	8·2632	0·1774
0·48	1·1536	0·1947	5·3690	18·617	18·588	0·2888	0·6069	0·6019	9·0362	0·1847
0·50	1·1961	0·1955	5·7468	19·218	19·188	0·2995	0·6645	0·6600	9·8487	0·1919
0·52	1·2385	0·1963	6·1365	19·820	19·788	0·3101	0·7252	0·7213	10·701	0·1990
0·54	1·2807	0·1970	6·5383	20·421	20·388	0·3207	0·7890	0·7859	11·595	0·2062
0·56	1·3228	0·1977	6·9521	21·022	20·988	0·3312	0·8561	0·8538	12·530	0·2133
0·58	1·3648	0·1983	7·3778	21·624	21·588	0·3418	0·9264	0·9252	13·507	0·2204
0·60	1·4066	0·1988	7·8156	22·225	22·188	0·3522	1·0000	1·0000	14·526	0·2275
0·64	1·4901	0·1998	8·7271	23·428	23·388	0·3731	1·1575	1·1604	16·694	0·2416
0·68	1·5731	0·2006	9·6866	24·630	24·588	0·3940	1·3288	1·3354	19·040	0·2557
0·72	1·6559	0·2014	10·694	25·833	25·788	0·4147	1·5145	1·5255	21·566	0·2697
0·76	1·7384	0·2020	11·750	27·036	26·988	0·4354	1·7149	1·7313	24·278	0·2837
0·80	1·8207	0·2026	12·853	28·238	28·188	0·4560	1·9305	1·9532	27·180	0·2976
0·84	1·9028	0·2030	14·005	29·441	29·388	0·4765	2·1617	2·1917	30·275	0·3115
0·88	1·9847	0·2035	15·204	30·644	30·588	0·4971	2·4088	2·4472	33·568	0·3253
0·92	2·0664	0·2038	16·452	31·846	31·788	0·5175	2·6723	2·7202	37·064	0·3391
0·96	2·1480	0·2042	17·747	33·049	32·988	0·5380	2·9525	3·0112	40·764	0·3529
1·00	2·2295	0·2045	19·091	34·252	34·188	0·5584	3·2498	3·3206	44·675	0·3666

C97 Triangular open channel (modified table)

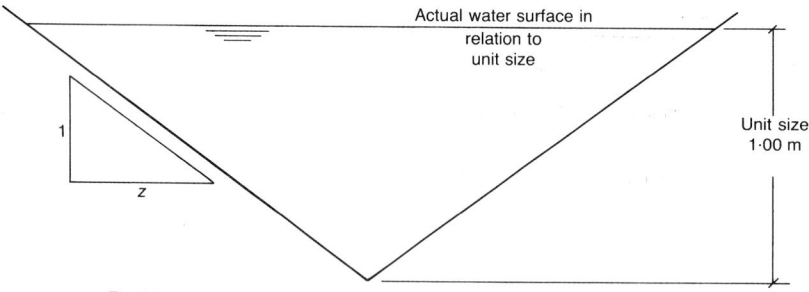

Actual water surface in relation to unit size

Unit size 1·00 m

Roughness size k_s 1.50 mm

1·00 m

Gradient S 0·001

Medial case for discharge ratio for Colebrook–White solutions $\Theta_{ep} = 474$

For this shape, the values in column 1 are of z (as above). Thus all data relates to triangular channels of specified side slopes which are full to unit depth

Side slope z	U.equ. dia. $D_{ep(u)}$ (m)	Equiv. disch. factor J	Unit sect. area A_u (m²)	Unit wetted perim. P_u (m)	Unit surf. brdth $B_{s(u)}$ (m)	Unit mean depth $y_{m(u)}$ (m)	Discharge ratios $Q/Q_{0.60}$ medial	Mann'g	U.crit. disch. $Q_{c(u)}$ (m³s⁻¹)	Unit depth cntrd $y_{d(u)}$ (m)
0·25	0·4851	0·7392	0·2500	2·0616	0·5000	0·5000	0·0550	0·0524	0·5536	0·3333
0·375	0·7022	1·0328	0·3750	2·1360	0·7500	0·5000	0·1045	0·1005	0·8304	0·3333
0·50	0·8944	1·2566	0·5000	2·2361	1·0000	0·5000	0·1623	0·1575	1·1072	0·3333
0·625	1·0600	1·4119	0·6250	2·3585	1·2500	0·5000	0·2257	0·2205	1·3840	0·3333
0·75	1·2000	1·5079	0·7500	2·5000	1·5000	0·5000	0·2927	0·2874	1·6608	0·3333
0·875	1·3170	1·5569	0·8750	2·6575	1·7500	0·5000	0·3619	0·3567	1·9376	0·3333
1·00	1·4142	1·5708	1·0000	2·8284	2·0000	0·5000	0·4323	0·4275	2·2143	0·3333
1·125	1·4948	1·5599	1·1250	3·0104	2·2500	0·5000	0·5034	0·4990	2·4911	0·3333
1·25	1·5617	1·5324	1·2500	3·2016	2·5000	0·5000	0·5747	0·5709	2·7679	0·3333
1·375	1·6175	1·4943	1·3750	3·4004	2·7500	0·5000	0·6460	0·6429	3·0447	0·3333
1·50	1·6641	1·4499	1·5000	3·6056	3·0000	0·5000	0·7172	0·7147	3·3215	0·3333
1·75	1·7365	1·3533	1·7500	4·0311	3·5000	0·5000	0·8591	0·8578	3·8751	0·3333
2·00	1·7889	1·2566	2·0000	4·4721	4·0000	0·5000	1·0000	1·0000	4·4287	0·3333
2·50	1·8570	1·0833	2·5000	5·3852	5·0000	0·5000	1·2792	1·2815	5·5359	0·3333
3·00	1·8974	0·9425	3·0000	6·3246	6·0000	0·5000	1·5555	1·5601	6·6430	0·3333
3·50	1·9230	0·8298	3·5000	7·2801	7·0000	0·5000	1·8299	1·8365	7·7502	0·3333
4·00	1·9403	0·7392	4·0000	8·2462	8·0000	0·5000	2·1029	2·1113	8·8574	0·3333
4·50	1·9524	0·6653	4·5000	9·2195	9·0000	0·5000	2·3748	2·3851	9·9646	0·3333
5·00	1·9612	0·6041	5·0000	10·198	10·000	0·5000	2·6460	2·6581	11·072	0·3333
6·00	1·9728	0·5094	6·0000	12·166	12·000	0·5000	3·1867	3·2023	13·286	0·3333
7·00	1·9799	0·4398	7·0000	14·142	14·000	0·5000	3·7261	3·7450	15·500	0·3333
8·00	1·9846	0·3866	8·0000	16·125	16·000	0·5000	4·2646	4·2867	17·715	0·3333
9·00	1·9878	0·3448	9·0000	18·111	18·000	0·5000	4·8024	4·8277	19·929	0·3333
10·00	1·9901	0·3110	10·000	20·100	20·000	0·5000	5·3398	5·3683	22·143	0·3333
12·00	1·9931	0·2600	12·000	24·083	24·000	0·5000	6·4138	6·4484	26·572	0·3333
14·00	1·9949	0·2233	14·000	28·071	28·000	0·5000	7·4870	7·5277	31·001	0·3333
16·00	1·9961	0·1956	16·000	32·062	32·000	0·5000	8·5597	8·6065	35·430	0·3333
18·00	1·9969	0·1740	18·000	36·056	36·000	0·5000	9·6320	9·6850	39·858	0·3333
20·00	1·9975	0·1567	20·000	40·050	40·000	0·5000	10·704	10·763	44·287	0·3333
25·00	1·9984	0·1255	25·000	50·040	50·000	0·5000	13·384	13·458	55·359	0·3333
30·00	1·9989	0·1046	30·000	60·033	60·000	0·5000	16·063	16·152	66·430	0·3333
35·00	1·9992	0·0897	35·000	70·029	70·000	0·5000	18·742	18·846	77·502	0·3333
40·00	1·9994	0·0785	40·000	80·025	80·000	0·5000	21·421	21·540	88·574	0·3333
45·00	1·9995	0·0698	45·000	90·022	90·000	0·5000	24·099	24·233	99·646	0·3333
50·00	1·9996	0·0628	50·000	100·02	100·00	0·5000	26·778	26·927	110·72	0·3333
60·00	1·9997	0·0523	60·000	120·02	120·00	0·5000	32·135	32·314	132·86	0·3333
70·00	1·9998	0·0449	70·000	140·01	140·00	0·5000	37·491	37·700	155·00	0·3333
80·00	1·9998	0·0393	80·000	160·01	160·00	0·5000	42·848	43·086	177·15	0·3333
90·00	1·9999	0·0349	90·000	180·01	180·00	0·5000	48·204	48·473	199·29	0·3333
100·0	1·9999	0·0314	100·00	200·01	200·00	0·5000	53·560	53·859	221·43	0·3333

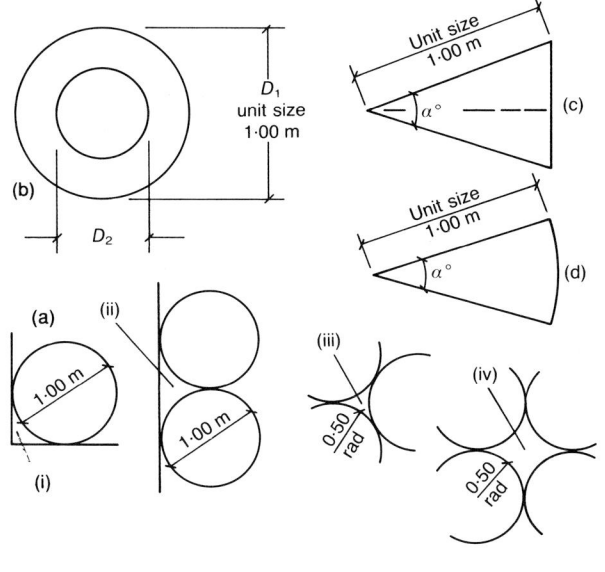

(a) Tube interstices
(1·00 m dia· unit size throughout)

Shape	U·equ· dia. $D_{ep(u)}$ (m)	Equiv. disch. factor J	Unit sect. area A_u (m²)	Unit wetted perim. P_u (m)
(i)	0·1202	0·2115	0·0537	1·7854
(ii)	0·1670	0·2040	0·1073	2·5708
(iii)	0·1027	0·2053	0·0403	1·5708
(iv)	0·2732	0·2732	0·2146	3·1416

(b) Annulus

D_2/D_1	U.equ. dia. $D_{ep(u)}$ (m)	Equiv. disch. factor J	Unit sect. area A_u (m²)	Unit wetted perim. P_u (m)
0·10	0·9000	0·8182	0·7775	3·4558
0·20	0·8000	0·6667	0·7540	3·7699
0·30	0·7000	0·5385	0·7147	4·0841
0·40	0·6000	0·4286	0·6597	4·3982
0·50	0·5000	0·3333	0·5890	4·7124
0·55	0·4500	0·2903	0·5478	4·8695
0·60	0·4000	0·2500	0·5027	5·0265
0·65	0·3500	0·2121	0·4536	5·1836
0·70	0·3000	0·1765	0·4006	5·3407
0·75	0·2500	0·1429	0·3436	5·4978
0·80	0·2000	0·1111	0·2827	5·6549
0·85	0·1500	0·0811	0·2179	5·8119
0·90	0·1000	0·0526	0·1492	5·9690
0·95	0·0500	0·0256	0·0756	6·1261
0·98	0·0200	0·0101	0·0311	6·2204

(c) Isosceles triangle

$\alpha°$	U·equ. dia. $D_{ep(u)}$ (m)	Equiv. disch. factor J	Unit sect. area A_u (m²)	Unit wetted perim. P_u (m)
10	0·1597	0·2308	0·0868	2·1743
20	0·2914	0·3900	0·1710	2·3473
30	0·3972	0·4956	0·2500	2·5176
40	0·4790	0·5606	0·3239	2·6840
50	0·5385	0·5946	0·3830	2·8452
60	0·5774	0·6046	0·4330	3·0000
70	0·5972	0·5961	0·4698	3·1472
80	0·5995	0·5732	0·4924	3·2856
90	0·5858	0·5390	0·5000	3·4142
100	0·5576	0·4960	0·4924	3·5321
110	0·5166	0·4460	0·4698	3·6383
120	0·4641	0·3907	0·4330	3·7321
130	0·4018	0·3311	0·3830	3·8126
140	0·3314	0·2687	0·3214	3·8794
150	0·2543	0·2032	0·2500	3·9319

(d) Sector of circle

$\alpha°$	U.equ. dia. $D_{ep(u)}$ (m)	Equiv. disch. factor J	Unit sect. area A_u (m²)	Unit wetted perim. P_u (m)
10	0·1605	0·2319	0·0873	2·1745
20	0·2972	0·3975	0·1745	2·3491
30	0·4150	0·5166	0·2618	2·5236
40	0·5175	0·6025	0·3491	2·6981
50	0·6076	0·6644	0·4363	2·8727
60	0·6873	0·7086	0·5236	3·0472
70	0·7584	0·7396	0·6109	3·2217
80	0·8222	0·7606	0·6981	3·3963
90	0·8798	0·7741	0·7854	3·5708
100	0·9320	0·7818	0·8727	3·7453
110	0·9796	0·7851	0·9599	3·9199
120	1·0231	0·7850	1·0472	4·0944
140	1·0998	0·7776	1·2217	4·4435
160	1·1654	0·7639	1·3963	4·7925
180	1·2220	0·7467	1·5708	5·1416

D1

m = Manning $n \times 100$
S = 0·00050 to 0·00280

Full bore conditions; Select m to divide mV and/or mQ.

ie hydraulic gradient =
1 in 2000 to 1 in 357

mV to give velocities in ms^{-1}
mQ to give discharges in litres/sec

Gradient	(Equivalent) Pipe diameters in mm												
	20	25	30	40	50	60	65	70	75	80	90	100	110
0·00050	0·065	0·076	0·086	0·104	0·120	0·136	0·143	0·151	0·158	0·165	0·178	0·191	0·204
1/ 2000	0·0205	0·0372	0·0606	0·1304	0·2365	0·3845	0·4760	0·5800	0·6972	0·8281	1·1337	1·5015	1·9360
0·00055	0·069	0·080	0·090	0·109	0·126	0·143	0·150	0·158	0·166	0·173	0·187	0·201	0·214
1/ 1818	0·0215	0·0391	0·0635	0·1368	0·2480	0·4033	0·4993	0·6084	0·7312	0·8686	1·1891	1·5748	2·0305
0·00060	0·072	0·083	0·094	0·114	0·132	0·149	0·157	0·165	0·173	0·180	0·195	0·209	0·223
1/ 1667	0·0225	0·0408	0·0663	0·1429	0·2590	0·4212	0·5215	0·6354	0·7638	0·9072	1·2420	1·6448	2·1208
0·00065	0·075	0·087	0·098	0·118	0·137	0·155	0·164	0·172	0·180	0·188	0·203	0·218	0·232
1/ 1538	0·0234	0·0425	0·0690	0·1487	0·2696	0·4384	0·5428	0·6614	0·7949	0·9442	1·2927	1·7120	2·2074
0·00070	0·077	0·090	0·101	0·123	0·143	0·161	0·170	0·178	0·187	0·195	0·211	0·226	0·241
1/ 1429	0·0243	0·0441	0·0717	0·1543	0·2798	0·4550	0·5632	0·6863	0·8250	0·9799	1·3415	1·7766	2·2908
0·00075	0·080	0·093	0·105	0·127	0·148	0·167	0·176	0·185	0·193	0·202	0·218	0·234	0·250
1/ 1333	0·0252	0·0456	0·0742	0·1597	0·2896	0·4710	0·5830	0·7104	0·8539	1·0143	1·3885	1·8390	2·3712
0·00080	0·083	0·096	0·108	0·131	0·152	0·172	0·181	0·191	0·200	0·208	0·225	0·242	0·258
1/ 1250	0·0260	0·0471	0·0766	0·1650	0·2991	0·4864	0·6021	0·7337	0·8819	1·0475	1·4341	1·8993	2·4489
0·00085	0·085	0·099	0·112	0·135	0·157	0·177	0·187	0·197	0·206	0·215	0·232	0·249	0·266
1/ 1176	0·0268	0·0486	0·0790	0·1701	0·3083	0·5014	0·6207	0·7563	0·9091	1·0798	1·4782	1·9578	2·5243
0·00090	0·088	0·102	0·115	0·139	0·162	0·182	0·192	0·202	0·212	0·221	0·239	0·256	0·273
1/ 1111	0·0276	0·0500	0·0813	0·1750	0·3173	0·5159	0·6387	0·7782	0·9354	1·1111	1·5211	2·0145	2·5975
0·00095	0·090	0·105	0·118	0·143	0·166	0·187	0·198	0·208	0·218	0·227	0·246	0·264	0·281
1/ 1053	0·0283	0·0513	0·0835	0·1798	0·3260	0·5300	0·6562	0·7995	0·9610	1·1415	1·5628	2·0697	2·6686
0·00100	0·092	0·107	0·121	0·147	0·170	0·192	0·203	0·213	0·223	0·233	0·252	0·270	0·288
1/ 1000	0·0290	0·0527	0·0856	0·1844	0·3344	0·5438	0·6732	0·8203	0·9860	1·1712	1·6034	2·1235	2·7380
0·00110	0·097	0·113	0·127	0·154	0·179	0·202	0·213	0·224	0·234	0·244	0·264	0·284	0·302
1/ 909	0·0305	0·0552	0·0898	0·1935	0·3508	0·5704	0·7061	0·8603	1·0341	1·2283	1·6816	2·2271	2·8716
0·00120	0·101	0·118	0·133	0·161	0·187	0·211	0·222	0·233	0·244	0·255	0·276	0·296	0·316
1/ 833	0·0318	0·0577	0·0938	0·2021	0·3663	0·5957	0·7375	0·8986	1·0801	1·2830	1·7564	2·3262	2·9993
0·00130	0·105	0·122	0·138	0·167	0·194	0·219	0·231	0·243	0·254	0·266	0·287	0·308	0·328
1/ 769	0·0331	0·0601	0·0977	0·2103	0·3813	0·6200	0·7676	0·9353	1·1242	1·3353	1·8281	2·4211	3·1218
0·00140	0·109	0·127	0·143	0·174	0·202	0·228	0·240	0·252	0·264	0·276	0·298	0·320	0·341
1/ 714	0·0344	0·0623	0·1013	0·2182	0·3957	0·6435	0·7966	0·9706	1·1667	1·3858	1·8971	2·5125	3·2396
0·00150	0·113	0·131	0·148	0·180	0·209	0·236	0·248	0·261	0·273	0·285	0·309	0·331	0·353
1/ 667	0·0356	0·0645	0·1049	0·2259	0·4096	0·6660	0·8245	1·0047	1·2076	1·4344	1·9637	2·6007	3·3533
0·00160	0·117	0·136	0·153	0·186	0·215	0·243	0·257	0·270	0·282	0·295	0·319	0·342	0·364
1/ 625	0·0367	0·0666	0·1083	0·2333	0·4230	0·6879	0·8516	1·0376	1·2472	1·4814	2·0281	2·6860	3·4633
0·00170	0·121	0·140	0·158	0·191	0·222	0·251	0·265	0·278	0·291	0·304	0·329	0·353	0·376
1/ 588	0·0379	0·0687	0·1117	0·2405	0·4360	0·7091	0·8778	1·0696	1·2856	1·5270	2·0905	2·7687	3·5699
0·00180	0·124	0·144	0·163	0·197	0·229	0·258	0·272	0·286	0·299	0·313	0·338	0·363	0·387
1/ 556	0·0390	0·0707	0·1149	0·2475	0·4487	0·7296	0·9032	1·1006	1·3229	1·5713	2·1511	2·8490	3·6734
0·00190	0·127	0·148	0·167	0·202	0·235	0·265	0·280	0·294	0·308	0·321	0·347	0·373	0·397
1/ 526	0·0400	0·0726	0·1181	0·2542	0·4610	0·7496	0·9280	1·1307	1·3591	1·6144	2·2101	2·9270	3·7740
0·00200	0·131	0·152	0·171	0·208	0·241	0·272	0·287	0·301	0·316	0·330	0·356	0·382	0·407
1/ 500	0·0411	0·0745	0·1211	0·2608	0·4730	0·7691	0·9521	1·1601	1·3944	1·6563	2·2675	3·0031	3·8721
0·00220	0·137	0·159	0·180	0·218	0·253	0·285	0·301	0·316	0·331	0·346	0·374	0·401	0·427
1/ 455	0·0431	0·0781	0·1270	0·2736	0·4960	0·8066	0·9985	1·2167	1·4625	1·7371	2·3782	3·1496	4·0611
0·00240	0·143	0·166	0·188	0·227	0·264	0·298	0·314	0·330	0·346	0·361	0·390	0·419	0·446
1/ 417	0·0450	0·0816	0·1327	0·2857	0·5181	0·8425	1·0429	1·2708	1·5275	1·8144	2·4839	3·2897	4·2417
0·00260	0·149	0·173	0·195	0·237	0·275	0·310	0·327	0·344	0·360	0·376	0·406	0·436	0·465
1/ 385	0·0468	0·0849	0·1381	0·2974	0·5392	0·8769	1·0855	1·3227	1·5899	1·8885	2·5853	3·4240	4·4149
0·00280	0·155	0·180	0·203	0·246	0·285	0·322	0·339	0·357	0·373	0·390	0·422	0·452	0·482
1/ 357	0·0486	0·0881	0·1433	0·3086	0·5596	0·9100	1·1265	1·3726	1·6499	1·9598	2·6829	3·5533	4·5815
	20	25	30	40	50	60	65	70	75	80	90	100	110

Gradient   (Equivalent) Pipe diameters in mm

S = 0·00050 to 0·00280

m = Manning $n \times 100$
S = 0.00300 to 0.01400

ie hydraulic gradient =
1 in 333 to 1 in 71

Full bore conditions; Select
m to divide mV and/or mQ.

mV to give velocities in ms^{-1}
mQ to give discharges in litres/sec

Gradient (Equivalent) Pipe diameters in mm

Gradient	20	25	30	40	50	60	65	70	75	80	90	100	110
0.00300	0.160	0.186	0.210	0.254	0.295	0.333	0.351	0.369	0.387	0.404	0.437	0.468	0.499
1/ 333	0.0503	0.0912	0.1483	0.3195	0.5792	0.9419	1.1660	1.4208	1.7078	2.0285	2.7771	3.6780	4.7423
0.00320	0.165	0.192	0.217	0.263	0.305	0.344	0.363	0.381	0.399	0.417	0.451	0.484	0.515
1/ 313	0.0520	0.0942	0.1532	0.3300	0.5982	0.9728	1.2043	1.4674	1.7638	2.0951	2.8682	3.7986	4.8978
0.00340	0.170	0.198	0.223	0.271	0.314	0.355	0.374	0.393	0.412	0.430	0.465	0.499	0.531
1/ 294	0.0536	0.0971	0.1579	0.3401	0.6167	1.0027	1.2413	1.5126	1.8181	2.1595	2.9564	3.9155	5.0486
0.00360	0.175	0.204	0.230	0.278	0.323	0.365	0.385	0.404	0.423	0.442	0.478	0.513	0.547
1/ 278	0.0551	0.0999	0.1625	0.3500	0.6345	1.0318	1.2773	1.5564	1.8708	2.2222	3.0421	4.0290	5.1949
0.00380	0.180	0.209	0.236	0.286	0.332	0.375	0.395	0.416	0.435	0.454	0.491	0.527	0.562
1/ 263	0.0566	0.1027	0.1670	0.3596	0.6519	1.0601	1.3123	1.5991	1.9221	2.2830	3.1255	4.1394	5.3373
0.00400	0.185	0.215	0.242	0.294	0.341	0.385	0.406	0.426	0.446	0.466	0.504	0.541	0.576
1/ 250	0.0581	0.1053	0.1713	0.3689	0.6689	1.0876	1.3464	1.6406	1.9720	2.3424	3.2067	4.2470	5.4760
0.00420	0.189	0.220	0.248	0.301	0.349	0.394	0.416	0.437	0.457	0.478	0.517	0.554	0.590
1/ 238	0.0595	0.1079	0.1755	0.3780	0.6854	1.1145	1.3797	1.6811	2.0207	2.4002	3.2859	4.3518	5.6112
0.00440	0.194	0.225	0.254	0.308	0.357	0.403	0.426	0.447	0.468	0.489	0.529	0.567	0.604
1/ 227	0.0609	0.1105	0.1797	0.3869	0.7015	1.1407	1.4121	1.7207	2.0683	2.4567	3.3632	4.4543	5.7432
0.00460	0.198	0.230	0.260	0.315	0.365	0.413	0.435	0.457	0.479	0.500	0.541	0.580	0.618
1/ 217	0.0623	0.1130	0.1837	0.3956	0.7173	1.1664	1.4439	1.7594	2.1147	2.5119	3.4388	4.5544	5.8723
0.00480	0.203	0.235	0.265	0.322	0.373	0.421	0.444	0.467	0.489	0.510	0.552	0.592	0.631
1/ 208	0.0636	0.1154	0.1876	0.4041	0.7327	1.1914	1.4749	1.7972	2.1602	2.5659	3.5128	4.6523	5.9986
0.00500	0.207	0.240	0.271	0.328	0.381	0.430	0.454	0.477	0.499	0.521	0.564	0.605	0.644
1/ 200	0.0650	0.1178	0.1915	0.4124	0.7478	1.2160	1.5053	1.8343	2.2048	2.6188	3.5852	4.7483	6.1223
0.00550	0.217	0.252	0.284	0.344	0.399	0.451	0.476	0.500	0.523	0.546	0.591	0.634	0.676
1/ 182	0.0681	0.1235	0.2009	0.4326	0.7843	1.2754	1.5788	1.9238	2.3124	2.7467	3.7602	4.9800	6.4211
0.00600	0.226	0.263	0.297	0.360	0.417	0.471	0.497	0.522	0.547	0.571	0.617	0.662	0.706
1/ 167	0.0712	0.1290	0.2098	0.4518	0.8192	1.3321	1.6490	2.0093	2.4152	2.8688	3.9274	5.2015	6.7066
0.00650	0.236	0.274	0.309	0.374	0.434	0.490	0.517	0.543	0.569	0.594	0.643	0.689	0.735
1/ 154	0.0741	0.1343	0.2184	0.4703	0.8526	1.3865	1.7164	2.0914	2.5138	2.9859	4.0878	5.4138	6.9805
0.00700	0.245	0.284	0.321	0.388	0.451	0.509	0.537	0.564	0.590	0.616	0.667	0.715	0.762
1/ 143	0.0769	0.1393	0.2266	0.4880	0.8848	1.4388	1.7811	2.1703	2.6087	3.0986	4.2421	5.6182	7.2440
0.00750	0.253	0.294	0.332	0.402	0.466	0.527	0.556	0.584	0.611	0.638	0.690	0.740	0.789
1/ 133	0.0796	0.1442	0.2346	0.5051	0.9159	1.4893	1.8437	2.2465	2.7003	3.2074	4.3910	5.8154	7.4983
0.00800	0.262	0.303	0.343	0.415	0.482	0.544	0.574	0.603	0.631	0.659	0.713	0.765	0.815
1/ 125	0.0822	0.1490	0.2422	0.5217	0.9459	1.5381	1.9041	2.3202	2.7888	3.3126	4.5350	6.0061	7.7442
0.00850	0.270	0.313	0.353	0.428	0.497	0.561	0.591	0.621	0.651	0.679	0.735	0.788	0.840
1/ 118	0.0847	0.1536	0.2497	0.5378	0.9750	1.5855	1.9627	2.3916	2.8747	3.4145	4.6745	6.1910	7.9825
0.00900	0.277	0.322	0.363	0.440	0.511	0.577	0.609	0.639	0.670	0.699	0.756	0.811	0.864
1/ 111	0.0871	0.1580	0.2569	0.5533	1.0033	1.6314	2.0196	2.4609	2.9580	3.5135	4.8101	6.3705	8.2139
0.00950	0.285	0.331	0.373	0.452	0.525	0.593	0.625	0.657	0.688	0.718	0.777	0.833	0.888
1/ 105	0.0895	0.1623	0.2640	0.5685	1.0308	1.6762	2.0750	2.5284	3.0391	3.6098	4.9419	6.5450	8.4390
0.01000	0.292	0.339	0.383	0.464	0.539	0.608	0.642	0.674	0.706	0.737	0.797	0.855	0.911
1/ 100	0.0919	0.1666	0.2708	0.5833	1.0576	1.7197	2.1289	2.5940	3.1180	3.7036	5.0702	6.7150	8.6582
0.01100	0.307	0.356	0.402	0.487	0.565	0.638	0.673	0.707	0.740	0.773	0.836	0.897	0.956
1/ 91	0.0963	0.1747	0.2841	0.6117	1.1092	1.8036	2.2328	2.7207	3.2702	3.8844	5.3177	7.0428	9.0808
0.01200	0.320	0.372	0.420	0.508	0.590	0.666	0.703	0.738	0.773	0.807	0.873	0.937	0.998
1/ 83	0.1006	0.1825	0.2967	0.6389	1.1585	1.8838	2.3321	2.8416	3.4156	4.0571	5.5542	7.3560	9.4846
0.01300	0.333	0.387	0.437	0.529	0.614	0.693	0.731	0.769	0.805	0.840	0.909	0.975	1.039
1/ 77	0.1047	0.1899	0.3088	0.6650	1.2058	1.9608	2.4273	2.9577	3.5551	4.2227	5.7810	7.6563	9.8719
0.01400	0.346	0.401	0.453	0.549	0.637	0.720	0.759	0.798	0.835	0.872	0.943	1.012	1.078
1/ 71	0.1087	0.1971	0.3205	0.6901	1.2513	2.0348	2.5189	3.0693	3.6893	4.3821	5.9992	7.9454	10.245
	20	25	30	40	50	60	65	70	75	80	90	100	110

Gradient (Equivalent) Pipe diameters in mm

S = 0.00300 to 0.01400

m = Manning $n \times 100$
S = 0·01500 to 0·07000

Full bore conditions; Select m to divide mV and/or mQ.

ie hydraulic gradient =
1 in 67 to 1 in 14·3

mV to give velocities in ms^{-1}
mQ to give discharges in litres/sec

Gradient (Equivalent) Pipe diameters in mm

Gradient	20	25	30	40	50	60	65	70	75	80	90	100	110
0·01500	0·358	0·416	0·469	0·568	0·660	0·745	0·786	0·826	0·864	0·902	0·976	1·047	1·116
1/ 67	0·1125	0·2040	0·3317	0·7144	1·2952	2·1062	2·6073	3·1770	3·8188	4·5359	6·2098	8·2242	10·604
0·01600	0·370	0·429	0·485	0·587	0·681	0·769	0·812	0·853	0·893	0·932	1·008	1·081	1·152
1/ 63	0·1162	0·2107	0·3426	0·7378	1·3377	2·1753	2·6928	3·2812	3·9440	4·6847	6·4134	8·4939	10·952
0·01700	0·381	0·442	0·500	0·605	0·702	0·793	0·836	0·879	0·920	0·961	1·039	1·115	1·188
1/ 59	0·1198	0·2172	0·3531	0·7605	1·3789	2·2422	2·7757	3·3822	4·0654	4·8289	6·6108	8·7554	11·289
0·01800	0·392	0·455	0·514	0·623	0·723	0·816	0·861	0·904	0·947	0·989	1·069	1·147	1·222
1/ 56	0·1232	0·2235	0·3634	0·7825	1·4189	2·3072	2·8562	3·4803	4·1833	4·9689	6·8025	9·0092	11·616
0·01900	0·403	0·468	0·528	0·640	0·742	0·838	0·884	0·929	0·973	1·016	1·099	1·179	1·256
1/ 53	0·1266	0·2296	0·3733	0·8040	1·4577	2·3704	2·9345	3·5756	4·2979	5·1050	6·9889	9·2561	11·935
0·02000	0·414	0·480	0·542	0·656	0·762	0·860	0·907	0·953	0·998	1·042	1·127	1·209	1·288
1/ 50	0·1299	0·2355	0·3830	0·8249	1·4956	2·4320	3·0107	3·6685	4·4095	5·2377	7·1704	9·4965	12·245
0·02200	0·434	0·503	0·568	0·688	0·799	0·902	0·952	1·000	1·047	1·093	1·182	1·268	1·351
1/ 45·5	0·1363	0·2470	0·4017	0·8651	1·5686	2·5507	3·1576	3·8476	4·6248	5·4933	7·5204	9·9600	12·842
0·02400	0·453	0·526	0·594	0·719	0·834	0·942	0·994	1·044	1·093	1·141	1·235	1·325	1·411
1/ 41·7	0·1423	0·2580	0·4196	0·9036	1·6384	2·6641	3·2980	4·0187	4·8304	5·7376	7·8548	10·403	13·413
0·02600	0·471	0·547	0·618	0·748	0·868	0·981	1·034	1·087	1·138	1·188	1·285	1·379	1·469
1/ 38·5	0·1481	0·2686	0·4367	0·9405	1·7053	2·7729	3·4327	4·1828	5·0277	5·9719	8·1755	10·828	13·961
0·02800	0·489	0·568	0·641	0·777	0·901	1·018	1·074	1·128	1·181	1·233	1·334	1·431	1·525
1/ 35·7	0·1537	0·2787	0·4532	0·9760	1·7696	2·8776	3·5623	4·3407	5·2174	6·1973	8·4841	11·236	14·488
0·03000	0·506	0·588	0·664	0·804	0·933	1·053	1·111	1·167	1·222	1·276	1·380	1·481	1·578
1/ 33·3	0·1591	0·2885	0·4691	1·0103	1·8317	2·9786	3·6873	4·4930	5·4006	6·4148	8·7819	11·631	14·997
0·03200	0·523	0·607	0·685	0·830	0·963	1·088	1·148	1·206	1·263	1·318	1·426	1·529	1·630
1/ 31·3	0·1643	0·2979	0·4845	1·0434	1·8918	3·0763	3·8083	4·6404	5·5777	6·6252	9·0699	12·012	15·488
0·03400	0·539	0·626	0·707	0·856	0·993	1·122	1·183	1·243	1·301	1·359	1·470	1·577	1·680
1/ 29·4	0·1694	0·3071	0·4994	1·0755	1·9500	3·1710	3·9255	4·7832	5·7493	6·8291	9·3491	12·382	15·965
0·03600	0·555	0·644	0·727	0·881	1·022	1·154	1·217	1·279	1·339	1·398	1·512	1·622	1·729
1/ 27·8	0·1743	0·3160	0·5139	1·1067	2·0066	3·2629	4·0393	4·9219	5·9160	7·0271	9·6201	12·741	16·428
0·03800	0·570	0·661	0·747	0·905	1·050	1·186	1·251	1·314	1·376	1·436	1·554	1·667	1·776
1/ 26·3	0·1791	0·3247	0·5280	1·1370	2·0616	3·3523	4·1500	5·0567	6·0781	7·2196	9·8837	13·090	16·878
0·04000	0·585	0·679	0·766	0·928	1·077	1·216	1·283	1·348	1·412	1·474	1·594	1·710	1·822
1/ 25·0	0·1837	0·3331	0·5417	1·1666	2·1151	3·4394	4·2578	5·1881	6·2360	7·4072	10·140	13·430	17·316
0·04200	0·599	0·695	0·785	0·951	1·104	1·246	1·315	1·381	1·446	1·510	1·633	1·752	1·867
1/ 23·8	0·1883	0·3413	0·5550	1·1954	2·1673	3·5243	4·3629	5·3162	6·3900	7·5901	10·391	13·762	17·744
0·04400	0·613	0·712	0·804	0·974	1·130	1·276	1·346	1·414	1·480	1·546	1·672	1·793	1·911
1/ 22·7	0·1927	0·3494	0·5681	1·2235	2·2183	3·6073	4·4656	5·4413	6·5404	7·7687	10·635	14·086	18·162
0·04600	0·627	0·728	0·822	0·996	1·155	1·304	1·376	1·446	1·514	1·580	1·709	1·834	1·954
1/ 21·7	0·1970	0·3572	0·5809	1·2510	2·2682	3·6883	4·5659	5·5636	6·6874	7·9433	10·874	14·402	18·570
0·04800	0·641	0·743	0·839	1·017	1·180	1·333	1·406	1·477	1·546	1·614	1·746	1·873	1·996
1/ 20·8	0·2013	0·3649	0·5934	1·2779	2·3170	3·7677	4·6641	5·6833	6·8312	8·1142	11·108	14·712	18·969
0·05000	0·654	0·759	0·857	1·038	1·204	1·360	1·435	1·507	1·578	1·648	1·782	1·912	2·037
1/ 20·0	0·2054	0·3724	0·6056	1·3042	2·3648	3·8454	4·7603	5·8005	6·9721	8·2815	11·337	15·015	19·360
0·05500	0·686	0·796	0·899	1·089	1·263	1·426	1·505	1·581	1·655	1·728	1·869	2·005	2·137
1/ 18·2	0·2154	0·3906	0·6352	1·3679	2·4802	4·0331	4·9927	6·0836	7·3124	8·6857	11·891	15·748	20·305
0·06000	0·716	0·831	0·939	1·137	1·319	1·490	1·571	1·651	1·729	1·805	1·952	2·094	2·232
1/ 16·7	0·2250	0·4080	0·6634	1·4287	2·5905	4·2124	5·2147	6·3541	7·6376	9·0719	12·420	16·448	21·208
0·06500	0·745	0·865	0·977	1·183	1·373	1·551	1·636	1·718	1·799	1·878	2·032	2·180	2·323
1/ 15·4	0·2342	0·4246	0·6905	1·4871	2·6962	4·3844	5·4276	6·6135	7·9494	9·4423	12·927	17·120	22·074
0·07000	0·774	0·898	1·014	1·228	1·425	1·609	1·697	1·783	1·867	1·949	2·109	2·262	2·410
1/ 14·3	0·2430	0·4407	0·7166	1·5432	2·7980	4·5499	5·6325	6·8632	8·2495	9·7988	13·415	17·766	22·908
	20	25	30	40	50	60	65	70	75	80	90	100	110

Gradient (Equivalent) Pipe diameters in mm

S = 0·01500 to 0·07000

m = Manning $n \times 100$
S = 0·07500 to 0·40000

Full bore conditions; Select m to divide mV and/or mQ.

D1
continued

ie hydraulic gradient =
1 in 13·3 to 1 in 2·5

mV to give velocities in ms^{-1}
mQ to give discharges in litres/sec

Gradient (Equivalent) Pipe diameters in mm

Gradient	20	25	30	40	50	60	65	70	75	80	90	100	110
0·07500	0·801	0·929	1·049	1·271	1·475	1·666	1·757	1·846	1·933	2·018	2·183	2·341	2·495
1/ 13·3	0·2516	0·4561	0·7417	1·5974	2·8962	4·7096	5·8302	7·1041	8·5391	10·143	13·885	18·390	23·712
0·08000	0·827	0·960	1·084	1·313	1·523	1·720	1·815	1·906	1·996	2·084	2·254	2·418	2·577
1/ 12·5	0·2598	0·4711	0·7660	1·6498	2·9912	4·8640	6·0214	7·3371	8·8191	10·475	14·341	18·993	24·489
0·08500	0·852	0·989	1·117	1·353	1·570	1·773	1·870	1·965	2·058	2·148	2·324	2·493	2·656
1/ 11·8	0·2678	0·4856	0·7896	1·7005	3·0833	5·0137	6·2067	7·5629	9·0905	10·798	14·782	19·578	25·243
0·09000	0·877	1·018	1·149	1·392	1·616	1·825	1·925	2·022	2·117	2·210	2·391	2·565	2·733
1/ 11·1	0·2756	0·4997	0·8125	1·7498	3·1727	5·1591	6·3866	7·7821	9·3541	11·111	15·211	20·145	25·975
0·09500	0·901	1·046	1·181	1·431	1·660	1·875	1·977	2·078	2·175	2·271	2·456	2·635	2·808
1/ 10·5	0·2831	0·5134	0·8348	1·7978	3·2596	5·3005	6·5617	7·9954	9·6104	11·415	15·628	20·697	26·686
0·10000	0·925	1·073	1·212	1·468	1·703	1·923	2·029	2·132	2·232	2·330	2·520	2·704	2·881
1/ 10·0	0·2905	0·5267	0·8565	1·8445	3·3443	5·4382	6·7321	8·2031	9·8601	11·712	16·034	21·235	27·380
0·11000	0·970	1·125	1·271	1·539	1·786	2·017	2·128	2·236	2·341	2·444	2·643	2·836	3·022
1/ 9·1	0·3047	0·5524	0·8983	1·9345	3·5075	5·7036	7·0607	8·6035	10·341	12·283	16·816	22·271	28·716
0·12000	1·013	1·175	1·327	1·608	1·866	2·107	2·222	2·335	2·445	2·552	2·761	2·962	3·156
1/ 8·3	0·3182	0·5770	0·9382	2·0205	3·6635	5·9572	7·3747	8·9860	10·801	12·830	17·564	23·262	29·993
0·13000	1·054	1·223	1·381	1·674	1·942	2·193	2·313	2·430	2·545	2·657	2·874	3·083	3·285
1/ 7·7	0·3312	0·6005	0·9765	2·1030	3·8131	6·2005	7·6758	9·3530	11·242	13·353	18·281	24·211	31·218
0·14000	1·094	1·270	1·434	1·737	2·015	2·276	2·400	2·522	2·641	2·757	2·982	3·199	3·409
1/ 7·1	0·3437	0·6232	1·0134	2·1824	3·9570	6·4345	7·9655	9·7060	11·667	13·858	18·971	25·125	32·396
0·15000	1·132	1·314	1·484	1·798	2·086	2·356	2·485	2·611	2·733	2·854	3·087	3·311	3·529
1/ 6·7	0·3558	0·6451	1·0489	2·2590	4·0959	6·6604	8·2451	10·047	12·076	14·344	19·637	26·007	33·533
0·16000	1·170	1·357	1·533	1·857	2·154	2·433	2·566	2·696	2·823	2·947	3·188	3·420	3·644
1/ 6·3	0·3674	0·6662	1·0833	2·3331	4·2302	6·8788	8·5155	10·376	12·472	14·814	20·281	26·860	34·633
0·17000	1·206	1·399	1·580	1·914	2·221	2·508	2·645	2·779	2·910	3·038	3·286	3·525	3·756
1/ 5·9	0·3788	0·6867	1·1167	2·4049	4·3604	7·0905	8·7776	10·696	12·856	15·270	20·905	27·687	35·699
0·18000	1·241	1·440	1·626	1·969	2·285	2·580	2·722	2·860	2·994	3·126	3·381	3·627	3·865
1/ 5·6	0·3897	0·7066	1·1491	2·4746	4·4868	7·2961	9·0321	11·006	13·229	15·713	21·511	28·490	36·734
0·19000	1·275	1·479	1·670	2·023	2·348	2·651	2·796	2·938	3·076	3·212	3·474	3·727	3·971
1/ 5·3	0·4004	0·7260	1·1805	2·5425	4·6098	7·4960	9·2796	11·307	13·591	16·144	22·101	29·270	37·740
0·20000	1·308	1·517	1·714	2·076	2·409	2·720	2·869	3·014	3·156	3·295	3·564	3·824	4·074
1/ 5·0	0·4108	0·7449	1·2112	2·6085	4·7295	7·6907	9·5206	11·601	13·944	16·563	22·675	30·031	38·721
0·22000	1·371	1·591	1·797	2·177	2·526	2·853	3·009	3·162	3·310	3·456	3·738	4·010	4·273
1/ 4·5	0·4309	0·7812	1·2703	2·7358	4·9604	8·0661	9·9853	12·167	14·625	17·371	23·782	31·496	40·611
0·24000	1·432	1·662	1·877	2·274	2·639	2·980	3·143	3·302	3·458	3·610	3·904	4·189	4·463
1/ 4·2	0·4500	0·8159	1·3268	2·8575	5·1809	8·4248	10·429	12·708	15·275	18·144	24·839	32·897	42·417
0·26000	1·491	1·730	1·954	2·367	2·746	3·101	3·271	3·437	3·599	3·757	4·064	4·360	4·646
1/ 3·8	0·4684	0·8493	1·3810	2·9741	5·3925	8·7688	10·855	13·227	15·899	18·885	25·853	34·240	44·149
0·28000	1·547	1·795	2·027	2·456	2·850	3·218	3·395	3·567	3·735	3·899	4·217	4·524	4·821
1/ 3·6	0·4861	0·8813	1·4331	3·0864	5·5960	9·0998	11·265	13·726	16·499	19·598	26·829	35·533	45·815
0·30000	1·602	1·858	2·099	2·542	2·950	3·331	3·514	3·692	3·866	4·036	4·365	4·683	4·990
1/ 3·3	0·5031	0·9123	1·4834	3·1947	5·7925	9·4192	11·660	14·208	17·078	20·285	27·771	36·780	47·423
0·32000	1·654	1·919	2·167	2·626	3·047	3·441	3·629	3·813	3·992	4·168	4·508	4·837	5·154
1/ 3·1	0·5196	0·9422	1·5321	3·2995	5·9824	9·7281	12·043	14·674	17·638	20·951	28·682	37·986	48·978
0·34000	1·705	1·978	2·234	2·706	3·141	3·547	3·741	3·930	4·115	4·296	4·647	4·985	5·312
1/ 2·9	0·5356	0·9712	1·5792	3·4011	6·1665	10·027	12·413	15·126	18·181	21·595	29·564	39·155	50·486
0·36000	1·754	2·036	2·299	2·785	3·232	3·649	3·849	4·044	4·235	4·421	4·782	5·130	5·466
1/ 2·8	0·5512	0·9993	1·6250	3·4997	6·3453	10·318	12·773	15·564	18·708	22·222	30·421	40·290	51·949
0·38000	1·802	2·092	2·362	2·861	3·320	3·749	3·955	4·155	4·351	4·542	4·913	5·270	5·616
1/ 2·6	0·5663	1·0267	1·6695	3·5956	6·5192	10·601	13·123	15·991	19·221	22·830	31·255	41·394	53·373
0·40000	1·848	2·146	2·423	2·936	3·406	3·847	4·058	4·263	4·464	4·660	5·041	5·407	5·762
1/ 2·5	0·5810	1·0538	1·7129	3·6890	6·6886	10·876	13·464	16·406	19·720	23·424	32·067	42·470	54·760
	20	25	30	40	50	60	65	70	75	80	90	100	110

Gradient (Equivalent) Pipe diameters in mm

S = 0·07500 to 0·40000

m = Manning $n \times 100$
S = 0.00015 to 0.00070

Full bore conditions; Select m to divide mV and/or mQ.

ie hydraulic gradient = 1 in 6667 to 1 in 1429

mV to give velocities in ms^{-1}
mQ to give discharges in litres/sec

Gradient — (Equivalent) Pipe diameters in mm

Gradient	110	125	135	150	175	200	225	250	275	300	315	330	350
0.00015	0.112	0.122	0.128	0.137	0.152	0.166	0.180	0.193	0.206	0.218	0.225	0.232	0.241
1/ 6667	1.0604	1.4912	1.8308	2.4248	3.6576	5.2221	7.1490	9.4682	12.208	15.396	17.536	19.852	23.224
0.00016	0.115	0.125	0.132	0.142	0.157	0.172	0.186	0.199	0.212	0.225	0.232	0.240	0.249
1/ 6250	1.0952	1.5401	1.8909	2.5043	3.7776	5.3933	7.3835	9.7787	12.608	15.901	18.111	20.503	23.986
0.00017	0.119	0.129	0.136	0.146	0.162	0.177	0.191	0.205	0.219	0.232	0.240	0.247	0.257
1/ 5882	1.1289	1.5875	1.9491	2.5814	3.8938	5.5593	7.6107	10.080	12.997	16.391	18.668	21.134	24.724
0.00018	0.122	0.133	0.140	0.150	0.167	0.182	0.197	0.211	0.225	0.239	0.246	0.254	0.264
1/ 5556	1.1616	1.6335	2.0056	2.6562	4.0067	5.7205	7.8314	10.372	13.373	16.866	19.209	21.746	25.441
0.00019	0.126	0.137	0.144	0.154	0.171	0.187	0.202	0.217	0.231	0.245	0.253	0.261	0.272
1/ 5263	1.1935	1.6782	2.0605	2.7290	4.1165	5.8772	8.0460	10.656	13.740	17.328	19.736	22.342	26.138
0.00020	0.129	0.140	0.148	0.158	0.176	0.192	0.208	0.223	0.237	0.252	0.260	0.268	0.279
1/ 5000	1.2245	1.7218	2.1141	2.7999	4.2234	6.0299	8.2550	10.933	14.097	17.778	20.248	22.923	26.817
0.00022	0.135	0.147	0.155	0.166	0.184	0.201	0.218	0.234	0.249	0.264	0.273	0.281	0.292
1/ 4545	1.2842	1.8059	2.2173	2.9365	4.4296	6.3242	8.6579	11.467	14.785	18.646	21.237	24.042	28.126
0.00024	0.141	0.154	0.162	0.174	0.192	0.210	0.227	0.244	0.260	0.276	0.285	0.294	0.305
1/ 4167	1.3413	1.8862	2.3159	3.0671	4.6265	6.6054	9.0429	11.976	15.442	19.475	22.181	25.111	29.377
0.00026	0.147	0.160	0.168	0.181	0.200	0.219	0.237	0.254	0.271	0.287	0.296	0.306	0.318
1/ 3846	1.3961	1.9632	2.4104	3.1924	4.8155	6.8752	9.4122	12.465	16.073	20.270	23.087	26.136	30.576
0.00028	0.152	0.166	0.175	0.187	0.208	0.227	0.246	0.264	0.281	0.298	0.307	0.317	0.330
1/ 3571	1.4488	2.0373	2.5014	3.3129	4.9972	7.1347	9.7675	12.936	16.679	21.035	23.958	27.123	31.730
0.00030	0.158	0.172	0.181	0.194	0.215	0.235	0.254	0.273	0.291	0.308	0.318	0.328	0.341
1/ 3333	1.4997	2.1088	2.5892	3.4292	5.1726	7.3851	10.110	13.390	17.265	21.774	24.799	28.075	32.844
0.00032	0.163	0.177	0.187	0.200	0.222	0.243	0.263	0.282	0.300	0.318	0.329	0.339	0.353
1/ 3125	1.5488	2.1780	2.6741	3.5416	5.3423	7.6273	10.442	13.829	17.831	22.488	25.613	28.995	33.921
0.00034	0.168	0.183	0.193	0.207	0.229	0.250	0.271	0.290	0.309	0.328	0.339	0.349	0.363
1/ 2941	1.5965	2.2450	2.7564	3.6506	5.5067	7.8620	10.763	14.255	18.380	23.180	26.401	29.888	34.965
0.00036	0.173	0.188	0.198	0.213	0.236	0.258	0.279	0.299	0.318	0.337	0.349	0.360	0.374
1/ 2778	1.6428	2.3101	2.8363	3.7564	5.6663	8.0900	11.075	14.668	18.913	23.852	27.166	30.754	35.979
0.00038	0.178	0.193	0.204	0.218	0.242	0.265	0.286	0.307	0.327	0.347	0.358	0.369	0.384
1/ 2632	1.6878	2.3734	2.9141	3.8594	5.8216	8.3117	11.379	15.070	19.431	24.506	27.911	31.597	36.965
0.00040	0.182	0.198	0.209	0.224	0.248	0.271	0.294	0.315	0.336	0.356	0.367	0.379	0.394
1/ 2500	1.7316	2.4350	2.9898	3.9596	5.9728	8.5276	11.674	15.462	19.936	25.142	28.636	32.418	37.925
0.00042	0.187	0.203	0.214	0.230	0.254	0.278	0.301	0.323	0.344	0.364	0.377	0.388	0.404
1/ 2381	1.7744	2.4952	3.0636	4.0574	6.1203	8.7382	11.963	15.843	20.428	25.763	29.343	33.218	38.862
0.00044	0.191	0.208	0.219	0.235	0.260	0.285	0.308	0.330	0.352	0.373	0.385	0.398	0.413
1/ 2273	1.8162	2.5539	3.1357	4.1529	6.2644	8.9438	12.244	16.216	20.909	26.369	30.033	34.000	39.776
0.00046	0.195	0.213	0.224	0.240	0.266	0.291	0.315	0.338	0.360	0.381	0.394	0.406	0.423
1/ 2174	1.8570	2.6113	3.2062	4.2462	6.4052	9.1448	12.519	16.581	21.379	26.962	30.708	34.764	40.670
0.00048	0.200	0.217	0.229	0.245	0.272	0.297	0.322	0.345	0.368	0.390	0.403	0.415	0.432
1/ 2083	1.8969	2.6675	3.2751	4.3376	6.5429	9.3415	12.789	16.937	21.839	27.542	31.369	35.512	41.545
0.00050	0.204	0.222	0.234	0.251	0.278	0.303	0.328	0.352	0.375	0.398	0.411	0.424	0.441
1/ 2000	1.9360	2.7225	3.3426	4.4270	6.6778	9.5341	13.052	17.287	22.289	28.110	32.016	36.244	42.402
0.00055	0.214	0.233	0.245	0.263	0.291	0.318	0.344	0.369	0.394	0.417	0.431	0.444	0.462
1/ 1818	2.0305	2.8553	3.5058	4.6431	7.0038	9.9995	13.689	18.130	23.377	29.482	33.578	38.013	44.471
0.00060	0.223	0.243	0.256	0.274	0.304	0.332	0.360	0.386	0.411	0.436	0.450	0.464	0.483
1/ 1667	2.1208	2.9823	3.6617	4.8496	7.3152	10.444	14.298	18.936	24.416	30.793	35.071	39.703	46.449
0.00065	0.232	0.253	0.266	0.286	0.317	0.346	0.374	0.402	0.428	0.453	0.468	0.483	0.502
1/ 1538	2.2074	3.1041	3.8112	5.0476	7.6139	10.871	14.882	19.710	25.413	32.050	36.503	41.325	48.345
0.00070	0.241	0.262	0.276	0.296	0.328	0.359	0.388	0.417	0.444	0.471	0.486	0.501	0.521
1/ 1429	2.2908	3.2213	3.9551	5.2381	7.9013	11.281	15.444	20.454	26.373	33.260	37.881	42.885	50.170
	110	125	135	150	175	200	225	250	275	300	315	330	350

Gradient — (Equivalent) Pipe diameters in mm

S = 0.00015 to 0.00070

m = Manning $n \times 100$
S = 0·00075 to 0·00380

ie hydraulic gradient =
1 in 1333 to 1 in 263

Full bore conditions; Select
m to divide mV and/or mQ.

mV to give velocities in ms^{-1}
mQ to give discharges in litres/sec

D2
continued

Gradient (Equivalent) Pipe diameters in mm

Gradient	110	125	135	150	175	200	225	250	275	300	315	330	350
0·00075	0·250	0·272	0·286	0·307	0·340	0·372	0·402	0·431	0·460	0·487	0·503	0·519	0·540
1/ 1333	2·3712	3·3343	4·0939	5·4220	8·1786	11·677	15·986	21·172	27·298	34·427	39·211	44·390	51·931
0·00080	0·258	0·281	0·295	0·317	0·351	0·384	0·415	0·445	0·475	0·503	0·520	0·536	0·557
1/ 1250	2·4489	3·4437	4·2282	5·5998	8·4469	12·060	16·510	21·866	28·193	35·556	40·497	45·846	53·634
0·00085	0·266	0·289	0·304	0·327	0·362	0·396	0·428	0·459	0·489	0·519	0·536	0·553	0·575
1/ 1176	2·5243	3·5496	4·3583	5·7721	8·7068	12·431	17·018	22·539	29·061	36·651	41·743	47·257	55·285
0·00090	0·273	0·298	0·313	0·336	0·372	0·407	0·440	0·472	0·503	0·534	0·551	0·569	0·591
1/ 1111	2·5975	3·6526	4·4846	5·9395	8·9593	12·791	17·512	23·192	29·904	37·713	42·954	48·627	56·888
0·00095	0·281	0·306	0·322	0·345	0·383	0·418	0·452	0·485	0·517	0·548	0·566	0·584	0·607
1/ 1053	2·6686	3·7526	4·6075	6·1022	9·2048	13·142	17·991	23·828	30·723	38·747	44·131	49·959	58·447
0·00100	0·288	0·314	0·330	0·354	0·393	0·429	0·464	0·498	0·531	0·562	0·581	0·599	0·623
1/ 1000	2·7380	3·8501	4·7272	6·2607	9·4439	13·483	18·459	24·447	31·521	39·753	45·277	51·257	59·965
0·00110	0·302	0·329	0·346	0·372	0·412	0·450	0·487	0·522	0·557	0·590	0·609	0·629	0·654
1/ 909	2·8716	4·0381	4·9580	6·5663	9·9048	14·141	19·360	25·640	33·060	41·694	47·487	53·759	62·892
0·00120	0·316	0·344	0·362	0·388	0·430	0·470	0·509	0·546	0·581	0·616	0·636	0·656	0·683
1/ 833	2·9993	4·2176	5·1784	6·8583	10·345	14·770	20·221	26·780	34·530	43·547	49·598	56·149	65·688
0·00130	0·328	0·358	0·377	0·404	0·448	0·489	0·529	0·568	0·605	0·641	0·662	0·683	0·711
1/ 769	3·1218	4·3898	5·3899	7·1383	10·768	15·373	21·046	27·874	35·940	45·326	51·624	58·442	68·371
0·00140	0·341	0·371	0·391	0·419	0·465	0·508	0·549	0·589	0·628	0·665	0·687	0·709	0·737
1/ 714	3·2396	4·5555	5·5933	7·4078	11·174	15·954	21·841	28·926	37·296	47·037	53·572	60·648	70·951
0·00150	0·353	0·384	0·404	0·434	0·481	0·526	0·569	0·610	0·650	0·689	0·712	0·734	0·763
1/ 667	3·3533	4·7154	5·7896	7·6678	11·566	16·514	22·607	29·941	38·605	48·688	55·453	62·777	73·442
0·00160	0·364	0·397	0·418	0·448	0·497	0·543	0·587	0·630	0·671	0·711	0·735	0·758	0·788
1/ 625	3·4633	4·8701	5·9795	7·9193	11·946	17·055	23·349	30·923	39·872	50·284	57·271	64·836	75·850
0·00170	0·376	0·409	0·431	0·462	0·512	0·560	0·605	0·649	0·692	0·733	0·758	0·781	0·813
1/ 588	3·5699	5·0200	6·1635	8·1630	12·313	17·580	24·067	31·875	41·099	51·832	59·034	66·831	78·185
0·00180	0·387	0·421	0·443	0·475	0·527	0·576	0·623	0·668	0·712	0·755	0·779	0·804	0·836
1/ 556	3·6734	5·1655	6·3422	8·3997	12·670	18·090	24·765	32·799	42·290	53·335	60·745	68·768	80·451
0·00190	0·397	0·432	0·455	0·488	0·541	0·592	0·640	0·686	0·732	0·775	0·801	0·826	0·859
1/ 526	3·7740	5·3070	6·5160	8·6298	13·017	18·585	25·444	33·698	43·449	54·796	62·410	70·653	82·656
0·00200	0·407	0·444	0·467	0·501	0·555	0·607	0·657	0·704	0·751	0·795	0·822	0·848	0·881
1/ 500	3·8721	5·4449	6·6853	8·8540	13·356	19·068	26·105	34·573	44·578	56·220	64·031	72·488	84·803
0·00220	0·427	0·465	0·490	0·525	0·582	0·637	0·689	0·739	0·787	0·834	0·862	0·889	0·924
1/ 455	4·0611	5·7107	7·0116	9·2862	14·008	19·999	27·379	36·261	46·754	58·964	67·157	76·026	88·942
0·00240	0·446	0·486	0·512	0·549	0·608	0·665	0·719	0·772	0·822	0·871	0·900	0·928	0·966
1/ 417	4·2417	5·9646	7·3234	9·6991	14·630	20·888	28·596	37·873	48·832	61·585	70·143	79·407	92·897
0·00260	0·465	0·506	0·533	0·571	0·633	0·692	0·749	0·803	0·856	0·907	0·937	0·966	1·005
1/ 385	4·4149	6·2082	7·6224	10·095	15·228	21·741	29·764	39·419	50·826	64·100	73·007	82·649	96·691
0·00280	0·482	0·525	0·553	0·593	0·657	0·718	0·777	0·833	0·888	0·941	0·972	1·003	1·043
1/ 357	4·5815	6·4425	7·9102	10·476	15·803	22·562	30·887	40·907	52·745	66·520	75·763	85·769	100·34
0·00300	0·499	0·543	0·572	0·614	0·680	0·743	0·804	0·863	0·919	0·974	1·006	1·038	1·080
1/ 333	4·7423	6·6686	8·1878	10·844	16·357	23·354	31·972	42·343	54·596	68·855	78·422	88·780	103·86
0·00320	0·515	0·561	0·591	0·634	0·702	0·768	0·830	0·891	0·949	1·006	1·039	1·072	1·115
1/ 313	4·8978	6·8873	8·4563	11·200	16·894	24·120	33·020	43·732	56·387	71·113	80·994	91·691	107·27
0·00340	0·531	0·579	0·609	0·653	0·724	0·791	0·856	0·918	0·979	1·037	1·071	1·105	1·149
1/ 294	5·0486	7·0993	8·7166	11·544	17·414	24·862	34·036	45·078	58·122	73·301	83·487	94·513	110·57
0·00360	0·547	0·595	0·627	0·672	0·745	0·814	0·881	0·945	1·007	1·067	1·102	1·137	1·183
1/ 278	5·1949	7·3051	8·9693	11·879	17·919	25·583	35·023	46·385	59·807	75·426	85·907	97·253	113·78
0·00380	0·562	0·612	0·644	0·691	0·765	0·837	0·905	0·971	1·035	1·096	1·133	1·168	1·215
1/ 263	5·3373	7·5053	9·2150	12·204	18·410	26·284	35·983	47·656	61·446	77·493	88·261	99·918	116·89
	110	125	135	150	175	200	225	250	275	300	315	330	350

Gradient (Equivalent) Pipe diameters in mm

S = 0·00075 to 0·00380

m = Manning $n \times 100$
S = 0·00400 to 0·01900

ie hydraulic gradient =
1 in 250   to 1 in 53

Full bore conditions; Select
m to divide mV and/or mQ.

mV to give velocities in ms^{-1}
mQ to give discharges in litres/sec

Gradient	(Equivalent) Pipe diameters in mm												
	110	125	135	150	175	200	225	250	275	300	315	330	350
0·00400	0·576	0·627	0·661	0·709	0·785	0·858	0·928	0·996	1·061	1·125	1·162	1·199	1·247
1/ 250	5·4760	7·7003	9·4544	12·521	18·888	26·967	36·918	48·894	63·042	79·506	90·554	102·51	119·93
0·00420	0·590	0·643	0·677	0·726	0·805	0·880	0·951	1·021	1·088	1·153	1·191	1·228	1·277
1/ 238	5·6112	7·8904	9·6879	12·831	19·354	27·633	37·829	50·101	64·599	81·470	92·790	105·05	122·89
0·00440	0·604	0·658	0·693	0·743	0·824	0·900	0·974	1·045	1·113	1·180	1·219	1·257	1·307
1/ 227	5·7432	8·0761	9·9159	13·133	19·810	28·283	38·719	51·280	66·119	83·387	94·974	107·52	125·78
0·00460	0·618	0·673	0·708	0·760	0·842	0·921	0·996	1·068	1·138	1·206	1·246	1·285	1·337
1/ 217	5·8723	8·2576	10·139	13·428	20·255	28·918	39·590	52·433	67·605	85·261	97·108	109·93	128·61
0·00480	0·631	0·687	0·724	0·776	0·860	0·940	1·017	1·091	1·163	1·232	1·273	1·313	1·366
1/ 208	5·9986	8·4352	10·357	13·717	20·691	29·540	40·441	53·560	69·060	87·095	99·197	112·30	131·38
0·00500	0·644	0·702	0·738	0·792	0·878	0·960	1·038	1·114	1·187	1·258	1·299	1·340	1·394
1/ 200	6·1223	8·6092	10·570	13·999	21·117	30·150	41·275	54·665	70·484	88·891	101·24	114·61	134·09
0·00550	0·676	0·736	0·775	0·831	0·921	1·007	1·089	1·168	1·245	1·319	1·363	1·405	1·462
1/ 182	6·4211	9·0294	11·086	14·683	22·148	31·621	43·290	57·333	73·924	93·230	106·18	120·21	140·63
0·00600	0·706	0·768	0·809	0·868	0·962	1·051	1·137	1·220	1·300	1·378	1·423	1·468	1·527
1/ 167	6·7066	9·4309	11·579	15·336	23·133	33·027	45·215	59·882	77·211	97·375	110·91	125·55	146·88
0·00650	0·735	0·800	0·842	0·903	1·001	1·094	1·184	1·270	1·353	1·434	1·481	1·528	1·589
1/ 154	6·9805	9·8160	12·052	15·962	24·077	34·376	47·061	62·327	80·364	101·35	115·43	130·68	152·88
0·00700	0·762	0·830	0·874	0·937	1·039	1·136	1·228	1·318	1·404	1·488	1·537	1·586	1·649
1/ 143	7·2440	10·186	12·507	16·564	24·986	35·673	48·837	64·680	83·397	105·18	119·79	135·61	158·65
0·00750	0·789	0·859	0·904	0·970	1·075	1·175	1·271	1·364	1·453	1·540	1·591	1·641	1·707
1/ 133	7·4983	10·544	12·946	17·146	25·863	36·926	50·551	66·950	86·324	108·87	124·00	140·37	164·22
0·00800	0·815	0·887	0·934	1·002	1·111	1·214	1·313	1·409	1·501	1·591	1·643	1·695	1·763
1/ 125	7·7442	10·890	13·371	17·708	26·711	38·137	52·209	69·146	89·155	112·44	128·06	144·98	169·61
0·00850	0·840	0·915	0·963	1·033	1·145	1·251	1·353	1·452	1·547	1·640	1·694	1·747	1·817
1/ 118	7·9825	11·225	13·782	18·253	27·533	39·310	53·816	71·274	91·899	115·90	132·00	149·44	174·83
0·00900	0·864	0·941	0·991	1·063	1·178	1·288	1·393	1·494	1·592	1·687	1·743	1·798	1·870
1/ 111	8·2139	11·550	14·182	18·782	28·332	40·450	55·376	73·340	94·564	119·26	135·83	153·77	179·89
0·00950	0·888	0·967	1·018	1·092	1·210	1·323	1·431	1·535	1·636	1·733	1·791	1·847	1·921
1/ 105	8·4390	11·867	14·570	19·297	29·108	41·558	56·894	75·350	97·155	122·53	139·55	157·98	184·82
0·01000	0·911	0·992	1·044	1·120	1·242	1·357	1·468	1·575	1·678	1·778	1·837	1·895	1·971
1/ 100	8·6582	12·175	14·949	19·798	29·864	42·638	58·372	77·308	99·679	125·71	143·18	162·09	189·63
0·01100	0·956	1·041	1·095	1·175	1·302	1·423	1·540	1·652	1·760	1·865	1·927	1·988	2·067
1/ 91	9·0808	12·769	15·678	20·765	31·322	44·719	61·221	81·081	104·54	131·85	150·17	170·00	198·88
0·01200	0·998	1·087	1·144	1·227	1·360	1·487	1·608	1·725	1·838	1·948	2·013	2·076	2·159
1/ 83	9·4846	13·337	16·376	21·688	32·715	46·707	63·943	84·686	109·19	137·71	156·84	177·56	207·72
0·01300	1·039	1·131	1·191	1·277	1·416	1·547	1·674	1·796	1·913	2·028	2·095	2·161	2·247
1/ 77	9·8719	13·882	17·044	22·573	34·050	48·615	66·554	88·144	113·65	143·33	163·25	184·81	216·21
0·01400	1·078	1·174	1·236	1·326	1·469	1·606	1·737	1·863	1·986	2·104	2·174	2·242	2·332
1/ 71	10·245	14·406	17·688	23·426	35·336	50·450	69·066	91·472	117·94	148·74	169·41	191·79	224·37
0·01500	1·116	1·215	1·279	1·372	1·521	1·662	1·798	1·929	2·055	2·178	2·250	2·321	2·414
1/ 67	10·604	14·912	18·308	24·248	36·576	52·221	71·490	94·682	122·08	153·96	175·36	198·52	232·24
0·01600	1·152	1·255	1·321	1·417	1·571	1·717	1·857	1·992	2·123	2·250	2·324	2·397	2·493
1/ 63	10·952	15·401	18·909	25·043	37·776	53·933	73·835	97·787	126·08	159·01	181·11	205·03	239·86
0·01700	1·188	1·294	1·362	1·461	1·619	1·770	1·914	2·053	2·188	2·319	2·395	2·471	2·570
1/ 59	11·289	15·875	19·491	25·814	38·938	55·593	76·107	100·80	129·97	163·91	186·68	211·34	247·24
0·01800	1·222	1·331	1·401	1·503	1·666	1·821	1·970	2·113	2·252	2·386	2·465	2·543	2·644
1/ 56	11·616	16·335	20·056	26·562	40·067	57·205	78·314	103·72	133·73	168·66	192·09	217·46	254·41
0·01900	1·256	1·368	1·440	1·544	1·711	1·871	2·024	2·171	2·313	2·451	2·532	2·612	2·717
1/ 53	11·935	16·782	20·605	27·290	41·165	58·772	80·460	106·56	137·40	173·28	197·36	223·42	261·38
	110	125	135	150	175	200	225	250	275	300	315	330	350

Gradient   (Equivalent) Pipe diameters in mm

S = 0·00400 to 0·01900

m = Manning $n \times 100$
S = 0·02000 to 0·10000

ie hydraulic gradient =
1 in 50 to 1 in 10·0

Full bore conditions; Select m to divide mV and/or mQ.

mV to give velocities in ms^{-1}
mQ to give discharges in litres/sec

Gradient — (Equivalent) Pipe diameters in mm

Gradient	110	125	135	150	175	200	225	250	275	300	315	330	350
0·02000	1·288	1·403	1·477	1·584	1·756	1·919	2·076	2·227	2·373	2·515	2·598	2·680	2·787
1/ 50	12·245	17·218	21·141	27·999	42·234	60·299	82·550	109·33	140·97	177·78	202·48	229·23	268·17
0·02200	1·351	1·472	1·549	1·662	1·842	2·013	2·178	2·336	2·489	2·638	2·725	2·811	2·923
1/ 45·5	12·842	18·059	22·173	29·365	44·296	63·242	86·579	114·67	147·85	186·46	212·37	240·42	281·26
0·02400	1·411	1·537	1·618	1·736	1·923	2·103	2·274	2·440	2·600	2·755	2·846	2·936	3·053
1/ 41·7	13·413	18·862	23·159	30·671	46·265	66·054	90·429	119·76	154·42	194·75	221·81	251·11	293·77
0·02600	1·469	1·600	1·684	1·807	2·002	2·188	2·367	2·539	2·706	2·868	2·962	3·056	3·178
1/ 38·5	13·961	19·632	24·104	31·924	48·155	68·752	94·122	124·65	160·73	202·70	230·87	261·36	305·76
0·02800	1·525	1·660	1·748	1·875	2·078	2·271	2·457	2·635	2·808	2·976	3·074	3·171	3·298
1/ 35·7	14·488	20·373	25·014	33·129	49·972	71·347	97·675	129·36	166·79	210·35	239·58	271·23	317·30
0·03000	1·578	1·718	1·809	1·941	2·151	2·351	2·543	2·728	2·907	3·080	3·182	3·282	3·414
1/ 33·3	14·997	21·088	25·892	34·292	51·726	73·851	101·10	133·90	172·65	217·74	247·99	280·75	328·44
0·03200	1·630	1·775	1·868	2·004	2·221	2·428	2·626	2·817	3·002	3·181	3·287	3·390	3·526
1/ 31·3	15·488	21·780	26·741	35·416	53·423	76·273	104·42	138·29	178·31	224·88	256·13	289·95	339·21
0·03400	1·680	1·829	1·926	2·066	2·289	2·503	2·707	2·904	3·094	3·279	3·388	3·494	3·634
1/ 29·4	15·965	22·450	27·564	36·506	55·067	78·620	107·63	142·55	183·80	231·80	264·01	298·88	349·65
0·03600	1·729	1·882	1·982	2·126	2·356	2·575	2·785	2·988	3·184	3·374	3·486	3·596	3·740
1/ 27·8	16·428	23·101	28·363	37·564	56·663	80·900	110·75	146·68	189·13	238·52	271·66	307·54	359·79
0·03800	1·776	1·934	2·036	2·184	2·420	2·646	2·862	3·070	3·271	3·467	3·581	3·694	3·842
1/ 26·3	16·878	23·734	29·141	38·594	58·216	83·117	113·79	150·70	194·31	245·06	279·11	315·97	369·65
0·04000	1·822	1·984	2·089	2·241	2·483	2·714	2·936	3·150	3·356	3·557	3·674	3·790	3·942
1/ 25·0	17·316	24·350	29·898	39·596	59·728	85·276	116·74	154·62	199·36	251·42	286·36	324·18	379·25
0·04200	1·867	2·033	2·140	2·296	2·545	2·781	3·009	3·228	3·439	3·645	3·765	3·884	4·039
1/ 23·8	17·744	24·952	30·636	40·574	61·203	87·382	119·63	158·43	204·28	257·63	293·43	332·18	388·62
0·04400	1·911	2·081	2·191	2·350	2·604	2·847	3·079	3·304	3·520	3·730	3·854	3·975	4·134
1/ 22·7	18·162	25·539	31·357	41·529	62·644	89·438	122·44	162·16	209·09	263·69	300·33	340·00	397·76
0·04600	1·954	2·128	2·240	2·403	2·663	2·911	3·149	3·378	3·599	3·814	3·940	4·065	4·227
1/ 21·7	18·570	26·113	32·062	42·462	64·052	91·448	125·19	165·81	213·79	269·62	307·08	347·64	406·70
0·04800	1·996	2·174	2·288	2·455	2·720	2·973	3·216	3·450	3·677	3·896	4·025	4·152	4·318
1/ 20·8	18·969	26·675	32·751	43·376	65·429	93·415	127·89	169·37	218·39	275·42	313·69	355·12	415·45
0·05000	2·037	2·218	2·335	2·505	2·776	3·035	3·283	3·522	3·753	3·977	4·108	4·238	4·407
1/ 20·0	19·360	27·225	33·426	44·270	66·778	95·341	130·52	172·87	222·89	281·10	320·16	362·44	424·02
0·05500	2·137	2·327	2·449	2·627	2·912	3·183	3·443	3·693	3·936	4·171	4·309	4·444	4·622
1/ 18·2	20·305	28·553	35·058	46·431	70·038	99·995	136·89	181·30	233·77	294·82	335·78	380·13	444·71
0·06000	2·232	2·430	2·558	2·744	3·041	3·324	3·596	3·858	4·111	4·356	4·500	4·642	4·828
1/ 16·7	21·208	29·823	36·617	48·496	73·152	104·44	142·98	189·36	244·16	307·93	350·71	397·03	464·49
0·06500	2·323	2·529	2·663	2·856	3·165	3·460	3·743	4·015	4·279	4·534	4·684	4·832	5·025
1/ 15·4	22·074	31·041	38·112	50·476	76·139	108·71	148·82	197·10	254·13	320·50	365·03	413·25	483·45
0·07000	2·410	2·625	2·763	2·964	3·285	3·591	3·884	4·167	4·440	4·705	4·861	5·014	5·215
1/ 14·3	22·908	32·213	39·551	52·381	79·013	112·81	154·44	204·54	263·73	332·60	378·81	428·85	501·70
0·07500	2·495	2·717	2·860	3·068	3·400	3·717	4·020	4·313	4·596	4·870	5·031	5·190	5·398
1/ 13·3	23·712	33·343	40·939	54·220	81·786	116·77	159·86	211·72	272·98	344·27	392·11	443·90	519·31
0·08000	2·577	2·806	2·954	3·169	3·512	3·839	4·152	4·454	4·747	5·030	5·197	5·360	5·575
1/ 12·5	24·489	34·437	42·282	55·998	84·469	120·60	165·10	218·66	281·93	355·56	404·97	458·46	536·34
0·08500	2·656	2·893	3·045	3·266	3·620	3·957	4·280	4·592	4·893	5·185	5·356	5·525	5·746
1/ 11·8	25·243	35·496	43·583	57·721	87·068	124·31	170·18	225·39	290·61	366·51	417·43	472·57	552·85
0·09000	2·733	2·976	3·133	3·361	3·725	4·072	4·404	4·725	5·035	5·335	5·512	5·685	5·913
1/ 11·1	25·975	36·526	44·846	59·395	89·593	127·91	175·12	231·92	299·04	377·13	429·54	486·27	568·88
0·09500	2·808	3·058	3·219	3·453	3·827	4·183	4·525	4·854	5·173	5·482	5·663	5·841	6·075
1/ 10·5	26·686	37·526	46·075	61·022	92·048	131·42	179·91	238·28	307·23	387·47	441·31	499·59	584·47
0·10000	2·881	3·137	3·303	3·543	3·926	4·292	4·642	4·980	5·307	5·624	5·810	5·993	6·233
1/ 10·0	27·380	38·501	47·272	62·607	94·439	134·83	184·59	244·47	315·21	397·53	452·77	512·57	599·65
	110	125	135	150	175	200	225	250	275	300	315	330	350

Gradient — (Equivalent) Pipe diameters in mm

S = 0·02000 to 0·10000

D3

m = Manning $n \times 100$
S = 0.00015 to 0.00070

ie hydraulic gradient =
1 in 6667 to 1 in 1429

Full bore conditions; Select
m to divide mV and/or mQ.

mV to give velocities in ms^{-1}
mQ to give discharges in litres/sec

Gradient	(Equivalent) Pipe diameters in mm												
	350	375	400	425	450	475	500	525	550	600	630	675	700
0.00015	0.241	0.253	0.264	0.275	0.285	0.296	0.306	0.316	0.326	0.346	0.357	0.374	0.383
1/ 6667	23.224	27.915	33.158	38.976	45.394	52.434	60.119	68.473	77.517	97.761	111.34	133.84	147.47
0.00016	0.249	0.261	0.273	0.284	0.295	0.306	0.316	0.327	0.337	0.357	0.369	0.386	0.396
1/ 6250	23.986	28.831	34.245	40.254	46.882	54.153	62.091	70.719	80.059	100.97	115.00	138.22	152.30
0.00017	0.257	0.269	0.281	0.292	0.304	0.315	0.326	0.337	0.347	0.368	0.380	0.398	0.408
1/ 5882	24.724	29.718	35.299	41.493	48.325	55.820	64.002	72.895	82.523	104.07	118.54	142.48	156.99
0.00018	0.264	0.277	0.289	0.301	0.313	0.324	0.335	0.346	0.357	0.379	0.391	0.410	0.420
1/ 5556	25.441	30.580	36.323	42.696	49.726	57.438	65.858	75.008	84.915	107.09	121.97	146.61	161.54
0.00019	0.272	0.284	0.297	0.309	0.321	0.333	0.345	0.356	0.367	0.389	0.402	0.421	0.431
1/ 5263	26.138	31.418	37.318	43.866	51.089	59.012	67.662	77.064	87.242	110.03	125.31	150.63	165.97
0.00020	0.279	0.292	0.305	0.317	0.330	0.342	0.354	0.365	0.377	0.399	0.412	0.432	0.442
1/ 5000	26.817	32.234	38.288	45.006	52.416	60.545	69.420	79.066	89.509	112.88	128.57	154.54	170.28
0.00022	0.292	0.306	0.320	0.333	0.346	0.358	0.371	0.383	0.395	0.419	0.433	0.453	0.464
1/ 4545	28.126	33.807	40.156	47.202	54.974	63.500	72.808	82.925	93.877	118.39	134.85	162.08	178.59
0.00024	0.305	0.320	0.334	0.348	0.361	0.374	0.387	0.400	0.413	0.437	0.452	0.473	0.485
1/ 4167	29.377	35.311	41.942	49.301	57.419	66.324	76.046	86.612	98.052	123.66	140.84	169.29	186.53
0.00026	0.318	0.333	0.347	0.362	0.376	0.390	0.403	0.416	0.430	0.455	0.470	0.492	0.504
1/ 3846	30.576	36.752	43.655	51.314	59.763	69.032	79.151	90.149	102.06	128.71	146.59	176.20	194.15
0.00028	0.330	0.345	0.361	0.375	0.390	0.404	0.418	0.432	0.446	0.472	0.488	0.511	0.524
1/ 3571	31.730	38.140	45.302	53.252	62.020	71.638	82.139	93.552	105.91	133.57	152.13	182.85	201.48
0.00030	0.341	0.357	0.373	0.389	0.404	0.418	0.433	0.447	0.461	0.489	0.505	0.529	0.542
1/ 3333	32.844	39.478	46.892	55.121	64.196	74.153	85.022	96.836	109.63	138.25	157.47	189.27	208.55
0.00032	0.353	0.369	0.385	0.401	0.417	0.432	0.447	0.462	0.477	0.505	0.522	0.546	0.560
1/ 3125	33.921	40.773	48.430	56.928	66.302	76.584	87.810	100.01	113.22	142.79	162.63	195.48	215.39
0.00034	0.363	0.381	0.397	0.414	0.430	0.445	0.461	0.476	0.491	0.521	0.538	0.563	0.577
1/ 2941	34.965	42.028	49.921	58.680	68.342	78.941	90.513	103.09	116.70	147.18	167.63	201.50	222.02
0.00036	0.374	0.392	0.409	0.426	0.442	0.458	0.474	0.490	0.505	0.536	0.553	0.579	0.594
1/ 2778	35.979	43.246	51.368	60.382	70.324	81.230	93.137	106.08	120.09	151.45	172.49	207.34	228.45
0.00038	0.384	0.402	0.420	0.437	0.454	0.471	0.487	0.503	0.519	0.550	0.569	0.595	0.610
1/ 2632	36.965	44.432	52.776	62.036	72.251	83.456	95.689	108.98	123.38	155.60	177.22	213.02	234.71
0.00040	0.394	0.413	0.431	0.449	0.466	0.483	0.500	0.517	0.533	0.565	0.583	0.611	0.626
1/ 2500	37.925	45.586	54.147	63.648	74.127	85.624	98.175	111.82	126.58	159.64	181.83	218.55	240.81
0.00042	0.404	0.423	0.442	0.460	0.478	0.495	0.512	0.529	0.546	0.579	0.598	0.626	0.641
1/ 2381	38.862	46.712	55.484	65.220	75.958	87.738	100.60	114.58	129.71	163.59	186.32	223.95	246.76
0.00044	0.413	0.433	0.452	0.471	0.489	0.507	0.524	0.542	0.559	0.592	0.612	0.641	0.656
1/ 2273	39.776	47.811	56.790	66.754	77.746	89.803	102.97	117.27	132.76	167.43	190.70	229.22	252.56
0.00046	0.423	0.443	0.462	0.481	0.500	0.518	0.536	0.554	0.571	0.605	0.626	0.655	0.671
1/ 2174	40.670	48.885	58.066	68.255	79.493	91.822	105.28	119.91	135.75	171.20	194.99	234.37	258.24
0.00048	0.432	0.452	0.472	0.491	0.511	0.529	0.548	0.566	0.584	0.619	0.639	0.669	0.685
1/ 2083	41.545	49.937	59.315	69.723	81.203	93.796	107.54	122.49	138.67	174.88	199.18	239.41	263.79
0.00050	0.441	0.461	0.482	0.502	0.521	0.540	0.559	0.577	0.596	0.631	0.652	0.683	0.700
1/ 2000	42.402	50.966	60.538	71.160	82.877	95.731	109.76	125.01	141.53	178.49	203.29	244.35	269.23
0.00055	0.462	0.484	0.505	0.526	0.547	0.567	0.586	0.606	0.625	0.662	0.684	0.716	0.734
1/ 1818	44.471	53.454	63.493	74.634	86.922	100.40	115.12	131.12	148.43	187.20	213.21	256.28	282.37
0.00060	0.483	0.506	0.528	0.549	0.571	0.592	0.612	0.633	0.653	0.692	0.714	0.748	0.766
1/ 1667	46.449	55.831	66.316	77.952	90.787	104.87	120.24	136.95	155.03	195.52	222.69	267.67	294.93
0.00065	0.502	0.526	0.549	0.572	0.594	0.616	0.637	0.658	0.679	0.720	0.744	0.779	0.798
1/ 1538	48.345	58.111	69.024	81.135	94.494	109.15	125.15	142.54	161.36	203.51	231.78	278.60	306.97
0.00070	0.521	0.546	0.570	0.594	0.617	0.639	0.661	0.683	0.705	0.747	0.772	0.808	0.828
1/ 1429	50.170	60.304	71.629	84.198	98.061	113.27	129.87	147.92	167.46	211.19	240.53	289.12	318.56
	350	375	400	425	450	475	500	525	550	600	630	675	700
Gradient	(Equivalent) Pipe diameters in mm												

S = 0.00015 to 0.00070

176

m = Manning $n \times 100$
S = 0·00075 to 0·00380

ie hydraulic gradient =
1 in 1333 to 1 in 263

Full bore conditions; Select
m to divide mV and/or mQ.

mV to give velocities in ms^{-1}
mQ to give discharges in litres/sec

Gradient   **(Equivalent) Pipe diameters in mm**

Gradient	350	375	400	425	450	475	500	525	550	600	630	675	700
0·00075	0·540	0·565	0·590	0·614	0·638	0·662	0·685	0·707	0·730	0·773	0·799	0·836	0·857
1/ 1333	51·931	62·421	74·144	87·153	101·50	117·25	134·43	153·11	173·33	218·60	248·97	299·27	329·74
0·00080	0·557	0·584	0·609	0·634	0·659	0·683	0·707	0·730	0·753	0·798	0·825	0·864	0·885
1/ 1250	53·634	64·468	76·575	90·012	104·83	121·09	138·84	158·13	179·02	225·77	257·14	309·08	340·56
0·00085	0·575	0·602	0·628	0·654	0·679	0·704	0·729	0·753	0·777	0·823	0·850	0·890	0·912
1/ 1176	55·285	66·452	78·932	92·782	108·06	124·82	143·11	163·00	184·53	232·72	265·05	318·59	351·04
0·00090	0·591	0·619	0·646	0·673	0·699	0·725	0·750	0·775	0·799	0·847	0·875	0·916	0·939
1/ 1111	56·888	68·379	81·220	95·472	111·19	128·44	147·26	167·72	189·88	239·46	272·74	327·83	361·21
0·00095	0·607	0·636	0·664	0·691	0·718	0·745	0·771	0·796	0·821	0·870	0·899	0·941	0·964
1/ 1053	58·447	70·252	83·446	98·088	114·24	131·96	151·30	172·32	195·08	246·03	280·21	336·81	371·11
0·00100	0·623	0·653	0·681	0·709	0·737	0·764	0·791	0·817	0·842	0·893	0·922	0·966	0·989
1/ 1000	59·965	72·077	85·614	100·64	117·21	135·38	155·23	176·80	200·15	252·42	287·49	345·56	380·75
0·00110	0·654	0·684	0·715	0·744	0·773	0·801	0·829	0·857	0·884	0·936	0·967	1·013	1·038
1/ 909	62·892	75·595	89·792	105·55	122·93	141·99	162·80	185·43	209·92	264·74	301·52	362·43	399·34
0·00120	0·683	0·715	0·746	0·777	0·807	0·837	0·866	0·895	0·923	0·978	1·010	1·058	1·084
1/ 833	65·688	78·957	93·785	110·24	128·39	148·31	170·04	193·67	219·25	276·51	314·93	378·54	417·09
0·00130	0·711	0·744	0·777	0·809	0·840	0·871	0·901	0·931	0·961	1·018	1·052	1·101	1·128
1/ 769	68·371	82·181	97·614	114·74	133·64	154·36	176·99	201·58	228·20	287·80	327·79	394·00	434·13
0·00140	0·737	0·772	0·806	0·839	0·872	0·904	0·935	0·966	0·997	1·056	1·091	1·143	1·171
1/ 714	70·951	85·283	101·30	119·07	138·68	160·19	183·67	209·19	236·82	298·66	340·16	408·87	450·51
0·00150	0·763	0·799	0·834	0·869	0·903	0·936	0·968	1·000	1·032	1·093	1·130	1·183	1·212
1/ 667	73·442	88·276	104·85	123·25	143·55	165·81	190·11	216·53	245·13	309·15	352·10	423·23	466·33
0·00160	0·788	0·825	0·862	0·897	0·932	0·966	1·000	1·033	1·066	1·129	1·167	1·221	1·251
1/ 625	75·850	91·172	108·29	127·30	148·25	171·25	196·35	223·63	253·17	319·29	363·65	437·11	481·62
0·00170	0·813	0·851	0·888	0·925	0·961	0·996	1·031	1·065	1·098	1·164	1·202	1·259	1·290
1/ 588	78·185	93·977	111·63	131·21	152·82	176·52	202·39	230·51	260·96	329·11	374·84	450·56	496·44
0·00180	0·836	0·876	0·914	0·952	0·989	1·025	1·061	1·096	1·130	1·198	1·237	1·296	1·327
1/ 556	80·451	96·702	114·86	135·02	157·25	181·64	208·26	237·20	268·53	338·65	385·71	463·62	510·83
0·00190	0·859	0·900	0·939	0·978	1·016	1·053	1·090	1·126	1·161	1·231	1·271	1·331	1·364
1/ 526	82·656	99·352	118·01	138·72	161·56	186·61	213·97	243·70	275·88	347·93	396·28	476·32	524·83
0·00200	0·881	0·923	0·963	1·003	1·042	1·080	1·118	1·155	1·191	1·263	1·304	1·366	1·399
1/ 500	84·803	101·93	121·08	142·32	165·75	191·46	219·53	250·03	283·05	356·97	406·57	488·70	538·47
0·00220	0·924	0·968	1·011	1·052	1·093	1·133	1·173	1·211	1·250	1·324	1·368	1·432	1·467
1/ 455	88·942	106·91	126·99	149·27	173·84	200·81	230·24	262·23	296·87	374·40	426·42	512·55	564·75
0·00240	0·966	1·011	1·055	1·099	1·142	1·184	1·225	1·265	1·305	1·383	1·429	1·496	1·533
1/ 417	92·897	111·66	132·63	155·90	181·57	209·74	240·48	273·89	310·07	391·04	445·38	535·34	589·86
0·00260	1·005	1·052	1·099	1·144	1·188	1·232	1·275	1·317	1·358	1·440	1·487	1·557	1·595
1/ 385	96·691	116·22	138·05	162·27	188·99	218·30	250·30	285·08	322·73	407·01	463·57	557·20	613·95
0·00280	1·043	1·092	1·140	1·187	1·233	1·278	1·323	1·367	1·410	1·494	1·543	1·616	1·656
1/ 357	100·34	120·61	143·26	168·40	196·12	226·54	259·75	295·84	334·91	422·38	481·06	578·24	637·12
0·00300	1·080	1·130	1·180	1·229	1·276	1·323	1·369	1·415	1·459	1·546	1·597	1·673	1·714
1/ 333	103·86	124·84	148·29	174·31	203·01	234·49	268·86	306·22	346·67	437·20	497·95	598·53	659·48
0·00320	1·115	1·167	1·219	1·269	1·318	1·367	1·414	1·461	1·507	1·597	1·650	1·727	1·770
1/ 313	107·27	128·94	153·15	180·02	209·66	242·18	277·68	316·26	358·03	451·54	514·28	618·16	681·11
0·00340	1·149	1·203	1·256	1·308	1·359	1·409	1·458	1·506	1·553	1·646	1·701	1·781	1·824
1/ 294	110·57	132·90	157·86	185·56	216·12	249·63	286·23	326·00	369·05	465·43	530·11	637·18	702·07
0·00360	1·183	1·238	1·293	1·346	1·398	1·450	1·500	1·550	1·598	1·694	1·750	1·832	1·877
1/ 278	113·78	136·76	162·44	190·94	222·38	256·87	294·52	335·45	379·75	478·93	545·48	655·66	722·43
0·00380	1·215	1·272	1·328	1·383	1·437	1·489	1·541	1·592	1·642	1·740	1·798	1·882	1·929
1/ 263	116·89	140·50	166·89	196·18	228·48	263·91	302·59	344·64	390·16	492·05	560·42	673·62	742·23
	350	375	400	425	450	475	500	525	550	600	630	675	700

Gradient   **(Equivalent) Pipe diameters in mm**

S = 0·00075 to 0·00380

D3
continued

m = Manning $n \times 100$
S = 0·00400 to 0·01900

ie hydraulic gradient =
1 in 250 to 1 in 53

Full bore conditions; Select
m to divide mV and/or mQ.

mV to give velocities in ms^{-1}
mQ to give discharges in litres/sec

Gradient	(Equivalent) Pipe diameters in mm												
	350	375	400	425	450	475	500	525	550	600	630	675	700
0·00400	1·247	1·305	1·363	1·419	1·474	1·528	1·581	1·633	1·685	1·785	1·845	1·931	1·979
1/ 250	119·93	144·15	171·23	201·27	234·41	270·77	310·46	353·59	400·29	504·83	574·98	691·12	761·51
0·00420	1·277	1·337	1·396	1·454	1·510	1·566	1·620	1·674	1·726	1·830	1·890	1·979	2·028
1/ 238	122·89	147·71	175·46	206·24	240·20	277·45	318·12	362·33	410·18	517·30	589·18	708·19	780·31
0·00440	1·307	1·369	1·429	1·488	1·546	1·603	1·658	1·713	1·767	1·873	1·935	2·026	2·075
1/ 227	125·78	151·19	179·58	211·10	245·85	283·98	325·61	370·85	419·83	529·48	603·05	724·86	798·68
0·00460	1·337	1·400	1·461	1·521	1·581	1·639	1·696	1·752	1·807	1·915	1·978	2·071	2·122
1/ 217	128·61	154·59	183·62	215·84	251·38	290·37	332·93	379·19	429·27	541·37	616·60	741·15	816·63
0·00480	1·366	1·430	1·493	1·554	1·615	1·674	1·732	1·789	1·846	1·956	2·021	2·116	2·168
1/ 208	131·38	157·91	187·57	220·48	256·79	296·61	340·09	387·34	438·50	553·02	629·86	757·09	834·19
0·00500	1·394	1·459	1·523	1·586	1·648	1·708	1·768	1·826	1·884	1·996	2·062	2·159	2·212
1/ 200	134·09	161·17	191·44	225·03	262·08	302·73	347·10	395·33	447·54	564·42	642·85	772·70	851·39
0·00550	1·462	1·530	1·598	1·664	1·728	1·792	1·854	1·915	1·976	2·094	2·163	2·265	2·320
1/ 182	140·63	169·04	200·78	236·01	274·87	317·50	364·04	414·62	469·39	591·97	674·23	810·41	892·95
0·00600	1·527	1·599	1·669	1·738	1·805	1·871	1·936	2·001	2·064	2·187	2·259	2·365	2·423
1/ 167	146·88	176·55	209·71	246·51	287·09	331·62	380·23	433·06	490·26	618·29	704·21	846·45	932·65
0·00650	1·589	1·664	1·737	1·809	1·879	1·948	2·016	2·082	2·148	2·276	2·351	2·462	2·522
1/ 154	152·88	183·76	218·27	256·57	298·82	345·16	395·75	450·74	510·28	643·54	732·96	881·01	970·73
0·00700	1·649	1·727	1·803	1·877	1·950	2·021	2·092	2·161	2·229	2·362	2·440	2·555	2·618
1/ 143	158·65	190·70	226·51	266·26	310·10	358·19	410·69	467·76	529·54	667·83	760·63	914·27	1007·4
0·00750	1·707	1·787	1·866	1·943	2·018	2·092	2·165	2·237	2·307	2·445	2·526	2·645	2·709
1/ 133	164·22	197·39	234·46	275·60	320·98	370·76	425·11	484·18	548·13	691·27	787·33	946·36	1042·7
0·00800	1·763	1·846	1·927	2·006	2·084	2·161	2·236	2·310	2·383	2·525	2·609	2·731	2·798
1/ 125	169·61	203·87	242·15	284·64	331·51	382·92	439·05	500·06	566·10	713·94	813·15	977·40	1076·9
0·00850	1·817	1·903	1·986	2·068	2·149	2·227	2·305	2·381	2·456	2·603	2·689	2·815	2·884
1/ 118	174·83	210·14	249·60	293·40	341·71	394·71	452·56	515·45	583·52	735·92	838·17	1007·5	1110·1
0·00900	1·870	1·958	2·044	2·128	2·211	2·292	2·372	2·450	2·527	2·678	2·767	2·897	2·968
1/ 111	179·89	216·23	256·84	301·91	351·62	406·15	465·68	530·39	600·44	757·25	862·47	1036·7	1142·3
0·00950	1·921	2·011	2·100	2·186	2·271	2·355	2·437	2·517	2·597	2·752	2·843	2·976	3·049
1/ 105	184·82	222·16	263·88	310·18	361·25	417·28	478·44	544·92	616·90	778·00	886·11	1065·1	1173·6
0·01000	1·971	2·064	2·154	2·243	2·330	2·416	2·500	2·583	2·664	2·823	2·916	3·054	3·129
1/ 100	189·63	227·93	270·73	318·24	370·64	428·12	490·87	559·08	632·92	798·21	909·13	1092·8	1204·0
0·01100	2·067	2·164	2·260	2·353	2·444	2·534	2·622	2·709	2·794	2·961	3·059	3·203	3·281
1/ 91	198·88	239·05	283·95	333·77	388·73	449·02	514·83	586·37	663·81	837·17	953·50	1146·1	1262·8
0·01200	2·159	2·261	2·360	2·457	2·553	2·647	2·739	2·829	2·918	3·093	3·195	3·345	3·427
1/ 83	207·72	249·68	296·57	348·61	406·01	468·98	537·72	612·44	693·33	874·40	995·90	1197·1	1319·0
0·01300	2·247	2·353	2·456	2·558	2·657	2·755	2·850	2·945	3·037	3·219	3·325	3·482	3·567
1/ 77	216·21	259·88	308·68	362·85	422·59	488·13	559·68	637·45	721·64	910·10	1036·6	1245·9	1372·8
0·01400	2·332	2·442	2·549	2·654	2·757	2·859	2·958	3·056	3·152	3·340	3·451	3·613	3·702
1/ 71	224·37	269·69	320·34	376·55	438·54	506·56	580·81	661·51	748·88	944·46	1075·7	1293·0	1424·6
0·01500	2·414	2·528	2·639	2·747	2·854	2·959	3·062	3·163	3·263	3·458	3·572	3·740	3·832
1/ 67	232·24	279·15	331·58	389·76	453·94	524·34	601·19	684·73	775·17	977·61	1113·4	1338·4	1474·7
0·01600	2·493	2·610	2·725	2·838	2·948	3·056	3·162	3·267	3·370	3·571	3·689	3·863	3·957
1/ 63	239·86	288·31	342·45	402·54	468·82	541·53	620·91	707·19	800·59	1009·7	1150·0	1382·2	1523·0
0·01700	2·570	2·691	2·809	2·925	3·038	3·150	3·260	3·367	3·473	3·681	3·803	3·982	4·079
1/ 59	247·24	297·18	352·99	414·93	483·25	558·20	640·02	728·95	825·23	1040·7	1185·4	1424·8	1569·9
0·01800	2·644	2·769	2·890	3·010	3·127	3·241	3·354	3·465	3·574	3·788	3·913	4·097	4·198
1/ 56	254·41	305·80	363·23	426·96	497·26	574·38	658·58	750·08	849·15	1070·9	1219·7	1466·1	1615·4
0·01900	2·717	2·845	2·970	3·092	3·212	3·330	3·446	3·560	3·672	3·891	4·020	4·209	4·313
1/ 53	261·38	314·18	373·18	438·66	510·89	590·12	676·62	770·64	872·42	1100·3	1253·1	1506·3	1659·7
	350	375	400	425	450	475	500	525	550	600	630	675	700
Gradient	(Equivalent) Pipe diameters in mm												

·S = 0·00400 to 0·01900

178

m = Manning $n \times 100$
S = 0·02000 to 0·10000

ie hydraulic gradient =
1 in 50 to 1 in 10·0

Full bore conditions; Select
m to divide mV and/or mQ.

mV to give velocities in ms^{-1}
mQ to give discharges in litres/sec

D3
continued

Gradient (Equivalent) Pipe diameters in mm

Gradient	350	375	400	425	450	475	500	525	550	600	630	675	700
0·02000	2·787	2·919	3·047	3·172	3·296	3·417	3·536	3·652	3·767	3·992	4·124	4·319	4·425
1/ 50	268·17	322·34	382·88	450·06	524·16	605·45	694·20	790·66	895·09	1128·8	1285·7	1545·4	1702·8
0·02200	2·923	3·061	3·196	3·327	3·457	3·583	3·708	3·831	3·951	4·187	4·326	4·529	4·641
1/ 45·5	281·26	338·07	401·56	472·02	549·74	635·00	728·08	829·25	938·77	1183·9	1348·5	1620·8	1785·9
0·02400	3·053	3·197	3·338	3·475	3·610	3·743	3·873	4·001	4·127	4·374	4·518	4·731	4·847
1/ 41·7	293·77	353·11	419·42	493·01	574·19	663·24	760·46	866·12	980·52	1236·6	1408·4	1692·9	1865·3
0·02600	3·178	3·328	3·474	3·617	3·758	3·896	4·031	4·164	4·296	4·552	4·703	4·924	5·045
1/ 38·5	305·76	367·52	436·55	513·14	597·63	690·32	791·51	901·49	1020·6	1287·1	1465·9	1762·0	1941·5
0·02800	3·298	3·453	3·605	3·754	3·900	4·043	4·183	4·322	4·458	4·724	4·880	5·110	5·235
1/ 35·7	317·30	381·40	453·02	532·52	620·20	716·38	821·39	935·52	1059·1	1335·7	1521·3	1828·5	2014·8
0·03000	3·414	3·574	3·732	3·885	4·036	4·185	4·330	4·473	4·614	4·890	5·051	5·289	5·419
1/ 33·3	328·44	394·78	468·92	551·21	641·96	741·53	850·22	968·36	1096·3	1382·5	1574·7	1892·7	2085·5
0·03200	3·526	3·692	3·854	4·013	4·169	4·322	4·472	4·620	4·766	5·050	5·217	5·463	5·597
1/ 31·3	339·21	407·73	484·30	569·28	663·02	765·84	878·10	1000·1	1132·2	1427·9	1626·3	1954·8	2153·9
0·03400	3·634	3·805	3·973	4·136	4·297	4·455	4·610	4·762	4·912	5·206	5·378	5·631	5·769
1/ 29·4	349·65	420·28	499·21	586·80	683·42	789·41	905·13	1030·9	1167·0	1471·8	1676·3	2015·0	2220·2
0·03600	3·740	3·916	4·088	4·256	4·422	4·584	4·743	4·900	5·055	5·356	5·534	5·794	5·936
1/ 27·8	359·79	432·46	513·68	603·82	703·24	812·30	931·37	1060·8	1200·9	1514·5	1724·9	2073·4	2284·5
0·03800	3·842	4·023	4·200	4·373	4·543	4·710	4·873	5·035	5·193	5·503	5·685	5·953	6·099
1/ 26·3	369·65	444·32	527·76	620·36	722·51	834·56	956·89	1089·8	1233·8	1556·0	1772·2	2130·2	2347·1
0·04000	3·942	4·127	4·309	4·487	4·661	4·832	5·000	5·165	5·328	5·646	5·833	6·107	6·257
1/ 25·0	379·25	455·86	541·47	636·48	741·27	856·24	981·75	1118·2	1265·8	1596·4	1818·3	2185·5	2408·1
0·04200	4·039	4·229	4·415	4·597	4·776	4·951	5·123	5·293	5·460	5·786	5·977	6·258	6·412
1/ 23·8	388·62	467·12	554·84	652·20	759·58	877·38	1006·0	1145·8	1297·1	1635·9	1863·2	2239·5	2467·6
0·04400	4·134	4·329	4·519	4·706	4·888	5·068	5·244	5·417	5·588	5·922	6·118	6·406	6·563
1/ 22·7	397·76	478·11	567·90	667·54	777·46	898·03	1029·7	1172·7	1327·6	1674·3	1907·0	2292·2	2525·6
0·04600	4·227	4·426	4·621	4·811	4·998	5·182	5·362	5·539	5·714	6·055	6·255	6·549	6·710
1/ 21·7	406·70	488·85	580·66	682·55	794·93	918·22	1052·8	1199·1	1357·5	1712·0	1949·9	2343·7	2582·4
0·04800	4·318	4·521	4·720	4·915	5·106	5·293	5·477	5·658	5·837	6·185	6·390	6·690	6·855
1/ 20·8	415·45	499·37	593·15	697·23	812·03	937·96	1075·4	1224·9	1386·7	1748·8	1991·8	2394·1	2637·9
0·05000	4·407	4·615	4·817	5·016	5·211	5·402	5·590	5·775	5·957	6·313	6·521	6·828	6·996
1/ 20·0	424·02	509·66	605·38	711·60	828·77	957·31	1097·6	1250·1	1415·3	1784·9	2032·9	2443·5	2692·3
0·05500	4·622	4·840	5·053	5·261	5·465	5·666	5·863	6·057	6·248	6·621	6·840	7·162	7·337
1/ 18·2	444·71	534·54	634·93	746·34	869·22	1004·0	1151·2	1311·2	1484·3	1872·0	2132·1	2562·8	2823·7
0·06000	4·828	5·055	5·277	5·495	5·708	5·918	6·124	6·326	6·525	6·915	7·144	7·480	7·664
1/ 16·7	464·49	558·31	663·16	779·52	907·87	1048·7	1202·4	1369·5	1550·3	1955·2	2226·9	2676·7	2949·3
0·06500	5·025	5·261	5·493	5·719	5·941	6·160	6·374	6·584	6·792	7·198	7·435	7·785	7·977
1/ 15·4	483·45	581·11	690·24	811·35	944·94	1091·5	1251·5	1425·4	1613·6	2035·1	2317·8	2786·0	3069·7
0·07000	5·215	5·460	5·700	5·935	6·166	6·392	6·614	6·833	7·048	7·469	7·716	8·079	8·278
1/ 14·3	501·70	603·04	716·29	841·98	980·61	1132·7	1298·7	1479·2	1674·6	2111·9	2405·3	2891·2	3185·6
0·07500	5·398	5·652	5·900	6·144	6·382	6·616	6·847	7·073	7·296	7·731	7·987	8·363	8·568
1/ 13·3	519·31	624·21	741·44	871·53	1015·0	1172·5	1344·3	1531·1	1733·3	2186·0	2489·7	2992·7	3297·4
0·08000	5·575	5·837	6·094	6·345	6·591	6·833	7·071	7·305	7·535	7·985	8·249	8·637	8·849
1/ 12·5	536·34	644·68	765·75	900·12	1048·3	1210·9	1388·4	1581·3	1790·2	2257·7	2571·4	3090·8	3405·6
0·08500	5·746	6·017	6·281	6·540	6·794	7·044	7·289	7·530	7·767	8·231	8·503	8·903	9·122
1/ 11·8	552·85	664·52	789·32	927·82	1080·6	1248·2	1431·1	1630·0	1845·3	2327·2	2650·5	3185·9	3510·4
0·09000	5·913	6·191	6·463	6·730	6·991	7·248	7·500	7·748	7·992	8·469	8·749	9·161	9·386
1/ 11·1	568·88	683·79	812·20	954·72	1111·9	1284·4	1472·6	1677·2	1898·8	2394·6	2727·4	3278·3	3612·1
0·09500	6·075	6·361	6·640	6·914	7·183	7·446	7·706	7·960	8·211	8·701	8·989	9·412	9·643
1/ 10·5	584·47	702·52	834·46	980·88	1142·4	1319·6	1513·0	1723·2	1950·8	2460·3	2802·1	3368·1	3711·1
0·10000	6·233	6·526	6·813	7·094	7·369	7·640	7·906	8·167	8·424	8·927	9·223	9·657	9·894
1/ 10·0	599·65	720·77	856·14	1006·4	1172·1	1353·8	1552·3	1768·0	2001·5	2524·2	2874·9	3455·6	3807·5
	350	375	400	425	450	475	500	525	550	600	630	675	700

Gradient (Equivalent) Pipe diameters in mm

S = 0·02000 to 0·10000

Tables D1 to D3 before this point show values of mV in black

to give mean velocities V in ms^{-1}

and values of mQ in green to give discharges Q in litres/sec

(i.e. Q in m^{3}s^{-1}/1000)

Tables D4 to D12 after this point show values of mV in black

to give mean velocities V in ms^{-1}

and values of mQ in blue to give discharges Q in m^{3}s^{-1}

Tables E follow Tables D

Tables E show values of m_C in black

to give Colebrook-White solutions from Tables D

and values of Ackers' parameter θ , for part-full pipes, in blue

m = Manning $n \times 100$
S = 0·00015 to 0·00070

Full bore conditions; Select m to divide mV and/or mQ.

ie hydraulic gradient =
1 in 6667 to 1 in 1429

mV to give velocities in ms⁻¹
mQ to give discharges in m³s⁻¹

Gradient (Equivalent) Pipe diameters in mm

Gradient	700	750	800	825	850	900	950	975	1000	1025	1050	1100	1125
0·00015	0·383	0·401	0·419	0·428	0·436	0·453	0·470	0·478	0·486	0·494	0·502	0·518	0·526
1/ 6667	0·1475	0·1773	0·2105	0·2285	0·2475	0·2882	0·3329	0·3568	0·3817	0·4077	0·4348	0·4922	0·5226
0·00016	0·396	0·414	0·433	0·442	0·450	0·468	0·485	0·494	0·502	0·510	0·519	0·535	0·543
1/ 6250	0·1523	0·1831	0·2174	0·2360	0·2556	0·2977	0·3439	0·3685	0·3943	0·4211	0·4490	0·5083	0·5397
0·00017	0·408	0·427	0·446	0·455	0·464	0·482	0·500	0·509	0·517	0·526	0·535	0·551	0·560
1/ 5882	0·1570	0·1887	0·2241	0·2433	0·2635	0·3068	0·3544	0·3799	0·4064	0·4340	0·4629	0·5240	0·5563
0·00018	0·420	0·440	0·459	0·468	0·478	0·496	0·515	0·524	0·532	0·541	0·550	0·567	0·576
1/ 5556	0·1615	0·1942	0·2306	0·2504	0·2711	0·3157	0·3647	0·3909	0·4182	0·4466	0·4763	0·5392	0·5725
0·00019	0·431	0·452	0·471	0·481	0·491	0·510	0·529	0·538	0·547	0·556	0·565	0·583	0·592
1/ 5263	0·1660	0·1995	0·2370	0·2572	0·2785	0·3244	0·3747	0·4016	0·4296	0·4589	0·4893	0·5540	0·5882
0·00020	0·442	0·463	0·484	0·494	0·504	0·523	0·542	0·552	0·561	0·571	0·580	0·598	0·607
1/ 5000	0·1703	0·2047	0·2431	0·2639	0·2858	0·3328	0·3844	0·4120	0·4408	0·4708	0·5020	0·5683	0·6034
0·00022	0·464	0·486	0·507	0·518	0·528	0·549	0·569	0·579	0·589	0·598	0·608	0·627	0·637
1/ 4545	0·1786	0·2147	0·2550	0·2768	0·2997	0·3491	0·4032	0·4321	0·4623	0·4938	0·5265	0·5961	0·6329
0·00024	0·485	0·508	0·530	0·541	0·552	0·573	0·594	0·605	0·615	0·625	0·635	0·655	0·665
1/ 4167	0·1865	0·2242	0·2663	0·2891	0·3130	0·3646	0·4211	0·4513	0·4829	0·5157	0·5500	0·6226	0·6610
0·00026	0·504	0·528	0·551	0·563	0·574	0·596	0·618	0·629	0·640	0·651	0·661	0·682	0·692
1/ 3846	0·1941	0·2334	0·2772	0·3009	0·3258	0·3795	0·4383	0·4698	0·5026	0·5368	0·5724	0·6480	0·6880
0·00028	0·524	0·548	0·572	0·584	0·596	0·619	0·642	0·653	0·664	0·675	0·686	0·708	0·718
1/ 3571	0·2015	0·2422	0·2877	0·3123	0·3381	0·3938	0·4549	0·4875	0·5215	0·5570	0·5940	0·6725	0·7140
0·00030	0·542	0·567	0·592	0·605	0·617	0·641	0·664	0·676	0·687	0·699	0·710	0·732	0·744
1/ 3333	0·2085	0·2507	0·2977	0·3232	0·3500	0·4076	0·4708	0·5046	0·5399	0·5766	0·6149	0·6961	0·7391
0·00032	0·560	0·586	0·612	0·624	0·637	0·662	0·686	0·698	0·710	0·722	0·733	0·756	0·768
1/ 3125	0·2154	0·2589	0·3075	0·3338	0·3615	0·4210	0·4863	0·5212	0·5576	0·5955	0·6350	0·7189	0·7633
0·00034	0·577	0·604	0·631	0·644	0·657	0·682	0·707	0·720	0·732	0·744	0·756	0·780	0·792
1/ 2941	0·2220	0·2669	0·3170	0·3441	0·3726	0·4339	0·5012	0·5372	0·5747	0·6138	0·6546	0·7410	0·7868
0·00036	0·594	0·622	0·649	0·662	0·676	0·702	0·728	0·740	0·753	0·765	0·778	0·802	0·814
1/ 2778	0·2285	0·2746	0·3262	0·3541	0·3834	0·4465	0·5158	0·5528	0·5914	0·6316	0·6736	0·7625	0·8096
0·00038	0·610	0·639	0·667	0·680	0·694	0·721	0·748	0·761	0·774	0·786	0·799	0·824	0·837
1/ 2632	0·2347	0·2821	0·3351	0·3638	0·3939	0·4588	0·5299	0·5679	0·6076	0·6489	0·6920	0·7834	0·8318
0·00040	0·626	0·655	0·684	0·698	0·712	0·740	0·767	0·780	0·794	0·807	0·820	0·846	0·859
1/ 2500	0·2408	0·2895	0·3438	0·3732	0·4041	0·4707	0·5437	0·5827	0·6234	0·6658	0·7100	0·8038	0·8534
0·00042	0·641	0·671	0·701	0·715	0·730	0·758	0·786	0·800	0·813	0·827	0·840	0·867	0·880
1/ 2381	0·2468	0·2966	0·3523	0·3824	0·4141	0·4823	0·5571	0·5971	0·6388	0·6822	0·7275	0·8236	0·8745
0·00044	0·656	0·687	0·717	0·732	0·747	0·776	0·804	0·819	0·832	0·846	0·860	0·887	0·900
1/ 2273	0·2526	0·3036	0·3606	0·3914	0·4239	0·4937	0·5702	0·6111	0·6538	0·6983	0·7446	0·8430	0·8951
0·00046	0·671	0·703	0·733	0·749	0·764	0·793	0·823	0·837	0·851	0·865	0·879	0·907	0·921
1/ 2174	0·2582	0·3104	0·3687	0·4002	0·4334	0·5047	0·5830	0·6248	0·6685	0·7140	0·7614	0·8619	0·9152
0·00048	0·685	0·718	0·749	0·765	0·780	0·810	0·840	0·855	0·869	0·884	0·898	0·926	0·940
1/ 2083	0·2638	0·3171	0·3766	0·4088	0·4427	0·5156	0·5956	0·6383	0·6829	0·7293	0·7778	0·8805	0·9349
0·00050	0·700	0·733	0·765	0·781	0·796	0·827	0·858	0·873	0·887	0·902	0·917	0·946	0·960
1/ 2000	0·2692	0·3236	0·3844	0·4173	0·4518	0·5262	0·6079	0·6514	0·6969	0·7444	0·7938	0·8986	0·9541
0·00055	0·734	0·768	0·802	0·819	0·835	0·868	0·899	0·915	0·931	0·946	0·961	0·992	1·007
1/ 1818	0·2824	0·3394	0·4032	0·4376	0·4739	0·5519	0·6375	0·6832	0·7310	0·7807	0·8325	0·9425	1·0007
0·00060	0·766	0·802	0·838	0·855	0·872	0·906	0·939	0·956	0·972	0·988	1·004	1·036	1·051
1/ 1667	0·2949	0·3545	0·4211	0·4571	0·4950	0·5765	0·6659	0·7136	0·7635	0·8154	0·8696	0·9844	1·0452
0·00065	0·798	0·835	0·872	0·890	0·908	0·943	0·978	0·995	1·012	1·029	1·045	1·078	1·094
1/ 1538	0·3070	0·3690	0·4383	0·4758	0·5152	0·6000	0·6931	0·7428	0·7946	0·8487	0·9051	1·0246	1·0879
0·00070	0·828	0·867	0·905	0·924	0·942	0·979	1·015	1·032	1·050	1·067	1·085	1·119	1·136
1/ 1429	0·3186	0·3829	0·4548	0·4937	0·5346	0·6227	0·7192	0·7708	0·8246	0·8808	0·9392	1·0633	1·1289
Gradient	700	750	800	825	850	900	950	975	1000	1025	1050	1100	1125

Gradient (Equivalent) Pipe diameters in mm

S = 0·00015 to 0·00070

m = Manning $n \times 100$
S = 0.00075 to 0.00380

ie hydraulic gradient = 1 in 1333 to 1 in 263

Full bore conditions; Select m to divide mV and/or mQ.

mV to give velocities in ms⁻¹
mQ to give discharges in m³s⁻¹

Gradient — (Equivalent) Pipe diameters in mm

Gradient	700	750	800	825	850	900	950	975	1000	1025	1050	1100	1125
0.00075	0.857	0.897	0.937	0.956	0.975	1.013	1.050	1.069	1.087	1.105	1.123	1.158	1.176
1/ 1333	0.3297	0.3963	0.4708	0.5110	0.5534	0.6445	0.7445	0.7979	0.8536	0.9117	0.9722	1.1006	1.1686
0.00080	0.885	0.927	0.967	0.987	1.007	1.046	1.085	1.104	1.122	1.141	1.160	1.196	1.214
1/ 1250	0.3406	0.4093	0.4862	0.5278	0.5715	0.6656	0.7689	0.8240	0.8816	0.9416	1.0041	1.1367	1.2069
0.00085	0.912	0.955	0.997	1.018	1.038	1.079	1.118	1.138	1.157	1.176	1.195	1.233	1.252
1/ 1176	0.3510	0.4219	0.5012	0.5440	0.5891	0.6861	0.7925	0.8494	0.9087	0.9706	1.0350	1.1717	1.2440
0.00090	0.939	0.983	1.026	1.047	1.068	1.110	1.151	1.171	1.191	1.210	1.230	1.269	1.288
1/ 1111	0.3612	0.4342	0.5157	0.5598	0.6062	0.7060	0.8155	0.8740	0.9351	0.9987	1.0650	1.2056	1.2801
0.00095	0.964	1.010	1.054	1.076	1.098	1.140	1.182	1.203	1.223	1.243	1.264	1.303	1.323
1/ 1053	0.3711	0.4461	0.5298	0.5752	0.6228	0.7254	0.8379	0.8980	0.9607	1.0261	1.0942	1.2387	1.3152
0.00100	0.989	1.036	1.081	1.104	1.126	1.170	1.213	1.234	1.255	1.276	1.296	1.337	1.357
1/ 1000	0.3808	0.4577	0.5436	0.5901	0.6390	0.7442	0.8596	0.9213	0.9856	1.0527	1.1226	1.2709	1.3493
0.00110	1.038	1.087	1.134	1.158	1.181	1.227	1.272	1.294	1.316	1.338	1.360	1.403	1.424
1/ 909	0.3993	0.4800	0.5701	0.6189	0.6702	0.7805	0.9016	0.9663	1.0337	1.1041	1.1774	1.3329	1.4152
0.00120	1.084	1.135	1.185	1.209	1.234	1.281	1.329	1.352	1.375	1.398	1.420	1.465	1.487
1/ 833	0.4171	0.5013	0.5955	0.6464	0.7000	0.8152	0.9417	1.0092	1.0797	1.1532	1.2297	1.3922	1.4781
0.00130	1.128	1.181	1.233	1.259	1.284	1.334	1.383	1.407	1.431	1.455	1.478	1.525	1.548
1/ 769	0.4341	0.5218	0.6198	0.6728	0.7286	0.8485	0.9801	1.0504	1.1238	1.2003	1.2799	1.4490	1.5385
0.00140	1.171	1.226	1.280	1.306	1.332	1.384	1.435	1.460	1.485	1.510	1.534	1.582	1.606
1/ 714	0.4505	0.5415	0.6432	0.6982	0.7561	0.8806	1.0171	1.0901	1.1662	1.2456	1.3283	1.5037	1.5966
0.00150	1.212	1.269	1.325	1.352	1.379	1.433	1.485	1.511	1.537	1.563	1.588	1.638	1.663
1/ 667	0.4663	0.5605	0.6658	0.7227	0.7826	0.9115	1.0528	1.1283	1.2072	1.2893	1.3749	1.5565	1.6526
0.00160	1.251	1.310	1.368	1.396	1.424	1.480	1.534	1.561	1.587	1.614	1.640	1.692	1.717
1/ 625	0.4816	0.5789	0.6876	0.7464	0.8083	0.9414	1.0874	1.1653	1.2467	1.3316	1.4200	1.6075	1.7068
0.00170	1.290	1.351	1.410	1.439	1.468	1.525	1.581	1.609	1.636	1.663	1.690	1.744	1.770
1/ 588	0.4964	0.5967	0.7088	0.7694	0.8332	0.9703	1.1208	1.2012	1.2851	1.3726	1.4637	1.6570	1.7593
0.00180	1.327	1.390	1.451	1.481	1.511	1.569	1.627	1.656	1.684	1.712	1.739	1.794	1.821
1/ 556	0.5108	0.6140	0.7293	0.7917	0.8573	0.9985	1.1533	1.2360	1.3224	1.4124	1.5061	1.7050	1.8103
0.00190	1.364	1.428	1.491	1.522	1.552	1.612	1.672	1.701	1.730	1.759	1.787	1.843	1.871
1/ 526	0.5248	0.6308	0.7493	0.8134	0.8808	1.0258	1.1849	1.2699	1.3586	1.4511	1.5474	1.7518	1.8599
0.00200	1.399	1.465	1.529	1.561	1.593	1.654	1.715	1.745	1.775	1.804	1.833	1.891	1.920
1/ 500	0.5385	0.6472	0.7688	0.8345	0.9037	1.0525	1.2157	1.3029	1.3939	1.4888	1.5876	1.7973	1.9083
0.00220	1.467	1.537	1.604	1.637	1.670	1.735	1.799	1.830	1.861	1.892	1.923	1.984	2.013
1/ 455	0.5647	0.6788	0.8063	0.8753	0.9478	1.1038	1.2750	1.3665	1.4619	1.5614	1.6651	1.8850	2.0014
0.00240	1.533	1.605	1.675	1.710	1.745	1.812	1.879	1.912	1.944	1.976	2.008	2.072	2.103
1/ 417	0.5899	0.7090	0.8422	0.9142	0.9899	1.1529	1.3317	1.4273	1.5269	1.6309	1.7391	1.9688	2.0904
0.00260	1.595	1.670	1.744	1.780	1.816	1.886	1.956	1.990	2.024	2.057	2.090	2.156	2.189
1/ 385	0.6139	0.7380	0.8765	0.9515	1.0304	1.2000	1.3861	1.4855	1.5893	1.6975	1.8101	2.0492	2.1758
0.00280	1.656	1.733	1.810	1.847	1.884	1.957	2.029	2.065	2.100	2.135	2.169	2.238	2.271
1/ 357	0.6371	0.7658	0.9096	0.9874	1.0692	1.2453	1.4384	1.5416	1.6493	1.7615	1.8785	2.1265	2.2579
0.00300	1.714	1.794	1.873	1.912	1.950	2.026	2.101	2.137	2.174	2.210	2.246	2.316	2.351
1/ 333	0.6595	0.7927	0.9416	1.0221	1.1068	1.2890	1.4889	1.5957	1.7072	1.8234	1.9444	2.2012	2.3371
0.00320	1.770	1.853	1.935	1.975	2.014	2.093	2.169	2.207	2.245	2.282	2.319	2.392	2.428
1/ 313	0.6811	0.8187	0.9724	1.0556	1.1431	1.3313	1.5378	1.6480	1.7632	1.8832	2.0081	2.2734	2.4138
0.00340	1.824	1.910	1.994	2.035	2.076	2.157	2.236	2.275	2.314	2.352	2.391	2.466	2.503
1/ 294	0.7021	0.8439	1.0024	1.0881	1.1783	1.3723	1.5851	1.6988	1.8174	1.9411	2.0700	2.3433	2.4881
0.00360	1.877	1.966	2.052	2.094	2.137	2.220	2.301	2.341	2.381	2.421	2.460	2.537	2.576
1/ 278	0.7224	0.8684	1.0314	1.1196	1.2124	1.4120	1.6310	1.7480	1.8701	1.9974	2.1300	2.4113	2.5602
0.00380	1.929	2.019	2.108	2.152	2.195	2.280	2.364	2.405	2.446	2.487	2.527	2.607	2.646
1/ 263	0.7422	0.8921	1.0597	1.1503	1.2456	1.4507	1.6757	1.7959	1.9214	2.0521	2.1883	2.4774	2.6304
	700	750	800	825	850	900	950	975	1000	1025	1050	1100	1125

Gradient — (Equivalent) Pipe diameters in mm

S = 0.00075 to 0.00380

D4
continued

m = Manning $n \times 100$
S = 0·00400 to 0·01900

ie hydraulic gradient =
1 in 250 to 1 in 53

Full bore conditions; Select
m to divide mV and/or mQ.

mV to give velocities in ms^{-1}
mQ to give discharges in m^{3}s^{-1}

Gradient	(Equivalent) Pipe diameters in mm												
	700	750	800	825	850	900	950	975	1000	1025	1050	1100	1125
0·00400	1·979	2·072	2·163	2·208	2·252	2·340	2·426	2·468	2·510	2·552	2·593	2·675	2·715
1/ 250	0·7615	0·9153	1·0872	1·1802	1·2780	1·4884	1·7193	1·8426	1·9713	2·1054	2·2452	2·5417	2·6987
0·00420	2·028	2·123	2·216	2·262	2·308	2·397	2·485	2·529	2·572	2·615	2·657	2·741	2·782
1/ 238	0·7803	0·9379	1·1141	1·2093	1·3096	1·5252	1·7617	1·8881	2·0199	2·1574	2·3006	2·6045	2·7653
0·00440	2·075	2·173	2·269	2·316	2·362	2·454	2·544	2·588	2·632	2·676	2·719	2·805	2·847
1/ 227	0·7987	0·9600	1·1403	1·2378	1·3404	1·5611	1·8032	1·9325	2·0675	2·2082	2·3548	2·6658	2·8304
0·00460	2·122	2·222	2·320	2·368	2·415	2·509	2·601	2·647	2·692	2·736	2·781	2·868	2·911
1/ 217	0·8166	0·9816	1·1659	1·2656	1·3705	1·5962	1·8437	1·9759	2·1140	2·2578	2·4077	2·7257	2·8940
0·00480	2·168	2·270	2·369	2·419	2·467	2·563	2·657	2·703	2·749	2·795	2·840	2·930	2·974
1/ 208	0·8342	1·0027	1·1910	1·2928	1·4000	1·6305	1·8834	2·0184	2·1594	2·3064	2·4595	2·7843	2·9563
0·00500	2·212	2·316	2·418	2·468	2·518	2·616	2·712	2·759	2·806	2·853	2·899	2·990	3·035
1/ 200	0·8514	1·0234	1·2156	1·3195	1·4288	1·6641	1·9222	2·0601	2·2039	2·3540	2·5102	2·8417	3·0172
0·00550	2·320	2·429	2·536	2·589	2·641	2·743	2·844	2·894	2·943	2·992	3·040	3·136	3·184
1/ 182	0·8929	1·0733	1·2749	1·3839	1·4986	1·7453	2·0160	2·1606	2·3115	2·4688	2·6327	2·9804	3·1645
0·00600	2·423	2·538	2·649	2·704	2·758	2·865	2·971	3·023	3·074	3·125	3·176	3·276	3·325
1/ 167	0·9327	1·1210	1·3316	1·4454	1·5652	1·8229	2·1057	2·2567	2·4143	2·5786	2·7498	3·1129	3·3052
0·00650	2·522	2·641	2·757	2·814	2·871	2·982	3·092	3·146	3·200	3·253	3·305	3·409	3·461
1/ 154	0·9707	1·1668	1·3859	1·5045	1·6291	1·8974	2·1916	2·3488	2·5129	2·6839	2·8621	3·2401	3·4402
0·00700	2·618	2·741	2·861	2·921	2·979	3·095	3·209	3·265	3·320	3·375	3·430	3·538	3·592
1/ 143	1·0074	1·2109	1·4383	1·5613	1·6906	1·9690	2·2744	2·4375	2·6077	2·7852	2·9701	3·3624	3·5700
0·00750	2·709	2·837	2·962	3·023	3·084	3·204	3·321	3·379	3·437	3·494	3·550	3·662	3·718
1/ 133	1·0427	1·2534	1·4887	1·6161	1·7500	2·0381	2·3542	2·5230	2·6993	2·8830	3·0743	3·4804	3·6953
0·00800	2·798	2·930	3·059	3·122	3·185	3·309	3·430	3·490	3·550	3·608	3·667	3·782	3·839
1/ 125	1·0769	1·2945	1·5376	1·6691	1·8074	2·1049	2·4314	2·6058	2·7878	2·9775	3·1752	3·5945	3·8165
0·00850	2·884	3·020	3·153	3·218	3·283	3·411	3·536	3·598	3·659	3·720	3·780	3·899	3·958
1/ 118	1·1101	1·3343	1·5849	1·7204	1·8630	2·1697	2·5062	2·6860	2·8736	3·0692	3·2729	3·7052	3·9340
0·00900	2·968	3·108	3·244	3·312	3·378	3·509	3·638	3·702	3·765	3·827	3·889	4·012	4·072
1/ 111	1·1423	1·3730	1·6308	1·7703	1·9170	2·2326	2·5789	2·7639	2·9569	3·1582	3·3678	3·8126	4·0480
0·00950	3·049	3·193	3·333	3·402	3·471	3·606	3·738	3·803	3·868	3·932	3·996	4·122	4·184
1/ 105	1·1736	1·4106	1·6755	1·8188	1·9695	2·2938	2·6496	2·8396	3·0379	3·2447	3·4601	3·9170	4·1590
0·01000	3·129	3·276	3·420	3·491	3·561	3·699	3·835	3·902	3·969	4·034	4·100	4·229	4·293
1/ 100	1·2040	1·4473	1·7191	1·8661	2·0207	2·3534	2·7184	2·9134	3·1169	3·3290	3·5499	4·0188	4·2670
0·01100	3·281	3·436	3·587	3·661	3·735	3·880	4·022	4·093	4·162	4·231	4·300	4·435	4·502
1/ 91	1·2628	1·5179	1·8030	1·9571	2·1193	2·4683	2·8511	3·0556	3·2690	3·4915	3·7232	4·2150	4·4753
0·01200	3·427	3·589	3·746	3·824	3·901	4·052	4·201	4·275	4·347	4·419	4·491	4·632	4·702
1/ 83	1·3190	1·5854	1·8831	2·0442	2·2136	2·5780	2·9778	3·1914	3·4143	3·6467	3·8888	4·4024	4·6743
0·01300	3·567	3·735	3·899	3·980	4·060	4·218	4·373	4·449	4·525	4·600	4·674	4·822	4·894
1/ 77	1·3728	1·6501	1·9600	2·1276	2·3039	2·6833	3·0994	3·3217	3·5538	3·7956	4·0476	4·5821	4·8651
0·01400	3·702	3·876	4·047	4·130	4·213	4·377	4·538	4·617	4·696	4·774	4·851	5·004	5·079
1/ 71	1·4246	1·7124	2·0340	2·2080	2·3909	2·7846	3·2164	3·4471	3·6879	3·9389	4·2003	4·7551	5·0488
0·01500	3·832	4·012	4·189	4·275	4·361	4·531	4·697	4·779	4·860	4·941	5·021	5·179	5·257
1/ 67	1·4747	1·7725	2·1054	2·2855	2·4748	2·8823	3·3293	3·5681	3·8173	4·0772	4·3478	4·9220	5·2260
0·01600	3·957	4·144	4·326	4·416	4·504	4·679	4·851	4·936	5·020	5·103	5·186	5·349	5·430
1/ 63	1·5230	1·8307	2·1744	2·3604	2·5560	2·9768	3·4385	3·6851	3·9425	4·2109	4·4904	5·0834	5·3974
0·01700	4·079	4·271	4·459	4·551	4·643	4·823	5·000	5·088	5·174	5·260	5·345	5·514	5·597
1/ 59	1·5699	1·8870	2·2414	2·4330	2·6347	3·0685	3·5444	3·7986	4·0639	4·3405	4·6286	5·2399	5·5635
0·01800	4·198	4·395	4·588	4·683	4·778	4·963	5·145	5·235	5·324	5·413	5·500	5·674	5·759
1/ 56	1·6154	1·9417	2·3064	2·5036	2·7110	3·1574	3·6471	3·9087	4·1817	4·4663	4·7627	5·3918	5·7248
0·01900	4·313	4·516	4·714	4·812	4·909	5·099	5·286	5·379	5·470	5·561	5·651	5·829	5·917
1/ 53	1·6597	1·9949	2·3696	2·5722	2·7853	3·2439	3·7470	4·0158	4·2963	4·5887	4·8933	5·5395	5·8817
	700	750	800	825	850	900	950	975	1000	1025	1050	1100	1125

Gradient (Equivalent) Pipe diameters in mm

S = 0·00400 to 0·01900

m = Manning $n \times 100$
S = 0.02000 to 0.10000

ie hydraulic gradient =
1 in 50 to 1 in 10.0

Full bore conditions; Select
m to divide mV and/or mQ.

mV to give velocities in ms^{-1}
mQ to give discharges in m^{3}s^{-1}

Gradient   (Equivalent) Pipe diameters in mm

Gradient	700	750	800	825	850	900	950	975	1000	1025	1050	1100	1125
0.02000	4.425	4.633	4.837	4.937	5.036	5.232	5.424	5.518	5.612	5.705	5.798	5.980	6.071
1/ 50	1.7028	2.0467	2.4311	2.6390	2.8577	3.3282	3.8444	4.1201	4.4079	4.7079	5.0204	5.6834	6.0344
0.02200	4.641	4.859	5.073	5.178	5.282	5.487	5.688	5.788	5.886	5.984	6.081	6.272	6.367
1/ 45.5	1.7859	2.1466	2.5498	2.7678	2.9972	3.4907	4.0320	4.3212	4.6230	4.9377	5.2654	5.9608	6.3290
0.02400	4.847	5.075	5.298	5.408	5.517	5.731	5.941	6.045	6.148	6.250	6.351	6.551	6.650
1/ 41.7	1.8653	2.2421	2.6631	2.8909	3.1304	3.6459	4.2113	4.5134	4.8286	5.1573	5.4995	6.2259	6.6104
0.02600	5.045	5.282	5.515	5.629	5.742	5.965	6.184	6.292	6.399	6.505	6.611	6.819	6.922
1/ 38.5	1.9415	2.3336	2.7719	3.0089	3.2583	3.7947	4.3833	4.6977	5.0258	5.3678	5.7241	6.4801	6.8803
0.02800	5.235	5.482	5.723	5.841	5.959	6.190	6.417	6.529	6.641	6.751	6.860	7.076	7.183
1/ 35.7	2.0148	2.4217	2.8765	3.1225	3.3813	3.9380	4.5487	4.8750	5.2155	5.5705	5.9402	6.7247	7.1401
0.03000	5.419	5.674	5.924	6.046	6.168	6.407	6.643	6.759	6.874	6.988	7.101	7.325	7.435
1/ 33.3	2.0855	2.5067	2.9775	3.2321	3.4999	4.0762	4.7084	5.0461	5.3985	5.7660	6.1487	6.9608	7.3907
0.03200	5.597	5.860	6.118	6.245	6.370	6.618	6.860	6.980	7.099	7.217	7.334	7.565	7.679
1/ 31.3	2.1539	2.5889	3.0751	3.3381	3.6147	4.2099	4.8628	5.2116	5.5756	5.9551	6.3503	7.1890	7.6330
0.03400	5.769	6.041	6.306	6.437	6.566	6.821	7.072	7.195	7.318	7.439	7.559	7.798	7.915
1/ 29.4	2.2202	2.6686	3.1698	3.4409	3.7260	4.3395	5.0125	5.3720	5.7472	6.1384	6.5458	7.4103	7.8680
0.03600	5.936	6.216	6.489	6.623	6.757	7.019	7.277	7.404	7.530	7.655	7.779	8.024	8.145
1/ 27.8	2.2845	2.7460	3.2617	3.5406	3.8340	4.4653	5.1578	5.5277	5.9138	6.3163	6.7355	7.6251	8.0961
0.03800	6.099	6.386	6.667	6.805	6.942	7.211	7.476	7.607	7.736	7.864	7.992	8.244	8.368
1/ 26.3	2.3471	2.8212	3.3511	3.6376	3.9391	4.5876	5.2991	5.6792	6.0759	6.4894	6.9201	7.8341	8.3179
0.04000	6.257	6.552	6.840	6.982	7.122	7.399	7.670	7.804	7.937	8.069	8.199	8.458	8.585
1/ 25.0	2.4081	2.8945	3.4381	3.7321	4.0414	4.7068	5.4368	5.8267	6.2337	6.6580	7.0999	8.0376	8.5340
0.04200	6.412	6.714	7.009	7.154	7.298	7.581	7.860	7.997	8.133	8.268	8.402	8.667	8.797
1/ 23.8	2.4676	2.9660	3.5230	3.8243	4.1412	4.8230	5.5710	5.9706	6.3876	6.8224	7.2752	8.2361	8.7447
0.04400	6.563	6.872	7.174	7.322	7.470	7.760	8.045	8.185	8.324	8.463	8.600	8.870	9.004
1/ 22.7	2.5256	3.0358	3.6059	3.9143	4.2386	4.9365	5.7021	6.1111	6.5380	6.9830	7.4464	8.4299	8.9505
0.04600	6.710	7.026	7.335	7.487	7.638	7.934	8.225	8.369	8.511	8.653	8.793	9.070	9.207
1/ 21.7	2.5824	3.1040	3.6870	4.0023	4.3339	5.0475	5.8303	6.2485	6.6849	7.1399	7.6138	8.6194	9.1517
0.04800	6.855	7.177	7.493	7.648	7.802	8.105	8.402	8.549	8.695	8.839	8.982	9.265	9.405
1/ 20.8	2.6379	3.1708	3.7663	4.0883	4.4271	5.1560	5.9557	6.3829	6.8287	7.2935	7.7775	8.8048	9.3485
0.05000	6.996	7.325	7.647	7.806	7.963	8.272	8.576	8.725	8.874	9.021	9.167	9.456	9.599
1/ 20.0	2.6923	3.2362	3.8439	4.1726	4.5184	5.2624	6.0785	6.5145	6.9695	7.4439	7.9379	8.9863	9.5413
0.05500	7.337	7.683	8.020	8.187	8.351	8.676	8.994	9.151	9.307	9.461	9.615	9.918	10.07
1/ 18.2	2.8237	3.3941	4.0315	4.3763	4.7389	5.5192	6.3752	6.8324	7.3097	7.8072	8.3253	9.4249	10.007
0.06000	7.664	8.024	8.377	8.551	8.723	9.061	9.394	9.558	9.721	9.882	10.04	10.36	10.51
1/ 16.7	2.9493	3.5450	4.2108	4.5709	4.9497	5.7646	6.6587	7.1363	7.6347	8.1543	8.6955	9.8440	10.452
0.06500	7.977	8.352	8.719	8.900	9.079	9.431	9.778	9.948	10.12	10.29	10.45	10.78	10.94
1/ 15.4	3.0697	3.6898	4.3827	4.7575	5.1518	6.0000	6.9306	7.4277	7.9464	8.4873	9.0506	10.246	10.879
0.07000	8.278	8.667	9.048	9.236	9.422	9.787	10.15	10.32	10.50	10.67	10.85	11.19	11.36
1/ 14.3	3.1856	3.8291	4.5482	4.9371	5.3462	6.2265	7.1922	7.7080	8.2464	8.8077	9.3923	10.633	11.289
0.07500	8.568	8.971	9.366	9.560	9.752	10.13	10.50	10.69	10.87	11.05	11.23	11.58	11.76
1/ 13.3	3.2974	3.9635	4.7078	5.1104	5.5339	6.4451	7.4446	7.9786	8.5358	9.1168	9.7219	11.006	11.686
0.08000	8.849	9.266	9.673	9.874	10.07	10.46	10.85	11.04	11.22	11.41	11.60	11.96	12.14
1/ 12.5	3.4056	4.0935	4.8622	5.2780	5.7154	6.6564	7.6888	8.2402	8.8158	9.4158	10.041	11.367	12.069
0.08500	9.122	9.551	9.971	10.18	10.38	10.79	11.18	11.38	11.57	11.76	11.95	12.33	12.52
1/ 11.8	3.5104	4.2194	5.0119	5.4405	5.8913	6.8613	7.9254	8.4938	9.0871	9.7056	10.350	11.717	12.440
0.09000	9.386	9.828	10.26	10.47	10.68	11.10	11.51	11.71	11.91	12.10	12.30	12.69	12.88
1/ 11.1	3.6121	4.3418	5.1572	5.5982	6.0621	7.0602	8.1552	8.7401	9.3506	9.9870	10.650	12.056	12.801
0.09500	9.643	10.10	10.54	10.76	10.98	11.40	11.82	12.03	12.23	12.43	12.64	13.03	13.23
1/ 10.5	3.7111	4.4607	5.2985	5.7516	6.2282	7.2537	8.3786	8.9796	9.6068	10.261	10.942	12.387	13.152
0.10000	9.894	10.36	10.81	11.04	11.26	11.70	12.13	12.34	12.55	12.76	12.96	13.37	13.57
1/ 10.0	3.8075	4.5766	5.4361	5.9010	6.3900	7.4421	8.5963	9.2129	9.8563	10.527	11.226	12.709	13.493
	700	750	800	825	850	900	950	975	1000	1025	1050	1100	1125

Gradient   (Equivalent) Pipe diameters in mm

S = 0.02000 to 0.10000

D5

m = Manning $n \times 100$
S = 0·00010 to 0·00048

ie hydraulic gradient =
1 in 10000 to 1 in 2083

Full bore conditions; Select m to divide mV and/or mQ.

mV to give velocities in ms^{-1}
mQ to give discharges in m^{3}s^{-1}

Gradient (Equivalent) Pipe diameters in mm

Gradient	1125	1150	1200	1250	1275	1300	1350	1400	1425	1450	1500	1550	1575
0·00010	0·429	0·436	0·448	0·461	0·467	0·473	0·485	0·497	0·503	0·508	0·520	0·532	0·537
1/10000	0·4267	0·4525	0·5068	0·5651	0·5958	0·6274	0·6939	0·7645	0·8015	0·8395	0·9190	1·0029	1·0466
0·00011	0·450	0·457	0·470	0·483	0·489	0·496	0·508	0·521	0·527	0·533	0·545	0·557	0·563
1/ 9091	0·4475	0·4745	0·5316	0·5927	0·6248	0·6581	0·7277	0·8018	0·8406	0·8805	0·9638	1·0519	1·0977
0·00012	0·470	0·477	0·491	0·504	0·511	0·518	0·531	0·544	0·551	0·557	0·570	0·582	0·588
1/ 8333	0·4674	0·4956	0·5552	0·6191	0·6526	0·6873	0·7601	0·8375	0·8780	0·9196	1·0067	1·0986	1·1465
0·00013	0·489	0·497	0·511	0·525	0·532	0·539	0·553	0·566	0·573	0·580	0·593	0·606	0·613
1/ 7692	0·4865	0·5159	0·5779	0·6443	0·6793	0·7154	0·7911	0·8717	0·9138	0·9572	1·0478	1·1435	1·1934
0·00014	0·508	0·515	0·530	0·545	0·552	0·559	0·574	0·588	0·595	0·602	0·615	0·629	0·636
1/ 7143	0·5049	0·5354	0·5997	0·6687	0·7049	0·7424	0·8210	0·9046	0·9483	0·9933	1·0873	1·1867	1·2384
0·00015	0·526	0·534	0·549	0·564	0·571	0·579	0·594	0·608	0·615	0·623	0·637	0·651	0·658
1/ 6667	0·5226	0·5541	0·6207	0·6921	0·7297	0·7684	0·8498	0·9363	0·9816	1·0282	1·1255	1·2283	1·2819
0·00016	0·543	0·551	0·567	0·582	0·590	0·598	0·613	0·628	0·636	0·643	0·658	0·672	0·680
1/ 6250	0·5397	0·5723	0·6411	0·7148	0·7536	0·7936	0·8777	0·9671	1·0138	1·0619	1·1624	1·2686	1·3239
0·00017	0·560	0·568	0·584	0·600	0·608	0·616	0·632	0·648	0·655	0·663	0·678	0·693	0·700
1/ 5882	0·5563	0·5899	0·6608	0·7368	0·7768	0·8181	0·9047	0·9968	1·0450	1·0946	1·1982	1·3077	1·3647
0·00018	0·576	0·584	0·601	0·618	0·626	0·634	0·650	0·666	0·674	0·682	0·698	0·713	0·721
1/ 5556	0·5725	0·6070	0·6800	0·7582	0·7993	0·8418	0·9309	1·0257	1·0753	1·1263	1·2329	1·3456	1·4042
0·00019	0·592	0·600	0·618	0·635	0·643	0·652	0·668	0·685	0·693	0·701	0·717	0·733	0·740
1/ 5263	0·5882	0·6237	0·6986	0·7790	0·8212	0·8649	0·9564	1·0538	1·1048	1·1572	1·2667	1·3824	1·4427
0·00020	0·607	0·616	0·634	0·651	0·660	0·669	0·686	0·702	0·711	0·719	0·735	0·752	0·760
1/ 5000	0·6034	0·6399	0·7168	0·7992	0·8425	0·8873	0·9813	1·0812	1·1335	1·1873	1·2996	1·4183	1·4802
0·00022	0·637	0·646	0·665	0·683	0·692	0·701	0·719	0·737	0·745	0·754	0·771	0·788	0·797
1/ 4545	0·6329	0·6711	0·7518	0·8382	0·8837	0·9306	1·0292	1·1340	1·1888	1·2452	1·3630	1·4876	1·5524
0·00024	0·665	0·675	0·694	0·713	0·723	0·732	0·751	0·769	0·779	0·788	0·806	0·823	0·832
1/ 4167	0·6610	0·7009	0·7852	0·8755	0·9230	0·9720	1·0749	1·1844	1·2416	1·3006	1·4236	1·5537	1·6214
0·00026	0·692	0·702	0·723	0·743	0·752	0·762	0·782	0·801	0·810	0·820	0·839	0·857	0·866
1/ 3846	0·6880	0·7296	0·8172	0·9112	0·9606	1·0117	1·1188	1·2328	1·2923	1·3537	1·4818	1·6172	1·6877
0·00028	0·718	0·729	0·750	0·771	0·781	0·791	0·811	0·831	0·841	0·851	0·870	0·889	0·899
1/ 3571	0·7140	0·7571	0·8481	0·9456	0·9969	1·0499	1·1611	1·2793	1·3411	1·4048	1·5377	1·6782	1·7514
0·00030	0·744	0·754	0·776	0·798	0·808	0·819	0·840	0·860	0·870	0·881	0·901	0·921	0·930
1/ 3333	0·7391	0·7837	0·8779	0·9788	1·0319	1·0867	1·2018	1·3242	1·3882	1·4541	1·5917	1·7371	1·8128
0·00032	0·768	0·779	0·802	0·824	0·835	0·846	0·867	0·888	0·899	0·909	0·930	0·951	0·961
1/ 3125	0·7633	0·8094	0·9067	1·0109	1·0657	1·1224	1·2412	1·3676	1·4337	1·5018	1·6439	1·7941	1·8723
0·00034	0·792	0·803	0·826	0·849	0·860	0·872	0·894	0·916	0·927	0·937	0·959	0·980	0·991
1/ 2941	0·7868	0·8343	0·9346	1·0420	1·0985	1·1569	1·2794	1·4097	1·4778	1·5480	1·6945	1·8493	1·9299
0·00036	0·814	0·827	0·850	0·874	0·885	0·897	0·920	0·942	0·954	0·965	0·987	1·008	1·019
1/ 2778	0·8096	0·8585	0·9617	1·0722	1·1304	1·1905	1·3165	1·4506	1·5207	1·5929	1·7436	1·9029	1·9859
0·00038	0·837	0·849	0·874	0·898	0·910	0·921	0·945	0·968	0·980	0·991	1·014	1·036	1·047
1/ 2632	0·8318	0·8820	0·9880	1·1016	1·1614	1·2231	1·3526	1·4903	1·5624	1·6365	1·7914	1·9551	2·0403
0·00040	0·859	0·871	0·896	0·921	0·933	0·945	0·969	0·993	1·005	1·017	1·040	1·063	1·074
1/ 2500	0·8534	0·9049	1·0137	1·1302	1·1915	1·2549	1·3877	1·5290	1·6029	1·6790	1·8379	2·0058	2·0933
0·00042	0·880	0·893	0·918	0·944	0·956	0·969	0·993	1·018	1·030	1·042	1·066	1·089	1·101
1/ 2381	0·8745	0·9273	1·0387	1·1582	1·2210	1·2858	1·4220	1·5668	1·6425	1·7205	1·8833	2·0554	2·1450
0·00044	0·900	0·914	0·940	0·966	0·979	0·992	1·017	1·042	1·054	1·066	1·091	1·115	1·127
1/ 2273	0·8951	0·9491	1·0631	1·1854	1·2497	1·3161	1·4555	1·6037	1·6812	1·7610	1·9276	2·1037	2·1955
0·00046	0·921	0·934	0·961	0·988	1·001	1·014	1·040	1·065	1·078	1·090	1·115	1·140	1·152
1/ 2174	0·9152	0·9704	1·0870	1·2121	1·2778	1·3457	1·4882	1·6397	1·7190	1·8006	1·9709	2·1510	2·2448
0·00048	0·940	0·954	0·982	1·009	1·022	1·036	1·062	1·088	1·101	1·114	1·139	1·164	1·177
1/ 2083	0·9349	0·9913	1·1104	1·2381	1·3053	1·3746	1·5202	1·6750	1·7559	1·8393	2·0133	2·1973	2·2931
	1125	1150	1200	1250	1275	1300	1350	1400	1425	1450	1500	1550	1575

Gradient (Equivalent) Pipe diameters in mm

S = 0·00010 to 0·00048

m = Manning $n \times 100$
S = 0·00050 to 0·00280

ie hydraulic gradient =
1 in 2000 to 1 in 357

Full bore conditions; Select m to divide mV and/or mQ.

mV to give velocities in ms^{-1}
mQ to give discharges in $m^3 s^{-1}$

Gradient — (Equivalent) Pipe diameters in mm

Gradient	1125	1150	1200	1250	1275	1300	1350	1400	1425	1450	1500	1550	1575
0·00050	0·960	0·974	1·002	1·030	1·043	1·057	1·084	1·111	1·124	1·137	1·163	1·189	1·201
1/ 2000	0·9541	1·0117	1·1333	1·2637	1·3322	1·4030	1·5515	1·7095	1·7921	1·8772	2·0548	2·2426	2·3404
0·00055	1·007	1·022	1·051	1·080	1·094	1·109	1·137	1·165	1·179	1·192	1·220	1·247	1·260
1/ 1818	1·0007	1·0611	1·1886	1·3253	1·3972	1·4715	1·6272	1·7930	1·8796	1·9688	2·1551	2·3521	2·4546
0·00060	1·051	1·067	1·098	1·128	1·143	1·158	1·187	1·217	1·231	1·245	1·274	1·302	1·316
1/ 1667	1·0452	1·1083	1·2415	1·3843	1·4593	1·5369	1·6996	1·8727	1·9632	2·0564	2·2510	2·4566	2·5637
0·00065	1·094	1·111	1·143	1·174	1·190	1·205	1·236	1·266	1·281	1·296	1·326	1·355	1·370
1/ 1538	1·0879	1·1535	1·2922	1·4408	1·5189	1·5996	1·7690	1·9492	2·0434	2·1404	2·3429	2·5570	2·6684
0·00070	1·136	1·152	1·186	1·218	1·235	1·251	1·283	1·314	1·330	1·345	1·376	1·406	1·421
1/ 1429	1·1289	1·1971	1·3410	1·4952	1·5763	1·6600	1·8358	2·0227	2·1205	2·2212	2·4313	2·6535	2·7691
0·00075	1·176	1·193	1·227	1·261	1·278	1·295	1·328	1·360	1·376	1·392	1·424	1·456	1·471
1/ 1333	1·1686	1·2391	1·3880	1·5477	1·6316	1·7183	1·9002	2·0937	2·1949	2·2991	2·5167	2·7466	2·8663
0·00080	1·214	1·232	1·268	1·303	1·320	1·337	1·371	1·405	1·421	1·438	1·471	1·503	1·519
1/ 1250	1·2069	1·2797	1·4335	1·5984	1·6851	1·7746	1·9625	2·1624	2·2669	2·3745	2·5992	2·8367	2·9603
0·00085	1·252	1·270	1·307	1·343	1·360	1·378	1·413	1·448	1·465	1·482	1·516	1·550	1·566
1/ 1176	1·2440	1·3191	1·4777	1·6476	1·7369	1·8293	2·0229	2·2289	2·3367	2·4476	2·6792	2·9240	3·0515
0·00090	1·288	1·307	1·344	1·382	1·400	1·418	1·454	1·490	1·508	1·525	1·560	1·595	1·612
1/ 1111	1·2801	1·3574	1·5205	1·6954	1·7873	1·8823	2·0816	2·2936	2·4044	2·5186	2·7569	3·0088	3·1399
0·00095	1·323	1·343	1·381	1·419	1·438	1·457	1·494	1·531	1·549	1·567	1·603	1·638	1·656
1/ 1053	1·3152	1·3946	1·5622	1·7418	1·8363	1·9339	2·1386	2·3564	2·4703	2·5876	2·8324	3·0912	3·2260
0·00100	1·357	1·378	1·417	1·456	1·476	1·495	1·533	1·571	1·589	1·608	1·644	1·681	1·699
1/ 1000	1·3493	1·4308	1·6028	1·7871	1·8840	1·9841	2·1942	2·4176	2·5345	2·6548	2·9060	3·1715	3·3098
0·00110	1·424	1·445	1·486	1·527	1·548	1·568	1·608	1·647	1·667	1·686	1·725	1·763	1·782
1/ 909	1·4152	1·5006	1·6810	1·8743	1·9759	2·0809	2·3013	2·5356	2·6582	2·7844	3·0478	3·3263	3·4713
0·00120	1·487	1·509	1·552	1·595	1·616	1·637	1·679	1·720	1·741	1·761	1·801	1·841	1·861
1/ 833	1·4781	1·5674	1·7557	1·9576	2·0638	2·1735	2·4036	2·6484	2·7764	2·9082	3·1833	3·4742	3·6257
0·00130	1·548	1·571	1·616	1·660	1·682	1·704	1·748	1·791	1·812	1·833	1·875	1·916	1·937
1/ 769	1·5385	1·6314	1·8274	2·0376	2·1481	2·2622	2·5018	2·7565	2·8898	3·0269	3·3133	3·6161	3·7737
0·00140	1·606	1·630	1·677	1·723	1·746	1·769	1·814	1·858	1·880	1·902	1·946	1·989	2·010
1/ 714	1·5966	1·6929	1·8964	2·1145	2·2292	2·3476	2·5962	2·8606	2·9988	3·1412	3·4384	3·7526	3·9162
0·00150	1·663	1·687	1·736	1·784	1·807	1·831	1·877	1·923	1·946	1·969	2·014	2·059	2·081
1/ 667	1·6526	1·7524	1·9630	2·1887	2·3074	2·4300	2·6873	2·9610	3·1041	3·2514	3·5591	3·8843	4·0536
0·00160	1·717	1·742	1·793	1·842	1·866	1·891	1·939	1·987	2·010	2·034	2·080	2·126	2·149
1/ 625	1·7068	1·8098	2·0273	2·2605	2·3831	2·5097	2·7754	3·0581	3·2059	3·3581	3·6758	4·0117	4·1866
0·00170	1·770	1·796	1·848	1·899	1·924	1·949	1·999	2·048	2·072	2·096	2·144	2·191	2·215
1/ 588	1·7593	1·8655	2·0897	2·3301	2·4564	2·5870	2·8609	3·1522	3·3046	3·4614	3·7889	4·1352	4·3154
0·00180	1·821	1·848	1·901	1·954	1·980	2·006	2·057	2·107	2·132	2·157	2·206	2·255	2·279
1/ 556	1·8103	1·9196	2·1503	2·3976	2·5276	2·6620	2·9438	3·2436	3·4004	3·5618	3·8988	4·2550	4·4405
0·00190	1·871	1·899	1·953	2·007	2·034	2·060	2·113	2·165	2·191	2·216	2·267	2·317	2·342
1/ 526	1·8599	1·9722	2·2092	2·4633	2·5969	2·7349	3·0245	3·3325	3·4935	3·6594	4·0056	4·3716	4·5622
0·00200	1·920	1·948	2·004	2·059	2·087	2·114	2·168	2·221	2·247	2·274	2·326	2·377	2·402
1/ 500	1·9083	2·0234	2·2666	2·5273	2·6644	2·8059	3·1030	3·4191	3·5843	3·7544	4·1097	4·4852	4·6807
0·00220	2·013	2·043	2·102	2·160	2·189	2·217	2·274	2·329	2·357	2·385	2·439	2·493	2·520
1/ 455	2·0014	2·1222	2·3773	2·6507	2·7944	2·9429	3·2545	3·5859	3·7592	3·9377	4·3103	4·7041	4·9092
0·00240	2·103	2·134	2·195	2·256	2·286	2·316	2·375	2·433	2·462	2·491	2·548	2·604	2·632
1/ 417	2·0904	2·2166	2·4830	2·7685	2·9187	3·0738	3·3992	3·7454	3·9264	4·1128	4·5019	4·9133	5·1275
0·00260	2·189	2·221	2·285	2·348	2·379	2·410	2·472	2·532	2·562	2·592	2·652	2·710	2·739
1/ 385	2·1758	2·3071	2·5844	2·8816	3·0378	3·1993	3·5380	3·8983	4·0867	4·2807	4·6858	5·1139	5·3368
0·00280	2·271	2·305	2·371	2·437	2·469	2·501	2·565	2·628	2·659	2·690	2·752	2·813	2·843
1/ 357	2·2579	2·3942	2·6819	2·9903	3·1525	3·3200	3·6716	4·0455	4·2410	4·4423	4·8626	5·3070	5·5383
	1125	**1150**	**1200**	**1250**	**1275**	**1300**	**1350**	**1400**	**1425**	**1450**	**1500**	**1550**	**1575**

Gradient — (Equivalent) Pipe diameters in mm

S = 0·00050 to 0·00280

D5
continued

m = Manning $n \times 100$
S = 0.00300 to 0.01400

Full bore conditions; Select m to divide mV and/or mQ.

ie hydraulic gradient = 1 in 333 to 1 in 71

mV to give velocities in ms^{-1}
mQ to give discharges in m^{3}s^{-1}

Gradient (Equivalent) Pipe diameters in mm

Gradient	1125	1150	1200	1250	1275	1300	1350	1400	1425	1450	1500	1550	1575
0.00300 1/ 333	2.351	2.386	2.455	2.522	2.556	2.589	2.655	2.720	2.753	2.785	2.848	2.911	2.942
	2.3371	2.4782	2.7760	3.0953	3.2632	3.4366	3.8004	4.1875	4.3898	4.5982	5.0333	5.4932	5.7327
0.00320 1/ 313	2.428	2.464	2.535	2.605	2.640	2.674	2.742	2.809	2.843	2.876	2.942	3.007	3.039
	2.4138	2.5595	2.8671	3.1968	3.3702	3.5493	3.9251	4.3248	4.5338	4.7490	5.1984	5.6734	5.9207
0.00340 1/ 294	2.503	2.540	2.613	2.685	2.721	2.756	2.827	2.896	2.930	2.964	3.032	3.099	3.132
	2.4881	2.6383	2.9553	3.2952	3.4739	3.6585	4.0459	4.4579	4.6734	4.8952	5.3584	5.8480	6.1029
0.00360 1/ 278	2.576	2.614	2.689	2.763	2.800	2.836	2.908	2.980	3.015	3.050	3.120	3.189	3.223
	2.5602	2.7147	3.0410	3.3907	3.5746	3.7646	4.1632	4.5871	4.8088	5.0371	5.5137	6.0175	6.2798
0.00380 1/ 263	2.646	2.685	2.763	2.839	2.876	2.914	2.988	3.062	3.098	3.134	3.206	3.276	3.312
	2.6304	2.7891	3.1243	3.4836	3.6726	3.8677	4.2772	4.7128	4.9406	5.1751	5.6648	6.1824	6.4519
0.00400 1/ 250	2.715	2.755	2.834	2.912	2.951	2.990	3.066	3.141	3.178	3.215	3.289	3.362	3.398
	2.6987	2.8616	3.2055	3.5741	3.7680	3.9682	4.3884	4.8353	5.0690	5.3096	5.8120	6.3430	6.6195
0.00420 1/ 238	2.782	2.823	2.904	2.984	3.024	3.063	3.142	3.219	3.257	3.295	3.370	3.445	3.482
	2.7653	2.9323	3.2847	3.6624	3.8610	4.0662	4.4967	4.9547	5.1941	5.4407	5.9555	6.4997	6.7830
0.00440 1/ 227	2.847	2.889	2.973	3.055	3.095	3.136	3.215	3.294	3.333	3.372	3.449	3.526	3.563
	2.8304	3.0013	3.3620	3.7486	3.9519	4.1619	4.6026	5.0713	5.3164	5.5687	6.0956	6.6526	6.9426
0.00460 1/ 217	2.911	2.954	3.039	3.123	3.165	3.206	3.288	3.368	3.408	3.448	3.527	3.605	3.644
	2.8940	3.0687	3.4375	3.8328	4.0407	4.2554	4.7060	5.1852	5.4359	5.6939	6.2326	6.8021	7.0987
0.00480 1/ 208	2.974	3.018	3.105	3.190	3.233	3.275	3.358	3.441	3.482	3.522	3.603	3.682	3.722
	2.9563	3.1347	3.5115	3.9153	4.1276	4.3470	4.8072	5.2968	5.5528	5.8164	6.3667	6.9484	7.2513
0.00500 1/ 200	3.035	3.080	3.169	3.256	3.300	3.343	3.428	3.512	3.553	3.595	3.677	3.758	3.799
	3.0172	3.1994	3.5839	3.9960	4.2127	4.4366	4.9063	5.4060	5.6673	5.9363	6.4980	7.0917	7.4009
0.00550 1/ 182	3.184	3.231	3.323	3.415	3.461	3.506	3.595	3.683	3.727	3.770	3.857	3.942	3.984
	3.1645	3.3555	3.7588	4.1911	4.4183	4.6531	5.1458	5.6699	5.9439	6.2260	6.8151	7.4379	7.7621
0.00600 1/ 167	3.325	3.374	3.471	3.567	3.614	3.662	3.755	3.847	3.893	3.938	4.028	4.117	4.161
	3.3052	3.5047	3.9259	4.3774	4.6148	4.8600	5.3746	5.9220	6.2082	6.5029	7.1182	7.7686	8.1072
0.00650 1/ 154	3.461	3.512	3.613	3.713	3.762	3.811	3.908	4.004	4.052	4.099	4.193	4.285	4.331
	3.4402	3.6478	4.0862	4.5562	4.8032	5.0585	5.5941	6.1638	6.4617	6.7684	7.4088	8.0858	8.4383
0.00700 1/ 143	3.592	3.645	3.749	3.853	3.904	3.955	4.056	4.155	4.205	4.254	4.351	4.447	4.495
	3.5700	3.7855	4.2405	4.7282	4.9845	5.2495	5.8053	6.3965	6.7056	7.0239	7.6885	8.3910	8.7568
0.00750 1/ 133	3.718	3.772	3.881	3.988	4.041	4.094	4.198	4.301	4.352	4.403	4.504	4.603	4.652
	3.6953	3.9184	4.3893	4.8941	5.1595	5.4337	6.0090	6.6210	6.9410	7.2704	7.9584	8.6856	9.0642
0.00800 1/ 125	3.839	3.896	4.008	4.119	4.174	4.228	4.336	4.442	4.495	4.547	4.651	4.754	4.805
	3.8165	4.0469	4.5333	5.0546	5.3287	5.6119	6.2061	6.8381	7.1686	7.5089	8.2194	8.9704	9.3614
0.00850 1/ 118	3.958	4.016	4.132	4.246	4.302	4.358	4.469	4.579	4.633	4.687	4.794	4.900	4.953
	3.9340	4.1714	4.6728	5.2102	5.4927	5.7846	6.3971	7.0486	7.3892	7.7400	8.4723	9.2465	9.6495
0.00900 1/ 111	4.072	4.133	4.251	4.369	4.427	4.484	4.599	4.712	4.768	4.823	4.933	5.042	5.096
	4.0480	4.2924	4.8083	5.3612	5.6519	5.9523	6.5825	7.2529	7.6034	7.9644	8.7179	9.5146	9.9293
0.00950 1/ 105	4.184	4.246	4.368	4.488	4.548	4.607	4.725	4.841	4.898	4.955	5.069	5.181	5.236
	4.1590	4.4100	4.9400	5.5081	5.8068	6.1154	6.7629	7.4517	7.8118	8.1826	8.9568	9.7753	10.201
0.01000 1/ 100	4.293	4.356	4.481	4.605	4.666	4.727	4.847	4.966	5.025	5.084	5.200	5.315	5.372
	4.2670	4.5246	5.0683	5.6512	5.9577	6.2743	6.9386	7.6452	8.0147	8.3952	9.1895	10.029	10.466
0.01100 1/ 91	4.502	4.569	4.700	4.830	4.894	4.958	5.084	5.209	5.271	5.332	5.454	5.575	5.634
	4.4753	4.7454	5.3157	5.9271	6.2485	6.5805	7.2773	8.0184	8.4059	8.8050	9.6380	10.519	10.977
0.01200 1/ 83	4.702	4.772	4.909	5.045	5.112	5.178	5.310	5.440	5.505	5.569	5.697	5.822	5.885
	4.6743	4.9564	5.5521	6.1906	6.5263	6.8731	7.6009	8.3749	8.7797	9.1965	10.067	10.986	11.465
0.01300 1/ 77	4.894	4.967	5.110	5.251	5.320	5.390	5.527	5.663	5.730	5.797	5.929	6.060	6.125
	4.8651	5.1588	5.7788	6.4434	6.7928	7.1538	7.9112	8.7169	9.1382	9.5720	10.478	11.435	11.934
0.01400 1/ 71	5.079	5.154	5.302	5.449	5.521	5.593	5.736	5.876	5.946	6.015	6.153	6.289	6.356
	5.0488	5.3535	5.9969	6.6866	7.0492	7.4238	8.2099	9.0460	9.4832	9.9333	10.873	11.867	12.384
Gradient	1125	1150	1200	1250	1275	1300	1350	1400	1425	1450	1500	1550	1575

Gradient (Equivalent) Pipe diameters in mm

S = 0.00300 to 0.01400

m = Manning $n \times 100$
S = 0·01500 to 0·07500

ie hydraulic gradient =
1 in 67   to   1 in 13·3

Full bore conditions; Select m to divide mV and/or mQ.

mV to give velocities in ms^{-1}
mQ to give discharges in m^{3}s^{-1}

Gradient (Equivalent) Pipe diameters in mm

Gradient	1125	1150	1200	1250	1275	1300	1350	1400	1425	1450	1500	1550	1575
0·01500	5·257	5·335	5·489	5·640	5·715	5·789	5·937	6·083	6·155	6·227	6·369	6·510	6·579
1/ 67	5·2260	5·5414	6·2074	6·9213	7·2966	7·6844	8·4980	9·3635	9·8160	10·282	11·255	12·283	12·819
0·01600	5·430	5·510	5·669	5·825	5·902	5·979	6·132	6·282	6·357	6·431	6·578	6·723	6·795
1/ 63	5·3974	5·7232	6·4110	7·1483	7·5359	7·9364	8·7767	9·6705	10·138	10·619	11·624	12·686	13·239
0·01700	5·597	5·680	5·843	6·004	6·084	6·163	6·320	6·475	6·552	6·629	6·780	6·930	7·004
1/ 59	5·5635	5·8993	6·6083	7·3683	7·7679	8·1807	9·0468	9·9682	10·450	10·946	11·982	13·077	13·647
0·01800	5·759	5·844	6·012	6·178	6·260	6·342	6·504	6·663	6·742	6·821	6·977	7·131	7·207
1/ 56	5·7248	6·0703	6·7999	7·5819	7·9931	8·4178	9·3091	10·257	10·753	11·263	12·329	13·456	14·042
0·01900	5·917	6·004	6·177	6·348	6·432	6·516	6·682	6·846	6·927	7·008	7·168	7·326	7·405
1/ 53	5·8817	6·2367	6·9862	7·7897	8·2121	8·6485	9·5642	10·538	11·048	11·572	12·667	13·824	14·427
0·02000	6·071	6·160	6·338	6·513	6·599	6·685	6·855	7·024	7·107	7·190	7·354	7·517	7·597
1/ 50	6·0344	6·3987	7·1677	7·9920	8·4254	8·8732	9·8127	10·812	11·335	11·873	12·996	14·183	14·802
0·02200	6·367	6·461	6·647	6·830	6·921	7·011	7·190	7·366	7·454	7·541	7·713	7·884	7·968
1/ 45·5	6·3290	6·7110	7·5176	8·3821	8·8367	9·3063	10·292	11·340	11·888	12·452	13·630	14·876	15·524
0·02400	6·650	6·748	6·943	7·134	7·229	7·323	7·510	7·694	7·785	7·876	8·056	8·234	8·322
1/ 41·7	6·6104	7·0094	7·8518	8·7548	9·2296	9·7201	10·749	11·844	12·416	13·006	14·236	15·537	16·214
0·02600	6·922	7·024	7·226	7·425	7·524	7·622	7·816	8·008	8·103	8·198	8·385	8·570	8·662
1/ 38·5	6·8803	7·2956	8·1725	9·1123	9·6065	10·117	11·188	12·328	12·923	13·537	14·818	16·172	16·877
0·02800	7·183	7·289	7·499	7·706	7·808	7·910	8·111	8·310	8·409	8·507	8·702	8·894	8·989
1/ 35·7	7·1401	7·5710	8·4810	9·4563	9·9691	10·499	11·611	12·793	13·411	14·048	15·377	16·782	17·514
0·03000	7·435	7·545	7·762	7·976	8·082	8·187	8·396	8·602	8·704	8·806	9·007	9·206	9·305
1/ 33·3	7·3907	7·8368	8·7786	9·7882	10·319	10·867	12·018	13·242	13·882	14·541	15·917	17·371	18·128
0·03200	7·679	7·792	8·017	8·238	8·347	8·456	8·671	8·884	8·990	9·095	9·302	9·508	9·610
1/ 31·3	7·6330	8·0938	9·0665	10·109	10·657	11·224	12·412	13·676	14·337	15·018	16·439	17·941	18·723
0·03400	7·915	8·032	8·263	8·491	8·604	8·716	8·938	9·158	9·266	9·374	9·589	9·801	9·906
1/ 29·4	7·8680	8·3429	9·3456	10·420	10·985	11·569	12·794	14·097	14·778	15·480	16·945	18·493	19·299
0·03600	8·145	8·265	8·503	8·737	8·854	8·969	9·197	9·423	9·535	9·646	9·867	10·08	10·19
1/ 27·8	8·0961	8·5848	9·6165	10·722	11·304	11·905	13·165	14·506	15·207	15·929	17·436	19·029	19·859
0·03800	8·368	8·491	8·736	8·977	9·096	9·215	9·449	9·681	9·796	9·911	10·14	10·36	10·47
1/ 26·3	8·3179	8·8200	9·8800	11·016	11·614	12·231	13·526	14·903	15·624	16·365	17·914	19·551	20·403
0·04000	8·585	8·712	8·963	9·210	9·332	9·454	9·695	9·933	10·05	10·17	10·40	10·63	10·74
1/ 25·0	8·5340	9·0491	10·137	11·302	11·915	12·549	13·877	15·290	16·029	16·790	18·379	20·058	20·933
0·04200	8·797	8·927	9·184	9·438	9·563	9·688	9·934	10·18	10·30	10·42	10·66	10·89	11·01
1/ 23·8	8·7447	9·2726	10·387	11·582	12·210	12·858	14·220	15·668	16·425	17·205	18·833	20·554	21·450
0·04400	9·004	9·137	9·400	9·660	9·788	9·916	10·17	10·42	10·54	10·66	10·91	11·15	11·27
1/ 22·7	8·9505	9·4908	10·631	11·854	12·497	13·161	14·555	16·037	16·812	17·610	19·276	21·037	21·955
0·04600	9·207	9·343	9·612	9·877	10·01	10·14	10·40	10·65	10·78	10·90	11·15	11·40	11·52
1/ 21·7	9·1517	9·7041	10·870	12·121	12·778	13·457	14·882	16·397	17·190	18·006	19·709	21·510	22·448
0·04800	9·405	9·544	9·818	10·09	10·22	10·36	10·62	10·88	11·01	11·14	11·39	11·64	11·77
1/ 20·8	9·3485	9·9128	11·104	12·381	13·053	13·746	15·202	16·750	17·559	18·393	20·133	21·973	22·931
0·05000	9·599	9·740	10·02	10·30	10·43	10·57	10·84	11·11	11·24	11·37	11·63	11·89	12·01
1/ 20·0	9·5413	10·117	11·333	12·637	13·322	14·030	15·515	17·095	17·921	18·772	20·548	22·426	23·404
0·05500	10·07	10·22	10·51	10·80	10·94	11·09	11·37	11·65	11·79	11·92	12·20	12·47	12·60
1/ 18·2	10·007	10·611	11·886	13·253	13·972	14·715	16·272	17·930	18·796	19·688	21·551	23·521	24·546
0·06000	10·51	10·67	10·98	11·28	11·43	11·58	11·87	12·17	12·31	12·45	12·74	13·02	13·16
1/ 16·7	10·452	11·083	12·415	13·843	14·593	15·369	16·996	18·727	19·632	20·564	22·510	24·566	25·637
0·06500	10·94	11·11	11·43	11·74	11·90	12·05	12·36	12·66	12·81	12·96	13·26	13·55	13·70
1/ 15·4	10·879	11·535	12·922	14·408	15·189	15·996	17·690	19·492	20·434	21·404	23·429	25·570	26·684
0·07000	11·36	11·52	11·86	12·18	12·35	12·51	12·83	13·14	13·30	13·45	13·76	14·06	14·21
1/ 14·3	11·289	11·971	13·410	14·952	15·763	16·600	18·358	20·227	21·205	22·212	24·313	26·535	27·691
0·07500	11·76	11·93	12·27	12·61	12·78	12·95	13·28	13·60	13·76	13·92	14·24	14·56	14·71
1/ 13·3	11·686	12·391	13·880	15·477	16·316	17·183	19·002	20·937	21·949	22·991	25·167	27·466	28·663
	1125	1150	1200	1250	1275	1300	1350	1400	1425	1450	1500	1550	1575

Gradient (Equivalent) Pipe diameters in mm

S = 0·01500 to 0·07500

m = Manning $n \times 100$
S = 0.00010 to 0.00048

ie hydraulic gradient =
1 in 10000 to 1 in 2083

Full bore conditions; Select
m to divide mV and/or mQ.

mV to give velocities in ms^{-1}
mQ to give discharges in m^3s^{-1}

Gradient   (Equivalent) Pipe diameters in mm

Gradient	1575	1600	1650	1700	1725	1750	1800	1850	1875	1900	1950	2000	2050
0.00010	0.537	0.543	0.554	0.565	0.571	0.576	0.587	0.598	0.603	0.609	0.619	0.630	0.640
1/10000	1.0466	1.0915	1.1849	1.2831	1.3340	1.3862	1.4943	1.6076	1.6662	1.7261	1.8499	1.9791	2.1138
0.00011	0.563	0.569	0.581	0.593	0.599	0.604	0.616	0.627	0.633	0.638	0.650	0.661	0.672
1/ 9091	1.0977	1.1448	1.2427	1.3457	1.3991	1.4538	1.5673	1.6860	1.7475	1.8103	1.9402	2.0757	2.2170
0.00012	0.588	0.595	0.607	0.619	0.625	0.631	0.643	0.655	0.661	0.667	0.679	0.690	0.702
1/ 8333	1.1465	1.1957	1.2980	1.4055	1.4613	1.5185	1.6369	1.7610	1.8252	1.8908	2.0264	2.1680	2.3155
0.00013	0.613	0.619	0.632	0.645	0.651	0.657	0.670	0.682	0.688	0.694	0.706	0.718	0.730
1/ 7692	1.1934	1.2445	1.3510	1.4629	1.5210	1.5805	1.7038	1.8329	1.8997	1.9680	2.1092	2.2565	2.4101
0.00014	0.636	0.642	0.656	0.669	0.675	0.682	0.695	0.708	0.714	0.720	0.733	0.745	0.758
1/ 7143	1.2384	1.2915	1.4020	1.5181	1.5784	1.6401	1.7681	1.9021	1.9714	2.0423	2.1888	2.3417	2.5011
0.00015	0.658	0.665	0.679	0.692	0.699	0.706	0.719	0.732	0.739	0.746	0.759	0.772	0.784
1/ 6667	1.2819	1.3368	1.4512	1.5714	1.6338	1.6977	1.8302	1.9689	2.0406	2.1140	2.2656	2.4239	2.5888
0.00016	0.680	0.687	0.701	0.715	0.722	0.729	0.743	0.756	0.763	0.770	0.784	0.797	0.810
1/ 6250	1.3239	1.3807	1.4988	1.6230	1.6874	1.7534	1.8902	2.0335	2.1076	2.1833	2.3399	2.5034	2.6737
0.00017	0.700	0.708	0.723	0.737	0.744	0.751	0.766	0.780	0.787	0.794	0.808	0.821	0.835
1/ 5882	1.3647	1.4232	1.5449	1.6729	1.7393	1.8073	1.9484	2.0960	2.1724	2.2505	2.4119	2.5804	2.7560
0.00018	0.721	0.728	0.743	0.758	0.766	0.773	0.788	0.802	0.810	0.817	0.831	0.845	0.859
1/ 5556	1.4042	1.4644	1.5897	1.7214	1.7897	1.8597	2.0048	2.1568	2.2354	2.3158	2.4819	2.6552	2.8359
0.00019	0.740	0.748	0.764	0.779	0.787	0.794	0.809	0.824	0.832	0.839	0.854	0.868	0.883
1/ 5263	1.4427	1.5046	1.6332	1.7686	1.8388	1.9107	2.0598	2.2159	2.2967	2.3792	2.5499	2.7280	2.9136
0.00020	0.760	0.768	0.784	0.799	0.807	0.815	0.830	0.846	0.853	0.861	0.876	0.891	0.906
1/ 5000	1.4802	1.5437	1.6757	1.8145	1.8866	1.9603	2.1133	2.2735	2.3563	2.4410	2.6161	2.7988	2.9893
0.00022	0.797	0.805	0.822	0.838	0.847	0.855	0.871	0.887	0.895	0.903	0.919	0.934	0.950
1/ 4545	1.5524	1.6190	1.7575	1.9031	1.9786	2.0560	2.2164	2.3844	2.4713	2.5602	2.7438	2.9354	3.1352
0.00024	0.832	0.841	0.858	0.876	0.884	0.893	0.910	0.927	0.935	0.943	0.960	0.976	0.992
1/ 4167	1.6214	1.6910	1.8356	1.9877	2.0666	2.1474	2.3150	2.4905	2.5812	2.6740	2.8658	3.0660	3.2747
0.00026	0.866	0.875	0.894	0.911	0.920	0.929	0.947	0.964	0.973	0.982	0.999	1.016	1.033
1/ 3846	1.6877	1.7600	1.9106	2.0689	2.1510	2.2351	2.4095	2.5922	2.6866	2.7832	2.9828	3.1912	3.4084
0.00028	0.899	0.908	0.927	0.946	0.955	0.964	0.983	1.001	1.010	1.019	1.036	1.054	1.072
1/ 3571	1.7514	1.8265	1.9827	2.1470	2.2322	2.3195	2.5005	2.6900	2.7880	2.8883	3.0954	3.3116	3.5370
0.00030	0.930	0.940	0.960	0.979	0.989	0.998	1.017	1.036	1.045	1.054	1.073	1.091	1.109
1/ 3333	1.8128	1.8906	2.0523	2.2223	2.3105	2.4009	2.5882	2.7844	2.8859	2.9896	3.2041	3.4279	3.6612
0.00032	0.961	0.971	0.991	1.011	1.021	1.031	1.050	1.070	1.079	1.089	1.108	1.127	1.146
1/ 3125	1.8723	1.9526	2.1196	2.2952	2.3863	2.4797	2.6731	2.8757	2.9805	3.0877	3.3092	3.5403	3.7812
0.00034	0.991	1.001	1.022	1.042	1.053	1.063	1.083	1.103	1.113	1.123	1.142	1.162	1.181
1/ 2941	1.9299	2.0127	2.1848	2.3658	2.4598	2.5560	2.7554	2.9642	3.0723	3.1827	3.4110	3.6492	3.8976
0.00036	1.019	1.030	1.051	1.073	1.083	1.093	1.114	1.135	1.145	1.155	1.175	1.195	1.215
1/ 2778	1.9859	2.0710	2.2481	2.4344	2.5311	2.6301	2.8353	3.0502	3.1613	3.2750	3.5099	3.7550	4.0106
0.00038	1.047	1.058	1.080	1.102	1.113	1.123	1.145	1.166	1.176	1.187	1.207	1.228	1.248
1/ 2632	2.0403	2.1278	2.3097	2.5011	2.6004	2.7021	2.9130	3.1338	3.2480	3.3647	3.6061	3.8579	4.1205
0.00040	1.074	1.086	1.108	1.131	1.142	1.153	1.174	1.196	1.207	1.218	1.239	1.260	1.281
1/ 2500	2.0933	2.1831	2.3698	2.5661	2.6680	2.7723	2.9886	3.2152	3.3323	3.4521	3.6997	3.9582	4.2276
0.00042	1.101	1.113	1.136	1.158	1.170	1.181	1.203	1.226	1.237	1.248	1.269	1.291	1.312
1/ 2381	2.1450	2.2370	2.4283	2.6295	2.7339	2.8408	3.0624	3.2946	3.4146	3.5374	3.7911	4.0559	4.3320
0.00044	1.127	1.139	1.162	1.186	1.197	1.209	1.232	1.254	1.266	1.277	1.299	1.321	1.343
1/ 2273	2.1955	2.2896	2.4854	2.6914	2.7982	2.9077	3.1345	3.3721	3.4950	3.6206	3.8803	4.1513	4.4339
0.00046	1.152	1.164	1.188	1.212	1.224	1.236	1.259	1.283	1.294	1.306	1.329	1.351	1.374
1/ 2174	2.2448	2.3411	2.5413	2.7519	2.8611	2.9730	3.2050	3.4479	3.5735	3.7020	3.9675	4.2446	4.5336
0.00048	1.177	1.189	1.214	1.238	1.251	1.263	1.287	1.310	1.322	1.334	1.357	1.380	1.403
1/ 2083	2.2931	2.3914	2.5959	2.8110	2.9226	3.0370	3.2739	3.5220	3.6504	3.7816	4.0529	4.3359	4.6311
	1575	1600	1650	1700	1725	1750	1800	1850	1875	1900	1950	2000	2050

Gradient   (Equivalent) Pipe diameters in mm

S = 0.00010 to 0.00048

m = Manning $n \times 100$
S = 0.00050 to 0.00280

Full bore conditions; Select m to divide mV and/or mQ.

ie hydraulic gradient =
1 in 2000 to 1 in 357

mV to give velocities in ms^{-1}
mQ to give discharges in m^3s^{-1}

Gradient — (Equivalent) Pipe diameters in mm

Gradient	1575	1600	1650	1700	1725	1750	1800	1850	1875	1900	1950	2000	2050
0.00050	1.201	1.214	1.239	1.264	1.276	1.289	1.313	1.337	1.349	1.361	1.385	1.409	1.432
1/ 2000	2.3404	2.4407	2.6495	2.8690	2.9829	3.0996	3.3414	3.5947	3.7257	3.8596	4.1364	4.4253	4.7266
0.00055	1.260	1.273	1.300	1.326	1.339	1.352	1.377	1.403	1.415	1.428	1.453	1.477	1.502
1/ 1818	2.4546	2.5599	2.7788	3.0090	3.1285	3.2509	3.5045	3.7701	3.9075	4.0480	4.3383	4.6413	4.9573
0.00060	1.316	1.330	1.357	1.385	1.398	1.412	1.438	1.465	1.478	1.491	1.517	1.543	1.569
1/ 1667	2.5637	2.6737	2.9023	3.1428	3.2676	3.3954	3.6603	3.9378	4.0813	4.2280	4.5312	4.8477	5.1777
0.00065	1.370	1.384	1.413	1.441	1.455	1.469	1.497	1.525	1.538	1.552	1.579	1.606	1.633
1/ 1538	2.6684	2.7829	3.0209	3.2712	3.4010	3.5341	3.8098	4.0986	4.2479	4.4006	4.7163	5.0457	5.3891
0.00070	1.421	1.436	1.466	1.496	1.510	1.525	1.554	1.582	1.597	1.611	1.639	1.667	1.694
1/ 1429	2.7691	2.8879	3.1349	3.3947	3.5294	3.6675	3.9536	4.2533	4.4083	4.5668	4.8943	5.2361	5.5925
0.00075	1.471	1.487	1.518	1.548	1.563	1.578	1.608	1.638	1.653	1.667	1.696	1.725	1.754
1/ 1333	2.8663	2.9893	3.2449	3.5138	3.6533	3.7962	4.0924	4.4026	4.5630	4.7270	5.0661	5.4199	5.7888
0.00080	1.519	1.536	1.567	1.599	1.614	1.630	1.661	1.692	1.707	1.722	1.752	1.782	1.811
1/ 1250	2.9603	3.0873	3.3513	3.6290	3.7731	3.9207	4.2266	4.5469	4.7126	4.8821	5.2322	5.5977	5.9787
0.00085	1.566	1.583	1.616	1.648	1.664	1.680	1.712	1.744	1.759	1.775	1.806	1.837	1.867
1/ 1176	3.0515	3.1823	3.4545	3.7407	3.8892	4.0414	4.3566	4.6869	4.8577	5.0323	5.3933	5.7700	6.1627
0.00090	1.612	1.629	1.662	1.696	1.712	1.729	1.762	1.794	1.810	1.826	1.858	1.890	1.921
1/ 1111	3.1399	3.2746	3.5546	3.8492	4.0020	4.1585	4.4829	4.8228	4.9985	5.1782	5.5496	5.9372	6.3413
0.00095	1.656	1.673	1.708	1.742	1.759	1.776	1.810	1.843	1.860	1.876	1.909	1.942	1.974
1/ 1053	3.2260	3.3643	3.6520	3.9547	4.1116	4.2725	4.6058	4.9549	5.1355	5.3201	5.7017	6.0999	6.5151
0.00100	1.699	1.717	1.752	1.788	1.805	1.822	1.857	1.891	1.908	1.925	1.959	1.992	2.025
1/ 1000	3.3098	3.4517	3.7469	4.0574	4.2185	4.3835	4.7254	5.0836	5.2689	5.4583	5.8498	6.2584	6.6844
0.00110	1.782	1.801	1.838	1.875	1.893	1.911	1.948	1.984	2.001	2.019	2.054	2.089	2.124
1/ 909	3.4713	3.6202	3.9298	4.2554	4.4244	4.5974	4.9561	5.3318	5.5261	5.7247	6.1353	6.5639	7.0106
0.00120	1.861	1.881	1.920	1.958	1.977	1.996	2.034	2.072	2.090	2.109	2.146	2.182	2.218
1/ 833	3.6257	3.7812	4.1045	4.4446	4.6211	4.8018	5.1765	5.5688	5.7718	5.9793	6.4081	6.8557	7.3223
0.00130	1.937	1.957	1.998	2.038	2.058	2.078	2.117	2.156	2.176	2.195	2.233	2.271	2.309
1/ 769	3.7737	3.9356	4.2721	4.6261	4.8098	4.9979	5.3878	5.7962	6.0075	6.2234	6.6698	7.1357	7.6213
0.00140	2.010	2.031	2.073	2.115	2.136	2.156	2.197	2.238	2.258	2.278	2.318	2.357	2.396
1/ 714	3.9162	4.0841	4.4334	4.8008	4.9913	5.1866	5.5912	6.0150	6.2342	6.4584	6.9216	7.4050	7.9090
0.00150	2.081	2.103	2.146	2.189	2.211	2.232	2.274	2.316	2.337	2.358	2.399	2.440	2.480
1/ 667	4.0536	4.2275	4.5890	4.9693	5.1665	5.3686	5.7875	6.2261	6.4530	6.6850	7.1645	7.6649	8.1866
0.00160	2.149	2.172	2.217	2.261	2.283	2.305	2.349	2.392	2.414	2.435	2.478	2.520	2.562
1/ 625	4.1866	4.3661	4.7395	5.1322	5.3360	5.5447	5.9773	6.4303	6.6647	6.9043	7.3995	7.9163	8.4551
0.00170	2.215	2.238	2.285	2.331	2.353	2.376	2.421	2.466	2.488	2.510	2.554	2.597	2.641
1/ 588	4.3154	4.5005	4.8854	5.2902	5.5002	5.7153	6.1612	6.6282	6.8698	7.1168	7.6272	8.1599	8.7153
0.00180	2.279	2.303	2.351	2.398	2.422	2.445	2.491	2.537	2.560	2.583	2.628	2.673	2.717
1/ 556	4.4405	4.6310	5.0270	5.4436	5.6597	5.8810	6.3398	6.8204	7.0690	7.3231	7.8483	8.3965	8.9680
0.00190	2.342	2.366	2.415	2.464	2.488	2.512	2.560	2.607	2.630	2.654	2.700	2.746	2.792
1/ 526	4.5622	4.7579	5.1648	5.5927	5.8147	6.0422	6.5136	7.0073	7.2627	7.5238	8.0634	8.6266	9.2137
0.00200	2.402	2.428	2.478	2.528	2.553	2.577	2.626	2.675	2.699	2.723	2.770	2.817	2.864
1/ 500	4.6807	4.8815	5.2989	5.7380	5.9658	6.1992	6.6828	7.1893	7.4513	7.7192	8.2729	8.8507	9.4531
0.00220	2.520	2.546	2.599	2.651	2.677	2.703	2.754	2.805	2.830	2.855	2.905	2.955	3.004
1/ 455	4.9092	5.1197	5.5576	6.0181	6.2570	6.5017	7.0090	7.5402	7.8150	8.0960	8.6767	9.2827	9.9145
0.00240	2.632	2.660	2.715	2.769	2.796	2.823	2.877	2.930	2.956	2.982	3.035	3.086	3.137
1/ 417	5.1275	5.3474	5.8047	6.2857	6.5352	6.7908	7.3206	7.8755	8.1625	8.4560	9.0625	9.6955	10.355
0.00260	2.739	2.768	2.826	2.882	2.911	2.939	2.994	3.049	3.077	3.104	3.158	3.212	3.265
1/ 385	5.3368	5.5657	6.0417	6.5423	6.8021	7.0681	7.6195	8.1971	8.4958	8.8013	9.4325	10.091	10.778
0.00280	2.843	2.873	2.932	2.991	3.020	3.050	3.107	3.165	3.193	3.221	3.278	3.333	3.389
1/ 357	5.5383	5.7758	6.2698	6.7893	7.0588	7.3349	7.9072	8.5065	8.8165	9.1335	9.7886	10.472	11.185
	1575	1600	1650	1700	1725	1750	1800	1850	1875	1900	1950	2000	2050

Gradient (Equivalent) Pipe diameters in mm

S = 0.00050 to 0.00280

m = Manning $n \times 100$
S = 0.00300 to 0.01400

ie hydraulic gradient =
1 in 333   to   1 in 71

Full bore conditions; Select
m to divide mV and/or mQ.

mV to give velocities in ms^{-1}
mQ to give discharges in m^{3}s^{-1}

Gradient (Equivalent) Pipe diameters in mm

Gradient	1575	1600	1650	1700	1725	1750	1800	1850	1875	1900	1950	2000	2050
0.00300	2.942	2.973	3.035	3.096	3.126	3.157	3.216	3.276	3.305	3.334	3.393	3.450	3.508
1/ 333	5.7327	5.9786	6.4898	7.0276	7.3066	7.5924	8.1847	8.8051	9.1260	9.4541	10.132	10.840	11.578
0.00320	3.039	3.071	3.135	3.198	3.229	3.260	3.322	3.383	3.414	3.444	3.504	3.564	3.623
1/ 313	5.9207	6.1746	6.7027	7.2581	7.5462	7.8414	8.4531	9.0939	9.4253	9.7641	10.464	11.195	11.957
0.00340	3.132	3.166	3.231	3.296	3.328	3.360	3.424	3.487	3.519	3.550	3.612	3.673	3.734
1/ 294	6.1029	6.3647	6.9090	7.4815	7.7784	8.0827	8.7133	9.3737	9.7154	10.065	10.787	11.540	12.325
0.00360	3.223	3.257	3.325	3.392	3.425	3.458	3.523	3.588	3.621	3.653	3.716	3.780	3.842
1/ 278	6.2798	6.5492	7.1093	7.6983	8.0040	8.3170	8.9659	9.6455	9.9970	10.356	11.099	11.874	12.683
0.00380	3.312	3.347	3.416	3.485	3.519	3.553	3.620	3.687	3.720	3.753	3.818	3.883	3.948
1/ 263	6.4519	6.7286	7.3041	7.9093	8.2233	8.5449	9.2116	9.9098	10.271	10.640	11.403	12.200	13.030
0.00400	3.398	3.433	3.505	3.575	3.610	3.645	3.714	3.782	3.816	3.850	3.918	3.984	4.050
1/ 250	6.6195	6.9034	7.4938	8.1148	8.4369	8.7669	9.4509	10.167	10.538	10.917	11.700	12.517	13.369
0.00420	3.482	3.518	3.591	3.663	3.699	3.735	3.806	3.876	3.911	3.945	4.014	4.083	4.150
1/ 238	6.7830	7.0739	7.6789	8.3152	8.6453	8.9834	9.6843	10.418	10.798	11.186	11.989	12.826	13.699
0.00440	3.563	3.601	3.676	3.750	3.786	3.823	3.895	3.967	4.003	4.038	4.109	4.179	4.248
1/ 227	6.9426	7.2404	7.8596	8.5108	8.8487	9.1948	9.9122	10.664	11.052	11.449	12.271	13.128	14.021
0.00460	3.644	3.682	3.758	3.834	3.871	3.909	3.983	4.056	4.093	4.129	4.201	4.273	4.344
1/ 217	7.0987	7.4031	8.0362	8.7021	9.0476	9.4015	10.135	10.903	11.301	11.707	12.546	13.423	14.336
0.00480	3.722	3.761	3.839	3.916	3.955	3.993	4.068	4.143	4.181	4.218	4.291	4.364	4.437
1/ 208	7.2513	7.5623	8.2091	8.8893	9.2422	9.6037	10.353	11.138	11.544	11.959	12.816	13.711	14.645
0.00500	3.799	3.839	3.918	3.997	4.036	4.075	4.152	4.229	4.267	4.305	4.380	4.454	4.528
1/ 200	7.4009	7.7183	8.3783	9.0726	9.4328	9.8017	10.566	11.367	11.782	12.205	13.081	13.994	14.947
0.00550	3.984	4.026	4.110	4.192	4.233	4.274	4.355	4.435	4.475	4.515	4.594	4.672	4.749
1/ 182	7.7621	8.0950	8.7873	9.5154	9.8932	10.280	11.082	11.922	12.357	12.801	13.719	14.677	15.676
0.00600	4.161	4.205	4.292	4.379	4.421	4.464	4.549	4.633	4.674	4.716	4.798	4.880	4.961
1/ 167	8.1072	8.4550	9.1780	9.9385	10.333	10.737	11.575	12.452	12.906	13.370	14.329	15.330	16.373
0.00650	4.331	4.377	4.468	4.557	4.602	4.646	4.734	4.822	4.865	4.908	4.994	5.079	5.163
1/ 154	8.4383	8.8002	9.5528	10.344	10.755	11.176	12.048	12.961	13.433	13.916	14.914	15.956	17.042
0.00700	4.495	4.542	4.636	4.729	4.776	4.822	4.913	5.004	5.049	5.093	5.182	5.271	5.358
1/ 143	8.7568	9.1324	9.9134	10.735	11.161	11.598	12.502	13.450	13.940	14.441	15.477	16.558	17.685
0.00750	4.652	4.702	4.799	4.895	4.943	4.991	5.086	5.179	5.226	5.272	5.364	5.456	5.546
1/ 133	9.0642	9.4529	10.261	11.112	11.553	12.005	12.941	13.922	14.429	14.948	16.020	17.139	18.306
0.00800	4.805	4.856	4.956	5.056	5.105	5.155	5.252	5.349	5.397	5.445	5.540	5.635	5.728
1/ 125	9.3614	9.7629	10.598	11.476	11.932	12.398	13.366	14.379	14.903	15.438	16.546	17.701	18.906
0.00850	4.953	5.005	5.109	5.212	5.263	5.313	5.414	5.514	5.563	5.613	5.711	5.808	5.904
1/ 118	9.6495	10.063	10.924	11.829	12.299	12.780	13.777	14.821	15.361	15.914	17.055	18.246	19.488
0.00900	5.096	5.150	5.257	5.363	5.415	5.467	5.571	5.674	5.725	5.775	5.876	5.976	6.076
1/ 111	9.9293	10.355	11.241	12.172	12.655	13.150	14.176	15.251	15.807	16.375	17.549	18.775	20.053
0.00950	5.236	5.291	5.401	5.510	5.563	5.617	5.724	5.829	5.882	5.934	6.037	6.140	6.242
1/ 105	10.201	10.639	11.549	12.506	13.002	13.511	14.565	15.669	16.240	16.824	18.030	19.290	20.603
0.01000	5.372	5.429	5.541	5.653	5.708	5.763	5.872	5.981	6.034	6.088	6.194	6.300	6.404
1/ 100	10.466	10.915	11.849	12.831	13.340	13.862	14.943	16.076	16.662	17.261	18.499	19.791	21.138
0.01100	5.634	5.694	5.812	5.929	5.987	6.044	6.159	6.272	6.329	6.385	6.496	6.607	6.717
1/ 91	10.977	11.448	12.427	13.457	13.991	14.538	15.673	16.860	17.475	18.103	19.402	20.757	22.170
0.01200	5.885	5.947	6.070	6.192	6.253	6.313	6.433	6.551	6.610	6.669	6.785	6.901	7.015
1/ 83	11.465	11.957	12.980	14.055	14.613	15.185	16.369	17.610	18.252	18.908	20.264	21.680	23.155
0.01300	6.125	6.190	6.318	6.445	6.508	6.571	6.695	6.819	6.880	6.941	7.062	7.183	7.302
1/ 77	11.934	12.445	13.510	14.629	15.210	15.805	17.038	18.329	18.997	19.680	21.092	22.565	24.101
0.01400	6.356	6.423	6.557	6.688	6.754	6.819	6.948	7.076	7.140	7.203	7.329	7.454	7.578
1/ 71	12.384	12.915	14.020	15.181	15.784	16.401	17.681	19.021	19.714	20.423	21.888	23.417	25.011
	1575	1600	1650	1700	1725	1750	1800	1850	1875	1900	1950	2000	2050

Gradient (Equivalent) Pipe diameters in mm

S = 0.00300 to 0.01400

m = Manning n × 100
S = 0·01500 to 0·07500

Full bore conditions; Select m to divide mV and/or mQ.

ie hydraulic gradient = 1 in 67 to 1 in 13·3

mV to give velocities in ms⁻¹
mQ to give discharges in m³s⁻¹

Gradient	(Equivalent) Pipe diameters in mm												
	1575	1600	1650	1700	1725	1750	1800	1850	1875	1900	1950	2000	2050
0·01500	6·579	6·649	6·787	6·923	6·991	7·058	7·192	7·325	7·390	7·456	7·586	7·715	7·843
1/ 67	12·819	13·368	14·512	15·714	16·338	16·977	18·302	19·689	20·406	21·140	22·656	24·239	25·888
0·01600	6·795	6·867	7·009	7·150	7·220	7·290	7·428	7·565	7·633	7·701	7·835	7·968	8·101
1/ 63	13·239	13·807	14·988	16·230	16·874	17·534	18·902	20·335	21·076	21·833	23·399	25·034	26·737
0·01700	7·004	7·078	7·225	7·370	7·442	7·514	7·657	7·798	7·868	7·938	8·076	8·214	8·350
1/ 59	13·647	14·232	15·449	16·729	17·393	18·073	19·484	20·960	21·724	22·505	24·119	25·804	27·560
0·01800	7·207	7·284	7·435	7·584	7·658	7·732	7·879	8·024	8·096	8·168	8·310	8·452	8·592
1/ 56	14·042	14·644	15·897	17·214	17·897	18·597	20·048	21·568	22·354	23·158	24·819	26·552	28·359
0·01900	7·405	7·483	7·638	7·792	7·868	7·944	8·094	8·244	8·318	8·391	8·538	8·683	8·828
1/ 53	14·427	15·046	16·332	17·686	18·388	19·107	20·598	22·159	22·967	23·792	25·499	27·280	29·136
0·02000	7·597	7·678	7·837	7·994	8·072	8·150	8·305	8·458	8·534	8·609	8·760	8·909	9·057
1/ 50	14·802	15·437	16·757	18·145	18·866	19·603	21·133	22·735	23·563	24·410	26·161	27·988	29·893
0·02200	7·968	8·052	8·219	8·384	8·466	8·548	8·710	8·871	8·950	9·030	9·187	9·344	9·499
1/ 45·5	15·524	16·190	17·575	19·031	19·786	20·560	22·164	23·844	24·713	25·602	27·438	29·354	31·352
0·02400	8·322	8·410	8·585	8·757	8·843	8·928	9·097	9·265	9·348	9·431	9·596	9·759	9·921
1/ 41·7	16·214	16·910	18·356	19·877	20·666	21·474	23·150	24·905	25·812	26·740	28·658	30·660	32·747
0·02600	8·662	8·754	8·935	9·115	9·204	9·293	9·469	9·643	9·730	9·816	9·988	10·16	10·33
1/ 38·5	16·877	17·600	19·106	20·689	21·510	22·351	24·095	25·922	26·866	27·832	29·828	31·912	34·084
0·02800	8·989	9·084	9·272	9·459	9·551	9·643	9·826	10·01	10·10	10·19	10·36	10·54	10·72
1/ 35·7	17·514	18·265	19·827	21·470	22·322	23·195	25·005	26·900	27·880	28·883	30·954	33·116	35·370
0·03000	9·305	9·403	9·598	9·791	9·887	9·982	10·17	10·36	10·45	10·54	10·73	10·91	11·09
1/ 33·3	18·128	18·906	20·523	22·223	23·105	24·009	25·882	27·844	28·859	29·896	32·041	34·279	36·612
0·03200	9·610	9·711	9·913	10·11	10·21	10·31	10·50	10·70	10·79	10·89	11·08	11·27	11·46
1/ 31·3	18·723	19·526	21·196	22·952	23·863	24·797	26·731	28·757	29·805	30·877	33·092	35·403	37·812
0·03400	9·906	10·01	10·22	10·42	10·53	10·63	10·83	11·03	11·13	11·23	11·42	11·62	11·81
1/ 29·4	19·299	20·127	21·848	23·658	24·598	25·560	27·554	29·642	30·723	31·827	34·110	36·492	38·976
0·03600	10·19	10·30	10·51	10·73	10·83	10·93	11·14	11·35	11·45	11·55	11·75	11·95	12·15
1/ 27·8	19·859	20·710	22·481	24·344	25·311	26·301	28·353	30·502	31·613	32·750	35·099	37·550	40·106
0·03800	10·47	10·58	10·80	11·02	11·13	11·23	11·45	11·66	11·76	11·87	12·07	12·28	12·48
1/ 26·3	20·403	21·278	23·097	25·011	26·004	27·021	29·130	31·338	32·480	33·647	36·061	38·579	41·205
0·04000	10·74	10·86	11·08	11·31	11·42	11·53	11·74	11·96	12·07	12·18	12·39	12·60	12·81
1/ 25·0	20·933	21·831	23·698	25·661	26·680	27·723	29·886	32·152	33·323	34·521	36·997	39·582	42·276
0·04200	11·01	11·13	11·36	11·58	11·70	11·81	12·03	12·26	12·37	12·48	12·69	12·91	13·12
1/ 23·8	21·450	22·370	24·283	26·295	27·339	28·408	30·624	32·946	34·146	35·374	37·911	40·559	43·320
0·04400	11·27	11·39	11·62	11·86	11·97	12·09	12·32	12·54	12·66	12·77	12·99	13·21	13·43
1/ 22·7	21·955	22·896	24·854	26·914	27·982	29·077	31·345	33·721	34·950	36·206	38·803	41·513	44·339
0·04600	11·52	11·64	11·88	12·12	12·24	12·36	12·59	12·83	12·94	13·06	13·29	13·51	13·74
1/ 21·7	22·448	23·411	25·413	27·519	28·611	29·730	32·050	34·479	35·735	37·020	39·675	42·446	45·336
0·04800	11·77	11·89	12·14	12·38	12·51	12·63	12·87	13·10	13·22	13·34	13·57	13·80	14·03
1/ 20·8	22·931	23·914	25·959	28·110	29·226	30·370	32·739	35·220	36·504	37·816	40·529	43·359	46·311
0·05000	12·01	12·14	12·39	12·64	12·76	12·89	13·13	13·37	13·49	13·61	13·85	14·09	14·32
1/ 20·0	23·404	24·407	26·495	28·690	29·829	30·996	33·414	35·947	37·257	38·596	41·364	44·253	47·266
0·05500	12·60	12·73	13·00	13·26	13·39	13·52	13·77	14·03	14·15	14·28	14·53	14·77	15·02
1/ 18·2	24·546	25·599	27·788	30·090	31·285	32·509	35·045	37·701	39·075	40·480	43·383	46·413	49·573
0·06000	13·16	13·30	13·57	13·85	13·98	14·12	14·38	14·65	14·78	14·91	15·17	15·43	15·69
1/ 16·7	25·637	26·737	29·023	31·428	32·676	33·954	36·603	39·378	40·813	42·280	45·312	48·477	51·777
0·06500	13·70	13·84	14·13	14·41	14·55	14·69	14·97	15·25	15·38	15·52	15·79	16·06	16·33
1/ 15·4	26·684	27·829	30·209	32·712	34·010	35·341	38·098	40·986	42·479	44·006	47·163	50·457	53·891
0·07000	14·21	14·36	14·66	14·96	15·10	15·25	15·54	15·82	15·97	16·11	16·39	16·67	16·94
1/ 14·3	27·691	28·879	31·349	33·947	35·294	36·675	39·536	42·533	44·083	45·668	48·943	52·361	55·925
0·07500	14·71	14·87	15·18	15·48	15·63	15·78	16·08	16·38	16·53	16·67	16·96	17·25	17·54
1/ 13·3	28·663	29·893	32·449	35·138	36·533	37·962	40·924	44·026	45·630	47·270	50·661	54·199	57·888
	1575	1600	1650	1700	1725	1750	1800	1850	1875	1900	1950	2000	2050

Gradient (Equivalent) Pipe diameters in mm

S = 0·01500 to 0·07500

m = Manning $n \times 100$
S = 0·00010 to 0·00048

ie hydraulic gradient =
1 in 10000 to 1 in 2083

Full bore conditions; Select
m to divide mV and/or mQ.

mV to give velocities in ms^{-1}
mQ to give discharges in m^{3}s^{-1}

Gradient — (Equivalent) Pipe diameters in mm

Gradient	2050	2100	2150	2200	2250	2300	2350	2400	2450	2500	2550	2600	2650
0·00010	0·640	0·651	0·661	0·671	0·681	0·691	0·701	0·711	0·721	0·731	0·741	0·750	0·760
1/10000	2·1138	2·2541	2·4000	2·5518	2·7094	2·8729	3·0425	3·2182	3·4001	3·5883	3·7829	3·9839	4·1915
0·00011	0·672	0·683	0·693	0·704	0·715	0·725	0·736	0·746	0·756	0·767	0·777	0·787	0·797
1/ 9091	2·2170	2·3641	2·5172	2·6763	2·8416	3·0131	3·1910	3·3753	3·5661	3·7634	3·9675	4·1784	4·3961
0·00012	0·702	0·713	0·724	0·735	0·746	0·757	0·768	0·779	0·790	0·801	0·811	0·822	0·832
1/ 8333	2·3155	2·4692	2·6291	2·7953	2·9680	3·1471	3·3329	3·5254	3·7246	3·9308	4·1439	4·3642	4·5916
0·00013	0·730	0·742	0·754	0·765	0·777	0·788	0·800	0·811	0·822	0·833	0·845	0·856	0·866
1/ 7692	2·4101	2·5700	2·7365	2·9095	3·0892	3·2756	3·4690	3·6693	3·8767	4·0913	4·3132	4·5424	4·7791
0·00014	0·758	0·770	0·782	0·794	0·806	0·818	0·830	0·842	0·853	0·865	0·876	0·888	0·899
1/ 7143	2·5011	2·6671	2·8398	3·0193	3·2058	3·3993	3·5999	3·8078	4·0231	4·2457	4·4760	4·7138	4·9595
0·00015	0·784	0·797	0·810	0·822	0·835	0·847	0·859	0·871	0·883	0·895	0·907	0·919	0·931
1/ 6667	2·5888	2·7607	2·9394	3·1253	3·3183	3·5186	3·7263	3·9415	4·1643	4·3948	4·6331	4·8793	5·1335
0·00016	0·810	0·823	0·836	0·849	0·862	0·875	0·887	0·900	0·912	0·925	0·937	0·949	0·961
1/ 6250	2·6737	2·8512	3·0358	3·2278	3·4271	3·6340	3·8485	4·0707	4·3008	4·5389	4·7850	5·0393	5·3019
0·00017	0·835	0·849	0·862	0·875	0·888	0·902	0·915	0·928	0·940	0·953	0·966	0·978	0·991
1/ 5882	2·7560	2·9389	3·1293	3·3271	3·5326	3·7458	3·9669	4·1960	4·4332	4·6786	4·9323	5·1944	5·4651
0·00018	0·859	0·873	0·887	0·901	0·914	0·928	0·941	0·954	0·968	0·981	0·994	1·007	1·020
1/ 5556	2·8359	3·0242	3·2200	3·4236	3·6350	3·8544	4·0819	4·3177	4·5617	4·8142	5·0753	5·3450	5·6235
0·00019	0·883	0·897	0·911	0·925	0·939	0·953	0·967	0·981	0·994	1·008	1·021	1·034	1·048
1/ 5263	2·9136	3·1070	3·3082	3·5174	3·7346	3·9600	4·1938	4·4360	4·6867	4·9461	5·2143	5·4915	5·7776
0·00020	0·906	0·920	0·935	0·949	0·964	0·978	0·992	1·006	1·020	1·034	1·048	1·061	1·075
1/ 5000	2·9893	3·1877	3·3942	3·6088	3·8316	4·0629	4·3027	4·5512	4·8085	5·0746	5·3498	5·6341	5·9277
0·00022	0·950	0·965	0·981	0·996	1·011	1·026	1·040	1·055	1·070	1·084	1·099	1·113	1·127
1/ 4545	3·1352	3·3433	3·5598	3·7849	4·0187	4·2612	4·5128	4·7734	5·0432	5·3223	5·6109	5·9091	6·2170
0·00024	0·992	1·008	1·024	1·040	1·056	1·071	1·087	1·102	1·117	1·132	1·148	1·162	1·177
1/ 4167	3·2747	3·4920	3·7181	3·9532	4·1973	4·4507	4·7134	4·9856	5·2674	5·5590	5·8604	6·1719	6·4935
0·00026	1·033	1·049	1·066	1·082	1·099	1·115	1·131	1·147	1·163	1·179	1·194	1·210	1·225
1/ 3846	3·4084	3·6346	3·8699	4·1146	4·3687	4·6324	4·9059	5·1892	5·4825	5·7860	6·0997	6·4239	6·7586
0·00028	1·072	1·089	1·106	1·123	1·140	1·157	1·174	1·190	1·207	1·223	1·239	1·256	1·272
1/ 3571	3·5370	3·7718	4·0160	4·2699	4·5337	4·8073	5·0911	5·3851	5·6895	6·0044	6·3300	6·6664	7·0138
0·00030	1·109	1·127	1·145	1·163	1·180	1·198	1·215	1·232	1·249	1·266	1·283	1·300	1·316
1/ 3333	3·6612	3·9042	4·1570	4·4198	4·6928	4·9760	5·2698	5·5741	5·8892	6·2151	6·5521	6·9004	7·2599
0·00032	1·146	1·164	1·183	1·201	1·219	1·237	1·255	1·273	1·290	1·308	1·325	1·342	1·359
1/ 3125	3·7812	4·0322	4·2933	4·5648	4·8467	5·1392	5·4426	5·7569	6·0823	6·4190	6·7670	7·1267	7·4980
0·00034	1·181	1·200	1·219	1·238	1·256	1·275	1·293	1·312	1·330	1·348	1·366	1·384	1·401
1/ 2941	3·8976	4·1563	4·4255	4·7052	4·9958	5·2974	5·6101	5·9341	6·2695	6·6165	6·9753	7·3460	7·7288
0·00036	1·215	1·235	1·254	1·274	1·293	1·312	1·331	1·350	1·368	1·387	1·405	1·424	1·442
1/ 2778	4·0106	4·2768	4·5538	4·8417	5·1407	5·4510	5·7727	6·1061	6·4512	6·8083	7·1775	7·5590	7·9529
0·00038	1·248	1·269	1·289	1·309	1·328	1·348	1·367	1·387	1·406	1·425	1·444	1·463	1·481
1/ 2632	4·1205	4·3940	4·6785	4·9743	5·2815	5·6004	5·9309	6·2734	6·6280	6·9949	7·3742	7·7661	8·1708
0·00040	1·281	1·302	1·322	1·343	1·363	1·383	1·403	1·423	1·442	1·462	1·481	1·501	1·520
1/ 2500	4·2276	4·5081	4·8001	5·1036	5·4188	5·7458	6·0850	6·4364	6·8002	7·1766	7·5658	7·9679	8·3830
0·00042	1·312	1·334	1·355	1·376	1·396	1·417	1·438	1·458	1·478	1·498	1·518	1·538	1·557
1/ 2381	4·3320	4·6195	4·9186	5·2296	5·5526	5·8877	6·2353	6·5953	6·9681	7·3538	7·7526	8·1646	8·5901
0·00044	1·343	1·365	1·387	1·408	1·429	1·450	1·471	1·492	1·513	1·533	1·554	1·574	1·594
1/ 2273	4·4339	4·7282	5·0344	5·3527	5·6832	6·0263	6·3820	6·7505	7·1321	7·5269	7·9350	8·3568	8·7922
0·00046	1·374	1·396	1·418	1·440	1·461	1·483	1·504	1·526	1·547	1·568	1·589	1·609	1·630
1/ 2174	4·5336	4·8344	5·1475	5·4730	5·8110	6·1617	6·5254	6·9023	7·2924	7·6961	8·1134	8·5446	8·9898
0·00048	1·403	1·426	1·448	1·471	1·493	1·515	1·537	1·559	1·580	1·602	1·623	1·644	1·665
1/ 2083	4·6311	4·9384	5·2582	5·5907	5·9359	6·2943	6·6658	7·0507	7·4493	7·8616	8·2879	8·7284	9·1832
Gradient	2050	2100	2150	2200	2250	2300	2350	2400	2450	2500	2550	2600	2650

Gradient — (Equivalent) Pipe diameters in mm

S = 0·00010 to 0·00048

m = Manning $n \times 100$
S = 0·00050 to 0·00280

ie hydraulic gradient =
1 in 2000 to 1 in 357

Full bore conditions; Select m to divide mV and/or mQ.

mV to give velocities in ms^{-1}
mQ to give discharges in m^{3}s^{-1}

Gradient (Equivalent) Pipe diameters in mm

Gradient	2050	2100	2150	2200	2250	2300	2350	2400	2450	2500	2550	2600	2650
0·00050	1·432	1·455	1·478	1·501	1·524	1·546	1·569	1·591	1·613	1·635	1·656	1·678	1·699
1/ 2000	4·7266	5·0403	5·3667	5·7060	6·0583	6·4240	6·8032	7·1961	7·6029	8·0237	8·4588	8·9083	9·3725
0·00055	1·502	1·526	1·550	1·574	1·598	1·622	1·645	1·668	1·691	1·714	1·737	1·760	1·782
1/ 1818	4·9573	5·2863	5·6286	5·9845	6·3541	6·7376	7·1353	7·5473	7·9740	8·4153	8·8717	9·3431	9·8300
0·00060	1·569	1·594	1·619	1·644	1·669	1·694	1·718	1·743	1·767	1·791	1·814	1·838	1·862
1/ 1667	5·1777	5·5213	5·8789	6·2506	6·6366	7·0372	7·4526	7·8829	8·3285	8·7895	9·2661	9·7586	10·267
0·00065	1·633	1·659	1·685	1·711	1·737	1·763	1·788	1·814	1·839	1·864	1·888	1·913	1·938
1/ 1538	5·3891	5·7468	6·1189	6·5058	6·9076	7·3245	7·7569	8·2048	8·6686	9·1484	9·6445	10·157	10·686
0·00070	1·694	1·722	1·749	1·776	1·803	1·829	1·856	1·882	1·908	1·934	1·960	1·985	2·011
1/ 1429	5·5925	5·9637	6·3499	6·7514	7·1683	7·6010	8·0497	8·5145	8·9958	9·4938	10·009	10·540	11·090
0·00075	1·754	1·782	1·810	1·838	1·866	1·894	1·921	1·948	1·975	2·002	2·029	2·055	2·081
1/ 1333	5·7888	6·1730	6·5728	6·9883	7·4199	7·8678	8·3322	8·8134	9·3116	9·8270	10·360	10·910	11·479
0·00080	1·811	1·841	1·870	1·899	1·927	1·956	1·984	2·012	2·040	2·068	2·095	2·122	2·149
1/ 1250	5·9787	6·3755	6·7883	7·2175	7·6633	8·1258	8·6055	9·1024	9·6169	10·149	10·700	11·268	11·855
0·00085	1·867	1·897	1·927	1·957	1·987	2·016	2·045	2·074	2·103	2·131	2·160	2·188	2·216
1/ 1176	6·1627	6·5717	6·9973	7·4397	7·8991	8·3759	8·8703	9·3826	9·9129	10·462	11·029	11·615	12·220
0·00090	1·921	1·952	1·983	2·014	2·044	2·074	2·104	2·134	2·164	2·193	2·222	2·251	2·280
1/ 1111	6·3413	6·7622	7·2001	7·6553	8·1281	8·6188	9·1275	9·6546	10·200	10·765	11·349	11·952	12·575
0·00095	1·974	2·006	2·038	2·069	2·100	2·131	2·162	2·193	2·223	2·253	2·283	2·313	2·342
1/ 1053	6·5151	6·9475	7·3974	7·8651	8·3509	8·8549	9·3776	9·9191	10·480	11·060	11·660	12·279	12·919
0·00100	2·025	2·058	2·091	2·123	2·155	2·187	2·218	2·250	2·281	2·312	2·342	2·373	2·403
1/ 1000	6·6844	7·1280	7·5896	8·0694	8·5678	9·0850	9·6212	10·177	10·752	11·347	11·963	12·598	13·255
0·00110	2·124	2·158	2·193	2·226	2·260	2·293	2·326	2·359	2·392	2·424	2·457	2·489	2·520
1/ 909	7·0106	7·4759	7·9600	8·4633	8·9860	9·5284	10·091	10·674	11·277	11·901	12·546	13·213	13·902
0·00120	2·218	2·254	2·290	2·325	2·361	2·395	2·430	2·464	2·498	2·532	2·566	2·599	2·633
1/ 833	7·3223	7·8083	8·3140	8·8396	9·3856	9·9521	10·540	11·148	11·778	12·430	13·104	13·801	14·520
0·00130	2·309	2·346	2·384	2·420	2·457	2·493	2·529	2·565	2·600	2·636	2·671	2·705	2·740
1/ 769	7·6213	8·1272	8·6535	9·2006	9·7688	10·358	10·970	11·603	12·259	12·938	13·639	14·364	15·113
0·00140	2·396	2·435	2·474	2·512	2·550	2·587	2·625	2·662	2·699	2·735	2·772	2·808	2·844
1/ 714	7·9090	8·4340	8·9801	9·5479	10·138	10·749	11·384	12·041	12·722	13·426	14·154	14·906	15·683
0·00150	2·480	2·520	2·560	2·600	2·639	2·678	2·717	2·755	2·793	2·831	2·869	2·906	2·943
1/ 667	8·1866	8·7300	9·2953	9·8830	10·493	11·127	11·784	12·464	13·169	13·897	14·651	15·430	16·234
0·00160	2·562	2·603	2·644	2·685	2·726	2·766	2·806	2·846	2·885	2·924	2·963	3·001	3·040
1/ 625	8·4551	9·0163	9·6002	10·207	10·838	11·492	12·170	12·873	13·600	14·353	15·132	15·936	16·766
0·00170	2·641	2·683	2·726	2·768	2·810	2·851	2·892	2·933	2·974	3·014	3·054	3·094	3·133
1/ 588	8·7153	9·2938	9·8956	10·521	11·171	11·845	12·545	13·269	14·019	14·795	15·597	16·426	17·282
0·00180	2·717	2·761	2·805	2·848	2·891	2·934	2·976	3·018	3·060	3·101	3·143	3·184	3·224
1/ 556	8·9680	9·5632	10·183	10·826	11·495	12·189	12·908	13·654	14·425	15·224	16·049	16·902	17·783
0·00190	2·792	2·837	2·882	2·926	2·970	3·014	3·058	3·101	3·144	3·186	3·229	3·271	3·313
1/ 526	9·2137	9·8253	10·462	11·123	11·810	12·523	13·262	14·028	14·821	15·641	16·489	17·366	18·270
0·00200	2·864	2·910	2·956	3·002	3·047	3·092	3·137	3·181	3·225	3·269	3·313	3·356	3·399
1/ 500	9·4531	10·081	10·733	11·412	12·117	12·848	13·606	14·392	15·206	16·047	16·918	17·817	18·745
0·00220	3·004	3·052	3·101	3·149	3·196	3·243	3·290	3·337	3·383	3·429	3·474	3·520	3·565
1/ 455	9·9145	10·573	11·257	11·969	12·708	13·475	14·271	15·095	15·948	16·831	17·743	18·686	19·660
0·00240	3·137	3·188	3·239	3·289	3·338	3·388	3·436	3·485	3·533	3·581	3·629	3·676	3·723
1/ 417	10·355	11·043	11·758	12·501	13·273	14·074	14·905	15·766	16·657	17·579	18·532	19·517	20·534
0·00260	3·265	3·318	3·371	3·423	3·475	3·526	3·577	3·627	3·678	3·727	3·777	3·826	3·875
1/ 385	10·778	11·494	12·238	13·012	13·815	14·649	15·514	16·410	17·337	18·297	19·289	20·314	21·373
0·00280	3·389	3·444	3·498	3·552	3·606	3·659	3·712	3·764	3·816	3·868	3·920	3·971	4·021
1/ 357	11·185	11·927	12·700	13·503	14·337	15·202	16·099	17·029	17·992	18·988	20·017	21·081	22·179
	2050	2100	2150	2200	2250	2300	2350	2400	2450	2500	2550	2600	2650

Gradient   (Equivalent) Pipe diameters in mm

S = 0·00050 to 0·00280

m = Manning $n \times 100$
S = 0·00300 to 0·01400

ie hydraulic gradient =
1 in 333 to 1 in 71

Full bore conditions; Select
m to divide mV and/or mQ.

mV to give velocities in ms^{-1}
mQ to give discharges in m^{3}s^{-1}

Gradient   (Equivalent) Pipe diameters in mm

Gradient	2050	2100	2150	2200	2250	2300	2350	2400	2450	2500	2550	2600	2650
0·00300	3·508	3·565	3·621	3·677	3·732	3·787	3·842	3·896	3·950	4·004	4·057	4·110	4·162
1/ 333	11·578	12·346	13·146	13·977	14·840	15·736	16·664	17·627	18·623	19·654	20·720	21·821	22·958
0·00320	3·623	3·681	3·740	3·797	3·855	3·912	3·968	4·024	4·080	4·135	4·190	4·245	4·299
1/ 313	11·957	12·751	13·577	14·435	15·327	16·252	17·211	18·205	19·234	20·299	21·399	22·537	23·711
0·00340	3·734	3·795	3·855	3·914	3·973	4·032	4·090	4·148	4·205	4·262	4·319	4·375	4·431
1/ 294	12·325	13·143	13·995	14·879	15·798	16·752	17·741	18·765	19·826	20·923	22·058	23·230	24·441
0·00360	3·842	3·905	3·966	4·028	4·089	4·149	4·209	4·268	4·327	4·386	4·444	4·502	4·560
1/ 278	12·683	13·524	14·400	15·311	16·256	17·238	18·255	19·309	20·401	21·530	22·697	23·904	25·149
0·00380	3·948	4·012	4·075	4·138	4·201	4·263	4·324	4·385	4·446	4·506	4·566	4·626	4·685
1/ 263	13·030	13·895	14·795	15·730	16·702	17·710	18·755	19·838	20·960	22·120	23·319	24·559	25·838
0·00400	4·050	4·116	4·181	4·246	4·310	4·373	4·436	4·499	4·561	4·623	4·685	4·746	4·806
1/ 250	13·369	14·256	15·179	16·139	17·136	18·170	19·242	20·354	21·504	22·694	23·925	25·197	26·510
0·00420	4·150	4·218	4·284	4·350	4·416	4·481	4·546	4·610	4·674	4·737	4·800	4·863	4·925
1/ 238	13·699	14·608	15·554	16·537	17·559	18·619	19·718	20·856	22·035	23·255	24·516	25·819	27·164
0·00440	4·248	4·317	4·385	4·453	4·520	4·587	4·653	4·719	4·784	4·849	4·913	4·977	5·041
1/ 227	14·021	14·952	15·920	16·927	17·972	19·057	20·182	21·347	22·554	23·802	25·093	26·426	27·803
0·00460	4·344	4·414	4·484	4·553	4·622	4·690	4·758	4·825	4·892	4·958	5·024	5·089	5·154
1/ 217	14·336	15·288	16·278	17·307	18·376	19·485	20·635	21·827	23·061	24·337	25·657	27·020	28·428
0·00480	4·437	4·509	4·580	4·651	4·721	4·791	4·860	4·929	4·997	5·065	5·132	5·199	5·265
1/ 208	14·645	15·617	16·628	17·679	18·771	19·904	21·079	22·296	23·557	24·861	26·209	27·601	29·040
0·00500	4·528	4·602	4·675	4·747	4·818	4·889	4·960	5·030	5·100	5·169	5·238	5·306	5·374
1/ 200	14·947	15·939	16·971	18·044	19·158	20·315	21·514	22·756	24·042	25·373	26·749	28·171	29·639
0·00550	4·749	4·826	4·903	4·978	5·054	5·128	5·202	5·276	5·349	5·421	5·493	5·565	5·636
1/ 182	15·676	16·717	17·799	18·925	20·093	21·306	22·564	23·867	25·216	26·612	28·055	29·546	31·085
0·00600	4·961	5·041	5·121	5·200	5·278	5·356	5·434	5·510	5·587	5·662	5·738	5·812	5·887
1/ 167	16·373	17·460	18·591	19·766	20·987	22·254	23·567	24·928	26·337	27·795	29·302	30·859	32·467
0·00650	5·163	5·247	5·330	5·412	5·494	5·575	5·655	5·735	5·815	5·894	5·972	6·050	6·127
1/ 154	17·042	18·173	19·350	20·573	21·844	23·162	24·529	25·946	27·413	28·930	30·499	32·119	33·793
0·00700	5·358	5·445	5·531	5·616	5·701	5·785	5·869	5·952	6·034	6·116	6·197	6·278	6·358
1/ 143	17·685	18·859	20·080	21·350	22·668	24·037	25·455	26·925	28·447	30·022	31·650	33·332	35·069
0·00750	5·546	5·636	5·725	5·814	5·901	5·988	6·075	6·161	6·246	6·331	6·415	6·498	6·581
1/ 133	18·306	19·521	20·785	22·099	23·464	24·880	26·349	27·870	29·446	31·076	32·761	34·502	36·300
0·00800	5·728	5·821	5·913	6·004	6·095	6·185	6·274	6·363	6·451	6·538	6·625	6·712	6·797
1/ 125	18·906	20·161	21·467	22·824	24·233	25·696	27·213	28·784	30·411	32·095	33·835	35·633	37·490
0·00850	5·904	6·000	6·095	6·189	6·282	6·375	6·467	6·559	6·649	6·740	6·829	6·918	7·006
1/ 118	19·488	20·781	22·127	23·526	24·979	26·487	28·050	29·670	31·347	33·083	34·876	36·730	38·644
0·00900	6·076	6·174	6·272	6·368	6·465	6·560	6·655	6·749	6·842	6·935	7·027	7·119	7·210
1/ 111	20·053	21·384	22·769	24·208	25·703	27·255	28·864	30·530	32·256	34·042	35·888	37·795	39·764
0·00950	6·242	6·343	6·443	6·543	6·642	6·740	6·837	6·934	7·030	7·125	7·220	7·314	7·407
1/ 105	20·603	21·970	23·393	24·872	26·408	28·002	29·655	31·367	33·140	34·974	36·871	38·831	40·854
0·01000	6·404	6·508	6·611	6·713	6·814	6·915	7·015	7·114	7·212	7·310	7·407	7·504	7·600
1/ 100	21·138	22·541	24·000	25·518	27·094	28·729	30·425	32·182	34·001	35·883	37·829	39·839	41·915
0·01100	6·717	6·826	6·933	7·041	7·147	7·252	7·357	7·461	7·564	7·667	7·769	7·870	7·971
1/ 91	22·170	23·641	25·172	26·763	28·416	30·131	31·910	33·753	35·661	37·634	39·675	41·784	43·961
0·01200	7·015	7·129	7·242	7·354	7·465	7·575	7·684	7·793	7·901	8·008	8·114	8·220	8·325
1/ 83	23·155	24·692	26·291	27·953	29·680	31·471	33·329	35·254	37·246	39·308	41·439	43·642	45·916
0·01300	7·302	7·420	7·537	7·654	7·769	7·884	7·998	8·111	8·223	8·335	8·445	8·556	8·665
1/ 77	24·101	25·700	27·365	29·095	30·892	32·756	34·690	36·693	38·767	40·913	43·132	45·424	47·791
0·01400	7·578	7·700	7·822	7·943	8·063	8·182	8·300	8·417	8·534	8·649	8·764	8·878	8·992
1/ 71	25·011	26·671	28·398	30·193	32·058	33·993	35·999	38·078	40·231	42·457	44·760	47·138	49·595
	2050	2100	2150	2200	2250	2300	2350	2400	2450	2500	2550	2600	2650

Gradient   (Equivalent) Pipe diameters in mm

S = 0·00300 to 0·01400

m = Manning $n \times 100$
S = 0·01500 to 0·07500

Full bore conditions; Select m to divide mV and/or mQ.

ie hydraulic gradient =
1 in 67 to 1 in 13·3

mV to give velocities in ms^{-1}
mQ to give discharges in m^3s^{-1}

Gradient (Equivalent) Pipe diameters in mm

Gradient	2050	2100	2150	2200	2250	2300	2350	2400	2450	2500	2550	2600	2650
0·01500	7·843	7·970	8·097	8·222	8·346	8·469	8·591	8·713	8·833	8·953	9·072	9·190	9·308
1/ 67	25·888	27·607	29·394	31·253	33·183	35·186	37·263	39·415	41·643	43·948	46·331	48·793	51·335
0·01600	8·101	8·232	8·362	8·491	8·619	8·747	8·873	8·998	9·123	9·247	9·369	9·492	9·613
1/ 63	26·737	28·512	30·358	32·278	34·271	36·340	38·485	40·707	43·008	45·389	47·850	50·393	53·019
0·01700	8·350	8·485	8·619	8·753	8·885	9·016	9·146	9·275	9·404	9·531	9·658	9·784	9·909
1/ 59	27·560	29·389	31·293	33·271	35·326	37·458	39·669	41·960	44·332	46·786	49·323	51·944	54·651
0·01800	8·592	8·731	8·869	9·006	9·142	9·277	9·411	9·544	9·676	9·807	9·938	10·07	10·20
1/ 56	28·359	30·242	32·200	34·236	36·350	38·544	40·819	43·177	45·617	48·142	50·753	53·450	56·235
0·01900	8·828	8·970	9·112	9·253	9·393	9·531	9·669	9·806	9·941	10·08	10·21	10·34	10·48
1/ 53	29·136	31·070	33·082	35·174	37·346	39·600	41·938	44·360	46·867	49·461	52·143	54·915	57·776
0·02000	9·057	9·204	9·349	9·493	9·637	9·779	9·920	10·06	10·20	10·34	10·48	10·61	10·75
1/ 50	29·893	31·877	33·942	36·088	38·316	40·629	43·027	45·512	48·085	50·746	53·498	56·341	59·277
0·02200	9·499	9·653	9·805	9·957	10·11	10·26	10·40	10·55	10·70	10·84	10·99	11·13	11·27
1/ 45·5	31·352	33·433	35·598	37·849	40·187	42·612	45·128	47·734	50·432	53·223	56·109	59·091	62·170
0·02400	9·921	10·08	10·24	10·40	10·56	10·71	10·87	11·02	11·17	11·32	11·48	11·62	11·77
1/ 41·7	32·747	34·920	37·181	39·532	41·973	44·507	47·134	49·856	52·674	55·590	58·604	61·719	64·935
0·02600	10·33	10·49	10·66	10·82	10·99	11·15	11·31	11·47	11·63	11·79	11·94	12·10	12·25
1/ 38·5	34·084	36·346	38·699	41·146	43·687	46·324	49·059	51·892	54·825	57·860	60·997	64·239	67·586
0·02800	10·72	10·89	11·06	11·23	11·40	11·57	11·74	11·90	12·07	12·23	12·39	12·56	12·72
1/ 35·7	35·370	37·718	40·160	42·699	45·337	48·073	50·911	53·851	56·895	60·044	63·300	66·664	70·138
0·03000	11·09	11·27	11·45	11·63	11·80	11·98	12·15	12·32	12·49	12·66	12·83	13·00	13·16
1/ 33·3	36·612	39·042	41·570	44·198	46·928	49·760	52·698	55·741	58·892	62·151	65·521	69·004	72·599
0·03200	11·46	11·64	11·83	12·01	12·19	12·37	12·55	12·73	12·90	13·08	13·25	13·42	13·59
1/ 31·3	37·812	40·322	42·933	45·648	48·467	51·392	54·426	57·569	60·823	64·190	67·670	71·267	74·980
0·03400	11·81	12·00	12·19	12·38	12·56	12·75	12·93	13·12	13·30	13·48	13·66	13·84	14·01
1/ 29·4	38·976	41·563	44·255	47·052	49·958	52·974	56·101	59·341	62·695	66·165	69·753	73·460	77·288
0·03600	12·15	12·35	12·54	12·74	12·93	13·12	13·31	13·50	13·68	13·87	14·05	14·24	14·42
1/ 27·8	40·106	42·768	45·538	48·417	51·407	54·510	57·727	61·061	64·512	68·083	71·775	75·590	79·529
0·03800	12·48	12·69	12·89	13·09	13·28	13·48	13·67	13·87	14·06	14·25	14·44	14·63	14·81
1/ 26·3	41·205	43·940	46·785	49·743	52·815	56·004	59·309	62·734	66·280	69·949	73·742	77·661	81·708
0·04000	12·81	13·02	13·22	13·43	13·63	13·83	14·03	14·23	14·42	14·62	14·81	15·01	15·20
1/ 25·0	42·276	45·081	48·001	51·036	54·188	57·458	60·850	64·364	68·002	71·766	75·658	79·679	83·830
0·04200	13·12	13·34	13·55	13·76	13·96	14·17	14·38	14·58	14·78	14·98	15·18	15·38	15·57
1/ 23·8	43·320	46·195	49·186	52·296	55·526	58·877	62·353	65·953	69·681	73·538	77·526	81·646	85·901
0·04400	13·43	13·65	13·87	14·08	14·29	14·50	14·71	14·92	15·13	15·33	15·54	15·74	15·94
1/ 22·7	44·339	47·282	50·344	53·527	56·832	60·263	63·820	67·505	71·321	75·269	79·350	83·568	87·922
0·04600	13·74	13·96	14·18	14·40	14·61	14·83	15·04	15·26	15·47	15·68	15·89	16·09	16·30
1/ 21·7	45·336	48·344	51·475	54·730	58·110	61·617	65·254	69·023	72·924	76·961	81·134	85·446	89·898
0·04800	14·03	14·26	14·48	14·71	14·93	15·15	15·37	15·59	15·80	16·02	16·23	16·44	16·65
1/ 20·8	46·311	49·384	52·582	55·907	59·359	62·943	66·658	70·507	74·493	78·616	82·879	87·284	91·832
0·05000	14·32	14·55	14·78	15·01	15·24	15·46	15·69	15·91	16·13	16·35	16·56	16·78	16·99
1/ 20·0	47·266	50·403	53·667	57·060	60·583	64·240	68·032	71·961	76·029	80·237	84·588	89·083	93·725
0·05500	15·02	15·26	15·50	15·74	15·98	16·22	16·45	16·68	16·91	17·14	17·37	17·60	17·82
1/ 18·2	49·573	52·863	56·286	59·845	63·541	67·376	71·353	75·473	79·740	84·153	88·717	93·431	98·300
0·06000	15·69	15·94	16·19	16·44	16·69	16·94	17·18	17·43	17·67	17·91	18·14	18·38	18·62
1/ 16·7	51·777	55·213	58·789	62·506	66·366	70·372	74·526	78·829	83·285	87·895	92·661	97·586	102·67
0·06500	16·33	16·59	16·85	17·11	17·37	17·63	17·88	18·14	18·39	18·64	18·88	19·13	19·38
1/ 15·4	53·891	57·468	61·189	65·058	69·076	73·245	77·569	82·048	86·686	91·484	96·445	101·57	106·86
0·07000	16·94	17·22	17·49	17·76	18·03	18·29	18·56	18·82	19·08	19·34	19·60	19·85	20·11
1/ 14·3	55·925	59·637	63·499	67·514	71·683	76·010	80·497	85·145	89·958	94·938	100·09	105·40	110·90
0·07500	17·54	17·82	18·10	18·38	18·66	18·94	19·21	19·48	19·75	20·02	20·29	20·55	20·81
1/ 13·3	57·888	61·730	65·728	69·883	74·199	78·678	83·322	88·134	93·116	98·270	103·60	109·10	114·79
	2050	2100	2150	2200	2250	2300	2350	2400	2450	2500	2550	2600	2650

Gradient (Equivalent) Pipe diameters in mm

S = 0·01500 to 0·07500

D8

m = Manning $n \times 100$
S = 0.00010 to 0.00048

ie hydraulic gradient =
1 in 10000 to 1 in 2083

Full bore conditions; Select
m to divide mV and/or mQ.

mV to give velocities in ms^{-1}
mQ to give discharges in m^{3}s^{-1}

Gradient (Equivalent) Pipe diameters in mm

Gradient	2650	2700	2750	2800	2850	2900	3000	3200	3400	3500	3600	3800	4000
0.00010	0.760	0.769	0.779	0.788	0.798	0.807	0.825	0.862	0.897	0.915	0.932	0.966	1.000
1/10000	4.1915	4.4057	4.6267	4.8544	5.0890	5.3306	5.8350	6.9308	8.1469	8.8016	9.4883	10.960	12.566
0.00011	0.797	0.807	0.817	0.827	0.837	0.846	0.866	0.904	0.941	0.959	0.978	1.014	1.049
1/ 9091	4.3961	4.6208	4.8525	5.0914	5.3374	5.5908	6.1198	7.2691	8.5446	9.2312	9.9514	11.495	13.180
0.00012	0.832	0.843	0.853	0.864	0.874	0.884	0.904	0.944	0.983	1.002	1.021	1.059	1.095
1/ 8333	4.5916	4.8263	5.0683	5.3178	5.5748	5.8394	6.3919	7.5923	8.9245	9.6417	10.394	12.006	13.766
0.00013	0.866	0.877	0.888	0.899	0.910	0.920	0.941	0.983	1.023	1.043	1.063	1.102	1.140
1/ 7692	4.7791	5.0233	5.2752	5.5349	5.8024	6.0778	6.6529	7.9023	9.2889	10.035	10.818	12.496	14.328
0.00014	0.899	0.910	0.922	0.933	0.944	0.955	0.977	1.020	1.062	1.082	1.103	1.143	1.183
1/ 7143	4.9595	5.2129	5.4744	5.7438	6.0214	6.3073	6.9040	8.2006	9.6396	10.414	11.227	12.968	14.869
0.00015	0.931	0.942	0.954	0.966	0.977	0.988	1.011	1.055	1.099	1.120	1.142	1.184	1.225
1/ 6667	5.1335	5.3959	5.6665	5.9454	6.2328	6.5286	7.1464	8.4884	9.9779	10.780	11.621	13.423	15.391
0.00016	0.961	0.973	0.985	0.997	1.009	1.021	1.044	1.090	1.135	1.157	1.179	1.222	1.265
1/ 6250	5.3019	5.5729	5.8523	6.1404	6.4372	6.7428	7.3807	8.7668	10.305	11.133	12.002	13.863	15.895
0.00017	0.991	1.003	1.016	1.028	1.040	1.052	1.076	1.124	1.170	1.193	1.215	1.260	1.304
1/ 5882	5.4651	5.7444	6.0325	6.3294	6.6353	6.9503	7.6079	9.0366	10.622	11.476	12.371	14.290	16.385
0.00018	1.020	1.032	1.045	1.058	1.070	1.083	1.107	1.156	1.204	1.227	1.251	1.297	1.342
1/ 5556	5.6235	5.9109	6.2073	6.5129	6.8277	7.1518	7.8284	9.2986	10.930	11.809	12.730	14.704	16.860
0.00019	1.048	1.061	1.074	1.087	1.100	1.112	1.138	1.188	1.237	1.261	1.285	1.332	1.378
1/ 5263	5.7776	6.0729	6.3774	6.6914	7.0148	7.3477	8.0430	9.5534	11.230	12.132	13.079	15.107	17.322
0.00020	1.075	1.088	1.102	1.115	1.128	1.141	1.167	1.219	1.269	1.294	1.318	1.367	1.414
1/ 5000	5.9277	6.2307	6.5431	6.8652	7.1970	7.5386	8.2519	9.8016	11.521	12.447	13.419	15.500	17.772
0.00022	1.127	1.141	1.155	1.169	1.183	1.197	1.224	1.278	1.331	1.357	1.383	1.433	1.483
1/ 4545	6.2170	6.5348	6.8625	7.2003	7.5483	7.9066	8.6547	10.280	12.084	13.055	14.073	16.256	18.639
0.00024	1.177	1.192	1.207	1.221	1.236	1.250	1.279	1.335	1.390	1.417	1.444	1.497	1.549
1/ 4167	6.4935	6.8254	7.1676	7.5204	7.8839	8.2582	9.0395	10.737	12.621	13.635	14.699	16.979	19.468
0.00026	1.225	1.241	1.256	1.271	1.286	1.301	1.331	1.390	1.447	1.475	1.503	1.558	1.612
1/ 3846	6.7586	7.1040	7.4603	7.8275	8.2058	8.5954	9.4086	11.176	13.137	14.192	15.299	17.672	20.263
0.00028	1.272	1.288	1.303	1.319	1.335	1.350	1.381	1.442	1.501	1.531	1.560	1.617	1.673
1/ 3571	7.0138	7.3722	7.7419	8.1230	8.5156	8.9198	9.7638	11.597	13.632	14.728	15.877	18.339	21.028
0.00030	1.316	1.333	1.349	1.366	1.382	1.398	1.430	1.493	1.554	1.585	1.615	1.674	1.732
1/ 3333	7.2599	7.6310	8.0137	8.4081	8.8145	9.2329	10.106	12.004	14.111	15.245	16.434	18.983	21.766
0.00032	1.359	1.377	1.393	1.410	1.427	1.444	1.477	1.542	1.605	1.636	1.668	1.729	1.789
1/ 3125	7.4980	7.8812	8.2765	8.6839	9.1035	9.5357	10.438	12.398	14.574	15.745	16.973	19.606	22.479
0.00034	1.401	1.419	1.436	1.454	1.471	1.488	1.522	1.589	1.655	1.687	1.719	1.782	1.844
1/ 2941	7.7288	8.1238	8.5312	8.9511	9.3837	9.8292	10.759	12.780	15.022	16.229	17.496	20.209	23.171
0.00036	1.442	1.460	1.478	1.496	1.514	1.531	1.566	1.635	1.703	1.736	1.769	1.834	1.897
1/ 2778	7.9529	8.3593	8.7785	9.2106	9.6558	10.114	11.071	13.150	15.458	16.700	18.003	20.795	23.843
0.00038	1.481	1.500	1.518	1.537	1.555	1.573	1.609	1.680	1.749	1.783	1.817	1.884	1.949
1/ 2632	8.1708	8.5884	9.0191	9.4630	9.9204	10.391	11.374	13.511	15.881	17.158	18.496	21.365	24.496
0.00040	1.520	1.539	1.558	1.577	1.595	1.614	1.651	1.724	1.795	1.830	1.864	1.933	2.000
1/ 2500	8.3830	8.8115	9.2534	9.7088	10.178	10.661	11.670	13.862	16.294	17.603	18.977	21.920	25.133
0.00042	1.557	1.577	1.596	1.616	1.635	1.654	1.692	1.766	1.839	1.875	1.910	1.980	2.049
1/ 2381	8.5901	9.0291	9.4819	9.9486	10.429	10.925	11.958	14.204	16.696	18.038	19.445	22.461	25.753
0.00044	1.594	1.614	1.634	1.654	1.673	1.693	1.732	1.808	1.882	1.919	1.955	2.027	2.098
1/ 2273	8.7922	9.2416	9.7050	10.183	10.675	11.182	12.240	14.538	17.089	18.462	19.903	22.990	26.359
0.00046	1.630	1.650	1.671	1.691	1.711	1.731	1.770	1.848	1.925	1.962	1.999	2.073	2.145
1/ 2174	8.9898	9.4493	9.9231	10.412	10.915	11.433	12.515	14.865	17.473	18.877	20.350	23.506	26.952
0.00048	1.665	1.686	1.707	1.727	1.748	1.768	1.809	1.888	1.966	2.004	2.042	2.117	2.191
1/ 2083	9.1832	9.6525	10.137	10.636	11.150	11.679	12.784	15.185	17.849	19.283	20.788	24.012	27.532
	2650	2700	2750	2800	2850	2900	3000	3200	3400	3500	3600	3800	4000

Gradient (Equivalent) Pipe diameters in mm

S = 0.00010 to 0.00048

m = Manning $n \times 100$
S = 0.00050 to 0.00280

ie hydraulic gradient =
1 in 2000 to 1 in 357

Full bore conditions; Select m to divide mV and/or mQ.

mV to give velocities in ms^{-1}
mQ to give discharges in m^{3}s^{-1}

Gradient (Equivalent) Pipe diameters in mm

Gradient	2650	2700	2750	2800	2850	2900	3000	3200	3400	3500	3600	3800	4000
0.00050	1.699	1.721	1.742	1.763	1.784	1.805	1.846	1.927	2.006	2.046	2.084	2.161	2.236
1/ 2000	9.3725	9.8515	10.346	10.855	11.379	11.920	13.047	15.498	18.217	19.681	21.217	24.507	28.099
0.00055	1.782	1.805	1.827	1.849	1.871	1.893	1.936	2.021	2.104	2.145	2.186	2.266	2.345
1/ 1818	9.8300	10.332	10.851	11.385	11.935	12.501	13.684	16.254	19.106	20.642	22.252	25.703	29.471
0.00060	1.862	1.885	1.908	1.931	1.954	1.977	2.022	2.111	2.198	2.241	2.283	2.367	2.449
1/ 1667	10.267	10.792	11.333	11.891	12.466	13.057	14.293	16.977	19.956	21.560	23.242	26.846	30.781
0.00065	1.938	1.962	1.986	2.010	2.034	2.058	2.105	2.197	2.288	2.332	2.377	2.464	2.550
1/ 1538	10.686	11.232	11.796	12.376	12.975	13.590	14.876	17.670	20.771	22.440	24.191	27.942	32.038
0.00070	2.011	2.036	2.061	2.086	2.111	2.135	2.184	2.280	2.374	2.420	2.466	2.557	2.646
1/ 1429	11.090	11.657	12.241	12.844	13.464	14.103	15.438	18.337	21.555	23.287	25.104	28.997	33.247
0.00075	2.081	2.107	2.133	2.159	2.185	2.210	2.261	2.360	2.457	2.505	2.553	2.647	2.739
1/ 1333	11.479	12.066	12.671	13.294	13.937	14.598	15.980	18.981	22.311	24.104	25.985	30.015	34.414
0.00080	2.149	2.176	2.203	2.230	2.256	2.283	2.335	2.437	2.538	2.588	2.637	2.733	2.828
1/ 1250	11.855	12.461	13.086	13.730	14.394	15.077	16.504	19.603	23.043	24.895	26.837	30.999	35.543
0.00085	2.216	2.243	2.271	2.298	2.326	2.353	2.407	2.512	2.616	2.667	2.718	2.817	2.915
1/ 1176	12.220	12.845	13.489	14.153	14.837	15.541	17.012	20.207	23.752	25.661	27.663	31.953	36.637
0.00090	2.280	2.308	2.337	2.365	2.393	2.421	2.476	2.585	2.692	2.744	2.797	2.899	3.000
1/ 1111	12.575	13.217	13.880	14.563	15.267	15.992	17.505	20.792	24.441	26.405	28.465	32.880	37.699
0.00095	2.342	2.372	2.401	2.430	2.459	2.487	2.544	2.656	2.766	2.820	2.873	2.979	3.082
1/ 1053	12.919	13.579	14.260	14.962	15.685	16.430	17.985	21.362	25.110	27.129	29.245	33.781	38.732
0.00100	2.403	2.433	2.463	2.493	2.523	2.552	2.610	2.725	2.838	2.893	2.948	3.056	3.162
1/ 1000	13.255	13.932	14.631	15.351	16.093	16.857	18.452	21.917	25.763	27.833	30.005	34.658	39.738
0.00110	2.520	2.552	2.584	2.615	2.646	2.677	2.738	2.858	2.976	3.034	3.092	3.205	3.317
1/ 909	13.902	14.612	15.345	16.100	16.878	17.680	19.352	22.987	27.020	29.192	31.469	36.350	41.678
0.00120	2.633	2.666	2.698	2.731	2.763	2.796	2.860	2.985	3.108	3.169	3.229	3.348	3.464
1/ 833	14.520	15.262	16.027	16.816	17.629	18.466	20.213	24.009	28.222	30.490	32.869	37.966	43.531
0.00130	2.740	2.774	2.809	2.843	2.876	2.910	2.976	3.107	3.235	3.298	3.361	3.484	3.606
1/ 769	15.113	15.885	16.682	17.503	18.349	19.220	21.038	24.989	29.374	31.735	34.211	39.516	45.309
0.00140	2.844	2.879	2.915	2.950	2.985	3.020	3.089	3.224	3.357	3.423	3.488	3.616	3.742
1/ 714	15.683	16.485	17.311	18.164	19.041	19.945	21.832	25.933	30.483	32.933	35.502	41.008	47.019
0.00150	2.943	2.980	3.017	3.053	3.090	3.126	3.197	3.338	3.475	3.543	3.610	3.743	3.873
1/ 667	16.234	17.063	17.919	18.801	19.710	20.645	22.599	26.843	31.553	34.089	36.748	42.447	48.669
0.00160	3.040	3.078	3.116	3.153	3.191	3.228	3.302	3.447	3.589	3.659	3.729	3.866	4.000
1/ 625	16.766	17.623	18.507	19.418	20.356	21.322	23.340	27.723	32.588	35.207	37.953	43.839	50.265
0.00170	3.133	3.173	3.212	3.251	3.289	3.327	3.404	3.553	3.700	3.772	3.843	3.984	4.123
1/ 588	17.282	18.165	19.076	20.015	20.983	21.979	24.058	28.576	33.591	36.290	39.121	45.189	51.812
0.00180	3.224	3.265	3.305	3.345	3.384	3.424	3.502	3.656	3.807	3.881	3.955	4.100	4.243
1/ 556	17.783	18.692	19.629	20.596	21.591	22.616	24.756	29.405	34.564	37.342	40.256	46.499	53.315
0.00190	3.313	3.354	3.395	3.436	3.477	3.518	3.598	3.756	3.911	3.988	4.063	4.212	4.359
1/ 526	18.270	19.204	20.167	21.160	22.183	23.236	25.434	30.211	35.512	38.366	41.359	47.773	54.775
0.00200	3.399	3.441	3.484	3.526	3.568	3.609	3.692	3.854	4.013	4.091	4.169	4.322	4.472
1/ 500	18.745	19.703	20.691	21.710	22.759	23.839	26.095	30.995	36.434	39.362	42.433	49.014	56.198
0.00220	3.565	3.609	3.654	3.698	3.742	3.785	3.872	4.042	4.209	4.291	4.372	4.533	4.690
1/ 455	19.660	20.665	21.701	22.769	23.870	25.003	27.368	32.508	38.212	41.283	44.504	51.406	58.941
0.00240	3.723	3.770	3.816	3.862	3.908	3.954	4.044	4.222	4.396	4.482	4.567	4.734	4.899
1/ 417	20.534	21.584	22.666	23.782	24.931	26.115	28.585	33.954	39.912	43.119	46.483	53.692	61.562
0.00260	3.875	3.924	3.972	4.020	4.068	4.115	4.209	4.394	4.575	4.665	4.753	4.928	5.099
1/ 385	21.373	22.465	23.592	24.753	25.949	27.181	29.753	35.340	41.541	44.880	48.381	55.885	64.076
0.00280	4.021	4.072	4.122	4.172	4.221	4.270	4.368	4.560	4.748	4.841	4.933	5.114	5.291
1/ 357	22.179	23.313	24.482	25.687	26.929	28.207	30.876	36.674	43.109	46.574	50.207	57.994	66.495
	2650	2700	2750	2800	2850	2900	3000	3200	3400	3500	3600	3800	4000

Gradient (Equivalent) Pipe diameters in mm

S = 0.00050 to 0.00280

m = Manning $n \times 100$

S = 0·00300 to 0·01400

ie hydraulic gradient = 1 in 333 to 1 in 71

Full bore conditions; Select m to divide mV and/or mQ.

mV to give velocities in ms^{-1}

mQ to give discharges in m^{3}s^{-1}

Gradient   (Equivalent) Pipe diameters in mm

Gradient	2650	2700	2750	2800	2850	2900	3000	3200	3400	3500	3600	3800	4000
0·00300	4·162	4·215	4·267	4·318	4·369	4·420	4·521	4·720	4·915	5·011	5·106	5·293	5·477
1/ 333	22·958	24·131	25·341	26·589	27·874	29·197	31·959	37·961	44·622	48·209	51·970	60·030	68·829
0·00320	4·299	4·353	4·406	4·460	4·513	4·565	4·670	4·875	5·076	5·175	5·273	5·467	5·657
1/ 313	23·711	24·923	26·172	27·461	28·788	30·155	33·008	39·206	46·086	49·790	53·674	61·998	71·086
0·00340	4·431	4·487	4·542	4·597	4·652	4·706	4·813	5·025	5·232	5·334	5·435	5·635	5·831
1/ 294	24·441	25·690	26·978	28·306	29·674	31·083	34·023	40·413	47·504	51·322	55·326	63·906	73·274
0·00360	4·560	4·617	4·674	4·730	4·786	4·842	4·953	5·171	5·384	5·489	5·593	5·798	6·000
1/ 278	25·149	26·434	27·760	29·127	30·534	31·984	35·010	41·585	48·881	52·810	56·930	65·759	75·398
0·00380	4·685	4·743	4·802	4·860	4·918	4·975	5·089	5·312	5·531	5·639	5·746	5·957	6·164
1/ 263	25·838	27·159	28·521	29·925	31·371	32·860	35·969	42·724	50·221	54·257	58·490	67·561	77·464
0·00400	4·806	4·867	4·927	4·986	5·045	5·104	5·221	5·450	5·675	5·786	5·896	6·112	6·325
1/ 250	26·510	27·864	29·262	30·702	32·186	33·714	36·904	43·834	51·526	55·667	60·009	69·316	79·477
0·00420	4·925	4·987	5·048	5·109	5·170	5·230	5·350	5·585	5·815	5·929	6·041	6·263	6·481
1/ 238	27·164	28·552	29·984	31·460	32·981	34·546	37·815	44·917	52·798	57·041	61·491	71·028	81·439
0·00440	5·041	5·104	5·167	5·229	5·292	5·353	5·476	5·716	5·952	6·068	6·183	6·410	6·633
1/ 227	27·803	29·224	30·690	32·201	33·757	35·359	38·705	45·974	54·041	58·384	62·938	72·700	83·356
0·00460	5·154	5·219	5·283	5·347	5·410	5·474	5·599	5·845	6·086	6·205	6·322	6·554	6·782
1/ 217	28·428	29·881	31·380	32·924	34·516	36·154	39·575	47·007	55·255	59·696	64·353	74·333	85·229
0·00480	5·265	5·331	5·397	5·462	5·527	5·591	5·719	5·971	6·217	6·338	6·458	6·695	6·928
1/ 208	29·040	30·524	32·055	33·632	35·258	36·932	40·426	48·018	56·443	60·980	65·737	75·932	87·062
0·00500	5·374	5·441	5·508	5·575	5·641	5·707	5·837	6·094	6·345	6·469	6·591	6·833	7·071
1/ 200	29·639	31·153	32·716	34·326	35·985	37·693	41·260	49·008	57·607	62·237	67·093	77·498	88·858
0·00550	5·636	5·707	5·777	5·847	5·916	5·985	6·122	6·391	6·655	6·785	6·913	7·167	7·416
1/ 182	31·085	32·674	34·312	36·001	37·741	39·533	43·273	51·400	60·419	65·275	70·367	81·281	93·195
0·00600	5·887	5·960	6·034	6·107	6·179	6·251	6·394	6·675	6·951	7·086	7·221	7·486	7·746
1/ 167	32·467	34·127	35·838	37·602	39·420	41·291	45·198	53·686	63·106	68·177	73·496	84·895	97·339
0·00650	6·127	6·204	6·280	6·356	6·432	6·507	6·655	6·948	7·234	7·376	7·515	7·791	8·062
1/ 154	33·793	35·520	37·302	39·138	41·029	42·977	47·043	55·878	65·683	70·961	76·497	88·361	101·31
0·00700	6·358	6·438	6·517	6·596	6·674	6·752	6·906	7·210	7·507	7·654	7·799	8·085	8·367
1/ 143	35·069	36·861	38·710	40·615	42·578	44·599	48·819	57·987	68·162	73·640	79·385	91·697	105·14
0·00750	6·581	6·664	6·746	6·828	6·909	6·989	7·149	7·463	7·771	7·923	8·073	8·369	8·660
1/ 133	36·300	38·155	40·068	42·041	44·072	46·164	50·532	60·022	70·554	76·225	82·171	94·915	108·83
0·00800	6·797	6·883	6·967	7·051	7·135	7·218	7·383	7·708	8·026	8·182	8·338	8·644	8·944
1/ 125	37·490	39·406	41·382	43·419	45·518	47·678	52·190	61·991	72·868	78·724	84·866	98·028	112·40
0·00850	7·006	7·094	7·182	7·268	7·355	7·440	7·611	7·945	8·273	8·434	8·594	8·910	9·220
1/ 118	38·644	40·619	42·656	44·756	46·919	49·146	53·796	63·899	75·111	81·147	87·478	101·05	115·86
0·00900	7·210	7·300	7·390	7·479	7·568	7·656	7·831	8·175	8·513	8·679	8·843	9·168	9·487
1/ 111	39·764	41·797	43·893	46·053	48·279	50·571	55·355	65·751	77·288	83·500	90·014	103·97	119·21
0·00950	7·407	7·500	7·592	7·684	7·775	7·866	8·046	8·400	8·746	8·917	9·086	9·419	9·747
1/ 105	40·854	42·942	45·095	47·315	49·602	51·956	56·872	67·553	79·406	85·788	92·481	106·82	122·48
0·01000	7·600	7·695	7·790	7·884	7·977	8·070	8·255	8·618	8·973	9·148	9·322	9·664	10·00
1/ 100	41·915	44·057	46·267	48·544	50·890	53·306	58·350	69·308	81·469	88·016	94·883	109·60	125·66
0·01100	7·971	8·070	8·170	8·269	8·367	8·464	8·658	9·038	9·411	9·595	9·777	10·14	10·49
1/ 91	43·961	46·208	48·525	50·914	53·374	55·908	61·198	72·691	85·446	92·312	99·514	114·95	131·80
0·01200	8·325	8·429	8·533	8·636	8·739	8·841	9·043	9·440	9·830	10·02	10·21	10·59	10·95
1/ 83	45·916	48·263	50·683	53·178	55·748	58·394	63·919	75·923	89·245	96·417	103·94	120·06	137·66
0·01300	8·665	8·774	8·882	8·989	9·096	9·202	9·412	9·826	10·23	10·43	10·63	11·02	11·40
1/ 77	47·791	50·233	52·752	55·349	58·024	60·778	66·529	79·023	92·889	100·35	108·18	124·96	143·28
0·01400	8·992	9·105	9·217	9·328	9·439	9·549	9·767	10·20	10·62	10·82	11·03	11·43	11·83
1/ 71	49·595	52·129	54·744	57·438	60·214	63·073	69·040	82·006	96·396	104·14	112·27	129·68	148·69
	2650	2700	2750	2800	2850	2900	3000	3200	3400	3500	3600	3800	4000

Gradient   (Equivalent) Pipe diameters in mm

S = 0·00300 to 0·01400

m = Manning n × 100
S = 0.01500 to 0.07500

Full bore conditions; Select m to divide mV and/or mQ.

ie hydraulic gradient = 1 in 67 to 1 in 13·3

mV to give velocities in ms⁻¹
mQ to give discharges in m³s⁻¹

D8 continued

Gradient   (Equivalent) Pipe diameters in mm

Gradient	2650	2700	2750	2800	2850	2900	3000	3200	3400	3500	3600	3800	4000
0·01500	9·308	9·424	9·540	9·656	9·770	9·884	10·11	10·55	10·99	11·20	11·42	11·84	12·25
1/ 67	51·335	53·959	56·665	59·454	62·328	65·286	71·464	84·884	99·779	107·80	116·21	134·23	153·91
0·01600	9·613	9·733	9·853	9·972	10·09	10·21	10·44	10·90	11·35	11·57	11·79	12·22	12·65
1/ 63	53·019	55·729	58·523	61·404	64·372	67·428	73·807	87·668	103·05	111·33	120·02	138·63	158·95
0·01700	9·909	10·03	10·16	10·28	10·40	10·52	10·76	11·24	11·70	11·93	12·15	12·60	13·04
1/ 59	54·651	57·444	60·325	63·294	66·353	69·503	76·079	90·366	106·22	114·76	123·71	142·90	163·85
0·01800	10·20	10·32	10·45	10·58	10·70	10·83	11·07	11·56	12·04	12·27	12·51	12·97	13·42
1/ 56	56·235	59·109	62·073	65·129	68·277	71·518	78·284	92·986	109·30	118·09	127·30	147·04	168·60
0·01900	10·48	10·61	10·74	10·87	11·00	11·12	11·38	11·88	12·37	12·61	12·85	13·32	13·78
1/ 53	57·776	60·729	63·774	66·914	70·148	73·477	80·430	95·534	112·30	121·32	130·79	151·07	173·22
0·02000	10·75	10·88	11·02	11·15	11·28	11·41	11·67	12·19	12·69	12·94	13·18	13·67	14·14
1/ 50	59·277	62·307	65·431	68·652	71·970	75·386	82·519	98·016	115·21	124·47	134·19	155·00	177·72
0·02200	11·27	11·41	11·55	11·69	11·83	11·97	12·24	12·78	13·31	13·57	13·83	14·33	14·83
1/ 45.5	62·170	65·348	68·625	72·003	75·483	79·066	86·547	102·80	120·84	130·55	140·73	162·56	186·39
0·02400	11·77	11·92	12·07	12·21	12·36	12·50	12·79	13·35	13·90	14·17	14·44	14·97	15·49
1/ 41.7	64·935	68·254	71·676	75·204	78·839	82·582	90·395	107·37	126·21	136·35	146·99	169·79	194·68
0·02600	12·25	12·41	12·56	12·71	12·86	13·01	13·31	13·90	14·47	14·75	15·03	15·58	16·12
1/ 38.5	67·586	71·040	74·603	78·275	82·058	85·954	94·086	111·76	131·37	141·92	152·99	176·72	202·63
0·02800	12·72	12·88	13·03	13·19	13·35	13·50	13·81	14·42	15·01	15·31	15·60	16·17	16·73
1/ 35.7	70·138	73·722	77·419	81·230	85·156	89·198	97·638	115·97	136·32	147·28	158·77	183·39	210·28
0·03000	13·16	13·33	13·49	13·66	13·82	13·98	14·30	14·93	15·54	15·85	16·15	16·74	17·32
1/ 33.3	72·599	76·310	80·137	84·081	88·145	92·329	101·06	120·04	141·11	152·45	164·34	189·83	217·66
0·03200	13·59	13·77	13·93	14·10	14·27	14·44	14·77	15·42	16·05	16·36	16·68	17·29	17·89
1/ 31.3	74·980	78·812	82·765	86·839	91·035	95·357	104·38	123·98	145·74	157·45	169·73	196·06	224·79
0·03400	14·01	14·19	14·36	14·54	14·71	14·88	15·22	15·89	16·55	16·87	17·19	17·82	18·44
1/ 29.4	77·288	81·238	85·312	89·511	93·837	98·292	107·59	127·80	150·22	162·29	174·96	202·09	231·71
0·03600	14·42	14·60	14·78	14·96	15·14	15·31	15·66	16·35	17·03	17·36	17·69	18·34	18·97
1/ 27.8	79·529	83·593	87·785	92·106	96·558	101·14	110·71	131·50	154·58	167·00	180·03	207·95	238·43
0·03800	14·81	15·00	15·18	15·37	15·55	15·73	16·09	16·80	17·49	17·83	18·17	18·84	19·49
1/ 26.3	81·708	85·884	90·191	94·630	99·204	103·91	113·74	135·11	158·81	171·58	184·96	213·65	244·96
0·04000	15·20	15·39	15·58	15·77	15·95	16·14	16·51	17·24	17·95	18·30	18·64	19·33	20·00
1/ 25.0	83·830	88·115	92·534	97·088	101·78	106·61	116·70	138·62	162·94	176·03	189·77	219·20	251·33
0·04200	15·57	15·77	15·96	16·16	16·35	16·54	16·92	17·66	18·39	18·75	19·10	19·80	20·49
1/ 23.8	85·901	90·291	94·819	99·486	104·29	109·25	119·58	142·04	166·96	180·38	194·45	224·61	257·53
0·04400	15·94	16·14	16·34	16·54	16·73	16·93	17·32	18·08	18·82	19·19	19·55	20·27	20·98
1/ 22.7	87·922	92·416	97·050	101·83	106·75	111·82	122·40	145·38	170·89	184·62	199·03	229·90	263·59
0·04600	16·30	16·50	16·71	16·91	17·11	17·31	17·70	18·48	19·25	19·62	19·99	20·73	21·45
1/ 21.7	89·898	94·493	99·231	104·12	109·15	114·33	125·15	148·65	174·73	188·77	203·50	235·06	269·52
0·04800	16·65	16·86	17·07	17·27	17·48	17·68	18·09	18·88	19·66	20·04	20·42	21·17	21·91
1/ 20.8	91·832	96·525	101·37	106·36	111·50	116·79	127·84	151·85	178·49	192·83	207·88	240·12	275·32
0·05000	16·99	17·21	17·42	17·63	17·84	18·05	18·46	19·27	20·06	20·46	20·84	21·61	22·36
1/ 20.0	93·725	98·515	103·46	108·55	113·79	119·20	130·47	154·98	182·17	196·81	212·17	245·07	280·99
0·05500	17·82	18·05	18·27	18·49	18·71	18·93	19·36	20·21	21·04	21·45	21·86	22·66	23·45
1/ 18.2	98·300	103·32	108·51	113·85	119·35	125·01	136·84	162·54	191·06	206·42	222·52	257·03	294·71
0·06000	18·62	18·85	19·08	19·31	19·54	19·77	20·22	21·11	21·98	22·41	22·83	23·67	24·49
1/ 16.7	102·67	107·92	113·33	118·91	124·66	130·57	142·93	169·77	199·56	215·60	232·42	268·46	307·81
0·06500	19·38	19·62	19·86	20·10	20·34	20·58	21·05	21·97	22·88	23·32	23·77	24·64	25·50
1/ 15.4	106·86	112·32	117·96	123·76	129·75	135·90	148·76	176·70	207·71	224·40	241·91	279·42	320·38
0·07000	20·11	20·36	20·61	20·86	21·11	21·35	21·84	22·80	23·74	24·20	24·66	25·57	26·46
1/ 14.3	110·90	116·57	122·41	128·44	134·64	141·03	154·38	183·37	215·55	232·87	251·04	289·97	332·47
0·07500	20·81	21·07	21·33	21·59	21·85	22·10	22·61	23·60	24·57	25·05	25·53	26·47	27·39
1/ 13.3	114·79	120·66	126·71	132·94	139·37	145·98	159·80	189·81	223·11	241·04	259·85	300·15	344·14
	2650	**2700**	**2750**	**2800**	**2850**	**2900**	**3000**	**3200**	**3400**	**3500**	**3600**	**3800**	**4000**

Gradient   (Equivalent) Pipe diameters in mm

S = 0.01500 to 0.07500

D9

m = Manning $n \times 100$
S = 0·00010 to 0·00048

Full bore conditions; Select m to divide mV and/or mQ.

ie hydraulic gradient =
1 in 10000 to 1 in 2083

mV to give velocities in ms^{-1}
mQ to give discharges in m^{3}s^{-1}

Gradient   (Equivalent) Pipe diameters in m

Gradient	4·000	4·250	4·500	4·750	5·000	5·250	5·500	5·750	6·000	6·250	6·500	7·000	7·500
0·00010	1·000	1·041	1·082	1·121	1·160	1·199	1·237	1·274	1·310	1·347	1·382	1·452	1·521
1/10000	12·566	14·771	17·203	19·872	22·784	25·950	29·378	33·075	37·050	41·311	45·865	55·887	67·176
0·00011	1·049	1·092	1·134	1·176	1·217	1·257	1·297	1·336	1·374	1·412	1·450	1·523	1·595
1/ 9091	13·180	15·492	18·043	20·841	23·896	27·217	30·811	34·689	38·858	43·327	48·104	58·615	70·455
0·00012	1·095	1·141	1·185	1·228	1·271	1·313	1·355	1·395	1·435	1·475	1·514	1·591	1·666
1/ 8333	13·766	16·181	18·845	21·768	24·959	28·427	32·182	36·232	40·586	45·254	50·243	61·221	73·587
0·00013	1·140	1·187	1·233	1·279	1·323	1·367	1·410	1·452	1·494	1·535	1·576	1·656	1·734
1/ 7692	14·328	16·842	19·615	22·657	25·978	29·588	33·496	37·711	42·243	47·101	52·295	63·721	76·592
0·00014	1·183	1·232	1·280	1·327	1·373	1·418	1·463	1·507	1·550	1·593	1·635	1·718	1·799
1/ 7143	14·869	17·478	20·355	23·512	26·959	30·705	34·760	39·135	43·838	48·879	54·269	66·126	79·483
0·00015	1·225	1·275	1·325	1·373	1·421	1·468	1·514	1·560	1·605	1·649	1·693	1·779	1·862
1/ 6667	15·391	18·091	21·070	24·338	27·905	31·782	35·980	40·508	45·377	50·595	56·173	68·447	82·273
0·00016	1·265	1·317	1·368	1·418	1·468	1·516	1·564	1·611	1·658	1·703	1·748	1·837	1·923
1/ 6250	15·895	18·684	21·761	25·136	28·820	32·825	37·160	41·837	46·865	52·254	58·016	70·692	84·971
0·00017	1·304	1·358	1·410	1·462	1·513	1·563	1·612	1·661	1·709	1·756	1·802	1·893	1·983
1/ 5882	16·385	19·259	22·431	25·909	29·707	33·835	38·304	43·124	48·307	53·863	59·801	72·868	87·586
0·00018	1·342	1·397	1·451	1·504	1·557	1·608	1·659	1·709	1·758	1·807	1·854	1·948	2·040
1/ 5556	16·860	19·818	23·081	26·661	30·568	34·816	39·414	44·374	49·708	55·424	61·535	74·980	90·126
0·00019	1·378	1·435	1·491	1·546	1·599	1·652	1·704	1·756	1·806	1·856	1·905	2·002	2·096
1/ 5263	17·322	20·361	23·713	27·391	31·406	35·770	40·494	45·590	51·070	56·943	63·221	77·035	92·595
0·00020	1·414	1·473	1·530	1·586	1·641	1·695	1·749	1·801	1·853	1·904	1·955	2·054	2·150
1/ 5000	17·772	20·890	24·329	28·103	32·222	36·699	41·546	46·775	52·396	58·422	64·863	79·036	95·001
0·00022	1·483	1·544	1·604	1·663	1·721	1·778	1·834	1·889	1·944	1·997	2·050	2·154	2·255
1/ 4545	18·639	21·909	25·517	29·474	33·795	38·490	43·574	49·058	54·954	61·274	68·029	82·894	99·638
0·00024	1·549	1·613	1·676	1·737	1·798	1·857	1·916	1·973	2·030	2·086	2·141	2·250	2·356
1/ 4167	19·468	22·884	26·651	30·785	35·297	40·202	45·512	51·239	57·397	63·998	71·054	86·580	104·07
0·00026	1·612	1·679	1·744	1·808	1·871	1·933	1·994	2·054	2·113	2·171	2·229	2·342	2·452
1/ 3846	20·263	23·818	27·740	32·042	36·739	41·843	47·370	53·331	59·741	66·612	73·956	90·115	108·32
0·00028	1·673	1·742	1·810	1·876	1·942	2·006	2·069	2·131	2·193	2·253	2·313	2·430	2·544
1/ 3571	21·028	24·717	28·787	33·252	38·125	43·423	49·158	55·345	61·996	69·126	76·747	93·517	112·41
0·00030	1·732	1·803	1·874	1·942	2·010	2·076	2·142	2·206	2·270	2·332	2·394	2·515	2·634
1/ 3333	21·766	25·585	29·797	34·419	39·464	44·947	50·884	57·287	64·172	71·552	79·441	96·799	116·35
0·00032	1·789	1·863	1·935	2·006	2·076	2·144	2·212	2·278	2·344	2·409	2·473	2·598	2·720
1/ 3125	22·479	26·424	30·774	35·547	40·758	46·421	52·552	59·166	66·277	73·899	82·046	99·974	120·17
0·00034	1·844	1·920	1·995	2·068	2·140	2·210	2·280	2·349	2·416	2·483	2·549	2·678	2·804
1/ 2941	23·171	27·237	31·722	36·641	42·012	47·850	54·170	60·987	68·316	76·173	84·572	103·05	123·87
0·00036	1·897	1·976	2·052	2·128	2·202	2·274	2·346	2·417	2·486	2·555	2·623	2·755	2·885
1/ 2778	23·843	28·027	32·641	37·704	43·230	49·237	55·740	62·755	70·297	78·382	87·023	106·04	127·46
0·00038	1·949	2·030	2·109	2·186	2·262	2·337	2·410	2·483	2·554	2·625	2·694	2·831	2·964
1/ 2632	24·496	28·795	33·536	38·737	44·415	50·586	57·268	64·475	72·223	80·529	89·408	108·94	130·95
0·00040	2·000	2·082	2·163	2·243	2·321	2·398	2·473	2·547	2·621	2·693	2·764	2·904	3·041
1/ 2500	25·133	29·543	34·407	39·743	45·569	51·900	58·755	66·150	74·100	82·621	91·731	111·77	134·35
0·00042	2·049	2·134	2·217	2·298	2·378	2·457	2·534	2·610	2·685	2·760	2·833	2·976	3·116
1/ 2381	25·753	30·272	35·257	40·725	46·694	53·182	60·206	67·783	75·930	84·662	93·996	114·53	137·67
0·00044	2·098	2·184	2·269	2·352	2·434	2·515	2·594	2·672	2·749	2·824	2·899	3·046	3·190
1/ 2273	26·359	30·985	36·086	41·683	47·793	54·434	61·623	69·378	77·716	86·654	96·208	117·23	140·91
0·00046	2·145	2·233	2·320	2·405	2·489	2·571	2·652	2·732	2·810	2·888	2·964	3·115	3·261
1/ 2174	26·952	31·681	36·897	42·620	48·867	55·657	63·008	70·937	79·463	88·602	98·370	119·86	144·08
0·00048	2·191	2·281	2·370	2·457	2·542	2·626	2·709	2·791	2·871	2·950	3·028	3·182	3·331
1/ 2083	27·532	32·362	37·691	43·536	49·918	56·854	64·363	72·463	81·172	90·507	100·49	122·44	147·17
	4·000	4·250	4·500	4·750	5·000	5·250	5·500	5·750	6·000	6·250	6·500	7·000	7·500

Gradient   (Equivalent) Pipe diameters in m

S = 0·00010 to 0·00048

m = Manning $n \times 100$
S = 0.00050 to 0.00280

ie hydraulic gradient =
1 in 2000 to 1 in 357

Full bore conditions; Select m to divide mV and/or mQ.

mV to give velocities in ms^{-1}
mQ to give discharges in m^{3}s^{-1}

Gradient	(Equivalent) Pipe diameters in m												
	4·000	4·250	4·500	4·750	5·000	5·250	5·500	5·750	6·000	6·250	6·500	7·000	7·500
0·00050	2·236	2·328	2·419	2·507	2·595	2·681	2·765	2·848	2·930	3·011	3·091	3·247	3·400
1/ 2000	28·099	33·030	38·468	44·434	50·947	58·026	65·690	73·957	82·846	92·374	102·56	124·97	150·21
0·00055	2·345	2·442	2·537	2·630	2·721	2·811	2·900	2·987	3·073	3·158	3·242	3·406	3·566
1/ 1818	29·471	34·642	40·346	46·603	53·434	60·859	68·897	77·567	86·890	96·882	107·56	131·07	157·54
0·00060	2·449	2·551	2·650	2·747	2·842	2·936	3·029	3·120	3·210	3·298	3·386	3·557	3·725
1/ 1667	30·781	36·182	42·140	48·675	55·810	63·565	71·960	81·016	90·753	101·19	112·35	136·89	164·55
0·00065	2·550	2·655	2·758	2·859	2·958	3·056	3·153	3·247	3·341	3·433	3·524	3·702	3·877
1/ 1538	32·038	37·660	43·860	50·663	58·089	66·160	74·899	84·324	94·459	105·32	116·93	142·48	171·27
0·00070	2·646	2·755	2·862	2·967	3·070	3·172	3·272	3·370	3·467	3·563	3·657	3·842	4·023
1/ 1429	33·247	39·081	45·516	52·575	60·282	68·658	77·726	87·508	98·025	109·30	121·35	147·86	177·73
0·00075	2·739	2·852	2·962	3·071	3·178	3·283	3·386	3·488	3·589	3·688	3·785	3·977	4·164
1/ 1333	34·414	40·453	47·114	54·421	62·397	71·068	80·454	90·579	101·47	113·13	125·61	153·05	183·97
0·00080	2·828	2·945	3·059	3·172	3·282	3·391	3·497	3·603	3·706	3·809	3·909	4·107	4·301
1/ 1250	35·543	41·780	48·659	56·205	64·444	73·398	83·092	93·550	104·79	116·84	129·73	158·07	190·00
0·00085	2·915	3·036	3·154	3·269	3·383	3·495	3·605	3·713	3·820	3·926	4·030	4·234	4·433
1/ 1176	36·637	43·065	50·156	57·935	66·427	75·657	85·650	96·429	108·02	120·44	133·72	162·94	195·85
0·00090	3·000	3·124	3·245	3·364	3·481	3·596	3·710	3·821	3·931	4·040	4·147	4·357	4·562
1/ 1111	37·699	44·314	51·610	59·615	68·353	77·851	88·133	99·224	111·15	123·93	137·60	167·66	201·53
0·00095	3·082	3·209	3·334	3·456	3·577	3·695	3·811	3·926	4·039	4·150	4·260	4·476	4·687
1/ 1053	38·732	45·528	53·025	61·248	70·226	79·984	90·548	101·94	114·20	127·33	141·37	172·26	207·05
0·00100	3·162	3·293	3·421	3·546	3·669	3·791	3·910	4·028	4·144	4·258	4·371	4·592	4·808
1/ 1000	39·738	46·711	54·402	62·839	72·050	82·062	92·900	104·59	117·16	130·64	145·04	176·73	212·43
0·00110	3·317	3·453	3·588	3·719	3·849	3·976	4·101	4·224	4·346	4·466	4·584	4·816	5·043
1/ 909	41·678	48·991	57·057	65·907	75·567	86·067	97·435	109·70	122·88	137·01	152·12	185·36	222·80
0·00120	3·464	3·607	3·747	3·885	4·020	4·153	4·283	4·412	4·539	4·664	4·788	5·031	5·267
1/ 833	43·531	51·169	59·595	68·837	78·927	89·894	101·77	114·57	128·34	143·10	158·88	193·60	232·70
0·00130	3·606	3·754	3·900	4·043	4·184	4·322	4·458	4·592	4·725	4·855	4·984	5·236	5·482
1/ 769	45·309	53·259	62·028	71·648	82·150	93·565	105·92	119·25	133·58	148·95	165·37	201·50	242·21
0·00140	3·742	3·896	4·047	4·196	4·342	4·485	4·627	4·766	4·903	5·038	5·172	5·434	5·689
1/ 714	47·019	55·269	64·369	74·353	85·251	97·097	109·92	123·75	138·63	154·57	171·61	209·11	251·35
0·00150	3·873	4·033	4·189	4·343	4·494	4·643	4·789	4·933	5·075	5·215	5·353	5·624	5·889
1/ 667	48·669	57·209	66·629	76·962	88·243	100·50	113·78	128·10	143·49	160·00	177·64	216·45	260·17
0·00160	4·000	4·165	4·327	4·486	4·642	4·795	4·946	5·095	5·241	5·386	5·529	5·809	6·082
1/ 625	50·265	59·085	68·814	79·486	91·137	103·80	117·51	132·30	148·20	165·24	183·46	223·55	268·70
0·00170	4·123	4·293	4·460	4·624	4·784	4·943	5·098	5·252	5·403	5·552	5·699	5·988	6·269
1/ 588	51·812	60·904	70·932	81·933	93·942	107·00	121·13	136·37	152·76	170·33	189·11	230·43	276·97
0·00180	4·243	4·418	4·589	4·758	4·923	5·086	5·246	5·404	5·559	5·713	5·864	6·161	6·451
1/ 556	53·315	62·669	72·988	84·308	96·666	110·10	124·64	140·32	157·19	175·27	194·59	237·11	285·00
0·00190	4·359	4·539	4·715	4·888	5·058	5·225	5·390	5·552	5·712	5·869	6·025	6·330	6·628
1/ 526	54·775	64·387	74·988	86·618	99·315	113·11	128·05	144·17	161·50	180·07	199·92	243·61	292·81
0·00200	4·472	4·657	4·837	5·015	5·189	5·361	5·530	5·696	5·860	6·022	6·181	6·494	6·800
1/ 500	56·198	66·059	76·936	88·868	101·89	116·05	131·38	147·91	165·69	184·75	205·12	249·93	300·42
0·00220	4·690	4·884	5·074	5·260	5·443	5·623	5·800	5·974	6·146	6·316	6·483	6·811	7·132
1/ 455	58·941	69·284	80·691	93·206	106·87	121·72	137·79	155·13	173·78	193·76	215·13	262·13	315·08
0·00240	4·899	5·101	5·299	5·494	5·685	5·873	6·058	6·240	6·419	6·597	6·771	7·114	7·449
1/ 417	61·562	72·364	84·279	97·350	111·62	127·13	143·92	162·03	181·51	202·38	224·69	273·79	329·09
0·00260	5·099	5·309	5·516	5·718	5·917	6·113	6·305	6·495	6·682	6·866	7·048	7·405	7·753
1/ 385	64·076	75·319	87·721	101·33	116·18	132·32	149·80	168·65	188·92	210·64	233·87	284·97	342·53
0·00280	5·291	5·510	5·724	5·934	6·140	6·343	6·543	6·740	6·934	7·125	7·314	7·684	8·046
1/ 357	66·495	78·163	91·032	105·15	120·56	137·32	155·45	175·02	196·05	218·60	242·70	295·73	355·46
	4·000	4·250	4·500	4·750	5·000	5·250	5·500	5·750	6·000	6·250	6·500	7·000	7·500

Gradient   (Equivalent) Pipe diameters in m

S = 0.00050 to 0.00280

D9
continued

m = Manning $n \times 100$
S = 0.00300 to 0.01400

ie hydraulic gradient =
1 in 333   to   1 in 71

Full bore conditions; Select
m to divide mV and/or mQ.

mV to give velocities in ms⁻¹
mQ to give discharges in m³s⁻¹

Gradient   (Equivalent) Pipe diameters in m

Gradient	4.000	4.250	4.500	4.750	5.000	5.250	5.500	5.750	6.000	6.250	6.500	7.000	7.500
0.00300	5.477	5.703	5.925	6.142	6.356	6.566	6.773	6.976	7.177	7.375	7.571	7.954	8.328
1/ 333	68.829	80.906	94.227	108.84	124.79	142.14	160.91	181.16	202.93	226.27	251.21	306.11	367.94
0.00320	5.657	5.890	6.119	6.344	6.564	6.781	6.995	7.205	7.413	7.617	7.819	8.215	8.602
1/ 313	71.086	83.559	97.318	112.41	128.89	146.80	166.18	187.10	209.59	233.69	259.45	316.14	380.00
0.00340	5.831	6.071	6.307	6.539	6.766	6.990	7.210	7.427	7.641	7.851	8.060	8.468	8.866
1/ 294	73.274	86.131	100.31	115.87	132.85	151.31	171.30	192.86	216.04	240.88	267.44	325.87	391.70
0.00360	6.000	6.247	6.490	6.728	6.962	7.193	7.419	7.642	7.862	8.079	8.293	8.713	9.123
1/ 278	75.398	88.628	103.22	119.23	136.71	155.70	176.27	198.45	222.30	247.86	275.19	335.32	403.05
0.00380	6.164	6.419	6.668	6.913	7.153	7.390	7.622	7.852	8.078	8.301	8.520	8.952	9.373
1/ 263	77.464	91.057	106.05	122.50	140.45	159.97	181.10	203.89	228.39	254.66	282.73	344.51	414.10
0.00400	6.325	6.585	6.841	7.092	7.339	7.582	7.820	8.056	8.288	8.516	8.742	9.184	9.617
1/ 250	79.477	93.422	108.80	125.68	144.10	164.12	185.80	209.18	234.32	261.27	290.08	353.46	424.86
0.00420	6.481	6.748	7.010	7.267	7.520	7.769	8.014	8.255	8.492	8.726	8.958	9.411	9.854
1/ 238	81.439	95.729	111.49	128.78	147.66	168.18	190.39	214.35	240.11	267.72	297.24	362.19	435.35
0.00440	6.633	6.907	7.175	7.438	7.697	7.952	8.202	8.449	8.692	8.932	9.168	9.633	10.09
1/ 227	83.356	97.982	114.11	131.81	151.13	172.13	194.87	219.39	245.76	274.02	304.24	370.71	445.59
0.00460	6.782	7.062	7.336	7.606	7.870	8.130	8.386	8.639	8.887	9.133	9.374	9.849	10.31
1/ 217	85.229	100.18	116.68	134.78	154.53	176.00	199.25	224.32	251.28	280.18	311.07	379.04	455.61
0.00480	6.928	7.214	7.494	7.769	8.039	8.305	8.567	8.825	9.079	9.329	9.576	10.06	10.53
1/ 208	87.062	102.34	119.19	137.67	157.85	179.79	203.53	229.15	256.69	286.21	317.76	387.20	465.41
0.00500	7.071	7.363	7.649	7.929	8.205	8.477	8.744	9.007	9.266	9.521	9.774	10.27	10.75
1/ 200	88.858	104.45	121.65	140.51	161.11	183.50	207.73	233.87	261.98	292.11	324.32	395.18	475.00
0.00550	7.416	7.722	8.022	8.316	8.606	8.890	9.170	9.446	9.718	9.986	10.25	10.77	11.28
1/ 182	93.195	109.55	127.58	147.37	168.97	192.45	217.87	245.29	274.77	306.37	340.15	414.47	498.19
0.00600	7.746	8.065	8.379	8.686	8.988	9.286	9.578	9.866	10.15	10.43	10.71	11.25	11.78
1/ 167	97.339	114.42	133.26	153.92	176.49	201.01	227.56	256.20	286.99	319.99	355.27	432.90	520.34
0.00650	8.062	8.395	8.721	9.041	9.355	9.665	9.969	10.27	10.56	10.86	11.14	11.71	12.26
1/ 154	101.31	119.09	138.70	160.21	183.69	209.22	236.85	266.66	298.71	333.06	369.78	450.58	541.59
0.00700	8.367	8.712	9.050	9.382	9.709	10.03	10.35	10.66	10.96	11.27	11.56	12.15	12.72
1/ 143	105.14	123.59	143.93	166.26	190.63	217.11	245.79	276.72	309.98	345.63	383.74	467.58	562.03
0.00750	8.660	9.017	9.368	9.711	10.05	10.38	10.71	11.03	11.35	11.66	11.97	12.58	13.17
1/ 133	108.83	127.92	148.99	172.09	197.32	224.74	254.42	286.44	320.86	357.76	397.21	484.00	581.76
0.00800	8.944	9.313	9.675	10.03	10.38	10.72	11.06	11.39	11.72	12.04	12.36	12.99	13.60
1/ 125	112.40	132.12	153.87	177.74	203.79	232.11	262.76	295.83	331.38	369.49	410.23	499.87	600.84
0.00850	9.220	9.600	9.973	10.34	10.70	11.05	11.40	11.74	12.08	12.41	12.74	13.39	14.02
1/ 118	115.86	136.18	158.61	183.21	210.06	239.25	270.85	304.93	341.58	380.87	422.86	515.25	619.33
0.00900	9.487	9.878	10.26	10.64	11.01	11.37	11.73	12.08	12.43	12.77	13.11	13.78	14.43
1/ 111	119.21	140.13	163.21	188.52	216.15	246.19	278.70	313.77	351.49	391.91	435.12	530.19	637.29
0.00950	9.747	10.15	10.54	10.93	11.31	11.68	12.05	12.41	12.77	13.12	13.47	14.15	14.82
1/ 105	122.48	143.97	167.68	193.68	222.07	252.93	286.34	322.37	361.12	402.65	447.04	544.72	654.75
0.01000	10.00	10.41	10.82	11.21	11.60	11.99	12.37	12.74	13.10	13.47	13.82	14.52	15.21
1/ 100	125.66	147.71	172.03	198.72	227.84	259.50	293.78	330.75	370.50	413.11	458.65	558.87	671.76
0.01100	10.49	10.92	11.34	11.76	12.17	12.57	12.97	13.36	13.74	14.12	14.50	15.23	15.95
1/ 91	131.80	154.92	180.43	208.41	238.96	272.17	308.11	346.89	388.58	433.27	481.04	586.15	704.55
0.01200	10.95	11.41	11.85	12.28	12.71	13.13	13.55	13.95	14.35	14.75	15.14	15.91	16.66
1/ 83	137.66	161.81	188.45	217.68	249.59	284.27	321.82	362.32	405.86	452.54	502.43	612.21	735.87
0.01300	11.40	11.87	12.33	12.79	13.23	13.67	14.10	14.52	14.94	15.35	15.76	16.56	17.34
1/ 77	143.28	168.42	196.15	226.57	259.78	295.88	334.96	377.11	422.43	471.01	522.95	637.21	765.92
0.01400	11.83	12.32	12.80	13.27	13.73	14.18	14.63	15.07	15.50	15.93	16.35	17.18	17.99
1/ 71	148.69	174.78	203.55	235.12	269.59	307.05	347.60	391.35	438.38	488.79	542.69	661.26	794.83
	4.000	4.250	4.500	4.750	5.000	5.250	5.500	5.750	6.000	6.250	6.500	7.000	7.500

Gradient   (Equivalent) Pipe diameters in m

S = 0.00300 to 0.01400

m = Manning $n \times 100$
S = 0·01500 to 0·07500

ie hydraulic gradient =
1 in 67   to   1 in 13·3

Full bore conditions; Select
m to divide mV and/or mQ.

mV to give velocities in ms^{-1}
mQ to give discharges in m^{3}s^{-1}

Gradient   (Equivalent) Pipe diameters in m

Gradient	4·000	4·250	4·500	4·750	5·000	5·250	5·500	5·750	6·000	6·250	6·500	7·000	7·500
0·01500	12·25	12·75	13·25	13·73	14·21	14·68	15·14	15·60	16·05	16·49	16·93	17·79	18·62
1/ 67	153·91	180·91	210·70	243·38	279·05	317·82	359·80	405·08	453·77	505·95	561·73	684·47	822·73
0·01600	12·65	13·17	13·68	14·18	14·68	15·16	15·64	16·11	16·58	17·03	17·48	18·37	19·23
1/ 63	158·95	186·84	217·61	251·36	288·20	328·25	371·60	418·37	468·65	522·54	580·16	706·92	849·71
0·01700	13·04	13·58	14·10	14·62	15·13	15·63	16·12	16·61	17·09	17·56	18·02	18·93	19·83
1/ 59	163·85	192·59	224·31	259·09	297·07	338·35	383·04	431·24	483·07	538·63	598·01	728·68	875·86
0·01800	13·42	13·97	14·51	15·04	15·57	16·08	16·59	17·09	17·58	18·07	18·54	19·48	20·40
1/ 56	168·60	198·18	230·81	266·61	305·68	348·16	394·14	443·74	497·08	554·24	615·35	749·80	901·26
0·01900	13·78	14·35	14·91	15·46	15·99	16·52	17·04	17·56	18·06	18·56	19·05	20·02	20·96
1/ 53	173·22	203·61	237·13	273·91	314·06	357·70	404·94	455·90	510·70	569·43	632·21	770·35	925·95
0·02000	14·14	14·73	15·30	15·86	16·41	16·95	17·49	18·01	18·53	19·04	19·55	20·54	21·50
1/ 50	177·72	208·90	243·29	281·03	322·22	366·99	415·46	467·75	523·96	584·22	648·63	790·36	950·01
0·02200	14·83	15·44	16·04	16·63	17·21	17·78	18·34	18·89	19·44	19·97	20·50	21·54	22·55
1/ 45·5	186·39	219·09	255·17	294·74	337·95	384·90	435·74	490·58	549·54	612·74	680·29	828·94	996·38
0·02400	15·49	16·13	16·76	17·37	17·98	18·57	19·16	19·73	20·30	20·86	21·41	22·50	23·56
1/ 41·7	194·68	228·84	266·51	307·85	352·97	402·02	455·12	512·39	573·97	639·98	710·54	865·80	1040·7
0·02600	16·12	16·79	17·44	18·08	18·71	19·33	19·94	20·54	21·13	21·71	22·29	23·42	24·52
1/ 38·5	202·63	238·18	277·40	320·42	367·39	418·43	473·70	533·31	597·41	666·12	739·56	901·15	1083·2
0·02800	16·73	17·42	18·10	18·76	19·42	20·06	20·69	21·31	21·93	22·53	23·13	24·30	25·44
1/ 35·7	210·28	247·17	287·87	332·52	381·25	434·23	491·58	553·45	619·96	691·26	767·47	935·17	1124·1
0·03000	17·32	18·03	18·74	19·42	20·10	20·76	21·42	22·06	22·70	23·32	23·94	25·15	26·34
1/ 33·3	217·66	255·85	297·97	344·19	394·64	449·47	508·84	572·87	641·72	715·52	794·41	967·99	1163·5
0·03200	17·89	18·63	19·35	20·06	20·76	21·44	22·12	22·78	23·44	24·09	24·73	25·98	27·20
1/ 31·3	224·79	264·24	307·74	355·47	407·58	464·21	525·52	591·66	662·77	738·99	820·46	999·74	1201·7
0·03400	18·44	19·20	19·95	20·68	21·40	22·10	22·80	23·49	24·16	24·83	25·49	26·78	28·04
1/ 29·4	231·71	272·37	317·22	366·41	420·12	478·50	541·70	609·87	683·16	761·73	845·72	1030·5	1238·7
0·03600	18·97	19·76	20·52	21·28	22·02	22·74	23·46	24·17	24·86	25·55	26·23	27·55	28·85
1/ 27·8	238·43	280·27	326·41	377·04	432·30	492·37	557·40	627·55	702·97	783·82	870·23	1060·4	1274·6
0·03800	19·49	20·30	21·09	21·86	22·62	23·37	24·10	24·83	25·54	26·25	26·94	28·31	29·64
1/ 26·3	244·96	287·95	335·36	387·37	444·15	505·86	572·68	644·75	722·23	805·29	894·08	1089·4	1309·5
0·04000	20·00	20·82	21·63	22·43	23·21	23·98	24·73	25·47	26·21	26·93	27·64	29·04	30·41
1/ 25·0	251·33	295·43	344·07	397·43	455·69	519·00	587·55	661·50	741·00	826·21	917·31	1117·7	1343·5
0·04200	20·49	21·34	22·17	22·98	23·78	24·57	25·34	26·10	26·85	27·60	28·33	29·76	31·16
1/ 23·8	257·53	302·72	352·57	407·25	466·94	531·82	602·06	677·83	759·30	846·62	939·96	1145·3	1376·7
0·04400	20·98	21·84	22·69	23·52	24·34	25·15	25·94	26·72	27·49	28·24	28·99	30·46	31·90
1/ 22·7	263·59	309·85	360·86	416·83	477·93	544·34	616·23	693·78	777·16	866·54	962·08	1172·3	1409·1
0·04600	21·45	22·33	23·20	24·05	24·89	25·71	26·52	27·32	28·10	28·88	29·64	31·15	32·61
1/ 21·7	269·52	316·81	368·97	426·20	488·67	556·57	630·08	709·37	794·63	886·02	983·70	1198·6	1440·8
0·04800	21·91	22·81	23·70	24·57	25·42	26·26	27·09	27·91	28·71	29·50	30·28	31·82	33·31
1/ 20·8	275·32	323·62	376·91	435·36	499·18	568·54	643·63	724·63	811·72	905·07	1004·9	1224·4	1471·7
0·05000	22·36	23·28	24·19	25·07	25·95	26·81	27·65	28·48	29·30	30·11	30·91	32·47	34·00
1/ 20·0	280·99	330·30	384·68	444·34	509·47	580·26	656·90	739·57	828·46	923·74	1025·6	1249·7	1502·1
0·05500	23·45	24·42	25·37	26·30	27·21	28·11	29·00	29·87	30·73	31·58	32·42	34·06	35·66
1/ 18·2	294·71	346·42	403·46	466·03	534·34	608·59	688·97	775·67	868·90	968·82	1075·6	1310·7	1575·4
0·06000	24·49	25·51	26·50	27·47	28·42	29·36	30·29	31·20	32·10	32·98	33·86	35·57	37·25
1/ 16·7	307·81	361·82	421·40	486·75	558·10	635·65	719·60	810·16	907·53	1011·9	1123·5	1368·9	1645·5
0·06500	25·50	26·55	27·58	28·59	29·58	30·56	31·53	32·47	33·41	34·33	35·24	37·02	38·77
1/ 15·4	320·38	376·60	438·60	506·63	580·89	661·60	748·99	843·24	944·59	1053·2	1169·3	1424·8	1712·7
0·07000	26·46	27·55	28·62	29·67	30·70	31·72	32·72	33·70	34·67	35·63	36·57	38·42	40·23
1/ 14·3	332·47	390·81	455·16	525·75	602·82	686·58	777·26	875·08	980·25	1093·0	1213·5	1478·6	1777·3
0·07500	27·39	28·52	29·62	30·71	31·78	32·83	33·86	34·88	35·89	36·88	37·85	39·77	41·64
1/ 13·3	344·14	404·53	471·14	544·21	623·97	710·68	804·54	905·79	1014·7	1131·3	1256·1	1530·5	1839·7
	4·000	4·250	4·500	4·750	5·000	5·250	5·500	5·750	6·000	6·250	6·500	7·000	7·500

Gradient   (Equivalent) Pipe diameters in m

S = 0·01500 to 0·07500

m = Manning $n \times 100$
S = 0.00010 to 0.00048

ie hydraulic gradient =
1 in 10000 to 1 in 2083

Full bore conditions; Select
m to divide mV and/or mQ.

mV to give velocities in ms^{-1}
mQ to give discharges in m^{3}s^{-1}

Gradient	(Equivalent) Pipe diameters in m												
	7.500	8.000	8.500	9.000	9.500	10.00	10.50	11.00	11.50	12.00	12.50	13.00	13.50
0.00010	1.521	1.587	1.653	1.717	1.780	1.842	1.903	1.963	2.022	2.080	2.137	2.194	2.250
1/10000	67.176	79.791	93.792	109.24	126.18	144.67	164.77	186.54	210.01	235.25	262.31	291.23	322.06
0.00011	1.595	1.665	1.734	1.801	1.867	1.932	1.996	2.059	2.121	2.182	2.242	2.301	2.360
1/ 9091	70.455	83.686	98.370	114.57	132.34	151.73	172.82	195.64	220.26	246.73	275.11	305.44	337.78
0.00012	1.666	1.739	1.811	1.881	1.950	2.018	2.085	2.150	2.215	2.279	2.341	2.404	2.465
1/ 8333	73.587	87.407	102.74	119.66	138.22	158.48	180.50	204.34	230.06	257.71	287.34	319.02	352.80
0.00013	1.734	1.810	1.885	1.958	2.030	2.100	2.170	2.238	2.305	2.372	2.437	2.502	2.565
1/ 7692	76.592	90.976	106.94	124.55	143.86	164.95	187.87	212.68	239.45	268.23	299.08	332.05	367.21
0.00014	1.799	1.878	1.956	2.032	2.106	2.180	2.252	2.322	2.392	2.461	2.529	2.596	2.662
1/ 7143	79.483	94.410	110.98	129.25	149.29	171.18	194.96	220.71	248.49	278.35	310.37	344.58	381.07
0.00015	1.862	1.944	2.024	2.103	2.180	2.256	2.331	2.404	2.476	2.548	2.618	2.687	2.756
1/ 6667	82.273	97.724	114.87	133.79	154.53	177.19	201.81	228.46	257.21	288.12	321.26	356.68	394.44
0.00016	1.923	2.008	2.091	2.172	2.252	2.330	2.407	2.483	2.558	2.631	2.704	2.775	2.846
1/ 6250	84.971	100.93	118.64	138.17	159.60	183.00	208.42	235.95	265.65	297.57	331.79	368.38	407.38
0.00017	1.983	2.070	2.155	2.239	2.321	2.402	2.481	2.559	2.636	2.712	2.787	2.861	2.934
1/ 5882	87.586	104.04	122.29	142.43	164.51	188.63	214.84	243.21	273.82	306.73	342.01	379.71	419.92
0.00018	2.040	2.130	2.218	2.304	2.388	2.471	2.553	2.633	2.713	2.791	2.868	2.944	3.019
1/ 5556	90.126	107.05	125.84	146.55	169.28	194.10	221.07	250.26	281.76	315.62	351.92	390.72	432.09
0.00019	2.096	2.188	2.278	2.367	2.454	2.539	2.623	2.706	2.787	2.867	2.946	3.024	3.101
1/ 5263	92.595	109.98	129.28	150.57	173.92	199.42	227.12	257.12	289.48	324.27	361.56	401.43	443.93
0.00020	2.150	2.245	2.338	2.428	2.517	2.605	2.691	2.776	2.859	2.942	3.023	3.103	3.182
1/ 5000	95.001	112.84	132.64	154.48	178.44	204.60	233.02	263.80	297.00	332.70	370.96	411.86	455.46
0.00022	2.255	2.354	2.452	2.547	2.640	2.732	2.822	2.911	2.999	3.085	3.170	3.254	3.337
1/ 4545	99.638	118.35	139.12	162.02	187.15	214.58	244.40	276.68	311.50	348.93	389.06	431.96	477.70
0.00024	2.356	2.459	2.561	2.660	2.758	2.854	2.948	3.041	3.132	3.222	3.311	3.399	3.486
1/ 4167	104.07	123.61	145.30	169.23	195.47	224.12	255.27	288.98	325.35	364.45	406.36	451.17	498.94
0.00026	2.452	2.560	2.665	2.769	2.870	2.970	3.068	3.165	3.260	3.354	3.447	3.538	3.628
1/ 3846	108.32	128.66	151.24	176.14	203.45	233.28	265.69	300.78	338.63	379.33	422.96	469.59	519.31
0.00028	2.544	2.656	2.766	2.873	2.979	3.082	3.184	3.284	3.383	3.481	3.577	3.671	3.765
1/ 3571	112.41	133.52	156.94	182.79	211.13	242.08	275.72	312.13	351.42	393.65	438.92	487.32	538.91
0.00030	2.634	2.749	2.863	2.974	3.083	3.190	3.296	3.400	3.502	3.603	3.702	3.800	3.897
1/ 3333	116.35	138.20	162.45	189.20	218.54	250.58	285.40	323.09	363.75	407.47	454.33	504.42	557.83
0.00032	2.720	2.840	2.957	3.072	3.184	3.295	3.404	3.511	3.617	3.721	3.824	3.925	4.025
1/ 3125	120.17	142.74	167.78	195.41	225.71	258.80	294.76	333.69	375.68	420.83	469.23	520.96	576.12
0.00034	2.804	2.927	3.048	3.166	3.282	3.397	3.509	3.619	3.728	3.835	3.941	4.046	4.149
1/ 2941	123.87	147.13	172.94	201.42	232.66	266.76	303.83	343.96	387.24	433.78	483.67	537.00	593.85
0.00036	2.885	3.012	3.136	3.258	3.377	3.495	3.611	3.724	3.836	3.947	4.056	4.163	4.269
1/ 2778	127.46	151.39	177.96	207.26	239.40	274.49	312.64	353.93	398.47	446.36	497.69	552.56	611.07
0.00038	2.964	3.094	3.222	3.347	3.470	3.591	3.709	3.826	3.941	4.055	4.167	4.277	4.386
1/ 2632	130.95	155.54	182.83	212.94	245.96	282.02	321.20	363.63	409.39	458.59	511.33	567.71	627.81
0.00040	3.041	3.175	3.306	3.434	3.560	3.684	3.806	3.926	4.044	4.160	4.275	4.388	4.500
1/ 2500	134.35	159.58	187.58	218.47	252.35	289.34	329.55	373.07	420.02	470.50	524.61	582.45	644.12
0.00042	3.116	3.253	3.387	3.519	3.648	3.775	3.900	4.023	4.144	4.263	4.381	4.497	4.611
1/ 2381	137.67	163.52	192.22	223.87	258.59	296.49	337.69	382.29	430.40	482.12	537.57	596.84	660.03
0.00044	3.190	3.330	3.467	3.602	3.734	3.864	3.992	4.117	4.241	4.363	4.484	4.602	4.720
1/ 2273	140.91	167.37	196.74	229.13	264.67	303.47	345.63	391.28	440.52	493.47	550.22	610.88	675.56
0.00046	3.261	3.405	3.545	3.683	3.818	3.951	4.081	4.210	4.336	4.461	4.584	4.706	4.826
1/ 2174	144.08	171.13	201.16	234.28	270.62	310.29	353.40	400.08	450.43	504.56	562.59	624.61	690.75
0.00048	3.331	3.478	3.621	3.762	3.900	4.036	4.169	4.300	4.430	4.557	4.683	4.807	4.930
1/ 2083	147.17	174.81	205.49	239.32	276.44	316.96	361.00	408.68	460.11	515.41	574.69	638.05	705.60
	7.500	8.000	8.500	9.000	9.500	10.00	10.50	11.00	11.50	12.00	12.50	13.00	13.50

Gradient (Equivalent) Pipe diameters in m

S = 0.00010 to 0.00048

m = Manning $n \times 100$
S = 0·00050 to 0·00280

ie hydraulic gradient =
1 in 2000 to 1 in 357

Full bore conditions; Select
m to divide mV and/or mQ.

mV to give velocities in ms^{-1}
mQ to give discharges in m^{3}s^{-1}

Gradient — (Equivalent) Pipe diameters in m

Gradient	7·500	8·000	8·500	9·000	9·500	10·00	10·50	11·00	11·50	12·00	12·50	13·00	13·50
0·00050	3·400	3·550	3·696	3·839	3·980	4·119	4·255	4·389	4·521	4·651	4·780	4·906	5·031
1/ 2000	150·21	178·42	209·73	244·26	282·14	323·50	368·44	417·11	469·60	526·04	586·54	651·20	720·15
0·00055	3·566	3·723	3·876	4·027	4·175	4·320	4·463	4·603	4·742	4·878	5·013	5·146	5·277
1/ 1818	157·54	187·13	219·96	256·18	295·91	339·28	386·43	437·47	492·52	551·71	615·16	682·99	755·30
0·00060	3·725	3·888	4·049	4·206	4·360	4·512	4·661	4·808	4·953	5·095	5·236	5·374	5·511
1/ 1667	164·55	195·45	229·74	267·57	309·07	354·37	403·61	456·92	514·42	576·25	642·52	713·36	788·89
0·00065	3·877	4·047	4·214	4·378	4·538	4·696	4·851	5·004	5·155	5·303	5·449	5·594	5·736
1/ 1538	171·27	203·43	239·12	278·50	321·69	368·84	420·09	475·58	535·43	599·78	668·75	742·49	821·10
0·00070	4·023	4·200	4·373	4·543	4·710	4·874	5·035	5·193	5·349	5·503	5·655	5·805	5·953
1/ 1429	177·73	211·11	248·15	289·01	333·83	382·76	435·95	493·53	555·64	622·42	694·00	770·51	852·10
0·00075	4·164	4·347	4·527	4·702	4·875	5·045	5·211	5·375	5·537	5·697	5·854	6·009	6·162
1/ 1333	183·97	218·52	256·86	299·15	345·55	396·20	451·25	510·85	575·14	644·26	718·36	797·56	882·00
0·00080	4·301	4·490	4·675	4·857	5·035	5·210	5·382	5·552	5·719	5·883	6·046	6·206	6·364
1/ 1250	190·00	225·68	265·28	308·96	356·88	409·19	466·05	527·60	594·00	665·39	741·92	823·71	910·93
0·00085	4·433	4·628	4·819	5·006	5·190	5·370	5·548	5·723	5·895	6·064	6·232	6·397	6·560
1/ 1176	195·85	232·63	273·45	318·47	367·86	421·79	480·39	543·84	612·28	685·87	764·75	849·07	938·96
0·00090	4·562	4·762	4·959	5·151	5·340	5·526	5·709	5·889	6·066	6·240	6·412	6·582	6·750
1/ 1111	201·53	239·37	281·38	327·71	378·53	434·01	494·32	559·61	630·04	705·75	786·92	873·68	966·19
0·00095	4·687	4·893	5·094	5·292	5·487	5·677	5·865	6·050	6·232	6·411	6·588	6·763	6·935
1/ 1053	207·05	245·93	289·09	336·69	388·90	445·91	507·87	574·94	647·30	725·09	808·48	897·62	992·66
0·00100	4·808	5·020	5·227	5·430	5·629	5·825	6·018	6·207	6·394	6·578	6·759	6·938	7·115
1/ 1000	212·43	252·32	296·60	345·43	399·01	457·49	521·06	589·88	664·12	743·93	829·49	920·94	1018·4
0·00110	5·043	5·265	5·482	5·695	5·904	6·109	6·311	6·510	6·706	6·899	7·089	7·277	7·462
1/ 909	222·80	264·64	311·07	362·29	418·48	479·82	546·49	618·67	696·53	780·24	869·97	965·89	1068·2
0·00120	5·267	5·499	5·726	5·948	6·166	6·381	6·592	6·800	7·004	7·206	7·404	7·601	7·794
1/ 833	232·70	276·41	324·91	378·40	437·09	501·16	570·79	646·18	727·50	814·94	908·66	1008·8	1115·7
0·00130	5·482	5·723	5·960	6·191	6·418	6·641	6·861	7·077	7·290	7·500	7·707	7·911	8·112
1/ 769	242·21	287·69	338·17	393·85	454·94	521·62	594·10	672·57	757·21	848·21	945·76	1050·0	1161·2
0·00140	5·689	5·940	6·184	6·425	6·660	6·892	7·120	7·344	7·565	7·783	7·998	8·210	8·419
1/ 714	251·35	298·55	350·94	408·72	472·11	541·31	616·53	697·95	785·79	880·23	981·46	1089·7	1205·0
0·00150	5·889	6·148	6·402	6·650	6·894	7·134	7·370	7·602	7·831	8·056	8·278	8·498	8·714
1/ 667	260·17	309·03	363·26	423·07	488·68	560·31	638·17	722·45	813·37	911·13	1015·9	1127·9	1247·3
0·00160	6·082	6·350	6·611	6·868	7·120	7·368	7·612	7·851	8·088	8·320	8·550	8·776	9·000
1/ 625	268·70	319·17	375·17	436·94	504·71	578·69	659·09	746·14	840·05	941·01	1049·2	1164·9	1288·2
0·00170	6·269	6·545	6·815	7·080	7·340	7·595	7·846	8·093	8·336	8·576	8·813	9·046	9·277
1/ 588	276·97	328·99	386·71	450·39	520·24	596·50	679·38	769·11	865·90	969·97	1081·5	1200·8	1327·9
0·00180	6·451	6·735	7·013	7·285	7·552	7·815	8·073	8·328	8·578	8·825	9·069	9·309	9·546
1/ 556	285·00	338·53	397·93	463·45	535·32	613·79	699·07	791·41	891·00	998·09	1112·9	1235·6	1366·4
0·00190	6·628	6·919	7·205	7·485	7·759	8·029	8·295	8·556	8·813	9·067	9·317	9·564	9·808
1/ 526	292·81	347·80	408·83	476·15	549·99	630·61	718·23	813·09	915·42	1025·4	1143·4	1269·4	1403·8
0·00200	6·800	7·099	7·392	7·679	7·961	8·238	8·510	8·778	9·042	9·302	9·559	9·812	10·06
1/ 500	300·42	356·84	419·45	488·51	564·28	646·99	736·89	834·22	939·20	1052·1	1173·1	1302·4	1440·3
0·00220	7·132	7·446	7·753	8·054	8·349	8·640	8·925	9·207	9·484	9·756	10·03	10·29	10·55
1/ 455	315·08	374·25	439·92	512·36	591·82	678·57	772·86	874·93	985·04	1103·4	1230·3	1366·0	1510·6
0·00240	7·449	7·777	8·097	8·412	8·721	9·024	9·322	9·616	9·905	10·19	10·47	10·75	11·02
1/ 417	329·09	390·90	459·49	535·14	618·14	708·74	807·22	913·84	1028·8	1152·5	1285·0	1426·7	1577·8
0·00260	7·753	8·094	8·428	8·755	9·077	9·392	9·703	10·01	10·31	10·61	10·90	11·19	11·47
1/ 385	342·53	406·86	478·25	556·99	643·38	737·68	840·18	951·15	1070·9	1199·6	1337·5	1485·0	1642·2
0·00280	8·046	8·400	8·746	9·086	9·419	9·747	10·07	10·39	10·70	11·01	11·31	11·61	11·91
1/ 357	355·46	422·22	496·30	578·02	667·66	765·53	871·90	987·06	1111·3	1244·8	1388·0	1541·0	1704·2
	7·500	8·000	8·500	9·000	9·500	10·00	10·50	11·00	11·50	12·00	12·50	13·00	13·50

Gradient — (Equivalent) Pipe diameters in m

S = 0·00050 to 0·00280

D10
continued

m = Manning $n \times 100$
S = 0·00300 to 0·01400

ie hydraulic gradient =
1 in 333   to   1 in 71

Full bore conditions; Select
m to divide mV and/or mQ.

mV to give velocities in ms⁻¹
mQ to give discharges in m³s⁻¹

Gradient   (Equivalent) Pipe diameters in m

Gradient	7·500	8·000	8·500	9·000	9·500	10·00	10·50	11·00	11·50	12·00	12·50	13·00	13·50
0·00300	8·328	8·695	9·053	9·405	9·750	10·09	10·42	10·75	11·07	11·39	11·71	12·02	12·32
1/ 333	367·94	437·04	513·72	598·31	691·10	792·40	902·50	1021·7	1150·3	1288·5	1436·7	1595·1	1764·0
0·00320	8·602	8·980	9·350	9·713	10·07	10·42	10·76	11·10	11·44	11·77	12·09	12·41	12·73
1/ 313	380·00	451·37	530·57	617·93	713·76	818·38	932·10	1055·2	1188·0	1330·8	1483·8	1647·4	1821·9
0·00340	8·866	9·256	9·638	10·01	10·38	10·74	11·10	11·45	11·79	12·13	12·46	12·79	13·12
1/ 294	391·70	465·26	546·90	636·95	735·73	843·57	960·79	1087·7	1224·6	1371·7	1529·5	1698·1	1877·9
0·00360	9·123	9·524	9·917	10·30	10·68	11·05	11·42	11·78	12·13	12·48	12·82	13·16	13·50
1/ 278	403·05	478·75	562·75	655·41	757·06	868·03	988·64	1119·2	1260·1	1411·5	1573·8	1747·4	1932·4
0·00380	9·373	9·785	10·19	10·58	10·97	11·35	11·73	12·10	12·46	12·82	13·18	13·53	13·87
1/ 263	414·10	491·87	578·17	673·37	777·81	891·81	1015·7	1149·9	1294·6	1450·2	1617·0	1795·2	1985·3
0·00400	9·617	10·04	10·45	10·86	11·26	11·65	12·04	12·41	12·79	13·16	13·52	13·88	14·23
1/ 250	424·86	504·64	593·19	690·86	798·01	914·98	1042·1	1179·8	1328·2	1487·9	1659·0	1841·9	2036·9
0·00420	9·854	10·29	10·71	11·13	11·54	11·94	12·33	12·72	13·10	13·48	13·85	14·22	14·58
1/ 238	435·35	517·11	607·84	707·93	817·72	937·58	1067·9	1208·9	1361·0	1524·6	1699·9	1887·4	2087·2
0·00440	10·09	10·53	10·96	11·39	11·81	12·22	12·62	13·02	13·41	13·80	14·18	14·55	14·92
1/ 227	445·59	529·28	622·15	724·58	836·96	959·64	1093·0	1237·3	1393·1	1560·5	1739·9	1931·8	2136·3
0·00460	10·31	10·77	11·21	11·65	12·07	12·49	12·91	13·31	13·71	14·11	14·50	14·88	15·26
1/ 217	455·61	541·17	636·13	740·87	855·77	981·21	1117·5	1265·2	1424·4	1595·6	1779·1	1975·2	2184·3
0·00480	10·53	11·00	11·45	11·90	12·33	12·76	13·18	13·60	14·01	14·41	14·81	15·20	15·59
1/ 208	465·41	552·81	649·81	756·80	874·18	1002·3	1141·6	1292·4	1455·0	1629·9	1817·3	2017·7	2231·3
0·00500	10·75	11·22	11·69	12·14	12·59	13·03	13·46	13·88	14·30	14·71	15·11	15·51	15·91
1/ 200	475·00	564·21	663·21	772·41	892·20	1023·0	1165·1	1319·0	1485·0	1663·5	1854·8	2059·3	2277·3
0·00550	11·28	11·77	12·26	12·73	13·20	13·66	14·11	14·56	14·99	15·43	15·85	16·27	16·69
1/ 182	498·19	591·75	695·58	810·11	935·75	1072·9	1222·0	1383·4	1557·5	1744·7	1945·3	2159·8	2388·5
0·00600	11·78	12·30	12·80	13·30	13·79	14·27	14·74	15·20	15·66	16·11	16·56	17·00	17·43
1/ 167	520·34	618·06	726·51	846·13	977·36	1120·6	1276·3	1444·9	1626·7	1822·3	2031·8	2255·8	2494·7
0·00650	12·26	12·80	13·33	13·84	14·35	14·85	15·34	15·83	16·30	16·77	17·23	17·69	18·14
1/ 154	541·59	643·30	756·18	880·68	1017·3	1166·4	1328·4	1503·9	1693·2	1896·7	2114·8	2347·9	2596·5
0·00700	12·72	13·28	13·83	14·37	14·89	15·41	15·92	16·42	16·92	17·40	17·88	18·36	18·82
1/ 143	562·03	667·58	784·72	913·93	1055·7	1210·4	1378·6	1560·7	1757·1	1968·3	2194·6	2436·6	2694·6
0·00750	13·17	13·75	14·31	14·87	15·42	15·95	16·48	17·00	17·51	18·01	18·51	19·00	19·49
1/ 133	581·76	691·01	812·26	946·00	1092·7	1252·9	1427·0	1615·5	1818·8	2037·3	2271·6	2522·1	2789·1
0·00800	13·60	14·20	14·78	15·36	15·92	16·48	17·02	17·56	18·08	18·60	19·12	19·62	20·12
1/ 125	600·84	713·68	838·90	977·03	1128·6	1294·0	1473·8	1668·4	1878·4	2104·2	2346·1	2604·8	2880·6
0·00850	14·02	14·64	15·24	15·83	16·41	16·98	17·54	18·10	18·64	19·18	19·71	20·23	20·74
1/ 118	619·33	735·64	864·72	1007·1	1163·3	1333·8	1519·1	1719·8	1936·2	2168·9	2418·3	2685·0	2969·3
0·00900	14·43	15·06	15·68	16·29	16·89	17·47	18·05	18·62	19·18	19·73	20·28	20·82	21·35
1/ 111	637·29	756·97	889·79	1036·3	1197·0	1372·5	1563·2	1769·6	1992·3	2231·8	2488·5	2762·8	3055·3
0·00950	14·82	15·47	16·11	16·74	17·35	17·95	18·55	19·13	19·71	20·27	20·83	21·39	21·93
1/ 105	654·75	777·71	914·17	1064·7	1229·8	1410·1	1606·0	1818·1	2046·9	2292·9	2556·6	2838·5	3139·1
0·01000	15·21	15·87	16·53	17·17	17·80	18·42	19·03	19·63	20·22	20·80	21·37	21·94	22·50
1/ 100	671·76	797·91	937·92	1092·4	1261·8	1446·7	1647·7	1865·4	2100·1	2352·5	2623·1	2912·3	3220·6
0·01100	15·95	16·65	17·34	18·01	18·67	19·32	19·96	20·59	21·21	21·82	22·42	23·01	23·60
1/ 91	704·55	836·86	983·70	1145·7	1323·4	1517·3	1728·2	1956·4	2202·6	2467·3	2751·1	3054·4	3377·8
0·01200	16·66	17·39	18·11	18·81	19·50	20·18	20·85	21·50	22·15	22·79	23·41	24·04	24·65
1/ 83	735·87	874·07	1027·4	1196·6	1382·2	1584·8	1805·0	2043·4	2300·6	2577·1	2873·4	3190·2	3528·0
0·01300	17·34	18·10	18·85	19·58	20·30	21·00	21·70	22·38	23·05	23·72	24·37	25·02	25·65
1/ 77	765·92	909·76	1069·4	1245·5	1438·6	1649·5	1878·7	2126·8	2394·5	2682·3	2990·8	3320·5	3672·1
0·01400	17·99	18·78	19·56	20·32	21·06	21·80	22·52	23·22	23·92	24·61	25·29	25·96	26·62
1/ 71	794·83	944·10	1109·8	1292·5	1492·9	1711·8	1949·6	2207·1	2484·9	2783·5	3103·7	3445·8	3810·7
	7·500	8·000	8·500	9·000	9·500	10·00	10·50	11·00	11·50	12·00	12·50	13·00	13·50

Gradient   (Equivalent) Pipe diameters in m

S = 0·00300 to 0·01400

m = Manning $n \times 100$
S = 0.01500 to 0.07500

ie hydraulic gradient =
1 in 67   to   1 in 13·3

Full bore conditions; Select m to divide mV and/or mQ.

mV to give velocities in ms^{-1}
mQ to give discharges in m^{3}s^{-1}

Gradient   (Equivalent) Pipe diameters in m

Gradient	7·500	8·000	8·500	9·000	9·500	10·00	10·50	11·00	11·50	12·00	12·50	13·00	13·50
0·01500	18·62	19·44	20·24	21·03	21·80	22·56	23·31	24·04	24·76	25·48	26·18	26·87	27·56
1/ 67	822·73	977·24	1148·7	1337·9	1545·3	1771·9	2018·1	2284·6	2572·1	2881·2	3212·6	3566·8	3944·4
0·01600	19·23	20·08	20·91	21·72	22·52	23·30	24·07	24·83	25·58	26·31	27·04	27·75	28·46
1/ 63	849·71	1009·3	1186·4	1381·7	1596·0	1830·0	2084·2	2359·5	2656·5	2975·7	3317·9	3683·8	4073·8
0·01700	19·83	20·70	21·55	22·39	23·21	24·02	24·81	25·59	26·36	27·12	27·87	28·61	29·34
1/ 59	875·86	1040·4	1222·9	1424·3	1645·1	1886·3	2148·4	2432·1	2738·2	3067·3	3420·1	3797·1	4199·2
0·01800	20·40	21·30	22·18	23·04	23·88	24·71	25·53	26·33	27·13	27·91	28·68	29·44	30·19
1/ 56	901·26	1070·5	1258·4	1465·5	1692·8	1941·0	2210·7	2502·6	2817·6	3156·2	3519·2	3907·2	4320·9
0·01900	20·96	21·88	22·78	23·67	24·54	25·39	26·23	27·06	27·87	28·67	29·46	30·24	31·01
1/ 53	925·95	1099·8	1292·8	1505·7	1739·2	1994·2	2271·2	2571·2	2894·8	3242·7	3615·6	4014·3	4439·3
0·02000	21·50	22·45	23·38	24·28	25·17	26·05	26·91	27·76	28·59	29·42	30·23	31·03	31·82
1/ 50	950·01	1128·4	1326·4	1544·8	1784·4	2046·0	2330·2	2638·0	2970·0	3327·0	3709·6	4118·6	4554·6
0·02200	22·55	23·54	24·52	25·47	26·40	27·32	28·22	29·11	29·99	30·85	31·70	32·54	33·37
1/ 45·5	996·38	1183·5	1391·2	1620·2	1871·5	2145·8	2444·0	2766·8	3115·0	3489·3	3890·6	4319·6	4777·0
0·02400	23·56	24·59	25·61	26·60	27·58	28·54	29·48	30·41	31·32	32·22	33·11	33·99	34·86
1/ 41·7	1040·7	1236·1	1453·0	1692·3	1954·7	2241·2	2552·7	2889·8	3253·5	3644·5	4063·6	4511·7	4989·4
0·02600	24·52	25·60	26·65	27·69	28·70	29·70	30·68	31·65	32·60	33·54	34·47	35·38	36·28
1/ 38·5	1083·2	1286·6	1512·4	1761·4	2034·5	2332·8	2656·9	3007·8	3386·3	3793·3	4229·6	4695·9	5193·1
0·02800	25·44	26·56	27·66	28·73	29·79	30·82	31·84	32·84	33·83	34·81	35·77	36·71	37·65
1/ 35·7	1124·1	1335·2	1569·4	1827·9	2111·3	2420·8	2757·2	3121·3	3514·2	3936·5	4389·2	4873·2	5389·1
0·03000	26·34	27·49	28·63	29·74	30·83	31·90	32·96	34·00	35·02	36·03	37·02	38·00	38·97
1/ 33·3	1163·5	1382·0	1624·5	1892·0	2185·4	2505·8	2854·0	3230·9	3637·5	4074·7	4543·3	5044·2	5578·3
0·03200	27·20	28·40	29·57	30·72	31·84	32·95	34·04	35·11	36·17	37·21	38·24	39·25	40·25
1/ 31·3	1201·7	1427·4	1677·8	1954·1	2257·1	2588·0	2947·6	3336·9	3756·8	4208·3	4692·3	5209·6	5761·2
0·03400	28·04	29·27	30·48	31·66	32·82	33·97	35·09	36·19	37·28	38·35	39·41	40·46	41·49
1/ 29·4	1238·7	1471·3	1729·4	2014·2	2326·6	2667·6	3038·3	3439·6	3872·4	4337·8	4836·7	5370·0	5938·5
0·03600	28·85	30·12	31·36	32·58	33·77	34·95	36·11	37·24	38·36	39·47	40·56	41·63	42·69
1/ 27·8	1274·6	1513·9	1779·6	2072·6	2394·0	2744·9	3126·4	3539·3	3984·7	4463·6	4976·9	5525·6	6110·7
0·03800	29·64	30·94	32·22	33·47	34·70	35·91	37·09	38·26	39·41	40·55	41·67	42·77	43·86
1/ 26·3	1309·5	1555·4	1828·3	2129·4	2459·6	2820·2	3212·0	3636·3	4093·9	4585·9	5113·3	5677·1	6278·1
0·04000	30·41	31·75	33·06	34·34	35·60	36·84	38·06	39·26	40·44	41·60	42·75	43·88	45·00
1/ 25·0	1343·5	1595·8	1875·8	2184·7	2523·5	2893·4	3295·5	3730·7	4200·2	4705·0	5246·1	5824·5	6441·2
0·04200	31·16	32·53	33·87	35·19	36·48	37·75	39·00	40·23	41·44	42·63	43·81	44·97	46·11
1/ 23·8	1376·7	1635·2	1922·2	2238·7	2585·9	2964·9	3376·9	3822·9	4304·0	4821·2	5375·7	5968·4	6600·3
0·04400	31·90	33·30	34·67	36·02	37·34	38·64	39·92	41·17	42·41	43·63	44·84	46·02	47·20
1/ 22·7	1409·1	1673·7	1967·4	2291·3	2646·7	3034·7	3456·3	3912·8	4405·2	4934·7	5502·2	6108·8	6755·6
0·04600	32·61	34·05	35·45	36·83	38·18	39·51	40·81	42·10	43·36	44·61	45·84	47·06	48·26
1/ 21·7	1440·8	1711·3	2011·6	2342·8	2706·2	3102·9	3534·0	4000·8	4504·3	5045·6	5625·9	6246·1	6907·5
0·04800	33·31	34·78	36·21	37·62	39·00	40·36	41·69	43·00	44·30	45·57	46·83	48·07	49·30
1/ 20·8	1471·7	1748·1	2054·9	2393·2	2764·4	3169·6	3610·0	4086·8	4601·1	5154·1	5746·9	6380·5	7056·0
0·05000	34·00	35·50	36·96	38·39	39·80	41·19	42·55	43·89	45·21	46·51	47·80	49·06	50·31
1/ 20·0	1502·1	1784·2	2097·3	2442·6	2821·4	3235·0	3684·4	4171·1	4696·0	5260·4	5865·4	6512·0	7201·5
0·05500	35·66	37·23	38·76	40·27	41·75	43·20	44·63	46·03	47·42	48·78	50·13	51·46	52·77
1/ 18·2	1575·4	1871·3	2199·6	2561·8	2959·1	3392·8	3864·3	4374·7	4925·2	5517·1	6151·6	6829·9	7553·0
0·06000	37·25	38·88	40·49	42·06	43·60	45·12	46·61	48·08	49·53	50·95	52·36	53·74	55·11
1/ 16·7	1645·5	1954·5	2297·4	2675·7	3090·7	3543·7	4036·1	4569·2	5144·2	5762·5	6425·2	7133·6	7888·9
0·06500	38·77	40·47	42·14	43·78	45·38	46·96	48·51	50·04	51·55	53·03	54·49	55·94	57·36
1/ 15·4	1712·7	2034·3	2391·2	2785·0	3216·9	3688·4	4200·9	4755·8	5354·3	5997·8	6687·5	7424·9	8211·0
0·07000	40·23	42·00	43·73	45·43	47·10	48·74	50·35	51·93	53·49	55·03	56·55	58·05	59·53
1/ 14·3	1777·3	2111·1	2481·5	2890·1	3338·3	3827·6	4359·5	4935·3	5556·4	6224·2	6940·0	7705·1	8521·0
0·07500	41·64	43·47	45·27	47·02	48·75	50·45	52·11	53·75	55·37	56·97	58·54	60·09	61·62
1/ 13·3	1839·7	2185·2	2568·6	2991·5	3455·5	3962·0	4512·5	5108·5	5751·4	6442·6	7183·6	7975·6	8820·0
	7·500	8·000	8·500	9·000	9·500	10·00	10·50	11·00	11·50	12·00	12·50	13·00	13·50

Gradient   (Equivalent) Pipe diameters in m

S = 0.01500 to 0.07500

m = Manning $n \times 100$
S = 0·000010 to 0·000048

ie hydraulic gradient =
1 in 100000 to 1 in 20833

Full bore conditions; Select
m to divide mV and/or mQ.

mV to give velocities in ms^{-1}
mQ to give discharges in m^{3}s^{-1}

Gradient **(Equivalent) Pipe diameters in m**

Gradient	13·50	14·00	15·00	16·00	17·00	18·00	19·00	20·00	21·00	22·00	24·00	26·00	28·00
0·000010	0·712	0·729	0·763	0·797	0·830	0·862	0·894	0·925	0·955	0·985	1·044	1·101	1·157
1/100000	101·84	112·22	134·88	160·21	188·33	219·34	253·35	290·49	330·85	374·55	472·37	584·76	712·53
0·000011	0·746	0·765	0·801	0·836	0·870	0·904	0·937	0·970	1·002	1·033	1·095	1·155	1·214
1/ 90909	106·82	117·69	141·47	168·03	197·52	230·04	265·72	304·67	347·00	392·83	495·42	613·30	747·31
0·000012	0·779	0·799	0·836	0·873	0·909	0·944	0·979	1·013	1·046	1·079	1·144	1·207	1·268
1/ 83333	111·57	122·93	147·76	175·51	206·30	240·27	277·53	318·21	362·43	410·30	517·45	640·57	780·54
0·000013	0·811	0·831	0·870	0·909	0·946	0·983	1·019	1·054	1·089	1·123	1·191	1·256	1·319
1/ 76923	116·12	127·95	153·79	182·67	214·73	250·08	288·87	331·21	377·23	427·05	538·58	666·73	812·41
0·000014	0·842	0·863	0·903	0·943	0·982	1·020	1·057	1·094	1·130	1·166	1·235	1·303	1·369
1/ 71429	120·50	132·78	159·60	189·57	222·83	259·52	299·77	343·71	391·47	443·17	558·91	691·90	843·08
0·000015	0·871	0·893	0·935	0·976	1·016	1·056	1·094	1·132	1·170	1·207	1·279	1·349	1·417
1/ 66667	124·73	137·44	165·20	196·22	230·65	268·63	310·29	355·77	405·21	458·73	578·53	716·18	872·67
0·000016	0·900	0·922	0·965	1·008	1·050	1·090	1·130	1·170	1·208	1·246	1·321	1·393	1·464
1/ 62500	128·82	141·94	170·62	202·66	238·22	277·44	320·47	367·44	418·50	473·77	597·50	739·67	901·29
0·000017	0·928	0·950	0·995	1·039	1·082	1·124	1·165	1·206	1·245	1·285	1·361	1·436	1·509
1/ 58824	132·79	146·31	175·87	208·89	245·55	285·98	330·33	378·75	431·38	488·35	615·89	762·43	929·03
0·000018	0·955	0·978	1·024	1·069	1·113	1·156	1·199	1·241	1·282	1·322	1·401	1·478	1·553
1/ 55556	136·64	150·55	180·97	214·95	252·67	294·27	339·91	389·73	443·88	502·51	633·75	784·54	955·96
0·000019	0·981	1·005	1·052	1·098	1·144	1·188	1·232	1·275	1·317	1·358	1·439	1·518	1·595
1/ 52632	140·38	154·68	185·92	220·84	259·59	302·33	349·22	400·41	456·05	516·28	651·11	806·04	982·16
0·000020	1·006	1·031	1·079	1·127	1·173	1·219	1·264	1·308	1·351	1·393	1·477	1·558	1·636
1/ 50000	144·03	158·70	190·75	226·58	266·33	310·19	358·29	410·81	467·90	529·69	668·03	826·98	1007·7
0·000022	1·055	1·081	1·132	1·182	1·231	1·278	1·325	1·371	1·417	1·461	1·549	1·634	1·716
1/ 45455	151·06	166·44	200·06	237·64	279·33	325·33	375·78	430·86	490·73	555·55	700·63	867·34	1056·9
0·000024	1·102	1·129	1·182	1·234	1·285	1·335	1·384	1·432	1·480	1·526	1·618	1·706	1·793
1/ 41667	157·78	173·85	208·96	248·20	291·76	339·79	392·49	450·02	512·55	580·25	731·79	905·91	1103·8
0·000026	1·147	1·175	1·231	1·285	1·338	1·390	1·441	1·491	1·540	1·589	1·684	1·776	1·866
1/ 38462	164·22	180·94	217·49	258·34	303·67	353·67	408·52	468·40	533·48	603·94	761·67	942·90	1148·9
0·000028	1·191	1·220	1·277	1·333	1·388	1·442	1·495	1·547	1·598	1·649	1·747	1·843	1·936
1/ 35714	170·42	187·77	225·70	268·09	315·13	367·02	423·94	486·08	553·62	626·74	790·42	978·49	1192·3
0·000030	1·232	1·263	1·322	1·380	1·437	1·493	1·548	1·602	1·654	1·707	1·809	1·908	2·004
1/ 33333	176·40	194·36	233·63	277·50	326·19	379·90	438·82	503·14	573·05	648·74	818·16	1012·8	1234·1
0·000032	1·273	1·304	1·365	1·425	1·484	1·542	1·598	1·654	1·709	1·763	1·868	1·970	2·070
1/ 31250	182·19	200·74	241·29	286·60	336·89	392·36	453·21	519·64	591·85	670·02	845·00	1046·1	1274·6
0·000034	1·312	1·344	1·407	1·469	1·530	1·589	1·648	1·705	1·761	1·817	1·925	2·031	2·134
1/ 29412	187·79	206·92	248·71	295·42	347·26	404·44	467·16	535·63	610·06	690·64	871·00	1078·2	1313·8
0·000036	1·350	1·383	1·448	1·512	1·574	1·635	1·695	1·754	1·812	1·870	1·981	2·090	2·196
1/ 27778	193·24	212·92	255·92	303·99	357·33	416·16	480·70	551·16	627·75	710·66	896·25	1109·5	1351·9
0·000038	1·387	1·421	1·488	1·553	1·617	1·680	1·742	1·802	1·862	1·921	2·035	2·147	2·256
1/ 26316	198·53	218·75	262·94	312·32	367·12	427·56	493·88	566·27	644·95	730·13	920·81	1139·9	1389·0
0·000040	1·423	1·458	1·527	1·594	1·659	1·724	1·787	1·849	1·910	1·971	2·088	2·203	2·314
1/ 25000	203·69	224·43	269·77	320·43	376·65	438·67	506·71	580·98	661·70	749·10	944·73	1169·5	1425·1
0·000042	1·458	1·494	1·564	1·633	1·700	1·766	1·831	1·895	1·958	2·019	2·140	2·257	2·371
1/ 23810	208·72	229·98	276·43	328·34	385·96	449·50	519·22	595·32	678·05	767·60	968·06	1198·4	1460·3
0·000044	1·492	1·529	1·601	1·671	1·740	1·808	1·874	1·940	2·004	2·067	2·190	2·310	2·427
1/ 22727	213·63	235·39	282·93	336·07	395·04	460·08	531·44	609·33	694·00	785·66	990·84	1226·6	1494·6
0·000046	1·526	1·563	1·637	1·709	1·780	1·849	1·916	1·983	2·049	2·113	2·239	2·362	2·482
1/ 21739	218·43	240·68	289·29	343·62	403·92	470·42	543·38	623·03	709·60	803·32	1013·1	1254·2	1528·2
0·000048	1·559	1·597	1·672	1·746	1·818	1·888	1·958	2·026	2·093	2·159	2·288	2·413	2·535
1/ 20833	223·13	245·85	295·52	351·01	412·60	480·54	555·07	636·43	724·86	820·60	1034·9	1281·1	1561·1
	13·50	14·00	15·00	16·00	17·00	18·00	19·00	20·00	21·00	22·00	24·00	26·00	28·00

Gradient **(Equivalent) Pipe diameters in m**

S = 0·000010 to 0·000048

m = Manning $n \times 100$
S = 0·000050 to 0·000280

ie hydraulic gradient =
1 in 20000 to 1 in 3571

Full bore conditions; Select m to divide mV and/or mQ.

mV to give velocities in ms⁻¹
mQ to give discharges in m³s⁻¹

Gradient   (Equivalent) Pipe diameters in m

Gradient	13·50	14·00	15·00	16·00	17·00	18·00	19·00	20·00	21·00	22·00	24·00	26·00	28·00
0·000050	1·591	1·630	1·707	1·782	1·855	1·927	1·998	2·068	2·136	2·203	2·335	2·463	2·588
1/ 20000	227·73	250·92	301·61	358·25	421·11	490·45	566·51	649·55	739·81	837·52	1056·2	1307·6	1593·3
0·000055	1·669	1·710	1·790	1·869	1·946	2·021	2·096	2·169	2·240	2·311	2·449	2·583	2·714
1/ 18182	238·85	263·17	316·33	375·74	441·67	514·39	594·16	681·26	775·92	878·40	1107·8	1371·4	1671·0
0·000060	1·743	1·786	1·870	1·952	2·032	2·111	2·189	2·265	2·340	2·414	2·558	2·698	2·834
1/ 16667	249·47	274·87	330·40	392·44	461·31	537·26	620·58	711·55	810·42	917·46	1157·1	1432·4	1745·3
0·000065	1·814	1·859	1·946	2·032	2·115	2·198	2·278	2·357	2·435	2·512	2·662	2·808	2·950
1/ 15385	259·65	286·10	343·89	408·47	480·14	559·20	645·93	740·60	843·51	954·92	1204·3	1490·9	1816·6
0·000070	1·882	1·929	2·019	2·108	2·195	2·280	2·364	2·446	2·527	2·607	2·763	2·914	3·062
1/ 14286	269·46	296·90	356·87	423·89	498·27	580·31	670·31	768·56	875·35	990·97	1249·8	1547·1	1885·2
0·000075	1·949	1·996	2·090	2·182	2·272	2·361	2·447	2·532	2·616	2·698	2·860	3·016	3·169
1/ 13333	278·91	307·32	369·39	438·77	515·76	600·68	693·84	795·54	906·08	1025·7	1293·6	1601·4	1951·3
0·000080	2·012	2·062	2·159	2·254	2·347	2·438	2·527	2·615	2·702	2·787	2·953	3·115	3·273
1/ 12500	288·06	317·40	381·51	453·16	532·67	620·38	716·59	821·63	935·79	1059·4	1336·1	1654·0	2015·3
0·000085	2·074	2·125	2·225	2·323	2·419	2·513	2·605	2·696	2·785	2·873	3·044	3·211	3·374
1/ 11765	296·93	327·17	393·25	467·10	549·06	639·47	738·64	846·91	964·59	1092·0	1377·2	1704·9	2077·4
0·000090	2·135	2·187	2·290	2·391	2·489	2·586	2·681	2·774	2·866	2·956	3·132	3·304	3·472
1/ 11111	305·53	336·65	404·65	480·64	564·98	658·01	760·06	871·47	992·56	1123·7	1417·1	1754·3	2137·6
0·000095	2·193	2·247	2·353	2·456	2·557	2·657	2·754	2·850	2·944	3·037	3·218	3·395	3·567
1/ 10526	313·91	345·88	415·74	493·82	580·46	676·04	780·89	895·35	1019·8	1154·4	1455·9	1802·4	2196·2
0·000100	2·250	2·305	2·414	2·520	2·624	2·726	2·826	2·924	3·021	3·116	3·302	3·483	3·659
1/ 10000	322·06	354·86	426·54	506·64	595·54	693·60	801·17	918·61	1046·2	1184·4	1493·8	1849·2	2253·2
0·000110	2·360	2·418	2·532	2·643	2·752	2·859	2·964	3·067	3·168	3·268	3·463	3·653	3·838
1/ 9091	337·78	372·18	447·36	531·37	624·61	727·45	840·28	963·44	1097·3	1242·2	1566·7	1939·4	2363·2
0·000120	2·465	2·525	2·644	2·760	2·874	2·986	3·095	3·203	3·309	3·413	3·617	3·815	4·009
1/ 8333	352·80	388·73	467·25	555·00	652·38	759·80	877·64	1006·3	1146·1	1297·5	1636·3	2025·7	2468·3
0·000130	2·565	2·628	2·752	2·873	2·992	3·108	3·222	3·334	3·444	3·553	3·765	3·971	4·172
1/ 7692	367·21	404·60	486·33	577·66	679·02	790·83	913·48	1047·4	1192·9	1350·5	1703·1	2108·4	2569·1
0·000140	2·662	2·728	2·856	2·982	3·104	3·225	3·343	3·460	3·574	3·687	3·907	4·121	4·330
1/ 7143	381·07	419·88	504·69	599·47	704·66	820·68	947·96	1086·9	1237·9	1401·4	1767·4	2188·0	2666·0
0·000150	2·756	2·823	2·956	3·086	3·213	3·338	3·461	3·581	3·700	3·816	4·044	4·266	4·482
1/ 6667	394·44	434·61	522·40	620·51	729·39	849·48	981·23	1125·1	1281·4	1450·6	1829·5	2264·8	2759·6
0·000160	2·846	2·916	3·053	3·187	3·319	3·448	3·574	3·699	3·821	3·941	4·177	4·406	4·629
1/ 6250	407·38	448·87	539·53	640·86	753·31	877·34	1013·4	1162·0	1323·4	1498·2	1889·5	2339·0	2850·1
0·000170	2·934	3·006	3·147	3·285	3·421	3·554	3·684	3·812	3·938	4·063	4·305	4·541	4·771
1/ 5882	419·92	462·68	556·14	660·58	776·49	904·34	1044·6	1197·7	1364·1	1544·3	1947·6	2411·0	2937·8
0·000180	3·019	3·093	3·238	3·381	3·520	3·657	3·791	3·923	4·053	4·180	4·430	4·673	4·909
1/ 5556	432·09	476·10	572·26	679·73	799·00	930·56	1074·9	1232·4	1403·7	1589·1	2004·1	2480·9	3023·0
0·000190	3·101	3·178	3·327	3·473	3·617	3·757	3·895	4·030	4·164	4·295	4·551	4·801	5·044
1/ 5263	443·93	489·14	587·94	698·36	820·90	956·06	1104·3	1266·2	1442·2	1632·6	2059·0	2548·9	3105·9
0·000200	3·182	3·260	3·414	3·564	3·711	3·855	3·996	4·135	4·272	4·406	4·670	4·926	5·175
1/ 5000	455·46	501·85	603·22	716·50	842·23	980·90	1133·0	1299·1	1479·6	1675·0	2112·5	2615·1	3186·5
0·000220	3·337	3·419	3·580	3·738	3·892	4·043	4·191	4·337	4·480	4·622	4·898	5·166	5·428
1/ 4545	477·70	526·34	632·66	751·47	883·33	1028·8	1188·3	1362·5	1551·8	1756·8	2215·6	2742·8	3342·1
0·000240	3·486	3·571	3·739	3·904	4·065	4·223	4·378	4·530	4·680	4·827	5·115	5·396	5·669
1/ 4167	498·94	549·75	660·79	784·89	922·61	1074·5	1241·2	1423·1	1620·8	1834·9	2314·1	2864·7	3490·7
0·000260	3·628	3·717	3·892	4·063	4·231	4·395	4·556	4·715	4·871	5·024	5·324	5·616	5·900
1/ 3846	519·31	572·20	687·77	816·94	960·28	1118·4	1291·9	1481·2	1687·0	1909·8	2408·6	2981·7	3633·2
0·000280	3·765	3·857	4·039	4·216	4·390	4·561	4·728	4·893	5·055	5·214	5·525	5·828	6·123
1/ 3571	538·91	593·79	713·74	847·78	996·53	1160·6	1340·6	1537·1	1750·7	1981·9	2499·5	3094·3	3770·4
	13·50	14·00	15·00	16·00	17·00	18·00	19·00	20·00	21·00	22·00	24·00	26·00	28·00

Gradient   (Equivalent) Pipe diameters in m

S = 0·000050 to 0·000280

m = Manning $n \times 100$
S = 0·00030 to 0·00140

ie hydraulic gradient =
1 in 3333 to 1 in 714

Full bore conditions; Select
m to divide mV and/or mQ.

mV to give velocities in ms^{-1}
mQ to give discharges in m^{3}s^{-1}

Gradient   (Equivalent) Pipe diameters in m

Gradient	13·50	14·00	15·00	16·00	17·00	18·00	19·00	20·00	21·00	22·00	24·00	26·00	28·00
0·00030	3·897	3·993	4·181	4·364	4·545	4·721	4·894	5·065	5·232	5·397	5·719	6·033	6·338
1/ 3333	557·83	614·64	738·79	877·53	1031·5	1201·4	1387·7	1591·1	1812·2	2051·5	2587·3	3202·9	3902·7
0·00032	4·025	4·124	4·318	4·508	4·694	4·876	5·055	5·231	5·404	5·574	5·907	6·230	6·546
1/ 3125	576·12	634·79	763·02	906·31	1065·3	1240·8	1433·2	1643·3	1871·6	2118·8	2672·1	3307·9	4030·7
0·00034	4·149	4·251	4·451	4·646	4·838	5·026	5·210	5·392	5·570	5·745	6·088	6·422	6·747
1/ 2941	593·85	654·33	786·50	934·20	1098·1	1278·9	1477·3	1693·8	1929·2	2184·0	2754·3	3409·7	4154·7
0·00036	4·269	4·374	4·580	4·781	4·978	5·172	5·361	5·548	5·731	5·912	6·265	6·608	6·943
1/ 2778	611·07	673·30	809·30	961·29	1130·0	1316·0	1520·1	1742·9	1985·1	2247·3	2834·2	3508·6	4275·2
0·00038	4·386	4·494	4·705	4·912	5·115	5·313	5·508	5·700	5·888	6·074	6·437	6·789	7·133
1/ 2632	627·81	691·75	831·48	987·63	1160·9	1352·1	1561·8	1790·7	2039·5	2308·9	2911·9	3604·7	4392·3
0·00040	4·500	4·610	4·827	5·040	5·248	5·451	5·651	5·848	6·041	6·232	6·604	6·966	7·319
1/ 2500	644·12	709·72	853·08	1013·3	1191·1	1387·2	1602·3	1837·2	2092·5	2368·9	2987·5	3698·4	4506·4
0·00042	4·611	4·724	4·947	5·164	5·377	5·586	5·791	5·992	6·191	6·386	6·767	7·138	7·499
1/ 2381	660·03	727·25	874·15	1038·3	1220·5	1421·5	1641·9	1882·6	2144·2	2427·4	3061·3	3789·7	4617·7
0·00044	4·720	4·835	5·063	5·286	5·504	5·717	5·927	6·133	6·336	6·536	6·926	7·306	7·676
1/ 2273	675·56	744·36	894·72	1062·7	1249·2	1454·9	1680·6	1926·9	2194·6	2484·5	3133·3	3878·9	4726·4
0·00046	4·826	4·944	5·177	5·404	5·627	5·846	6·060	6·271	6·479	6·683	7·082	7·470	7·848
1/ 2174	690·75	761·09	914·83	1086·6	1277·3	1487·6	1718·3	1970·2	2243·9	2540·3	3203·7	3966·0	4832·6
0·00048	4·930	5·050	5·288	5·521	5·748	5·972	6·191	6·406	6·618	6·826	7·234	7·631	8·017
1/ 2083	705·60	777·46	934·50	1110·0	1304·8	1519·6	1755·3	2012·6	2292·2	2595·0	3272·7	4051·3	4936·6
0·00050	5·031	5·155	5·397	5·635	5·867	6·095	6·318	6·538	6·754	6·967	7·383	7·788	8·182
1/ 2000	720·15	793·49	953·77	1132·9	1331·7	1550·9	1791·5	2054·1	2339·5	2648·5	3340·1	4134·9	5038·4
0·00055	5·277	5·406	5·661	5·910	6·153	6·392	6·627	6·857	7·084	7·307	7·744	8·168	8·582
1/ 1818	755·30	832·22	1000·3	1188·2	1396·7	1626·6	1878·9	2154·3	2453·7	2777·7	3503·2	4336·7	5284·3
0·00060	5·511	5·647	5·912	6·172	6·427	6·677	6·922	7·162	7·399	7·632	8·088	8·531	8·963
1/ 1667	788·89	869·23	1044·8	1241·0	1458·8	1699·0	1962·5	2250·1	2562·8	2901·3	3658·9	4529·5	5519·2
0·00065	5·736	5·877	6·154	6·424	6·689	6·949	7·204	7·455	7·701	7·944	8·418	8·880	9·329
1/ 1538	821·10	904·72	1087·5	1291·7	1518·3	1768·3	2042·6	2342·0	2667·4	3019·7	3808·3	4714·5	5744·6
0·00070	5·953	6·099	6·386	6·667	6·942	7·211	7·476	7·736	7·992	8·244	8·736	9·215	9·682
1/ 1429	852·10	938·87	1128·5	1340·5	1575·7	1835·1	2119·7	2430·4	2768·1	3133·7	3952·1	4892·5	5961·5
0·00075	6·162	6·313	6·610	6·901	7·185	7·465	7·739	8·008	8·272	8·533	9·043	9·538	10·02
1/ 1333	882·00	971·82	1168·1	1387·5	1631·0	1899·5	2194·1	2515·7	2865·3	3243·7	4090·8	5064·2	6170·7
0·00080	6·364	6·520	6·827	7·127	7·421	7·709	7·992	8·270	8·544	8·813	9·339	9·851	10·35
1/ 1250	910·93	1003·7	1206·4	1433·0	1684·5	1961·8	2266·1	2598·2	2959·2	3350·1	4225·0	5230·3	6373·1
0·00085	6·560	6·721	7·037	7·347	7·650	7·947	8·238	8·525	8·807	9·084	9·627	10·15	10·67
1/ 1176	938·96	1034·6	1243·6	1477·1	1736·3	2022·2	2335·8	2678·2	3050·3	3453·2	4355·0	5391·2	6569·2
0·00090	6·750	6·916	7·241	7·560	7·871	8·177	8·477	8·772	9·062	9·348	9·906	10·45	10·98
1/ 1111	966·19	1064·6	1279·6	1519·9	1786·6	2080·8	2403·5	2755·8	3138·7	3553·3	4481·3	5547·5	6759·7
0·00095	6·935	7·105	7·440	7·767	8·087	8·401	8·709	9·012	9·310	9·604	10·18	10·74	11·28
1/ 1053	992·66	1093·8	1314·7	1561·6	1835·6	2137·8	2469·4	2831·3	3224·7	3650·7	4604·1	5699·5	6944·9
0·00100	7·115	7·290	7·633	7·968	8·297	8·619	8·936	9·247	9·552	9·853	10·44	11·01	11·57
1/ 1000	1018·4	1122·2	1348·8	1602·1	1883·3	2193·4	2533·5	2904·9	3308·5	3745·5	4723·7	5847·6	7125·3
0·00110	7·462	7·646	8·005	8·357	8·702	9·040	9·372	9·698	10·02	10·33	10·95	11·55	12·14
1/ 909	1068·2	1176·9	1414·7	1680·3	1975·2	2300·4	2657·2	3046·7	3470·0	3928·3	4954·2	6133·0	7473·1
0·00120	7·794	7·986	8·361	8·729	9·089	9·442	9·789	10·13	10·46	10·79	11·44	12·07	12·68
1/ 833	1115·7	1229·3	1477·6	1755·1	2063·0	2402·7	2775·3	3182·1	3624·3	4103·0	5174·5	6405·7	7805·4
0·00130	8·112	8·312	8·703	9·085	9·460	9·828	10·19	10·54	10·89	11·23	11·91	12·56	13·19
1/ 769	1161·2	1279·5	1537·9	1826·7	2147·3	2500·8	2888·7	3312·1	3772·3	4270·5	5385·8	6667·3	8124·1
0·00140	8·419	8·625	9·031	9·428	9·817	10·20	10·57	10·94	11·30	11·66	12·35	13·03	13·69
1/ 714	1205·0	1327·8	1596·0	1895·7	2228·3	2595·2	2997·7	3437·1	3914·7	4431·7	5589·1	6919·0	8430·8
	13·50	14·00	15·00	16·00	17·00	18·00	19·00	20·00	21·00	22·00	24·00	26·00	28·00

Gradient   (Equivalent) Pipe diameters in m

S = 0·00030 to 0·00140

m = Manning $n \times 100$ Full bore conditions; Select   **D11**
S = 0.00150 to 0.00700 m to divide mV and/or mQ.   continued

ie hydraulic gradient = mV to give velocities in ms^{-1}
1 in 667 to 1 in 143 mQ to give discharges in m^{3}s^{-1}

Gradient   (Equivalent) Pipe diameters in m

	13·50	14·00	15·00	16·00	17·00	18·00	19·00	20·00	21·00	22·00	24·00	26·00	28·00
0·00150	8·714	8·928	9·348	9·759	10·16	10·56	10·94	11·32	11·70	12·07	12·79	13·49	14·17
1/ 667	1247·3	1374·4	1652·0	1962·2	2306·5	2686·3	3102·9	3557·7	4052·1	4587·3	5785·3	7161·8	8726·7
0·00160	9·000	9·221	9·655	10·08	10·50	10·90	11·30	11·70	12·08	12·46	13·21	13·93	14·64
1/ 625	1288·2	1419·4	1706·2	2026·6	2382·2	2774·4	3204·7	3674·4	4185·0	4737·7	5975·0	7396·7	9012·9
0·00170	9·277	9·505	9·952	10·39	10·82	11·24	11·65	12·06	12·45	12·85	13·61	14·36	15·09
1/ 588	1327·9	1463·1	1758·7	2088·9	2455·5	2859·8	3303·3	3787·5	4313·8	4883·5	6158·9	7624·3	9290·3
0·00180	9·546	9·780	10·24	10·69	11·13	11·56	11·99	12·41	12·82	13·22	14·01	14·78	15·53
1/ 556	1366·4	1505·5	1809·7	2149·5	2526·7	2942·7	3399·1	3897·3	4438·8	5025·1	6337·5	7845·4	9559·6
0·00190	9·808	10·05	10·52	10·98	11·44	11·88	12·32	12·75	13·17	13·58	14·39	15·18	15·95
1/ 526	1403·8	1546·8	1859·2	2208·4	2595·9	3023·3	3492·2	4004·1	4560·5	5162·8	6511·1	8060·4	9821·6
0·00200	10·06	10·31	10·79	11·27	11·73	12·19	12·64	13·08	13·51	13·93	14·77	15·58	16·36
1/ 500	1440·3	1587·0	1907·5	2265·8	2663·3	3101·9	3582·9	4108·1	4679·0	5296·9	6680·3	8269·8	10077
0·00220	10·55	10·81	11·32	11·82	12·31	12·78	13·25	13·71	14·17	14·61	15·49	16·34	17·16
1/ 455	1510·6	1664·4	2000·6	2376·4	2793·3	3253·3	3757·8	4308·6	4907·3	5555·5	7006·3	8673·4	10569
0·00240	11·02	11·29	11·82	12·34	12·85	13·35	13·84	14·32	14·80	15·26	16·18	17·06	17·93
1/ 417	1577·8	1738·5	2089·6	2482·0	2917·6	3397·9	3924·9	4500·2	5125·5	5802·5	7317·9	9059·1	11038
0·00260	11·47	11·75	12·31	12·85	13·38	13·90	14·41	14·91	15·40	15·89	16·84	17·76	18·66
1/ 385	1642·2	1809·4	2174·9	2583·4	3036·7	3536·7	4085·2	4684·0	5334·8	6039·4	7616·7	9429·0	11489
0·00280	11·91	12·20	12·77	13·33	13·88	14·42	14·95	15·47	15·98	16·49	17·47	18·43	19·36
1/ 357	1704·2	1877·7	2257·0	2680·9	3151·3	3670·2	4239·4	4860·8	5536·2	6267·4	7904·2	9784·9	11923
0·00300	12·32	12·63	13·22	13·80	14·37	14·93	15·48	16·02	16·54	17·07	18·09	19·08	20·04
1/ 333	1764·0	1943·6	2336·3	2775·0	3261·9	3799·0	4388·2	5031·4	5730·5	6487·4	8181·6	10128	12341
0·00320	12·73	13·04	13·65	14·25	14·84	15·42	15·98	16·54	17·09	17·63	18·68	19·70	20·70
1/ 313	1821·9	2007·4	2412·9	2866·0	3368·9	3923·6	4532·1	5196·4	5918·5	6700·2	8450·0	10461	12746
0·00340	13·12	13·44	14·07	14·69	15·30	15·89	16·48	17·05	17·61	18·17	19·25	20·31	21·34
1/ 294	1877·9	2069·2	2487·1	2954·2	3472·6	4044·4	4671·6	5356·3	6100·6	6906·4	8710·0	10782	13138
0·00360	13·50	13·83	14·48	15·12	15·74	16·35	16·95	17·54	18·12	18·70	19·81	20·90	21·96
1/ 278	1932·4	2129·2	2559·2	3039·9	3573·3	4161·6	4807·0	5511·6	6277·5	7106·6	8962·5	11095	13519
0·00380	13·87	14·21	14·88	15·53	16·17	16·80	17·42	18·02	18·62	19·21	20·35	21·47	22·56
1/ 263	1985·3	2187·5	2629·4	3123·2	3671·2	4275·6	4938·8	5662·7	6449·5	7301·3	9208·1	11399	13890
0·00400	14·23	14·58	15·27	15·94	16·59	17·24	17·87	18·49	19·10	19·71	20·88	22·03	23·14
1/ 250	2036·9	2244·3	2697·7	3204·3	3766·5	4386·7	5067·1	5809·8	6617·0	7491·0	9447·3	11695	14251
0·00420	14·58	14·94	15·64	16·33	17·00	17·66	18·31	18·95	19·58	20·19	21·40	22·57	23·71
1/ 238	2087·2	2299·8	2764·3	3283·4	3859·6	4495·0	5192·2	5953·2	6780·5	7676·0	9680·6	11984	14603
0·00440	14·92	15·29	16·01	16·71	17·40	18·08	18·74	19·40	20·04	20·67	21·90	23·10	24·27
1/ 227	2136·3	2353·9	2829·3	3360·7	3950·4	4600·8	5314·4	6093·3	6940·0	7856·6	9908·4	12266	14946
0·00460	15·26	15·63	16·37	17·09	17·80	18·49	19·16	19·83	20·49	21·13	22·39	23·62	24·82
1/ 217	2184·3	2406·8	2892·9	3436·2	4039·2	4704·2	5433·8	6230·3	7096·0	8033·2	10131	12542	15282
0·00480	15·59	15·97	16·72	17·46	18·18	18·88	19·58	20·26	20·93	21·59	22·88	24·13	25·35
1/ 208	2231·3	2458·5	2955·2	3510·1	4126·0	4805·4	5550·7	6364·3	7248·6	8206·0	10349	12811	15611
0·00500	15·91	16·30	17·07	17·82	18·55	19·27	19·98	20·68	21·36	22·03	23·35	24·63	25·88
1/ 200	2277·3	2509·2	3016·1	3582·5	4211·1	4904·5	5665·1	6495·5	7398·1	8375·2	10562	13076	15933
0·00550	16·69	17·10	17·90	18·69	19·46	20·21	20·96	21·69	22·40	23·11	24·49	25·83	27·14
1/ 182	2388·5	2631·7	3163·3	3757·4	4416·7	5143·9	5941·6	6812·6	7759·2	8784·0	11078	13714	16710
0·00600	17·43	17·86	18·70	19·52	20·32	21·11	21·89	22·65	23·40	24·14	25·58	26·98	28·34
1/ 167	2494·7	2748·7	3304·0	3924·4	4613·1	5372·6	6205·8	7115·5	8104·2	9174·6	11571	14324	17453
0·00650	18·14	18·59	19·46	20·32	21·15	21·98	22·78	23·57	24·35	25·12	26·62	28·08	29·50
1/ 154	2596·5	2861·0	3438·9	4084·7	4801·4	5592·0	6459·3	7406·0	8435·1	9549·2	12043	14909	18166
0·00700	18·82	19·29	20·19	21·08	21·95	22·80	23·64	24·46	25·27	26·07	27·63	29·14	30·62
1/ 143	2694·6	2969·0	3568·7	4238·9	4982·7	5803·1	6703·1	7685·6	8753·5	9909·7	12498	15471	18852
	13·50	14·00	15·00	16·00	17·00	18·00	19·00	20·00	21·00	22·00	24·00	26·00	28·00

Gradient   (Equivalent) Pipe diameters in m

S = 0.00150 to 0.00700

m = Manning $n \times 100$
S = 0.07500 to 0.04000

Full bore conditions; Select m to divide mV and/or mQ.

ie hydraulic gradient = 1 in 133 to 1 in 25.0

mV to give velocities in ms^{-1}
mQ to give discharges in m^{3}s^{-1}

Gradient   (Equivalent) Pipe diameters in m

Gradient	13.50	14.00	15.00	16.00	17.00	18.00	19.00	20.00	21.00	22.00	24.00	26.00	28.00
0.00750	19.49	19.96	20.90	21.82	22.72	23.61	24.47	25.32	26.16	26.98	28.60	30.16	31.69
1/ 133	2789.1	3073.2	3693.9	4387.7	5157.6	6006.8	6938.4	7955.4	9060.8	10257	12936	16014	19513
0.00800	20.12	20.62	21.59	22.54	23.47	24.38	25.27	26.15	27.02	27.87	29.53	31.15	32.73
1/ 125	2880.6	3174.0	3815.1	4531.6	5326.7	6203.8	7165.9	8216.3	9357.9	10594	13361	16540	20153
0.00850	20.74	21.25	22.25	23.23	24.19	25.13	26.05	26.96	27.85	28.73	30.44	32.11	33.74
1/ 118	2969.3	3271.7	3932.5	4671.0	5490.6	6394.7	7386.4	8469.1	9645.9	10920	13772	17049	20774
0.00900	21.35	21.87	22.90	23.91	24.89	25.86	26.81	27.74	28.66	29.56	31.32	33.04	34.72
1/ 111	3055.3	3366.5	4046.5	4806.4	5649.8	6580.1	7600.6	8714.7	9925.6	11237	14171	17543	21376
0.00950	21.93	22.47	23.53	24.56	25.57	26.57	27.54	28.50	29.44	30.37	32.18	33.95	35.67
1/ 105	3139.1	3458.8	4157.4	4938.2	5804.6	6760.4	7808.9	8953.5	10198	11544	14559	18024	21962
0.01000	22.50	23.05	24.14	25.20	26.24	27.26	28.26	29.24	30.21	31.16	33.02	34.83	36.59
1/ 100	3220.6	3548.6	4265.4	5066.4	5955.4	6936.0	8011.7	9186.1	10462	11844	14938	18492	22532
0.01100	23.60	24.18	25.32	26.43	27.52	28.59	29.64	30.67	31.68	32.68	34.63	36.53	38.38
1/ 91	3377.8	3721.8	4473.6	5313.7	6246.1	7274.5	8402.8	9634.4	10973	12422	15667	19394	23632
0.01200	24.65	25.25	26.44	27.60	28.74	29.86	30.95	32.03	33.09	34.13	36.17	38.15	40.09
1/ 83	3528.0	3887.3	4672.5	5550.0	6523.8	7598.0	8776.4	10063	11461	12975	16363	20257	24683
0.01300	25.65	26.28	27.52	28.73	29.92	31.08	32.22	33.34	34.44	35.53	37.65	39.71	41.72
1/ 77	3672.1	4046.0	4863.3	5776.6	6790.2	7908.3	9134.8	10474	11929	13505	17031	21084	25691
0.01400	26.62	27.28	28.56	29.82	31.04	32.25	33.43	34.60	35.74	36.87	39.07	41.21	43.30
1/ 71	3810.7	4198.8	5046.9	5994.7	7046.6	8206.8	9479.6	10869	12379	14014	17674	21880	26660
0.01500	27.56	28.23	29.56	30.86	32.13	33.38	34.61	35.81	37.00	38.16	40.44	42.66	44.82
1/ 67	3944.4	4346.1	5224.0	6205.1	7293.9	8494.8	9812.3	11251	12814	14506	18295	22648	27596
0.01600	28.46	29.16	30.53	31.87	33.19	34.48	35.74	36.99	38.21	39.41	41.77	44.06	46.29
1/ 63	4073.8	4488.7	5395.3	6408.6	7533.1	8773.4	10134	11620	13234	14982	18895	23390	28501
0.01700	29.34	30.06	31.47	32.85	34.21	35.54	36.84	38.12	39.38	40.63	43.05	45.41	47.71
1/ 59	4199.2	4626.8	5561.4	6605.8	7764.9	9043.4	10446	11977	13641	15443	19476	24110	29378
0.01800	30.19	30.93	32.38	33.81	35.20	36.57	37.91	39.23	40.53	41.80	44.30	46.73	49.09
1/ 56	4320.9	4761.0	5722.6	6797.3	7990.0	9305.6	10749	12324	14037	15891	20041	24809	30230
0.01900	31.01	31.78	33.27	34.73	36.17	37.57	38.95	40.30	41.64	42.95	45.51	48.01	50.44
1/ 53	4439.3	4891.4	5879.4	6983.6	8209.0	9560.6	11043	12662	14422	16326	20590	25489	31059
0.02000	31.82	32.60	34.14	35.64	37.11	38.55	39.96	41.35	42.72	44.06	46.70	49.26	51.75
1/ 50	4554.6	5018.5	6032.2	7165.0	8422.3	9809.0	11330	12991	14796	16750	21125	26151	31865
0.02200	33.37	34.19	35.80	37.38	38.92	40.43	41.91	43.37	44.80	46.22	48.98	51.66	54.28
1/ 45.5	4777.0	5263.4	6326.6	7514.7	8833.3	10288	11883	13625	15518	17568	22156	27428	33421
0.02400	34.86	35.71	37.39	39.04	40.65	42.23	43.78	45.30	46.80	48.27	51.15	53.96	56.69
1/ 41.7	4989.4	5497.5	6607.9	7848.9	9226.1	10745	12412	14231	16208	18349	23141	28647	34907
0.02600	36.28	37.17	38.92	40.63	42.31	43.95	45.56	47.15	48.71	50.24	53.24	56.16	59.00
1/ 38.5	5193.1	5722.0	6877.7	8169.4	9602.8	11184	12919	14812	16870	19098	24086	29817	36332
0.02800	37.65	38.57	40.39	42.16	43.90	45.61	47.28	48.93	50.55	52.14	55.25	58.28	61.23
1/ 35.7	5389.1	5937.9	7137.4	8477.8	9965.3	11606	13406	15371	17507	19819	24995	30943	37704
0.03000	38.97	39.93	41.81	43.64	45.45	47.21	48.94	50.65	52.32	53.97	57.19	60.33	63.38
1/ 33.3	5578.3	6146.4	7387.9	8775.3	10315	12014	13877	15911	18122	20515	25873	32029	39027
0.03200	40.25	41.24	43.18	45.08	46.94	48.76	50.55	52.31	54.04	55.74	59.07	62.30	65.46
1/ 31.3	5761.2	6347.9	7630.2	9063.1	10653	12408	14332	16433	18716	21188	26721	33079	40307
0.03400	41.49	42.51	44.51	46.46	48.38	50.26	52.10	53.92	55.70	57.45	60.88	64.22	67.47
1/ 29.4	5938.5	6543.3	7865.0	9342.0	10981	12789	14773	16938	19292	21840	27543	34097	41547
0.03600	42.69	43.74	45.80	47.81	49.78	51.72	53.61	55.48	57.31	59.12	62.65	66.08	69.43
1/ 27.8	6110.7	6733.0	8093.0	9612.9	11300	13160	15201	17429	19851	22473	28342	35086	42752
0.03800	43.86	44.94	47.05	49.12	51.15	53.13	55.08	57.00	58.88	60.74	64.37	67.89	71.33
1/ 26.3	6278.1	6917.5	8314.8	9876.3	11609	13521	15618	17907	20395	23089	29119	36047	43923
0.04000	45.00	46.10	48.27	50.40	52.48	54.51	56.51	58.48	60.41	62.32	66.04	69.66	73.19
1/ 25.0	6441.2	7097.2	8530.8	10133	11911	13872	16023	18372	20925	23689	29875	36984	45064
	13.50	14.00	15.00	16.00	17.00	18.00	19.00	20.00	21.00	22.00	24.00	26.00	28.00

Gradient   (Equivalent) Pipe diameters in m

S = 0.07500 to 0.04000

m = Manning $n \times 100$
S = 0·000010 to 0·000048

Full bore conditions; Select
m to divide mV and/or mQ.

D12

ie hydraulic gradient =
1 in 100000 to 1 in 20833

mV to give velocities in ms^{-1}
mQ to give discharges in m^{3}s^{-1}

Gradient   (Equivalent) Pipe diameters in m

Gradient	28.00	30.00	32.00	34.00	36.00	38.00	40.00	42.50	45.00	47.50	50.00	55.00	60.00
0·000010	1·157	1·212	1·265	1·317	1·368	1·418	1·468	1·528	1·588	1·646	1·703	1·815	1·923
1/100000	712·53	856·46	1017·3	1195·8	1392·7	1608·7	1844·5	2168·1	2525·1	2916·7	3344·3	4312·0	5438·2
0·000011	1·214	1·271	1·327	1·381	1·435	1·488	1·539	1·603	1·665	1·726	1·786	1·904	2·017
1/ 90909	747·31	898·26	1067·0	1254·2	1460·7	1687·2	1934·5	2274·0	2648·4	3059·1	3507·5	4522·5	5703·6
0·000012	1·268	1·327	1·386	1·443	1·499	1·554	1·608	1·674	1·739	1·803	1·866	1·988	2·107
1/ 83333	780·54	938·20	1114·4	1309·9	1525·6	1762·2	2020·5	2375·1	2766·1	3195·1	3663·5	4723·6	5957·2
0·000013	1·319	1·381	1·442	1·502	1·560	1·617	1·674	1·743	1·810	1·877	1·942	2·069	2·193
1/ 76923	812·41	976·51	1159·9	1363·4	1587·9	1834·2	2103·0	2472·1	2879·1	3325·6	3813·1	4916·5	6200·5
0·000014	1·369	1·434	1·497	1·558	1·619	1·678	1·737	1·808	1·879	1·948	2·015	2·147	2·276
1/ 71429	843·08	1013·4	1203·7	1414·9	1647·9	1903·4	2182·4	2565·4	2987·8	3451·1	3957·0	5102·1	6434·5
0·000015	1·417	1·484	1·549	1·613	1·676	1·737	1·798	1·872	1·945	2·016	2·086	2·223	2·356
1/ 66667	872·67	1048·9	1245·9	1464·6	1705·7	1970·2	2259·0	2655·4	3092·6	3572·3	4095·9	5281·2	6660·4
0·000016	1·464	1·533	1·600	1·666	1·731	1·794	1·857	1·933	2·008	2·082	2·154	2·296	2·433
1/ 62500	901·29	1083·3	1286·8	1512·6	1761·6	2034·8	2333·1	2742·5	3194·1	3689·4	4230·2	5454·4	6878·8
0·000017	1·509	1·580	1·649	1·717	1·784	1·849	1·914	1·993	2·070	2·146	2·221	2·366	2·508
1/ 58824	929·03	1116·7	1326·4	1559·1	1815·9	2097·5	2404·9	2826·9	3292·4	3803·0	4360·4	5622·2	7090·5
0·000018	1·553	1·626	1·697	1·767	1·836	1·903	1·969	2·050	2·130	2·208	2·285	2·435	2·580
1/ 55556	955·96	1149·1	1364·9	1604·3	1868·5	2158·3	2474·6	2908·9	3387·8	3913·2	4486·8	5785·2	7296·1
0·000019	1·595	1·670	1·744	1·815	1·886	1·955	2·023	2·107	2·188	2·269	2·348	2·502	2·651
1/ 52632	982·16	1180·5	1402·3	1648·3	1919·7	2217·4	2542·5	2988·6	3480·6	4020·5	4609·8	5943·7	7496·0
0·000020	1·636	1·714	1·789	1·863	1·935	2·006	2·076	2·161	2·245	2·328	2·409	2·567	2·720
1/ 50000	1007·7	1211·2	1438·7	1691·1	1969·6	2275·0	2608·5	3066·2	3571·1	4124·9	4729·5	6098·2	7690·7
0·000022	1·716	1·797	1·876	1·954	2·029	2·104	2·177	2·267	2·355	2·441	2·526	2·692	2·853
1/ 45455	1056·9	1270·3	1508·9	1773·7	2065·7	2386·1	2735·8	3215·9	3745·4	4326·2	4960·4	6395·8	8066·1
0·000024	1·793	1·877	1·960	2·040	2·120	2·197	2·274	2·368	2·460	2·550	2·639	2·812	2·980
1/ 41667	1103·8	1326·8	1576·0	1852·5	2157·6	2492·2	2857·5	3358·9	3911·9	4518·6	5180·9	6680·2	8424·8
0·000026	1·866	1·954	2·040	2·124	2·206	2·287	2·367	2·464	2·560	2·654	2·746	2·927	3·101
1/ 38462	1148·9	1381·0	1640·3	1928·2	2245·7	2593·9	2974·1	3496·0	4071·6	4703·1	5392·5	6953·0	8768·8
0·000028	1·936	2·027	2·117	2·204	2·289	2·374	2·456	2·557	2·657	2·754	2·850	3·037	3·218
1/ 35714	1192·3	1433·1	1702·3	2001·0	2330·4	2691·9	3086·4	3628·0	4225·3	4880·7	5596·1	7215·4	9099·8
0·000030	2·004	2·099	2·191	2·281	2·370	2·457	2·542	2·647	2·750	2·851	2·950	3·144	3·331
1/ 33333	1234·1	1483·4	1762·0	2071·2	2412·2	2786·3	3194·7	3755·3	4373·6	5052·0	5792·5	7468·7	9419·2
0·000032	2·070	2·167	2·263	2·356	2·448	2·537	2·626	2·734	2·840	2·944	3·047	3·247	3·441
1/ 31250	1274·6	1532·1	1819·8	2139·1	2491·3	2877·7	3299·5	3878·5	4517·1	5217·6	5982·4	7713·6	9728·1
0·000034	2·134	2·234	2·332	2·429	2·523	2·616	2·706	2·818	2·928	3·035	3·141	3·347	3·547
1/ 29412	1313·8	1579·2	1875·8	2205·0	2568·0	2966·3	3401·1	3997·8	4656·1	5378·2	6166·5	7951·0	10027
0·000036	2·196	2·299	2·400	2·499	2·596	2·691	2·785	2·900	3·012	3·123	3·232	3·444	3·649
1/ 27778	1351·9	1625·0	1930·2	2268·9	2642·5	3052·3	3499·7	4113·7	4791·1	5534·1	6345·3	8181·5	10318
0·000038	2·256	2·362	2·466	2·567	2·667	2·765	2·861	2·979	3·095	3·209	3·320	3·538	3·749
1/ 26316	1389·0	1669·5	1983·1	2331·1	2714·9	3135·9	3595·6	4226·5	4922·4	5685·8	6519·2	8405·7	10601
0·000040	2·314	2·423	2·530	2·634	2·736	2·837	2·936	3·057	3·175	3·292	3·406	3·630	3·847
1/ 25000	1425·1	1712·9	2034·6	2391·6	2785·4	3217·4	3689·0	4336·3	5050·2	5833·5	6688·6	8624·1	10876
0·000042	2·371	2·483	2·592	2·699	2·804	2·907	3·008	3·132	3·254	3·373	3·491	3·720	3·942
1/ 23810	1460·3	1755·2	2084·8	2450·7	2854·2	3296·8	3780·1	4443·4	5175·0	5977·6	6853·7	8837·1	11145
0·000044	2·427	2·542	2·653	2·763	2·870	2·975	3·079	3·206	3·330	3·453	3·573	3·807	4·034
1/ 22727	1494·6	1796·5	2133·9	2508·3	2921·3	3374·4	3869·0	4547·9	5296·7	6118·2	7015·0	9045·0	11407
0·000046	2·482	2·599	2·713	2·825	2·935	3·042	3·148	3·278	3·405	3·530	3·653	3·893	4·125
1/ 21739	1528·2	1836·9	2181·9	2564·7	2987·0	3450·3	3956·0	4650·1	5415·8	6255·7	7172·7	9248·3	11664
0·000048	2·535	2·655	2·771	2·886	2·998	3·108	3·216	3·348	3·478	3·606	3·732	3·976	4·214
1/ 20833	1561·1	1876·4	2228·8	2619·9	3051·2	3524·5	4041·1	4750·1	5532·3	6390·3	7327·0	9447·2	11914
	28·00	30·00	32·00	34·00	36·00	38·00	40·00	42·50	45·00	47·50	50·00	55·00	60·00

Gradient   (Equivalent) Pipe diameters in m

S = 0·000010 to 0·000048

D12
continued

m = Manning $n \times 100$
S = 0.000050 to 0.000280

ie hydraulic gradient =
1 in 20000 to 1 in 3571

Full bore conditions; Select
m to divide mV and/or mQ.

mV to give velocities in ms^{-1}
mQ to give discharges in m^{3}s^{-1}

Gradient	(Equivalent) Pipe diameters in m												
	28.00	30.00	32.00	34.00	36.00	38.00	40.00	42.50	45.00	47.50	50.00	55.00	60.00
0.000050 1/20000	2.588	2.709	2.828	2.945	3.059	3.172	3.282	3.417	3.550	3.681	3.809	4.058	4.301
	1593.3	1915.1	2274.8	2673.9	3114.2	3597.1	4124.4	4848.1	5646.3	6522.0	7478.0	9642.0	12160
0.000055 1/18182	2.714	2.842	2.966	3.089	3.209	3.327	3.442	3.584	3.723	3.860	3.994	4.256	4.511
	1671.0	2008.6	2385.8	2804.4	3266.2	3772.7	4325.7	5084.7	5921.9	6840.4	7843.0	10113	12754
0.000060 1/16667	2.834	2.968	3.098	3.226	3.351	3.474	3.595	3.744	3.889	4.032	4.172	4.446	4.711
	1745.3	2097.9	2491.9	2929.1	3411.4	3940.5	4518.1	5310.8	6185.3	7144.5	8191.8	10562	13321
0.000065 1/15385	2.950	3.089	3.225	3.358	3.488	3.616	3.742	3.897	4.048	4.196	4.342	4.627	4.904
	1816.6	2183.5	2593.6	3048.7	3550.7	4101.4	4702.5	5527.7	6437.8	7436.3	8526.3	10994	13865
0.000070 1/14286	3.062	3.206	3.347	3.485	3.620	3.753	3.883	4.044	4.201	4.355	4.506	4.802	5.089
	1885.2	2266.0	2691.5	3163.8	3684.7	4256.2	4880.1	5736.3	6680.9	7717.0	8848.1	11409	14388
0.000075 1/13333	3.169	3.318	3.464	3.607	3.747	3.885	4.020	4.186	4.348	4.508	4.664	4.970	5.267
	1951.3	2345.5	2786.0	3274.8	3814.1	4405.6	5051.3	5937.7	6915.3	7987.8	9158.7	11809	14893
0.000080 1/12500	3.273	3.427	3.578	3.725	3.870	4.012	4.152	4.323	4.491	4.656	4.817	5.133	5.440
	2015.3	2422.4	2877.4	3382.2	3939.1	4550.1	5217.0	6132.4	7142.1	8249.8	9459.1	12196	15381
0.000085 1/11765	3.374	3.533	3.688	3.840	3.989	4.135	4.279	4.456	4.629	4.799	4.966	5.291	5.608
	2077.4	2497.0	2965.9	3486.3	4060.4	4690.1	5377.6	6321.1	7361.9	8503.7	9750.2	12572	15855
0.000090 1/11111	3.472	3.635	3.795	3.951	4.105	4.255	4.403	4.585	4.763	4.938	5.110	5.445	5.770
	2137.6	2569.4	3051.9	3587.4	4178.1	4826.1	5533.5	6504.4	7575.4	8750.2	10033	12936	16315
0.000095 1/10526	3.567	3.735	3.899	4.060	4.217	4.372	4.524	4.711	4.894	5.073	5.250	5.594	5.928
	2196.2	2639.8	3135.5	3685.7	4292.6	4958.3	5685.1	6682.6	7783.0	8990.0	10308	13291	16762
0.000100 1/10000	3.659	3.832	4.000	4.165	4.327	4.486	4.642	4.833	5.021	5.205	5.386	5.739	6.082
	2253.2	2708.4	3217.0	3781.5	4404.1	5087.1	5832.8	6856.2	7985.1	9223.6	10576	13636	17197
0.000110 1/9091	3.838	4.019	4.195	4.368	4.538	4.704	4.868	5.069	5.266	5.459	5.649	6.020	6.379
	2363.2	2840.5	3374.0	3966.0	4619.0	5335.4	6117.5	7190.9	8374.9	9673.8	11092	14301	18036
0.000120 1/8333	4.009	4.197	4.382	4.563	4.740	4.914	5.085	5.294	5.500	5.702	5.900	6.287	6.663
	2468.3	2966.9	3524.0	4142.4	4824.4	5572.7	6389.5	7510.6	8747.3	10104	11585	14937	18838
0.000130 1/7692	4.172	4.369	4.561	4.749	4.933	5.114	5.292	5.510	5.725	5.935	6.141	6.544	6.935
	2569.1	3088.0	3667.9	4311.5	5021.4	5800.2	6650.4	7817.3	9104.5	10516	12058	15547	19608
0.000140 1/7143	4.330	4.534	4.733	4.928	5.119	5.307	5.492	5.719	5.941	6.159	6.373	6.791	7.197
	2666.0	3204.6	3806.4	4474.3	5211.0	6019.2	6901.4	8112.4	9448.1	10913	12513	16134	20348
0.000150 1/6667	4.482	4.693	4.899	5.101	5.299	5.494	5.685	5.919	6.149	6.375	6.597	7.029	7.449
	2759.6	3317.0	3940.0	4631.3	5393.9	6230.4	7143.7	8397.2	9779.8	11297	12952	16700	21062
0.000160 1/6250	4.629	4.847	5.060	5.268	5.473	5.674	5.871	6.113	6.351	6.584	6.813	7.260	7.693
	2850.1	3425.8	4069.2	4783.2	5570.8	6434.8	7378.0	8672.5	10100	11667	13377	17248	21753
0.000170 1/5882	4.771	4.996	5.215	5.430	5.641	5.848	6.052	6.301	6.546	6.787	7.023	7.483	7.930
	2937.8	3531.3	4194.4	4930.4	5742.2	6632.8	7605.0	8939.5	10411	12026	13789	17779	22422
0.000180 1/5556	4.909	5.141	5.367	5.588	5.805	6.018	6.227	6.484	6.736	6.983	7.226	7.700	8.160
	3023.0	3633.6	4316.0	5073.4	5908.7	6825.1	7825.5	9198.6	10713	12375	14189	18294	23072
0.000190 1/5263	5.044	5.281	5.514	5.741	5.964	6.183	6.398	6.662	6.921	7.175	7.424	7.911	8.384
	3105.9	3733.2	4434.3	5212.4	6070.6	7012.1	8039.9	9450.7	11007	12714	14577	18796	23704
0.000200 1/5000	5.175	5.419	5.657	5.890	6.119	6.344	6.564	6.835	7.100	7.361	7.617	8.117	8.602
	3186.5	3830.2	4549.5	5347.8	6228.3	7194.3	8248.8	9696.2	11293	13044	14956	19284	24320
0.000220 1/4545	5.428	5.683	5.933	6.178	6.418	6.653	6.885	7.169	7.447	7.720	7.989	8.513	9.021
	3342.1	4017.1	4771.6	5608.8	6532.3	7545.4	8651.4	10169	11844	13681	15686	20225	25507
0.000240 1/4167	5.669	5.936	6.197	6.452	6.703	6.949	7.191	7.487	7.778	8.064	8.344	8.891	9.423
	3490.7	4195.8	4983.7	5858.2	6822.8	7880.9	9036.1	10622	12371	14289	16384	21125	26641
0.000260 1/3846	5.900	6.178	6.450	6.716	6.977	7.233	7.484	7.793	8.096	8.393	8.685	9.255	9.807
	3633.2	4367.1	5187.2	6097.4	7101.4	8202.7	9405.1	11055	12876	14873	17053	21987	27729
0.000280 1/3571	6.123	6.411	6.693	6.969	7.240	7.506	7.767	8.087	8.401	8.710	9.013	9.604	10.18
	3770.4	4531.9	5383.0	6327.6	7369.4	8512.4	9760.1	11473	13362	15434	17696	22817	28776
	28.00	30.00	32.00	34.00	36.00	38.00	40.00	42.50	45.00	47.50	50.00	55.00	60.00

Gradient   (Equivalent) Pipe diameters in m

S = 0.000050 to 0.000280

m = Manning $n \times 100$
S = 0·00030 to 0·00140

ie hydraulic gradient =
1 in 3333   to   1 in 714

Full bore conditions; Select m to divide mV and/or mQ.

mV to give velocities in ms^{-1}
mQ to give discharges in m^3s^{-1}

Gradient	(Equivalent) Pipe diameters in m												
	28·00	30·00	32·00	34·00	36·00	38·00	40·00	42·50	45·00	47·50	50·00	55·00	60·00
0·00030	6·338	6·636	6·928	7·214	7·494	7·769	8·039	8·371	8·696	9·015	9·329	9·941	10·53
1/ 3333	3902·7	4691·0	5572·0	6549·7	7628·1	8811·2	10103	11875	13831	15976	18317	23618	29786
0·00032	6·546	6·854	7·155	7·451	7·740	8·024	8·303	8·646	8·981	9·311	9·635	10·27	10·88
1/ 3125	4030·7	4844·9	5754·7	6764·5	7878·3	9100·1	10434	12265	14284	16500	18918	24393	30763
0·00034	6·747	7·065	7·376	7·680	7·978	8·271	8·559	8·912	9·258	9·598	9·931	10·58	11·22
1/ 2941	4154·7	4994·0	5931·8	6972·7	8120·7	9380·2	10755	12642	14724	17007	19500	25143	31710
0·00036	6·943	7·270	7·589	7·902	8·209	8·511	8·807	9·170	9·526	9·876	10·22	10·89	11·54
1/ 2778	4275·2	5138·7	6103·8	7174·8	8356·2	9652·1	11067	13009	15151	17500	20066	25872	32629
0·00038	7·133	7·469	7·797	8·119	8·434	8·744	9·048	9·421	9·787	10·15	10·50	11·19	11·86
1/ 2632	4392·3	5279·6	6271·1	7371·4	8585·1	9916·6	11370	13365	15566	17980	20616	26581	33523
0·00040	7·319	7·663	8·000	8·330	8·653	8·971	9·283	9·666	10·04	10·41	10·77	11·48	12·16
1/ 2500	4506·4	5416·7	6434·0	7562·9	8808·2	10174	11666	13712	15970	18447	21151	27272	34394
0·00042	7·499	7·852	8·198	8·536	8·867	9·193	9·512	9·905	10·29	10·67	11·04	11·76	12·46
1/ 2381	4617·7	5550·5	6592·9	7749·7	9025·7	10426	11954	14051	16365	18903	21673	27945	35243
0·00044	7·676	8·037	8·390	8·737	9·076	9·409	9·736	10·14	10·53	10·92	11·30	12·04	12·76
1/ 2273	4726·4	5681·1	6748·0	7932·1	9238·1	10671	12235	14382	16750	19348	22183	28603	36073
0·00046	7·848	8·218	8·579	8·933	9·280	9·620	9·955	10·37	10·77	11·16	11·55	12·31	13·04
1/ 2174	4832·6	5808·8	6899·7	8110·3	9445·7	10911	12510	14705	17126	19782	22682	29246	36883
0·00048	8·017	8·394	8·764	9·125	9·479	9·827	10·17	10·59	11·00	11·40	11·80	12·57	13·33
1/ 2083	4936·6	5933·7	7048·1	8284·8	9648·9	11145	12779	15021	17495	20208	23170	29875	37677
0·00050	8·182	8·568	8·944	9·313	9·675	10·03	10·38	10·81	11·23	11·64	12·04	12·83	13·60
1/ 2000	5038·4	6056·1	7193·4	8455·6	9847·8	11375	13043	15331	17855	20625	23648	30491	38454
0·00055	8·582	8·986	9·381	9·768	10·15	10·52	10·89	11·33	11·77	12·21	12·63	13·46	14·26
1/ 1818	5284·3	6351·7	7544·5	8868·3	10329	11930	13679	16079	18727	21631	24802	31979	40331
0·00060	8·963	9·385	9·798	10·20	10·60	10·99	11·37	11·84	12·30	12·75	13·19	14·06	14·90
1/ 1667	5519·2	6634·1	7880·0	9262·7	10788	12461	14287	16794	19560	22593	25905	33401	42124
0·00065	9·329	9·769	10·20	10·62	11·03	11·44	11·83	12·32	12·80	13·27	13·73	14·63	15·51
1/ 1538	5744·6	6905·0	8201·7	9640·9	11228	12970	14871	17480	20358	23516	26962	34765	43844
0·00070	9·682	10·14	10·58	11·02	11·45	11·87	12·28	12·79	13·28	13·77	14·25	15·19	16·09
1/ 1429	5961·5	7165·6	8511·3	10005	11652	13459	15432	18140	21127	24403	27980	36077	45499
0·00075	10·02	10·49	10·95	11·41	11·85	12·28	12·71	13·24	13·75	14·25	14·75	15·72	16·66
1/ 1333	6170·7	7417·1	8810·1	10356	12061	13932	15974	18777	21868	25260	28962	37343	47096
0·00080	10·35	10·84	11·31	11·78	12·24	12·69	13·13	13·67	14·20	14·72	15·23	16·23	17·20
1/ 1250	6373·1	7660·4	9099·0	10696	12457	14389	16498	19392	22585	26088	29912	38568	48640
0·00085	10·67	11·17	11·66	12·14	12·61	13·08	13·53	14·09	14·64	15·18	15·70	16·73	17·73
1/ 1176	6569·2	7896·2	9379·0	11025	12840	14831	17005	19989	23280	26891	30833	39755	50137
0·00090	10·98	11·49	12·00	12·49	12·98	13·46	13·92	14·50	15·06	15·62	16·16	17·22	18·25
1/ 1111	6759·7	8125·1	9651·0	11344	13212	15261	17498	20569	23955	27671	31727	40908	51591
0·00095	11·28	11·81	12·33	12·84	13·34	13·83	14·31	14·90	15·47	16·04	16·60	17·69	18·75
1/ 1053	6944·9	8347·7	9915·4	11655	13574	15680	17978	21132	24612	28429	32596	42029	53005
0·00100	11·57	12·12	12·65	13·17	13·68	14·18	14·68	15·28	15·88	16·46	17·03	18·15	19·23
1/ 1000	7125·3	8564·6	10173	11958	13927	16087	18445	21681	25251	29167	33443	43120	54382
0·00110	12·14	12·71	13·27	13·81	14·35	14·88	15·39	16·03	16·65	17·26	17·86	19·04	20·17
1/ 909	7473·1	8982·6	10670	12542	14607	16872	19345	22740	26484	30591	35075	45225	57036
0·00120	12·68	13·27	13·86	14·43	14·99	15·54	16·08	16·74	17·39	18·03	18·66	19·88	21·07
1/ 833	7805·4	9382·0	11144	13099	15256	17622	20205	23751	27661	31951	36635	47236	59572
0·00130	13·19	13·81	14·42	15·02	15·60	16·17	16·74	17·43	18·10	18·77	19·42	20·69	21·93
1/ 769	8124·1	9765·1	11599	13634	15879	18342	21030	24721	28791	33256	38131	49165	62005
0·00140	13·69	14·34	14·97	15·58	16·19	16·78	17·37	18·08	18·79	19·48	20·15	21·47	22·76
1/ 714	8430·8	10134	12037	14149	16479	19034	21824	25654	29878	34511	39570	51021	64345
	28·00	30·00	32·00	34·00	36·00	38·00	40·00	42·50	45·00	47·50	50·00	55·00	60·00

Gradient   (Equivalent) Pipe diameters in m

S = 0·00030 to 0·00140

m = Manning $n \times 100$
S = 0·00150 to 0·00700

Full bore conditions; Select m to divide mV and/or mQ.

ie hydraulic gradient = 1 in 667 to 1 in 143

mV to give velocities in ms^{-1}
mQ to give discharges in m^{3}s^{-1}

Gradient	28.00	30.00	32.00	34.00	36.00	38.00	40.00	42.50	45.00	47.50	50.00	55.00	60.00
0·00150	14·17	14·84	15·49	16·13	16·76	17·37	17·98	18·72	19·45	20·16	20·86	22·23	23·56
1/ 667	8726·7	10489	12459	14646	17057	19702	22590	26554	30926	35723	40959	52812	66604
0·00160	14·64	15·33	16·00	16·66	17·31	17·94	18·57	19·33	20·08	20·82	21·54	22·96	24·33
1/ 625	9012·9	10833	12868	15126	17616	20348	23331	27425	31941	36894	42302	54544	68788
0·00170	15·09	15·80	16·49	17·17	17·84	18·49	19·14	19·93	20·70	21·46	22·21	23·66	25·08
1/ 588	9290·3	11167	13264	15591	18159	20975	24049	28269	32924	38030	43604	56222	70905
0·00180	15·53	16·26	16·97	17·67	18·36	19·03	19·69	20·50	21·30	22·08	22·85	24·35	25·80
1/ 556	9559·6	11491	13649	16043	18685	21583	24746	29089	33878	39132	44868	57852	72961
0·00190	15·95	16·70	17·44	18·15	18·86	19·55	20·23	21·07	21·88	22·69	23·48	25·02	26·51
1/ 526	9821·6	11805	14023	16483	19197	22174	25425	29886	34806	40205	46098	59437	74960
0·00200	16·36	17·14	17·89	18·63	19·35	20·06	20·76	21·61	22·45	23·28	24·09	25·67	27·20
1/ 500	10077	12112	14387	16911	19696	22750	26085	30662	35711	41249	47295	60982	76907
0·00220	17·16	17·97	18·76	19·54	20·29	21·04	21·77	22·67	23·55	24·41	25·26	26·92	28·53
1/ 455	10569	12703	15089	17737	20657	23861	27358	32159	37454	43262	49604	63958	80661
0·00240	17·93	18·77	19·60	20·40	21·20	21·97	22·74	23·68	24·60	25·50	26·39	28·12	29·80
1/ 417	11038	13268	15760	18525	21576	24922	28575	33589	39119	45186	51809	66802	84248
0·00260	18·66	19·54	20·40	21·24	22·06	22·87	23·67	24·64	25·60	26·54	27·46	29·27	31·01
1/ 385	11489	13810	16403	19282	22457	25939	29741	34960	40716	47031	53925	69530	87688
0·00280	19·36	20·27	21·17	22·04	22·89	23·74	24·56	25·57	26·57	27·54	28·50	30·37	32·18
1/ 357	11923	14331	17023	20010	23304	26919	30864	36280	42253	48807	55961	72154	90998
0·00300	20·04	20·99	21·91	22·81	23·70	24·57	25·42	26·47	27·50	28·51	29·50	31·44	33·31
1/ 333	12341	14834	17620	20712	24122	27863	31947	37553	43736	50520	57925	74687	94192
0·00320	20·70	21·67	22·63	23·56	24·48	25·37	26·26	27·34	28·40	29·44	30·47	32·47	34·41
1/ 313	12746	15321	18198	21391	24913	28777	32995	38785	45171	52176	59824	77136	97281
0·00340	21·34	22·34	23·32	24·29	25·23	26·16	27·06	28·18	29·28	30·35	31·41	33·47	35·47
1/ 294	13138	15792	18758	22050	25680	29663	34011	39978	46561	53782	61665	79510	100275
0·00360	21·96	22·99	24·00	24·99	25·96	26·91	27·85	29·00	30·12	31·23	32·32	34·44	36·49
1/ 278	13519	16250	19302	22689	26425	30523	34997	41137	47911	55341	63453	81815	103182
0·00380	22·56	23·62	24·66	25·67	26·67	27·65	28·61	29·79	30·95	32·09	33·20	35·38	37·49
1/ 263	13890	16695	19831	23311	27149	31359	35956	42265	49224	56858	65192	84057	106009
0·00400	23·14	24·23	25·30	26·34	27·36	28·37	29·36	30·57	31·75	32·92	34·06	36·30	38·47
1/ 250	14251	17129	20346	23916	27854	32174	36890	43363	50502	58335	66886	86241	108763
0·00420	23·71	24·83	25·92	26·99	28·04	29·07	30·08	31·32	32·54	33·73	34·91	37·20	39·42
1/ 238	14603	17552	20848	24507	28542	32968	37801	44434	51750	59776	68537	88371	111449
0·00440	24·27	25·42	26·53	27·63	28·70	29·75	30·79	32·06	33·30	34·53	35·73	38·07	40·34
1/ 227	14946	17965	21339	25083	29213	33744	38690	45479	52967	61182	70150	90450	114072
0·00460	24·82	25·99	27·13	28·25	29·35	30·42	31·48	32·78	34·05	35·30	36·53	38·93	41·25
1/ 217	15282	18369	21819	25647	29870	34503	39560	46501	54158	62557	71727	92483	116636
0·00480	25·35	26·55	27·71	28·86	29·98	31·08	32·16	33·48	34·78	36·06	37·32	39·76	42·14
1/ 208	15611	18764	22288	26199	30512	35245	40411	47501	55323	63903	73270	94472	119144
0·00500	25·88	27·09	28·28	29·45	30·59	31·72	32·82	34·17	35·50	36·81	38·09	40·58	43·01
1/ 200	15933	19151	22748	26739	31142	35971	41244	48481	56463	65220	74780	96420	121601
0·00550	27·14	28·42	29·66	30·89	32·09	33·27	34·42	35·84	37·23	38·60	39·94	42·56	45·11
1/ 182	16710	20086	23858	28044	32662	37727	43257	50847	59219	68404	78430	101126	127536
0·00600	28·34	29·68	30·98	32·26	33·51	34·74	35·95	37·44	38·89	40·32	41·72	44·46	47·11
1/ 167	17453	20979	24919	29291	34114	39405	45181	53108	61853	71445	81918	105623	133207
0·00650	29·50	30·89	32·25	33·58	34·88	36·16	37·42	38·97	40·48	41·96	43·42	46·27	49·04
1/ 154	18166	21835	25936	30487	35507	41014	47025	55277	64378	74363	85263	109936	138647
0·00700	30·62	32·06	33·47	34·85	36·20	37·53	38·83	40·44	42·01	43·55	45·06	48·02	50·89
1/ 143	18852	22660	26915	31638	36847	42562	48801	57363	66809	77170	88481	114086	143880
	28.00	30.00	32.00	34.00	36.00	38.00	40.00	42.50	45.00	47.50	50.00	55.00	60.00

Gradient (Equivalent) Pipe diameters in m

S = 0·00150 to 0·00700

m = Manning $n \times 100$
S = 0.07500 to 0.04000

ie hydraulic gradient =
1 in 133   to   1 in 25.0

Full bore conditions; Select m to divide mV and/or mQ.

mV to give velocities in ms^{-1}
mQ to give discharges in m^{3}s^{-1}

Gradient   (Equivalent) Pipe diameters in m

Gradient	28.00	30.00	32.00	34.00	36.00	38.00	40.00	42.50	45.00	47.50	50.00	55.00	60.00
0.00750	31.69	33.18	34.64	36.07	37.47	38.85	40.20	41.86	43.48	45.08	46.64	49.70	52.67
1/ 133	19513	23455	27860	32748	38141	44056	50513	59377	69153	79878	91587	118090	148930
0.00800	32.73	34.27	35.78	37.25	38.70	40.12	41.52	43.23	44.91	46.56	48.17	51.33	54.40
1/ 125	20153	24224	28774	33822	39391	45501	52170	61324	71421	82498	94591	121963	153815
0.00850	33.74	35.33	36.88	38.40	39.89	41.35	42.79	44.56	46.29	47.99	49.66	52.91	56.08
1/ 118	20774	24970	29659	34863	40604	46901	53776	63211	73619	85037	97502	125717	158549
0.00900	34.72	36.35	37.95	39.51	41.05	42.55	44.03	45.85	47.63	49.38	51.10	54.45	57.70
1/ 111	21376	25694	30519	35874	41781	48261	55335	65044	75754	87502	100328	129361	163145
0.00950	35.67	37.35	38.99	40.60	42.17	43.72	45.24	47.11	48.94	50.73	52.50	55.94	59.28
1/ 105	21962	26398	31355	36857	42926	49583	56851	66826	77830	89900	103078	132906	167616
0.01000	36.59	38.32	40.00	41.65	43.27	44.86	46.42	48.33	50.21	52.05	53.86	57.39	60.82
1/ 100	22532	27084	32170	37815	44041	50871	58328	68562	79851	92236	105755	136359	171970
0.01100	38.38	40.19	41.95	43.68	45.38	47.04	48.68	50.69	52.66	54.59	56.49	60.20	63.79
1/ 91	23632	28405	33740	39660	46190	53354	61175	71909	83749	96738	110917	143014	180364
0.01200	40.09	41.97	43.82	45.63	47.40	49.14	50.85	52.94	55.00	57.02	59.00	62.87	66.63
1/ 83	24683	29669	35240	41424	48244	55727	63895	75106	87473	101039	115849	149374	188384
0.01300	41.72	43.69	45.61	47.49	49.33	51.14	52.92	55.10	57.25	59.35	61.41	65.44	69.35
1/ 77	25691	30880	36679	43115	50214	58002	66504	78173	91045	105165	120580	155473	196076
0.01400	43.30	45.34	47.33	49.28	51.19	53.07	54.92	57.19	59.41	61.59	63.73	67.91	71.97
1/ 71	26660	32046	38064	44743	52110	60192	69014	81124	94481	109135	125132	161342	203478
0.01500	44.82	46.93	48.99	51.01	52.99	54.94	56.85	59.19	61.49	63.75	65.97	70.29	74.49
1/ 67	27596	33170	39400	46313	53939	62304	71437	83972	97798	112965	129523	167005	210619
0.01600	46.29	48.47	50.60	52.68	54.73	56.74	58.71	61.13	63.51	65.84	68.13	72.60	76.93
1/ 63	28501	34258	40692	47832	55708	64348	73780	86725	101005	116670	133771	172482	217527
0.01700	47.71	49.96	52.15	54.30	56.41	58.48	60.52	63.01	65.46	67.87	70.23	74.83	79.30
1/ 59	29378	35313	41944	49304	57422	66328	76050	89395	104114	120261	137888	177790	224221
0.01800	49.09	51.41	53.67	55.88	58.05	60.18	62.27	64.84	67.36	69.83	72.26	77.00	81.60
1/ 56	30230	36336	43160	50734	59087	68251	78255	91986	107132	123747	141886	182945	230722
0.01900	50.44	52.81	55.14	57.41	59.64	61.83	63.98	66.62	69.21	71.75	74.24	79.11	83.84
1/ 53	31059	37332	44343	52124	60706	70121	80399	94507	110068	127138	145774	187958	237044
0.02000	51.75	54.19	56.57	58.90	61.19	63.44	65.64	68.35	71.00	73.61	76.17	81.17	86.02
1/ 50	31865	38302	45495	53478	62283	71943	82488	96962	112927	130441	149561	192840	243202
0.02200	54.28	56.83	59.33	61.78	64.18	66.53	68.85	71.69	74.47	77.20	79.89	85.13	90.21
1/ 45.5	33421	40171	47716	56088	65323	75454	86514	101695	118439	136808	156861	202253	255073
0.02400	56.69	59.36	61.97	64.52	67.03	69.49	71.91	74.87	77.78	80.64	83.44	88.91	94.23
1/ 41.7	34907	41958	49837	58582	68228	78809	90361	106217	123705	142891	163836	211246	266415
0.02600	59.00	61.78	64.50	67.16	69.77	72.33	74.84	77.93	80.96	83.93	86.85	92.55	98.07
1/ 38.5	36332	43671	51872	60974	71014	82027	94051	110554	128757	148726	170526	219872	277293
0.02800	61.23	64.11	66.93	69.69	72.40	75.06	77.67	80.87	84.01	87.10	90.13	96.04	101.8
1/ 35.7	37704	45319	53830	63276	73694	85124	97601	114727	133617	154340	176963	228172	287761
0.03000	63.38	66.36	69.28	72.14	74.94	77.69	80.39	83.71	86.96	90.15	93.29	99.41	105.3
1/ 33.3	39027	46910	55720	65497	76281	88112	101027	118754	138307	159757	183174	236180	297861
0.03200	65.46	68.54	71.55	74.51	77.40	80.24	83.03	86.46	89.81	93.11	96.35	102.7	108.8
1/ 31.3	40307	48449	57547	67645	78783	91001	104340	122648	142843	164996	189181	243926	307629
0.03400	67.47	70.65	73.76	76.80	79.78	82.71	85.59	89.12	92.58	95.98	99.31	105.8	112.2
1/ 29.4	41547	49940	59318	69727	81207	93802	107551	126423	147239	170074	195003	251433	317097
0.03600	69.43	72.70	75.89	79.02	82.09	85.11	88.07	91.70	95.26	98.76	102.2	108.9	115.4
1/ 27.8	42752	51387	61038	71748	83562	96521	110669	130088	151507	175005	200657	258723	326290
0.03800	71.33	74.69	77.97	81.19	84.34	87.44	90.48	94.21	97.87	101.5	105.0	111.9	118.6
1/ 26.3	43923	52796	62711	73714	85851	99166	113702	133653	155659	179800	206155	265812	335231
0.04000	73.19	76.63	80.00	83.30	86.53	89.71	92.83	96.66	100.4	104.1	107.7	114.8	121.6
1/ 25.0	45064	54167	64340	75629	88082	101742	116656	137125	159703	184471	211511	272718	343940
	28.00	30.00	32.00	34.00	36.00	38.00	40.00	42.50	45.00	47.50	50.00	55.00	60.00

Gradient   (Equivalent) Pipe diameters in m

S = 0.07500 to 0.04000

E1

$k_s = 0.003$ mm
$S = 0.00010$ to 0.60000

Water (or sewage) at 15°C; full bore conditions.
This table shows values of m for use with Tables D

ie hydraulic gradient =
1 in 10000 to 1 in 1·7

m_C for Colebrook-White solutions; or m_P for laminar flow
θ for use with part-full circular pipes (turbulent flows)

Gradient — (Equivalent) Pipe diameters in mm

Gradient		20	30	40	60	80	100	150	200	300	400	600	800	1000
0·00010	m	2·722	1·585	1·080	*	1·157	1·136	1·107	1·092	1·078	1·072	1·068	1·069	1·071
1/10000	θ	-	-	-		13	17	26	34	50	65	100	130	170
0·00015	m	2·222	1·294	*	*	1·118	1·100	1·075	1·062	1·051	1·046	1·045	1·046	1·049
1/ 6667	θ	-	-			15	19	28	38	55	75	110	150	190
0·00020	m	1·925	1·121	*	1·119	1·092	1·076	1·053	1·042	1·032	1·029	1·028	1·031	1·034
1/ 5000	θ	-	-		13	17	22	32	42	65	85	130	170	210
0·00030	m	1·571	0·915	*	1·081	1·058	1·044	1·024	1·015	1·008	1·005	1·007	1·010	1·014
1/ 3333	θ	-	-		14	19	24	36	48	70	95	140	190	240
0·00040	m	1·361	*	*	1·056	1·035	1·022	1·005	0·997	0·991	0·990	0·992	0·996	1·000
1/ 2500	θ	-			16	22	26	40	55	80	110	160	210	270
0·00060	m	1·111	*	*	1·022	1·004	0·993	0·979	0·972	0·968	0·968	0·972	0·976	0·981
1/ 1667	θ	-			18	24	30	46	60	90	120	180	240	300
0·00080	m	0·962	*	1·031	1·000	0·983	0·973	0·961	0·956	0·953	0·953	0·958	0·963	0·969
1/ 1250	θ	-		13	20	26	34	50	65	100	130	200	270	330
0·00100	m	0·861	*	1·012	0·983	0·968	0·959	0·948	0·943	0·941	0·942	0·947	0·953	0·959
1/ 1000	θ	-		14	22	28	36	55	70	110	140	220	290	360
0·00150	m	*	*	0·979	0·954	0·941	0·933	0·924	0·921	0·921	0·923	0·929	0·936	0·942
1/ 667	θ			16	24	32	42	60	80	120	160	250	330	410
0·00200	m	*	0·979	0·957	0·934	0·923	0·916	0·909	0·906	0·907	0·910	0·917	0·924	0·930
1/ 500	θ		14	18	28	36	46	70	90	140	180	270	360	450
0·00300	m	*	0·947	0·927	0·908	0·898	0·893	0·887	0·886	0·888	0·892	0·900	0·908	0·915
1/ 333	θ		16	20	32	42	50	80	100	160	210	310	410	500
0·00400	m	*	0·925	0·908	0·890	0·882	0·877	0·873	0·873	0·875	0·880	0·888	0·896	0·904
1/ 250	θ		17	22	34	46	55	85	110	170	230	340	460	550
0·00600	m	0·925	0·896	0·881	0·866	0·860	0·856	0·853	0·854	0·858	0·863	0·873	0·882	0·889
1/ 167	θ	13	20	26	40	50	65	100	130	200	260	390	500	650
0·00800	m	0·903	0·877	0·864	0·850	0·845	0·842	0·840	0·842	0·846	0·852	0·862	0·871	0·879
1/ 125	θ	14	22	28	44	55	70	110	140	220	290	430	550	700
0·01000	m	0·886	0·862	0·850	0·838	0·833	0·831	0·830	0·832	0·838	0·844	0·854	0·864	0·872
1/ 100	θ	15	24	30	46	60	75	120	150	230	310	460	600	750
0·01500	m	0·858	0·838	0·827	0·818	0·814	0·813	0·813	0·816	0·822	0·829	0·840	0·850	0·859
1/ 67	θ	18	26	36	55	70	90	130	180	270	350	550	700	900
0·02000	m	0·839	0·821	0·812	0·804	0·801	0·800	0·801	0·805	0·812	0·819	0·831	0·841	0·850
1/ 50	θ	19	30	38	60	80	95	150	190	290	390	600	800	950
0·03000	m	0·814	0·799	0·791	0·785	0·783	0·783	0·786	0·790	0·798	0·805	0·818	0·829	0·839
1/ 33.3	θ	22	34	44	65	90	110	170	220	330	450	650	900	1100
0·04000	m	0·798	0·784	0·777	0·772	0·771	0·771	0·775	0·779	0·788	0·796	0·810	0·821	0·831
1/ 25.0	θ	24	36	50	75	100	120	180	250	370	490	750	1000	1200
0·06000	m	0·775	0·763	0·758	0·755	0·755	0·756	0·761	0·766	0·776	0·784	0·798	0·810	0·820
1/ 16.7	θ	28	42	55	85	110	140	210	280	420	550	850	1100	1400
0·08000	m	0·760	0·750	0·746	0·743	0·744	0·745	0·751	0·757	0·767	0·776	0·791	0·803	0·813
1/ 12.5	θ	30	46	60	95	120	150	230	310	460	600	950	1200	1500
0·10000	m	0·749	0·740	0·736	0·735	0·736	0·738	0·744	0·750	0·761	0·770	0·785	0·798	0·808
1/ 10.0	θ	34	50	65	100	130	170	250	330	500	650	1000	1300	1700
0·15000	m	0·729	0·722	0·720	0·720	0·722	0·724	0·731	0·738	0·750	0·759	0·775	0·788	0·799
1/ 6.7	θ	38	55	75	110	150	190	290	380	550	750	1100	1500	1900
0·20000	m	0·716	0·710	0·709	0·709	0·712	0·715	0·723	0·730	0·742	0·752	0·769	0·782	0·793
1/ 5.0	θ	42	65	85	130	170	210	310	420	650	850	1300	1700	2100
0·30000	m	0·699	0·695	0·694	0·696	0·699	0·703	0·712	0·719	0·732	0·743	0·761	0·774	0·786
1/ 3.3	θ	48	70	95	140	190	240	360	480	700	950	1400	1900	2400
0·40000	m	0·687	0·684	0·684	0·687	0·691	0·695	0·704	0·712	0·726	0·737	0·755	0·769	0·781
1/ 2.5	θ	55	80	110	160	210	260	390	550	800	1100	1600	2100	2600
0·60000	m	0·672	0·670	0·671	0·675	0·680	0·684	0·694	0·703	0·717	0·729	0·748	0·762	0·775
1/ 1.7	θ	60	90	120	180	240	300	450	600	900	1200	1800	2400	3000
		20	30	40	60	80	100	150	200	300	400	600	800	1000

Gradient — (Equivalent) Pipe diameters in mm

$k_s = 0.003$ mm $S = 0.00010$ to 0.60000

k_s = 0·006 mm
S = 0·00010 to 0·60000

ie hydraulic gradient =
1 in 10000 to 1 in 1·7

Water (or sewage) at 15°C; full bore conditions.
This table shows values of m for use with Tables D

m_C for Colebrook-White solutions; or m_P for laminar flow
θ for use with part-full circular pipes (turbulent flows)

E2

(Equivalent) Pipe diameters in mm

Gradient	20	30	40	60	80	100	150	200	300	400	600	800	1000
0·00010	2·722	1·585	1·080	*	1·157	1·136	1·107	1·093	1·078	1·073	1·069	1·070	1·072
1/10000	-	-	-		13	17	26	34	50	65	100	130	170
0·00015	2·222	1·294	*	*	1·119	1·101	1·076	1·063	1·052	1·047	1·046	1·048	1·051
1/ 6667	-				15	19	28	38	55	75	110	150	190
0·00020	1·925	1·121	*	1·120	1·093	1·077	1·054	1·043	1·033	1·030	1·030	1·033	1·036
1/ 5000	-	-		13	17	22	32	42	65	85	130	170	210
0·00030	1·571	0·915	*	1·082	1·059	1·045	1·025	1·016	1·009	1·007	1·008	1·012	1·016
1/ 3333	-	-		14	19	24	36	48	70	95	140	190	240
0·00040	1·361	*	*	1·057	1·036	1·023	1·006	0·998	0·992	0·991	0·994	0·998	1·003
1/ 2500	-			16	22	26	40	55	80	110	160	210	260
0·00060	1·111	*	*	1·023	1·005	0·994	0·980	0·974	0·970	0·970	0·974	0·979	0·984
1/ 1667	-			18	24	30	46	60	90	120	180	240	300
0·00080	0·962	*	1·032	1·001	0·985	0·975	0·962	0·957	0·955	0·956	0·960	0·966	0·972
1/ 1250	-		13	20	26	34	50	65	100	130	200	270	330
0·00100	0·861	*	1·013	0·984	0·969	0·960	0·949	0·945	0·943	0·945	0·950	0·956	0·962
1/ 1000	-		14	22	28	36	55	70	110	140	220	290	360
0·00150	*	*	0·981	0·955	0·943	0·935	0·926	0·923	0·923	0·926	0·932	0·939	0·946
1/ 667			16	24	32	42	60	80	120	160	250	330	410
0·00200	*	0·981	0·959	0·936	0·925	0·918	0·911	0·909	0·910	0·913	0·920	0·928	0·935
1/ 500		14	18	28	36	46	70	90	140	180	270	360	450
0·00300	*	0·948	0·929	0·910	0·901	0·895	0·890	0·889	0·891	0·895	0·904	0·912	0·920
1/ 333		16	20	32	42	50	80	100	160	210	310	410	500
0·00400	*	0·927	0·910	0·892	0·884	0·880	0·876	0·876	0·879	0·884	0·893	0·902	0·909
1/ 250		17	22	34	46	55	85	110	170	230	340	460	550
0·00600	0·927	0·898	0·884	0·869	0·863	0·859	0·857	0·858	0·862	0·868	0·878	0·888	0·896
1/ 167	13	20	26	40	50	65	100	130	200	260	390	500	650
0·00800	0·905	0·879	0·866	0·853	0·848	0·845	0·844	0·846	0·851	0·857	0·868	0·878	0·887
1/ 125	14	22	28	44	55	70	110	140	220	290	430	550	700
0·01000	0·889	0·865	0·853	0·842	0·837	0·835	0·835	0·837	0·843	0·849	0·861	0·871	0·880
1/ 100	15	24	30	46	60	75	120	150	230	310	460	600	750
0·01500	0·861	0·841	0·831	0·821	0·818	0·817	0·818	0·821	0·828	0·835	0·848	0·859	0·868
1/ 67	18	26	36	55	70	90	130	180	260	350	550	700	900
0·02000	0·843	0·824	0·815	0·808	0·805	0·805	0·807	0·810	0·819	0·826	0·839	0·851	0·860
1/ 50	19	30	38	60	80	95	150	190	290	390	600	800	950
0·03000	0·818	0·803	0·795	0·789	0·788	0·788	0·792	0·796	0·805	0·814	0·828	0·840	0·850
1/ 33·3	22	34	44	65	90	110	170	220	330	440	650	900	1100
0·04000	0·802	0·788	0·782	0·777	0·777	0·777	0·782	0·787	0·797	0·806	0·820	0·833	0·843
1/ 25·0	24	36	48	75	100	120	180	240	370	490	750	1000	1200
0·06000	0·780	0·768	0·764	0·761	0·761	0·763	0·768	0·774	0·785	0·795	0·810	0·823	0·834
1/ 16·7	28	42	55	85	110	140	210	280	420	550	850	1100	1400
0·08000	0·765	0·755	0·752	0·750	0·751	0·753	0·760	0·766	0·778	0·788	0·804	0·817	0·828
1/ 12·5	30	46	60	90	120	150	230	310	460	600	900	1200	1500
0·10000	0·754	0·746	0·743	0·742	0·743	0·746	0·753	0·760	0·772	0·782	0·799	0·813	0·824
1/ 10·0	34	50	65	100	130	170	250	330	500	650	1000	1300	1700
0·15000	0·736	0·729	0·727	0·728	0·730	0·734	0·742	0·749	0·762	0·773	0·791	0·805	0·817
1/ 6·7	38	55	75	110	150	190	280	380	550	750	1100	1500	1900
0·20000	0·723	0·718	0·717	0·718	0·722	0·725	0·734	0·742	0·756	0·768	0·786	0·800	0·813
1/ 5·0	42	60	85	120	170	210	310	420	600	850	1200	1700	2100
0·30000	0·707	0·703	0·703	0·706	0·710	0·715	0·725	0·733	0·748	0·760	0·779	0·794	0·807
1/ 3·3	48	70	95	140	190	240	360	480	700	950	1400	1900	2400
0·40000	0·696	0·694	0·694	0·698	0·703	0·708	0·718	0·728	0·743	0·755	0·775	0·790	0·803
1/ 2·5	50	80	100	160	210	260	390	500	800	1000	1600	2100	2600
0·60000	0·682	0·681	0·682	0·688	0·693	0·698	0·710	0·720	0·736	0·749	0·769	0·785	0·799
1/ 1·7	60	90	120	180	240	300	450	600	900	1200	1800	2400	3000
	20	30	40	60	80	100	150	200	300	400	600	800	1000

Gradient   (Equivalent) Pipe diameters in mm

S = 0·00010 to 0·60000 k_s = 0·006 mm

E3

$k_s = 0.015$ mm
$S = 0.00010$ to 0.60000

Water (or sewage) at 15°C; full bore conditions.
This table shows values of m for use with Tables D

ie hydraulic gradient =
1 in 10000 to 1 in 1·7

m_C for Colebrook-White solutions; or m_P for laminar flow
θ   for use with part-full circular pipes (turbulent flows)

Gradient	\(Equivalent\) Pipe diameters in mm												
	20	30	40	60	80	100	150	200	300	400	600	800	1000
0·00010	2·722	1·585	1·080	*	1·160	1·139	1·110	1·095	1·081	1·076	1·073	1·074	1·076
1/10000	-	-	-		13	17	26	34	50	65	100	130	170
0·00015	2·222	1·294	*	*	1·121	1·103	1·078	1·066	1·055	1·051	1·050	1·052	1·055
1/ 6667	-	-			15	19	28	38	55	75	110	150	190
0·00020	1·925	1·121	*	1·122	1·096	1·079	1·057	1·046	1·037	1·034	1·034	1·037	1·041
1/ 5000	-	-		13	17	20	32	42	65	85	130	170	210
0·00030	1·571	0·915	*	1·085	1·062	1·048	1·029	1·020	1·013	1·011	1·013	1·018	1·022
1/ 3333	-	-		14	19	24	36	48	70	95	140	190	240
0·00040	1·361	*	*	1·060	1·039	1·027	1·010	1·002	0·997	0·996	0·999	1·004	1·009
1/ 2500	-			16	22	26	40	55	80	110	160	210	260
0·00060	1·111	*	*	1·027	1·009	0·998	0·984	0·979	0·975	0·976	0·980	0·986	0·992
1/ 1667	-			18	24	30	46	60	90	120	180	240	300
0·00080	0·962	*	1·036	1·005	0·989	0·979	0·967	0·963	0·961	0·962	0·968	0·974	0·980
1/ 1250	-		13	20	26	34	50	65	100	130	200	270	330
0·00100	0·861	*	1·017	0·988	0·974	0·965	0·955	0·951	0·950	0·952	0·958	0·965	0·972
1/ 1000	-		14	22	28	36	55	70	110	140	210	290	360
0·00150	*	*	0·985	0·960	0·948	0·941	0·932	0·930	0·931	0·934	0·942	0·949	0·957
1/ 667			16	24	32	40	60	80	120	160	250	330	410
0·00200	*	0·985	0·964	0·941	0·930	0·924	0·918	0·916	0·918	0·922	0·931	0·939	0·947
1/ 500		14	18	28	36	46	70	90	140	180	270	360	450
0·00300	*	0·954	0·935	0·916	0·907	0·902	0·898	0·898	0·901	0·906	0·916	0·925	0·933
1/ 333		15	20	30	42	50	75	100	150	210	310	410	500
0·00400	*	0·933	0·916	0·899	0·892	0·888	0·884	0·885	0·890	0·895	0·906	0·916	0·925
1/ 250		17	22	34	46	55	85	110	170	230	340	450	550
0·00600	0·933	0·905	0·891	0·877	0·871	0·868	0·867	0·869	0·875	0·881	0·893	0·904	0·913
1/ 167	13	19	26	38	50	65	95	130	190	260	390	500	650
0·00800	0·912	0·887	0·874	0·862	0·857	0·855	0·855	0·858	0·864	0·872	0·885	0·896	0·906
1/ 125	14	22	28	42	55	70	110	140	210	280	430	550	700
0·01000	0·896	0·873	0·862	0·851	0·847	0·845	0·846	0·849	0·857	0·865	0·878	0·890	0·900
1/ 100	15	24	30	46	60	75	120	150	230	310	460	600	750
0·01500	0·870	0·850	0·840	0·832	0·829	0·829	0·831	0·835	0·844	0·853	0·868	0·880	0·891
1/ 67	18	26	36	55	70	90	130	180	260	350	550	700	900
0·02000	0·852	0·834	0·826	0·819	0·818	0·818	0·821	0·826	0·836	0·845	0·861	0·874	0·885
1/ 50	19	28	38	60	75	95	140	190	290	390	600	750	950
0·03000	0·828	0·814	0·807	0·803	0·802	0·803	0·808	0·814	0·825	0·835	0·852	0·865	0·877
1/ 33·3	22	34	44	65	90	110	170	220	330	440	650	900	1100
0·04000	0·813	0·800	0·795	0·792	0·792	0·794	0·800	0·806	0·818	0·829	0·846	0·860	0·872
1/ 25·0	24	36	48	75	95	120	180	240	360	480	750	950	1200
0·06000	0·793	0·782	0·779	0·777	0·779	0·781	0·789	0·796	0·809	0·820	0·839	0·853	0·866
1/ 16·7	28	42	55	85	110	140	210	280	410	550	850	1100	1400
0·08000	0·779	0·771	0·768	0·768	0·770	0·773	0·782	0·790	0·803	0·815	0·834	0·849	0·862
1/ 12·5	30	46	60	90	120	150	230	300	450	600	900	1200	1500
0·10000	0·769	0·762	0·760	0·761	0·764	0·767	0·776	0·785	0·799	0·811	0·831	0·846	0·859
1/ 10·0	32	48	65	100	130	160	240	330	490	650	1000	1300	1600
0·15000	0·753	0·748	0·747	0·749	0·753	0·757	0·768	0·777	0·792	0·805	0·825	0·841	0·855
1/ 6·7	38	55	75	110	150	190	280	370	550	750	1100	1500	1900
0·20000	0·742	0·738	0·738	0·742	0·746	0·751	0·762	0·772	0·788	0·801	0·822	0·838	0·852
1/ 5·0	40	60	80	120	160	200	310	410	600	800	1200	1600	2000
0·30000	0·728	0·726	0·727	0·732	0·737	0·743	0·755	0·765	0·782	0·796	0·818	0·834	0·848
1/ 3·3	46	70	95	140	190	230	350	470	700	950	1400	1900	2300
0·40000	0·718	0·718	0·720	0·726	0·732	0·738	0·751	0·761	0·779	0·793	0·815	0·832	0·846
1/ 2·5	50	75	100	150	200	260	380	500	750	1000	1500	2000	2600
0·60000	0·707	0·708	0·711	0·718	0·725	0·731	0·745	0·756	0·775	0·789	0·812	0·829	0·843
1/ 1·7	60	85	120	170	230	290	440	600	850	1200	1700	2300	2900
	20	30	40	60	80	100	150	200	300	400	600	800	1000

Gradient   (Equivalent) Pipe diameters in mm

$k_s = 0.015$ mm $S = 0.00010$ to 0.60000

k_s = 0·015 mm
S = 0·000020 to 0·10000

ie hydraulic gradient =
1 in 50000 to 1 in 10·0

Water (or sewage) at 15°C; full bore conditions.
This table shows values of m for use with Tables D

m_C for Colebrook-White solutions;
θ for use with part-full circular pipes

(Equivalent) Pipe diameters in mm to 4000 mm, then in m

Gradient	1000	1500	2000	3000	4000	6·00	8·00	10·00	15·00	20·00	30·00	40·00	60·00
0·00002	1·170	1·172	1·177	1·186	1·196	1·213	1·228	1·240	1·266	1·286	1·318	1·342	1·380
1/50000	100	150	200	290	390	600	800	1000	1500	2000	2900	3900	6000
0·00003	1·145	1·148	1·154	1·165	1·176	1·194	1·209	1·222	1·248	1·269	1·301	1·327	1·365
1/33333	110	170	220	340	450	650	900	1100	1700	2200	3400	4500	6500
0·00004	1·128	1·132	1·138	1·150	1·161	1·180	1·196	1·209	1·236	1·257	1·290	1·316	1·355
1/25000	120	180	250	370	490	750	1000	1200	1800	2500	3700	4900	7500
0·00006	1·104	1·110	1·117	1·131	1·142	1·162	1·178	1·192	1·220	1·242	1·276	1·302	1·342
1/16667	140	210	280	420	550	850	1100	1400	2100	2800	4200	5500	8500
0·00008	1·088	1·095	1·103	1·117	1·129	1·150	1·167	1·181	1·209	1·231	1·266	1·292	1·333
1/12500	150	230	310	460	600	950	1200	1500	2300	3100	4600	6000	9500
0·00010	1·076	1·084	1·092	1·107	1·120	1·141	1·158	1·172	1·201	1·223	1·258	1·285	1·326
1/10000	170	250	330	500	650	1000	1300	1700	2500	3300	5000	6500	10000
0·00015	1·055	1·065	1·074	1·089	1·103	1·125	1·142	1·157	1·187	1·210	1·246	1·273	1·315
1/ 6667	190	290	380	550	750	1100	1500	1900	2900	3800	5500	7500	11000
0·00020	1·041	1·051	1·061	1·078	1·091	1·114	1·132	1·147	1·177	1·201	1·237	1·265	1·307
1/ 5000	210	310	420	650	850	1300	1700	2100	3100	4200	6500	8500	13000
0·00030	1·022	1·034	1·044	1·062	1·076	1·100	1·118	1·134	1·165	1·189	1·226	1·255	1·298
1/ 3333	240	360	480	700	950	1400	1900	2400	3600	4800	7000	9500	14000
0·00040	1·009	1·022	1·032	1·051	1·066	1·090	1·109	1·125	1·157	1·181	1·219	1·248	1·292
1/ 2500	260	400	550	800	1100	1600	2100	2600	4000	5500	8000	11000	16000
0·00060	0·992	1·005	1·017	1·036	1·052	1·077	1·097	1·113	1·146	1·171	1·210	1·239	1·284
1/ 1667	300	450	600	900	1200	1800	2400	3000	4500	6000	9000	12000	18000
0·00080	0·980	0·995	1·007	1·027	1·043	1·069	1·089	1·106	1·139	1·164	1·204	1·233	1·279
1/ 1250	330	500	650	1000	1300	2000	2700	3300	5000	6500	10000	13000	20000
0·00100	0·972	0·987	0·999	1·020	1·036	1·062	1·083	1·100	1·134	1·160	1·199	1·229	1·275
1/ 1000	360	550	700	1100	1400	2100	2900	3600	5500	7000	11000	14000	21000
0·00150	0·957	0·973	0·986	1·007	1·025	1·052	1·073	1·091	1·125	1·152	1·192	1·223	1·269
1/ 667	410	600	800	1200	1600	2500	3300	4100	6000	8000	12000	16000	25000
0·00200	0·947	0·963	0·977	0·999	1·017	1·045	1·066	1·084	1·119	1·146	1·187	1·218	1·265
1/ 500	450	700	900	1400	1800	2700	3600	4500	7000	9000	14000	18000	27000
0·00300	0·933	0·951	0·966	0·989	1·007	1·036	1·058	1·076	1·112	1·140	1·181	1·213	1·261
1/ 333	500	750	1000	1500	2100	3100	4100	5000	7500	10000	15000	21000	31000
0·00400	0·925	0·943	0·958	0·982	1·001	1·030	1·053	1·071	1·108	1·136	1·178	1·210	1·258
1/ 250	550	850	1100	1700	2300	3400	4500	5500	8500	11000	17000	23000	34000
0·00600	0·913	0·933	0·948	0·973	0·993	1·023	1·046	1·065	1·102	1·131	1·173	1·206	1·254
1/ 167	650	950	1300	1900	2600	3900	5000	6500	9500	13000	19000	26000	39000
0·00800	0·906	0·926	0·942	0·967	0·987	1·018	1·042	1·061	1·099	1·127	1·170	1·203	1·252
1/ 125	700	1100	1400	2100	2800	4300	5500	7000	11000	14000	21000	28000	43000
0·01000	0·900	0·921	0·937	0·963	0·983	1·015	1·038	1·058	1·096	1·125	1·168	1·201	1·250
1/ 100	750	1200	1500	2300	3100	4600	6000	7500	12000	15000	23000	31000	46000
0·01500	0·891	0·912	0·930	0·956	0·977	1·009	1·033	1·053	1·092	1·121	1·165	1·198	1·248
1/ 67	900	1300	1800	2600	3500	5500	7000	9000	13000	18000	26000	35000	53000
0·02000	0·885	0·907	0·925	0·952	0·973	1·006	1·030	1·050	1·090	1·119	1·163	1·197	1·246
1/ 50	950	1400	1900	2900	3900	6000	7500	9500	14000	19000	29000	39000	58000
0·03000	0·877	0·900	0·918	0·947	0·968	1·001	1·026	1·047	1·086	1·116	1·161	1·195	1·245
1/ 33·3	1100	1700	2200	3300	4400	6500	9000	11000	17000	22000	33000	44000	66000
0·04000	0·872	0·896	0·914	0·943	0·965	0·999	1·024	1·045	1·085	1·115	1·160	1·193	1·244
1/ 25·0	1200	1800	2400	3600	4800	7500	9500	12000	18000	24000	36000	48000	73000
0·06000	0·866	0·890	0·910	0·939	0·961	0·995	1·021	1·042	1·082	1·113	1·158	1·192	1·242
1/ 16·7	1400	2100	2800	4100	5500	8500	11000	14000	21000	28000	41000	55000	83000
0·08000	0·862	0·887	0·906	0·936	0·959	0·993	1·019	1·040	1·081	1·111	1·157	1·191	1·241
1/ 12·5	1500	2300	3000	4500	6000	9000	12000	15000	23000	30000	45000	61000	91000
0·10000	0·859	0·885	0·904	0·934	0·957	0·992	1·018	1·039	1·080	1·110	1·156	1·190	1·241
1/ 10·0	1600	2400	3300	4900	6500	10000	13000	16000	24000	33000	49000	65000	98000
	1000	1500	2000	3000	4000	6·00	8·00	10·00	15·00	20·00	30·00	40·00	60·00

Gradient (Equivalent) Pipe diameters in mm to 4000 mm, then in m

S = 0·000020 to 0·10000 k_s = 0·015 mm

E4

k_s = 0·030 mm
S = 0·00010 to 0·60000

Water (or sewage) at 15°C; full bore conditions.
This table shows values of m for use with Tables D

ie hydraulic gradient =
1 in 10000 to 1 in 1·7

m_C for Colebrook-White solutions; or m_P for laminar flow
θ for use with part-full circular pipes (turbulent flows)

Gradient	(Equivalent) Pipe diameters in mm												
	20	30	40	60	80	100	150	200	300	400	600	800	1000
0·00010	2·722	1·585	1·080	*	1·163	1·142	1·114	1·099	1·086	1·081	1·078	1·080	1·083
1/10000	-	-	-		13	17	24	34	50	65	100	130	170
0·00015	2·222	1·294	*	*	1·125	1·107	1·083	1·071	1·060	1·056	1·056	1·059	1·063
1/ 6667	-	-			15	19	28	38	55	75	110	150	190
0·00020	1·925	1·121	*	1·126	1·100	1·084	1·062	1·052	1·043	1·040	1·041	1·045	1·050
1/ 5000	-	-		13	17	20	32	42	65	85	130	170	210
0·00030	1·571	0·915	*	1·090	1·067	1·053	1·034	1·026	1·020	1·019	1·022	1·026	1·032
1/ 3333	-	-		14	19	24	36	48	70	95	140	190	240
0·00040	1·361	*	*	1·065	1·045	1·032	1·016	1·009	1·004	1·004	1·008	1·014	1·020
1/ 2500	-			16	22	26	40	55	80	110	160	210	260
0·00060	1·111	*	*	1·033	1·015	1·005	0·992	0·986	0·984	0·985	0·991	0·997	1·004
1/ 1667	-			18	24	30	46	60	90	120	180	240	300
0·00080	0·962	*	1·042	1·011	0·996	0·986	0·975	0·971	0·970	0·972	0·979	0·986	0·994
1/ 1250	-		13	20	26	34	50	65	100	130	200	260	330
0·00100	0·861	*	1·024	0·995	0·981	0·973	0·963	0·960	0·960	0·963	0·970	0·978	0·986
1/ 1000	-		14	22	28	36	55	70	110	140	210	280	360
0·00150	*	*	0·992	0·968	0·956	0·949	0·942	0·940	0·942	0·946	0·956	0·965	0·973
1/ 667			16	24	32	40	60	80	120	160	240	330	410
0·00200	*	0·993	0·972	0·950	0·939	0·934	0·928	0·927	0·931	0·935	0·946	0·955	0·964
1/ 500		13	18	26	36	44	65	90	130	180	270	360	450
0·00300	*	0·962	0·944	0·926	0·917	0·913	0·910	0·910	0·915	0·921	0·933	0·944	0·953
1/ 333		15	20	30	40	50	75	100	150	200	310	410	500
0·00400	*	0·942	0·926	0·910	0·903	0·900	0·898	0·899	0·905	0·912	0·925	0·936	0·946
1/ 250		17	22	34	44	55	85	110	170	220	340	450	550
0·00600	0·943	0·916	0·902	0·889	0·884	0·882	0·882	0·885	0·892	0·900	0·914	0·926	0·937
1/ 167	13	19	26	38	50	65	95	130	190	260	380	500	650
0·00800	0·923	0·899	0·886	0·875	0·871	0·870	0·871	0·875	0·884	0·892	0·907	0·920	0·931
1/ 125	14	22	28	42	55	70	110	140	210	280	420	550	700
0·01000	0·908	0·886	0·875	0·865	0·862	0·861	0·864	0·868	0·877	0·886	0·902	0·915	0·926
1/ 100	15	22	30	46	60	75	110	150	230	300	450	600	750
0·01500	0·883	0·864	0·855	0·848	0·846	0·847	0·851	0·856	0·867	0·877	0·894	0·907	0·919
1/ 67	17	26	34	50	70	85	130	170	260	350	500	700	850
0·02000	0·866	0·850	0·842	0·837	0·836	0·837	0·842	0·848	0·860	0·871	0·888	0·903	0·915
1/ 50	19	28	38	55	75	95	140	190	280	380	550	750	950
0·03000	0·845	0·831	0·825	0·822	0·823	0·825	0·832	0·839	0·852	0·863	0·882	0·897	0·909
1/ 33·3	22	32	44	65	85	110	160	220	320	430	650	850	1100
0·04000	0·831	0·819	0·814	0·813	0·814	0·817	0·825	0·832	0·846	0·858	0·877	0·893	0·906
1/ 25·0	24	36	48	70	95	120	180	240	360	470	700	950	1200
0·06000	0·812	0·803	0·800	0·800	0·803	0·807	0·816	0·825	0·840	0·852	0·872	0·888	0·902
1/ 16·7	28	40	55	80	110	140	200	270	410	550	800	1100	1400
0·08000	0·800	0·793	0·791	0·793	0·796	0·800	0·810	0·820	0·835	0·848	0·869	0·885	0·899
1/ 12·5	30	44	60	90	120	150	220	300	440	600	900	1200	1500
0·10000	0·792	0·786	0·785	0·787	0·791	0·796	0·806	0·816	0·832	0·846	0·867	0·883	0·897
1/ 10·0	32	48	65	95	130	160	240	320	480	650	950	1300	1600
0·15000	0·778	0·774	0·774	0·778	0·783	0·788	0·800	0·810	0·827	0·841	0·863	0·880	0·894
1/ 6·7	36	55	70	110	140	180	270	360	550	700	1100	1400	1800
0·20000	0·768	0·766	0·767	0·772	0·778	0·783	0·796	0·807	0·824	0·839	0·861	0·878	0·892
1/ 5·0	40	60	80	120	160	200	300	400	600	800	1200	1600	2000
0·30000	0·757	0·756	0·758	0·764	0·771	0·777	0·791	0·802	0·821	0·835	0·858	0·876	0·890
1/ 3·3	44	65	90	130	180	220	340	450	650	900	1300	1800	2200
0·40000	0·749	0·750	0·753	0·760	0·767	0·774	0·788	0·800	0·818	0·833	0·856	0·874	0·889
1/ 2·5	50	75	100	150	200	250	370	490	750	1000	1500	2000	2500
0·60000	0·740	0·742	0·746	0·754	0·762	0·769	0·784	0·796	0·816	0·831	0·854	0·872	0·887
1/ 1·7	55	85	110	170	220	280	420	550	850	1100	1700	2200	2800
	20	30	40	60	80	100	150	200	300	400	600	800	1000

Gradient (Equivalent) Pipe diameters in mm

k_s = 0·030 mm S = 0·00010 to 0·60000

k_s = 0·030 mm
S = 0·000020 to 0·10000

ie hydraulic gradient =
1 in 50000 to 1 in 10·0

Water (or sewage) at 15°C; full bore conditions.
This table shows values of m for use with Tables D

m_C for Colebrook-White solutions;
θ for use with part-full circular pipes

Gradient	(Equivalent) Pipe diameters in mm to 4000 mm, then in m												
	1000	1500	2000	3000	4000	6·00	8·00	10·00	15·00	20·00	30·00	40·00	60·00
0·00002	1·174	1·176	1·181	1·192	1·202	1·220	1·235	1·248	1·275	1·296	1·330	1·356	1·396
1/50000	95	150	190	290	390	600	800	950	1500	1900	2900	3900	6000
0·00003	1·149	1·153	1·159	1·171	1·182	1·201	1·217	1·231	1·259	1·281	1·315	1·342	1·383
1/33333	110	170	220	330	450	650	900	1100	1700	2200	3300	4500	6500
0·00004	1·132	1·137	1·144	1·157	1·169	1·189	1·205	1·219	1·248	1·270	1·305	1·333	1·374
1/25000	120	180	250	370	490	750	1000	1200	1800	2500	3700	4900	7500
0·00006	1·110	1·116	1·124	1·138	1·151	1·172	1·189	1·204	1·234	1·257	1·293	1·321	1·363
1/16667	140	210	280	420	550	850	1100	1400	2100	2800	4200	5500	8500
0·00008	1·094	1·102	1·111	1·126	1·139	1·161	1·179	1·194	1·224	1·248	1·284	1·313	1·356
1/12500	150	230	310	460	600	950	1200	1500	2300	3100	4600	6000	9500
0·00010	1·083	1·092	1·100	1·116	1·130	1·152	1·171	1·186	1·217	1·241	1·278	1·307	1·351
1/10000	170	250	330	500	650	1000	1300	1700	2500	3300	5000	6500	10000
0·00015	1·063	1·073	1·083	1·100	1·115	1·138	1·157	1·173	1·205	1·230	1·268	1·297	1·342
1/ 6667	190	290	380	550	750	1100	1500	1900	2900	3800	5500	7500	11000
0·00020	1·050	1·061	1·071	1·089	1·104	1·129	1·148	1·165	1·197	1·222	1·261	1·291	1·337
1/ 5000	210	310	420	650	850	1300	1700	2100	3100	4200	6500	8500	13000
0·00030	1·032	1·044	1·056	1·075	1·091	1·116	1·137	1·153	1·187	1·213	1·253	1·283	1·330
1/ 3333	240	360	480	700	950	1400	1900	2400	3600	4800	7000	9500	14000
0·00040	1·020	1·033	1·045	1·066	1·082	1·108	1·129	1·146	1·181	1·207	1·247	1·278	1·325
1/ 2500	260	390	550	800	1100	1600	2100	2600	3900	5500	8000	11000	16000
0·00060	1·004	1·019	1·032	1·053	1·070	1·098	1·119	1·137	1·172	1·199	1·241	1·272	1·320
1/ 1667	300	450	600	900	1200	1800	2400	3000	4500	6000	9000	12000	18000
0·00080	0·994	1·009	1·023	1·045	1·063	1·091	1·113	1·131	1·167	1·194	1·236	1·268	1·316
1/ 1250	330	500	650	1000	1300	2000	2600	3300	5000	6500	10000	13000	20000
0·00100	0·986	1·002	1·016	1·039	1·057	1·086	1·108	1·127	1·163	1·191	1·233	1·265	1·314
1/ 1000	360	550	700	1100	1400	2100	2800	3600	5500	7000	11000	14000	21000
0·00150	0·973	0·991	1·005	1·029	1·048	1·078	1·101	1·120	1·157	1·185	1·228	1·261	1·310
1/ 667	410	600	800	1200	1600	2400	3300	4100	6000	8000	12000	16000	24000
0·00200	0·964	0·983	0·998	1·023	1·042	1·072	1·096	1·115	1·153	1·182	1·225	1·258	1·307
1/ 500	450	650	900	1300	1800	2700	3600	4500	6500	9000	13000	18000	27000
0·00300	0·953	0·973	0·989	1·014	1·035	1·066	1·090	1·109	1·148	1·177	1·221	1·255	1·305
1/ 333	500	750	1000	1500	2000	3100	4100	5000	7500	10000	15000	20000	31000
0·00400	0·946	0·966	0·983	1·009	1·030	1·062	1·086	1·106	1·145	1·174	1·219	1·253	1·303
1/ 250	550	850	1100	1700	2200	3400	4500	5500	8500	11000	17000	22000	34000
0·00600	0·937	0·958	0·976	1·003	1·024	1·056	1·081	1·102	1·141	1·171	1·216	1·250	1·301
1/ 167	650	950	1300	1900	2600	3800	5000	6500	9500	13000	19000	26000	38000
0·00800	0·931	0·953	0·971	0·999	1·020	1·053	1·078	1·099	1·139	1·169	1·214	1·248	1·299
1/ 125	700	1100	1400	2100	2800	4200	5500	7000	11000	14000	21000	28000	42000
0·01000	0·926	0·949	0·967	0·996	1·017	1·051	1·076	1·097	1·137	1·168	1·213	1·247	1·298
1/ 100	750	1100	1500	2300	3000	4500	6000	7500	11000	15000	23000	30000	45000
0·01500	0·919	0·943	0·962	0·991	1·013	1·047	1·073	1·094	1·135	1·165	1·211	1·246	1·297
1/ 67	850	1300	1700	2600	3500	5000	7000	8500	13000	17000	26000	35000	52000
0·02000	0·915	0·939	0·958	0·988	1·011	1·045	1·071	1·092	1·133	1·164	1·210	1·245	1·296
1/ 50	950	1400	1900	2800	3800	5500	7500	9500	14000	19000	28000	38000	57000
0·03000	0·909	0·935	0·954	0·984	1·007	1·042	1·069	1·090	1·131	1·162	1·209	1·243	1·295
1/ 33.3	1100	1600	2200	3200	4300	6500	8500	11000	16000	22000	32000	43000	65000
0·04000	0·906	0·932	0·952	0·982	1·005	1·040	1·067	1·089	1·130	1·161	1·208	1·243	1·294
1/ 25.0	1200	1800	2400	3600	4700	7000	9500	12000	18000	24000	36000	47000	71000
0·06000	0·902	0·928	0·948	0·979	1·003	1·038	1·065	1·087	1·129	1·160	1·207	1·242	1·294
1/ 16.7	1400	2000	2700	4100	5500	8000	11000	14000	20000	27000	41000	54000	81000
0·08000	0·899	0·926	0·946	0·978	1·001	1·037	1·064	1·086	1·128	1·159	1·206	1·241	1·293
1/ 12.5	1500	2200	3000	4400	6000	9000	12000	15000	22000	30000	44000	59000	89000
0·10000	0·897	0·924	0·945	0·977	1·000	1·036	1·063	1·085	1·127	1·159	1·206	1·241	1·293
1/ 10.0	1600	2400	3200	4800	6500	9500	13000	16000	24000	32000	48000	64000	95000
	1000	1500	2000	3000	4000	6·00	8·00	10·00	15·00	20·00	30·00	40·00	60·00

Gradient   (Equivalent) Pipe diameters in mm to 4000 mm, then in m

S = 0·000020 to 0·10000 k_s = 0·030 mm

E5

$k_s = 0.060$ mm
$S = 0.00010$ to 0.60000

ie hydraulic gradient =
1 in 10000 to 1 in 1·7

Water (or sewage) at 15°C; full bore conditions.
This table shows values of m for use with Tables D

m_C for Colebrook-White solutions; or m_P for laminar flow
θ for use with part-full circular pipes (turbulent flows)

Gradient	20	30	40	60	80	100	150	200	300	400	600	800	1000
0·00010	2·722	1·585	1·080	*	1·170	1·149	1·121	1·107	1·095	1·090	1·089	1·091	1·095
1/10000	-	-	-		13	17	24	34	50	65	100	130	170
0·00015	2·222	1·294	*	*	1·133	1·115	1·091	1·080	1·070	1·067	1·068	1·072	1·077
1/ 6667	-	-			15	19	28	38	55	75	110	150	190
0·00020	1·925	1·121	*	*	1·109	1·093	1·071	1·062	1·054	1·052	1·055	1·059	1·065
1/ 5000	-	-			17	20	32	42	60	85	120	170	210
0·00030	1·571	0·915	*	1·099	1·076	1·063	1·045	1·037	1·032	1·032	1·037	1·043	1·049
1/ 3333	-	-		14	19	24	36	48	70	95	140	190	240
0·00040	1·361	*	*	1·075	1·055	1·043	1·028	1·021	1·018	1·019	1·025	1·032	1·039
1/ 2500	-			16	20	26	40	50	80	100	160	210	260
0·00060	1·111	*	*	1·044	1·027	1·017	1·005	1·000	0·999	1·002	1·009	1·017	1·025
1/ 1667	-			18	24	30	44	60	90	120	180	240	300
0·00080	0·962	*	†·054	1·024	1·008	1·000	0·990	0·987	0·987	0·990	0·999	1·008	1·016
1/ 1250	-		13	20	26	32	50	65	100	130	200	260	330
0·00100	0·861	*	1·036	1·009	0·995	0·987	0·979	0·976	0·978	0·982	0·992	1·001	1·010
1/ 1000	-		14	22	28	36	55	70	110	140	210	280	350
0·00150	*	*	1·006	0·983	0·972	0·966	0·960	0·959	0·963	0·968	0·979	0·990	0·999
1/ 667			16	24	32	40	60	80	120	160	240	320	400
0·00200	*	1·008	0·987	0·966	0·956	0·951	0·947	0·948	0·953	0·959	0·971	0·983	0·993
1/ 500		13	18	26	36	44	65	90	130	180	260	350	440
0·00300	*	0·979	0·961	0·944	0·937	0·933	0·931	0·933	0·940	0·947	0·961	0·973	0·984
1/ 333		15	20	30	40	50	75	100	150	200	300	400	500
0·00400	*	0·960	0·945	0·930	0·924	0·921	0·921	0·924	0·932	0·940	0·955	0·967	0·979
1/ 250		17	22	34	44	55	85	110	170	220	330	440	550
0·00600	0·963	0·936	0·923	0·911	0·907	0·906	0·907	0·912	0·921	0·930	0·946	0·960	0·972
1/ 167	13	19	26	38	50	65	95	130	190	250	380	500	650
0·00800	0·944	0·920	0·909	0·899	0·896	0·896	0·899	0·904	0·914	0·924	0·941	0·955	0·968
1/ 125	14	20	28	42	55	70	100	140	210	280	410	550	700
0·01000	0·930	0·909	0·898	0·890	0·888	0·888	0·892	0·898	0·910	0·920	0·938	0·952	0·965
1/ 100	15	22	30	44	60	75	110	150	220	300	440	600	750
0·01500	0·907	0·889	0·881	0·876	0·875	0·876	0·882	0·889	0·902	0·913	0·932	0·947	0·960
1/ 67	17	26	34	50	65	85	130	170	250	340	500	650	850
0·02000	0·893	0·877	0·870	0·866	0·867	0·869	0·876	0·883	0·897	0·909	0·928	0·944	0·957
1/ 50	18	28	36	55	75	90	140	180	280	370	550	750	900
0·03000	0·874	0·861	0·856	0·854	0·856	0·859	0·868	0·876	0·891	0·903	0·923	0·940	0·953
1/ 33·3	20	32	42	65	85	100	160	210	310	420	650	850	1000
0·04000	0·862	0·851	0·847	0·847	0·849	0·853	0·863	0·871	0·887	0·900	0·921	0·937	0·951
1/ 25·0	22	34	46	70	90	110	170	230	340	460	700	900	1100
0·06000	0·846	0·838	0·836	0·837	0·841	0·845	0·856	0·866	0·882	0·896	0·917	0·934	0·948
1/ 16·7	26	38	50	80	100	130	190	260	390	500	800	1000	1300
0·08000	0·836	0·830	0·829	0·831	0·836	0·841	0·852	0·862	0·879	0·893	0·915	0·932	0·946
1/ 12·5	28	42	55	85	110	140	210	280	430	550	850	1100	1400
0·10000	0·829	0·824	0·824	0·827	0·832	0·837	0·849	0·860	0·877	0·891	0·914	0·931	0·945
1/ 10·0	30	46	60	90	120	150	230	300	460	600	900	1200	1500
0·15000	0·818	0·815	0·816	0·821	0·826	0·832	0·845	0·856	0·874	0·889	0·911	0·929	0·944
1/ 6·7	34	50	70	100	140	170	260	340	500	700	1000	1400	1700
0·20000	0·811	0·809	0·810	0·816	0·823	0·829	0·842	0·854	0·872	0·887	0·910	0·928	0·942
1/ 5·0	38	55	75	110	150	190	280	370	550	750	1100	1500	1900
0·30000	0·802	0·802	0·804	0·811	0·818	0·825	0·839	0·851	0·870	0·885	0·908	0·926	0·941
1/ 3·3	42	65	85	130	170	210	320	420	650	850	1300	1700	2100
0·40000	0·796	0·797	0·800	0·808	0·815	0·822	0·837	0·849	0·868	0·884	0·907	0·925	0·940
1/ 2·5	46	70	90	140	180	230	340	460	700	900	1400	1800	2300
0·60000	0·789	0·792	0·796	0·804	0·812	0·819	0·835	0·847	0·867	0·882	0·906	0·924	0·939
1/ 1·7	50	75	100	150	210	260	390	500	750	1000	1500	2100	2600
	20	30	40	60	80	100	150	200	300	400	600	800	1000

Gradient   (Equivalent) Pipe diameters in mm

$k_s = 0.060$ mm $S = 0.00010$ to 0.60000

k_s = 0·060 mm
S = 0·000020 to 0·10000

Water (or sewage) at 15°C; full bore conditions.
This table shows values of m for use with Tables D

ie hydraulic gradient =
1 in 50000 to 1 in 10·0

m_C for Colebrook-White solutions;
θ for use with part-full circular pipes

Gradient	(Equivalent) Pipe diameters in mm to 4000 mm, then in m												
	1000	1500	2000	3000	4000	6·00	8·00	10·00	15·00	20·00	30·00	40·00	60·00
0·00002	1·181	1·184	1·190	1·201	1·213	1·232	1·249	1·263	1·292	1·315	1·351	1·379	1·422
1/50000	95	150	190	290	390	600	800	950	1500	1900	2900	3900	6000
0·00003	1·157	1·162	1·169	1·182	1·195	1·215	1·233	1·248	1·278	1·301	1·339	1·367	1·411
1/33333	110	170	220	330	440	650	900	1100	1700	2200	3300	4400	6500
0·00004	1·141	1·147	1·155	1·170	1·182	1·204	1·222	1·238	1·268	1·293	1·331	1·360	1·405
1/25000	120	180	240	370	490	750	1000	1200	1800	2400	3700	4900	7500
0·00006	1·120	1·128	1·137	1·153	1·167	1·190	1·208	1·224	1·257	1·282	1·321	1·351	1·397
1/16667	140	210	280	420	550	850	1100	1400	2100	2800	4200	5500	8500
0·00008	1·105	1·115	1·124	1·141	1·156	1·180	1·199	1·216	1·249	1·275	1·314	1·345	1·391
1/12500	150	230	310	460	600	900	1200	1500	2300	3100	4600	6000	9000
0·00010	1·095	1·105	1·115	1·133	1·148	1·173	1·193	1·210	1·243	1·269	1·310	1·341	1·388
1/10000	170	250	330	500	650	1000	1300	1700	2500	3300	5000	6500	10000
0·00015	1·077	1·089	1·100	1·119	1·135	1·161	1·182	1·199	1·234	1·261	1·302	1·334	1·382
1/ 6667	190	280	380	550	750	1100	1500	1900	2800	3800	5500	7500	11000
0·00020	1·065	1·078	1·090	1·110	1·127	1·153	1·175	1·193	1·228	1·256	1·297	1·329	1·378
1/ 5000	210	310	420	600	850	1200	1700	2100	3100	4200	6000	8500	12000
0·00030	1·049	1·064	1·076	1·098	1·116	1·144	1·166	1·184	1·221	1·249	1·291	1·324	1·373
1/ 3333	240	360	480	700	950	1400	1900	2400	3600	4800	7000	9500	14000
0·00040	1·039	1·054	1·068	1·090	1·109	1·137	1·160	1·179	1·216	1·244	1·288	1·321	1·370
1/ 2500	260	390	500	800	1000	1600	2100	2600	3900	5000	8000	10000	16000
0·00060	1·025	1·042	1·057	1·080	1·100	1·129	1·153	1·172	1·210	1·239	1·283	1·317	1·367
1/ 1667	300	450	600	900	1200	1800	2400	3000	4500	6000	9000	12000	18000
0·00080	1·016	1·034	1·050	1·074	1·094	1·124	1·148	1·168	1·206	1·236	1·280	1·314	1·365
1/ 1250	330	490	650	1000	1300	2000	2600	3300	4900	6500	10000	13000	20000
0·00100	1·010	1·029	1·045	1·070	1·090	1·121	1·145	1·165	1·204	1·234	1·278	1·312	1·363
1/ 1000	350	550	700	1100	1400	2100	2800	3500	5500	7000	11000	14000	21000
0·00150	0·999	1·020	1·036	1·062	1·083	1·115	1·140	1·160	1·200	1·230	1·275	1·309	1·361
1/ 667	400	600	800	1200	1600	2400	3200	4000	6000	8000	12000	16000	24000
0·00200	0·993	1·014	1·031	1·058	1·079	1·111	1·137	1·157	1·197	1·228	1·273	1·308	1·359
1/ 500	440	650	900	1300	1800	2600	3500	4400	6500	9000	13000	18000	26000
0·00300	0·984	1·006	1·024	1·052	1·074	1·107	1·133	1·154	1·194	1·225	1·271	1·306	1·358
1/ 333	500	750	1000	1500	2000	3000	4000	5000	7500	10000	15000	20000	30000
0·00400	0·979	1·001	1·020	1·048	1·070	1·104	1·130	1·151	1·192	1·223	1·270	1·304	1·356
1/ 250	550	850	1100	1700	2200	3300	4400	5500	8500	11000	17000	22000	33000
0·00600	0·972	0·996	1·015	1·044	1·066	1·101	1·127	1·149	1·190	1·221	1·268	1·303	1·355
1/ 167	650	950	1300	1900	2500	3800	5000	6500	9500	13000	19000	25000	38000
0·00800	0·968	0·992	1·011	1·041	1·064	1·099	1·125	1·147	1·189	1·220	1·267	1·302	1·354
1/ 125	700	1000	1400	2100	2800	4100	5500	7000	10000	14000	21000	28000	41000
0·01000	0·965	0·989	1·009	1·039	1·062	1·097	1·124	1·146	1·188	1·219	1·266	1·301	1·354
1/ 100	750	1100	1500	2200	3000	4400	6000	7500	11000	15000	22000	30000	44000
0·01500	0·960	0·985	1·005	1·036	1·060	1·095	1·122	1·144	1·186	1·218	1·265	1·300	1·353
1/ 67	850	1300	1700	2500	3400	5000	6500	8500	13000	17000	25000	34000	51000
0·02000	0·957	0·983	1·003	1·034	1·058	1·094	1·121	1·143	1·185	1·217	1·264	1·300	1·353
1/ 50	900	1400	1800	2800	3700	5500	7500	9000	14000	18000	28000	37000	55000
0·03000	0·953	0·980	1·001	1·032	1·056	1·092	1·119	1·142	1·184	1·216	1·264	1·299	1·352
1/ 33·3	1000	1600	2100	3100	4200	6500	8500	10000	16000	21000	31000	42000	63000
0·04000	0·951	0·978	0·999	1·031	1·055	1·091	1·119	1·141	1·183	1·215	1·263	1·299	1·352
1/ 25·0	1100	1700	2300	3400	4600	7000	9000	11000	17000	23000	34000	46000	69000
0·06000	0·948	0·976	0·997	1·029	1·053	1·090	1·118	1·140	1·183	1·215	1·263	1·298	1·351
1/ 16·7	1300	1900	2600	3900	5000	8000	10000	13000	19000	26000	39000	52000	78000
0·08000	0·946	0·974	0·996	1·028	1·052	1·089	1·117	1·139	1·182	1·214	1·262	1·298	1·351
1/ 12·5	1400	2100	2800	4300	5500	8500	11000	14000	21000	28000	43000	57000	85000
0·10000	0·945	0·973	0·995	1·027	1·052	1·089	1·116	1·139	1·182	1·214	1·262	1·298	1·351
1/ 10·0	1500	2300	3000	4600	6000	9000	12000	15000	23000	30000	46000	61000	91000
	1000	1500	2000	3000	4000	6·00	8·00	10·00	15·00	20·00	30·00	40·00	60·00

Gradient   (Equivalent) Pipe diameters in mm to 4000 mm, then in m

S = 0·000020 to 0·10000 k_s = 0·060 mm

k_s = 0·150 mm
S = 0·00010 to 0·60000

Water (or sewage) at 15°C; full bore conditions.
This table shows values of *m* for use with Tables D

ie hydraulic gradient =
1 in 10000 to 1 in 1·7

m_C for Colebrook-White solutions; or m_P for laminar flow
θ for use with part-full circular pipes (turbulent flows)

Gradient	20	30	40	60	80	100	150	200	300	400	600	800	1000
0·00010	2·722	1·585	1·080	*	1·189	1·169	1·142	1·129	1·119	1·115	1·117	1·121	1·126
1/10000	-	-	-		13	16	24	32	48	65	100	130	160
0·00015	2·222	1·294	*	*	1·155	1·138	1·115	1·105	1·097	1·096	1·100	1·105	1·112
1/ 6667	-	-			15	19	28	38	55	75	110	150	190
0·00020	1·925	1·121	*	*	1·132	1·117	1·097	1·089	1·083	1·083	1·088	1·095	1·102
1/ 5000	-	-			16	20	30	40	60	80	120	160	200
0·00030	1·571	0·915	*	1·125	1·103	1·090	1·074	1·068	1·065	1·067	1·074	1·082	1·090
1/ 3333	-	-		14	19	24	34	46	70	95	140	190	230
0·00040	1·361	*	*	1·103	1·084	1·073	1·059	1·055	1·054	1·056	1·065	1·074	1·083
1/ 2500	-			15	20	26	38	50	75	100	150	200	260
0·00060	1·111	*	*	1·075	1·059	1·050	1·040	1·037	1·039	1·043	1·053	1·064	1·073
1/ 1667	-			17	24	30	44	60	85	120	170	230	290
0·00080	0·962	*	1·086	1·057	1·043	1·035	1·028	1·026	1·029	1·034	1·046	1·057	1·067
1/ 1250	-		13	19	26	32	48	65	95	130	190	250	320
0·00100	0·861	*	1·070	1·044	1·032	1·025	1·019	1·018	1·022	1·028	1·041	1·052	1·063
1/ 1000	-		14	20	28	34	50	70	100	140	200	270	340
0·00150	*	*	1·044	1·022	1·012	1·007	1·004	1·005	1·011	1·018	1·032	1·045	1·056
1/ 667			16	24	32	38	60	80	120	160	230	310	390
0·00200	*	1·048	1·028	1·008	1·000	0·996	0·994	0·996	1·004	1·012	1·027	1·040	1·052
1/ 500		13	17	26	34	42	65	85	130	170	250	340	420
0·00300	*	1·023	1·006	0·990	0·984	0·982	0·982	0·986	0·995	1·004	1·020	1·034	1·046
1/ 333		14	19	28	38	48	70	95	140	190	290	390	480
0·00400	*	1·007	0·992	0·979	0·974	0·973	0·975	0·979	0·989	0·999	1·016	1·031	1·043
1/ 250		16	22	32	42	55	80	110	160	210	320	420	550
0·00600	*	0·987	0·975	0·965	0·962	0·961	0·965	0·971	0·982	0·993	1·011	1·026	1·039
1/ 167		18	24	36	48	60	90	120	180	240	360	480	600
0·00800	0·998	0·975	0·964	0·956	0·954	0·954	0·959	0·966	0·978	0·989	1·008	1·024	1·037
1/ 125	13	19	26	38	50	65	95	130	190	260	390	500	650
0·01000	0·987	0·966	0·956	0·949	0·948	0·949	0·955	0·962	0·975	0·987	1·006	1·022	1·035
1/ 100	14	20	28	42	55	70	100	140	210	280	420	550	700
0·01500	0·969	0·951	0·944	0·939	0·939	0·941	0·949	0·956	0·970	0·983	1·003	1·019	1·032
1/ 67	16	24	32	48	65	80	120	160	240	310	470	650	800
0·02000	0·957	0·942	0·936	0·932	0·934	0·936	0·944	0·953	0·968	0·980	1·001	1·017	1·031
1/ 50	17	26	34	50	70	85	130	170	260	340	500	700	850
0·03000	0·943	0·930	0·926	0·924	0·927	0·930	0·939	0·948	0·964	0·977	0·998	1·015	1·029
1/ 33·3	19	28	38	55	75	95	140	190	290	380	550	750	950
0·04000	0·934	0·923	0·920	0·920	0·922	0·926	0·936	0·946	0·962	0·975	0·997	1·014	1·028
1/ 25·0	20	32	42	60	85	100	160	210	310	420	600	850	1000
0·06000	0·923	0·914	0·912	0·914	0·917	0·922	0·933	0·943	0·959	0·973	0·995	1·012	1·026
1/ 16·7	24	34	46	70	95	120	170	230	350	470	700	950	1200
0·08000	0·916	0·909	0·908	0·910	0·914	0·919	0·930	0·941	0·958	0·972	0·994	1·011	1·025
1/ 12·5	26	38	50	75	100	130	190	250	380	500	750	1000	1300
0·10000	0·911	0·905	0·904	0·907	0·912	0·917	0·929	0·939	0·957	0·971	0·993	1·010	1·025
1/ 10·0	26	40	55	80	110	130	200	270	400	550	800	1100	1300
0·15000	0·904	0·899	0·899	0·903	0·909	0·914	0·926	0·937	0·955	0·969	0·992	1·009	1·024
1/ 6·7	30	44	60	90	120	150	220	300	450	600	900	1200	1500
0·20000	0·899	0·896	0·896	0·901	0·906	0·912	0·925	0·936	0·954	0·968	0·991	1·009	1·023
1/ 5·0	32	48	65	95	130	160	240	320	480	650	950	1300	1600
0·30000	0·893	0·891	0·892	0·898	0·904	0·910	0·923	0·934	0·953	0·967	0·990	1·008	1·023
1/ 3·3	36	55	70	110	140	180	270	350	550	700	1100	1400	1800
0·40000	0·890	0·888	0·890	0·896	0·902	0·909	0·922	0·933	0·952	0·967	0·990	1·008	1·022
1/ 2·5	38	55	75	110	150	190	280	380	550	750	1100	1500	1900
0·60000	0·886	0·885	0·887	0·894	0·901	0·907	0·921	0·932	0·951	0·966	0·989	1·007	1·022
1/ 1·7	42	65	85	130	170	210	310	420	650	850	1300	1700	2100
Gradient	20	30	40	60	80	100	150	200	300	400	600	800	1000

(Equivalent) Pipe diameters in mm

k_s = 0·150 mm   S = 0·00010 to 0·60000

k_s = 0·150 mm
S = 0·000020 to 0·10000

Water (or sewage) at 15°C; full bore conditions.
This table shows values of *m* for use with Tables D

ie hydraulic gradient =
1 in 50000 to 1 in 10·0

m_C for Colebrook-White solutions;
θ for use with part-full circular pipes

Gradient — (Equivalent) Pipe diameters in mm to 4000 mm, then in m

Gradient	1000	1500	2000	3000	4000	6·00	8·00	10·00	15·00	20·00	30·00	40·00	60·00
0·00002	1·200	1·206	1·213	1·227	1·241	1·263	1·282	1·298	1·331	1·357	1·397	1·428	1·476
1/50000	95	140	190	290	390	600	750	950	1400	1900	2900	3900	6000
0·00003	1·179	1·187	1·195	1·212	1·226	1·250	1·270	1·287	1·321	1·348	1·389	1·421	1·470
1/33333	110	170	220	330	440	650	900	1100	1700	2200	3300	4400	6500
0·00004	1·165	1·174	1·184	1·201	1·216	1·242	1·262	1·280	1·315	1·342	1·384	1·417	1·466
1/25000	120	180	240	360	480	750	950	1200	1800	2400	3600	4800	7500
0·00006	1·147	1·158	1·169	1·188	1·204	1·231	1·253	1·271	1·307	1·335	1·378	1·411	1·461
1/16667	140	210	280	410	550	850	1100	1400	2100	2800	4100	5500	8500
0·00008	1·135	1·147	1·159	1·179	1·196	1·224	1·246	1·265	1·302	1·330	1·374	1·407	1·458
1/12500	150	230	300	450	600	900	1200	1500	2300	3000	4500	6000	9000
0·00010	1·126	1·140	1·152	1·173	1·191	1·219	1·242	1·261	1·298	1·327	1·371	1·405	1·456
1/10000	160	240	330	490	650	1000	1300	1600	2400	3300	4900	6500	10000
0·00015	1·112	1·127	1·140	1·163	1·182	1·211	1·235	1·254	1·293	1·322	1·367	1·401	1·452
1/ 6667	190	280	370	550	750	1100	1500	1900	2800	3700	5500	7500	11000
0·00020	1·102	1·119	1·133	1·157	1·176	1·206	1·230	1·250	1·289	1·319	1·364	1·399	1·450
1/ 5000	200	310	410	600	800	1200	1600	2000	3100	4100	6000	8000	12000
0·00030	1·090	1·108	1·124	1·149	1·169	1·200	1·225	1·245	1·285	1·315	1·361	1·396	1·448
1/ 3333	230	350	470	700	950	1400	1900	2300	3500	4700	7000	9500	14000
0·00040	1·083	1·102	1·118	1·143	1·164	1·196	1·221	1·242	1·282	1·313	1·359	1·394	1·446
1/ 2500	260	380	500	750	1000	1500	2000	2600	3800	5000	7500	10000	15000
0·00060	1·073	1·094	1·110	1·137	1·159	1·192	1·217	1·238	1·279	1·310	1·357	1·392	1·445
1/ 1667	290	440	600	850	1200	1700	2300	2900	4400	6000	8500	12000	17000
0·00080	1·067	1·088	1·106	1·133	1·155	1·189	1·214	1·236	1·277	1·308	1·355	1·391	1·444
1/ 1250	320	480	650	950	1300	1900	2500	3200	4800	6500	9500	13000	19000
0·00100	1·063	1·085	1·103	1·131	1·153	1·187	1·213	1·234	1·276	1·307	1·354	1·390	1·443
1/ 1000	340	500	700	1000	1400	2000	2700	3400	5000	7000	10000	14000	20000
0·00150	1·056	1·079	1·097	1·126	1·149	1·183	1·210	1·232	1·273	1·305	1·353	1·388	1·442
1/ 667	390	600	800	1200	1600	2300	3100	3900	6000	8000	12000	16000	23000
0·00200	1·052	1·075	1·094	1·124	1·146	1·181	1·208	1·230	1·272	1·304	1·352	1·388	1·441
1/ 500	420	650	850	1300	1700	2500	3400	4200	6500	8500	13000	17000	25000
0·00300	1·046	1·071	1·090	1·120	1·144	1·179	1·206	1·228	1·271	1·303	1·351	1·387	1·440
1/ 333	480	700	950	1400	1900	2900	3900	4800	7000	9500	14000	19000	29000
0·00400	1·043	1·068	1·088	1·118	1·142	1·178	1·205	1·227	1·270	1·302	1·350	1·386	1·440
1/ 250	550	800	1100	1600	2100	3200	4200	5500	8000	11000	16000	21000	32000
0·00600	1·039	1·065	1·085	1·116	1·140	1·176	1·203	1·226	1·269	1·301	1·349	1·385	1·439
1/ 167	600	900	1200	1800	2400	3600	4800	6000	9000	12000	18000	24000	36000
0·00800	1·037	1·063	1·083	1·115	1·139	1·175	1·202	1·225	1·268	1·300	1·349	1·385	1·439
1/ 125	650	950	1300	1900	2600	3900	5000	6500	9500	13000	19000	26000	39000
0·01000	1·035	1·061	1·082	1·114	1·138	1·174	1·202	1·224	1·267	1·300	1·348	1·385	1·439
1/ 100	700	1000	1400	2100	2800	4200	5500	7000	10000	14000	21000	28000	42000
0·01500	1·032	1·059	1·080	1·112	1·136	1·173	1·201	1·223	1·267	1·299	1·348	1·384	1·438
1/ 67	800	1200	1600	2400	3100	4700	6500	8000	12000	16000	24000	31000	47000
0·02000	1·031	1·058	1·079	1·111	1·135	1·172	1·200	1·223	1·266	1·299	1·347	1·384	1·438
1/ 50	850	1300	1700	2600	3400	5000	7000	8500	13000	17000	26000	34000	51000
0·03000	1·029	1·056	1·078	1·110	1·135	1·172	1·200	1·222	1·266	1·298	1·347	1·383	1·438
1/ 33·3	950	1400	1900	2900	3800	5500	7500	9500	14000	19000	29000	38000	57000
0·04000	1·028	1·055	1·077	1·109	1·134	1·171	1·199	1·222	1·265	1·298	1·347	1·383	1·437
1/ 25·0	1000	1600	2100	3100	4200	6000	8500	10000	16000	21000	31000	42000	62000
0·06000	1·026	1·054	1·076	1·108	1·133	1·170	1·199	1·221	1·265	1·298	1·347	1·383	1·437
1/ 16·7	1200	1700	2300	3500	4700	7000	9500	12000	17000	23000	35000	47000	70000
0·08000	1·025	1·054	1·075	1·108	1·133	1·170	1·198	1·221	1·265	1·298	1·346	1·383	1·437
1/ 12·5	1300	1900	2500	3800	5000	7500	10000	13000	19000	25000	38000	50000	76000
0·10000	1·025	1·053	1·075	1·108	1·133	1·170	1·198	1·221	1·265	1·297	1·346	1·383	1·437
1/ 10·0	1300	2000	2700	4000	5500	8000	11000	13000	20000	27000	40000	53000	80000
	1000	1500	2000	3000	4000	6·00	8·00	10·00	15·00	20·00	30·00	40·00	60·00

Gradient — (Equivalent) Pipe diameters in mm to 4000 mm, then in m

S = 0·000020 to 0·10000   **k_s = 0·150 mm**

E7

k_s = 0.30 mm
S = 0.00010 to 0.60000

Water (or sewage) at 15°C; full bore conditions.
This table shows values of m for use with Tables D

ie hydraulic gradient =
1 in 10000 to 1 in 1.7

m_C for Colebrook-White solutions; or m_P for laminar flow
θ   for use with part-full circular pipes (turbulent flows)

Gradient	(Equivalent) Pipe diameters in mm												
	20	30	40	60	80	100	150	200	300	400	600	800	1000
0.00010	2.722	1.585	1.080	*	1.219	1.200	1.174	1.162	1.153	1.152	1.155	1.161	1.168
1/10000	-	-	-		13	16	24	32	48	65	95	130	160
0.00015	2.222	1.294	*	*	1.188	1.171	1.150	1.141	1.135	1.136	1.141	1.149	1.157
1/ 6667	-	-			14	18	28	36	55	70	110	140	180
0.00020	1.925	1.121	*	*	1.168	1.153	1.135	1.128	1.124	1.126	1.133	1.141	1.150
1/ 5000	-	-			16	20	30	40	60	80	120	160	200
0.00030	1.571	0.915	*	1.163	1.142	1.130	1.116	1.111	1.110	1.113	1.122	1.132	1.141
1/ 3333	-	-		13	18	22	34	44	65	90	130	180	220
0.00040	1.361	*	*	1.144	1.126	1.115	1.103	1.100	1.100	1.105	1.115	1.126	1.135
1/ 2500	-			15	20	24	36	50	75	100	150	200	250
0.00060	1.111	*	*	1.120	1.105	1.096	1.088	1.086	1.089	1.095	1.107	1.118	1.129
1/ 1667	-			17	22	28	42	55	85	110	170	220	280
0.00080	0.962	*	1.133	1.105	1.092	1.084	1.078	1.077	1.082	1.088	1.102	1.114	1.125
1/ 1250	-		12	18	24	30	46	60	90	120	180	240	300
0.00100	0.861	*	1.120	1.094	1.082	1.076	1.071	1.071	1.077	1.084	1.098	1.111	1.122
1/ 1000	-		13	19	26	32	48	65	95	130	190	260	320
0.00150	*	*	1.098	1.076	1.067	1.062	1.059	1.061	1.069	1.077	1.092	1.105	1.117
1/ 667			15	22	30	36	55	75	110	150	220	290	370
0.00200	*	1.104	1.084	1.065	1.057	1.053	1.052	1.055	1.064	1.072	1.089	1.102	1.114
1/ 500		12	16	24	32	40	60	80	120	160	240	320	400
0.00300	*	1.084	1.067	1.051	1.045	1.043	1.044	1.048	1.057	1.067	1.084	1.099	1.111
1/ 333		13	18	26	36	44	65	90	130	180	270	360	450
0.00400	*	1.071	1.056	1.042	1.037	1.036	1.038	1.043	1.054	1.064	1.081	1.096	1.109
1/ 250		15	20	30	40	48	75	100	150	200	290	390	490
0.00600	*	1.055	1.042	1.032	1.028	1.028	1.032	1.037	1.049	1.060	1.078	1.093	1.106
1/ 167		16	22	32	44	55	80	110	160	220	330	440	550
0.00800	1.071	1.046	1.034	1.025	1.023	1.023	1.028	1.034	1.046	1.057	1.076	1.092	1.105
1/ 125	12	18	24	36	48	60	90	120	180	240	360	470	600
0.01000	1.062	1.039	1.028	1.020	1.019	1.019	1.025	1.031	1.044	1.056	1.075	1.091	1.104
1/ 100	13	19	26	38	50	65	95	130	190	250	380	500	650
0.01500	1.048	1.028	1.019	1.013	1.012	1.014	1.020	1.028	1.041	1.053	1.073	1.089	1.102
1/ 67	14	22	28	42	55	70	110	140	210	280	420	550	700
0.02000	1.039	1.021	1.013	1.008	1.009	1.010	1.018	1.025	1.039	1.052	1.072	1.088	1.101
1/ 50	15	22	30	46	60	75	110	150	230	300	450	600	750
0.03000	1.028	1.012	1.006	1.003	1.004	1.006	1.014	1.023	1.037	1.050	1.070	1.086	1.100
1/ 33.3	17	26	34	50	65	85	130	170	250	340	500	650	850
0.04000	1.022	1.007	1.002	1.000	1.001	1.004	1.012	1.021	1.036	1.049	1.069	1.086	1.099
1/ 25.0	18	26	36	55	70	90	130	180	270	360	550	700	900
0.06000	1.014	1.001	0.997	0.995	0.998	1.001	1.010	1.019	1.034	1.047	1.068	1.085	1.099
1/ 16.7	20	30	40	60	80	100	150	200	300	400	600	800	1000
0.08000	1.009	0.998	0.994	0.993	0.996	0.999	1.009	1.018	1.033	1.046	1.067	1.084	1.098
1/ 12.5	22	32	42	65	85	110	160	210	320	420	650	850	1100
0.10000	1.005	0.995	0.992	0.991	0.994	0.998	1.008	1.017	1.033	1.046	1.067	1.084	1.098
1/ 10.0	22	34	44	65	90	110	170	220	330	450	650	900	1100
0.15000	1.000	0.991	0.988	0.989	0.992	0.996	1.006	1.016	1.032	1.045	1.066	1.083	1.097
1/ 6.7	24	36	48	75	95	120	180	240	360	490	750	950	1200
0.20000	0.997	0.988	0.986	0.987	0.991	0.995	1.005	1.015	1.031	1.044	1.066	1.083	1.097
1/ 5.0	26	38	50	75	100	130	190	260	390	500	750	1000	1300
0.30000	0.993	0.986	0.984	0.985	0.989	0.993	1.004	1.014	1.030	1.044	1.065	1.082	1.097
1/ 3.3	28	42	55	85	110	140	210	280	420	550	850	1100	1400
0.40000	0.991	0.984	0.982	0.984	0.988	0.993	1.004	1.013	1.030	1.043	1.065	1.082	1.096
1/ 2.5	30	44	60	90	120	150	220	300	440	600	900	1200	1500
0.60000	0.988	0.982	0.980	0.983	0.987	0.992	1.003	1.013	1.029	1.043	1.065	1.082	1.096
1/ 1.7	32	48	65	95	130	160	240	320	480	650	950	1300	1600
	20	30	40	60	80	100	150	200	300	400	600	800	1000

Gradient   (Equivalent) Pipe diameters in mm

k_s = 0.30 mm S = 0.00010 to 0.60000

$k_s = 0.30$ mm
$S = 0.000020$ to 0.10000

Water (or sewage) at 15°C; full bore conditions.
This table shows values of m for use with Tables D

ie hydraulic gradient =
1 in 50000 to 1 in 10·0

m_C for Colebrook-White solutions;
θ for use with part-full circular pipes

Gradient	(Equivalent) Pipe diameters in mm to 4000 mm, then in m												
	1000	1500	2000	3000	4000	6·00	8·00	10·00	15·00	20·00	30·00	40·00	60·00
0·00002	1·229	1·236	1·245	1·263	1·278	1·304	1·325	1·343	1·379	1·407	1·450	1·483	1·534
1/50000	95	140	190	280	380	550	750	950	1400	1900	2800	3800	5500
0·00003	1·211	1·221	1·231	1·250	1·267	1·294	1·316	1·335	1·372	1·401	1·445	1·479	1·530
1/33333	110	160	220	320	430	650	850	1100	1600	2200	3200	4300	6500
0·00004	1·199	1·210	1·222	1·242	1·260	1·288	1·311	1·330	1·367	1·397	1·441	1·476	1·527
1/25000	120	180	240	360	470	700	950	1200	1800	2400	3600	4700	7000
0·00006	1·184	1·198	1·210	1·232	1·251	1·280	1·304	1·323	1·362	1·392	1·437	1·472	1·524
1/16667	140	200	270	410	550	800	1100	1400	2000	2700	4100	5500	8000
0·00008	1·174	1·189	1·203	1·226	1·245	1·275	1·299	1·319	1·359	1·389	1·435	1·470	1·522
1/12500	150	220	300	440	600	900	1200	1500	2200	3000	4400	6000	9000
0·00010	1·168	1·184	1·198	1·222	1·241	1·272	1·297	1·317	1·357	1·387	1·433	1·468	1·521
1/10000	160	240	320	480	650	950	1300	1600	2400	3200	4800	6500	9500
0·00015	1·157	1·174	1·189	1·214	1·235	1·267	1·292	1·312	1·353	1·384	1·431	1·466	1·519
1/ 6667	180	270	360	550	700	1100	1400	1800	2700	3600	5500	7000	11000
0·00020	1·150	1·168	1·184	1·210	1·231	1·263	1·289	1·310	1·351	1·382	1·429	1·465	1·518
1/ 5000	200	300	400	600	800	1200	1600	2000	3000	4000	6000	8000	12000
0·00030	1·141	1·161	1·178	1·205	1·226	1·259	1·285	1·307	1·348	1·380	1·427	1·463	1·517
1/ 3333	220	340	450	650	900	1300	1800	2200	3400	4500	6500	9000	13000
0·00040	1·135	1·156	1·174	1·201	1·223	1·257	1·283	1·305	1·346	1·378	1·426	1·462	1·516
1/ 2500	250	370	490	750	1000	1500	2000	2500	3700	4900	7500	10000	15000
0·00060	1·129	1·151	1·169	1·197	1·220	1·254	1·281	1·302	1·345	1·377	1·425	1·461	1·515
1/ 1667	280	420	550	850	1100	1700	2200	2800	4200	5500	8500	11000	17000
0·00080	1·125	1·147	1·166	1·195	1·217	1·252	1·279	1·301	1·343	1·375	1·424	1·460	1·514
1/ 1250	300	460	600	900	1200	1800	2400	3000	4600	6000	9000	12000	18000
0·00100	1·122	1·145	1·164	1·193	1·216	1·251	1·278	1·300	1·343	1·375	1·423	1·459	1·514
1/ 1000	320	490	650	950	1300	1900	2600	3200	4900	6500	9500	13000	19000
0·00150	1·117	1·141	1·160	1·190	1·214	1·249	1·276	1·298	1·341	1·374	1·422	1·459	1·513
1/ 667	370	550	750	1100	1500	2200	2900	3700	5500	7500	11000	15000	22000
0·00200	1·114	1·139	1·158	1·189	1·212	1·248	1·275	1·298	1·341	1·373	1·422	1·458	1·513
1/ 500	400	600	800	1200	1600	2400	3200	4000	6000	8000	12000	16000	24000
0·00300	1·111	1·136	1·156	1·187	1·210	1·247	1·274	1·296	1·340	1·372	1·421	1·458	1·512
1/ 333	450	650	900	1300	1800	2700	3600	4500	6500	9000	13000	18000	27000
0·00400	1·109	1·134	1·155	1·186	1·209	1·246	1·273	1·296	1·339	1·372	1·421	1·457	1·512
1/ 250	490	750	1000	1500	2000	2900	3900	4900	7500	10000	15000	20000	29000
0·00600	1·106	1·132	1·153	1·184	1·208	1·245	1·272	1·295	1·338	1·371	1·420	1·457	1·512
1/ 167	550	800	1100	1600	2200	3300	4400	5500	8000	11000	16000	22000	33000
0·00800	1·105	1·131	1·152	1·183	1·207	1·244	1·272	1·295	1·338	1·371	1·420	1·457	1·511
1/ 125	600	900	1200	1800	2400	3600	4700	6000	9000	12000	18000	24000	36000
0·01000	1·104	1·130	1·151	1·183	1·207	1·244	1·272	1·294	1·338	1·371	1·420	1·456	1·511
1/ 100	650	950	1300	1900	2500	3800	5000	6500	9500	13000	19000	25000	38000
0·01500	1·102	1·129	1·150	1·182	1·206	1·243	1·271	1·294	1·337	1·370	1·419	1·456	1·511
1/ 67	700	1100	1400	2100	2800	4200	5500	7000	11000	14000	21000	28000	42000
0·02000	1·101	1·128	1·149	1·181	1·206	1·243	1·271	1·293	1·337	1·370	1·419	1·456	1·511
1/ 50	750	1100	1500	2300	3000	4500	6000	7500	11000	15000	23000	30000	45000
0·03000	1·100	1·127	1·148	1·181	1·205	1·242	1·270	1·293	1·337	1·370	1·419	1·456	1·511
1/ 33.3	850	1300	1700	2500	3400	5000	6500	8500	13000	17000	25000	34000	50000
0·04000	1·099	1·127	1·148	1·180	1·205	1·242	1·270	1·293	1·337	1·370	1·419	1·456	1·511
1/ 25.0	900	1300	1800	2700	3600	5500	7000	9000	13000	18000	27000	36000	54000
0·06000	1·099	1·126	1·147	1·180	1·204	1·242	1·270	1·293	1·336	1·369	1·419	1·456	1·511
1/ 16.7	1000	1500	2000	3000	4000	6000	8000	10000	15000	20000	30000	40000	59000
0·08000	1·098	1·126	1·147	1·179	1·204	1·241	1·270	1·292	1·336	1·369	1·419	1·456	1·511
1/ 12.5	1100	1600	2100	3200	4200	6500	8500	11000	16000	21000	32000	42000	64000
0·10000	1·098	1·125	1·147	1·179	1·204	1·241	1·269	1·292	1·336	1·369	1·419	1·456	1·510
1/ 10.0	1100	1700	2200	3300	4500	6500	9000	11000	17000	22000	33000	45000	67000
	1000	1500	2000	3000	4000	6·00	8·00	10·00	15·00	20·00	30·00	40·00	60·00

Gradient   (Equivalent) Pipe diameters in mm to 4000 mm, then in m

$S = 0.000020$ to 0.10000 $k_s = 0.30$ mm

E8

$k_s = 0.60$ mm
$S = 0.00010$ to 0.60000

Water (or sewage) at 15°C; full bore conditions.
This table shows values of m for use with Tables D

ie hydraulic gradient =
1 in 10000 to 1 in 1.7

m_C for Colebrook-White solutions; or m_P for laminar flow
θ for use with part-full circular pipes (turbulent flows)

Gradient	20	30	40	60	80	100	150	200	300	400	600	800	1000
0.00010	2.722	1.585	1.080	*	1.272	1.253	1.228	1.217	1.210	1.209	1.214	1.222	1.229
1/10000	-	-	-		12	15	22	30	46	60	90	120	150
0.00015	2.222	1.294	*	*	1.246	1.229	1.209	1.200	1.196	1.197	1.204	1.213	1.221
1/ 6667	-	-			14	17	26	34	50	70	100	140	170
0.00020	1.925	1.121	*	*	1.229	1.214	1.197	1.190	1.187	1.190	1.198	1.208	1.216
1/ 5000	-	-			15	19	28	38	55	75	110	150	190
0.00030	1.571	0.915	*	1.229	1.208	1.196	1.181	1.177	1.177	1.180	1.191	1.201	1.211
1/ 3333	-	-		13	17	22	32	42	65	85	130	170	210
0.00040	1.361	*	*	1.213	1.195	1.184	1.172	1.169	1.170	1.175	1.186	1.197	1.207
1/ 2500	-			14	18	22	34	46	70	90	140	180	230
0.00060	1.111	*	*	1.194	1.178	1.169	1.160	1.159	1.162	1.168	1.180	1.192	1.203
1/ 1667	-			15	20	26	38	50	75	100	150	210	260
0.00080	0.962	*	*	1.182	1.168	1.160	1.153	1.152	1.157	1.163	1.177	1.189	1.200
1/ 1250	-			17	22	28	42	55	85	110	170	220	280
0.00100	0.861	*	1.202	1.174	1.161	1.154	1.148	1.148	1.153	1.160	1.174	1.187	1.198
1/ 1000	-		12	18	24	30	44	60	90	120	180	240	300
0.00150	*	*	1.184	1.160	1.149	1.144	1.140	1.141	1.148	1.156	1.171	1.184	1.195
1/ 667			13	20	26	34	50	65	100	130	200	260	330
0.00200	*	*	1.173	1.152	1.142	1.138	1.135	1.137	1.145	1.153	1.168	1.182	1.194
1/ 500			14	22	28	36	55	70	110	140	210	290	360
0.00300	*	1.181	1.160	1.141	1.134	1.130	1.129	1.132	1.141	1.149	1.166	1.179	1.191
1/ 333		12	16	24	32	40	60	80	120	160	240	320	400
0.00400	*	1.171	1.152	1.135	1.128	1.126	1.126	1.129	1.138	1.147	1.164	1.178	1.190
1/ 250		13	17	26	34	42	65	85	130	170	260	340	430
0.00600	*	1.159	1.142	1.128	1.122	1.120	1.121	1.125	1.135	1.145	1.162	1.176	1.189
1/ 167		14	19	28	38	46	70	95	140	190	280	380	470
0.00800	*	1.152	1.136	1.123	1.118	1.117	1.119	1.123	1.133	1.143	1.161	1.175	1.188
1/ 125		15	20	30	40	50	75	100	150	200	300	400	500
0.01000	1.177	1.147	1.132	1.120	1.115	1.114	1.117	1.122	1.132	1.142	1.160	1.175	1.187
1/ 100	11	16	22	32	42	55	80	110	160	210	320	420	550
0.01500	1.166	1.139	1.126	1.115	1.111	1.111	1.114	1.119	1.130	1.141	1.159	1.173	1.186
1/ 67	12	17	24	34	46	60	85	120	170	230	350	460	600
0.02000	1.160	1.134	1.122	1.112	1.109	1.108	1.112	1.118	1.129	1.140	1.158	1.173	1.186
1/ 50	12	18	24	36	50	60	90	120	180	250	370	490	600
0.03000	1.152	1.128	1.117	1.108	1.106	1.106	1.110	1.116	1.128	1.139	1.157	1.172	1.185
1/ 33.3	13	20	26	40	55	65	100	130	200	270	400	550	650
0.04000	1.147	1.125	1.114	1.106	1.104	1.104	1.109	1.115	1.127	1.138	1.156	1.172	1.184
1/ 25.0	14	22	28	42	55	70	110	140	210	280	420	550	700
0.06000	1.142	1.120	1.111	1.103	1.102	1.102	1.107	1.114	1.126	1.137	1.156	1.171	1.184
1/ 16.7	15	22	30	46	60	75	110	150	230	310	460	600	750
0.08000	1.138	1.118	1.109	1.102	1.100	1.101	1.106	1.113	1.125	1.137	1.155	1.171	1.184
1/ 12.5	16	24	32	48	65	80	120	160	240	320	480	650	800
0.10000	1.136	1.116	1.107	1.100	1.099	1.100	1.106	1.112	1.125	1.136	1.155	1.170	1.183
1/ 10.0	17	26	34	50	65	85	130	170	250	330	500	650	850
0.15000	1.132	1.113	1.105	1.099	1.098	1.099	1.105	1.112	1.124	1.136	1.155	1.170	1.183
1/ 6.7	18	26	36	55	70	90	130	180	270	360	550	700	900
0.20000	1.130	1.112	1.104	1.098	1.097	1.098	1.104	1.111	1.124	1.135	1.154	1.170	1.183
1/ 5.0	19	28	38	55	75	95	140	190	280	370	550	750	950
0.30000	1.127	1.110	1.102	1.097	1.096	1.098	1.104	1.111	1.124	1.135	1.154	1.170	1.183
1/ 3.3	20	30	40	60	80	100	150	200	300	390	600	800	1000
0.40000	1.126	1.109	1.101	1.096	1.095	1.097	1.103	1.110	1.123	1.135	1.154	1.169	1.183
1/ 2.5	20	30	40	60	80	100	150	200	310	410	600	800	1000
0.60000	1.124	1.107	1.100	1.095	1.095	1.096	1.103	1.110	1.123	1.135	1.154	1.169	1.182
1/ 1.7	22	32	44	65	85	110	160	220	320	430	650	850	1100
	20	30	40	60	80	100	150	200	300	400	600	800	1000

Gradient   (Equivalent) Pipe diameters in mm

$k_s = 0.60$ mm $S = 0.00010$ to 0.60000

$k_s = 0.60$ mm
$S = 0.000020$ to 0.10000

Water (or sewage) at 15°C; full bore conditions.
This table shows values of m for use with Tables D

ie hydraulic gradient =
1 in 50000 to 1 in 10.0

m_C for Colebrook-White solutions;
θ for use with part-full circular pipes

Gradient	(Equivalent) Pipe diameters in mm to 4000 mm, then in m												
	1000	1500	2000	3000	4000	6.00	8.00	10.00	15.00	20.00	30.00	40.00	60.00
0.00002	1.275	1.285	1.296	1.316	1.334	1.362	1.385	1.404	1.443	1.472	1.518	1.553	1.605
1/50000	90	140	180	280	370	550	750	900	1400	1800	2800	3700	5500
0.00003	1.261	1.273	1.286	1.307	1.326	1.355	1.379	1.399	1.438	1.469	1.515	1.550	1.603
1/33333	100	160	210	310	420	650	850	1000	1600	2100	3100	4200	6500
0.00004	1.252	1.266	1.279	1.302	1.321	1.351	1.375	1.396	1.435	1.466	1.513	1.548	1.602
1/25000	110	170	230	340	460	700	900	1100	1700	2300	3400	4600	7000
0.00006	1.241	1.257	1.271	1.295	1.315	1.346	1.371	1.392	1.432	1.463	1.510	1.546	1.600
1/16667	130	190	260	390	500	800	1000	1300	1900	2600	3900	5000	8000
0.00008	1.234	1.251	1.266	1.291	1.311	1.343	1.368	1.389	1.430	1.461	1.509	1.545	1.599
1/12500	140	210	280	430	550	850	1100	1400	2100	2800	4300	5500	8500
0.00010	1.229	1.247	1.262	1.288	1.308	1.341	1.367	1.388	1.429	1.460	1.508	1.544	1.598
1/10000	150	230	300	460	600	900	1200	1500	2300	3000	4600	6000	9000
0.00015	1.221	1.240	1.257	1.283	1.304	1.338	1.364	1.385	1.427	1.458	1.506	1.543	1.597
1/ 6667	170	260	340	500	700	1000	1400	1700	2600	3400	5000	7000	10000
0.00020	1.216	1.236	1.253	1.280	1.302	1.336	1.362	1.383	1.425	1.457	1.505	1.542	1.596
1/ 5000	190	280	370	550	750	1100	1500	1900	2800	3700	5500	7500	11000
0.00030	1.211	1.232	1.249	1.277	1.299	1.333	1.360	1.381	1.424	1.456	1.504	1.541	1.595
1/ 3333	210	320	420	650	850	1300	1700	2100	3200	4200	6500	8500	13000
0.00040	1.207	1.229	1.246	1.275	1.297	1.332	1.358	1.380	1.423	1.455	1.504	1.540	1.595
1/ 2500	230	340	460	700	900	1400	1800	2300	3400	4600	7000	9000	14000
0.00060	1.203	1.225	1.243	1.272	1.295	1.330	1.357	1.379	1.422	1.454	1.503	1.539	1.594
1/ 1667	260	390	500	750	1000	1500	2100	2600	3900	5000	7500	10000	15000
0.00080	1.200	1.223	1.241	1.271	1.294	1.329	1.356	1.378	1.421	1.453	1.502	1.539	1.594
1/ 1250	280	420	550	850	1100	1700	2200	2800	4200	5500	8500	11000	17000
0.00100	1.198	1.221	1.240	1.270	1.293	1.328	1.355	1.377	1.420	1.453	1.502	1.539	1.594
1/ 1000	300	440	600	900	1200	1800	2400	3000	4400	6000	9000	12000	18000
0.00150	1.195	1.219	1.238	1.268	1.291	1.327	1.354	1.377	1.420	1.452	1.501	1.538	1.593
1/ 667	330	500	650	1000	1300	2000	2600	3300	5000	6500	10000	13000	20000
0.00200	1.194	1.218	1.237	1.267	1.290	1.326	1.354	1.376	1.419	1.452	1.501	1.538	1.593
1/ 500	360	550	700	1100	1400	2100	2900	3600	5500	7000	11000	14000	21000
0.00300	1.191	1.216	1.236	1.266	1.289	1.325	1.353	1.375	1.419	1.452	1.501	1.538	1.593
1/ 333	400	600	800	1200	1600	2400	3200	4000	6000	8000	12000	16000	24000
0.00400	1.190	1.215	1.235	1.265	1.289	1.325	1.353	1.375	1.419	1.451	1.501	1.538	1.593
1/ 250	430	650	850	1300	1700	2600	3400	4300	6500	8500	13000	17000	26000
0.00600	1.189	1.214	1.234	1.264	1.288	1.324	1.352	1.375	1.418	1.451	1.500	1.537	1.593
1/ 167	470	700	950	1400	1900	2800	3800	4700	7000	9500	14000	19000	28000
0.00800	1.188	1.213	1.233	1.264	1.288	1.324	1.352	1.374	1.418	1.451	1.500	1.537	1.592
1/ 125	500	750	1000	1500	2000	3000	4000	5000	7500	10000	15000	20000	30000
0.01000	1.187	1.213	1.233	1.264	1.287	1.324	1.352	1.374	1.418	1.451	1.500	1.537	1.592
1/ 100	550	800	1100	1600	2100	3200	4200	5500	8000	11000	16000	21000	32000
0.01500	1.186	1.212	1.232	1.263	1.287	1.323	1.351	1.374	1.418	1.451	1.500	1.537	1.592
1/ 67	600	850	1200	1700	2300	3500	4600	6000	8500	12000	17000	23000	35000
0.02000	1.186	1.211	1.232	1.263	1.287	1.323	1.351	1.374	1.417	1.450	1.500	1.537	1.592
1/ 50	600	900	1200	1800	2500	3700	4900	6000	9000	12000	18000	25000	37000
0.03000	1.185	1.211	1.231	1.262	1.286	1.323	1.351	1.373	1.417	1.450	1.500	1.537	1.592
1/ 33.3	650	1000	1300	2000	2700	4000	5500	6500	10000	13000	20000	27000	40000
0.04000	1.184	1.210	1.231	1.262	1.286	1.323	1.351	1.373	1.417	1.450	1.500	1.537	1.592
1/ 25.0	700	1100	1400	2100	2800	4200	5500	7000	11000	14000	21000	28000	42000
0.06000	1.184	1.210	1.230	1.262	1.286	1.323	1.351	1.373	1.417	1.450	1.499	1.537	1.592
1/ 16.7	750	1100	1500	2300	3100	4600	6000	7500	11000	15000	23000	31000	46000
0.08000	1.184	1.210	1.230	1.262	1.286	1.322	1.350	1.373	1.417	1.450	1.499	1.537	1.592
1/ 12.5	800	1200	1600	2400	3200	4800	6500	8000	12000	16000	24000	32000	48000
0.10000	1.183	1.210	1.230	1.262	1.286	1.322	1.350	1.373	1.417	1.450	1.499	1.537	1.592
1/ 10.0	850	1300	1700	2500	3300	5000	6500	8500	13000	17000	25000	33000	50000
	1000	1500	2000	3000	4000	6.00	8.00	10.00	15.00	20.00	30.00	40.00	60.00

Gradient   (Equivalent) Pipe diameters in mm to 4000 mm, then in m

$S = 0.000020$ to 0.10000 $k_s = 0.60$ mm

E9

k_s = 1·50 mm
S = 0·00010 to 0·60000

ie hydraulic gradient =
1 in 10000 to 1 in 1·7

Water (or sewage) at 15°C; full bore conditions.
This table shows values of m for use with Tables D

m_C for Colebrook-White solutions; or m_P for laminar flow
θ for use with part-full circular pipes (turbulent flows)

Gradient	\(Equivalent\) Pipe diameters in mm												
	20	30	40	60	80	100	150	200	300	400	600	800	1000
0·00010	2·722	1·585	1·080	*	*	1·378	1·350	1·338	1·329	1·327	1·332	1·339	1·346
1/10000	-	-	-			13	20	26	40	55	80	110	130
0·00015	2·222	1·294	*	*	1·380	1·361	1·337	1·326	1·320	1·320	1·326	1·334	1·341
1/ 6667	-	-			12	15	22	30	44	60	90	120	150
0·00020	1·925	1·121	*	*	1·368	1·350	1·329	1·320	1·314	1·315	1·322	1·330	1·339
1/ 5000	-	-			13	16	24	32	48	65	95	130	160
0·00030	1·571	0·915	*	1·379	1·353	1·338	1·319	1·311	1·308	1·310	1·318	1·327	1·335
1/ 3333	-	-		11	14	18	26	36	55	70	110	140	180
0·00040	1·361	*	*	1·368	1·344	1·330	1·313	1·306	1·304	1·307	1·315	1·325	1·334
1/ 2500	-			11	15	19	28	38	55	75	110	150	190
0·00060	1·111	*	*	1·355	1·333	1·320	1·305	1·300	1·299	1·303	1·312	1·322	1·331
1/ 1667	-			13	17	20	32	42	65	85	130	170	210
0·00080	0·962	*	*	1·347	1·326	1·315	1·301	1·297	1·296	1·300	1·310	1·320	1·330
1/ 1250	-			13	18	22	34	44	65	90	130	180	220
0·00100	0·861	*	*	1·341	1·322	1·311	1·298	1·294	1·294	1·298	1·309	1·319	1·329
1/ 1000	-			14	19	24	36	46	70	95	140	190	230
0·00150	*	*	1·369	1·333	1·315	1·304	1·293	1·290	1·291	1·296	1·307	1·317	1·327
1/ 667			10	15	20	26	38	50	75	100	150	200	250
0·00200	*	*	1·362	1·327	1·310	1·301	1·290	1·288	1·289	1·294	1·306	1·316	1·326
1/ 500			11	16	22	26	40	55	80	110	160	220	270
0·00300	*	*	1·353	1·321	1·305	1·296	1·287	1·285	1·287	1·292	1·304	1·315	1·325
1/ 333			12	18	24	30	44	60	90	120	180	230	290
0·00400	*	1·379	1·348	1·317	1·302	1·294	1·285	1·283	1·286	1·291	1·303	1·314	1·325
1/ 250		9	12	18	24	30	46	60	90	120	180	230	290
0·00600	*	1·371	1·342	1·312	1·298	1·290	1·282	1·281	1·284	1·290	1·302	1·314	1·324
1/ 167		10	13	20	26	34	50	65	100	130	200	250	310
0·00800	*	1·366	1·338	1·309	1·296	1·288	1·281	1·280	1·283	1·289	1·302	1·313	1·323
1/ 125		10	14	20	28	34	50	70	100	140	210	280	350
0·01000	*	1·363	1·336	1·308	1·294	1·287	1·280	1·279	1·283	1·289	1·301	1·313	1·323
1/ 100		11	14	22	28	36	55	70	110	140	220	290	360
0·01500	1·410	1·358	1·331	1·305	1·292	1·285	1·278	1·278	1·282	1·288	1·300	1·312	1·323
1/ 67	8	11	15	22	30	38	55	75	110	150	230	300	380
0·02000	1·405	1·355	1·329	1·303	1·290	1·284	1·277	1·277	1·281	1·287	1·300	1·312	1·322
1/ 50	8	12	16	24	32	40	60	80	120	160	240	320	400
0·03000	1·400	1·352	1·326	1·301	1·289	1·282	1·276	1·276	1·280	1·287	1·300	1·311	1·322
1/ 33·3	8	13	17	26	34	42	65	85	130	170	250	330	420
0·04000	1·397	1·349	1·324	1·299	1·288	1·281	1·275	1·275	1·280	1·286	1·299	1·311	1·322
1/ 25·0	9	13	17	26	34	44	65	85	130	170	260	350	430
0·06000	1·394	1·347	1·322	1·298	1·286	1·280	1·275	1·275	1·279	1·286	1·299	1·311	1·321
1/ 16·7	9	14	18	28	36	46	70	90	140	180	270	360	450
0·08000	1·392	1·345	1·321	1·297	1·286	1·280	1·274	1·274	1·279	1·286	1·299	1·311	1·321
1/ 12·5	9	14	19	28	38	46	70	95	140	190	280	370	470
0·10000	1·390	1·344	1·320	1·296	1·285	1·279	1·274	1·274	1·279	1·285	1·299	1·311	1·321
1/ 10·0	10	14	19	28	38	48	70	95	140	190	290	380	480
0·15000	1·388	1·343	1·319	1·295	1·284	1·278	1·273	1·274	1·279	1·285	1·298	1·310	1·321
1/ 6·7	10	15	20	30	40	50	75	100	150	200	300	380	480
0·20000	1·387	1·342	1·318	1·295	1·284	1·278	1·273	1·273	1·278	1·285	1·298	1·310	1·321
1/ 5·0	10	15	20	30	40	50	75	100	150	200	300	400	490
0·30000	1·385	1·340	1·317	1·294	1·283	1·278	1·273	1·273	1·278	1·285	1·298	1·310	1·321
1/ 3·3	10	16	20	32	42	50	80	100	160	210	310	420	500
0·40000	1·384	1·340	1·317	1·294	1·283	1·277	1·272	1·273	1·278	1·285	1·298	1·310	1·321
1/ 2·5	11	16	22	32	42	55	80	110	160	210	320	430	550
0·60000	1·383	1·339	1·316	1·293	1·283	1·277	1·272	1·273	1·278	1·285	1·298	1·310	1·321
1/ 1·7	11	16	22	32	44	55	80	110	160	220	330	440	550
	20	30	40	60	80	100	150	200	300	400	600	800	1000

Gradient   \(Equivalent\) Pipe diameters in mm

k_s = 1·50 mm S = 0·00010 to 0·60000

$k_s = 1.50$ mm
$S = 0.000020$ to 0.10000

ie hydraulic gradient =
1 in 50000 to 1 in 10.0

Water (or sewage) at 15°C; full bore conditions.
This table shows values of m for use with Tables D

m_C for Colebrook-White solutions;
θ for use with part-full circular pipes

Gradient — (Equivalent) Pipe diameters in mm to 4000 mm, then in m

Gradient	1000	1500	2000	3000	4000	6.00	8.00	10.00	15.00	20.00	30.00	40.00	60.00
0.00002	1.374	1.386	1.398	1.420	1.439	1.469	1.493	1.513	1.553	1.584	1.631	1.667	1.721
1/50000	85	130	170	260	340	500	700	850	1300	1700	2600	3400	5000
0.00003	1.365	1.379	1.392	1.415	1.434	1.465	1.490	1.510	1.551	1.582	1.629	1.665	1.720
1/33333	95	140	190	290	380	550	750	950	1400	1900	2900	3800	5500
0.00004	1.360	1.374	1.388	1.412	1.431	1.463	1.488	1.508	1.549	1.581	1.628	1.664	1.719
1/25000	100	160	210	310	420	600	850	1000	1600	2100	3100	4200	6000
0.00006	1.353	1.369	1.383	1.408	1.428	1.460	1.485	1.506	1.547	1.579	1.627	1.663	1.718
1/16667	120	170	230	350	470	700	950	1200	1700	2300	3500	4700	7000
0.00008	1.349	1.366	1.381	1.406	1.426	1.459	1.484	1.505	1.546	1.578	1.626	1.663	1.718
1/12500	130	190	250	380	500	750	1000	1300	1900	2500	3800	5000	7500
0.00010	1.346	1.363	1.379	1.404	1.425	1.457	1.483	1.504	1.546	1.578	1.626	1.662	1.717
1/10000	130	200	270	400	550	800	1100	1300	2000	2700	4000	5500	8000
0.00015	1.341	1.360	1.376	1.402	1.423	1.456	1.482	1.503	1.545	1.577	1.625	1.662	1.717
1/6667	150	220	300	450	600	900	1200	1500	2200	3000	4500	6000	9000
0.00020	1.339	1.358	1.374	1.400	1.421	1.455	1.481	1.502	1.544	1.576	1.625	1.661	1.716
1/5000	160	240	320	480	650	950	1300	1600	2400	3200	4800	6500	9500
0.00030	1.335	1.355	1.372	1.398	1.420	1.453	1.480	1.501	1.543	1.575	1.624	1.661	1.716
1/3333	180	270	350	550	700	1100	1400	1800	2700	3500	5500	7000	11000
0.00040	1.334	1.353	1.370	1.397	1.419	1.453	1.479	1.501	1.543	1.575	1.624	1.661	1.716
1/2500	190	280	380	550	750	1100	1500	1900	2800	3800	5500	7500	11000
0.00060	1.331	1.352	1.369	1.396	1.418	1.452	1.478	1.500	1.542	1.575	1.623	1.660	1.715
1/1667	210	310	420	650	850	1300	1700	2100	3100	4200	6500	8500	13000
0.00080	1.330	1.350	1.368	1.395	1.417	1.451	1.478	1.500	1.542	1.574	1.623	1.660	1.715
1/1250	220	330	450	650	900	1300	1800	2200	3300	4500	6500	9000	13000
0.00100	1.329	1.350	1.367	1.395	1.417	1.451	1.477	1.499	1.542	1.574	1.623	1.660	1.715
1/1000	230	350	470	700	950	1400	1900	2300	3500	4700	7000	9500	14000
0.00150	1.327	1.348	1.366	1.394	1.416	1.450	1.477	1.499	1.541	1.574	1.623	1.660	1.715
1/667	250	380	500	750	1000	1500	2000	2500	3800	5000	7500	10000	15000
0.00200	1.326	1.348	1.365	1.393	1.416	1.450	1.477	1.499	1.541	1.574	1.623	1.660	1.715
1/500	270	400	550	800	1100	1600	2200	2700	4000	5500	8000	11000	16000
0.00300	1.325	1.347	1.365	1.393	1.415	1.450	1.476	1.498	1.541	1.573	1.622	1.659	1.715
1/333	290	440	600	900	1200	1800	2300	2900	4400	6000	9000	12000	18000
0.00400	1.325	1.346	1.364	1.392	1.415	1.449	1.476	1.498	1.541	1.573	1.622	1.659	1.715
1/250	310	460	600	900	1200	1800	2500	3100	4600	6000	9000	12000	18000
0.00600	1.324	1.346	1.364	1.392	1.414	1.449	1.476	1.498	1.541	1.573	1.622	1.659	1.715
1/167	330	500	650	1000	1300	2000	2600	3300	5000	6500	10000	13000	20000
0.00800	1.323	1.345	1.363	1.392	1.414	1.449	1.476	1.498	1.541	1.573	1.622	1.659	1.715
1/125	350	500	700	1000	1400	2100	2800	3500	5000	7000	10000	14000	21000
0.01000	1.323	1.345	1.363	1.392	1.414	1.449	1.476	1.498	1.540	1.573	1.622	1.659	1.715
1/100	360	550	700	1100	1400	2200	2900	3600	5500	7000	11000	14000	22000
0.01500	1.323	1.345	1.363	1.391	1.414	1.449	1.475	1.498	1.540	1.573	1.622	1.659	1.714
1/67	380	550	750	1100	1500	2300	3000	3800	5500	7500	11000	15000	23000
0.02000	1.322	1.345	1.363	1.391	1.414	1.449	1.475	1.497	1.540	1.573	1.622	1.659	1.714
1/50	400	600	800	1200	1600	2400	3200	4000	6000	8000	12000	16000	24000
0.03000	1.322	1.344	1.362	1.391	1.414	1.448	1.475	1.497	1.540	1.573	1.622	1.659	1.714
1/33.3	420	650	850	1300	1700	2500	3300	4200	6500	8500	13000	17000	25000
0.04000	1.322	1.344	1.362	1.391	1.414	1.448	1.475	1.497	1.540	1.573	1.622	1.659	1.714
1/25.0	430	650	850	1300	1700	2600	3500	4300	6500	8500	13000	17000	26000
0.06000	1.321	1.344	1.362	1.391	1.413	1.448	1.475	1.497	1.540	1.573	1.622	1.659	1.714
1/16.7	450	700	900	1400	1800	2700	3600	4500	7000	9000	14000	18000	27000
0.08000	1.321	1.344	1.362	1.391	1.413	1.448	1.475	1.497	1.540	1.573	1.622	1.659	1.714
1/12.5	470	700	950	1400	1900	2800	3700	4700	7000	9500	14000	19000	28000
0.10000	1.321	1.344	1.362	1.391	1.413	1.448	1.475	1.497	1.540	1.573	1.622	1.659	1.714
1/10.0	480	700	950	1400	1900	2900	3800	4800	7000	9500	14000	19000	29000
	1000	1500	2000	3000	4000	6.00	8.00	10.00	15.00	20.00	30.00	40.00	60.00

Gradient — (Equivalent) Pipe diameters in mm to 4000 mm, then in m

$S = 0.000020$ to 0.10000 $k_s = 1.50$ mm

k_s = 3·0 mm
S = 0.00010 to 0.60000

Water (or sewage) at 15°C; full bore conditions.
This table shows values of m for use with Tables D

ie hydraulic gradient =
1 in 10000 to 1 in 1·7

m_C for Colebrook-White solutions; or m_P for laminar flow
θ for use with part-full circular pipes (turbulent flows)

Gradient — (Equivalent) Pipe diameters in mm

Gradient		20	30	40	60	80	100	150	200	300	400	600	800	1000
0·00010 1/10000	m	2·722	1·585	1·080	*	*	1·531	1·493	1·476	1·460	1·455	1·455	1·459	1·465
	θ	-	-	-			11	17	22	34	44	65	90	110
0·00015 1/ 6667	m	2·222	1·294	*	*	1·544	1·518	1·484	1·468	1·454	1·450	1·451	1·456	1·462
	θ	-	-			10	12	18	24	36	48	75	95	120
0·00020 1/ 5000	m	1·925	1·121	*	*	1·535	1·511	1·478	1·463	1·451	1·447	1·449	1·454	1·460
	θ	-	-			10	13	19	26	38	50	75	100	130
0·00030 1/ 3333	m	1·571	0·915	*	*	1·524	1·501	1·472	1·458	1·447	1·444	1·446	1·452	1·458
	θ	-	-			11	14	20	28	42	55	85	110	140
0·00040 1/ 2500	m	1·361	*	*	1·553	1·518	1·496	1·467	1·454	1·444	1·442	1·445	1·450	1·457
	θ	-			9	12	15	22	30	44	60	90	120	150
0·00060 1/ 1667	m	1·111	*	*	1·544	1·510	1·489	1·462	1·450	1·441	1·439	1·443	1·449	1·455
	θ	-			10	13	16	24	32	48	65	95	130	160
0·00080 1/ 1250	m	0·962	*	*	1·538	1·505	1·485	1·459	1·448	1·439	1·438	1·441	1·448	1·455
	θ	-			10	13	17	26	34	50	65	100	130	170
0·00100 1/ 1000	m	0·861	*	*	1·534	1·502	1·483	1·457	1·446	1·438	1·436	1·441	1·447	1·454
	θ	-			10	14	17	26	34	50	70	100	140	170
0·00150 1/ 667	m	*	*	*	1·528	1·497	1·478	1·454	1·443	1·436	1·435	1·439	1·446	1·453
	θ				11	15	18	28	36	55	75	110	150	180
0·00200 1/ 500	m	*	*	1·581	1·524	1·494	1·476	1·452	1·442	1·435	1·434	1·438	1·445	1·453
	θ			8	12	15	19	28	38	60	75	120	150	190
0·00300 1/ 333	m	*	*	1·575	1·519	1·490	1·473	1·450	1·440	1·433	1·433	1·438	1·445	1·452
	θ			8	12	16	20	30	40	60	80	120	160	200
0·00400 1/ 250	m	*	*	1·571	1·517	1·488	1·471	1·449	1·439	1·432	1·432	1·437	1·444	1·451
	θ			8	13	17	22	32	42	65	85	130	170	210
0·00600 1/ 167	m	*	1·617	1·567	1·513	1·486	1·469	1·447	1·438	1·431	1·431	1·436	1·444	1·451
	θ		7	9	13	18	22	34	44	65	90	130	180	220
0·00800 1/ 125	m	*	1·614	1·564	1·511	1·484	1·467	1·446	1·437	1·431	1·431	1·436	1·443	1·451
	θ		7	9	14	18	22	34	46	70	90	140	180	230
0·01000 1/ 100	m	*	1·612	1·562	1·510	1·483	1·467	1·445	1·436	1·430	1·430	1·436	1·443	1·451
	θ		7	9	14	19	24	34	46	70	95	140	190	230
0·01500 1/ 67	m	*	1·608	1·559	1·508	1·481	1·465	1·444	1·435	1·430	1·430	1·435	1·443	1·450
	θ		7	10	15	19	24	36	48	75	95	150	190	240
0·02000 1/ 50	m	1·696	1·606	1·558	1·507	1·480	1·464	1·444	1·435	1·429	1·429	1·435	1·443	1·450
	θ	5	7	10	15	20	24	38	50	75	100	150	200	250
0·03000 1/ 33·3	m	1·692	1·603	1·556	1·505	1·479	1·463	1·443	1·434	1·429	1·429	1·435	1·442	1·450
	θ	5	8	10	15	20	26	38	50	75	100	150	210	260
0·04000 1/ 25·0	m	1·690	1·602	1·554	1·504	1·478	1·463	1·443	1·434	1·429	1·429	1·435	1·442	1·450
	θ	5	8	10	16	20	26	40	50	80	100	160	210	260
0·06000 1/ 16·7	m	1·687	1·600	1·553	1·503	1·478	1·462	1·442	1·434	1·428	1·429	1·434	1·442	1·450
	θ	5	8	11	16	22	26	40	55	80	110	160	220	270
0·08000 1/ 12·5	m	1·686	1·599	1·552	1·503	1·477	1·462	1·442	1·433	1·428	1·428	1·434	1·442	1·450
	θ	5	8	11	16	22	28	42	55	80	110	160	220	270
0·10000 1/ 10·0	m	1·685	1·598	1·551	1·502	1·477	1·461	1·442	1·433	1·428	1·428	1·434	1·442	1·449
	θ	6	8	11	17	22	28	42	55	85	110	170	220	280
0·15000 1/ 6·7	m	1·683	1·597	1·551	1·502	1·476	1·461	1·441	1·433	1·428	1·428	1·434	1·442	1·449
	θ	6	9	11	17	22	28	42	55	85	110	170	230	280
0·20000 1/ 5·0	m	1·682	1·596	1·550	1·501	1·476	1·461	1·441	1·433	1·428	1·428	1·434	1·442	1·449
	θ	6	9	12	17	24	28	44	60	85	120	170	230	290
0·30000 1/ 3·3	m	1·681	1·595	1·549	1·501	1·476	1·460	1·441	1·432	1·427	1·428	1·434	1·442	1·449
	θ	6	9	12	18	24	30	44	60	90	120	180	230	290
0·40000 1/ 2·5	m	1·680	1·595	1·549	1·501	1·475	1·460	1·441	1·432	1·427	1·428	1·434	1·441	1·449
	θ	6	9	12	18	24	30	44	60	90	120	180	240	300
0·60000 1/ 1·7	m	1·679	1·594	1·549	1·500	1·475	1·460	1·441	1·432	1·427	1·428	1·434	1·441	1·449
	θ	6	9	12	18	24	30	46	60	90	120	180	240	300
		20	30	40	60	80	100	150	200	300	400	600	800	1000

Gradient — (Equivalent) Pipe diameters in mm

k_s = 3·0 mm S = 0.00010 to 0.60000

k_s = 3·0 mm
S = 0.000020 to 0·10000

ie hydraulic gradient =
1 in 50000 to 1 in 10·0

Water (or sewage) at 15°C; full bore conditions.
This table shows values of m for use with Tables D

m_C for Colebrook-White solutions;
θ for use with part-full circular pipes

Gradient	(Equivalent) Pipe diameters in mm to 4000 mm, then in m												
	1000	1500	2000	3000	4000	6·00	8·00	10·00	15·00	20·00	30·00	40·00	60·00
0·00002 1/50000	1·483 75	1·494 110	1·505 150	1·526 230	1·544 300	1·573 450	1·597 600	1·616 750	1·656 1100	1·687 1500	1·734 2300	1·770 3000	1·824 4500
0·00003 1/33333	1·477 85	1·489 130	1·501 170	1·522 250	1·541 340	1·571 500	1·595 650	1·615 850	1·655 1300	1·686 1700	1·733 2500	1·769 3400	1·823 5000
0·00004 1/25000	1·474 90	1·486 130	1·499 180	1·521 270	1·539 360	1·569 550	1·593 700	1·614 900	1·654 1300	1·685 1800	1·732 2700	1·768 3600	1·823 5500
0·00006 1/16667	1·469 100	1·483 150	1·496 200	1·518 300	1·537 400	1·568 600	1·592 800	1·613 1000	1·653 1500	1·684 2000	1·732 3000	1·768 4000	1·822 6000
0·00008 1/12500	1·467 110	1·481 160	1·494 210	1·517 320	1·536 420	1·567 650	1·591 850	1·612 1100	1·652 1600	1·684 2100	1·731 3200	1·767 4200	1·822 6500
0·00010 1/10000	1·465 110	1·479 170	1·493 220	1·516 330	1·535 450	1·566 650	1·591 900	1·611 1100	1·652 1700	1·683 2200	1·731 3300	1·767 4500	1·822 6500
0·00015 1/ 6667	1·462 120	1·477 180	1·491 240	1·514 360	1·534 490	1·565 750	1·590 950	1·611 1200	1·651 1800	1·683 2400	1·730 3600	1·767 4900	1·822 7500
0·00020 1/ 5000	1·460 130	1·476 190	1·490 260	1·513 390	1·533 500	1·564 750	1·589 1000	1·610 1300	1·651 1900	1·682 2600	1·730 3900	1·767 5000	1·821 7500
0·00030 1/ 3333	1·458 140	1·474 210	1·488 280	1·512 420	1·532 550	1·564 850	1·589 1100	1·610 1400	1·650 2100	1·682 2800	1·730 4200	1·766 5500	1·821 8500
0·00040 1/ 2500	1·457 150	1·473 220	1·487 300	1·512 440	1·532 600	1·563 900	1·588 1200	1·609 1500	1·650 2200	1·682 3000	1·730 4400	1·766 6000	1·821 9000
0·00060 1/ 1667	1·455 160	1·472 240	1·486 320	1·511 480	1·531 650	1·563 950	1·588 1300	1·609 1600	1·650 2400	1·681 3200	1·729 4800	1·766 6500	1·821 9500
0·00080 1/ 1250	1·455 170	1·471 250	1·486 330	1·510 500	1·531 650	1·562 1000	1·588 1300	1·609 1700	1·650 2500	1·681 3300	1·729 5000	1·766 6500	1·821 10000
0·00100 1/ 1000	1·454 170	1·471 260	1·485 350	1·510 500	1·530 700	1·562 1000	1·587 1400	1·608 1700	1·650 2600	1·681 3500	1·729 5000	1·766 7000	1·821 10000
0·00150 1/ 667	1·453 180	1·470 280	1·485 370	1·510 550	1·530 750	1·562 1100	1·587 1500	1·608 1800	1·649 2800	1·681 3700	1·729 5500	1·766 7500	1·821 11000
0·00200 1/ 500	1·453 190	1·470 290	1·484 380	1·509 600	1·530 750	1·562 1200	1·587 1500	1·608 1900	1·649 2900	1·681 3800	1·729 6000	1·766 7500	1·821 12000
0·00300 1/ 333	1·452 200	1·469 300	1·484 410	1·509 600	1·529 800	1·562 1200	1·587 1600	1·608 2000	1·649 3000	1·681 4100	1·729 6000	1·765 8000	1·820 12000
0·00400 1/ 250	1·451 210	1·469 320	1·484 420	1·509 650	1·529 850	1·561 1300	1·587 1700	1·608 2100	1·649 3200	1·681 4200	1·729 6500	1·765 8500	1·820 13000
0·00600 1/ 167	1·451 220	1·468 330	1·483 440	1·509 650	1·529 900	1·561 1300	1·587 1800	1·608 2200	1·649 3300	1·681 4400	1·729 6500	1·765 9000	1·820 13000
0·00800 1/ 125	1·451 230	1·468 340	1·483 460	1·508 700	1·529 900	1·561 1400	1·586 1800	1·608 2300	1·649 3400	1·681 4600	1·729 7000	1·765 9000	1·820 14000
0·01000 1/ 100	1·451 230	1·468 350	1·483 470	1·508 700	1·529 950	1·561 1400	1·586 1900	1·607 2300	1·649 3500	1·681 4700	1·729 7000	1·765 9500	1·820 14000
0·01500 1/ 67	1·450 240	1·468 360	1·483 480	1·508 750	1·529 950	1·561 1500	1·586 1900	1·607 2400	1·649 3600	1·680 4800	1·729 7500	1·765 9500	1·820 15000
0·02000 1/ 50	1·450 250	1·468 370	1·483 500	1·508 750	1·529 1000	1·561 1500	1·586 2000	1·607 2500	1·649 3700	1·680 5000	1·729 7500	1·765 10000	1·820 15000
0·03000 1/ 33·3	1·450 260	1·467 390	1·483 500	1·508 750	1·528 1000	1·561 1500	1·586 2100	1·607 2600	1·649 3900	1·680 5000	1·729 7500	1·765 10000	1·820 15000
0·04000 1/ 25·0	1·450 260	1·467 390	1·483 500	1·508 800	1·528 1000	1·561 1600	1·586 2100	1·607 2600	1·649 3900	1·680 5000	1·729 8000	1·765 10000	1·820 16000
0·06000 1/ 16·7	1·450 270	1·467 400	1·483 550	1·508 800	1·528 1100	1·561 1600	1·586 2200	1·607 2700	1·649 4000	1·680 5500	1·729 8000	1·765 11000	1·820 16000
0·08000 1/ 12·5	1·450 270	1·467 410	1·482 550	1·508 800	1·528 1100	1·561 1600	1·586 2200	1·607 2700	1·649 4100	1·680 5500	1·729 8000	1·765 11000	1·820 16000
0·10000 1/ 10·0	1·449 280	1·467 420	1·482 550	1·508 850	1·528 1100	1·561 1700	1·586 2200	1·607 2800	1·649 4200	1·680 5500	1·729 8500	1·765 11000	1·820 17000
	1000	1500	2000	3000	4000	6·00	8·00	10·00	15·00	20·00	30·00	40·00	60·00

Gradient   (Equivalent) Pipe diameters in mm to 4000 mm, then in m

S = 0·000020 to 0·10000 k_s = 3·0 mm

$k_s = 6.0$ mm
$S = 0.00010$ to 0.60000

ie hydraulic gradient =
1 in 10000 to 1 in 1.7

Water (or sewage) at 15°C; full bore conditions.
This table shows values of m for use with Tables D

m_C for Colebrook-White solutions; or m_P for laminar flow
θ for use with part-full circular pipes (turbulent flows)

Gradient — (Equivalent) Pipe diameters in mm

Gradient		20	30	40	60	80	100	150	200	300	400	600	800	1000
0.00010	m		1.585	1.080	*	*	1.755	1.697	1.667	1.638	1.625	1.615	1.614	1.615
1/10000	θ		-	-			8	13	17	26	34	50	65	85
0.00015	m		1.294	*	*	*	1.746	1.690	1.662	1.634	1.622	1.613	1.612	1.613
1/ 6667	θ		-				9	13	18	26	36	55	70	90
0.00020	m		1.121	*	*	1.779	1.740	1.686	1.659	1.632	1.620	1.611	1.610	1.612
1/ 5000	θ		-			7	9	14	19	28	38	55	75	95
0.00030	m		0.915	*	*	1.771	1.734	1.682	1.655	1.629	1.617	1.609	1.609	1.611
1/ 3333	θ		-			8	10	15	20	30	40	60	80	100
0.00040	m		*	*	*	1.767	1.730	1.679	1.653	1.627	1.616	1.608	1.608	1.610
1/ 2500	θ					8	10	15	20	30	40	60	80	100
0.00060	m		*	*	1.819	1.761	1.725	1.675	1.650	1.625	1.614	1.607	1.607	1.609
1/ 1667	θ				6	9	11	16	22	32	44	65	85	110
0.00080	m		*	*	1.814	1.758	1.722	1.673	1.648	1.624	1.613	1.606	1.606	1.609
1/ 1250	θ				7	9	11	17	22	34	44	65	90	110
0.00100	m		*	*	1.812	1.756	1.720	1.672	1.647	1.623	1.613	1.606	1.606	1.608
1/ 1000	θ				7	9	11	17	22	34	46	70	90	110
0.00150	m		*	*	1.807	1.752	1.718	1.670	1.645	1.622	1.612	1.605	1.605	1.608
1/ 667	θ				7	9	12	18	24	36	48	70	95	120
0.00200	m		*	*	1.804	1.750	1.716	1.668	1.644	1.621	1.611	1.605	1.605	1.607
1/ 500	θ				7	10	12	18	24	36	48	75	100	120
0.00300	m		*	1.902	1.801	1.747	1.714	1.667	1.643	1.620	1.610	1.604	1.604	1.607
1/ 333	θ			5	8	10	13	19	26	38	50	75	100	130
0.00400	m		*	1.899	1.799	1.746	1.712	1.666	1.642	1.620	1.610	1.604	1.604	1.607
1/ 250	θ			5	8	10	13	19	26	38	50	75	100	130
0.00600	m		*	1.896	1.797	1.744	1.711	1.665	1.642	1.619	1.609	1.603	1.604	1.606
1/ 167	θ			5	8	11	13	20	26	40	55	80	110	130
0.00800	m		1.986	1.894	1.795	1.743	1.710	1.664	1.641	1.619	1.609	1.603	1.604	1.606
1/ 125	θ		4	5	8	11	14	20	28	40	55	80	110	140
0.01000	m		1.985	1.892	1.794	1.742	1.709	1.664	1.641	1.618	1.609	1.603	1.603	1.606
1/ 100	θ		4	5	8	11	14	20	28	42	55	80	110	140
0.01500	m		1.982	1.890	1.793	1.741	1.708	1.663	1.640	1.618	1.608	1.603	1.603	1.606
1/ 67	θ		4	6	8	11	14	22	28	42	55	85	110	140
0.02000	m		1.980	1.889	1.792	1.740	1.708	1.663	1.640	1.618	1.608	1.602	1.603	1.606
1/ 50	θ		4	6	9	11	14	22	28	42	55	85	110	140
0.03000	m		1.978	1.887	1.791	1.739	1.707	1.662	1.639	1.617	1.608	1.602	1.603	1.606
1/ 33.3	θ		4	6	9	12	15	22	30	44	60	85	120	150
0.04000	m		1.977	1.886	1.790	1.739	1.707	1.662	1.639	1.617	1.608	1.602	1.603	1.606
1/ 25.0	θ		4	6	9	12	15	22	30	44	60	90	120	150
0.06000	m		1.975	1.885	1.790	1.738	1.706	1.662	1.639	1.617	1.608	1.602	1.603	1.606
1/ 16.7	θ		4	6	9	12	15	22	30	44	60	90	120	150
0.08000	m		1.975	1.885	1.789	1.738	1.706	1.661	1.639	1.617	1.608	1.602	1.603	1.606
1/ 12.5	θ		5	6	9	12	15	22	30	46	60	90	120	150
0.10000	m		1.974	1.884	1.789	1.738	1.706	1.661	1.639	1.617	1.608	1.602	1.603	1.606
1/ 10.0	θ		5	6	9	12	15	22	30	46	60	90	120	150
0.15000	m		1.973	1.884	1.788	1.737	1.705	1.661	1.638	1.617	1.607	1.602	1.603	1.605
1/ 6.7	θ		5	6	9	12	15	22	30	46	60	90	120	150
0.20000	m		1.973	1.883	1.788	1.737	1.705	1.661	1.638	1.617	1.607	1.602	1.602	1.605
1/ 5.0	θ		5	6	9	12	15	24	30	46	60	95	120	150
0.30000	m		1.972	1.883	1.788	1.737	1.705	1.661	1.638	1.617	1.607	1.602	1.602	1.605
1/ 3.3	θ		5	6	9	12	16	24	32	46	60	95	120	160
0.40000	m		1.972	1.882	1.788	1.737	1.705	1.661	1.638	1.616	1.607	1.602	1.602	1.605
1/ 2.5	θ		5	6	9	13	16	24	32	48	65	95	130	160
0.60000	m		1.971	1.882	1.787	1.737	1.705	1.661	1.638	1.616	1.607	1.602	1.602	1.605
1/ 1.7	θ		5	6	9	13	16	24	32	48	65	95	130	160
		20	30	40	60	80	100	150	200	300	400	600	800	1000

Gradient — (Equivalent) Pipe diameters in mm

$k_s = 6.0$ mm $S = 0.00010$ to 0.60000

k_s = 6·0 mm
S = 0·000020 to 0·10000

Water (or sewage) at 15°C; full bore conditions.
This table shows values of m for use with Tables D

ie hydraulic gradient =
1 in 50000 to 1 in 10·0

m_C for Colebrook-White solutions;
θ for use with part-full circular pipes

Gradient	(Equivalent) Pipe diameters in mm to 4000 mm, then in m												
	1000	1500	2000	3000	4000	6·00	8·00	10·00	15·00	20·00	30·00	40·00	60·00
0·00002	1·627	1·632	1·640	1·657	1·673	1·699	1·721	1·740	1·778	1·808	1·853	1·889	1·942
1/50000	60	90	120	180	250	370	490	600	900	1200	1800	2500	3700
0·00003	1·623	1·629	1·638	1·655	1·671	1·698	1·720	1·739	1·777	1·807	1·853	1·888	1·942
1/33333	65	100	130	200	270	400	550	650	1000	1300	2000	2700	4000
0·00004	1·621	1·627	1·636	1·654	1·670	1·697	1·720	1·739	1·777	1·807	1·852	1·888	1·942
1/25000	70	110	140	210	280	420	550	700	1100	1400	2100	2800	4200
0·00006	1·618	1·625	1·635	1·653	1·669	1·696	1·719	1·738	1·776	1·806	1·852	1·888	1·941
1/16667	75	110	150	230	310	460	600	750	1100	1500	2300	3100	4600
0·00008	1·616	1·624	1·634	1·652	1·668	1·696	1·718	1·737	1·776	1·806	1·852	1·887	1·941
1/12500	80	120	160	240	320	480	650	800	1200	1600	2400	3200	4800
0·00010	1·615	1·623	1·633	1·651	1·668	1·695	1·718	1·737	1·776	1·806	1·852	1·887	1·941
1/10000	85	130	170	250	330	500	650	850	1300	1700	2500	3300	5000
0·00015	1·613	1·622	1·632	1·650	1·667	1·695	1·717	1·737	1·775	1·805	1·851	1·887	1·941
1/ 6667	90	130	180	270	360	550	700	900	1300	1800	2700	3600	5500
0·00020	1·612	1·621	1·631	1·650	1·667	1·694	1·717	1·736	1·775	1·805	1·851	1·887	1·941
1/ 5000	95	140	190	280	370	550	750	950	1400	1900	2800	3700	5500
0·00030	1·611	1·620	1·630	1·649	1·666	1·694	1·717	1·736	1·775	1·805	1·851	1·887	1·941
1/ 3333	100	150	200	300	390	600	800	1000	1500	2000	3000	3900	6000
0·00040	1·610	1·619	1·630	1·649	1·666	1·694	1·717	1·736	1·775	1·805	1·851	1·887	1·941
1/ 2500	100	150	200	310	410	600	800	1000	1500	2000	3100	4100	6000
0·00060	1·609	1·619	1·629	1·648	1·665	1·693	1·716	1·736	1·774	1·805	1·851	1·886	1·940
1/ 1667	110	160	220	320	430	650	850	1100	1600	2200	3200	4300	6500
0·00080	1·609	1·618	1·629	1·648	1·665	1·693	1·716	1·736	1·774	1·804	1·851	1·886	1·940
1/ 1250	110	170	220	330	440	650	900	1100	1700	2200	3300	4400	6500
0·00100	1·608	1·618	1·628	1·648	1·665	1·693	1·716	1·735	1·774	1·804	1·851	1·886	1·940
1/ 1000	110	170	230	340	460	700	900	1100	1700	2300	3400	4600	7000
0·00150	1·608	1·617	1·628	1·648	1·665	1·693	1·716	1·735	1·774	1·804	1·851	1·886	1·940
1/ 667	120	180	240	360	470	700	950	1200	1800	2400	3600	4700	7000
0·00200	1·607	1·617	1·628	1·647	1·665	1·693	1·716	1·735	1·774	1·804	1·851	1·886	1·940
1/ 500	120	180	240	370	490	750	1000	1200	1800	2400	3700	4900	7500
0·00300	1·607	1·617	1·628	1·647	1·664	1·693	1·716	1·735	1·774	1·804	1·851	1·886	1·940
1/ 333	130	190	250	380	500	750	1000	1300	1900	2500	3800	5000	7500
0·00400	1·607	1·617	1·627	1·647	1·664	1·693	1·716	1·735	1·774	1·804	1·851	1·886	1·940
1/ 250	130	190	260	390	500	750	1000	1300	1900	2600	3900	5000	7500
0·00600	1·606	1·616	1·627	1·647	1·664	1·693	1·716	1·735	1·774	1·804	1·851	1·886	1·940
1/ 167	130	200	270	400	550	800	1100	1300	2000	2700	4000	5500	8000
0·00800	1·606	1·616	1·627	1·647	1·664	1·692	1·715	1·735	1·774	1·804	1·850	1·886	1·940
1/ 125	140	200	270	410	550	800	1100	1400	2000	2700	4100	5500	8000
0·01000	1·606	1·616	1·627	1·647	1·664	1·692	1·715	1·735	1·774	1·804	1·850	1·886	1·940
1/ 100	140	210	270	410	550	800	1100	1400	2100	2700	4100	5500	8000
0·01500	1·606	1·616	1·627	1·647	1·664	1·692	1·715	1·735	1·774	1·804	1·850	1·886	1·940
1/ 67	140	210	280	420	550	850	1100	1400	2100	2800	4200	5500	8500
0·02000	1·606	1·616	1·627	1·647	1·664	1·692	1·715	1·735	1·774	1·804	1·850	1·886	1·940
1/ 50	140	210	280	430	550	850	1100	1400	2100	2800	4300	5500	8500
0·03000	1·606	1·616	1·627	1·647	1·664	1·692	1·715	1·735	1·774	1·804	1·850	1·886	1·940
1/ 33·3	150	220	290	440	600	850	1200	1500	2200	2900	4400	6000	8500
0·04000	1·606	1·616	1·627	1·647	1·664	1·692	1·715	1·735	1·774	1·804	1·850	1·886	1·940
1/ 25·0	150	220	290	440	600	900	1200	1500	2200	2900	4400	6000	9000
0·06000	1·606	1·616	1·627	1·647	1·664	1·692	1·715	1·735	1·774	1·804	1·850	1·886	1·940
1/ 16·7	150	220	300	450	600	900	1200	1500	2200	3000	4500	6000	9000
0·08000	1·606	1·616	1·627	1·646	1·664	1·692	1·715	1·735	1·774	1·804	1·850	1·886	1·940
1/ 12·5	150	230	300	450	600	900	1200	1500	2300	3000	4500	6000	9000
0·10000	1·606	1·616	1·627	1·646	1·664	1·692	1·715	1·735	1·774	1·804	1·850	1·886	1·940
1/ 10·0	150	230	300	450	600	900	1200	1500	2300	3000	4500	6000	9000
	1000	1500	2000	3000	4000	6·00	8·00	10·00	15·00	20·00	30·00	40·00	60·00
Gradient	(Equivalent) Pipe diameters in mm to 4000 mm, then in m												

S = 0·000020 to 0·10000 k_s = 6·0 mm

E12

k_s = 15·0 mm
S = 0·00010 to 0·60000

Water (or sewage) at 15°C; full bore conditions.
This table shows values of m for use with Tables D

ie hydraulic gradient =
1 in 10000 to 1 in 1·7

m_C for Colebrook-White solutions;
θ for use with part-full circular pipes (turbulent flows)

Gradient	(Equivalent) Pipe diameters in mm												
	20	30	40	60	80	100	150	200	300	400	600	800	1000
0·00010 1/10000					*	*	2·105	2·041	1·973	1·938	1·903	1·886	1·877
							7	10	14	19	28	38	48
0·00015 1/ 6667					*	2·219	2·100	2·037	1·971	1·936	1·901	1·885	1·877
						5	7	10	15	20	30	40	50
0·00020 1/ 5000					*	2·215	2·098	2·035	1·969	1·935	1·900	1·884	1·876
						5	8	10	15	20	30	40	50
0·00030 1/ 3333					*	2·211	2·095	2·033	1·968	1·934	1·899	1·884	1·875
						5	8	10	16	20	32	42	50
0·00040 1/ 2500					2·290	2·208	2·093	2·031	1·967	1·933	1·899	1·883	1·875
					4	5	8	11	16	22	32	42	55
0·00060 1/ 1667					2·286	2·205	2·091	2·030	1·965	1·932	1·898	1·882	1·874
					4	5	8	11	16	22	32	44	55
0·00080 1/ 1250					2·284	2·203	2·089	2·029	1·965	1·931	1·898	1·882	1·874
					4	6	8	11	17	22	34	44	55
0·00100 1/ 1000					2·282	2·202	2·089	2·028	1·964	1·931	1·897	1·882	1·874
					4	6	8	11	17	22	34	44	55
0·00150 1/ 667					2·280	2·200	2·087	2·027	1·963	1·930	1·897	1·882	1·873
					5	6	9	11	17	22	34	46	55
0·00200 1/ 500					2·278	2·199	2·086	2·026	1·963	1·930	1·897	1·881	1·873
					5	6	9	12	17	24	34	46	60
0·00300 1/ 333					2·277	2·197	2·085	2·026	1·962	1·929	1·896	1·881	1·873
					5	6	9	12	18	24	36	48	60
0·00400 1/ 250					2·275	2·196	2·085	2·025	1·962	1·929	1·896	1·881	1·873
					5	6	9	12	18	24	36	48	60
0·00600 1/ 167					2·274	2·195	2·084	2·025	1·962	1·929	1·896	1·881	1·873
					5	6	9	12	18	24	36	48	60
0·00800 1/ 125					2·273	2·195	2·084	2·024	1·961	1·929	1·896	1·881	1·873
					5	6	9	12	18	24	36	48	60
0·01000 1/ 100					2·273	2·194	2·083	2·024	1·961	1·928	1·896	1·881	1·873
					5	6	9	12	18	24	36	50	60
0·01500 1/ 67					2·272	2·194	2·083	2·024	1·961	1·928	1·896	1·880	1·873
					5	6	9	12	19	24	38	50	60
0·02000 1/ 50					2·272	2·193	2·083	2·024	1·961	1·928	1·896	1·880	1·872
					5	6	9	12	19	24	38	50	60
0·03000 1/ 33·3					2·271	2·193	2·082	2·023	1·961	1·928	1·895	1·880	1·872
					5	6	9	13	19	26	38	50	65
0·04000 1/ 25·0					2·271	2·193	2·082	2·023	1·961	1·928	1·895	1·880	1·872
					5	6	9	13	19	26	38	50	65
0·06000 1/ 16·7					2·270	2·192	2·082	2·023	1·960	1·928	1·895	1·880	1·872
					5	6	10	13	19	26	38	50	65
0·08000 1/ 12·5					2·270	2·192	2·082	2·023	1·960	1·928	1·895	1·880	1·872
					5	6	10	13	19	26	38	50	65
0·10000 1/ 10·0					2·270	2·192	2·082	2·023	1·960	1·928	1·895	1·880	1·872
					5	6	10	13	19	26	38	50	65
0·15000 1/ 6·7					2·270	2·192	2·082	2·023	1·960	1·928	1·895	1·880	1·872
					5	6	10	13	19	26	38	50	65
0·20000 1/ 5·0					2·270	2·192	2·082	2·023	1·960	1·928	1·895	1·880	1·872
					5	6	10	13	19	26	38	50	65
0·30000 1/ 3·3					2·269	2·191	2·081	2·023	1·960	1·928	1·895	1·880	1·872
					5	6	10	13	19	26	38	50	65
0·40000 1/ 2·5					2·269	2·191	2·081	2·023	1·960	1·928	1·895	1·880	1·872
					5	7	10	13	20	26	40	50	65
0·60000 1/ 1·7					2·269	2·191	2·081	2·022	1·960	1·928	1·895	1·880	1·872
					5	7	10	13	20	26	40	50	65
	20	30	40	60	80	100	150	200	300	400	600	800	1000

Gradient   (Equivalent) Pipe diameters in mm

k_s = 15·0 mm S = 0·00010 to 0·60000

k_s = 15·0 mm
S = 0.000020 to 0·10000

ie hydraulic gradient =
1 in 50000 to 1 in 10·0

Water (or sewage) at 15°C; full bore conditions.
This table shows values of m for use with Tables D

m_C for Colebrook-White solutions;
θ for use with part-full circular pipes

Gradient	1000	1500	2000	3000	4000	6·00	8·00	10·00	15·00	20·00	30·00	40·00	60·00
0·00002	1·884	1·875	1·874	1·880	1·889	1·908	1·925	1·941	1·973	2·000	2·042	2·075	2·126
1/50000	40	60	80	120	160	240	320	400	600	800	1200	1600	2400
0·00003	1·882	1·873	1·873	1·879	1·889	1·907	1·925	1·940	1·973	2·000	2·042	2·075	2·126
1/33333	42	65	85	130	170	250	330	420	650	850	1300	1700	2500
0·00004	1·881	1·872	1·872	1·879	1·888	1·907	1·924	1·940	1·973	1·999	2·042	2·075	2·126
1/25000	44	65	85	130	170	260	350	430	650	850	1300	1700	2600
0·00006	1·879	1·871	1·871	1·878	1·887	1·907	1·924	1·940	1·973	1·999	2·041	2·075	2·126
1/16667	46	70	90	140	180	270	360	450	700	900	1400	1800	2700
0·00008	1·878	1·870	1·870	1·877	1·887	1·906	1·924	1·939	1·972	1·999	2·041	2·074	2·126
1/12500	46	70	95	140	190	280	370	470	700	950	1400	1900	2800
0·00010	1·877	1·870	1·870	1·877	1·887	1·906	1·924	1·939	1·972	1·999	2·041	2·074	2·126
1/10000	48	70	95	140	190	290	380	480	700	950	1400	1900	2900
0·00015	1·877	1·869	1·869	1·877	1·886	1·906	1·923	1·939	1·972	1·999	2·041	2·074	2·126
1/ 6667	50	75	100	150	200	300	400	490	750	1000	1500	2000	3000
0·00020	1·876	1·869	1·869	1·876	1·886	1·906	1·923	1·939	1·972	1·999	2·041	2·074	2·125
1/ 5000	50	75	100	150	200	300	410	500	750	1000	1500	2000	3000
0·00030	1·875	1·868	1·869	1·876	1·886	1·905	1·923	1·939	1·972	1·999	2·041	2·074	2·125
1/ 3333	50	80	100	160	210	310	420	500	800	1000	1600	2100	3100
0·00040	1·875	1·868	1·868	1·876	1·886	1·905	1·923	1·939	1·972	1·999	2·041	2·074	2·125
1/ 2500	55	80	110	160	210	320	430	550	800	1100	1600	2100	3200
0·00060	1·874	1·867	1·868	1·876	1·885	1·905	1·923	1·939	1·972	1·998	2·041	2·074	2·125
1/ 1667	55	80	110	160	220	330	440	550	800	1100	1600	2200	3300
0·00080	1·874	1·867	1·868	1·875	1·885	1·905	1·923	1·939	1·972	1·998	2·041	2·074	2·125
1/ 1250	55	85	110	170	220	330	440	550	850	1100	1700	2200	3300
0·00100	1·874	1·867	1·868	1·875	1·885	1·905	1·923	1·938	1·972	1·998	2·041	2·074	2·125
1/ 1000	55	85	110	170	220	340	450	550	850	1100	1700	2200	3400
0·00150	1·873	1·867	1·867	1·875	1·885	1·905	1·923	1·938	1·972	1·998	2·041	2·074	2·125
1/ 667	55	85	110	170	230	340	460	550	850	1100	1700	2300	3400
0·00200	1·873	1·867	1·867	1·875	1·885	1·905	1·923	1·938	1·971	1·998	2·041	2·074	2·125
1/ 500	60	85	120	170	230	350	460	600	850	1200	1700	2300	3500
0·00300	1·873	1·866	1·867	1·875	1·885	1·905	1·922	1·938	1·971	1·998	2·041	2·074	2·125
1/ 333	60	90	120	180	240	350	470	600	900	1200	1800	2400	3500
0·00400	1·873	1·866	1·867	1·875	1·885	1·905	1·922	1·938	1·971	1·998	2·041	2·074	2·125
1/ 250	60	90	120	180	240	360	480	600	900	1200	1800	2400	3600
0·00600	1·873	1·866	1·867	1·875	1·885	1·905	1·922	1·938	1·971	1·998	2·041	2·074	2·125
1/ 167	60	90	120	180	240	360	480	600	900	1200	1800	2400	3600
0·00800	1·873	1·866	1·867	1·875	1·885	1·905	1·922	1·938	1·971	1·998	2·041	2·074	2·125
1/ 125	60	90	120	180	240	370	490	600	900	1200	1800	2400	3700
0·01000	1·873	1·866	1·867	1·875	1·885	1·905	1·922	1·938	1·971	1·998	2·041	2·074	2·125
1/ 100	60	90	120	180	250	370	490	600	900	1200	1800	2500	3700
0·01500	1·873	1·866	1·867	1·875	1·885	1·905	1·922	1·938	1·971	1·998	2·041	2·074	2·125
1/ 67	60	95	120	190	250	370	500	600	950	1200	1900	2500	3700
0·02000	1·872	1·866	1·867	1·875	1·885	1·905	1·922	1·938	1·971	1·998	2·041	2·074	2·125
1/ 50	60	95	120	190	250	370	500	600	950	1200	1900	2500	3700
0·03000	1·872	1·866	1·867	1·875	1·885	1·904	1·922	1·938	1·971	1·998	2·041	2·074	2·125
1/ 33·3	65	95	130	190	250	380	500	650	950	1300	1900	2500	3800
0·04000	1·872	1·866	1·867	1·875	1·885	1·904	1·922	1·938	1·971	1·998	2·041	2·074	2·125
1/ 25·0	65	95	130	190	250	380	500	650	950	1300	1900	2500	3800
0·06000	1·872	1·866	1·867	1·875	1·885	1·904	1·922	1·938	1·971	1·998	2·041	2·074	2·125
1/ 16·7	65	95	130	190	250	380	500	650	950	1300	1900	2500	3800
0·08000	1·872	1·866	1·867	1·875	1·885	1·904	1·922	1·938	1·971	1·998	2·041	2·074	2·125
1/ 12·5	65	95	130	190	260	380	500	650	950	1300	1900	2600	3800
0·10000	1·872	1·866	1·867	1·875	1·885	1·904	1·922	1·938	1·971	1·998	2·041	2·074	2·125
1/ 10·0	65	95	130	190	260	380	500	650	950	1300	1900	2600	3800
	1000	1500	2000	3000	4000	6·00	8·00	10·00	15·00	20·00	30·00	40·00	60·00

Gradient (Equivalent) Pipe diameters in mm to 4000 mm, then in m

S = 0.000020 to 0·10000   **k_s = 15·0 mm**

E13

k_s = roughness size in mm

For all values of S or hydraulic gradient

Water (or sewage) at 15°C; full bore conditions.
This table shows values of m for use with Tables D

m_C for Colebrook-White solutions;
θ for use with part-full circular pipes (turbulent flows)

(Equivalent) Pipe diameters in mm

k_s	20	30	40	60	80	100	150	200	300	400	600	800	1000
22·5							2·35 / 6	2·26 / 8	2·17 / 12	2·12 / 16	2·06 / 24	2·04 / 32	2·02 / 40
30·0							2·58 / 5	2·47 / 6	2·34 / 9	2·27 / 12	2·20 / 18	2·17 / 24	2·14 / 30
45·0									2·63 / 6	2·54 / 8	2·43 / 13	2·37 / 17	2·34 / 21
60·0									2·89 / 5	2·76 / 6	2·62 / 10	2·55 / 13	2·50 / 16
100											3·05 / 6	2·93 / 8	2·86 / 10
150												3·33 / 5	3·22 / 7

(Equivalent) Pipe diameters in mm — k_s

(Equivalent) Pipe diameters in mm to 4000 mm, then in m

k_s	1000	1500	2000	3000	4000	6·00	8·00	10·00	15·00	20·00	30·00	40·00	60·00
22·5	2·02 / 40	2·00 / 60	2·00 / 80	2·00 / 120	2·00 / 160	2·02 / 240	2·03 / 320	2·04 / 400	2·07 / 600	2·10 / 800	2·14 / 1200	2·17 / 1600	2·22 / 2400
30·0	2·15 / 30	2·12 / 46	2·10 / 60	2·10 / 90	2·10 / 120	2·11 / 180	2·12 / 240	2·13 / 310	2·15 / 460	2·18 / 600	2·21 / 900	2·24 / 1200	2·29 / 1800
45·0	2·34 / 21	2·30 / 31	2·27 / 42	2·25 / 60	2·24 / 85	2·24 / 130	2·25 / 170	2·26 / 210	2·28 / 310	2·29 / 420	2·33 / 650	2·36 / 850	2·40 / 1250
60·0	2·50 / 16	2·44 / 24	2·41 / 32	2·37 / 48	2·36 / 65	2·35 / 95	2·35 / 130	2·36 / 160	2·37 / 240	2·39 / 320	2·42 / 480	2·44 / 650	2·48 / 950
100	2·86 / 10	2·75 / 15	2·69 / 20	2·63 / 30	2·60 / 40	2·57 / 60	2·56 / 80	2·56 / 100	2·56 / 150	2·57 / 200	2·59 / 300	2·61 / 390	2·65 / 600
150	3·22 / 7	3·06 / 10	2·97 / 13	2·88 / 20	2·83 / 26	2·78 / 40	2·76 / 50	2·75 / 65	2·74 / 100	2·74 / 130	2·75 / 200	2·77 / 260	2·80 / 390
225		3·44 / 7	3·31 / 9	3·18 / 13	3·10 / 18	3·03 / 26	2·99 / 35	2·97 / 44	2·94 / 65	2·93 / 90	2·93 / 130	2·94 / 180	2·96 / 260
300		3·78 / 5	3·61 / 7	3·43 / 10	3·33 / 13	3·23 / 20	3·18 / 26	3·14 / 34	3·10 / 50	3·08 / 65	3·07 / 100	3·08 / 130	3·09 / 200
450				3·86 / 7	3·72 / 9	3·56 / 13	3·48 / 18	3·43 / 22	3·36 / 33	3·33 / 44	3·30 / 65	3·29 / 90	3·29 / 130
600				4·24 / 5	4·05 / 7	3·85 / 10	3·74 / 13	3·67 / 17	3·58 / 24	3·53 / 34	3·48 / 50	3·46 / 65	3·45 / 100
1000						4·48 / 6	4·30 / 8	4·19 / 10	4·03 / 15	3·95 / 20	3·86 / 30	3·82 / 40	3·78 / 60
1500							4·89 / 5	4·72 / 7	4·48 / 10	4·36 / 13	4·22 / 20	4·15 / 27	4·08 / 40
2250									5·05 / 7	4·86 / 9	4·66 / 13	4·55 / 18	4·44 / 27
3000									5·55 / 6	5·30 / 7	5·03 / 10	4·89 / 13	4·74 / 20
4500											5·67 / 7	5·46 / 9	5·23 / 13
6000											6·23 / 5	5·95 / 7	5·65 / 10
10000													6·58 / 6

(Equivalent) Pipe diameters in mm to 4000 mm, then in m — k_s

ANNEXURE - Hazen-Williams solutions from Tables D

The Hazen-Williams[1] formula for the mean velocity V of turbulent flow of water in a circular pipe, with basic SI units, is

$$V = 0.849\, C R^{0.63} S^{0.54}$$

where C is a coefficient depending on the pipe surface and applying to both basic SI and basic FPS units. The constant 0.849 allows for the use of basic SI units and thus conforms here with V being in ms^{-1} and hydraulic mean depth R (hydraulic radius) being in m, with S indicating piezometric gradient.

Together with the Manning formula adjusted to use $m = 100\,n$, and with both formula adjusted to $D = 4R$, the Hazen-Williams formula gives, for selected combinations of values of D, S and C, values of m_H to be applied as divisor to values of mV and/or mQ in Tables D. This then gives mean velocity V and/or discharge Q according to the Hazen-Williams formula, given pipe diameter D and gradient S. or, in the former case, equivalent diameter D_{ep} then leading to mean velocity and/or discharge Q_{ep} in the equivalent pipe.

There follow Tables H1 to H19 giving values of m_H against diameter and gradient for values of C from 140 to 40. Commonly accepted values[2,3] of C for pipes are as follows, but the main intention is for users to draw on past experience in selecting C.

Extremely smooth pipes	140	Old cast iron, brick	100
Very smooth pipes	130	Old riveted steel	95
Asbestos cement, concrete	120	Old iron pipe	60 - 80
New riveted steel, clay	110	Small, badly tuberculated	40 - 50

1. WILLIAMS, G.S. and HAZEN, A.H. *Hydraulic tables*, 3rd ed., Wiley, N.Y., 1933.

2. KING, H.O. *Handbook of hydraulics*, 3rd ed., McGraw-Hill, N.Y., 1939.

3. VENNARD, J.K. and STREET, R.L. *Elementary fluid mechanics*, 6th ed., SI version, Wiley, N.Y., 1982.

C = 140
$S = 0.00010$ to 0.40000

ie hydraulic gradient =
1 in 10000 to 1 in 2.5

Water (or sewage) at normal temperature;
full bore conditions.

This table shows values of m_H, the Hazen-Williams based divisor of mV and/or mQ, for use with Tables D

H1

Gradient		(Equivalent) Pipe diameters in mm										
		40	60	80	100	150	200	300	400	600	800	1000	1500
0.00010	(1/10000)	1.027	1.042	1.054	1.062	1.078	1.090	1.106	1.118	1.134	1.146	1.156	1.173
0.00015	(1/ 6667)	1.011	1.026	1.037	1.045	1.061	1.072	1.088	1.100	1.116	1.128	1.137	1.154
0.00020	(1/ 5000)	0.999	1.014	1.025	1.033	1.049	1.060	1.076	1.087	1.103	1.115	1.124	1.141
0.00030	(1/ 3333)	0.983	0.998	1.008	1.017	1.032	1.043	1.058	1.070	1.086	1.097	1.106	1.123
0.00040	(1/ 2500)	0.972	0.986	0.997	1.005	1.020	1.031	1.046	1.057	1.073	1.085	1.093	1.110
0.00060	(1/ 1667)	0.956	0.970	0.981	0.989	1.004	1.014	1.029	1.040	1.056	1.067	1.076	1.092
0.00080	(1/ 1250)	0.945	0.959	0.969	0.977	0.992	1.003	1.018	1.028	1.044	1.055	1.064	1.080
0.00100	(1/ 1000)	0.937	0.951	0.961	0.969	0.983	0.994	1.009	1.019	1.035	1.046	1.054	1.070
0.00150	(1/ 667)	0.922	0.935	0.945	0.953	0.967	0.978	0.992	1.003	1.018	1.029	1.037	1.053
0.00200	(1/ 500)	0.911	0.925	0.935	0.942	0.956	0.967	0.981	0.991	1.006	1.017	1.025	1.041
0.00300	(1/ 333)	0.896	0.910	0.920	0.927	0.941	0.951	0.965	0.975	0.990	1.001	1.009	1.024
0.00400	(1/ 250)	0.886	0.899	0.909	0.916	0.930	0.940	0.954	0.964	0.979	0.989	0.997	1.012
0.00600	(1/ 167)	0.872	0.885	0.894	0.902	0.915	0.925	0.939	0.949	0.963	0.973	0.981	0.996
0.00800	(1/ 125)	0.862	0.875	0.884	0.891	0.905	0.914	0.928	0.938	0.952	0.962	0.970	0.985
0.01000	(1/ 100)	0.854	0.867	0.876	0.883	0.897	0.906	0.920	0.930	0.944	0.954	0.961	0.976
0.01500	(1/ 67)	0.841	0.853	0.862	0.869	0.882	0.892	0.905	0.915	0.928	0.938	0.946	0.960
0.02000	(1/ 50)	0.831	0.843	0.852	0.859	0.872	0.881	0.895	0.904	0.918	0.927	0.935	0.949
0.03000	(1/ 33.3)	0.818	0.830	0.839	0.845	0.858	0.867	0.880	0.890	0.903	0.913	0.920	0.934
0.04000	(1/ 25.0)	0.808	0.820	0.829	0.836	0.848	0.857	0.870	0.879	0.893	0.902	0.910	0.923
0.06000	(1/ 16.7)	0.795	0.807	0.816	0.822	0.835	0.844	0.856	0.865	0.878	0.888	0.895	0.908
0.08000	(1/ 12.5)	0.786	0.798	0.806	0.813	0.825	0.834	0.846	0.855	0.868	0.877	0.885	0.898
0.10000	(1/ 10.0)	0.779	0.791	0.799	0.806	0.818	0.827	0.839	0.848	0.861	0.870	0.877	0.890
0.15000	(1/ 6.7)	0.767	0.778	0.786	0.793	0.805	0.813	0.825	0.834	0.847	0.856	0.863	0.876
0.20000	(1/ 5.0)	0.758	0.769	0.777	0.784	0.795	0.804	0.816	0.825	0.837	0.846	0.853	0.866
0.30000	(1/ 3.3)	0.746	0.757	0.765	0.771	0.783	0.791	0.803	0.811	0.824	0.832	0.839	0.852
0.40000	(1/ 2.5)	0.737	0.748	0.756	0.762	0.774	0.782	0.794	0.802	0.814	0.823	0.829	0.842
		40	60	80	100	150	200	300	400	600	800	1000	1500

Gradient (Equivalent) Pipe diameters in mm (C = 140)

243

H2

C = 135
S = 0.00010 to 0.40000

ie hydraulic gradient =
1 in 10000 to 1 in 2.5

Water (or sewage) at normal temperature; full bore conditions.

This table shows values of m_H, the Hazen-Williams based divisor of mV and/or mQ, for use with Tables D

Gradient — (Equivalent) Pipe diameters in mm

Gradient		40	60	80	100	150	200	300	400	600	800	1000	1500
0.00010	(1/10000)	1.065	1.081	1.093	1.102	1.118	1.130	1.147	1.159	1.176	1.189	1.199	1.217
0.00015	(1/ 6667)	1.048	1.064	1.075	1.084	1.100	1.112	1.128	1.140	1.157	1.170	1.179	1.197
0.00020	(1/ 5000)	1.036	1.051	1.063	1.071	1.087	1.099	1.115	1.127	1.144	1.156	1.166	1.183
0.00030	(1/ 3333)	1.019	1.035	1.046	1.054	1.070	1.081	1.098	1.109	1.126	1.138	1.147	1.164
0.00040	(1/ 2500)	1.008	1.023	1.034	1.042	1.058	1.069	1.085	1.096	1.113	1.125	1.134	1.151
0.00060	(1/ 1667)	0.991	1.006	1.017	1.025	1.041	1.052	1.068	1.079	1.095	1.107	1.116	1.132
0.00080	(1/ 1250)	0.980	0.995	1.005	1.014	1.029	1.040	1.055	1.067	1.082	1.094	1.103	1.120
0.00100	(1/ 1000)	0.971	0.986	0.996	1.005	1.020	1.030	1.046	1.057	1.073	1.084	1.093	1.110
0.00150	(1/ 667)	0.956	0.970	0.980	0.988	1.003	1.014	1.029	1.040	1.056	1.067	1.076	1.092
0.00200	(1/ 500)	0.945	0.959	0.969	0.977	0.992	1.002	1.017	1.028	1.044	1.055	1.063	1.079
0.00300	(1/ 333)	0.930	0.944	0.954	0.961	0.976	0.986	1.001	1.012	1.027	1.038	1.046	1.062
0.00400	(1/ 250)	0.919	0.933	0.943	0.950	0.965	0.975	0.990	1.000	1.015	1.026	1.034	1.050
0.00600	(1/ 167)	0.904	0.918	0.927	0.935	0.949	0.959	0.974	0.984	0.999	1.009	1.018	1.033
0.00800	(1/ 125)	0.894	0.907	0.917	0.924	0.938	0.948	0.962	0.973	0.987	0.998	1.006	1.021
0.01000	(1/ 100)	0.886	0.899	0.909	0.916	0.930	0.940	0.954	0.964	0.978	0.989	0.997	1.012
0.01500	(1/ 67)	0.872	0.885	0.894	0.901	0.915	0.925	0.939	0.949	0.963	0.973	0.981	0.996
0.02000	(1/ 50)	0.862	0.875	0.884	0.891	0.905	0.914	0.928	0.938	0.952	0.962	0.970	0.984
0.03000	(1/ 33.3)	0.848	0.861	0.870	0.877	0.890	0.899	0.913	0.923	0.936	0.946	0.954	0.968
0.04000	(1/ 25.0)	0.838	0.851	0.860	0.867	0.880	0.889	0.902	0.912	0.926	0.936	0.943	0.957
0.06000	(1/ 16.7)	0.825	0.837	0.846	0.853	0.866	0.875	0.888	0.897	0.911	0.920	0.928	0.942
0.08000	(1/ 12.5)	0.815	0.827	0.836	0.843	0.856	0.865	0.878	0.887	0.900	0.910	0.917	0.931
0.10000	(1/ 10.0)	0.808	0.820	0.829	0.836	0.848	0.857	0.870	0.879	0.892	0.902	0.909	0.923
0.15000	(1/ 6.7)	0.795	0.807	0.815	0.822	0.834	0.843	0.856	0.865	0.878	0.887	0.895	0.908
0.20000	(1/ 5.0)	0.786	0.798	0.806	0.813	0.825	0.834	0.846	0.855	0.868	0.877	0.884	0.898
0.30000	(1/ 3.3)	0.773	0.785	0.793	0.800	0.812	0.820	0.833	0.841	0.854	0.863	0.870	0.883
0.40000	(1/ 2.5)	0.764	0.776	0.784	0.791	0.802	0.811	0.823	0.832	0.844	0.853	0.860	0.873
		40	**60**	**80**	**100**	**150**	**200**	**300**	**400**	**600**	**800**	**1000**	**1500**

Gradient — (Equivalent) Pipe diameters in mm **(C = 135)**

H3

C = 130
S = 0.00010 to 0.40000

ie hydraulic gradient =
1 in 10000 to 1 in 2.5

Water (or sewage) at normal temperature; full bore conditions.

This table shows values of m_H, the Hazen-Williams based divisor of mV and/or mQ, for use with Tables D

Gradient — (Equivalent) Pipe diameters in mm

Gradient		40	60	80	100	150	200	300	400	600	800	1000	1500
0.00010	(1/10000)	1.106	1.123	1.135	1.144	1.161	1.173	1.191	1.204	1.222	1.235	1.245	1.263
0.00015	(1/ 6667)	1.088	1.105	1.116	1.125	1.142	1.154	1.172	1.184	1.202	1.215	1.225	1.243
0.00020	(1/ 5000)	1.076	1.092	1.104	1.113	1.129	1.141	1.158	1.171	1.188	1.201	1.211	1.229
0.00030	(1/ 3333)	1.059	1.074	1.086	1.095	1.111	1.123	1.140	1.152	1.169	1.182	1.191	1.209
0.00040	(1/ 2500)	1.046	1.062	1.073	1.082	1.098	1.110	1.127	1.139	1.156	1.168	1.178	1.195
0.00060	(1/ 1667)	1.030	1.045	1.056	1.065	1.081	1.092	1.109	1.120	1.137	1.149	1.159	1.176
0.00080	(1/ 1250)	1.018	1.033	1.044	1.053	1.068	1.080	1.096	1.108	1.124	1.136	1.145	1.163
0.00100	(1/ 1000)	1.009	1.024	1.035	1.043	1.059	1.070	1.086	1.098	1.114	1.126	1.135	1.152
0.00150	(1/ 667)	0.993	1.007	1.018	1.026	1.042	1.053	1.069	1.080	1.096	1.108	1.117	1.134
0.00200	(1/ 500)	0.981	0.996	1.006	1.015	1.030	1.041	1.056	1.068	1.084	1.095	1.104	1.121
0.00300	(1/ 333)	0.965	0.980	0.990	0.998	1.013	1.024	1.039	1.050	1.066	1.078	1.086	1.103
0.00400	(1/ 250)	0.954	0.969	0.979	0.987	1.002	1.012	1.028	1.038	1.054	1.065	1.074	1.090
0.00600	(1/ 167)	0.939	0.953	0.963	0.971	0.986	0.996	1.011	1.022	1.037	1.048	1.057	1.073
0.00800	(1/ 125)	0.928	0.942	0.952	0.960	0.974	0.985	0.999	1.010	1.025	1.036	1.045	1.060
0.01000	(1/ 100)	0.920	0.934	0.944	0.951	0.966	0.976	0.991	1.001	1.016	1.027	1.035	1.051
0.01500	(1/ 67)	0.905	0.919	0.928	0.936	0.950	0.960	0.975	0.985	1.000	1.010	1.019	1.034
0.02000	(1/ 50)	0.895	0.908	0.918	0.925	0.939	0.949	0.963	0.974	0.988	0.999	1.007	1.022
0.03000	(1/ 33.3)	0.880	0.894	0.903	0.911	0.924	0.934	0.948	0.958	0.972	0.983	0.991	1.006
0.04000	(1/ 25.0)	0.870	0.883	0.893	0.900	0.914	0.923	0.937	0.947	0.961	0.971	0.979	0.994
0.06000	(1/ 16.7)	0.856	0.869	0.878	0.886	0.899	0.908	0.922	0.932	0.946	0.956	0.964	0.978
0.08000	(1/ 12.5)	0.847	0.859	0.868	0.876	0.889	0.898	0.912	0.921	0.935	0.945	0.953	0.967
0.10000	(1/ 10.0)	0.839	0.852	0.861	0.868	0.881	0.890	0.903	0.913	0.927	0.937	0.944	0.958
0.15000	(1/ 6.7)	0.826	0.838	0.847	0.854	0.867	0.876	0.889	0.898	0.912	0.921	0.929	0.943
0.20000	(1/ 5.0)	0.816	0.828	0.837	0.844	0.857	0.866	0.879	0.888	0.901	0.911	0.918	0.932
0.30000	(1/ 3.3)	0.803	0.815	0.824	0.830	0.843	0.852	0.865	0.874	0.887	0.896	0.904	0.917
0.40000	(1/ 2.5)	0.794	0.806	0.814	0.821	0.833	0.842	0.855	0.864	0.877	0.886	0.893	0.907
		40	**60**	**80**	**100**	**150**	**200**	**300**	**400**	**600**	**800**	**1000**	**1500**

Gradient — (Equivalent) Pipe diameters in mm **(C = 130)**

C = 125
S = 0·00010 to 0·40000

ie hydraulic gradient =
1 in 10000 to 1 in 2·5

Water (or sewage) at normal temperature;
full bore conditions.

This table shows values of m_H, the Hazen-Williams
based divisor of mV and/or mQ, for use with Tables D

Gradient		40	60	80	100	150	200	300	400	600	800	1000	1500
0·00010	(1/10000)	1·150	1·168	1·180	1·190	1·207	1·220	1·239	1·252	1·270	1·284	1·295	1·314
0·00015	(1/ 6667)	1·132	1·149	1·161	1·170	1·188	1·201	1·219	1·232	1·250	1·263	1·274	1·293
0·00020	(1/ 5000)	1·119	1·136	1·148	1·157	1·174	1·187	1·205	1·217	1·236	1·249	1·259	1·278
0·00030	(1/ 3333)	1·101	1·117	1·129	1·138	1·156	1·168	1·185	1·198	1·216	1·229	1·239	1·257
0·00040	(1/ 2500)	1·088	1·105	1·116	1·125	1·142	1·154	1·172	1·184	1·202	1·215	1·225	1·243
0·00060	(1/ 1667)	1·071	1·087	1·098	1·107	1·124	1·136	1·153	1·165	1·183	1·195	1·205	1·223
0·00080	(1/ 1250)	1·058	1·074	1·086	1·095	1·111	1·123	1·140	1·152	1·169	1·182	1·191	1·209
0·00100	(1/ 1000)	1·049	1·065	1·076	1·085	1·101	1·113	1·130	1·142	1·159	1·171	1·181	1·198
0·00150	(1/ 667)	1·032	1·048	1·059	1·068	1·084	1·095	1·111	1·123	1·140	1·152	1·162	1·179
0·00200	(1/ 500)	1·020	1·036	1·047	1·055	1·071	1·082	1·099	1·110	1·127	1·139	1·148	1·166
0·00300	(1/ 333)	1·004	1·019	1·030	1·038	1·054	1·065	1·081	1·093	1·109	1·121	1·130	1·147
0·00400	(1/ 250)	0·992	1·007	1·018	1·026	1·042	1·053	1·069	1·080	1·096	1·108	1·117	1·134
0·00600	(1/ 167)	0·977	0·991	1·002	1·010	1·025	1·036	1·051	1·063	1·079	1·090	1·099	1·115
0·00800	(1/ 125)	0·965	0·980	0·990	0·998	1·013	1·024	1·039	1·050	1·066	1·078	1·086	1·103
0·01000	(1/ 100)	0·957	0·971	0·981	0·989	1·004	1·015	1·030	1·041	1·057	1·068	1·077	1·093
0·01500	(1/ 67)	0·941	0·955	0·966	0·974	0·988	0·999	1·014	1·024	1·040	1·051	1·059	1·075
0·02000	(1/ 50)	0·931	0·945	0·955	0·962	0·977	0·987	1·002	1·013	1·028	1·039	1·047	1·063
0·03000	(1/ 33·3)	0·916	0·929	0·939	0·947	0·961	0·971	0·986	0·996	1·011	1·022	1·030	1·046
0·04000	(1/ 25·0)	0·905	0·919	0·928	0·936	0·950	0·960	0·975	0·985	1·000	1·010	1·019	1·034
0·06000	(1/ 16·7)	0·891	0·904	0·914	0·921	0·935	0·945	0·959	0·969	0·984	0·994	1·002	1·017
0·08000	(1/ 12·5)	0·880	0·894	0·903	0·911	0·924	0·934	0·948	0·958	0·972	0·983	0·991	1·006
0·10000	(1/ 10·0)	0·873	0·886	0·895	0·902	0·916	0·926	0·940	0·950	0·964	0·974	0·982	0·997
0·15000	(1/ 6·7)	0·859	0·871	0·881	0·888	0·901	0·911	0·924	0·934	0·948	0·958	0·966	0·981
0·20000	(1/ 5·0)	0·849	0·861	0·871	0·878	0·891	0·900	0·914	0·924	0·937	0·947	0·955	0·969
0·30000	(1/ 3·3)	0·835	0·848	0·857	0·864	0·877	0·886	0·899	0·909	0·922	0·932	0·940	0·954
0·40000	(1/ 2·5)	0·826	0·838	0·847	0·854	0·867	0·876	0·889	0·898	0·912	0·921	0·929	0·943
		40	60	80	100	150	200	300	400	600	800	1000	1500

Gradient — (Equivalent) Pipe diameters in mm — (C = 125)

C = 120
S = 0·00010 to 0·40000

ie hydraulic gradient =
1 in 10000 to 1 in 2·5

Water (or sewage) at normal temperature;
full bore conditions.

This table shows values of m_H, the Hazen-Williams
based divisor of mV and/or mQ, for use with Tables D

Gradient		40	60	80	100	150	200	300	400	600	800	1000	1500
0·00010	(1/10000)	1·198	1·216	1·229	1·239	1·258	1·271	1·290	1·304	1·323	1·337	1·348	1·369
0·00015	(1/ 6667)	1·179	1·197	1·209	1·219	1·238	1·251	1·269	1·283	1·302	1·316	1·327	1·347
0·00020	(1/ 5000)	1·165	1·183	1·195	1·205	1·223	1·236	1·255	1·268	1·287	1·301	1·312	1·331
0·00030	(1/ 3333)	1·147	1·164	1·176	1·186	1·204	1·216	1·235	1·248	1·267	1·280	1·290	1·310
0·00040	(1/ 2500)	1·134	1·151	1·163	1·172	1·190	1·203	1·221	1·234	1·252	1·265	1·276	1·295
0·00060	(1/ 1667)	1·115	1·132	1·144	1·153	1·171	1·183	1·201	1·214	1·232	1·245	1·255	1·274
0·00080	(1/ 1250)	1·103	1·119	1·131	1·140	1·157	1·170	1·187	1·200	1·218	1·231	1·241	1·259
0·00100	(1/ 1000)	1·093	1·109	1·121	1·130	1·147	1·159	1·177	1·189	1·207	1·220	1·230	1·248
0·00150	(1/ 667)	1·075	1·091	1·103	1·112	1·129	1·141	1·158	1·170	1·188	1·200	1·210	1·228
0·00200	(1/ 500)	1·063	1·079	1·090	1·099	1·116	1·128	1·144	1·157	1·174	1·186	1·196	1·214
0·00300	(1/ 333)	1·046	1·061	1·073	1·082	1·098	1·109	1·126	1·138	1·155	1·167	1·177	1·195
0·00400	(1/ 250)	1·034	1·049	1·060	1·069	1·085	1·097	1·113	1·125	1·142	1·154	1·163	1·181
0·00600	(1/ 167)	1·017	1·032	1·043	1·052	1·068	1·079	1·095	1·107	1·124	1·135	1·145	1·162
0·00800	(1/ 125)	1·006	1·021	1·031	1·040	1·056	1·067	1·083	1·094	1·111	1·122	1·132	1·149
0·01000	(1/ 100)	0·997	1·012	1·022	1·031	1·046	1·057	1·073	1·085	1·101	1·112	1·122	1·138
0·01500	(1/ 67)	0·981	0·995	1·006	1·014	1·029	1·040	1·056	1·067	1·083	1·095	1·104	1·120
0·02000	(1/ 50)	0·969	0·984	0·994	1·003	1·018	1·028	1·044	1·055	1·071	1·082	1·091	1·107
0·03000	(1/ 33·3)	0·954	0·968	0·978	0·986	1·001	1·012	1·027	1·038	1·053	1·065	1·073	1·089
0·04000	(1/ 25·0)	0·943	0·957	0·967	0·975	0·990	1·000	1·015	1·026	1·041	1·052	1·061	1·077
0·06000	(1/ 16·7)	0·928	0·942	0·952	0·959	0·974	0·984	0·999	1·010	1·025	1·036	1·044	1·060
0·08000	(1/ 12·5)	0·917	0·931	0·941	0·948	0·963	0·973	0·987	0·998	1·013	1·024	1·032	1·048
0·10000	(1/ 10·0)	0·909	0·923	0·932	0·940	0·954	0·964	0·979	0·989	1·004	1·015	1·023	1·038
0·15000	(1/ 6·7)	0·894	0·908	0·917	0·925	0·939	0·949	0·963	0·973	0·988	0·998	1·006	1·022
0·20000	(1/ 5·0)	0·884	0·897	0·907	0·914	0·928	0·938	0·952	0·962	0·976	0·987	0·995	1·010
0·30000	(1/ 3·3)	0·870	0·883	0·892	0·900	0·913	0·923	0·937	0·947	0·961	0·971	0·979	0·994
0·40000	(1/ 2·5)	0·860	0·873	0·882	0·889	0·903	0·912	0·926	0·936	0·950	0·960	0·968	0·982
		40	60	80	100	150	200	300	400	600	800	1000	1500

Gradient — (Equivalent) Pipe diameters in mm — (C = 120)

245

H6

C = 115
S = 0·00010 to 0·40000

ie hydraulic gradient =
1 in 10000 to 1 in 2·5

Water (or sewage) at normal temperature; full bore conditions.

This table shows values of m_H, the Hazen-Williams based divisor of mV and/or mQ, for use with Tables D

Gradient		40	60	80	100	150	200	300	400	600	800	1000	1500
0·00010	(1/10000)	1·250	1·269	1·283	1·293	1·312	1·326	1·346	1·361	1·381	1·396	1·407	1·428
0·00015	(1/ 6667)	1·230	1·249	1·262	1·272	1·291	1·305	1·325	1·339	1·359	1·373	1·384	1·405
0·00020	(1/ 5000)	1·216	1·234	1·247	1·258	1·277	1·290	1·309	1·323	1·343	1·357	1·369	1·389
0·00030	(1/ 3333)	1·197	1·215	1·227	1·237	1·256	1·269	1·288	1·302	1·322	1·336	1·347	1·367
0·00040	(1/ 2500)	1·183	1·201	1·213	1·223	1·242	1·255	1·274	1·287	1·306	1·320	1·331	1·351
0·00060	(1/ 1667)	1·164	1·181	1·194	1·204	1·222	1·235	1·253	1·266	1·285	1·299	1·310	1·329
0·00080	(1/ 1250)	1·151	1·168	1·180	1·190	1·208	1·221	1·239	1·252	1·271	1·284	1·295	1·314
0·00100	(1/ 1000)	1·140	1·157	1·170	1·179	1·197	1·210	1·228	1·241	1·259	1·273	1·283	1·303
0·00150	(1/ 667)	1·122	1·139	1·151	1·160	1·178	1·190	1·208	1·221	1·239	1·252	1·263	1·282
0·00200	(1/ 500)	1·109	1·126	1·138	1·147	1·164	1·177	1·194	1·207	1·225	1·238	1·248	1·267
0·00300	(1/ 333)	1·091	1·108	1·119	1·129	1·146	1·158	1·175	1·188	1·205	1·218	1·228	1·247
0·00400	(1/ 250)	1·079	1·095	1·107	1·116	1·132	1·144	1·162	1·174	1·192	1·204	1·214	1·232
0·00600	(1/ 167)	1·061	1·077	1·089	1·098	1·114	1·126	1·143	1·155	1·172	1·185	1·195	1·212
0·00800	(1/ 125)	1·049	1·065	1·076	1·085	1·101	1·113	1·130	1·142	1·159	1·171	1·181	1·199
0·01000	(1/ 100)	1·040	1·056	1·067	1·076	1·092	1·103	1·120	1·132	1·149	1·161	1·170	1·188
0·01500	(1/ 67)	1·023	1·039	1·050	1·058	1·074	1·086	1·102	1·113	1·130	1·142	1·152	1·169
0·02000	(1/ 50)	1·012	1·027	1·038	1·046	1·062	1·073	1·089	1·101	1·117	1·129	1·138	1·155
0·03000	(1/ 33·3)	0·995	1·010	1·021	1·029	1·045	1·056	1·072	1·083	1·099	1·111	1·120	1·137
0·04000	(1/ 25·0)	0·984	0·999	1·009	1·018	1·033	1·044	1·059	1·071	1·087	1·098	1·107	1·124
0·06000	(1/ 16·7)	0·968	0·983	0·993	1·001	1·016	1·027	1·042	1·053	1·069	1·081	1·089	1·106
0·08000	(1/ 12·5)	0·957	0·971	0·982	0·990	1·005	1·015	1·030	1·041	1·057	1·068	1·077	1·093
0·10000	(1/ 10·0)	0·948	0·963	0·973	0·981	0·996	1·006	1·021	1·032	1·048	1·059	1·067	1·083
0·15000	(1/ 6·7)	0·933	0·947	0·957	0·965	0·980	0·990	1·005	1·015	1·031	1·042	1·050	1·066
0·20000	(1/ 5·0)	0·923	0·936	0·946	0·954	0·968	0·979	0·993	1·004	1·019	1·030	1·038	1·054
0·30000	(1/ 3·3)	0·908	0·921	0·931	0·939	0·953	0·963	0·977	0·988	1·003	1·013	1·022	1·037
0·40000	(1/ 2·5)	0·897	0·911	0·920	0·928	0·942	0·952	0·966	0·976	0·991	1·002	1·010	1·025
Gradient		**40**	**60**	**80**	**100**	**150**	**200**	**300**	**400**	**600**	**800**	**1000**	**1500**

(Equivalent) Pipe diameters in mm

(C = 115)

H7

C = 110
S = 0·00010 to 0·40000

ie hydraulic gradient =
1 in 10000 to 1 in 2·5

Water (or sewage) at normal temperature; full bore conditions.

This table shows values of m_H, the Hazen-Williams based divisor of mV and/or mQ, for use with Tables D

Gradient		40	60	80	100	150	200	300	400	600	800	1000	1500
0·00010	(1/10000)	1·307	1·327	1·341	1·352	1·372	1·387	1·407	1·422	1·444	1·459	1·471	1·493
0·00015	(1/ 6667)	1·286	1·305	1·319	1·330	1·350	1·364	1·385	1·400	1·421	1·436	1·447	1·469
0·00020	(1/ 5000)	1·271	1·290	1·304	1·315	1·335	1·349	1·369	1·384	1·404	1·419	1·431	1·452
0·00030	(1/ 3333)	1·251	1·270	1·283	1·294	1·313	1·327	1·347	1·361	1·382	1·396	1·408	1·429
0·00040	(1/ 2500)	1·237	1·255	1·269	1·279	1·298	1·312	1·332	1·346	1·366	1·380	1·392	1·413
0·00060	(1/ 1667)	1·217	1·235	1·248	1·258	1·277	1·291	1·310	1·324	1·344	1·358	1·369	1·390
0·00080	(1/ 1250)	1·203	1·221	1·234	1·244	1·263	1·276	1·295	1·309	1·329	1·343	1·354	1·374
0·00100	(1/ 1000)	1·192	1·210	1·223	1·233	1·251	1·265	1·284	1·297	1·317	1·331	1·342	1·362
0·00150	(1/ 667)	1·173	1·191	1·203	1·213	1·231	1·244	1·263	1·276	1·296	1·309	1·320	1·340
0·00200	(1/ 500)	1·160	1·177	1·189	1·199	1·217	1·230	1·249	1·262	1·281	1·294	1·305	1·325
0·00300	(1/ 333)	1·141	1·158	1·170	1·180	1·198	1·210	1·228	1·241	1·260	1·273	1·284	1·303
0·00400	(1/ 250)	1·128	1·145	1·157	1·166	1·184	1·196	1·214	1·227	1·246	1·259	1·269	1·288
0·00600	(1/ 167)	1·110	1·126	1·138	1·148	1·165	1·177	1·195	1·208	1·226	1·239	1·249	1·268
0·00800	(1/ 125)	1·097	1·113	1·125	1·135	1·152	1·164	1·181	1·194	1·212	1·224	1·235	1·253
0·01000	(1/ 100)	1·087	1·104	1·115	1·124	1·141	1·153	1·171	1·183	1·201	1·214	1·224	1·242
0·01500	(1/ 67)	1·070	1·086	1·097	1·106	1·123	1·135	1·152	1·164	1·182	1·194	1·204	1·222
0·02000	(1/ 50)	1·058	1·073	1·085	1·094	1·110	1·122	1·139	1·151	1·168	1·180	1·190	1·208
0·03000	(1/ 33·3)	1·041	1·056	1·067	1·076	1·092	1·104	1·120	1·132	1·149	1·161	1·171	1·189
0·04000	(1/ 25·0)	1·029	1·044	1·055	1·064	1·080	1·091	1·108	1·119	1·136	1·148	1·158	1·175
0·06000	(1/ 16·7)	1·012	1·027	1·038	1·047	1·062	1·074	1·090	1·101	1·118	1·130	1·139	1·156
0·08000	(1/ 12·5)	1·000	1·015	1·026	1·035	1·050	1·061	1·077	1·089	1·105	1·117	1·126	1·143
0·10000	(1/ 10·0)	0·992	1·006	1·017	1·025	1·041	1·052	1·068	1·079	1·095	1·107	1·116	1·133
0·15000	(1/ 6·7)	0·976	0·990	1·001	1·009	1·024	1·035	1·051	1·062	1·078	1·089	1·098	1·114
0·20000	(1/ 5·0)	0·964	0·979	0·989	0·997	1·012	1·023	1·038	1·050	1·065	1·077	1·085	1·102
0·30000	(1/ 3·3)	0·949	0·963	0·973	0·981	0·996	1·007	1·022	1·033	1·048	1·059	1·068	1·084
0·40000	(1/ 2·5)	0·938	0·952	0·962	0·970	0·985	0·995	1·010	1·021	1·036	1·047	1·056	1·072
Gradient		**40**	**60**	**80**	**100**	**150**	**200**	**300**	**400**	**600**	**800**	**1000**	**1500**

(Equivalent) Pipe diameters in mm

(C = 110)

C = 105
S = 0·00010 to 0·40000

ie hydraulic gradient =
1 in 10000 to 1 in 2·5

Water (or sewage) at normal temperature; full bore conditions.

This table shows values of m_H, the Hazen-Williams based divisor of mV and/or mQ, for use with Tables D

Gradient		(Equivalent) Pipe diameters in mm											
		40	60	80	100	150	200	300	400	600	800	1000	1500
0·00010	(1/10000)	1·369	1·390	1·405	1·416	1·437	1·453	1·475	1·490	1·512	1·529	1·541	1·564
0·00015	(1/ 6667)	1·347	1·368	1·382	1·393	1·414	1·429	1·451	1·466	1·488	1·504	1·516	1·539
0·00020	(1/ 5000)	1·332	1·352	1·366	1·378	1·398	1·413	1·434	1·449	1·471	1·487	1·499	1·521
0·00030	(1/ 3333)	1·311	1·330	1·344	1·355	1·376	1·390	1·411	1·426	1·447	1·463	1·475	1·497
0·00040	(1/ 2500)	1·296	1·315	1·329	1·340	1·360	1·374	1·395	1·410	1·431	1·446	1·458	1·480
0·00060	(1/ 1667)	1·275	1·294	1·308	1·318	1·338	1·352	1·373	1·387	1·408	1·423	1·435	1·456
0·00080	(1/ 1250)	1·260	1·279	1·293	1·303	1·323	1·337	1·357	1·371	1·392	1·407	1·418	1·439
0·00100	(1/ 1000)	1·249	1·268	1·281	1·292	1·311	1·325	1·345	1·359	1·379	1·394	1·406	1·427
0·00150	(1/ 667)	1·229	1·247	1·260	1·271	1·290	1·304	1·323	1·337	1·357	1·372	1·383	1·404
0·00200	(1/ 500)	1·215	1·233	1·246	1·256	1·275	1·289	1·308	1·322	1·342	1·356	1·367	1·388
0·00300	(1/ 333)	1·195	1·213	1·226	1·236	1·255	1·268	1·287	1·301	1·320	1·334	1·345	1·365
0·00400	(1/ 250)	1·182	1·199	1·212	1·222	1·240	1·253	1·272	1·286	1·305	1·319	1·330	1·350
0·00600	(1/ 167)	1·163	1·180	1·192	1·202	1·220	1·233	1·252	1·265	1·284	1·298	1·308	1·328
0·00800	(1/ 125)	1·149	1·166	1·179	1·189	1·206	1·219	1·237	1·251	1·269	1·283	1·293	1·313
0·01000	(1/ 100)	1·139	1·156	1·168	1·178	1·196	1·208	1·226	1·239	1·258	1·271	1·282	1·301
0·01500	(1/ 67)	1·121	1·137	1·150	1·159	1·176	1·189	1·207	1·220	1·238	1·251	1·261	1·280
0·02000	(1/ 50)	1·108	1·124	1·136	1·146	1·163	1·175	1·193	1·206	1·224	1·237	1·247	1·265
0·03000	(1/ 33·3)	1·090	1·106	1·118	1·127	1·144	1·156	1·174	1·186	1·204	1·217	1·227	1·245
0·04000	(1/ 25·0)	1·078	1·094	1·105	1·114	1·131	1·143	1·160	1·173	1·190	1·203	1·213	1·231
0·06000	(1/ 16·7)	1·060	1·076	1·088	1·097	1·113	1·125	1·142	1·154	1·171	1·183	1·193	1·211
0·08000	(1/ 12·5)	1·048	1·064	1·075	1·084	1·100	1·112	1·129	1·141	1·158	1·170	1·180	1·197
0·10000	(1/ 10·0)	1·039	1·054	1·066	1·074	1·090	1·102	1·119	1·130	1·147	1·160	1·169	1·187
0·15000	(1/ 6·7)	1·022	1·037	1·048	1·057	1·073	1·084	1·101	1·112	1·129	1·141	1·150	1·167
0·20000	(1/ 5·0)	1·010	1·026	1·036	1·045	1·061	1·072	1·088	1·099	1·116	1·128	1·137	1·154
0·30000	(1/ 3·3)	0·994	1·009	1·020	1·028	1·044	1·055	1·070	1·082	1·098	1·110	1·119	1·136
0·40000	(1/ 2·5)	0·983	0·997	1·008	1·016	1·032	1·043	1·058	1·069	1·085	1·097	1·106	1·123
		40	60	80	100	150	200	300	400	600	800	1000	1500

Gradient (Equivalent) Pipe diameters in mm (C = 105)

C = 100
S = 0·00010 to 0·40000

ie hydraulic gradient =
1 in 10000 to 1 in 2·5

Water (or sewage) at normal temperature; full bore conditions.

This table shows values of m_H, the Hazen-Williams based divisor of mV and/or mQ, for use with Tables D

Gradient		(Equivalent) Pipe diameters in mm											
		40	60	80	100	150	200	300	400	600	800	1000	1500
0·00010	(1/10000)	1·438	1·459	1·475	1·487	1·509	1·525	1·548	1·565	1·588	1·605	1·618	1·642
0·00015	(1/ 6667)	1·415	1·436	1·451	1·463	1·485	1·501	1·523	1·539	1·563	1·579	1·592	1·616
0·00020	(1/ 5000)	1·399	1·420	1·435	1·446	1·468	1·484	1·506	1·522	1·545	1·561	1·574	1·598
0·00030	(1/ 3333)	1·376	1·397	1·412	1·423	1·444	1·460	1·482	1·497	1·520	1·536	1·549	1·572
0·00040	(1/ 2500)	1·360	1·381	1·395	1·407	1·428	1·443	1·465	1·480	1·502	1·518	1·531	1·554
0·00060	(1/ 1667)	1·338	1·358	1·373	1·384	1·405	1·420	1·441	1·456	1·478	1·494	1·506	1·529
0·00080	(1/ 1250)	1·323	1·343	1·357	1·368	1·389	1·404	1·425	1·440	1·461	1·477	1·489	1·511
0·00100	(1/ 1000)	1·311	1·331	1·345	1·356	1·377	1·391	1·412	1·427	1·448	1·464	1·476	1·498
0·00150	(1/ 667)	1·290	1·310	1·323	1·334	1·354	1·369	1·389	1·404	1·425	1·440	1·452	1·474
0·00200	(1/ 500)	1·275	1·295	1·308	1·319	1·339	1·353	1·373	1·388	1·409	1·424	1·435	1·457
0·00300	(1/ 333)	1·255	1·274	1·287	1·298	1·317	1·331	1·351	1·366	1·386	1·401	1·412	1·434
0·00400	(1/ 250)	1·241	1·259	1·273	1·283	1·302	1·316	1·336	1·350	1·370	1·385	1·396	1·417
0·00600	(1/ 167)	1·221	1·239	1·252	1·262	1·281	1·295	1·314	1·328	1·348	1·363	1·374	1·394
0·00800	(1/ 125)	1·207	1·225	1·238	1·248	1·267	1·280	1·299	1·313	1·333	1·347	1·358	1·378
0·01000	(1/ 100)	1·196	1·214	1·227	1·237	1·255	1·269	1·288	1·301	1·321	1·335	1·346	1·366
0·01500	(1/ 67)	1·177	1·194	1·207	1·217	1·235	1·248	1·267	1·280	1·300	1·313	1·324	1·344
0·02000	(1/ 50)	1·163	1·181	1·193	1·203	1·221	1·234	1·253	1·266	1·285	1·298	1·309	1·329
0·03000	(1/ 33·3)	1·145	1·162	1·174	1·184	1·201	1·214	1·232	1·245	1·264	1·278	1·288	1·307
0·04000	(1/ 25·0)	1·131	1·148	1·161	1·170	1·188	1·200	1·218	1·231	1·250	1·263	1·273	1·292
0·06000	(1/ 16·7)	1·113	1·130	1·142	1·151	1·169	1·181	1·199	1·211	1·230	1·243	1·253	1·272
0·08000	(1/ 12·5)	1·101	1·117	1·129	1·138	1·155	1·167	1·185	1·198	1·216	1·228	1·239	1·257
0·10000	(1/ 10·0)	1·091	1·107	1·119	1·128	1·145	1·157	1·174	1·187	1·205	1·217	1·228	1·246
0·15000	(1/ 6·7)	1·073	1·089	1·101	1·110	1·127	1·138	1·156	1·168	1·185	1·198	1·208	1·226
0·20000	(1/ 5·0)	1·061	1·077	1·088	1·097	1·114	1·125	1·142	1·154	1·172	1·184	1·194	1·212
0·30000	(1/ 3·3)	1·044	1·059	1·071	1·080	1·096	1·107	1·124	1·136	1·153	1·165	1·175	1·192
0·40000	(1/ 2·5)	1·032	1·047	1·058	1·067	1·083	1·095	1·111	1·123	1·140	1·152	1·161	1·179
		40	60	80	100	150	200	300	400	600	800	1000	1500

Gradient (Equivalent) Pipe diameters in mm (C = 100)

C = 95
S = 0·00010 to 0·40000

ie hydraulic gradient =
1 in 10000 to 1 in 2·5

Water (or sewage) at normal temperature;
full bore conditions.

This table shows values of m_H, the Hazen-Williams based divisor of mV and/or mQ, for use with Tables D

Gradient		40	60	80	100	150	200	300	400	600	800	1000	1500
0·00010	(1/10000)	1·514	1·536	1·553	1·565	1·589	1·606	1·630	1·647	1·672	1·689	1·703	1·729
0·00015	(1/ 6667)	1·489	1·512	1·528	1·540	1·563	1·580	1·603	1·621	1·645	1·662	1·676	1·701
0·00020	(1/ 5000)	1·472	1·494	1·510	1·523	1·545	1·562	1·585	1·602	1·626	1·643	1·657	1·682
0·00030	(1/ 3333)	1·448	1·470	1·486	1·498	1·520	1·537	1·560	1·576	1·600	1·617	1·630	1·655
0·00040	(1/ 2500)	1·432	1·453	1·469	1·481	1·503	1·519	1·542	1·558	1·582	1·598	1·611	1·636
0·00060	(1/ 1667)	1·409	1·430	1·445	1·457	1·479	1·495	1·517	1·533	1·556	1·573	1·586	1·609
0·00080	(1/ 1250)	1·393	1·414	1·429	1·440	1·462	1·477	1·500	1·516	1·538	1·555	1·567	1·591
0·00100	(1/ 1000)	1·380	1·401	1·416	1·428	1·449	1·464	1·486	1·502	1·525	1·541	1·553	1·577
0·00150	(1/ 667)	1·358	1·379	1·393	1·405	1·426	1·441	1·462	1·478	1·500	1·516	1·528	1·551
0·00200	(1/ 500)	1·343	1·363	1·377	1·389	1·409	1·424	1·446	1·461	1·483	1·499	1·511	1·534
0·00300	(1/ 333)	1·321	1·341	1·355	1·366	1·387	1·401	1·422	1·438	1·459	1·475	1·487	1·509
0·00400	(1/ 250)	1·306	1·325	1·340	1·351	1·371	1·385	1·406	1·421	1·442	1·458	1·470	1·492
0·00600	(1/ 167)	1·285	1·304	1·318	1·329	1·349	1·363	1·384	1·398	1·419	1·434	1·446	1·468
0·00800	(1/ 125)	1·270	1·289	1·303	1·314	1·333	1·347	1·368	1·382	1·403	1·418	1·429	1·451
0·01000	(1/ 100)	1·259	1·278	1·291	1·302	1·321	1·336	1·356	1·370	1·390	1·405	1·417	1·438
0·01500	(1/ 67)	1·239	1·257	1·271	1·281	1·300	1·314	1·334	1·348	1·368	1·383	1·394	1·415
0·02000	(1/ 50)	1·224	1·243	1·256	1·266	1·285	1·299	1·318	1·332	1·352	1·367	1·378	1·399
0·03000	(1/ 33·3)	1·205	1·223	1·236	1·246	1·265	1·278	1·297	1·311	1·331	1·345	1·356	1·376
0·04000	(1/ 25·0)	1·191	1·209	1·222	1·232	1·250	1·263	1·282	1·296	1·315	1·329	1·340	1·360
0·06000	(1/ 16·7)	1·172	1·189	1·202	1·212	1·230	1·243	1·262	1·275	1·294	1·308	1·319	1·339
0·08000	(1/ 12·5)	1·158	1·176	1·188	1·198	1·216	1·229	1·247	1·261	1·279	1·293	1·304	1·323
0·10000	(1/ 10·0)	1·148	1·165	1·178	1·187	1·205	1·218	1·236	1·249	1·268	1·282	1·292	1·311
0·15000	(1/ 6·7)	1·130	1·147	1·159	1·168	1·186	1·198	1·216	1·229	1·248	1·261	1·271	1·290
0·20000	(1/ 5·0)	1·117	1·133	1·146	1·155	1·172	1·185	1·202	1·215	1·233	1·247	1·257	1·276
0·30000	(1/ 3·3)	1·099	1·115	1·127	1·136	1·153	1·166	1·183	1·196	1·214	1·226	1·237	1·255
0·40000	(1/ 2·5)	1·086	1·102	1·114	1·123	1·140	1·152	1·170	1·182	1·200	1·212	1·222	1·241
Gradient		40	60	80	100	150	200	300	400	600	800	1000	1500

(Equivalent) Pipe diameters in mm

(C = 95)

C = 90
S = 0·00010 to 0·40000

ie hydraulic gradient =
1 in 10000 to 1 in 2·5

Water (or sewage) at normal temperature;
full bore conditions.

This table shows values of m_H, the Hazen-Williams based divisor of mV and/or mQ, for use with Tables D

Gradient		40	60	80	100	150	200	300	400	600	800	1000	1500
0·00010	(1/10000)	1·598	1·622	1·639	1·652	1·677	1·695	1·720	1·739	1·765	1·783	1·798	1·825
0·00015	(1/ 6667)	1·572	1·595	1·612	1·626	1·650	1·668	1·693	1·711	1·736	1·755	1·769	1·796
0·00020	(1/ 5000)	1·554	1·577	1·594	1·607	1·631	1·648	1·673	1·691	1·716	1·735	1·749	1·775
0·00030	(1/ 3333)	1·529	1·552	1·568	1·581	1·605	1·622	1·646	1·664	1·689	1·707	1·721	1·746
0·00040	(1/ 2500)	1·511	1·534	1·550	1·563	1·587	1·603	1·627	1·645	1·669	1·687	1·701	1·726
0·00060	(1/ 1667)	1·487	1·509	1·525	1·538	1·561	1·578	1·601	1·618	1·643	1·660	1·674	1·699
0·00080	(1/ 1250)	1·470	1·492	1·508	1·520	1·543	1·560	1·583	1·600	1·624	1·641	1·654	1·679
0·00100	(1/ 1000)	1·457	1·479	1·495	1·507	1·529	1·546	1·569	1·586	1·609	1·626	1·640	1·664
0·00150	(1/ 667)	1·434	1·455	1·471	1·483	1·505	1·521	1·544	1·560	1·583	1·600	1·613	1·638
0·00200	(1/ 500)	1·417	1·438	1·454	1·466	1·488	1·503	1·526	1·542	1·565	1·582	1·595	1·619
0·00300	(1/ 333)	1·394	1·415	1·430	1·442	1·464	1·479	1·501	1·517	1·540	1·556	1·569	1·593
0·00400	(1/ 250)	1·378	1·399	1·414	1·426	1·447	1·462	1·484	1·500	1·523	1·539	1·551	1·575
0·00600	(1/ 167)	1·356	1·377	1·391	1·403	1·424	1·439	1·460	1·476	1·498	1·514	1·526	1·549
0·00800	(1/ 125)	1·341	1·361	1·375	1·387	1·407	1·422	1·444	1·459	1·481	1·497	1·509	1·532
0·01000	(1/ 100)	1·329	1·349	1·363	1·374	1·395	1·410	1·431	1·446	1·468	1·483	1·495	1·518
0·01500	(1/ 67)	1·307	1·327	1·341	1·352	1·372	1·387	1·408	1·423	1·444	1·459	1·471	1·493
0·02000	(1/ 50)	1·293	1·312	1·326	1·337	1·357	1·371	1·392	1·406	1·428	1·443	1·455	1·476
0·03000	(1/ 33·3)	1·272	1·291	1·304	1·315	1·335	1·349	1·369	1·384	1·405	1·420	1·431	1·453
0·04000	(1/ 25·0)	1·257	1·276	1·290	1·300	1·320	1·334	1·354	1·368	1·389	1·403	1·415	1·436
0·06000	(1/ 16·7)	1·237	1·255	1·269	1·279	1·298	1·312	1·332	1·346	1·366	1·381	1·392	1·413
0·08000	(1/ 12·5)	1·223	1·241	1·254	1·265	1·284	1·297	1·317	1·331	1·351	1·365	1·376	1·397
0·10000	(1/ 10·0)	1·212	1·230	1·243	1·253	1·272	1·286	1·305	1·319	1·339	1·353	1·364	1·384
0·15000	(1/ 6·7)	1·192	1·210	1·223	1·233	1·252	1·265	1·284	1·298	1·317	1·331	1·342	1·362
0·20000	(1/ 5·0)	1·179	1·196	1·209	1·219	1·237	1·251	1·269	1·283	1·302	1·316	1·327	1·346
0·30000	(1/ 3·3)	1·160	1·177	1·190	1·199	1·217	1·230	1·249	1·262	1·281	1·295	1·305	1·325
0·40000	(1/ 2·5)	1·147	1·164	1·176	1·186	1·204	1·216	1·235	1·248	1·266	1·280	1·290	1·310
Gradient		40	60	80	100	150	200	300	400	600	800	1000	1500

(Equivalent) Pipe diameters in mm

(C = 90)

C = 85
S = 0·00010 to 0·40000

ie hydraulic gradient =
1 in 10000 to 1 in 2·5

Water (or sewage) at normal temperature;
full bore conditions.

This table shows values of m_H, the Hazen-Williams
based divisor of mV and/or mQ, for use with Tables D

Gradient		40	60	80	100	150	200	300	400	600	800	1000	1500
		(Equivalent) Pipe diameters in mm											
0·00010	(1/10000)	1·692	1·717	1·735	1·749	1·776	1·795	1·821	1·841	1·868	1·888	1·904	1·932
0·00015	(1/ 6667)	1·664	1·689	1·707	1·721	1·747	1·766	1·792	1·811	1·838	1·858	1·873	1·901
0·00020	(1/ 5000)	1·645	1·670	1·688	1·702	1·727	1·745	1·772	1·790	1·817	1·837	1·852	1·879
0·00030	(1/ 3333)	1·619	1·643	1·661	1·674	1·699	1·717	1·743	1·762	1·788	1·807	1·822	1·849
0·00040	(1/ 2500)	1·600	1·624	1·642	1·655	1·680	1·698	1·723	1·741	1·768	1·786	1·801	1·828
0·00060	(1/ 1667)	1·575	1·598	1·615	1·628	1·653	1·670	1·695	1·713	1·739	1·758	1·772	1·799
0·00080	(1/ 1250)	1·557	1·580	1·597	1·610	1·634	1·651	1·676	1·694	1·719	1·738	1·752	1·778
0·00100	(1/ 1000)	1·543	1·566	1·583	1·596	1·619	1·637	1·661	1·679	1·704	1·722	1·736	1·762
0·00150	(1/ 667)	1·518	1·541	1·557	1·570	1·593	1·610	1·634	1·652	1·677	1·694	1·708	1·734
0·00200	(1/ 500)	1·501	1·523	1·539	1·552	1·575	1·592	1·616	1·633	1·657	1·675	1·689	1·714
0·00300	(1/ 333)	1·476	1·499	1·514	1·527	1·550	1·566	1·590	1·607	1·631	1·648	1·662	1·686
0·00400	(1/ 250)	1·460	1·481	1·497	1·509	1·532	1·548	1·572	1·588	1·612	1·629	1·643	1·667
0·00600	(1/ 167)	1·436	1·458	1·473	1·485	1·507	1·523	1·546	1·563	1·586	1·603	1·616	1·640
0·00800	(1/ 125)	1·420	1·441	1·456	1·468	1·490	1·506	1·529	1·545	1·568	1·585	1·598	1·622
0·01000	(1/ 100)	1·407	1·428	1·443	1·455	1·477	1·493	1·515	1·531	1·554	1·571	1·583	1·607
0·01500	(1/ 67)	1·384	1·405	1·420	1·432	1·453	1·469	1·491	1·506	1·529	1·545	1·558	1·581
0·02000	(1/ 50)	1·369	1·389	1·404	1·415	1·437	1·452	1·474	1·489	1·512	1·528	1·540	1·563
0·03000	(1/ 33·3)	1·347	1·367	1·381	1·393	1·413	1·428	1·450	1·465	1·487	1·503	1·515	1·538
0·04000	(1/ 25·0)	1·331	1·351	1·365	1·377	1·397	1·412	1·433	1·448	1·470	1·486	1·498	1·520
0·06000	(1/ 16·7)	1·310	1·329	1·343	1·355	1·375	1·389	1·410	1·425	1·447	1·462	1·474	1·496
0·08000	(1/ 12·5)	1·295	1·314	1·328	1·339	1·359	1·374	1·394	1·409	1·430	1·445	1·457	1·479
0·10000	(1/ 10·0)	1·283	1·302	1·316	1·327	1·347	1·361	1·382	1·396	1·417	1·432	1·444	1·466
0·15000	(1/ 6·7)	1·263	1·282	1·295	1·306	1·325	1·339	1·359	1·374	1·395	1·409	1·421	1·442
0·20000	(1/ 5·0)	1·248	1·267	1·280	1·291	1·310	1·324	1·344	1·358	1·379	1·393	1·405	1·426
0·30000	(1/ 3·3)	1·228	1·246	1·260	1·270	1·289	1·303	1·322	1·336	1·356	1·371	1·382	1·403
0·40000	(1/ 2·5)	1·214	1·232	1·245	1·256	1·274	1·288	1·307	1·321	1·341	1·355	1·366	1·387
		40	60	80	100	150	200	300	400	600	800	1000	1500

Gradient (Equivalent) Pipe diameters in mm **(C = 85)**

C = 80
S = 0·00010 to 0·40000

ie hydraulic gradient =
1 in 10000 to 1 in 2·5

Water (or sewage) at normal temperature;
full bore conditions.

This table shows values of m_H, the Hazen-Williams
based divisor of mV and/or mQ, for use with Tables D

Gradient		40	60	80	100	150	200	300	400	600	800	1000	1500
		(Equivalent) Pipe diameters in mm											
0·00010	(1/10000)	1·797	1·824	1·844	1·859	1·887	1·907	1·935	1·956	1·985	2·006	2·023	2·053
0·00015	(1/ 6667)	1·768	1·795	1·814	1·829	1·856	1·876	1·904	1·924	1·953	1·974	1·990	2·020
0·00020	(1/ 5000)	1·748	1·774	1·793	1·808	1·835	1·855	1·882	1·902	1·931	1·951	1·967	1·997
0·00030	(1/ 3333)	1·720	1·746	1·764	1·779	1·806	1·825	1·852	1·872	1·900	1·920	1·936	1·965
0·00040	(1/ 2500)	1·700	1·726	1·744	1·759	1·785	1·804	1·831	1·850	1·878	1·898	1·914	1·942
0·00060	(1/ 1667)	1·673	1·698	1·716	1·730	1·756	1·775	1·801	1·821	1·848	1·867	1·883	1·911
0·00080	(1/ 1250)	1·654	1·679	1·697	1·710	1·736	1·755	1·781	1·800	1·827	1·846	1·861	1·889
0·00100	(1/ 1000)	1·639	1·664	1·681	1·695	1·721	1·739	1·765	1·784	1·810	1·830	1·845	1·872
0·00150	(1/ 667)	1·613	1·637	1·654	1·668	1·693	1·711	1·737	1·755	1·781	1·800	1·815	1·842
0·00200	(1/ 500)	1·594	1·618	1·635	1·649	1·674	1·691	1·717	1·735	1·761	1·780	1·794	1·821
0·00300	(1/ 333)	1·569	1·592	1·609	1·622	1·647	1·664	1·689	1·707	1·733	1·751	1·765	1·792
0·00400	(1/ 250)	1·551	1·574	1·591	1·604	1·628	1·645	1·670	1·688	1·713	1·731	1·745	1·771
0·00600	(1/ 167)	1·526	1·549	1·565	1·578	1·602	1·619	1·643	1·660	1·685	1·703	1·717	1·743
0·00800	(1/ 125)	1·508	1·531	1·547	1·560	1·583	1·600	1·624	1·641	1·666	1·684	1·698	1·723
0·01000	(1/ 100)	1·495	1·517	1·533	1·546	1·569	1·586	1·610	1·627	1·651	1·669	1·682	1·708
0·01500	(1/ 67)	1·471	1·493	1·509	1·521	1·544	1·560	1·584	1·601	1·625	1·642	1·655	1·680
0·02000	(1/ 50)	1·454	1·476	1·492	1·504	1·526	1·543	1·566	1·582	1·606	1·623	1·636	1·661
0·03000	(1/ 33·3)	1·431	1·452	1·468	1·480	1·502	1·518	1·540	1·557	1·580	1·597	1·610	1·634
0·04000	(1/ 25·0)	1·414	1·436	1·451	1·463	1·485	1·500	1·523	1·539	1·562	1·579	1·592	1·616
0·06000	(1/ 16·7)	1·392	1·412	1·427	1·439	1·461	1·476	1·498	1·514	1·537	1·553	1·566	1·590
0·08000	(1/ 12·5)	1·376	1·396	1·411	1·423	1·444	1·459	1·481	1·497	1·519	1·536	1·548	1·571
0·10000	(1/ 10·0)	1·363	1·384	1·399	1·410	1·431	1·446	1·468	1·484	1·506	1·522	1·534	1·557
0·15000	(1/ 6·7)	1·341	1·362	1·376	1·387	1·408	1·423	1·444	1·460	1·482	1·497	1·510	1·532
0·20000	(1/ 5·0)	1·326	1·346	1·360	1·371	1·392	1·407	1·428	1·443	1·465	1·480	1·492	1·515
0·30000	(1/ 3·3)	1·305	1·324	1·338	1·349	1·370	1·384	1·405	1·420	1·441	1·456	1·468	1·490
0·40000	(1/ 2·5)	1·290	1·309	1·323	1·334	1·354	1·368	1·389	1·404	1·425	1·440	1·452	1·473
		40	60	80	100	150	200	300	400	600	800	1000	1500

Gradient (Equivalent) Pipe diameters in mm **(C = 80)**

H14

C = 75
S = 0·00010 to 0·40000

ie hydraulic gradient =
1 in 10000 to 1 in 2·5

Water (or sewage) at normal temperature;
full bore conditions.

This table shows values of m_H, the Hazen-Williams based divisor of mV and/or mQ, for use with Tables D

Gradient   (Equivalent) Pipe diameters in mm

Gradient		40	60	80	100	150	200	300	400	600	800	1000	1500
0·00010	(1/10000)	1·917	1·946	1·967	1·983	2·012	2·034	2·064	2·086	2·117	2·140	2·158	2·190
0·00015	(1/ 6667)	1·886	1·915	1·935	1·951	1·980	2·001	2·031	2·053	2·083	2·106	2·123	2·155
0·00020	(1/ 5000)	1·865	1·893	1·913	1·929	1·957	1·978	2·008	2·029	2·060	2·081	2·099	2·130
0·00030	(1/ 3333)	1·835	1·862	1·882	1·897	1·926	1·946	1·976	1·997	2·026	2·048	2·065	2·096
0·00040	(1/ 2500)	1·814	1·841	1·860	1·876	1·904	1·924	1·953	1·974	2·003	2·025	2·041	2·072
0·00060	(1/ 1667)	1·785	1·811	1·831	1·846	1·873	1·893	1·922	1·942	1·971	1·992	2·008	2·038
0·00080	(1/ 1250)	1·764	1·791	1·810	1·824	1·852	1·871	1·900	1·920	1·948	1·969	1·985	2·015
0·00100	(1/ 1000)	1·748	1·775	1·794	1·808	1·835	1·855	1·883	1·903	1·931	1·952	1·968	1·997
0·00150	(1/ 667)	1·720	1·746	1·765	1·779	1·806	1·825	1·852	1·872	1·900	1·920	1·936	1·965
0·00200	(1/ 500)	1·701	1·726	1·744	1·759	1·785	1·804	1·831	1·851	1·878	1·898	1·914	1·943
0·00300	(1/ 333)	1·673	1·698	1·716	1·731	1·756	1·775	1·802	1·821	1·848	1·868	1·883	1·911
0·00400	(1/ 250)	1·654	1·679	1·697	1·711	1·736	1·755	1·781	1·800	1·827	1·846	1·862	1·889
0·00600	(1/ 167)	1·628	1·652	1·669	1·683	1·708	1·727	1·752	1·771	1·798	1·817	1·832	1·859
0·00800	(1/ 125)	1·609	1·633	1·650	1·664	1·689	1·707	1·732	1·751	1·777	1·796	1·811	1·838
0·01000	(1/ 100)	1·595	1·619	1·636	1·649	1·674	1·692	1·717	1·735	1·761	1·780	1·795	1·821
0·01500	(1/ 67)	1·569	1·592	1·609	1·623	1·647	1·664	1·689	1·707	1·733	1·751	1·766	1·792
0·02000	(1/ 50)	1·551	1·574	1·591	1·604	1·628	1·645	1·670	1·688	1·713	1·731	1·746	1·772
0·03000	(1/ 33·3)	1·526	1·549	1·565	1·578	1·602	1·619	1·643	1·661	1·686	1·703	1·717	1·743
0·04000	(1/ 25·0)	1·509	1·531	1·547	1·560	1·584	1·600	1·624	1·642	1·666	1·684	1·698	1·723
0·06000	(1/ 16·7)	1·484	1·507	1·523	1·535	1·558	1·575	1·598	1·615	1·639	1·657	1·670	1·696
0·08000	(1/ 12·5)	1·467	1·489	1·505	1·518	1·540	1·557	1·580	1·597	1·621	1·638	1·651	1·676
0·10000	(1/ 10·0)	1·454	1·476	1·492	1·504	1·527	1·543	1·566	1·583	1·606	1·623	1·637	1·661
0·15000	(1/ 6·7)	1·431	1·452	1·468	1·480	1·502	1·518	1·541	1·557	1·580	1·597	1·610	1·634
0·20000	(1/ 5·0)	1·415	1·436	1·451	1·463	1·485	1·501	1·523	1·539	1·562	1·579	1·592	1·616
0·30000	(1/ 3·3)	1·392	1·413	1·428	1·439	1·461	1·476	1·499	1·515	1·537	1·554	1·566	1·590
0·40000	(1/ 2·5)	1·376	1·396	1·411	1·423	1·444	1·460	1·481	1·497	1·520	1·536	1·548	1·572
		40	60	80	100	150	200	300	400	600	800	1000	1500

Gradient   (Equivalent) Pipe diameters in mm   **(C = 75)**

H15

C = 70
S = 0·00010 to 0·40000

ie hydraulic gradient =
1 in 10000 to 1 in 2·5

Water (or sewage) at normal temperature;
full bore conditions.

This table shows values of m_H, the Hazen-Williams based divisor of mV and/or mQ, for use with Tables D

Gradient   (Equivalent) Pipe diameters in mm

Gradient		40	60	80	100	150	200	300	400	600	800	1000	1500
0·00010	(1/10000)	2·054	2·085	2·107	2·124	2·156	2·179	2·212	2·235	2·269	2·293	2·312	2·346
0·00015	(1/ 6667)	2·021	2·051	2·073	2·090	2·122	2·144	2·176	2·199	2·232	2·256	2·274	2·309
0·00020	(1/ 5000)	1·998	2·028	2·049	2·066	2·097	2·119	2·151	2·174	2·207	2·230	2·248	2·282
0·00030	(1/ 3333)	1·966	1·995	2·016	2·033	2·063	2·085	2·117	2·139	2·171	2·194	2·212	2·245
0·00040	(1/ 2500)	1·943	1·972	1·993	2·010	2·040	2·062	2·092	2·115	2·146	2·169	2·187	2·220
0·00060	(1/ 1667)	1·912	1·941	1·961	1·977	2·007	2·028	2·059	2·081	2·112	2·134	2·152	2·184
0·00080	(1/ 1250)	1·890	1·918	1·939	1·955	1·984	2·005	2·035	2·057	2·088	2·110	2·127	2·159
0·00100	(1/ 1000)	1·873	1·901	1·922	1·937	1·966	1·987	2·017	2·039	2·069	2·091	2·108	2·140
0·00150	(1/ 667)	1·843	1·871	1·891	1·906	1·935	1·955	1·985	2·006	2·036	2·057	2·074	2·105
0·00200	(1/ 500)	1·822	1·849	1·869	1·884	1·913	1·933	1·962	1·983	2·013	2·034	2·051	2·081
0·00300	(1/ 333)	1·793	1·820	1·839	1·854	1·882	1·902	1·930	1·951	1·980	2·001	2·018	2·048
0·00400	(1/ 250)	1·772	1·799	1·818	1·833	1·860	1·880	1·908	1·929	1·958	1·978	1·995	2·024
0·00600	(1/ 167)	1·744	1·770	1·789	1·803	1·830	1·850	1·878	1·898	1·926	1·946	1·962	1·992
0·00800	(1/ 125)	1·724	1·750	1·768	1·783	1·810	1·829	1·856	1·876	1·904	1·924	1·940	1·969
0·01000	(1/ 100)	1·709	1·734	1·753	1·767	1·793	1·812	1·840	1·859	1·887	1·907	1·923	1·952
0·01500	(1/ 67)	1·681	1·706	1·724	1·739	1·765	1·783	1·810	1·829	1·857	1·876	1·892	1·920
0·02000	(1/ 50)	1·662	1·687	1·705	1·719	1·744	1·763	1·789	1·808	1·835	1·855	1·870	1·898
0·03000	(1/ 33·3)	1·635	1·660	1·677	1·691	1·716	1·735	1·761	1·779	1·806	1·825	1·840	1·868
0·04000	(1/ 25·0)	1·616	1·641	1·658	1·672	1·697	1·715	1·740	1·759	1·785	1·804	1·819	1·846
0·06000	(1/ 16·7)	1·590	1·614	1·631	1·645	1·669	1·687	1·712	1·731	1·757	1·775	1·790	1·817
0·08000	(1/ 12·5)	1·572	1·596	1·613	1·626	1·650	1·668	1·693	1·711	1·736	1·755	1·769	1·796
0·10000	(1/ 10·0)	1·558	1·582	1·598	1·611	1·636	1·653	1·678	1·696	1·721	1·739	1·754	1·780
0·15000	(1/ 6·7)	1·533	1·556	1·573	1·586	1·609	1·626	1·651	1·668	1·693	1·711	1·725	1·751
0·20000	(1/ 5·0)	1·516	1·538	1·555	1·567	1·591	1·608	1·632	1·649	1·674	1·692	1·706	1·731
0·30000	(1/ 3·3)	1·491	1·514	1·530	1·542	1·565	1·582	1·606	1·623	1·647	1·665	1·678	1·703
0·40000	(1/ 2·5)	1·474	1·496	1·512	1·525	1·547	1·564	1·587	1·604	1·628	1·645	1·659	1·684
		40	60	80	100	150	200	300	400	600	800	1000	1500

Gradient   (Equivalent) Pipe diameters in mm   **(C = 70)**

C = 65
S = 0·00010 to 0·40000

ie hydraulic gradient =
1 in 10000 to 1 in 2·5

Water (or sewage) at normal temperature; full bore conditions.

This table shows values of m_H, the Hazen-Williams based divisor of mV and/or mQ, for use with Tables D

Gradient		(Equivalent) Pipe diameters in mm											
		40	60	80	100	150	200	300	400	600	800	1000	1500
0·00010	(1/10000)	2·212	2·245	2·269	2·288	2·322	2·347	2·382	2·407	2·443	2·469	2·489	2·527
0·00015	(1/ 6667)	2·177	2·209	2·233	2·251	2·285	2·309	2·344	2·368	2·404	2·429	2·449	2·486
0·00020	(1/ 5000)	2·152	2·184	2·207	2·225	2·259	2·283	2·317	2·341	2·376	2·402	2·421	2·458
0·00030	(1/ 3333)	2·117	2·149	2·172	2·189	2·222	2·246	2·279	2·304	2·338	2·363	2·382	2·418
0·00040	(1/ 2500)	2·093	2·124	2·147	2·164	2·197	2·220	2·253	2·277	2·311	2·336	2·355	2·391
0·00060	(1/ 1667)	2·059	2·090	2·112	2·130	2·161	2·184	2·217	2·241	2·274	2·298	2·317	2·352
0·00080	(1/ 1250)	2·036	2·066	2·088	2·105	2·137	2·159	2·192	2·215	2·248	2·272	2·291	2·325
0·00100	(1/ 1000)	2·017	2·048	2·069	2·086	2·118	2·140	2·172	2·195	2·228	2·252	2·270	2·304
0·00150	(1/ 667)	1·985	2·015	2·036	2·053	2·084	2·106	2·137	2·160	2·192	2·216	2·234	2·267
0·00200	(1/ 500)	1·962	1·992	2·013	2·029	2·060	2·082	2·113	2·135	2·167	2·190	2·208	2·241
0·00300	(1/ 333)	1·931	1·960	1·980	1·997	2·027	2·048	2·079	2·101	2·132	2·155	2·173	2·205
0·00400	(1/ 250)	1·909	1·937	1·958	1·974	2·004	2·025	2·055	2·077	2·108	2·130	2·148	2·180
0·00600	(1/ 167)	1·878	1·906	1·926	1·942	1·971	1·992	2·022	2·044	2·074	2·096	2·113	2·145
0·00800	(1/ 125)	1·856	1·884	1·904	1·920	1·949	1·969	1·999	2·020	2·050	2·072	2·089	2·121
0·01000	(1/ 100)	1·840	1·868	1·887	1·903	1·931	1·952	1·981	2·002	2·032	2·054	2·071	2·102
0·01500	(1/ 67)	1·810	1·837	1·857	1·872	1·900	1·921	1·949	1·970	2·000	2·021	2·037	2·068
0·02000	(1/ 50)	1·790	1·816	1·836	1·851	1·879	1·899	1·927	1·947	1·977	1·998	2·014	2·044
0·03000	(1/ 33·3)	1·761	1·787	1·806	1·821	1·848	1·868	1·896	1·916	1·945	1·965	1·982	2·011
0·04000	(1/ 25·0)	1·741	1·767	1·786	1·800	1·827	1·847	1·874	1·894	1·923	1·943	1·959	1·988
0·06000	(1/ 16·7)	1·713	1·738	1·757	1·771	1·798	1·817	1·844	1·864	1·892	1·912	1·927	1·956
0·08000	(1/ 12·5)	1·693	1·718	1·737	1·751	1·777	1·796	1·823	1·842	1·870	1·890	1·905	1·934
0·10000	(1/ 10·0)	1·678	1·703	1·721	1·735	1·761	1·780	1·807	1·826	1·853	1·873	1·888	1·917
0·15000	(1/ 6·7)	1·651	1·676	1·694	1·708	1·733	1·752	1·778	1·797	1·824	1·843	1·858	1·886
0·20000	(1/ 5·0)	1·632	1·657	1·674	1·688	1·713	1·731	1·757	1·776	1·803	1·822	1·837	1·864
0·30000	(1/ 3·3)	1·606	1·630	1·647	1·661	1·686	1·704	1·729	1·748	1·774	1·793	1·807	1·834
0·40000	(1/ 2·5)	1·588	1·611	1·628	1·642	1·666	1·684	1·709	1·728	1·753	1·772	1·787	1·813
		40	60	80	100	150	200	300	400	600	800	1000	1500

Gradient — (Equivalent) Pipe diameters in mm — **(C = 65)**

C = 60
S = 0·00010 to 0·40000

ie hydraulic gradient =
1 in 10000 to 1 in 2·5

Water (or sewage) at normal temperature; full bore conditions.

This table shows values of m_H, the Hazen-Williams based divisor of mV and/or mQ, for use with Tables D

Gradient		(Equivalent) Pipe diameters in mm											
		40	60	80	100	150	200	300	400	600	800	1000	1500
0·00010	(1/10000)	2·396	2·432	2·458	2·478	2·516	2·542	2·580	2·608	2·647	2·675	2·697	2·737
0·00015	(1/ 6667)	2·358	2·393	2·419	2·439	2·475	2·501	2·539	2·566	2·604	2·632	2·654	2·693
0·00020	(1/ 5000)	2·331	2·366	2·391	2·411	2·447	2·473	2·510	2·536	2·574	2·602	2·623	2·663
0·00030	(1/ 3333)	2·293	2·328	2·353	2·372	2·407	2·433	2·469	2·496	2·533	2·560	2·581	2·620
0·00040	(1/ 2500)	2·267	2·301	2·326	2·345	2·380	2·405	2·441	2·467	2·504	2·531	2·551	2·590
0·00060	(1/ 1667)	2·231	2·264	2·288	2·307	2·342	2·366	2·402	2·427	2·464	2·490	2·510	2·548
0·00080	(1/ 1250)	2·205	2·238	2·262	2·281	2·315	2·339	2·374	2·400	2·436	2·461	2·482	2·519
0·00100	(1/ 1000)	2·186	2·218	2·242	2·260	2·294	2·319	2·353	2·378	2·414	2·440	2·460	2·497
0·00150	(1/ 667)	2·150	2·183	2·206	2·224	2·257	2·281	2·315	2·340	2·375	2·400	2·420	2·456
0·00200	(1/ 500)	2·126	2·158	2·181	2·199	2·231	2·255	2·289	2·313	2·348	2·373	2·392	2·428
0·00300	(1/ 333)	2·092	2·123	2·146	2·163	2·196	2·219	2·252	2·276	2·310	2·335	2·354	2·389
0·00400	(1/ 250)	2·068	2·099	2·121	2·138	2·170	2·194	2·226	2·250	2·284	2·308	2·327	2·362
0·00600	(1/ 167)	2·034	2·065	2·087	2·104	2·136	2·158	2·191	2·214	2·247	2·271	2·290	2·324
0·00800	(1/ 125)	2·011	2·041	2·063	2·080	2·111	2·134	2·166	2·188	2·221	2·245	2·263	2·297
0·01000	(1/ 100)	1·993	2·023	2·045	2·061	2·092	2·115	2·146	2·169	2·202	2·225	2·243	2·277
0·01500	(1/ 67)	1·961	1·991	2·012	2·028	2·059	2·081	2·112	2·134	2·166	2·189	2·207	2·240
0·02000	(1/ 50)	1·939	1·968	1·989	2·005	2·035	2·057	2·088	2·110	2·141	2·164	2·182	2·215
0·03000	(1/ 33·3)	1·908	1·936	1·957	1·973	2·002	2·024	2·054	2·076	2·107	2·129	2·147	2·179
0·04000	(1/ 25·0)	1·886	1·914	1·934	1·950	1·979	2·000	2·030	2·052	2·083	2·105	2·122	2·154
0·06000	(1/ 16·7)	1·855	1·883	1·903	1·919	1·948	1·968	1·998	2·019	2·049	2·071	2·088	2·119
0·08000	(1/ 12·5)	1·834	1·862	1·881	1·897	1·925	1·946	1·975	1·996	2·026	2·047	2·064	2·095
0·10000	(1/ 10·0)	1·818	1·845	1·865	1·880	1·908	1·929	1·957	1·978	2·008	2·029	2·046	2·077
0·15000	(1/ 6·7)	1·789	1·815	1·835	1·850	1·878	1·897	1·926	1·946	1·976	1·997	2·013	2·043
0·20000	(1/ 5·0)	1·768	1·795	1·814	1·829	1·856	1·876	1·904	1·924	1·953	1·974	1·990	2·020
0·30000	(1/ 3·3)	1·740	1·766	1·785	1·799	1·826	1·846	1·873	1·893	1·922	1·942	1·958	1·987
0·40000	(1/ 2·5)	1·720	1·746	1·764	1·779	1·805	1·824	1·852	1·871	1·900	1·920	1·935	1·965
		40	60	80	100	150	200	300	400	600	800	1000	1500

Gradient — (Equivalent) Pipe diameters in mm — **(C = 60)**

H18 C = 50
S = 0·00010 to 0·40000

Water (or sewage) at normal temperature;
full bore conditions.

ie hydraulic gradient =
1 in 10000 to 1 in 2·5

This table shows values of m_H, the Hazen-Williams
based divisor of mV and/or mQ, for use with Tables D

Gradient		(Equivalent) Pipe diameters in mm											
		40	60	80	100	150	200	300	400	600	800	1000	1500
0·00010	(1/10000)	2·876	2·919	2·950	2·974	3·019	3·051	3·096	3·129	3·176	3·210	3·236	3·285
0·00015	(1/ 6667)	2·829	2·872	2·902	2·926	2·970	3·002	3·047	3·079	3·125	3·158	3·184	3·232
0·00020	(1/ 5000)	2·797	2·839	2·869	2·893	2·936	2·967	3·012	3·044	3·089	3·122	3·148	3·195
0·00030	(1/ 3333)	2·752	2·793	2·823	2·846	2·889	2·920	2·963	2·995	3·040	3·072	3·097	3·144
0·00040	(1/ 2500)	2·721	2·761	2·791	2·814	2·856	2·886	2·929	2·961	3·005	3·037	3·062	3·108
0·00060	(1/ 1667)	2·677	2·717	2·746	2·768	2·810	2·840	2·882	2·913	2·957	2·988	3·013	3·058
0·00080	(1/ 1250)	2·646	2·686	2·714	2·737	2·778	2·807	2·849	2·880	2·923	2·954	2·978	3·023
0·00100	(1/ 1000)	2·623	2·662	2·690	2·712	2·753	2·782	2·824	2·854	2·897	2·928	2·952	2·996
0·00150	(1/ 667)	2·581	2·619	2·647	2·669	2·709	2·738	2·779	2·808	2·850	2·880	2·904	2·948
0·00200	(1/ 500)	2·551	2·589	2·617	2·638	2·678	2·706	2·747	2·776	2·818	2·847	2·871	2·914
0·00300	(1/ 333)	2·510	2·548	2·575	2·596	2·635	2·663	2·703	2·731	2·772	2·802	2·825	2·867
0·00400	(1/ 250)	2·481	2·518	2·545	2·566	2·605	2·632	2·672	2·700	2·741	2·770	2·792	2·834
0·00600	(1/ 167)	2·441	2·478	2·504	2·525	2·563	2·590	2·629	2·657	2·696	2·725	2·747	2·789
0·00800	(1/ 125)	2·413	2·450	2·476	2·496	2·533	2·560	2·599	2·626	2·666	2·694	2·716	2·757
0·01000	(1/ 100)	2·392	2·428	2·454	2·474	2·511	2·537	2·576	2·603	2·642	2·670	2·692	2·732
0·01500	(1/ 67)	2·353	2·389	2·414	2·434	2·470	2·497	2·534	2·561	2·599	2·627	2·649	2·688
0·02000	(1/ 50)	2·327	2·361	2·386	2·406	2·442	2·468	2·505	2·532	2·570	2·597	2·618	2·658
0·03000	(1/ 33·3)	2·289	2·323	2·348	2·367	2·403	2·428	2·465	2·491	2·528	2·555	2·576	2·615
0·04000	(1/ 25·0)	2·263	2·297	2·321	2·340	2·375	2·401	2·437	2·462	2·499	2·526	2·547	2·585
0·06000	(1/ 16·7)	2·227	2·260	2·284	2·303	2·337	2·362	2·397	2·423	2·459	2·485	2·506	2·543
0·08000	(1/ 12·5)	2·201	2·234	2·258	2·276	2·310	2·335	2·370	2·395	2·431	2·457	2·477	2·514
0·10000	(1/ 10·0)	2·181	2·214	2·238	2·256	2·290	2·314	2·349	2·374	2·409	2·435	2·455	2·492
0·15000	(1/ 6·7)	2·146	2·179	2·202	2·220	2·253	2·277	2·311	2·336	2·371	2·396	2·416	2·452
0·20000	(1/ 5·0)	2·122	2·154	2·176	2·194	2·227	2·251	2·285	2·309	2·344	2·368	2·388	2·424
0·30000	(1/ 3·3)	2·088	2·119	2·141	2·159	2·191	2·215	2·248	2·272	2·306	2·330	2·349	2·385
0·40000	(1/ 2·5)	2·064	2·095	2·117	2·134	2·166	2·189	2·222	2·246	2·279	2·304	2·323	2·357
		40	60	80	100	150	200	300	400	600	800	1000	1500

Gradient   (Equivalent) Pipe diameters in mm   (C = 50)

H19 C = 40
S = 0·00010 to 0·40000

Water (or sewage) at normal temperature;
full bore conditions.

ie hydraulic gradient =
1 in 10000 to 1 in 2·5

This table shows values of m_H, the Hazen-Williams
based divisor of mV and/or mQ, for use with Tables D

Gradient		(Equivalent) Pipe diameters in mm											
		40	60	80	100	150	200	300	400	600	800	1000	1500
0·00010	(1/10000)	3·595	3·649	3·687	3·718	3·773	3·813	3·871	3·912	3·970	4·012	4·045	4·106
0·00015	(1/ 6667)	3·537	3·590	3·628	3·658	3·713	3·752	3·808	3·849	3·906	3·948	3·980	4·040
0·00020	(1/ 5000)	3·496	3·549	3·586	3·616	3·670	3·709	3·765	3·805	3·862	3·903	3·935	3·994
0·00030	(1/ 3333)	3·440	3·492	3·529	3·558	3·611	3·649	3·704	3·743	3·800	3·840	3·871	3·930
0·00040	(1/ 2500)	3·401	3·452	3·488	3·517	3·570	3·608	3·662	3·701	3·756	3·796	3·827	3·885
0·00060	(1/ 1667)	3·346	3·396	3·432	3·460	3·512	3·550	3·603	3·641	3·696	3·735	3·766	3·822
0·00080	(1/ 1250)	3·308	3·357	3·393	3·421	3·472	3·509	3·562	3·599	3·653	3·692	3·723	3·778
0·00100	(1/ 1000)	3·278	3·328	3·363	3·391	3·441	3·478	3·530	3·567	3·621	3·659	3·689	3·745
0·00150	(1/ 667)	3·226	3·274	3·309	3·336	3·386	3·422	3·473	3·510	3·563	3·601	3·630	3·685
0·00200	(1/ 500)	3·189	3·237	3·271	3·298	3·347	3·383	3·433	3·470	3·522	3·559	3·589	3·642
0·00300	(1/ 333)	3·137	3·184	3·218	3·245	3·293	3·328	3·378	3·414	3·465	3·502	3·531	3·584
0·00400	(1/ 250)	3·102	3·148	3·181	3·208	3·256	3·290	3·340	3·375	3·426	3·462	3·490	3·543
0·00600	(1/ 167)	3·052	3·097	3·130	3·156	3·203	3·237	3·286	3·321	3·371	3·406	3·434	3·486
0·00800	(1/ 125)	3·017	3·062	3·094	3·120	3·167	3·200	3·248	3·283	3·332	3·367	3·395	3·446
0·01000	(1/ 100)	2·990	3·035	3·067	3·092	3·139	3·172	3·219	3·254	3·302	3·337	3·365	3·415
0·01500	(1/ 67)	2·942	2·986	3·018	3·042	3·088	3·121	3·168	3·201	3·249	3·284	3·311	3·360
0·02000	(1/ 50)	2·908	2·952	2·983	3·008	3·053	3·085	3·131	3·165	3·212	3·246	3·273	3·322
0·03000	(1/ 33·3)	2·861	2·904	2·935	2·959	3·004	3·035	3·081	3·114	3·160	3·194	3·220	3·268
0·04000	(1/ 25·0)	2·829	2·871	2·902	2·925	2·969	3·001	3·046	3·078	3·124	3·157	3·183	3·231
0·06000	(1/ 16·7)	2·783	2·825	2·855	2·878	2·921	2·952	2·997	3·029	3·074	3·107	3·132	3·179
0·08000	(1/ 12·5)	2·751	2·793	2·822	2·845	2·888	2·919	2·962	2·994	3·039	3·071	3·096	3·143
0·10000	(1/ 10·0)	2·727	2·768	2·797	2·820	2·862	2·893	2·936	2·967	3·012	3·044	3·069	3·115
0·15000	(1/ 6·7)	2·683	2·723	2·752	2·775	2·816	2·846	2·889	2·920	2·963	2·995	3·019	3·065
0·20000	(1/ 5·0)	2·652	2·692	2·721	2·743	2·784	2·814	2·856	2·886	2·929	2·961	2·985	3·030
0·30000	(1/ 3·3)	2·610	2·649	2·677	2·699	2·739	2·768	2·810	2·840	2·882	2·913	2·937	2·981
0·40000	(1/ 2·5)	2·580	2·618	2·646	2·668	2·708	2·737	2·778	2·807	2·849	2·880	2·903	2·947
		40	60	80	100	150	200	300	400	600	800	1000	1500

Gradient   (Equivalent) Pipe diameters in mm   (C = 40)